- The earth's life-support systems can take much stress and abuse, but there are limits (law of limits).

- In nature we can never do just one thing; everything we do creates effects that are often unpredictable (first law of ecology, or principle of ecological backlash).

- Everything is connected to and intermingled with everything else; we are all in this together (second law of ecology, or principle of interdependence).

- Any chemical that we produce should not interfere with the earth's natural biogeochemical cycles in ways that degrade the earth's life support systems for us or other species (third law of ecology).

- We can't expect to reduce the dangers from most hazards to zero, but the risks can be greatly reduced (principle of risk-benefit analysis).

- Any system that depends on fallible humans for its design, operation, and maintenance will sooner or later fail (limitation-of-risk-benefit-analysis principle).

- Nature is not only more complex than we think but also more complex than we can ever think (principle of complexity).

Economics and Politics

- The market price of anything should include all present and future costs of any pollution, environmental degradation, or other harmful effects connected with it that are passed on to society and the environment (principle of internalizing all external costs).

- Some forms of economic growth are harmful; don't produce harmful goods (principle of economic cancer).

- Short-term greed leads to long-term economic and environmental grief; don't deplete the earth's natural capital and mortgage the future (don't-live-off-earth-capital principle).

- The more things you own, the more you are owned by things (principle of overconsumption and thing tyranny).

- Don't give people subsidies and tax breaks to produce harmful goods and unnecessarily waste resources; either eliminate all resource subsidies or reward only producers who reduce resource waste, pollution, and environmental degradation (principle of economic and ecological wisdom).

- Put the poor and their environment first not last; help the poor sustain themselves and their local environment (eliminate-the-poverty-trap principle).

- Change earth degrading and depleting manufacturing processes, products, and businesses into earth-sustaining ones (break-it-and-fix-it-better principle).

- We cannot have a healthy economy in a sick environment (economics-as-if-the-earth-mattered principle).

- The earth can get along without us, but we can't get along without the earth; an exhausted earth is an exhausted economy (respect-your-roots, or earth-first, principle).

- Anticipating and preventing problems is cheaper and more effective than reacting to and trying to cure them; an ounce of prevention is worth a pound of cure (prevention, or input-control, principle).

- History shows that the most important changes come from the bottom up, not the top down (individuals-matter principle).

- Every crisis is an opportunity for change (bad-news-can-be-good-news principle).

- Think globally, act locally (principle of change).

Worldview and Ethics

- The earth does not belong to us; we belong to the earth (principle of humility).

- Our role is to understand and work with the rest of nature, not conquer it (principle of cooperation).

- Every living species has a right to live, or at least to struggle to live, simply because it exists; this right is not dependent on its actual or potential use to us (respect-for-nature principle).

- The best things in life aren't things (principle of love, caring, and joy).

- Something is right when it tends to maintain the earth's life-support systems for us and other species and wrong when it tends otherwise; the bottom line is that the earth is the bottom line (principle of sustainability and ecocentrism).

- It is wrong for humans to cause the premature extinction of any wild species and the elimination and degradation of their habitats (preservation-of-wildlife-and-biodiversity principle).

- When we alter nature to meet what we consider to be basic or nonbasic needs, we should choose the method that does the least possible harm to other living things; in minimizing harm it is generally worse to harm a species than an individual organism, and still worse to harm a biotic community (principle of minimum wrong).

- It is wrong to treat people and other living things primarily as factors of production, whose value is expressed only in economic terms (economics-is-not-everything principle).

- We should leave the earth in as good condition as we found it, if not better (rights-of-the-unborn principle).

- All people should be held responsible for their own pollution and environmental degradation (responsibility-of-the-born principle).

- No individual, corporation, or nation has a right to an ever-increasing share of the earth's finite resources; don't let need slide into greed (principle of enoughness).

- We should protect the earth's remaining wild ecosystems from our activities, rehabilitate or restore ecosystems we have degraded, use ecosystems only on a sustainable basis, and allow many of the ecosystems we have occupied and abused to return to a wild state (principle of ecosystem protection and healing).

- In protecting and sustaining nature, go farther than the law requires (ethics-often-exceed-legality principle).

- To prevent excessive deaths of people and other species, people must prevent excessive births (birth-control-is-better-than-death-control principle).

- Don't do anything that depletes the earth's physical, chemical, and biological capital, which supports all life and human economic activities; the earth deficit is the ultimate deficit (balanced-earth-budget principle).

- To love, cherish, and understand the earth and yourself, take time to experience and sense the air, water, soil, plants, animals, bacteria, and other parts of the earth directly; learning about the earth indirectly from books, TV images, and ideas is not enough (direct-experience-is-the-best-teacher principle).

- Learn about and love your local environment, and live gently within that place; walk lightly on the earth (love-your-neighborhood principle).

FOUR LEVELS OF ENVIRONMENTAL AWARENESS

First Level of Awareness

Pollution and Environmental Degradation

Environmental problems are seen as essentially pollution problems that threaten human health and welfare. Each environmental problem can be solved in isolation by waiting until it reaches a crisis level and then dealing with it by the use of legal, technological, and economic methods, mostly temporary output approaches. There are three major problems with staying at this awareness level. First, it is exclusively a human-centered view, not a life-centered view. Second, individuals see their own impacts as too tiny to matter, not realizing that billions of individual impacts acting together threaten the life-support systems for us and other species. Third, this approach seduces people into thinking that environmental and resource problems can be solved by quick technological solutions.

Second Level of Awareness

Consumption Overpopulation

We recognize that the causes of pollution, environmental degradation, and resource depletion are a combination of people overpopulation in poor countries and consumption overpopulation in rich countries, with the most environmentally damaging populations living in industrialized societies devoted to very high rates of resource consumption and waste production. At this level the answers seem obvious. Stabilize and then reduce population sizes in all countries. Reduce wasteful consumption of matter and energy resources—especially in the affluent countries that consume 80% of the world's resources.

At this level there is little emphasis on transforming political and economic systems in ways that help sustain the earth and in setting aside or restoring much larger areas of the earth's natural systems as wilderness areas, parks, and wildlife preserves. There is little awareness that most protected natural areas are too small to sustain their natural diversity of organisms and are being rapidly overwhelmed and biologically impoverished by the pollutants and unsustainable use of resources and the pollutants produced by technological, growth-oriented societies. This second level of awareness still views humans as above or outside nature and as more important than other species.

Third Level of Awareness

New Age/Spaceship Earth

The goal at this level is to use technology and existing economic and political systems to control population growth, pollution, and resource depletion to prevent environmental overload. The earth is viewed as a spaceship—a machine that we have the capacity and the duty to control and dominate by using advanced technology. If the earth becomes too crowded, we will build stations in space for the excess population. If the earth becomes depleted of mineral resources, we will mine other planets. Genetic engineering will be used to control the evolution of life forms and develop organisms that produce more food, clean up oil spills and toxic wastes, and satisfy more of our unlimited wants. Because of our ingenuity and power over the rest of nature, there will always be more.

This view of the earth as a spaceship is a sophisticated expression of the idea that through technology and human ingenuity we can control nature and create artificial environments and life forms to avoid environmental overload. Instead of novelty, spontaneity, joy, freedom, and biological and cultural diversity, the spaceship model is based on cultural sameness, social regimentation (ground control), artificiality, monotony, and gadgetry. This approach can also cause environmental overload and resource depletion in the long run because it is based on the false ideas that we understand how nature works and that there are no limits to the earth's resources and our ability to overcome any problem with technological innovations.

This view calls for sustainable economic development and sustainable societies for humans. *Sustainable* has become the buzzword for governments and businesses. But careful analysis reveals that some of the proposals now being made under the guise of sustainability are in the long run unsustainable. The human-centered spaceship worldview is inadequate for dealing with an overpopulated, environmentally stressed, and globally interconnected world that lives by depleting and degrading the earth's natural capital.

Fourth Level of Awareness

Sustainable Earth

The first three levels of understanding are human-centered views in which we shape the world to meet our needs. They do not recognize that the solution to our problems lies in giving up our destructive fantasies of omnipotence. Instead we must develop an earth-centered or life-centered worldview based on the principles summarized on the preceding two pages.

We cannot have sustainable or any form of economic development or sustainable societies unless we help sustain the entire earth by working with the earth's natural processes. We must do this not only because it helps ensure our survival but also because it is wrong to do otherwise.

PREFACE: TO THE INSTRUCTOR

Goals This book is designed to be used in introductory courses on environmental science. My goals for the book are to

- cover the diverse materials of an introductory course on environmental science in an accurate, balanced, and interesting way without the use of mathematics or complex scientific information

- help your students discover that dealing with environmental and resource issues is fun, interesting, and important to their lives

- allow you to use the material in a flexible manner depending on course length and what you believe are the important topics

- introduce students to key concepts and principles that govern how nature works and apply them to possible solutions to environmental and resource problems

- show how environmental and resource problems are interrelated and must be understood and dealt with in an integrated manner on a local, regional, national, and global basis

- give a realistic but hopeful view of how much has been done and what remains to be done in sustaining the earth for us and other species

- indicate what individuals can do in their personal lives to help sustain rather than degrade the earth's life-support systems

Three Different Textbooks Available This book is one of a series of three textbooks designed for different introductory courses on environmental science and resource conservation:

- *Living in the Environment* (6th ed., Wadsworth, 1990, 620 pages, 479 illustrations) gives broad and fairly detailed discussions of environmental and resource issues.

- *Resource Conservation and Management* (Wadsworth, 1990, 546 pages, 406 illustrations) has a different organization. It provides less detailed discussions of ecological concepts, population, and pollution than does *Living in the Environment*, but offers expanded coverage of renewable resources, including seven separate chapters on food, fishery, rangeland, forest (two chapters), and wildlife (two chapters) resources and their management

- This book, *Environmental Science* (3rd ed., Wadsworth, 1991, 465 pages, 392 illustrations), is the

briefest book and is designed for those instructors who want less detail than is available in *Living in the Environment*. It also has a different organization and combines some of the features of *Living in the Environment* and *Resource Conservation and Management*

Major Changes in This Edition *This new edition is a thorough and comprehensive revision.* I have rewritten every chapter extensively to eliminate unnecessary detail, for those instructors who want a briefer book. This has also allowed me to add 3 new chapters and 25 additional topics with only a minimal increase in the length of the book. The book now covers all the important topics. Major changes include:

- Updating and revising material throughout the book.

- Improving readability by reducing sentence and paragraph length, omitting unnecessary details in every chapter, and writing in a more personal style.

- Combining several chapters to give an integrated approach to key resources, their management, and the pollution associated with their use. These new integrated chapters are "Air Resources and Air Pollution" (Chapter 9), "Water Resources and Water Pollution" (Chapter 11), "Soil Resources and Hazardous Waste" (Chapter 12), and "Nonrenewable Mineral Resources and Solid Waste" (Chapter 19). Those wishing to teach use and pollution of these resources separately can do so by assigning the first half of these chapters (covering use) as a unit and the last half of these same chapters (covering pollution) as a unit.

- Adding three new chapters on "Cultural Changes, Worldviews, Ethics, and Environment" (Chapter 2), "Environmental Economics and Politics" (Chapter 7), and "Climate, Global Warming, Ozone Depletion, and Nuclear War: Ultimate Problems" (Chapter 10).

- Adding 42 *Case Studies*, 12 *Pro/Con* discussions of controversial issues, 17 *Individuals Matter* at the end of chapters (one for almost every chapter), and 8 *Guest Essays*. Many of the boxed *Spotlights* are also new.

- Adding or expanding coverage of many topics, including pollution prevention, waste reduction, loss of biodiversity, importance of wetlands, importance of sharks as ocean predators, global warming, depletion of the ozone layer, tropical deforestation, nature's secrets for sustainable living, Brazil's environmental problems, immigration and the United States, population regulation in Indonesia, water rights, flooding in Bangladesh, the Aral Sea ecological disaster, the environmental impact of agriculture, sustainable agriculture, sustainable development, debt-for-nature swaps, ecological protection and restoration in Costa Rica, Arctic National Wildlife Refuge, rock cycle, recycling of plastics, nuclear power as a solution to global warming, the international toxic waste trade, and the Valdez oil spill.

■ Reducing the number of chapters from 21 to 19.

■ Increasing the length of the basic text from 407 pages to 465 pages to accommodate the new chapters, Guest Essays, and 65 additional figures and photos. This still makes it one of the shortest textbooks available in this field. Those wanting an expanded textbook can use *Living in the Environment*, sixth edition (620 pages), or *Resource Conservation and Management* (546 pages).

■ Revising the entire art program. All of the illustrations and photographs (with the exception of a few portraits) are in full color. These 392 illustrations and photos make this the most graphically exciting and teachable edition of this book.

■ Increasing the number of maps from 22 to 37 to give students a better geographic perspective. A new feature is a series of "Where is _____?" maps that show students where areas being discussed are located.

■ Adding a two-page summary of key concepts and ideas inside the front cover.

See pp. vi–vii for a more detailed summary of major changes for each chapter.

Key Features

■ *Concept centered*: Uses basic principles and concepts to help students understand environmental and resource problems and possible solutions to these problems in an integrated manner.

■ *Global, national, and local* treatment of issues and solutions.

■ *Flexibility*: The book is divided into five major parts (see Brief Contents). After covering all or most of Parts One and Two, the rest of the book can be used in almost any order. In addition, most chapters and many sections within these chapters can be moved around or omitted to accommodate courses with different lengths and emphases.

■ *Comprehensive review of the professional literature*: Use of more than 10,000 research sources; several key readings are listed at the end of each chapter.

■ *Extensive manuscript review* by 221 experts and teachers (see list on pp. viii–ix) to help make the material accurate and up-to-date.

■ *Guest Essays* (8) to provide more information and to expose the reader to various points of view.

■ *Pro/Con boxes* (12) offering opposing views on controversial environmental and resource issues.

■ *Case Study boxes* (42) to give in-depth information about key issues and to apply concepts.

■ *Spotlight boxes* (28) to highlight and give further insights into environmental and resource problems.

■ *Individuals Matter boxes* (17) appear at the end of almost all chapters, giving examples of what individuals can do to help sustain the earth.

■ *Full-color diagrams* (225) to illustrate complex ideas simply; carefully selected *color photographs* (167) to show how the book's topics relate to the real world.

■ *Maps* (37) to give students a geographic perspective.

■ *Summary of key ideas* inside the front cover.

■ *General questions and issues* summarized at the beginning of each chapter.

■ *Key terms* shown in **boldface** type.

■ *Discussion topics* at the end of each chapter, with emphasis on encouraging students to think critically about and apply what they have learned.

■ *Glossary* of all key terms.

■ *Measurements* expressed in metric units and followed by their U.S. equivalents in parentheses.

Help Me Improve This Book To minimize errors, I have had all or parts of the manuscript reviewed by a large number of teachers and experts, but some errors inevitably creep in during the complex process of publishing a book. If you find any errors, please write them down and send them to me. Most errors can be corrected in subsequent printings of this edition.

Let me know how you think this book can be improved, and encourage your students to evaluate the book and send me their suggestions for improvement. Send any errors you find and your suggestions to Jack Carey, Science Editor, Wadsworth Publishing Company, 10 Davis Drive, Belmont, CA 94002. He will send them on to me.

Supplementary Materials Dr. David Cotter at Georgia College has written an excellent Instructor's Manual and Test Items Booklet for use with this text. It contains sample multiple-choice test questions with answers, suggested projects, field trips, experiments, and a list of topics suitable for term papers and reports for each chapter. Approximately 400 transparency masters are also available—one for every illustration in *Environmental Science, Living in the Environment*, and *Resource Conservation and Management*.

Also available is STELLA II software. For years, the laboratory has been a place where students have been able to apply the skills they've learned in class. STELLA II is a software tool that not only allows students to apply their existing skills, but to develop new skills—critical thinking skills. STELLA II can help students understand the hows and whys underlying biological processes ranging from cellular respiration to ecosystem dynamics and genetic evolution.

STELLA II provides a set of simple building blocks that can be used to piece together the relationships that govern a biological system. Students control the learning process as they incorporate assumptions into their model and then, through simulation and animation, discover the dynamic implications of their assumptions. Alternately, the instructor can lead the class in the construction of a model, providing students with immediate feedback based on their assumptions. STELLA II's exceptional sensitivity analysis capabilities enable students and instructors to explore a full range of variation of key variables within the system under investigation.

Annenberg/CPB Television Course This textbook is being offered as part of the Annenberg/CPB Project television series *Race to Save the Planet*, broadcast on PBS.

Race to Save the Planet is a ten-part public television series and a college-level television course examining the major environmental questions facing the world today, ranging from population growth to soil erosion, from the destruction of forests to climate changes induced by human activity. The series takes into account the wide spectrum of opinion about what constitutes an environmental problem, as well as the controversies about appropriate remedial measures. It analyzes problems and emphasizes the successful search for solutions. The course develops a number of key themes that cut across a broad range of environmental issues, including sustainability, the interconnection of the economy and the ecosystem, short-term versus long-term gains, and the tradeoffs involved in balancing problems and solutions.

In addition to my books *Environmental Science* and *Living in the Environment* and the video programs, the course includes a study guide and faculty guide available from Wadsworth Publishing Company that integrate the telecourse and my texts. The television program was developed as part of the Annenberg/CPB collection.

For further information about available television course licenses and off-air taping licenses, contact: PBS Adult Learning Service, 1320 Braddock Place, Alexandria, VA 22314-1698, 1-800-ALS-ALS-8.

For information about purchasing videocassettes, off-air taping licenses, duplication licenses, and print material, contact The Annenberg/CPB project, P.O. Box 1922, Santa Barbara, CA 93166-1922, 1-800-LEARNER.

Acknowledgments I wish to thank the many students and teachers who responded so favorably to the six editions of *Living in the Environment* and the two editions of *Environmental Science* and offered many helpful suggestions for improvement.

I am also deeply indebted to the many reviewers who pointed out errors and suggested important improvements for this and earlier editions, and to those who wrote Guest Essays for this edition. I am especially indebted to Kenneth J. Van Dellen, Macomb Community College, for his detailed and very helpful review of the entire manuscript. Any errors and deficiencies left are mine, not theirs.

My thanks to Wadsworth's talented production team, whose members have also made important contributions. My thanks also to Wadsworth's dedicated sales staff.

Special thanks to Jack Carey, Science Editor at Wadsworth, for his encouragement, help, friendship, and superb reviewing system. It helps immensely to work with the best and most experienced editor in college textbook publishing.

Finally, I wish to thank Peggy Sue O'Neal, my earthmate, spouse, and best friend, for her love and support of me and the earth. I dedicate this book to her and to the earth that sustains us all.

G. Tyler Miller, Jr.

SOME CHANGES IN THE THIRD EDITION

PART ONE
HUMANS AND NATURE: AN OVERVIEW

Chapter 1 Population, Resources, Environmental Degradation, and Pollution

4 color photos; 2 improved diagrams; Spotlight on The Importance of Biological Diversity and the Biodiversity Crisis; Case Study on Relative Oil Scarcity in the 1970s; table giving a health report summary of the earth's vital resources; expanded discussion of effects of pollution, input (prevention) and output (cleanup) methods for controlling pollution and managing resources, and summary of the major causes of major environmental, resource, and social problems; Guest Essay by Gus Speth; Individuals Matter box (also added to most other chapters).

Chapter 2 Cultural Changes, Worldviews, Ethics, and Environment

New chapter; includes 3 diagrams, 6 color photos, Spotlight on Listening to the Earth and Ourselves, and Case Study on the Chipko Movement.

PART TWO
BASIC CONCEPTS

Chapter 3 Matter and Energy Resources: Types and Concepts

Material on matter quality; 7 new diagrams; 1 color photo; Spotlight on Waste: An Outmoded and Dangerous Concept; Guest Essay by Amory B. Lovins.

Chapter 4 Ecosystems: What Are They and How Do They Work?

15 color photos; 8 new diagrams; revised order of chapter topics; expanded coverage of ecosystem components; new material on the four major types of species that can be found in ecosystems; Case Studies on Ecological Importance of the American Alligator and Sharks: The Oceans' Most Important Predator.

Chapter 5 Ecosystems: What Are the Major Types and What Can Happen to Them?

25 color photos; 6 new diagrams; 1 improved diagram; 3 Spotlights on Importance of the Coastal Zone, Importance of Inland Wetlands, and Nature's Secrets for Sustainable Living; table on Changes Affecting Ecosystems; Guest Essay by Edward J. Kormondy.

Chapter 6 Human Population Dynamics: Growth, Urbanization, and Regulation

Condensed coverage from 2 chapters to 1 chapter; 5 color photos; 9 new diagrams; 7 Case Studies on U.S. Population Stabilization, Teenage Pregnancy in the United States, Brazil, Mexico City, Sustainable Living in Davis, California, Immigration and Population Growth in the United States, and Indonesia; 2 Pro/Con boxes on Advantages and Disadvantages of the Automobile and Is Population Growth Good or Bad?; expanded discussion of resource and environmental problems of urban areas; table on effects of noise; Guest Essay by Garrett Hardin.

Chapter 7 Environmental Economics and Politics

New chapter; includes 3 color photos, 6 diagrams, 4 Spotlights, 1 Case Study, 2 Pro/Cons, and 1 Individuals Matter.

Chapter 8 Hazards, Risk, and Human Health

5 color photos; 4 new diagrams; 1 new table; Case Study on Smoking Risks; Pro/Con on Genetic Engineering.

PART THREE
AIR, WATER, AND SOIL RESOURCES

Chapter 9 Air Resources and Air Pollution

Integrates discussion of air resources and air pollution; 6 color photos; 2 new diagrams; Spotlight on The World's Most Polluted City; expanded discussion of air resources; new discussions on controlling toxic emissions and indoor air pollution.

Chapter 10   Climate, Global Warming, Ozone Depletion, and Nuclear War: Ultimate Problems

New chapter that greatly expands the discussion of these important topics; contains an introduction to climate, 1 color photo, 12 diagrams, 2 Spotlights, 2 Individuals Matter, and a Guest Essay by Kenneth E. Boulding.

Chapter 11   Water Resources and Water Pollution

Integrates discussion of water resources and water pollution; 7 color photos; 6 new diagrams; 5 improved figures; 5 Case Studies on Flooding in Bangladesh, Conflict over Water Supply in California, The Aral Sea Ecological Disaster, The Valdez Oil Spill, and Working with Nature to Purify Sewage; Spotlight on Water Rights in the United States.

Chapter 12   Soil Resources and Hazardous Waste

Integrates discussion of soil resources and hazardous waste; 10 color photos; 3 new diagrams; Spotlight on The International Hazardous-Waste Trade; Guest Essay by David Pimentel.

PART FOUR
LIVING RESOURCES

Chapter 13   Food Resources

2 color photos; 2 new diagrams; expanded discussions of the environmental impacts of agriculture and sustainable-earth agriculture; 2 Spotlights on Marasmus and Kwashiorkor and Major Environmental Impacts of Industrialized and Subsistence Agriculture; Case Study on Collapse of the Peruvian Anchovy Fishery; Pro/Con on Is Food Relief Helpful or Harmful?

Chapter 14   Protecting Food Resources: Pesticides and Pest Control

9 color photos; 2 Spotlights on Biological Amplification of Pesticides and Federal Regulation of Pesticides in the United States; Case Study on The Bhopal Tragedy; Pro/Con on Should Food Be Irradiated?; expanded discussion of biological pest control.

Chapter 15   Land Resources: Forests, Rangelands, Parks, and Wilderness

15 color photos; 3 new diagrams; expanded discussion of tropical deforestation and wilderness preservation; 2 Case Studies on Debt-for-Nature Swap in Bolivia and Ecological Protection and Restoration in Costa Rica; 3 Pro/Cons on How Much of the Amazon Basin Should Be Developed?, How Much Timber Should Be Harvested from National Forests?, and Grazing Fees on Public Rangeland.

Chapter 16   Wild Plant and Animal Resources

19 color photos; 3 new diagrams; expanded discussion of wildlife and fishery management; 3 Case Studies on Near Extinction of the American Bison, The Water Hyacinth, and Near Extinction of the Blue Whale; Spotlight on The Endangered Species Act; Pro/Con on Oil and Gas Development in the Arctic National Wildlife Refuge; Guest Essay by Norman Myers.

PART FIVE
ENERGY AND MINERAL RESOURCES

Chapter 17   Perpetual and Renewable Energy Resources

Condensed evaluation of energy resources from 3 chapters to 2 chapters; 10 color photos; 3 new figures; 2 Spotlights on The World's Biggest Energy User and Waster and Working with Nature: A Personal Progress Report.

Chapter 18   Nonrenewable Energy Resources

5 color photos; Case Study on The Search for a Radioactive Waste Depository; Pro/Con on Should More Nuclear Power Plants Be Built in the United States?; Guest Essay by Alvin M. Weinberg.

Chapter 19   Nonrenewable Mineral Resources and Solid Waste

Integrated discussion on nonrenewable mineral resources and solid wastes; 5 color photos; 1 new figure; expanded discussion of waste reduction; Spotlight on Mega-Landfills; 2 Case Studies on Recycling Aluminum and What Should We Do About Plastics?

Epilogue: Achieving a Sustainable-Earth Society

Added discussions on avoiding some common traps and 12 things individuals can do to become earth-sustaining citizens.

Appendixes

Added list of environmental organizations and government agencies.

AUTHORS OF GUEST ESSAYS

Kenneth E. Boulding, Institute of Behavioral Sciences, University of Colorado, Boulder; Garrett Hardin, Professor Emeritus of Human Ecology, University of California, Santa Barbara; Edward J. Kormondy, Chancellor and Professor of Biology, University of Hawaii-Hilo/West Oahu College; Amory B. Lovins, energy policy consultant and Director of Research, Rocky Mountain Institute; Norman Myers, consultant in environment and development; David Pimentel, Professor of Entomology, Cornell University; Gus Speth, President, World Resources Institute; Alvin M. Weinberg, Distinguished Fellow, Institute of Energy Analysis

REVIEWERS

Barbara J. Abraham, Hampton University; Donald D. Adams, Plattsburgh State University of New York; Larry G. Allen, California State University, Northridge; James R. Anderson, U.S. Geological Survey; Kenneth B. Armitage, University of Kansas; Gary J. Atchison, Iowa State University; Marvin W. Baker, Jr., University of Oklahoma; Virgil R. Baker, Arizona State University; Ian G. Barbour, Carleton College; Albert J. Beck, California State University, Chico; Keith L. Bildstein, Winthrop College; Jeff Bland, University of Puget Sound; Roger G. Bland, Central Michigan University; Georg Borgstrom, Michigan State University; Arthur C. Borror, University of New Hampshire; John H. Bounds, Sam Houston State University; Leon F. Bouvier, Population Reference Bureau; Michael F. Brewer, Resources for the Future, Inc.; Mark M. Brinson, East Carolina University; Patrick E. Brunelle, Contra Costa College; Terrence J. Burgess, Saddleback College North; David Byman, Pennsylvania State University, Worthington Scranton; Lynton K. Caldwell, Indiana University; Faith Thompson Campbell, Natural Resources Defense Council, Inc.; Ray Canterbery, Florida State University; Ted J. Case, University of San Diego; Ann Causey, Auburn University; Richard A. Cellarius, Evergreen State University; William U. Chandler, Worldwatch Institute; F. Christman, University of

North Carolina, Chapel Hill; Preston Cloud, University of California, Santa Barbara; Bernard C. Cohen, University of Pittsburgh; Richard A. Cooley, University of California, Santa Cruz; Dennis J. Corrigan; George Cox, San Diego State University; John D. Cunningham, Keene State College; Herman E. Daly, The World Bank; Raymond F. Dasmann, University of California, Santa Cruz; Kingsley Davis, Hoover Institution; Edward E. DeMartini, University of California, Santa Barbara; Thomas R. Detwyler, University of Wisconsin; Peter H. Diage, University of California, Riverside; Lon D. Drake, University of Iowa; T. Edmonson, University of Washington; Thomas Eisner, Cornell University; David E. Fairbrothers, Rutgers University; Paul P. Feeny, Cornell University; Nancy Field, Bellevue Community College; Allan Fitzsimmons, University of Kentucky; George L. Fouke, St. Andrews Presbyterian College; Kenneth O. Fulgham, Humboldt State University; Lowell L. Getz, University of Illinois at Urbana–Champaign; Frederick F. Gilbert, Washington State University; Jay Glassman, Los Angeles Valley College; Harold Goetz, North Dakota State University; Jeffery J. Gordon, Bowling Green State University; Eville Gorham, University of Minnesota; Michael Gough, Resources for the Future, Inc.; Ernest M. Gould, Jr., Harvard University; Peter Green, Golden West College; Katharine B. Gregg, West Virginia Wesleyan College; Peter Gregs, Golden West College; Paul K. Grogger, University of Colorado at Colorado Springs; L. Guernsey, Indiana State University; Ralph Guzman, University of California, Santa Cruz; Raymond Hames, University of Nebraska, Lincoln; Raymond E. Hampton, Central Michigan University; Ted L. Hanes, California State University, Fullerton; William S. Hardenbergh, Southern Illinois University, Carbondale; John P. Harley, Eastern Kentucky University; Neil A. Harriman, University of Wisconsin-Oshkosh; Grant A. Harris, Washington State University; Harry S. Hass, San Jose City College; Arthur N. Haupt, Population Reference Bureau; Denis A. Hayes, environmental consultant; John G. Hewston, Humboldt State University; David L. Hicks, Whitworth College; Eric Hirst, Oak Ridge National Laboratory; S. Holling, University of British Columbia; Donald Holtgrieve, California State University, Hayward; Michael H. Horn, California State University, Fullerton; Mark A. Hornberger, Bloomsberg University; Marilyn Houck, Pennsylvania State University; Richard D. Houk, Winthrop College; Robert J. Huggett, College of William and Mary; Donald Huisingh, North Carolina State University; Marlene K. Hutt, IBM; David R. Inglis, University of Massachusetts; Robert Janiskee, University of South Carolina; Hugo H. John, University of Connecticut; Brian A. Johnson, University of Pennsylvania, Bloomsburg; David I. Johnson, Michigan State University; Agnes Kadar, Nassau Community

College; Thomas L. Keefe, Eastern Kentucky University; Nathan Keyfitz, Harvard University; David Kidd, University of New Mexico; Edward J. Kormondy, University of Hawaii-Hilo/West Oahu College; John V. Krutilla, Resources for the Future, Inc.; Judith Kunofsky, Sierra Club; E. Kurtz; Theodore Kury, State University of New York at Buffalo; Steve Ladochy, University of Winnipeg; Mark B. Lapping, Kansas State University; Tom Leege, Idaho Department of Fish and Game; William S. Lindsay, Monterey Peninsula College; E. S. Lindstrom, Pennsylvania State University; M. Lippiman, New York University Medical Center; Valerie A. Liston, University of Minnesota; Dennis Livingston, Rensselaer Polytechnic Institute; James P. Lodge, air pollution consultant; Raymond C. Loehr, University of Texas at Austin; Ruth Logan, Santa Monica City College; Robert D. Loring, DePauw University; Paul F. Love, Angelo State University; Thomas Lovering, University of California, Santa Barbara; Amory B. Lovins, Rocky Mountain Institute; Hunter Lovins, Rocky Mountain Institute; Gene A. Lucas, Drake University; David Lynn; Timothy F. Lyon, Ball State University; Melvin G. Marcus, Arizona State University; Stuart A. Marks, St. Andrews Presbyterian College; Gordon E. Matzke, Oregon State University; Parker Mauldin, Rockefeller Foundation; Theodore R. McDowell, California State University, San Bernardino; Vincent E. McKelvey, U.S. Geological Survey; John G. Merriam, Bowling Green State University; A. Steven Messenger, Northern Illinois University; John Meyers, Middlesex Community College; Raymond W. Miller, Utah State University; Rolf Monteen, California Polytechnic State University; Ralph Morris, Brock University, St. Catherines, Ontario, Canada; William W. Murdoch, University of California, Santa Barbara; Norman Myers, environmental consultant; Brian C. Myres, Cypress College; A. Neale, Illinois State University; Duane Nellis, Kansas State University; Jan Newhouse, University of Hawaii, Manoa; John E. Oliver, Indiana State University; Eric Pallant, Allegheny College; Charles F. Park, Stanford University; Richard J. Pedersen, U.S. Department of Agriculture, Forest Service; Robert A. Pedigo, Callaway Gardens; David Pelliam, Bureau of Land Management, U.S. Department of Interior; Harry Perry, Library of Congress; Rodney Peterson, Colorado State University; William S. Pierce, Case Western Reserve University; David Pimentel, Cornell University; Peter Pizor, Northwest Community College; Robert B. Platt, Emory University; Mark D. Plunkett, Bellevue Community College; Grace L. Powell, University of Akron; James H. Price, Oklahoma College; Marian E. Reeve, Merritt College; Carl H. Reidel, University of Vermont; Roger Revelle, California State University, San Diego; L. Reynolds, University of Central Arkansas; Ronald R. Rhein, Kutztown University of Pennsylvania; Charles Rhyne, Jackson State University; Robert A. Richardson, University of Wisconsin; Benjamin F. Richason III, St. Cloud State University; Ronald Robberecht, University of Idaho; William Van B. Robertson, School of Medicine, Stanford University; C. Lee Rockett, Bowling Green State University; Terry D. Roelofs, Humboldt State University; Richard G. Rose, West Valley College; Stephen T. Ross, University of Southern Mississippi; Robert E. Roth, The Ohio State University; David Satterthwaite, I.E.E.D., London; Stephen W. Sawyer, University of Maryland; Arnold Schecter, State University of New York at Syracuse; William H. Schlesinger, Ecological Society of America; Stephen H. Schneider, National Center for Atmospheric Research; Clarence A. Schoenfeld, University of Wisconsin, Madison; Henry A. Schroeder, Dartmouth Medical School; Lauren A. Schroeder, Youngstown State University; Norman B. Schwartz, University of Delaware; George Sessions, Sierra College; David J. Severn, Clement Associates; Paul Shepard, Pitzer College and Claremont Graduate School; Frank Shiavo, San Jose State University; Michael P. Shields, Southern Illinois University at Carbondale; Kenneth Shiovitz; F. Siewert, Ball State University; E. K. Silbergold, Environmental Defense Fund; Joseph L. Simon, University of South Florida; William E. Sloey, University of Wisconsin–Oshkosh; Robert L. Smith, West Virginia University; Howard M. Smolkin, U.S. Environmental Protection Agency; Patricia M. Sparks, Glassboro State College; John E. Stanley, University of Virginia; Mel Stanley, California State Polytechnic University, Pomona; Norman R. Stewart, University of Wisconsin–Milwaukee; Frank E. Studnicka, University of Wisconsin–Platteville; William L. Thomas, California State University, Hayward; Kenneth J. Van Dellen, Macomb Community College; Tinco E. A. van Hylckama, Texas Tech University; Robert R. Van Kirk, Humboldt State University; Donald E. Van Meter, Ball State University; John D. Vitek, Oklahoma State University; Lee B. Waian, Saddleback College; C. Walker, Stephen F. Austin State University; Thomas D. Warner, South Dakota State University; Kenneth E. F. Watt, University of California, Davis; Alvin M. Weinberg, Institute of Energy Analysis, Oak Ridge Associated Universities; Brian Weiss; Raymond White, San Francisco City College; Douglas Wickum, University of Wisconsin–Stout; Charles G. Wilber, Colorado State University; Nancy Lee Wilkinson, San Francisco State University; John C. Williams, College of San Mateo; Ray Williams, Whittier College; Samuel J. Williamson, New York University; Ted L. Willrich, Oregon State University; James Winsor, Pennsylvania State University; Fred Witzig, University of Minnesota–Duluth; George M. Woodwell, Woods Hole Research Center; Robert Yoerg, Belmont Hills Hospital; Hideo Yonenaka, San Francisco State University; Malcolm J. Zwolinski, University of Arizona.

PREFACE: TO THE STUDENT

Why Study About Environmental and Resource Issues? The course you are taking is an introduction to how nature works, how the environment has been and is being used and abused, and what you can do to protect and improve it for yourself and other people, future generations, and other living things. I am convinced that nothing else deserves more of your energy, time, concern, and personal involvement.

What Is the Purpose of Learning? *You may be surprised to learn that the purpose of education is to learn as little as you can.* The goal of education is to learn how to sift through mountains of information and ideas and find the small number that are really useful and worth knowing.

This book is full of facts and numbers, but remember two things. First, they are merely stepping stones to ideas, concepts, and connections. Facts by themselves are useless and confusing. Second, most statistics and facts are human beings with the tears wiped off and living things whose lives we are threatening.

Inside the front cover of this book you will find a list of key principles that summarizes what I have learned so far about how the world works and what my role in it should be. In effect, it is a two-page summary of the key ideas in this book. I use these principles to evaluate other ideas and to make decisions about what to buy or not to buy and how to live my life with increased joy.

Learning is a never-ending, wonderful adventure, so I am also constantly striving to improve this list by modifying or removing some ideas and adding new ones. As you draw up your own list, please send me any ideas you have and suggest modifications to my list. We are all in this together, and we need all the help we can get.

How I Became Involved In 1966, when what we now know as the environmental movement began in the United States, I heard a scientist give a lecture on the problems of overpopulation and environmental abuse. Afterward I went to him and said, "If even a fraction of what you have said is true, I will feel ethically obligated to give up my present scientific research on the corrosion of metals and devote the rest of my life to environmental issues. Frankly, I don't want to believe a word you have said and change my life around, and I'm going into the liter-

ature to try to prove what you have said is either untrue or grossly distorted."

After six months of study I was convinced of the seriousness of these problems. Since then I have been studying, teaching, and writing about them. I have also attempted to live my life in an environmentally sound way—with varying degrees of success—by treading as lightly as possible on the earth (see p. 404 for a summary of my own progress in attempting to work with nature).

Readability Students often complain that textbooks are difficult and boring. I have tried to overcome this problem by writing this book in a clear, interesting, and informal style. My goal is to communicate with you, not confuse you. Let me know how to do this better.

I also relate the information in the book to the real world and to your own lives, in the main text and in boxed *Spotlights, Case Studies, Pro/Con* discussions of issues, and *Individuals Matters* (which suggest things you can do to help sustain the earth) sprinkled throughout the book.

A Realistic but Hopeful Local, National, and Global Outlook We face many interlocking environmental and resource problems. But any problem is an opportunity for change.

In this book I offer a realistic but hopeful view of the future. Much has been done since the mid-1960s, when many people first became aware of the resource and environmental problems we face. But much more needs to be done to protect the earth, which keeps us and other forms of life healthy and alive. The 1960s, 1970s, and 1980s were merely a dress rehearsal for the much more urgent and difficult work we must do in the 1990s and beyond. This book suggests ways you can help sustain the earth.

You will also learn that most environmental and resource problems and their possible solutions are interrelated. Treating them in isolation is a recipe for disaster. I point out many of these connections and give cross references to page numbers relating ideas discussed in various parts of this book. Environmental and resource problems must also be considered on a local, national, and global scale—as this book does.

How the Book Is Organized Take a look at the Brief Contents on p. xii to get an overview of the five major parts of this book and the major topics covered in each part. Before studying each chapter, I suggest you look over the major headings listed in the Detailed Contents on pp. xiii–xvii. This gives you a road map of where you will be going.

I have designed the book so that it can be used in courses with different lengths, emphases, and ordering of topics. So don't be concerned if your instructor skips around and omits material.

General Questions and Issues and Vocabulary
Each chapter begins with a few general questions to give you an idea of what you will be learning. After you finish a chapter, you can go back and try to answer these questions as a general review of what you have learned.

Each chapter will introduce new terms, whose meanings you need to know and understand. When a term is introduced and defined, it is printed in **boldface** type. There is also a glossary of all key terms at the end of the book.

Visual Aids To make this book graphically exciting, I have developed a number of four-color diagrams to illustrate concepts and complex ideas simply. I have also used a number of carefully selected color photos to give you a better picture of how key topics relate to the real world.

Discussion Topics Each chapter ends with a set of discussion questions designed to encourage you to think critically and apply what you have learned to your own life. They also ask you to take sides on controversial issues and to back up your conclusions and beliefs. I have not provided questions that test your recall of facts. This important but mechanical task is left to you and your instructor.

Further Readings If you become especially interested in some of the topics in this book, a few suggested readings are given at the end of each chapter. In Appendix 1 you will find a list of publications to help keep up to date on the book's material and a list of some key environmental organizations.

Interact with the Book When I read something, I interact with it. I mark sentences and paragraphs with a highlighter or pen. I put an asterisk in the margin next to something I think is important and double asterisks next to something that I think is really important. I write comments in the margins, such as *Beautiful, Confusing, Bull, Wrong,* and so on.

I fold down the top corner of pages with highlighted passages and the top and bottom corners of especially important pages. This way, I can flip through a book and quickly review the key passages. I hope you will interact in such ways with this book. You will learn more and have more fun. I hope you will often disagree with what I have written, take the time to think about or write down why, and send your thoughts to me.

Save This Book After you finish this course, you may be tempted to discard this book or resell it to the bookstore. But learning is a lifelong process, and you will have to deal with the vital issues discussed here for the rest of your life. Therefore, I hope you will keep this book in your personal library for future use. Or at least pass it on free to someone whom you want to learn about sustaining the earth.

Help Me Improve the Book Writing and publishing a book is such an incredibly complex process that this or any other book is likely to have some typographical and factual errors. If you find what you believe to be an error, write it down and send it to me.

I would also appreciate learning from you what you like and dislike about the book. This information helps make the book better in future editions. Some of the things you will read here were suggested by students like you.

Send any errors you find and any suggestions for improvement to Jack Carey, Science Editor, Wadsworth Publishing Company, 10 Davis Drive, Belmont, CA 94002. He will send them on to me. Your input helps me, students who take this course in the future, and the earth.

And Now Relax and enjoy yourself as you learn more about the exciting and challenging issues we all face in sustaining the earth for us and other forms of life.

G. Tyler Miller, Jr.

BRIEF CONTENTS

PART FOUR
LIVING RESOURCES 293

PART FIVE
ENERGY AND MINERAL
RESOURCES 377

Chapter 19

Nonrenewable Mineral Resources and Solid Waste 441

Earth Day celebration in Washington, D.C.,
on April 22, 1990.

*It is only in the most recent, and brief, period of their tenure that
human beings have developed in sufficient numbers, and acquired
enough power, to become one of the most potentially dangerous
organisms that the planet has ever hosted.*

JOHN MCHALE

The environmental crisis is an outward manifestation of a crisis of mind and spirit. There could be no greater misconception of its meaning than to believe it is concerned only with endangered wildlife, human-made ugliness, and pollution. These are part of it, but more importantly, the crisis is concerned with the kind of creatures we are and what we must become in order to survive.

LYNTON K. CALDWELL

CHAPTER ONE

POPULATION, RESOURCES, ENVIRONMENTAL DEGRADATION, AND POLLUTION

General Questions and Issues

1. How rapidly is the human population increasing?
2. What are the earth's major types of resources? How can they be depleted or degraded?
3. What are the major types of pollution?
4. What are the relationships among human population size, resource use, technology, environmental degradation, and pollution?

We must stop mortgaging the future to the present. We must stop destroying the air we breathe, the water we drink, the food we eat, and the forests that inspire awe in our hearts. . . . We need to prevent pollution at the source, not try to clean it up later. . . . It's time to remember that conservation is the cheapest and least polluting form of energy. . . . We need to come together and choose a new direction. We need to transform our society into one in which people live in true harmony—harmony among nations, harmony among the races of humankind, and harmony with nature. . . . We will either reduce, reuse, recycle, and restore—or we will perish.

REV. JESSE JACKSON

W E FACE A COMPLEX mix of interlocking problems that are reaching crisis levels on the beautiful blue and green planet that is the only home for us and a rich diversity of other life forms. One is population growth. World population has more than doubled, from 2.5 billion in 1950 to 5.3 billion in 1990. Unless death rates rise sharply, the world's population is projected to more than double to 10.8 billion by 2045 and almost triple to 14 billion before levelling off by the end of the next century. This will put severe stress on the earth's already strained life-support systems.

Each year more of the world's forests, grasslands, and wetlands disappear, and deserts grow in size as people increase their use of the earth's surface. Vital topsoil is washed or blown away from farmland and clogs rivers, lakes, and reservoirs with sediment (Figure 1-1). Water is withdrawn from many underground deposits faster than it is replenished. Each day several of the earth's wild species are driven to permanent extinction by our activities.

Burning of one-time deposits of fossil fuels to support affluent lifestyles is expected to change global climate and disrupt food and water supplies. This burning is also the major source of air pollution that threatens trees, lakes, and people and also causes extensive water pollution and land disruption. The oil that is used to run cars, heat homes, and produce food and most of the products we use will probably be depleted in your lifetime.

Toxic wastes produced by factories and homes are accumulating and poisoning the air, water, and soil. Agricultural pesticides contaminate the groundwater that many of us drink and some of the food we eat. Chemicals we have been adding to the atmosphere are drifting into the upper atmosphere and depleting ozone gas, which protects you and most other forms of life by filtering the sun's harmful ultraviolet radiation.

The bad news is that we are depleting and degrading the air, water, soil, wildlife, minerals, fossil fuels, and other things that make up the earth's resource capital. This inheritance for all living things provides enough for life to go on indefinitely as long as we live off the interest provided by the sun and the earth and don't deplete this natural capital.

But we are living in ways that are unsustainable. According to environmentalists, the increasing number of fishless lakes and streams, dying forests, eroded lands, extinct species, and millions of environmental refugees, whose homelands can no longer keep them alive, are clear signals that nature's bill for overexploitation of the earth's resource base is coming due and that we must drastically change our ways now (see Guest Essay on p. 18). Common sense tells us that we cannot continue to spend earth capital, which took billions of years to accumulate, at a rapid and increasing rate without going bankrupt.

Paying the massive bills we have run up on the earth's credit card will not be easy, and we will have to make some painful choices. But not doing this will be much more painful for us and for future generations who will inherit the earth deficits we have run up.

The good news is that we can help sustain the earth for human beings and other species for generations to come. To do this we will have to make major changes in the way we view the earth and in the ways we live, as summarized in the quote that opens this chapter. The best news is that living sustainably and in touch with the earth is one of the most meaningful and joyful things we can do.

Figure 1-1 Severe soil erosion on a hillside in Spain. Removal of forest cover leads to this type of erosion. Wildlife habitat is lost and eroded, and streams and lakes on the land below can be polluted with eroded sediment.

1-1 HUMAN POPULATION GROWTH

Rates of Change: Linear and Exponential Growth To understand and deal with a problem we need to know how fast it is growing. A car without brakes picks up speed as it goes down a mountain road and is very difficult to stop. The same car without brakes on a flat road is not accelerating and is easier to stop.

Things such as car speed, population size, resource use, and pollution can increase in two major ways: linearly (arithmetic growth) or exponentially (geometric growth). With **linear growth** a quantity increases by some fixed amount during each unit of time. An example is a quantity that increases during each unit of time by one unit: 1, 2, 3, 4, 5, and so on. For example, suppose you start your car and accelerate it by 1.6 kilometers (1 mile) an hour every second. After 60 seconds of such linear growth you would be travelling at 97 kilometers (60 miles) per hour. After two minutes your speed would be 193 kilometers (120 miles) per hour.

With **exponential growth** some quantity—such as population size or economic output—increases by some constant percentage of the quantity over a given time period. With exponential growth a quantity increases by doubling: 2, 4, 8, 16, 32, and so on. For example, if you doubled the speed of a supercharged car every second then it would take you only five seconds to reach a speed of 103 kilometers (64 miles) per hour and one second later you would be travelling 206 kilometers (128 miles) per hour. If you had a magic motor, after 30 seconds you would be travelling at 1.6 billion kilometers (1 billion miles) per hour. After 44 seconds your speed would be 27 trillion kilometers (17 trillion miles) per hour.

From this example we can understand some of the properties of exponential growth. It is deceptive because it starts off slowly. But a few doublings lead quickly to enormous numbers. Why? Because after the second doubling each additional doubling is more than the total of all preceding growth. One of the biggest challenges we face is that most of the environmental problems we face—population growth, resource use, wildlife extinction, and pollution—are growing exponentially.

The J-Shaped Curve of Human Population Growth Plotting the estimated number of people on earth over time gives us a curve with the shape of the letter J (Figure 1-2). This increase in the size of the human population is an example of exponential growth.

The slow, early phase of exponential growth is represented by the long horizontal part of the curve plotted in Figure 1-2. Then, as the base of people undergoing growth increases, the numbers rise sharply as the curve of population growth rounds the bend of the J and heads almost straight up from the horizontal axis (Figure 1-2).

This means that it has taken less time to add each new billion people. It took 2 million years to add the first billion people; 130 years to add the second billion; 30 years to add the third billion; 15 years to add the fourth billion; and only 12 years to add the fifth billion. With present growth rates, the sixth billion will be added during the 10-year period between 1987 and 1997, and the seventh billion is expected to be added 9 years later in 2006.

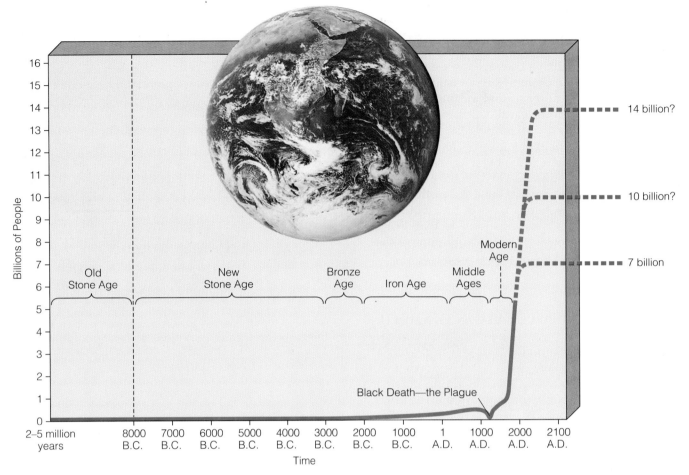

Figure 1-2 J-shaped curve of past exponential world population growth with projections to 2100. (Data from World Bank and United Nations)

In 1990, the world's population of 5.3 billion people was growing exponentially at a rate of 1.8%. This means that by 1991 there will be 95 million more people to feed, clothe, and house (5.3 billion people × 0.018 = 95 million). This amounted to an average increase of 1.8 million people a week, 260,000 a day, or 10,800 an hour. At this rate, it takes about

- 5 days to replace people equal in number to the Americans killed in all U.S. wars

- 9 months to add 75 million people—the number killed in the bubonic plague epidemic of the fourteenth century, the world's greatest disaster

- 21 months to add 165 million people—the number of people killed in all wars fought during the past 200 years

- 32 months to add 250 million people—the population of the United States in 1990

- 11.5 years to add 1.1 billion people—the population of China in 1990

These figures give you an idea of what it means to go around the bend of the J curve of exponential growth. This massive increase in population is happening when:

- one out of five people is hungry or malnourished (Figure 1-3)

- one out of five lacks clean drinking water and bathes in water contaminated with deadly disease-causing organisms

- one out of six does not have adequate housing and at least 100 million are homeless (Figure 1-4)

- one out of three does not have adequate health care or enough fuel to keep warm and cook food

- more than half of humanity lacks sanitary toilets

- one out of four adults cannot read or write

Population Growth in the More Developed and Less Developed Countries The world's 183 countries can be divided into two groups based on the average annual per capita (per person) gross national product (GNP)—the average market value of all goods and services for final use produced per year per person in each country (Figure 1-5).

The world's 33 **more developed countries (MDCs)** are highly industrialized and have high average GNPs per person. Most of these countries have generally

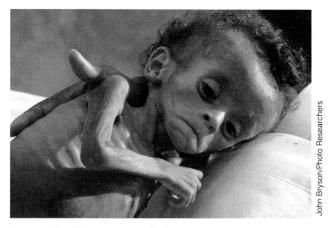

Figure 1-3 This Brazilian child is one of the estimated 1 billion people on earth who suffer from malnutrition caused by a diet without enough protein and other nutrients needed for good health.

Figure 1-4 One-sixth of the people in the world don't have adequate housing or have no housing at all. These homeless people in Calcutta, India, are forced to sleep on the street.

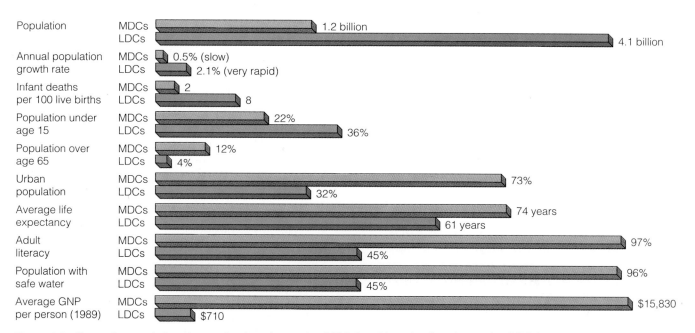

Population	MDCs	1.2 billion
	LDCs	4.1 billion
Annual population growth rate	MDCs	0.5% (slow)
	LDCs	2.1% (very rapid)
Infant deaths per 100 live births	MDCs	2
	LDCs	8
Population under age 15	MDCs	22%
	LDCs	36%
Population over age 65	MDCs	12%
	LDCs	4%
Urban population	MDCs	73%
	LDCs	32%
Average life expectancy	MDCs	74 years
	LDCs	61 years
Adult literacy	MDCs	97%
	LDCs	45%
Population with safe water	MDCs	96%
	LDCs	45%
Average GNP per person (1989)	MDCs	$15,830
	LDCs	$710

Figure 1-5 Some characteristics of more developed countries (MDCs) and less developed countries (LDCs) in 1990. (Data from United Nations and Population Reference Bureau)

favorable climates and fertile soils. MDCs include the United States, Canada, Japan, the Soviet Union, Australia, New Zealand, and all countries in western Europe. These MDCs, with 1.2 billion people (23% of the world's population), use about 80% of the world's mineral and energy resources.

The 150 **less developed countries (LDCs)** have low to moderate industrialization and low to moderate average GNPs per person. Most are located in the Southern Hemisphere in Africa, Asia, and Latin America. Many of these countries have less favorable climates and less fertile soils than most MDCs. The LDCs contain 4.1 billion people, or 77% of the world's population, but use only about 20% of the world's

mineral and energy resources. The LDCs, where 1 million people are added every 4.5 days, account for nine of every ten babies born and 98% of all infant and childhood deaths.

Most of the projected increase in world population will take place in the LDCs (Figure 1-6). An estimated three-fourths of these people will be living in mushrooming cities already facing an urban crisis from environmental degradation, deteriorating services and infrastructure, increased poverty, and neighborhood collapse. How the earth will accommodate the rapid exponential growth in population and resource use now taking place is one of the most important questions we face.

1-2 RESOURCES AND ENVIRONMENTAL DEGRADATION

Types of Resources A **resource** is anything we get from the physical environment to meet our needs and wants. Some resources are directly available for use. Examples are fresh air, fresh water in rivers and lakes, fertile soil, and naturally growing edible plants.

Most resources, such as oil, iron, groundwater (water found in underground resources), and modern crops, aren't directly available. They become resources only when we find ways to make them

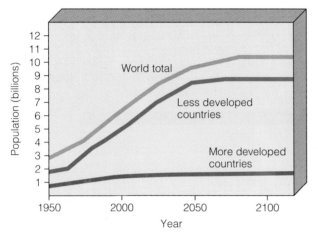

Figure 1-6 Past and projected population size for MDCs, LDCs, and the world, 1950–2120. (Data from United Nations)

available at affordable prices. Groundwater wasn't a resource until we developed the technology for drilling a well and installing pumps to bring it to the surface. Petroleum was a mysterious fluid until we learned how to find it, extract it, and refine it into gasoline, home heating oil, road tar, and other products at affordable prices.

Resources can be classified as perpetual, nonrenewable, and renewable (Figure 1-7). A **perpetual resource,** such as solar energy, is virtually inexhaustible on a human time scale. **Nonrenewable, or exhaustible, resources,** such as copper, aluminum, coal, and oil, exist in a fixed amount (stock) in various places in the earth's crust. They can be used up completely or **economically depleted** to the point where it costs too much to get what is left (typically when 80% of its total estimated supply has been removed and used).

Some nonrenewable resources can be recycled or reused to extend supplies—copper, aluminum, iron, and glass, for example. **Recycling** involves collecting and remelting or reprocessing a resource so that it can be made into new products. For example, aluminum beverage cans can be collected, melted, and converted into new beverage cans or other aluminum products. Also, glass bottles can be crushed and melted to make new glass bottles or other glass items. **Reuse** involves using a resource over and over in the same form. For example, glass bottles can be collected, washed, and refilled.

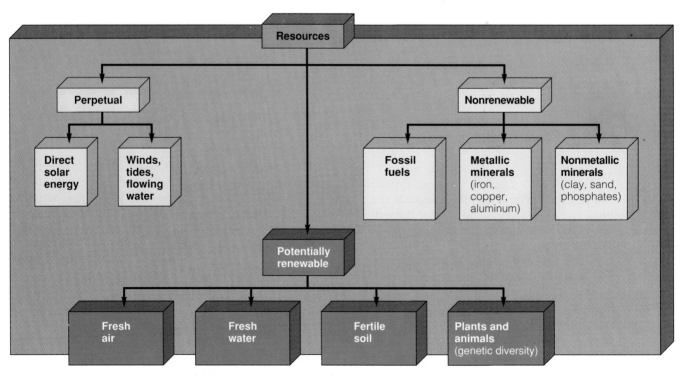

Figure 1-7 Major types of resources. This scheme, however, isn't fixed; potentially renewable resources can be converted to nonrenewable resources if used for a prolonged time faster than they are renewed by natural processes.

Other nonrenewable resources, such as fossil fuels (coal, oil, and natural gas), can't be recycled or reused. When burned, the high-quality, useful energy in these fuels is converted to low-quality waste heat and exhaust gases that pollute the atmosphere.

Since 1950, cheap oil has fueled remarkable economic growth in many countries. Oil is the key resource we have used to help us dominate the earth. Low oil prices have also encouraged waste of oil and discouraged the search for other sources of energy. In one year we use up oil that nature took 1 million years to produce.

Most resource analysts expect the price of oil to rise sharply some time between 1995 and 2010, when the world's oil demand exceeds the rate at which remaining supplies can be extracted. However, this projected oil-based energy crisis could be delayed a decade or two by greatly increased efforts to waste less oil and to find replacements for oil (see Case Study on p. 8).

Many analysts believe that over the next 50 years the world must gradually withdraw from its present heavy dependence on oil and find replacements—or face economic turmoil. They also believe that conventional and nuclear wars will become more likely as countries fight for control over the world's dwindling oil.

Often we can find a replacement for a scarce or expensive nonrenewable resource. But substitution isn't always possible. Some materials have properties that can't easily be matched. In other cases, replacements may be inferior, too costly, or too scarce.

A **potentially renewable resource** is one that can theoretically last forever because it is replaced through natural processes. Examples are trees in forests, grasses in grasslands, wild animals, fresh surface water in lakes and rivers, most deposits of groundwater, fresh air, and fertile soil. The planet's most valuable resource is its diversity of potentially renewable forms of life (see Spotlight below).

SPOTLIGHT   The Importance of Biological Diversity and the Biodiversity Crisis

Over billions of years the combination of the formation of new species and the extinction of species that could not adapt to changing environmental conditions has produced the planet's most valuable resource: **biological diversity,** or **biodiversity.** It is made up of three related concepts: genetic diversity, species diversity, and ecological diversity.

Genetic diversity is variability in the genetic makeup among individuals within a single species. **Species diversity** is the variety of species or types of organisms in different parts of the planet such as forests, grasslands, deserts, lakes, or oceans. **Ecological diversity** is the variety of forests, deserts, grasslands, oceans, rivers, lakes, and other biological communities that interact with one another and with their nonliving environments through various ecological processes.

Biodiversity is a critical part of the natural capital we have inherited from the earth. The priceless diversity within and among species has provided us with food, wood, fibers, energy, raw materials, industrial chemicals, and

medicines and has contributed hundreds of billions of dollars yearly to the world economy. Biological diversity has made the development of agriculture possible, enabled crops to adapt to new situations, and allowed us to develop new crop strains through crossbreeding and genetic engineering.

This vast genetic library also helps provide us and other species with free resource recycling and purification services. Every species here today represents stored genetic information that allows the species to adapt to certain changes in environmental conditions. We can think of biodiversity as nature's "insurance policy" against disasters.

Extinction is a natural process, but since agriculture began about 10,000 years ago, the rate of species extinction has increased sharply as human settlements have expanded worldwide. There is evidence that we are bringing about the greatest mass extinction since the end of the age of dinosaurs 65 million years ago. Already about 100 species per day are becoming extinct forever be-

cause of our activities. By early in the next century this extinction rate could easily rise several-fold and climb even more for several decades.

Biologists warn that if deforestation (especially of tropical forests), desertification, and destruction of wetlands and coral reefs continue at their present rates, at least 1 million of the earth's estimated 5 to 30 million species are likely to disappear over the next three or four decades.

This massive loss of biological diversity cannot be balanced by formation of new species because it takes between 2,000 and 100,000 generations for a new species to evolve. Genetic engineering is not a solution to this biological holocaust. Genetic engineers do not create new genes, they transfer genes from one organism to another. Thus, genetic engineering depends on natural biodiversity for its raw material.

Prematurely eliminating many of the earth's species for our own short-term economic gain is not only shortsighted—it is wrong. What do you think should be done to protect the earth's precious biodiversity from us?

Classifying something as a renewable resource, however, doesn't mean that it can't be depleted and that it will always be renewable. The highest rate at which a renewable resource can be used without decreasing its potential for renewal throughout the world or in a particular area is called its natural replacement rate or **sustainable yield.** If this yield is exceeded, the base supply of a renewable resource begins to shrink—a process known as **environmental** **degradation** (see Spotlight on p. 10). Overuse converts renewable resources to nonrenewable or nonexistent (extinct) ones on a human time scale by using them faster than they can be replenished. Pollution of renewable resources can make them unusable.

Types of Resource Scarcity Resource scarcity can be absolute or relative. History has shown that resource

---

CASE STUDY Relative Oil Scarcity in the 1970s

The relative scarcity of oil between 1973 and 1979 was caused by several factors. One was rapid economic growth during the 1960s, stimulated by low oil prices. Another factor was the growing dependence of the United States and many other MDCs on imported oil (Figure 1-8).

A third factor was that between 1973 and 1979 the Organization of Petroleum Exporting Countries (OPEC)* was able to control the

---

* OPEC was formed in 1960 so that LDCs with much of the world's known and projected oil supplies could get a higher price for this resource and stretch remaining supplies by forcing the world to reduce oil use and waste. Today its 13 members are Algeria, Ecuador, Gabon, Indonesia, Iran, Iraq, Kuwait, Libya, Nigeria, Qatar, Saudi Arabia, United Arab Emirates, and Venezuela.

---

world's supply, distribution, and price of oil. About 63% of the world's proven oil reserves are in the OPEC countries, compared to only 3% in the United States. In 1973 OPEC produced 56% of the world's oil and supplied about 84% of all oil imported by other countries.

This dependence of most MDCs on OPEC countries for oil set the stage for the two phases of the relative oil scarcity crisis of the 1970s. First, in 1973 Arab members of OPEC reduced oil exports to Western industrial countries and banned all shipments of their oil to the United States because of its support of Israel in its 18-day war with Egypt and Syria.

This embargo lasted until March 1974 and caused a fivefold increase in the average world

price of crude oil (Figure 1-9). The increase contributed to double-digit inflation in the United States and many other countries, high interest rates, soaring international debt, and a global economic recession. Americans, accustomed to cheap and plentiful fuel, waited for hours to buy gasoline and turned down thermostats in homes and offices.

Despite the sharp price increase, U.S. dependence on imported oil increased from 30% to 48% between 1973 and 1977. OPEC imports increased from 48% to 67% during the same period (Figure 1-8). This increasing dependence was caused mostly by the government's failure to lift controls that kept oil prices artificially low and encouraged energy waste.

The artificially low prices sent a false message to consumers and set the stage for the second phase of the oil distribution crisis. Available world oil supplies decreased when the 1979 revolution in Iran shut down most of that country's production. Gasoline waiting lines became even longer, and by 1981 the average world price of crude oil rose to about $35 a barrel.

A combination of energy conservation, substitution of other energy sources for oil, and increased oil production by non-OPEC countries led to a drop in world oil consumption between 1979 and 1988. The drop in demand and the inability of OPEC countries to reduce their oil production enough

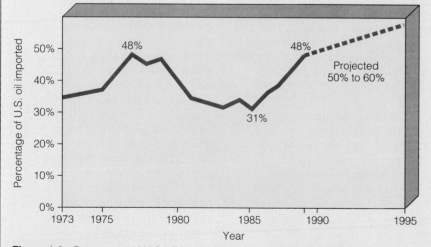

Figure 1-8 Percentage of U.S. oil imported between 1973 and 1989 with projections to 1995. (Data from U.S. Department of Energy and Spears and Associates, Tulsa, Oklahoma)

scarcity stimulates technological innovations to make resources last longer and to develop substitutes. But resource scarcity can also cause economic and social chaos, and has been a key factor in many wars.

Absolute resource scarcity occurs when there aren't enough actual or affordable supplies of a resource left to meet present or future demand. For example, the world's affordable supplies of nonrenewable oil may be used up within your lifetime. The period of absolute scarcity and increasing cost of oil may begin between 1995 and 2010.

Relative resource scarcity occurs when enough of a resource is still available to meet the demand but its distribution is unbalanced. For example, between 1973 and 1979, the world had enough oil to meet demand. But not enough oil was produced and distributed to meet the needs and wants of the United States, Japan, and many western European countries.

to sustain relative resource scarcity and high prices led to a glut of oil.

Because supply exceeded demand, the price of oil dropped from $35 to around $18 per barrel between 1981 and 1989. This meant that the inflation-adjusted price of crude oil in 1989 was about the same as it was in 1974 (Figure 1-9).

The oil glut has had some good effects for MDCs such as the United States and for LDCs heavily dependent on imported oil. It has stimulated economic growth and created new jobs (except in the oil industry), and it has reduced the rate of inflation. It has also reduced the percentage of world oil production by OPEC from 59% in 1980 to 47% in 1989.

At the same time, the price drop has had a number of undesirable effects:

- A sharp decrease in the search for new oil in the United States and most other countries.

- Economic chaos in many oil-producing countries, especially those with large international debts (such as Mexico), and in major oil-producing states (such as Texas, Oklahoma, Louisiana, and Alaska).

- Loss of many jobs in oil and related industries.

- Failure or near-failure of many U.S. banks with massive outstanding loans to oil companies and oil-producing LDCs such as Mexico.

- Decreased rate of improvement in energy efficiency and decreased development of energy alternatives to replace oil.

- Resumption of growth in energy consumption. After dropping at an annual rate of 1.9% between 1979 and 1983, average energy consumption in the United States rose by 1.8% a year between 1984 and 1989.

- Increased dependence on imported oil from a low of 31% in 1985 to 48% in 1989, draining over $64 billion a year from the U.S. economy (Figure 1-8).

Most energy analysts believe that the oil glut of the 1980s is only temporary. They expect that sometime between 1995 and 2010 the world will enter a period of increasing absolute scarcity of oil. They project that OPEC countries will increase their share of the world's oil market to 60% some-time between 1995 and 2010. Then OPEC will dominate world oil markets and raise prices even more than in the 1970s.

The Department of Energy and most major oil companies project that by 1995 or 2000 the United States could be dependent on imported oil for 60% to 70% of its oil consumption—much higher than in 1977 (Figure 1-8). This would drain the already debt-ridden United States of vast amounts of money, leading to severe inflation and widespread economic recession, perhaps even a major depression. This situation would also increase the chances of war as the world's MDCs compete for greater control over dwindling oil supplies to avoid economic collapse. Iraq's invasion of Kuwait in August 1990 showed how instability in this area can affect oil supplies and prices. What do you think should be done?

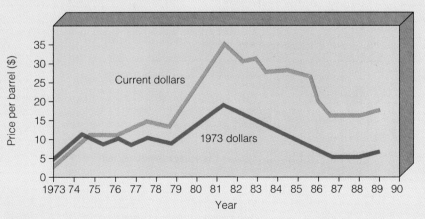

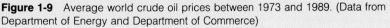

Figure 1-9 Average world crude oil prices between 1973 and 1989. (Data from Department of Energy and Department of Commerce)

Several types of environmental degradation can change potentially renewable resources into nonrenewable, permanently extinct, or unusable resources:

- Covering productive land with water, silt, concrete, asphalt, or buildings to such an extent that crop growth declines and places for wildlife to live (habitats) are lost.

- Cultivating land without proper soil management so that crop growth is reduced by soil erosion and depletion of plant nutrients. Each year the world's farmers must feed 95 million more people with 24 billion fewer metric tons (26 billion tons) of topsoil.

- Irrigating cropland without sufficient drainage so that excessive buildup of water (waterlogging) or salts (salinization) in the soil decreases crop growth.

- Removing water from underground reservoirs (aquifers) and surface waters (rivers and lakes) faster than it is replaced by natural processes. Water scarcity is emerging on every continent.

- Removing trees from large areas (deforestation) without adequate replanting so that wildlife habitats are destroyed and long-term timber growth is decreased. Every minute a piece of tropical forest the size of 20 city blocks disappears.

- Depletion of grass by livestock (overgrazing) so that soil is eroded to the point where productive grasslands are converted into unproductive land and deserts (desertification).

- Killing various forms (species) of wild plant and animal life through destruction of habitat, commercial hunting, pest control, and pollution to the point where these species no longer exist (extinction) (see Spotlight on p. 7).

Table 1-1 Health Report for Some of the Earth's Vital Resources

Land

Productive Land	About 8.1 million square kilometers (3.1 million square miles) of once-productive land (cropland, forests, grasslands) have become desert in the last 50 years. Each year almost 61,000 square kilometers (23,500 square miles) of new desert are formed.
Cropland Topsoil	Topsoil is eroding faster than it forms on about 35% of the world's cropland—a loss of about 24 billion metric tons (26 billion tons) of topsoil a year (see photo on p. 3). Crop productivity on one-third of the earth's irrigated cropland has been reduced by salt buildup in topsoil. Topsoil waterlogging has reduced productivity on at least one-tenth of the world's cropland.
Forest Cover	Almost half of the world's original expanse of tropical forests has been cleared. Each year about 202,000 square kilometers (78,000 square miles) of tropical forest are destroyed and another 202,000 square kilometers (78,000 square miles) are degraded. Within 30 to 50 years there may be little of these forests left. One-third of the people on earth cannot get enough fuelwood to meet their basic needs, and many are forced to meet their needs by cutting trees faster than they are being replenished. In MDCs 312,000 square kilometers (120,400 square miles) of forest have been damaged by air pollution. Also, many remaining areas of diverse, ancient forests are being cleared and replaced with more vulnerable tree farms that greatly reduce wildlife habitats and biodiversity.
Grasslands	Millions of hectares of grasslands have been overgrazed; some, especially in Africa and the Middle East, have been converted to desert. Almost two-thirds of U.S. rangeland is in fair to poor condition.

Water

Coastal and Inland Wetlands	Between 25% and 50% of the world's wetlands have been drained, built upon, or seriously polluted. Worldwide, millions of hectares of wetlands are lost each year. The United States has lost 56% of its wetlands in the lower 48 states, and each year loses another 150,000 hectares (371,000 acres).
Oceans	Most of the wastes we dump into the air, water, and land eventually end up in the oceans. Oil slicks, floating plastic debris, polluted estuaries and beaches, and contaminated fish and shellfish are visible signs that we are using the oceans as the world's largest trash dump.
Lakes	Thousands of lakes in eastern North America and in Scandinavia have become so acidic that they contain no fish; thousands of other lakes are dying; thousands are depleted of much of their oxygen because of inputs of various chemicals produced by human activities.

Data from Worldwatch Institute and World Resources Institute

Table 1-1 summarizes the status of key life-sustaining resources.

One situation that can cause environmental degradation is the use of **common-property resources**. These are resources that it is difficult to exclude people from using and for which each user depletes or degrades the available supply. Most are potentially renewable. Examples are clean air, fish in parts of the ocean not under the control of a coastal country, migratory birds, Antarctica, gases of the lower atmosphere, the ozone content of the upper atmosphere, and outer space.

Because excluding people from using the resources that make up the global commons is difficult, they can be polluted or over-harvested and converted from renewable to slowly renewable, nonrenewable, or unusable resources. Abuse or depletion of common-property resources is called the **tragedy of the commons**. It occurs because each user reasons, "If I don't use this resource, someone else will. The little bit I use or the little bit of pollution I create is not enough to matter."

When the number of users is small, there is no problem. But eventually the cumulative effect of many people trying to maximize their use of a common-property resource depletes or degrades the usable supply. Then no one can make a profit or otherwise benefit from the resource.

There are two ways out of this dilemma. One is for users of a common-property resource to agree voluntarily to limit their use to help sustain the resource. In some cases this has been successful at the local level where peer group pressure among neighbors and friends can play an important role. However, this is difficult at the regional and global levels. The second approach is for various local, regional, or national governments to agree to cooperate in regulating access to common-property resources.

Water (continued)

Drinking Water	In LDCs 61% of the people living in rural areas and 26% of urban dwellers do not have access to safe drinking water. Each year 5 million die from preventable waterborne diseases. In parts of China, India, Africa, and North America, water is withdrawn from underground reservoirs (aquifers) faster than it is replenished by precipitation. In the United States, one-fourth of the groundwater withdrawn each year is not replenished. Pesticides contaminate some groundwater deposits in 38 states. In MDCs hundreds of thousands of industrial and municipal landfills and settling ponds, several million underground storage tanks for gasoline and other chemicals, and thousands of abandoned toxic waste dumps threaten groundwater supplies.

Air

Climate	Emissions of carbon dioxide and other gases into the atmosphere from fossil fuel burning and other human activities may raise the average temperature of the earth's lower atmosphere several degrees between now and 2050. This would disrupt food production and flood low-lying coastal cities and croplands.
Atmosphere	Chlorofluorocarbons and halons released into the lower atmosphere are drifting into the upper atmosphere and reacting with and gradually depleting ozone faster than it is being formed. The thinner ozone layer will let in more ultraviolet radiation from the sun. This will cause increases in skin cancer and eye cataracts, and our immune-system defenses against many infectious diseases will be weakened. Levels of eye-burning smog, damaging ozone gas, and acid rain in the lower atmosphere will increase, and yields of some important food crops will decrease.

Biodiversity

Wildlife	An estimated 36,500 species of plants and animals become extinct each year, mostly because of human activities; if deforestation (especially of tropical forests), desertification, and destruction of wetlands and coral reefs continue at present rates, at least 500,000 and perhaps 1 million species will become extinct over the next 20 years.

People

Environmental Refugees	Today more than 10 million people worldwide have lost their homes and land because of environmental degradation. These people are now the world's single largest class of refugees.
Poverty	At least 1.2 billion people—more than one of every four—live in absolute poverty. During the 1980s this group increased by 200 million people.

During this period of relative resource scarcity, the price of oil rose from $3 to $35 a barrel (see Case Study on p. 8).

1-3 POLLUTION

What Is Pollution? Any undesirable change in the characteristics of the air, water, soil, or food that can adversely affect the health, survival, or activities of humans or other living organisms is called **pollution.** Most pollutants are unwanted solid, liquid, or gaseous chemicals produced as by-products or wastes when a resource is extracted, processed, made into products, and used. Pollution can also take the form of unwanted energy emissions such as excessive heat, noise, or radiation.

A major problem is that people differ in what they consider an acceptable level of pollution, especially if they have to choose between pollution control and losing their jobs. As philosopher Georg Hegel pointed out, the nature of tragedy is not the conflict between right and wrong but between right and right.

Sources of Pollution Pollutants can enter the environment naturally (for example, from volcanic eruptions) or through human activities (burning coal). Most natural pollution is dispersed over a large area and is often diluted or broken down to harmless levels by natural processes.

In contrast, most serious pollution from human activities occurs in or near urban and industrial areas, where large amounts of pollutants are concentrated in small volumes of air, water, and soil. Some pollutants contaminate the areas where they are produced. Others are carried by winds or flowing water to other areas. Pollution does not respect the state and national boundaries we draw on maps.

Some of the pollutants we add to the environment come from single, identifiable sources such as the smokestack of a power plant, the sewer pipe of a meat-packing plant, the chimney of a house, or the exhaust pipe of an automobile. These are called **point sources.** Other pollutants enter the air, water, or soil from dispersed and often hard-to-identify sources called **nonpoint sources.** Examples are the runoff of fertilizers and pesticides from farmlands and of numerous types of chemicals from urban areas and suburban lawns into lakes and streams. It is much easier and cheaper to control pollution from fixed point sources than from widely dispersed nonpoint sources.

Effects of Pollution Pollution can have a number of unwanted effects:

- *nuisance and aesthetic insult*—unpleasant smells and tastes, reduced atmospheric visibility, and soiling of buildings and monuments

- *property damage*—corrosion of metals, weathering or dissolution of building and monument materials, and soiling of clothes, buildings, and monuments

- *damage to plant and nonhuman animal life*—decreased tree and crop production, harmful health effects on animals, and extinction

- *damage to human health*—spread of infectious diseases, respiratory system irritation and diseases, genetic and reproductive harm, and cancers

- *disruption of natural life-support systems at local, regional, and global levels*—climate change and decreased natural recycling of chemicals, energy inputs, and biodiversity needed for good health and survival of people and other forms of life

Three factors determine how severe the effects of a pollutant will be. One is its *chemical nature*—how active and harmful it is to specific types of living organisms. Another is its *concentration*—the amount per volume unit of air, water, soil, or body weight. One way to reduce the concentration of a pollutant is to dilute it by adding it to a large volume of air or water. Until we started overwhelming the air and waterways with massive inputs of pollution, dilution was the solution to pollution. Now it is only a partial solution.

A third factor is a pollutant's *persistence*—how long it stays in the air, water, soil, or our bodies. **Degradable** or **nonpersistent** pollutants are broken down completely or reduced to acceptable levels by natural processes. Those broken down by living organisms are called **biodegradable pollutants.** Human sewage added to a river or the soil is biodegraded fairly quickly as long as it is not added faster than it can be degraded. Other pollutants, such as radioactive substances, are degraded by nonbiological processes.

A major problem is that many of the substances and products that we have made and introduced into the environment in large quantities often take decades or longer to degrade. Examples are the insecticide DDT, most plastics, aluminum cans, and chlorofluorocarbons (CFCs)—chemicals widely used as coolants in refrigerators and air conditioners, spray propellants, and foaming agents for making plastics such as Styrofoam. Chemists deliberately designed these substances to perform certain tasks without breaking down rapidly in the environment. The problem is that many of these products are being emitted or dumped into the environment instead of being recycled or reused. Because of a lack of oxygen and water, the paper, plastic, and most other items we bury in modern landfills take decades, even hundreds of years, to biodegrade.

Nondegradable pollutants are not broken down by natural processes. Examples are the toxic elements lead and mercury. The only way to deal with nondegradable pollutants is to recycle them, remove them from contaminated air, water, or soil (an expensive process), or not release them into the environment at all.

Pollution Control We can control pollution in two ways: input control and output control. **Input pollution control** prevents potential pollutants from entering the environment or sharply reduces the amounts released. For example, sulfur impurities can be removed from coal before it's burned. This stops or sharply reduces emissions of the air pollutant sulfur dioxide, a chemical that damages plants, buildings, metals, and our respiratory systems.

Reducing unnecessary use and waste of matter and energy resources is another major way to reduce harmful inputs of chemicals and excessive heat into the environment. We can also recycle or reuse chemical outputs from human activities instead of discarding them.

So far most attempts to control pollution have been output approaches, based on treating rather than preventing the problem. **Output pollution control** deals with wastes after they have entered the environment. The problem is that output approaches often remove a pollutant from one part of the environment and cause pollution in another part. For example, we can remove substances from polluted water at a sewage treatment plant, but this produces a gooey, often toxic, sludge. The sludge must either be burned (producing some air pollution), buried (possibly contaminating underground water supplies used for drinking water), or cleaned up and applied to the land as fertilizer.

Output approaches buy us some time. But eventually these temporary bandages are overwhelmed by the exponential growth of the human population, resource use, pollution, and environmental degradation. For example, adding catalytic converters to cars has helped reduce air pollution. But as the number of cars has increased, this output approach is being overwhelmed.

Often both input and output controls are needed. But environmentalists urge that we must place primary emphasis on prevention or input approaches because in the long run they work better and are cheaper. This is based on recognizing that environmental pollution is an incurable disease as long as population and resource use continue to increase. The only effective cure is prevention, not temporary

treatment of symptoms. As Benjamin Franklin reminded us long ago: "An ounce of prevention is worth a pound of cure."

Throughout this book you will see possible solutions to various environmental and resource problems divided into output approaches and input approaches. As you make decisions about what things to buy and what solutions to support, ask yourself "Is this an input (prevention) or an output (temporary) approach?" Our motto should be: *Output approaches are better than doing nothing, but input approaches are the best way to walk gently on the earth.*

1-4 RELATIONSHIPS AMONG POPULATION, RESOURCE USE, TECHNOLOGY, ENVIRONMENTAL DEGRADATION, AND POLLUTION

One Model of Environmental Degradation and Pollution According to one simple model, the total environmental degradation and pollution—that is, the environmental impact of population—in a given area depend on three factors: **(1)** the number of people, **(2)** the average number of units of resources each person uses, and **(3)** the amount of environmental degradation and pollution generated when each unit of resource is used (Figure 1-10).

Overpopulation occurs when the people in a country, a region, or the world use resources to such an extent that the resulting degradation or depletion of the resource base and pollution of the air, water, and soil are damaging their life-support systems. Overpopulation is a result of growing numbers of people, growing affluence (resource consumption), or both. The data in Table 1-1 suggest that the planet is already overpopulated.

Differences in the importance of the factors shown in Figure 1-10 lead to two types of overpopulation: people overpopulation and consumption overpopulation (Figure 1-11). **People overpopulation** exists where there are more people than the available supplies of food, water, and other important resources can support. In this type of overpopulation, population size and the resulting degradation of potentially renewable soil, grasslands, forests, and wildlife tend to be the key factors determining total environmental impact (Figure 1-11). In the world's poorest LDCs, people overpopulation causes premature death for at least 20 million, and perhaps 40 million, people each year and absolute poverty for 1.2 billion people.

Many analysts fear the plight of the rapidly growing number of people in these countries will get worse. The severity of this tragedy can be reduced by greatly increased efforts and funding by LDCs and MDCs for slowing population growth and environ-

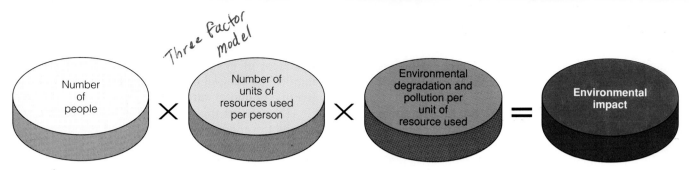

Figure 1-10 Simplified model of how three factors affect overall environmental degradation and pollution, or the environmental impact of population.

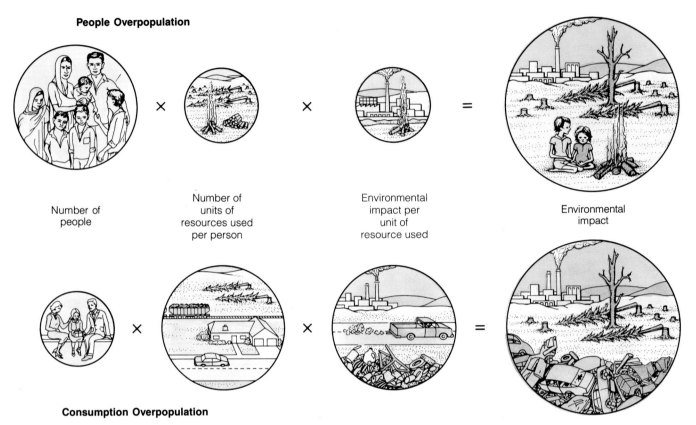

Figure 1-11 Two types of overpopulation based on the relative importance of the factors in the model shown in Figure 1-10. Circle size shows relative importance of each factor. People overpopulation is caused mostly by growing numbers of people. Consumption overpopulation is caused mostly by growing affluence (resource consumption).

mental degradation, reducing poverty, and restoring of degraded renewable resources. This must involve encouraging and helping LDCs undergo sustainable economic development that does not deplete the earth's capital upon which their present and future economic growth depends.

Industrialized countries have the second type of overpopulation: **consumption overpopulation.** It exists when a small number of people use resources at such a high rate that significant pollution, environmental degradation, and resource depletion occur. With this type of overpopulation, high rates of resource use per person and the resulting high levels of pollution and environmental degradation per person are the key factors determining overall environmental impact (Figure 1-11).

By controlling at least 80% of the world's wealth and material resources, people in MDCs presently enjoy an average standard of living at least 18 times that in LDCs. But this high standard of living is the major cause of the world's pollution and environmental degradation. This means that during your lifetime, the 9 million babies added to the population of MDCs in 1990 may do at least as much damage to the earth's life support systems as the 86 million babies added in LDCs.

According to the model in Figure 1-11, the United States has the world's highest level of consumption overpopulation. With only 4.8% of the world's population, it produces about 21% of all goods and services, uses about one-third of the world's processed nonrenewable energy and mineral resources, and produces at least one-third of the world's pollution and trash.

According to biologist Paul Ehrlich: "While overpopulation in poor nations tends to keep them poverty stricken, overpopulation in rich nations tends to undermine the life-support capacity of the entire planet. . . . A baby born in the United States will damage the planet 20 to 100 times more in a lifetime than a baby born into a poor family in an LDC. Each rich person in the United States does 1,000 times more damage than a poor person in an LDC." Thus, it is urgent that the United States and other MDCs switch from their present forms of earth-depleting, unsustainable economic development to earth-sustaining forms of economic development.

Multiple-Factor Model The three-factor model shown in Figure 1-10, though useful, is too simple. The major causes of the environmental, resource, and social problems we face are more complex. They include:

- *People overpopulation and consumption overpopulation* (Figure 1-11).

- *Population distribution*—the population implosion or urban crisis. The most severe air and water pollution problems occur when large numbers of people and industrial activities are concentrated in an urban area. Winds blow many of these pollutants over other cities and rural areas.

- *Overconsumption and wasteful patterns of resource use, especially in industrialized countries*—throwaway mentality, planned obsolescence, producing unnecessary and harmful items, and very little recycling and reuse of essential resources.

- *Belief that technology will solve our problems*—failure to distinguish between forms of technology that reduce pollution and unnecessary resource waste and help sustain the earth's life-support systems, and those that, without proper control, can degrade the earth's life-support systems.

- *Poverty*—failure of the world's economic and political systems to achieve a more equitable distribution of the world's economic output and wealth. Since 1950 the gap in average income between the world's rich people and poor people has been widening, a process that accelerated in the 1980s.

- *Oversimplification of the earth's life-support systems*—excessive reduction of the diversity of plant and animal life in forests, oceans, grasslands, and other parts of the earth's life-support system (see Spotlight on p. 7), resulting in increased soil erosion, flooding, accelerated extinction of wild species, and damage to crops from insects and diseases (Table 1-1).

- *Crisis in political and economic management*—overemphasis on all types of economic growth instead of encouraging sustainable forms of economic growth such as pollution control, recycling, and resource conservation, and discouraging polluting and resource-wasting forms of economic growth; a short-term outlook that leads governments to go from crisis to crisis instead of trying to anticipate problems and prevent them from reaching crisis levels; short-term bandages (output approaches) instead of long-term cures (input approaches).

- *Failure to have market prices represent the overall environmental cost of an economic good or service to society and the earth's life-support systems*—not knowing the harmful effects of the products we buy because most of the costs of pollution, environmental degradation, and resource depletion caused by their production and use are not included in their market prices; using economic systems to reward activities that degrade the earth instead of rewarding activities that help sustain the earth.

- *Human-centered (anthropocentric) worldview and behavior instead of earth-centered (biocentric) worldview and behavior*—tragedy of the commons (see Spotlight on p. 10); thinking that we are above and in charge of nature instead of just another part of nature; attempting to dominate and alter nature to suit our purposes rather than working with nature by walking gently on the earth; thinking that our cleverness and technology will allow us to escape the physical, chemical, and biological processes that govern the sustainability of the earth's life-support systems.

These and other factors interact in complex and largely unknown ways to produce the major environmental, resource, and social problems the world faces (Figure 1-12). This means that the population, energy, poverty, pollution, urban, and environmental degradation crises we face are interlocking parts of an overall crisis. The only way out is for us to act together to formulate interdisciplinary, integrated approaches to all these problems at the local, national, and global levels.

Everything You Do Matters *The most important message of this book is that we can deal with the problems we face and begin to turn things around within your lifetime.* It will not be easy, painless, or without

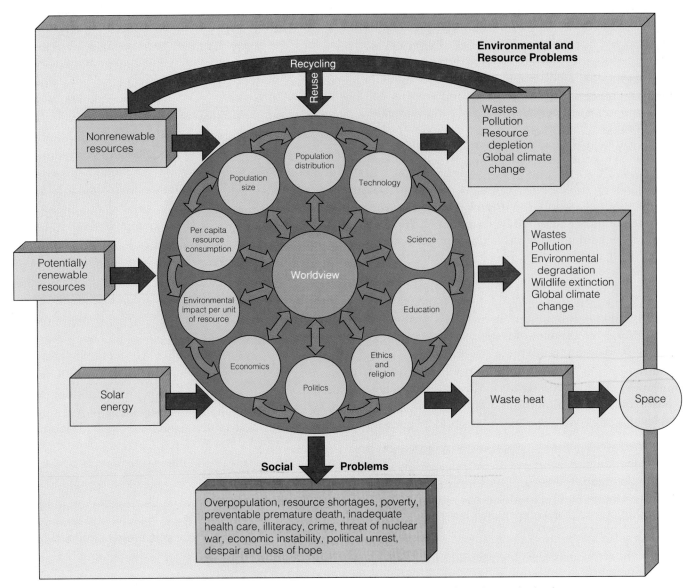

Figure 1-12 Environmental, resource, and social problems are caused by a complex, poorly understood mix of interacting factors, as illustrated by this simplified model.

controversy. But it can be done and environmentalists believe that it must be done.

The key to dealing with these problems is recognizing that *individuals matter*. Billions of individual actions contribute to the environmental and resource problems we face and to the solutions to these problems (see Individuals Matter on p. 17). Inside the back cover and throughout this book you will find a number of boxes titled Individuals Matter that suggest what you can do to help sustain the earth.

Every environmental and resource problem we face is an opportunity to make changes in the way we view the world and act in it. The choice is ours. We can walk hard on the earth or we can walk gently on the earth.

What's the use of a house if you don't have a decent planet to put it on?

HENRY DAVID THOREAU

DISCUSSION TOPICS

1. Do you favor instituting policies designed to reduce population growth and stabilize **(a)** the size of the world's population as soon as possible and **(b)** the size of the U.S. population as soon as possible? Explain. Would you favor providing tax breaks and other incentives for people who have one, two, or no children and imposing economic penalties for those who have more than one or two children? Explain. How many children do you plan to have?

- What are the major resource and environmental problems in the area where you live?

- List the following in order of decreasing importance as environmental threats to you: indoor air pollution, outdoor air pollution, depletion of the ozone layer, global warming, toxic chemicals in the air, drinking water contamination, leaking hazardous waste sites, acid rain, air pollution from automobiles, nuclear energy, coal-burning power plants, people overpopulation in LDCs, consumption overpopulation in MDCs, tropical deforestation, soil erosion, pesticides in your food and drinking water.

- How many people live in your community? Is it growing or shrinking in size?

- What have been the major beneficial and harmful effects of the growth or shrinkage of the population in your community?

- What major types of vegetation grow naturally in your area? What are the major forms of wildlife in or near your area? Are there many species of plants and animals, or only a few? What effects have people and activities had on local biodiversity during the last ten years?

- Are there major rivers, lakes, wetlands, or oceans nearby? How would you rate the quality of the water and wildlife in these bodies of water?

- Is there much agricultural land near where you live? Are there serious soil erosion problems on cropland, urban developments, and grazing land in your area?

- Does your community have a land-use plan? Does your campus have a land-use plan? If so, what are the major goals of these plans (e.g., encourage-ment of growth and economic development, slow growth and development, preservation of farmlands, forests, wetlands, and other natural areas)?

- Is your area prone to natural hazards such as earthquakes, hurricanes, tornadoes, droughts, or floods?

- Are there any power plants in your community? If so, what type are they, how close are they to where you live, and what are their safety records?

- What major industrial plants are in your community? How close are they to where you live? What are the major hazardous wastes generated in your community? What happens to these materials?

- What proportion of your diet comes from animals? What proportion comes from plants? How much of your food is grown locally? How much is imported? How much of your food is produced organically without the use of pesticides, commercial inorganic fertilizers, growth hormones, or antibiotics?

- How would you rate the quality of the air you breathe: good, fair, or poor? What are the major types and sources of air pollution where you live? How much of the pollution in the air you breathe comes from distant upwind sources? What is the quality of the indoor air where you live and work? Have the buildings in which you live or work been tested for the presence of asbestos and radioactive radon gas?

- Where does the water you drink and use for other purposes come from? How is the water you drink purified? How would you rate its quality? Where does the sewage you create go? How is it treated?

- About how much water do you use a day? In what ways do you conserve water? (See Individuals Matter inside the back cover.)

- How much solid waste do you generate each week? What are the major components of this trash? What happens to this waste? How much of it is buried? How much is burned? How much is recycled? Is there a recycling program in your community or on your campus? Do you separate cans, bottles, and paper for recycling for collection or take these materials to recycling centers? What products do you reuse?

- How is your house or dormitory heated and cooled? What energy source is used to supply the hot water you use? At what temperature do you keep your room or house in the winter? In the summer? How well insulated are your walls, ceiling, and floor? How airtight is your room or house?

- In what ways do you try to conserve energy? (See Individuals Matter inside the back cover.)

- Are you a member of a national, state, or local environmental organization? Do you regularly read an environmental publication? Have you participated in any groups or activities to improve the quality of your local environment? Have you written to or talked with a local, state, or federal elected official concerning some environmental issue within the last year? (See Appendix 1.)

Gus Speth

Since 1982 Gus Speth has served as president of the World Resources Institute, a center for policy research on global resource and environmental issues. He served as head of the President's Council on Environmental Quality (CEQ) between 1979 and 1981, after serving as a member of the council from 1977 to 1979. In 1980 he chaired the President's Task Force on Global Resources and Environment. Before his appointment to the CEQ, he was a senior attorney for the Natural Resources Defense Council, an environmental organization he helped found in 1970.

Writing recently in *Foreign Affairs*, George F. Kennan observed that "our world is at present faced with two unprecedented and supreme dangers": any major war at all among great industrial powers and "the devastating effect of modern industrialization and overpopulation on the world's natural environment."

The deterioration of the global environment to which Kennan refers has a scale that encompasses the great life-supporting systems of the planet's biosphere. It includes the alteration of the earth's climate and biogeochemical cycles, the accumulation of wastes, the exhaustion of soils, loss of forests, and the decline of ecological communities (Table 1-1, p. 10).

Since World War II, growth in human population and economic activity has been unprecedented. The world's population has doubled and now exceeds 5 billion, and another billion will be added by 1997. The gross world product has increased fourfold since 1950. With these increases in population and economic activity have come large increases in both pollution and pressure on natural resources.

Air pollution today poses problems for all countries. As use of fossil fuels has increased, so have emissions of sulfur and nitrogen oxides and other harmful gases. Acid rain, ozone, and other ills born of this pollution are now damaging public health and harming forests, fish, and crops over large areas of the globe.

Another gas emitted when fossil fuels burn is carbon dioxide, the chief culprit among the greenhouse gases, which trap heat in the atmosphere. If the buildup of greenhouse gases in the atmosphere is not halted, the global warming now apparently under way will bring about major climate changes. Regional impacts are difficult to predict accurately, but rainfall and monsoon patterns could shift, disrupting agriculture in many areas. Sea levels could rise, flooding coastal areas. Ocean currents could shift, further altering the climate and fisheries. Fewer plant and animal species could survive as favorable habitats are reduced. Heat waves, droughts, hurricanes, and other weather anomalies could harm susceptible people, crops, and forests.

Depletion of the stratosphere's ozone layer also threatens human health and natural systems. In 1987 an international treaty was negotiated to address this problem by reducing the use of chlorofluorocarbons (CFCs). This treaty, however, is already considered inadequate since scientists recently discovered more ozone depletion than expected.

These interrelated atmospheric issues constitute the most serious pollution threat in history. Simultaneous and gradual, their effects will be hard to reverse. Because pollutants react with other substances, with each other, and with the sun's energy, a well-planned response has to take all these factors into account. These air pollution issues are also linked to the use of fossil fuels. In the future, energy policy and environmental policy should be made together.

The United States can take some pride in actions to improve air quality. But the country still emits about 15% of the world's sulfur dioxide, about 25% of all nitrogen oxides, and 25% of the carbon dioxide, and it still manufactures about 30% of all CFCs.

Improvements in U.S. energy efficiency have been considerable. Per capita energy use dropped by 12% between 1973 and 1985—a period when per capita gross domestic product grew 17%. Yet the United States is still consuming one-fourth of the world's energy annually and producing only half as much GNP per unit of energy as its world market competitors such as West Germany and Japan.

Our national concern for the atmosphere must be matched by a growing awareness of the steady deteri-

2.   Explain why you agree or disagree with the following proposition: High levels of resource use by the United States and other MDCs is beneficial. MDCs stimulate the economic growth of LDCs by buying their raw materials. High levels of resource use also stimulate economic growth in MDCs. Economic growth provides money for more financial aid to LDCs and for reducing pollution, environmental degradation, and poverty.

3.   Do you believe that all automobiles, vans, and light trucks sold in the United States should be required to

oration of forests, soils, and water in much of the developing world. The UN Food and Agriculture Organization predicts that without corrective action, rainfed crops in the Third World will become 30% less productive by the end of this century because the soil is depleted or eroded.

In developing countries, 10 trees are cut down for every one replanted—30 trees for one in Africa—and every minute about 22 hectares (54 acres) of tropical forests disappear, as do uncounted species that inhabit them. Fuelwood shortages affect an estimated 1.5 billion people in 63 countries. Most people in developing countries lack access to basic sanitary facilities, and 80% of all illness is due to unsafe water supplies. People in LDCs now rank high among those exposed to toxic chemicals—from lead in Mexico to DDT in China.

In 1988 the World Commission on Environment and Development described a new consensus, supported by nations north and south. The old notion that environmental loss was the price of economic progress was rejected. Far from bringing about broad-based development, overexploitation or mismanagement of natural resources has contributed to famines and floods, dam reservoirs that fill with silt within a decade, irrigation schemes that salt the soil, and conversion of grasslands and tropical forests into unproductive wastelands. The commission's report, *Our Common Future*, stated: "Many forms of development erode the natural resources upon which they must be based, and environmental degradation can undermine economic development. Poverty is a major cause and effect of global environmental problems."

Fighting poverty requires diffusing the underlying pressures on the world's resource base. While many complex factors are involved, the LDCs must deal with:

- *Rapid population growth.* Of the 1 billion people to be added to the world's population between 1987 and 1997, nine out of ten will be born in developing countries.

- *Shortsighted economic policies pursued by governments of both industrial and developing countries.* These include direct and indirect subsidies that encourage the wasteful use of energy, water, and forests, and policies that favor city dwellers over the rural poor.

- *Misguided development and aid programs.* Many large-scale development projects have neglected environmental factors and local needs.

The United States is directly affected by these LDC concerns. Twenty percent of the carbon dioxide contributing to the greenhouse effect is estimated to come from tropical deforestation. A wide range of biological resources—species yet to be analyzed for their agricultural, industrial, or pharmaceutical value—is being lost.

The growth of developing economies expands global U.S. trade and job opportunities at home; already more than a third of U.S. trade is with LDCs. With economic recovery, the developing countries could absorb up to half of all U.S. exports by the year 2000. But sustained growth in much of the developing world requires better management of natural resources.

"Sustainable development" is the widely accepted answer—development that meets today's needs without compromising the ability of future generations to meet theirs. U.S. leadership in applying this approach requires both vigorous evaluation of environmental consequences of development assistance programs and also support for national development strategies that conserve and restore the land's productive capacity.

It means helping developing countries invest in reforestation, agroforestry (growing crops and trees together), water conservation, and energy efficiency. It also means reducing debt and other pressures that force LDCs to cash in their natural resources to earn foreign exchange. Family planning, primary health services, and better sanitation all deserve a high priority since they reduce child mortality and slow birth rates.

In the 1990s industrialized and developing countries will have to face these challenges together. Nations north and south, east and west, must act in concert to sustain the earth and its people.

Twenty years ago, the United States responded vigorously to the serious environmental concerns then emerging. New national policies were declared, new agencies created, and major pollution cleanup and resource management initiatives launched.

Today, as we enter the 1990s, the Bush administration and the members of Congress face a new agenda

(continued)

achieve at least 21 kilometers per liter (50 miles per gallon) by 1999? Explain. Do you believe that all houses and buildings should be required to meet stringent insulation, heating efficiency, lighting, and other standards designed to greatly reduce unnecessary energy waste? Explain.

4. Do you believe that all households and businesses should be required to separate trash into separate categories for recycling? Explain. Do you believe that a high disposal tax should be placed on all throwaway items to discourage their use? Explain.

of environmental concerns that are more serious and challenging than the problems of the 1970s.

These concerns are not the ones to which the United States addressed itself when environmental concerns emerged forcefully 20 years ago. They present us with new policy challenges that are more global in scope and international in implication.

The 1990s will be the crucial decade for action on these pressing concerns. If major national and international efforts are not pursued in this period, irreparable damage will be done to the world's environment, and the problems will prove increasingly intractable, expensive, and dominated by crises.

If the United States and other countries do respond, however, tropical deforestation can be arrested and disappearing species saved; poverty can be alleviated and human populations stabilized; soils can be conserved and more food provided; projected climate change can be slowed; regional and global pollution can be reduced.

These and other things can be done with means within our grasp. But success hinges on concerted effort made with some urgency to change many current policies, to strengthen and multiply successful programs, and to launch bold initiatives where they are needed.

There are ample grounds in the experience of the last two decades for both optimistic and pessimistic assumptions about the future. The gaps between success and failure in addressing resource, environmental, and population problems have been enormous. The good news is that these divergent outcomes are primarily the result of differences in policies and programs pursued by governments, the private sector, and others. In short, leadership and new initiatives can make a world of difference.

Guest Essay Discussion

1. How serious are the problems cited in this essay and throughout this chapter? How do these problems affect your life and lifestyle, now and in the future?

2. What is *sustainable development*? Is it being practiced in the United States? How could it be implemented in MDCs and LDCs?

5. Would you support a sharp increase in taxes if you could be sure the money was used to help improve environmental quality?

6. Would you support greatly increasing the amount of land protected from development as wilderness, even if the land contained valuable minerals, oil, natural gas, timber, or other resources?

FURTHER READINGS*

Brown, Lester R., et al. Annual. *State of the World*. New York: W. W. Norton.

Coalition of Environmental Groups. 1989. *Blueprint for the Environment*. Salt Lake City, Utah: Howe Brothers Press.

Commoner, Barry. 1990. *Making Peace with the Planet*. New York: Pantheon.

Dahlberg, Kenneth A., et al. 1985. *Environment and the Global Arena*. Durham, N.C.: Duke University Press.

Ehrlich, Anne H., and Paul R. Ehrlich. 1987. *Earth*. New York: Franklin Watts.

Ehrlich, Anne H., and Paul R. Ehrlich. 1990. *The Population Explosion*. New York: Doubleday.

Global Tomorrow Coalition. 1990. *The Global Ecology Handbook: What You Can Do About the Environmental Crisis*. Boston: Beacon Press.

Goldfarb, Theodore, ed. 1990. *Taking Sides: Clashing Views on Controversial Environmental Issues*. Guilford, Conn.: Duskin Publishing Group.

Hardin, Garrett. 1968. "The Tragedy of the Commons," *Science*, vol. 162, 1243–48.

Myers, Norman, ed. 1984. *Gaia: An Atlas of Planet Management*. Garden City, N.Y.: Anchor Press/Doubleday.

Office of Technology Assessment. 1987. *U.S. Oil Production: The Effect of Low Oil Prices*. Washington, D.C.: Government Printing Office.

Repetto, Robert. 1986. *World Enough and Time: Successful Strategies for Resource Management*. New Haven, Conn.: World Resources Institute.

Silver, Cheryl S., and Ruth S. Defries. 1990. *One Earth, One Future: Our Changing Global Environment*. Washington, D.C.: National Academy Press.

Weiner, Jonathan. 1990. *The Next One Hundred Years: Shaping the Fate of Our Living Earth*. New York: Bantam.

World Commission on Environment and Development. 1987. *Our Common Future*. New York: Oxford University Press.

World Resources Institute and International Institute for Environment and Development. Published every two years. *World Resources*. New York: Basic Books.

* Because of limited space only a brief list of key references is given for each chapter. For a more comprehensive set of references see Further Readings in the expanded version of this book: *Living in the Environment*, 6th ed. (Belmont, Calif.: Wadsworth, 1990).

CHAPTER TWO

CULTURAL CHANGES, WORLDVIEWS, ETHICS, AND ENVIRONMENT

General Questions and Issues

1. What major cultural changes have taken place during the 40,000 years that our species, *Homo sapiens sapiens*, has lived on earth, and what effects have these had on the earth's life support systems?

2. What worldview leads to the throwaway societies found in most of today's industrialized countries?

3. What worldview leads to a sustainable-earth society, and how can we achieve such a worldview?

A continent ages quickly once we come.
ERNEST HEMINGWAY

 OSSIL EVIDENCE SUGGESTS that humans and their extinct close relatives have lived on the earth for 4 to 6 million years. Such evidence also indicates that the most recent form of our species, *Homo sapiens sapiens,* has lived on earth for only about 40,000 years, a brief instant in the earth's estimated 4.6-billion-year existence.

During three-fourths of the 40,000 years our species has been around we survived as mostly nomadic hunter-gatherers. Since then there have been two major cultural shifts: the *Agricultural Revolution,* which began about 10,000 to 12,000 years ago, and the *Industrial Revolution,* which began about 275 years ago.

These cultural revolutions have given us much more energy and new technologies to alter and control increasingly larger parts of the earth to meet our basic needs and a rapidly expanding list of wants (Figure 2-1). By expanding food supplies, increasing average life spans, and improving average living standards, each of these shifts led to sharp increases in the size of the human population (Figure 2-2). But these cultural shifts have also led to the J-shaped curves of exponentially increasing resource use, pollution, and environmental degradation we are experiencing today (Chapter 1).

During the past thirty years a global environmental movement that began in the United States in the 1960s has questioned whether we can continue expanding in numbers and present forms of economic growth without disrupting the earth's life support systems for us and other species. Members of this rapidly growing movement believe there is an urgent need for us to make a new cultural change before we are overwhelmed by exponential growth in people,

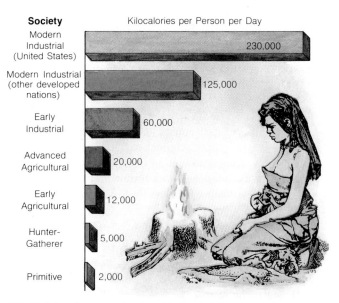

Figure 2-1 Average direct and indirect daily energy use per person at various stages of human cultural development.

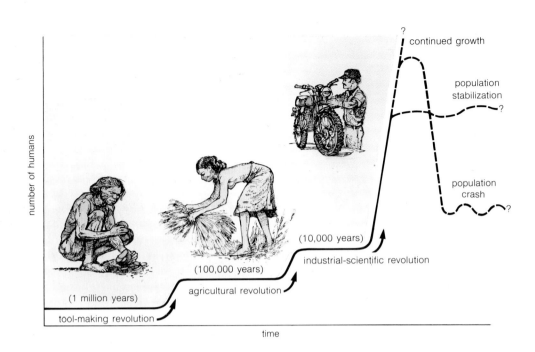

Figure 2-2 Human beings have expanded the earth's carrying capacity for their species through technological innovation, leading to several major cultural changes. Dashed lines represent possible future changes in human population size: continued growth, population stabilization, and continued growth followed by a crash and stabilization at a much lower level. This generalized curve is plotted by a different mathematical method (a plot of the logarithm of population size versus the logarithm of time) from the method (a plot of population size versus time) used in Figure 1-2, p. 4.

pollution, and environmental degradation. At this new turning point, called a **sustainable-earth revolution,** we are called upon to alter our lifestyles, political and economic systems, and the way we view and treat the earth to help sustain the earth for us and other species.

2-1 MAJOR CULTURAL CHANGES

Hunting and Gathering Societies Archaeological evidence suggests that during about three-fourths of our 40,000-year existence of our species we have been **hunter-gatherers** who survived by gathering wild plants and hunting and killing wild animals (including seafood) from the nearby environment (Figure 2-3). Evidence indicates that our hunter-gatherer ancestors lived in small groups of rarely more than 50 people. If food became scarce, they picked up their few possessions and moved to another area.

Our hunter-gatherer ancestors survived only by having expert knowledge about their natural surroundings. Their knowledge of nature enabled them to predict the weather and find water even in the desert. They discovered a variety of plants and animals that could be eaten and used as medicines. By using stones to chip sticks, other stones, and animal bones, they made primitive weapons for killing animals, and tools for cutting plants and scraping hides for clothing and shelter. These dwellers in nature had only two energy sources: sunlight captured by plants, which also served as food for the wild animals they hunted, and their own muscular power (Figure 2-1).

Although women typically gave birth to four or five children, usually only one or two survived to adulthood. Infant deaths from infectious diseases and infanticide (killing the newborn) led to an average life expectancy of about 30 years.

Groups made conscious efforts to keep their population size in balance with available food supplies. Population control practices varied with different cultures but included abstention from sexual intercourse, infanticide, abortion, late marriage, and feeding infants with breast milk as long as possible (a practice which provides some degree of birth control). This prevented the world's hunter-gatherers from undergoing rapid population growth (Figure 2-2).

Archaeological evidence indicates that hunter-gatherers gradually developed improved tools and hunting weapons. Some people learned to work together to hunt herds of reindeer, woolly mammoths, European bison, and other big game. They used fire to flush game from thickets toward hunters lying in wait and to stampede herds of animals into traps or over cliffs. Some also learned to burn vegetation to promote the growth of food plants and plants favored by the animals they hunted.

Advanced hunter-gatherers had a greater impact on their environment than early hunter-gatherers, especially in using fire to convert forests into grasslands. There is also evidence that they contributed to the extinction of some large game animals in different parts of the world.

But because of their small numbers, mobility, and dependence mostly on their own muscle power to modify the environment, the environmental impact of early and advanced hunter-gatherers was fairly

Figure 2-3 Most people who have lived on earth have survived by hunting wild game and gathering wild plants. These !Kung bushmen in Africa (left) are going hunting. This man (right) is digging for roots in a tropical forest in the Amazon Basin of Brazil.

small and localized. Both early and advanced hunter-gatherers were examples of *people in nature*, who survived by being immersed in and learning to work with nature.

The Agricultural Revolution About 10,000 to 12,000 years ago a cultural shift, known as the **Agricultural Revolution,** began taking place. It involved a gradual shift from small mobile hunting and gathering bands to settled agricultural communities where people survived by learning how to breed wild animals and to cultivate wild plants near where they lived.

Archaeological evidence indicates that plant cultivation, which we now call horticulture, probably began in tropical forest areas. People discovered that they could grow various wild food plants by digging holes with a stick (a primitive hoe) and placing roots or tubers of these plants in the holes.

To prepare for planting, they cleared small patches of forests by **slash-and-burn cultivation**—cutting down trees and other vegetation, leaving the cut vegetation on the ground to dry, and then burning it (Figures 2-4 and 2-5). The ashes that were left added plant nutrients to the nutrient-poor soils found in most tropical forest areas. Roots and tubers were then planted in holes dug between tree stumps.

These early growers also used **shifting cultivation** as part of their horticultural system (Figure 2-4). After a plot had been planted and harvested for a few years, few if any crops could be grown. By then either the soil was depleted of nutrients or the patch had been invaded by a dense growth of vegetation from the surrounding forest. When yields dropped, the growers shifted to a new area of forest and cleared a new plot. The growers learned that each abandoned patch had to be left fallow (unplanted) for 10 to 30 years before the soil became fertile enough to grow crops again. By doing this they practiced sustainable agriculture.

These growers practiced **subsistence farming,** growing only enough food to feed their families. Their dependence on human muscle power and crude stone or stick tools meant that they could cultivate only small plots; thus, they had relatively little impact on their environment.

True agriculture (as opposed to horticulture) began about 7,000 years ago with the invention of the metal plow, pulled by domesticated animals and steered by the farmer. Animal-pulled plows allowed farmers to cultivate larger plots of land and to break up fertile grassland soils, which previously couldn't be cultivated because of their thick and widespread root systems. This greatly increased the amount of land that could be cultivated and allowed population to increase (Figure 2-2).

The gradual shift from mobile hunting and gathering societies to settled agricultural communities had several major effects:

▪ Using domesticated animals to haul loads and do other tasks increased the average energy use per person (Figure 2-1).

▪ Population increased because of a larger, more constant supply of food (Figure 2-2).

▪ People cleared increasingly larger areas of land and began to control and shape the surface of the earth to meet their needs.

▪ People began accumulating material goods. Nomadic hunter-gatherers travelled with few possessions, but farmers living in one place can accumulate as much as they can afford.

▪ Urbanization—the formation of cities—began because a small number of farmers could produce enough food to feed their families plus a surplus that could be traded to other people. Many former farmers moved into permanent villages. Some villages gradually grew into

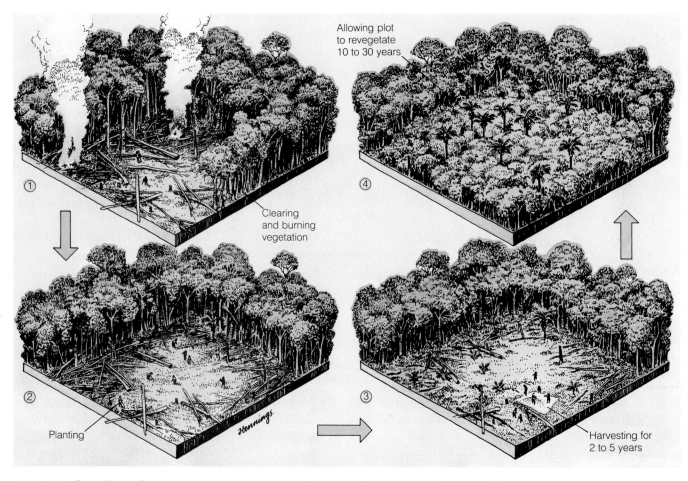

Allowing plot
to revegetate
10 to 30 years

Clearing
and burning
vegetation

Planting

Hennings.

Harvesting for
2 to 5 years

Figure 2-4 Probably the first crop-growing technique was a combination of slash-and-burn and shifting cultivation in tropical forests. This is a sustainable method if only a small portion of the forest is cleared. Soil fertility will not be restored unless each abandoned plot is left unplanted for 10 to 30 years. This form of agriculture can be sustained indefinitely if population levels are low. Today more than one out of five people are practicing this form of agriculture. Because of their numbers, they are rapidly depleting large areas of tropical forests and destroying the natural resource base that sustains them. This destruction and degradation of most of the world's remaining tropical forests within your lifetime is also being caused by clearing forests for fuelwood, timber (mostly for export to MDCs), grazing livestock, and mining. Such unsustainable clearing will eventually decrease food supplies, eliminate perhaps one-fifth of the earth's biodiversity, and contribute to projected global warming and climate change.

towns and cities, which served as centers for trade, government, and religion.

■ Specialized occupations and long-distance trade developed as former farmers in villages and towns learned crafts such as weaving, toolmaking, and pottery to produce handmade goods that could be exchanged for food.

■ Conflict increased as ownership of land and water rights became a valuable economic resource. Armies and their leaders rose to power and took over large areas of land. These rulers forced powerless people—slaves and landless peasants—to do the hard, disagreeable work of producing food and constructing irrigation systems, temples, and other projects.

The growing populations of these emerging civilizations needed more food and more wood for fuel

and buildings. To meet these needs, people cut down vast areas of forest and plowed up large areas of grasslands. Such massive land clearing destroyed and degraded the habitats of many forms of plant and animal wildlife, causing or hastening their extinction.

Poor management of many of the cleared areas led to greatly increased deforestation, soil erosion, and overgrazing of grasslands by huge herds of sheep, goats, and cattle. These unsustainable practices helped convert fertile land to desert. The topsoil that washed off these barren areas polluted streams, rivers, lakes, and irrigation canals, making them useless. The gradual degradation of the vital resource base of soil, water, forests, grazing land, and wildlife was a major factor in the downfall of many great civilizations.

For example, since 3000 B.C. much of the land that now makes up Iran and Iraq has been barren

Jack Swenson/Tom Stack & Associates

Figure 2-5 Slash-and-burn subsistence farming in a small patch of cleared tropical rain forest in Costa Rica. This family will be able to grow crops on this patch for a few years before the plant nutrients in the nutrient-poor soil are depleted. Then the family will move to another part of the forest and repeat this process. This shifting cultivation can be sustained indefinitely if population levels in the forest are low and the farmers allow each depleted patch to renew its soil fertility for several decades.

desert. But 4,000 years earlier in 7000 B.C. this land, the sites of the great Sumerian and Babylonian civilizations, was covered with productive forests and grasslands. With each generation, the elaborate network of irrigation canals that supported these civilizations became filled with more silt from deforestation, soil erosion, and overgrazing until the land became desert. These people squandered their natural capital by gradually depleting the soil that supported their civilizations. There is evidence that we may be repeating this mistake on a much larger scale (see Table 1-1, p. 10).

The gradual spread of agriculture meant that most of the earth's population shifted from hunter-gatherers *in nature* to shepherds, farmers, and urban dwellers *against nature* who viewed their role as learning to tame and control wild nature. Many analysts believe that this change in how people viewed their relationship to nature and used nature is a major cause of today's resource and environmental problems.

The Industrial Revolution The next major cultural change, known as the **Industrial Revolution,** began in England in the mid-1700s and spread to the United States in the 1800s. It changed our relationship with the earth even more than the Agricultural Revolution. It greatly increased the average amount of energy and thus other resources people in early and later more advanced industrial societies have to alter and shape the earth to meet basic needs and fuel economic growth (Figure 2-1). This led to greatly increased production, trade, and distribution of goods.

The Industrial Revolution arose as a response to absolute resource scarcity in England caused by the overuse and depletion of wood for fuel and construc-

tion. People began burning surface deposits of coal as a substitute for wood. The availability of coal led to the invention of coal-powered steam engines to pump water and perform other tasks.

People invented an increasing array of new machines powered by coal and later by oil and natural gas. Thus, the Industrial Revolution represented a shift from dependence on renewable wood and flowing water as major sources of energy to nonrenewable fossil fuels—first coal, and later oil and natural gas.

These new fuels and machines led to a shift from small-scale production of goods by hand to large-scale production of goods by machines in factories. The use of coal as an energy source meant that manufacturing plants could be located in cities rather than dispersed throughout the countryside to take advantage of wood supplies and power sources such as flowing water.

Increased agricultural production, plus the concentration of factories in cities, freed farm workers to move to the cities for work. With more income and a more reliable supply of food, the size of the human population began the sharp exponential increase we are still experiencing today (Figure 2-2 and Figure 1-2, p. 4).

After World War I (1914–18), more efficient machines and mass production techniques were developed, forming the basis of today's advanced industrial societies in the United States, Canada, Japan, and western Europe. These societies are characterized by

- greatly increased production and consumption of goods

- greatly increased dependence on nonrenewable resources such as oil, natural gas, coal, and various metals

- a shift from dependence on natural materials, which are broken down and recycled by natural processes, to synthetic materials, many of which break down slowly in the environment

- a sharp rise in the amount of energy used per person for transportation, manufacturing, agriculture, lighting, heating, and cooling (Figure 2-1)

Advanced industrial societies benefit most people living in them. These benefits include

- creation and mass production of many useful and economically affordable products

- a sharp increase in average agricultural productivity per person because of advanced industrialized agriculture, in which a small number of farmers produce large amounts of food

- a sharp rise in average life expectancy from improvements in sanitation, hygiene, nutrition, medicine, and birth control

- a gradual decline in the exponential rate of population growth because of improvements in health, birth control, education, average income, and old-age security

Along with their many benefits, advanced industrialized societies have intensified many existing resource and environmental problems and created new ones (Figure 1-11, p. 14, and Table 1-1, p. 10). The key factor responsible for rapid economic growth and today's environmental problems is the greatly increased use of cheap fossil fuels that supports industrialization and urbanization. Burning these fuels gives us enormous amounts of energy to alter the earth's surface to provide large quantities of resources for people with affluent lifestyles. Burning fossil fuels is also responsible for most of the world's air pollution and much of its water pollution.

This confronts us with a challenge. Nonrenewable fossil fuels made the Industrial Revolution possible. But they now cause so many problems and are being depleted so rapidly (especially oil) that we must learn to live differently by relying much more on perpetual and renewable energy from the sun, wind, flowing water, plants (biomass), and the earth's interior heat.

Making a New Cultural Change Most environmentalists believe that there is an urgent need to begin shifting from our present array of industrialized and partially industrialized societies to a series of *sustainable-earth societies* throughout the world within your lifetime. The 1960s and the 1970s were dress rehearsals for the real work to be done in the 1990s and beyond.

This shift would involve adopting practices from all the world's cultures, and using existing and new forms of appropriate technology to help sustain the earth's life-support systems for humans and other species. It would involve allowing parts of the world we have damaged to heal, helping restore severely damaged areas, and protecting remaining wild areas from any form of destructive development. Emphasis would be placed on making efficient use of locally available resources in sustainable ways. Existing economic and political systems would be used to reward earth-sustaining economic activities and discourage those that harm the earth.

Individuals, especially in MDCs, will have to change their lifestyles to reduce resource waste, environmental degradation, and pollution. Emphasis will have to shift from pollution control to pollution prevention, from waste disposal to waste prevention and reduction, from species protection to habitat protection, and from increased resource use to increased resource conservation.

Governments will have to cooperate to deal with a host of global and regional problems. These include global warming, ozone depletion, ocean pollution, acid deposition, rapid population growth, tropical deforestation, and the shift from an era of fossil fuels dependence to an era of energy efficiency and greatly increased dependence on perpetual and renewable energy.

On April 22, 1990, the twentieth Earth Day took place and an estimated 200 million people around the world participated—the largest global demonstration in history (see photo on p. 1). Its purpose was to get ordinary citizens, the media, elected officials, and leaders involved in changing the way we view the world and act in it.

2-2 THE THROWAWAY WORLDVIEW IN INDUSTRIAL SOCIETIES

Cultural Changes, Worldviews, and Ethics Your decisions and actions are built around your **worldview**—how you think the world works and what you think your role in the world should be—and your **ethics**—what you believe to be right or wrong behavior. Regardless of what you say you believe, how you act in the world reveals your true beliefs. According to E. F. Schumacher, "Environmental deterioration does not stem from science or technology, or from a lack of information, trained people, or money for research. It stems from the lifestyle of the modern world, which in turn arises from its basic beliefs."

Each culture has one or more worldviews based on survival experiences, religious beliefs, and the use of technology. Hunter-gatherers and people practicing subsistence agriculture live directly off the land. They have a sustainable worldview and must know how to live sustainably by not depleting the renewable capital provided by natural processes. To do otherwise is to face starvation.

Figure 2-6 Evidence of the throwaway worldview in which we believe that the earth plus our technological cleverness will produce a virtually unlimited supply of resources. If we believe there will always be more, it is not surprising that we dump, burn, or bury the solid wastes we produce, instead of seeing them as wasted resources that could be recycled, reused, or in many cases not produced in the first place. This is a sanitary landfill in Prince Georges County, Maryland.

Soil Conservation Service

When European settlers moved to North America they found a vast continent with seemingly unlimited resources. It is not surprising that they had a **frontier worldview** in which they viewed their role as expanding their use of the continent's resources. Resource conservation was not important because there was always more.

This worldview was in sharp contrast to that of many of the Native Americans whose land was taken over and whose cultures were fragmented or destroyed as European settlers spread across the continent. Although there were exceptions, most Native Americans had cultures based on a deep respect for the land and its animals. This way of viewing the earth is revealed by a Wintu Indian woman:

The white people never cared for the land or deer or bear. When the Indians kill meat, we eat it all up. When we dig roots we make little holes. When we build houses we make little holes. . . . We don't chop down trees. We only use dead wood. But the white people plow up the ground, pull down the trees, kill everything. The tree says: "Don't. I am sore. Don't hurt me." But they chop it down and cut it up. The spirit of the land hates them. . . . The white people destroy all. They blast rocks and scatter them on the ground. The rock says: "Don't. You are hurting me." But the white people pay no attention. . . . How can the spirit of the earth like the white man? . . . Everywhere the white man has touched the earth it is sore.

The Throwaway Worldview The important successes of the Industrial Revolution have given most people in MDCs the idea that there are no limits to human ingenuity, the earth's resources, and the ability of the earth's air, water, and soil to absorb our wastes.

Most people in today's industrialized societies have a **throwaway worldview** based on several beliefs:

- We are apart from nature.
- We are superior to other species.
- Our role is to conquer and subdue wild nature to further our goals by humanizing the earth's surface.
- Resources are unlimited because of our ingenuity in making them available or in finding substitutes—there is always more.
- The more we produce and consume, the better off we are.
- The most important nation is the one that can command and use the largest fraction of the world's resources.

You may not accept all these statements, but most people in today's industrialized societies act as if they did. And that's what counts.

This worldview has led to the throwaway lifestyles found in most MDCs (Figure 2-6). If there is always more, why go to the trouble and expense of picking up, recycling, or reusing what we dump into the environment? If the earth's resources, coupled with our ingenuity, are unlimited, why attempt to regulate population growth or discourage the production and consumption of anything people are willing to buy? If resources become scarce or a substitute can't be found, we can get materials from the moon and asteroids in the "new frontier" of space. If life will always get better because of our ingenuity, why should we make sacrifices now for future generations whose lives will be better anyway?

If the air, water, and soil can handle all the wastes we dump into them, why worry about pollution? Even if we do pollute an area, we can invent a technology to clean it up, move somewhere else, or live in space or on another planet. We can also use space as the ultimate waste dump. So don't worry, don't get involved, be happy. We will always be able to use technology to save us from ourselves.

Those who say there are limits to economic growth are sometimes derided as gloom-and-doom pessimists, but many analysts fear that continuing devotion to this seductive throwaway or there-will-always-be-more worldview will turn out to be a fatal attraction. They fear that our parasitic relationship to nature will squander the natural capital that keeps us and other species alive. Catholic theologian Thomas Berry calls the industrial-consumer society built upon the throwaway worldview the "supreme pathology of all history."

We can break the mountains apart; we can drain the rivers and flood the valleys. We can turn the most luxuriant forests into throwaway paper products. We can tear apart the great grass cover of the western plains and pour toxic chemicals into the soil and pesticides onto the fields until the soil is dead and blows away in the wind. We can pollute the air with acids, the rivers with sewage, the seas with oil—all this in a kind of intoxication with our power for devastation. . . . We can invent computers capable of processing ten million calculations per second. And why? To increase the volume and speed with which we move natural resources through the consumer economy to the junk pile or the waste heap. Our managerial skills are measured by our ability to accelerate this process. If in these activities the topography of the planet is damaged, if the environment is made inhospitable for a multitude of living species, then so be it. We are, supposedly, creating a technological wonderworld. . . . But our supposed progress toward an ever-improving human situation is bringing us to a wasteworld instead of a wonderworld.

2-3 A SUSTAINABLE-EARTH WORLDVIEW

Working with the Earth A small but growing number of people urge us to adopt a **sustainable-earth worldview.*** They believe that the earth does not have infinite resources and that ever-increasing production and consumption will put severe stress on the natural processes that renew and maintain the air, water, and soil and support the earth's variety of potentially renewable plant and animal life. This worldview is based on the following beliefs:

- We are part of nature (*principle of oneness*).

- We are a valuable species, but we are not superior to other species; in the words of Aldo Leopold, each of us is "to be a plain member and citizen of nature" (*principle of humility*).

- Our role is to understand and work with the rest of nature, not conquer it (*principle of cooperation*).

- Every living thing has a right to live, or at least to struggle to live, simply because it exists; this right is not dependent on its actual or potential use to us (*respect-for-nature principle*).

- Something is right when it tends to maintain the earth's life-support systems for us and all other species and wrong when it tends otherwise; the bottom line is that the earth is the bottom line (*principle of sustainability*).

- The best things in life aren't things (*principle of love, caring, and joy*).

- It is wrong for humans to cause the premature extinction of any wild species and the elimination and degradation of their habitats (*preservation of wildlife and biodiversity principle*).

- When we alter nature to meet what we consider to be basic or nonbasic needs, we should choose the method that does the least possible harm to other living things; in minimizing harm it is in general worse to harm a species than an individual organism, and still worse to harm a community of living organisms (*principle of minimum wrong*).

- Resources are limited and must not be wasted (*principle of limits*).

- No individual, corporation, or nation has a right to an ever-increasing share of the earth's finite resources. As the Indian philosopher and social activist Mahatma Gandhi said, "The earth provides enough to satisfy every person's need but not every person's greed" (*principle of enoughness*).

- We can never completely "do our own thing"; everything we do has mostly unpredictable present and future effects on other people and other species (*first law of ecology*).

- It is wrong to treat people and other living things primarily as factors of production, whose value is expressed only in economic terms (*economics-is-not-everything principle*).

- Everything we have or will have ultimately comes from the sun and the earth; the earth can get along without us, but we can't get along without the earth; an exhausted planet is an exhausted economy (*respect-your-roots or earth-first principle*).

- Don't do anything that depletes the physical, chemical, and biological capital that supports all life and human economic activities; the earth deficit is the ultimate deficit (*balanced earth budget principle*).

- We must leave the earth in as good a shape as we found it, if not better (*rights-of-the-unborn principle*).

---

* Others have used the terms *sustainable worldview* and *conserver worldview* to describe this idea. I add the word *earth* to make clear that it's all the earth's life-support systems and life, not just human beings and their societies, that must be sustained.

The essence, rhythms, and pulse of the earth within and around us can only be experienced at the deepest level by our senses and feelings—our emotions. We must tune our senses into the flow of air and water into our bodies that are absolute needs provided for us by nature at no charge.

We must listen to the soft, magnificent symphony of billions of organisms expressing their interdependency. We must pick up a handful of soil and try to sense its teeming microscopic life forms that keep us alive. We must look at a tree, a mountain, a rock, a bee and try to sense how they are a part of us and we are a part of them.

Michael J. Cohen urges each of us to recognize who we really are by saying,

I am a desire for water, air, food, love, warmth, beauty, freedom, sensations, life, community, place, and spirit in the natural world. These pulsating feelings are the Planet Earth, alive and well within me. I have two mothers: my human mother and my planet mother, Earth. The planet is my womb of life.

If we think of nature as separate from us and made up of disjointed parts to be manipulated by us, then we will tend to become people whose main motivation with regard to each other and to nature is manipulation and control. This is an unsatisfying, empty, and joyless way to live.

We need to understand that formal education is important but is not enough. Much of it is designed to socialize and homogenize us so that we will accept and participate in the worldview that our role is to conquer nature and to suppress and deny the deep feelings of guilt we have about doing thus.

The way to break out of this mental straitjacket is to experience nature directly so that you truly feel that you are part of nature and it is part of you. Find a *sense of place*—a river, mountain, or piece of the earth that you feel truly at one with. It can be a place where you live or where you occasionally visit and experience in your inner being. When you become part of a place it becomes a part of you. Then you are driven to defend it against damage and to heal its wounds.

Experiencing nature allows you to get in touch with your deepest self that has sensed from birth that when you destroy and degrade natural systems to insulate yourself from nature, you are attacking yourself. Then you will love the earth as an inseparable part of yourself and live your life in ways that sustain and replenish the earth and thus yourself and other living things.

Theologian Thomas Berry summarizes what we must do:

So long as we are under the illusion that we know best what is good for the earth and for ourselves, then we will continue our present course, with its devastating consequences on the entire earth community . . . . We need not a human answer to an earth problem, but an earth answer to an earth problem. . . . We need only listen to what the earth is telling us. . . . The time has come when we will listen, or we will die.

Thus, the most important frontier that we must explore and understand is ourselves. Discovering and understanding our true selves means experiencing and understanding the earth with our heart. This is true progress. This is living life at its fullest.

■ All people must be held responsible for their own pollution and environmental degradation *(responsibility-of-the-born principle)*.

■ We must protect the earth's remaining wild systems from our activities, rehabilitate or restore natural systems we have degraded, use natural systems only on a sustainable basis, and allow many of the systems we have occupied and abused to return to a wild state *(principle of earth protection and healing)*.

■ In protecting and sustaining nature, go further than the law requires *(ethics-often-exceeds-legality principle)*.

■ To prevent excessive deaths of people and other species, people must prevent excessive births *(birth-control-is-better-than-death-control principle)*.

■ To love, cherish, celebrate, and understand the earth and yourself, take time to experience and sense the air, water, soil, trees, animals, bacteria, and other parts and rhythms of the earth directly *(experience-is-the-best-teacher principle)*.

■ Learn about and love your local environment and live gently within that place; walk lightly on the earth *(love your neighborhood principle)*.

Levels of Environmental Awareness Achieving such a sustainable-earth worldview involves working our way through four levels of environmental awareness summarized inside the front cover, opposite the title page.

Sustaining life on earth requires not only a new way of thinking, but also a new way of feeling. We must work with our minds and our hearts (see Spotlight above).

Everyone wants to breathe clean air and drink uncontaminated water. But how much are you willing to pay in taxes and increased costs of some items to protect the environment and conserve resources? What changes in your lifestyle are you willing to make?

These are crucial questions. Why? Because each of us leading resource-intensive, industrialized lifestyles must recognize the truth in Pogo's statement, "We have met the enemy, and it is us," and begin to change our lifestyles accordingly.

Recent measurements have shown that the air inside most of today's houses, offices, and stores—where people spend most of their time—is more polluted and hazardous than the outside air. Would you be willing to support laws requiring annual testing of homes and buildings for indoor air pollution and being forced to meet certain indoor air pollution standards?

Would you favor much stricter twice-a-year inspections of air pollution control equipment on all motor vehicles and tough fines for not keeping these systems in good working order? In the United States almost 60% of such equipment on cars and trucks has either been dismantled or is not working properly.

Would you support requiring all cars, vans, and light trucks to get at least 21 kilometers per liter (50 miles per gallon) of gasoline

within the next ten years? This would probably mean the end of large luxury cars, fast cars, and perhaps vans.

Would you favor $2 a gallon tax on gasoline and heating oil to help reduce consumption, extend oil supplies, delay global warming, reduce air pollution, and stimulate improvements in energy efficiency and the search for less harmful energy sources for vehicles and furnaces? Such taxes are why gasoline in Tokyo costs $4.16 a gallon and sells for $3.00 to $3.89 a gallon in most western European countries. Would you support requiring all existing and new buildings to meet strict energy conservation standards?

Would you be willing to set your thermostat at 18°C (65°F) during the winter months to decrease the amount of carbon dioxide and other pollutants entering the atmosphere, prolong fossil fuel supplies, and save money? During hot weather would you be willing to set your thermostat at 26°C (78°F)? Would you be willing to do without air conditioning in cars and buildings if no acceptable substitutes are found for the chlorofluorocarbons that are depleting the ozone in the upper atmosphere and causing one-fourth of the projected global warming?

Would you support sharp increases in your monthly water bills to discourage water waste? Would you favor requiring expensive annual testing for all homes getting their water from under-

ground wells and installation of purification systems when such water is found to be contaminated?

Would you support laws that require you to separate your trash into paper, bottles, and cans for recycling and food wastes for composting? Would you support laws that ban throwaway bottles and cans and require that all beverage and food containers be reusable?

You want proper management of radioactive nuclear waste and industrial hazardous waste produced in supplying the electricity, products, and services you use. But most people want them placed in someone else's backyard, in another community, in another state, in another country, or perhaps in space. Would you be willing to take responsibility for some of these wastes you help create by having a nuclear waste dump, a hazardous waste incinerator, a landfill, or a recycling plant near where you live? If everyone had to accept more responsibility for the wastes they produce, pollution prevention, recycling, reuse, and waste reduction would become the planet's major forms of profitable businesses.

Would you support candidates for political office who promised to do these things? These are only a few of the tough and controversial issues we face if we want to sustain the earth.

Becoming Earth Citizens Our species is now at its most important turning point since the Agricultural Revolution. Sustaining the earth means that each of us, especially if we have an industrialized lifestyle, must change our individual consumption habits and lifestyles to reduce our environmental impact (see Individuals Matter above). This means adopting an earth-caring lifestyle. Doing this requires maintaining a positive attitude. Helen and Scott Nearing offer the following ten tips for doing this:

1. Do the best you can, whatever arises.

2. Be at peace with yourself.

3. Find a job you enjoy.

4. Simplify your life. Live in simple conditions: housing, food, clothing.

5. Contact nature every day. Feel the earth under your feet.

6. Exercise physically through hard work, gardening, or walking.

Figure 2-7 A growing number of people are pledging allegiance to the planet that keeps them alive. Some display the Earth Flag as a symbol of their commitment to sustaining the earth. (Courtesy of Earth Flag Co., 33 Roberts Road, Cambridge, MA 02138)

7. Don't worry. Live one day at a time.

8. Share something every day with someone.

9. Take time to wonder at life and the world. See some humor in life where you can.

10. Be kind to all creatures, and observe the one life in all things.

Many people now see themselves as members of a global community with ultimate loyalty to the planet, not merely a particular country (Figure 2-7). These earth citizens urge individual citizens to think globally and act locally to sustain the earth (see Case Study on p. 32). They are the earth's true heroes. They are guided by historian Arnold Toynbee's observation: "If you make the world ever so little better, you will have done splendidly, and your life will have been worthwhile," and by George Bernard Shaw's reminder that "indifference is the essence of inhumanity."

Can we do it? Yes. If we care enough to make the necessary commitment and in the process discover that caring for the earth is a never-ending source of joy and inner peace. We must become *earth conservers*, not *earth degraders*. We have an opportunity to participate in a sustainable-earth revolution that makes every day an Earth Day. No goal is more important, more urgent, and more worthy of our time, energy, creativity, and money.

The main ingredients of an environmental ethic are caring about the planet and all of its inhabitants, allowing unselfishness to control the immediate self-interest that harms others, and living each day so as to leave the lightest possible footprints on the planet.

ROBERT CAHN

In the late fifteenth century, Jambeshwar, the son of the leader of a village in northern India, renounced his inheritance and set out to teach people to care for their health and the environment. He developed 29 principles for living and founded a Hindu sect, called the Bishnois, based on a religious duty to protect trees and wild animals.

In 1730 when the Maharajah of Jodpur in northern India ordered that the few trees left in the area be cut down the Bishnois forbade it. Women rushed in and hugged the trees to protect them, but the Maharajah's minister ordered the work to proceed anyway. According to legend, 363 Bishnois women died on that day.

In the 1960s people in the villages located in the Himalayan mountain ranges of northern India faced disaster from severe deforestation, torrential floods, and landslides. Over two decades the slopes of many of the mountains had been cleared of most trees.

This deforestation occurred because of intensive commercial logging combined with the needs of a growing population for firewood and land for cultivation. Men were being forced to leave the village to find work in other parts of the country. Women were having to travel further every day in search of firewood.

In 1973 some women in the Himalayan village of Gopeshwar started the modern Chipko (an Indian word for "hug" or "cling to") movement to protect the remaining trees in a nearby forest from being cut down to make tennis rackets for export. It began when Chandi Prasad Bhatt, a village leader, urged villagers to "hold fast" or "chipko" to protect the trees from the loggers, thus carrying on the tradition started several hundred years earlier by the Bishnois (Figure 2-8).

Women and children went into the forests, flung their arms around trees and sat around them to prevent commercial loggers from cutting them down. This successful movement spread rapidly as their action inspired the women of other Himalayan villages to protect their forests. As a result, the commercial cutting of timber in the hills of the Indian state of Uttar Pradesh has been banned.

Since 1973 the Chipko movement has widened its efforts from embracing trees to embracing mountains and water threatened with destructive forms of development. Today, the women, who still guard trees from loggers, also plant trees, prepare village forestry plans, and build walls to stop soil erosion. This inspired Ghanshyam "Shaliana," a Chipko poet, to write: *Embrace the life of living trees and streams to your hearts. Resist the digging of mountains which kill our forests and streams.*

The spreading actions of such earth citizens should inspire us to help protect the earth. Sustaining the earth will come mostly from the bottom up, not from the top down. It will happen because of the daily actions of ordinary people who collectively force leaders to get on the bandwagon or lose their power.

Figure 2-8   Leader of the Chipko movement protecting a tree in northern India from being cut down for export to industrialized countries. In many cases the efforts of these earth heroes to help sustain the earth have been successful.

Robert Hutchison

DISCUSSION TOPICS

1. Do you think we would be better off if agriculture had never been discovered and we were still hunter-gatherers today? Explain.

2. Make a list of the major benefits and drawbacks of an advanced industrial society such as the United States. Do you feel that the benefits outweigh the drawbacks? Explain. What are the alternatives?

3. Governments are in the process of virtually eliminating the world's remaining hunter-gatherers and other indigenous peoples, who have lived gently on the land for centuries and who can teach us much about how to live sustainably. This cultural extinction is being done in the name of "progress" by taking over their lands for crop growing, timber cutting, cattle grazing, mining, building hydroelectric dams and reservoirs, and creating other forms of economic development. Some believe that the world's remaining indigenous people should be given title to the land they and their ancestors have lived on for centuries, a decisive voice in formulating policies about resource development in their areas, and the right to be left alone by modern civilization. We have created protected reserves for endangered wild species, so why not create reserves for these endangered human cultures? What do you think? Explain.

4. What is your worldview?

5. Do you agree with the principles and guidelines of the sustainable-earth worldview given in Section 2–3? Explain. Can you add others? Which ones do you try to follow?

FURTHER READINGS

Berry, Thomas. 1988. *The Dream of the Earth.* San Francisco: Sierra Club Books.

Berry, Wendell. 1990. *What Are People For?* Berkeley, Calif.: North Point Press.

Botkin, Daniel. 1990. *Discordant Harmonies: A New Ecology for the Twenty-first Century.* New York: Oxford University Press.

Bronowski, Jacob. 1974. *The Ascent of Man.* Boston: Little, Brown.

Cahn, Robert. 1978. *Footprints on the Planet: A Search for an Environmental Ethic.* New York: Universe Books.

Caplan, Ruth. 1990. *Our Earth, Ourselves.* Boston: Little, Brown.

Cohen, Michael J. 1988. *How Nature Works: Regenerating Kinship with Planet Earth.* Waldpole, N.H.: Stillpoint.

Devall, Bill, and George Sessions. 1985. *Deep Ecology: Living As If Nature Mattered.* Salt Lake City, Utah: Gibbs M. Smith.

Earth Works Group. 1990. *50 Simple Things You Can Do to Save the Earth.* Berkeley, Calif.: Earth Works Press.

Fritsch, Albert J. 1980. *Environmental Ethics: Choices for Concerned Citizens.* New York: Anchor Books.

Hardin, Garrett. 1978. *Exploring New Ethics for Survival.* 2nd ed. New York: Viking.

Hargrove, Eugene C. 1989. *Foundations of Environmental Ethics.* Englewood Cliffs, N.J.: Prentice-Hall.

Hollander, Jeffrey. 1990. *How to Make the World a Better Place.* New York: Quill.

Hynes, H. Patricia. 1990. *EarthRight: Every Citizen's Guide.* New York: Prima (St. Martin's Press).

Johnson, Warren. 1978. *Muddling Toward Frugality.* San Francisco: Sierra Club Books.

Leopold, Aldo. 1949. *A Sand County Almanac.* New York: Oxford University Press.

MacEachern, Diane. 1990. *Save Our Planet: 750 Everyday Ways You Can Help Clean Up the Earth.* New York: Dell.

McCormick, John. 1989. *Reclaiming Paradise: The Global Environmental Movement.* Bloomington: Indiana University Press.

Naar, Jon. 1990. *Design for a Liveable Planet.* New York: Harper & Row.

Naess, Arne. 1989. *Ecology, Community, and Lifestyle.* New York: Cambridge University Press.

Nash, Roderick. 1988. *The Rights of Nature: A History of Environmental Ethics.* Madison: University of Wisconsin Press.

Nearing, Helen, and Scott Nearing. 1970. *Living the Good Life.* New York: Schocken.

Rifkin, Jeremy, ed. 1990. *The Green Lifestyle Handbook: 1001 Ways You Can Heal the Earth.* New York: Henry Holt.

Rolston, Holmes, III. 1988. *Environmental Ethics: Duties to and Values in the Natural World.* Philadelphia: Temple University Press.

Roszak, Theodore. 1978. *Person/Planet.* Garden City, N.Y.: Doubleday.

Sale, Kirkpatrick. 1990. *Conquest of Paradise.* New York: Knopf.

Sargoff, Mark. 1988. *The Economy of the Earth.* New York: Cambridge University Press.

Schumacher, E. F. 1973. *Small Is Beautiful: Economics As If People Mattered.* New York: Harper & Row.

Simmons, I. G. 1989. *Changing the Face of the Earth: Culture, Environment, and History.* London: Basil Blackwell.

Sombke, Laurence. 1990. *The Solution to Pollution: 101 Things You Can Do to Clean Up Your Environment.* New York: MasterMedia.

Steager, Will, and John Bowermaster. 1990. *Saving the Earth.* New York: Knopf.

Taylor, Paul W. 1986. *Respect For Nature: A Theory of Environmental Ethics.* Lawrenceville, N.J.: Princeton University Press.

World Resources Institute. 1989. *The Crucial Decade: The 1990s and the Global Environmental Challenge.* Washington, D.C.: World Resources Institute.

Los Angeles, California, and other cities survive by using resources from around the world, producing enormous amounts of wastes.

Animal and vegetable life is too complicated a problem for human intelligence to solve, and we can never know how wide a circle of disturbance we produce in the harmonies of nature when we throw the smallest pebble into the ocean of organic life.

GEORGE PERKINS MARSH

CHAPTER THREE

MATTER AND ENERGY RESOURCES: TYPES AND CONCEPTS

General Questions and Issues

1. What are the major forms of matter? What is matter made of? What makes matter useful to us as a resource?

2. What are the major forms of energy? What major energy resources do we rely on? What makes energy useful to us as a resource?

3. What are physical and chemical changes, and what scientific law governs changes of matter from one physical or chemical form to another?

4. What are the three major types of nuclear changes that matter can undergo?

5. What two scientific laws govern changes of energy from one form to another?

6. How can we waste less energy, and how much useful energy is available from different energy resources?

7. How are the scientific laws governing changes of matter and energy from one form to another related to resource use and environmental disruption?

The laws of thermodynamics control the rise and fall of political systems, the freedom or bondage of nations, the movements of commerce and industry, the origins of wealth and poverty, and the general physical welfare of the human race.

FREDERICK SODDY (*Nobel Laureate, Chemistry*)

THIS BOOK, YOUR HAND, the water you drink, and the air you breathe are all samples of *matter*—the stuff all things are made of. The light and heat streaming from a burning lump of coal are examples of *energy*. It is what you and all living things use to move matter around, change its form, or cause a heat transfer between two objects at different temperatures. Cities are the ultimate expression of the use of the earth's matter and energy resources (see photo on p. 35).

This chapter is a brief introduction to what is going on in the world from a physical and chemical standpoint. It describes the major types of matter and energy and the scientific laws governing all changes of matter and energy from one form to another. The next two chapters are an introduction to what is going on in the world from an ecological standpoint, based on how key physical and chemical processes are integrated into the biological systems we call life.

3-1 MATTER: FORMS, STRUCTURE, AND QUALITY

Physical and Chemical Forms of Matter Anything that has mass (or weight on the earth's surface) and takes up space is **matter. Mass** is the amount of material in an object. All matter found in nature can be viewed as being organized in identifiable patterns, or *levels of organization*, according to size and function (Figure 3-1).

This section is devoted to a discussion of the three lowest levels of organization of matter—subatomic particles, atoms, and molecules—which make up the basic components of all higher levels. Chapter 4 discusses the five higher levels of organization of matter—organisms, populations, communities, ecosystems, and the ecosphere—the major concerns of ecology.

Any matter, such as water, can be found in three *physical forms:* solid (ice), liquid (liquid water), and gas (water vapor). All matter also consists of *chemical forms:* elements, compounds, or mixtures of elements and compounds.

Elements The 92 **elements** that occur naturally on earth are distinctive forms of matter that make up every material substance. Another 15 elements have been artificially synthesized in laboratories. Examples of these basic building blocks of all matter include hydrogen (represented by the symbol H), carbon (C), oxygen (O), nitrogen (N), phosphorus (P), sulfur (S), chlorine (Cl), fluorine (F), bromine (Br), sodium (Na), calcium (Ca), and uranium (U).

A sample of any element is composed of an incredibly large number of particular types of minute particles called **atoms.** Two or more atoms of the same or different elements can combine to form **molecules.** Some elements are found in nature as

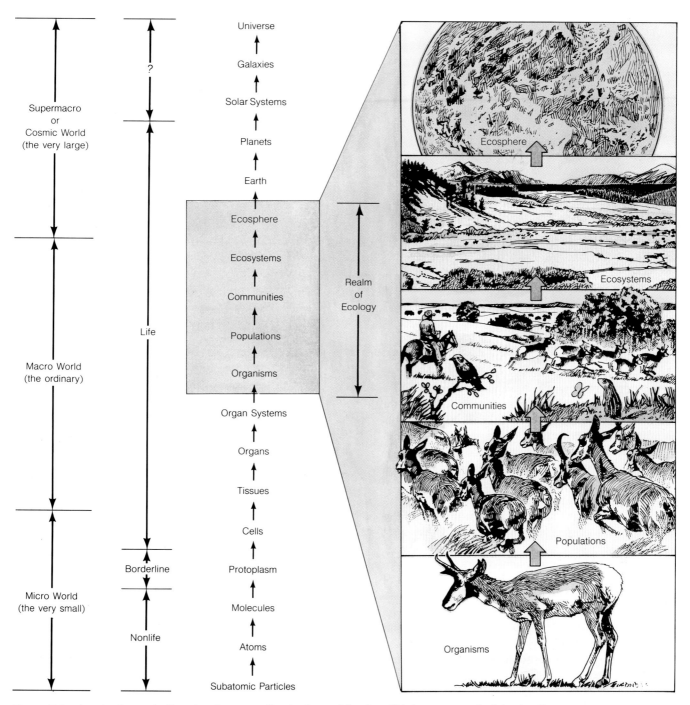

Figure 3-1 Levels of organization of matter, according to size and function. This is one way scientists classify patterns of matter found in nature.

molecules. Examples are the nitrogen and oxygen gases, making up about 99% of the volume of air we breathe. Two atoms of nitrogen (N) combine to form a nitrogen gas molecule with the shorthand formula N_2 (read as "N-two"). The subscript after the symbol of the element gives the number of atoms of that element in a molecule. Similarly, most of the oxygen gas in the atmosphere exists as O_2 (read as "O-two") molecules. A small amount of oxygen, found mostly in the upper atmosphere (stratosphere), exists as

ozone molecules with the formula O_3 (read as "O-three").

All atoms are made up of even smaller **subatomic particles:** protons, neutrons, and electrons. Each atom of an element has a tiny center, or **nucleus,** which contains almost all of an atom's mass. Atoms of a particular element have a unique number of **protons** (represented by the symbol p), each with a positive (+) electrical charge, in their nuclei. The nuclei of atoms of an element also contain uncharged **neutrons**

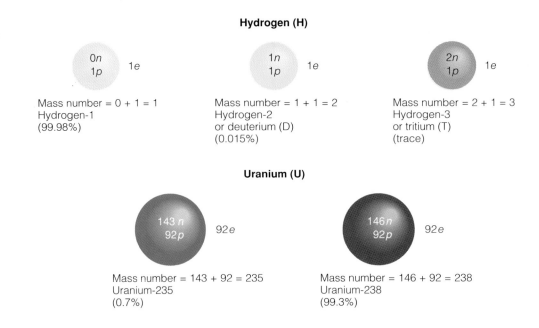

Figure 3-2 Isotopes of hydrogen and uranium. Figures in parentheses show the percent abundance by weight of each isotope in a natural sample of each element.

Hydrogen (H)

$0n$
$1p$ $1e$

Mass number = 0 + 1 = 1
Hydrogen-1
(99.98%)

$1n$
$1p$ $1e$

Mass number = 1 + 1 = 2
Hydrogen-2
or deuterium (D)
(0.015%)

$2n$
$1p$ $1e$

Mass number = 2 + 1 = 3
Hydrogen-3
or tritium (T)
(trace)

Uranium (U)

$143 n$
$92 p$ $92 e$

Mass number = 143 + 92 = 235
Uranium-235
(0.7%)

$146 n$
$92 p$ $92 e$

Mass number = 146 + 92 = 238
Uranium-238
(99.3%)

(n), varying in number from 0 (hydrogen-1 in Figure 3-2) to well over 100. Because an electron has an almost negligible mass (or weight) compared to a proton or a neutron, the approximate mass of an atom is determined by the number of neutrons plus the number of protons in its nucleus. This number is called its **mass number.**

All atoms of an element must have the same number of protons and electrons. However, they may have different numbers of uncharged neutrons in their nuclei, and thus different mass numbers. These different forms of an element are called **isotopes** of that element.

Isotopes of the same element are identified by attaching the mass number to the name or symbol of the element: hydrogen-1, or H-1; hydrogen-2, or H-2 (common name, deuterium); and hydrogen-3, or H-3 (common name, tritium). A natural sample of an element contains a mixture of its isotopes in a fixed proportion or percent abundance by weight (Figure 3-2).

One or more **electrons** (e), each with a negative (−) electrical charge, whiz around somewhere outside each nucleus. Each atom of any element always has an equal number of positively charged protons inside its nucleus and negatively charged electrons outside its nucleus. For example, each atom of the lightest element, hydrogen, has one positively charged proton in its nucleus and one negatively charged electron outside. Each atom of uranium, a much heavier element, has 92 protons and 92 electrons (Figure 3-2). Due to this electrical balance, each atom as a whole has no net electrical charge.

Atoms of some elements can lose or gain one or more electrons to form **ions:** atoms or groups of atoms with one or more net positive (+) or negative (−) electrical charges. The number of positive or negative charges is shown as a superscript after the symbol for an atom or group of atoms. Examples of positive ions are sodium ions (Na^+), calcium ions (Ca^{2+}), and ammonium ions (NH_4^+). Common negative ions are chloride ions (Cl^-), nitrate ions (NO_3^-), and phosphate ions (PO_4^{3-}).

Compounds Most matter exists as **compounds**—combinations of atoms, or oppositely charged ions, of two or more different elements held together in fixed proportions by attractive forces called chemical bonds. Water, for example, is made up of H_2O (read as "H-two-O") molecules, each consisting of two hydrogen atoms chemically bonded to an oxygen atom. Sodium chloride, or table salt, consists of a network of oppositely charged ions (Na^+ and Cl^-) held together by the forces of attraction that exist between opposite electric charges.

Table sugar, vitamins, oil, natural gas, plastics, detergents, aspirin, penicillin, and many other materials important to you and your lifestyle have one thing in common. They all are **organic compounds,** containing atoms of the element carbon, usually combined with each other and with atoms of one or more other elements such as hydrogen, oxygen, nitrogen, sulfur, phosphorus, chlorine, bromine, and fluorine.

The following are examples of the more than 7 million known organic compounds:

- *Hydrocarbons*—compounds of carbon and hydrogen atoms. An example is methane (CH_4), the major component of natural gas.

- *Chlorinated hydrocarbons*—compounds of carbon, hydrogen, and chlorine atoms. Examples are DDT ($C_{14}H_9Cl_5$), an insecticide, and toxic PCBs (such as $C_{12}H_5Cl_5$), used as insulating materials in electric transformers.

- *Chlorofluorocarbons (CFCs)*—compounds of carbon, chlorine, and fluorine atoms. An example is Freon-12 (CCl_2F_2), used as a coolant in refrigerators and air conditioners, aerosol propellants, and foaming agents for making plastics.

- *Simple sugars*—certain types of compounds of carbon, hydrogen, and oxygen atoms. An example is glucose ($C_6H_{12}O_6$), which most plants and animals break down in their cells to obtain energy.

All other compounds are **inorganic compounds.** Some inorganic compounds you will encounter in this book are sodium chloride (NaCl), water (H_2O), nitric oxide (NO), carbon monoxide (CO), carbon dioxide (CO_2), nitrogen dioxide (NO_2), sulfur dioxide (SO_2), ammonia (NH_3), sulfuric acid (H_2SO_4), and nitric acid (HNO_3).

Matter Quality **Matter quality** is a measure of how useful a matter resource is, based on its availability and concentration (Figure 3-3). **High-quality matter** is organized, concentrated, and usually found near the earth's surface. It has great potential for use as a matter resource. **Low-quality matter** is disorganized, dilute, or dispersed, and often found deep underground. It usually has little potential for use as a matter resource.

An aluminum can is a more concentrated, higher-quality form of aluminum than aluminum ore with the same amount of aluminum. That is why it usually takes less energy, water, and money to recycle an aluminum can than to get new aluminum from ore.

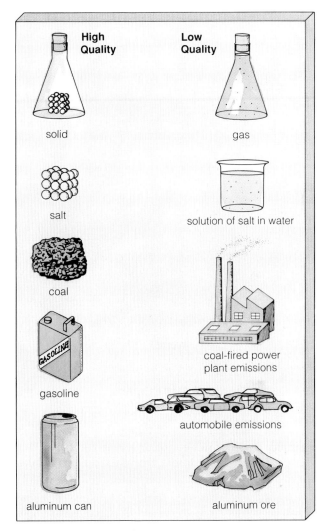

Figure 3-3 Examples of differences in matter quality. High-quality matter is fairly easy to get and is concentrated. Low-quality matter is hard to get and is dispersed.

3-2 ENERGY: TYPES, FORMS, AND QUALITY

Types of Energy Energy, not money, is the real "currency" of the world. We depend on it to grow our food, keep us and other organisms alive, and to warm and cool our bodies and the buildings where we work and live. We also use it to move people and other forms of matter from one place to another and to change matter from one physical or chemical form to another.

Energy is the ability to do work or to cause a heat transfer between two objects at different temperatures. **Work** is what happens when a force is used to push, pull, pump, or lift matter, such as this book, over some distance. Everything going on in and around us is based on work in which one form of energy is transformed into one or more other forms of energy.

Scientists classify energy as either potential or kinetic. **Kinetic energy** is the energy that matter has because of its motion and mass. Examples include a moving car, a falling rock, a speeding bullet, and the flow of water or charged particles (electrical energy). **Potential energy** is stored energy that is potentially available for use. A rock held in your hand, a stick of dynamite, still water behind a dam, and nuclear energy stored in the nuclei of atoms all have potential energy. Other examples are the chemical energy stored in molecules of gasoline and in the carbohydrates, proteins, and fats of the food you eat.

Heat refers to the total kinetic energy of all the moving atoms or molecules in a given substance. **Temperature** is a measure of the average speed of motion of the atoms or molecules in a substance at a given moment. A substance can have a high heat content (much mass and many moving atoms or molecules) but a low temperature (low average molecular speed). For example, the total heat content of a lake or an ocean is enormous, but its average temperature is low. Other samples of matter can have a low heat content and a high temperature. For

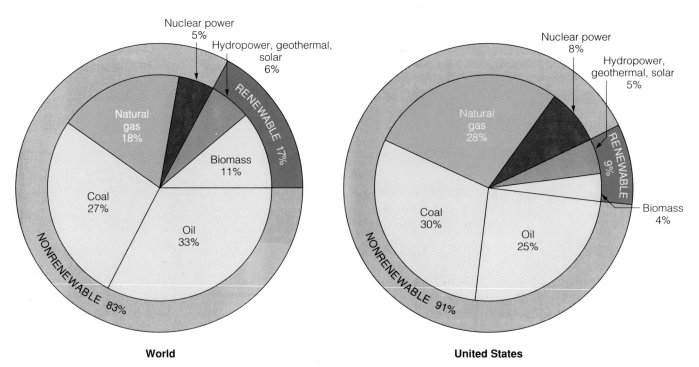

World **United States**

Figure 3-4 Commercial energy use by source in 1988 in the world (left) and in the United States (right). This amounts to only 1% of the energy used in the world. The other 99% of the energy used to heat the earth comes from the sun and is not sold in the marketplace. (Data from U.S. Department of Energy, British Petroleum, and Worldwatch Institute)

example, a cup of hot coffee or a burning match has a much lower heat content than an ocean or a lake but has a much higher temperature.

Energy Resources Used by People *The direct input of essentially inexhaustible solar energy alone supplies 99% of the energy used to heat the earth and all buildings.* Were it not for this direct input of energy from the sun, the average temperature would be −240°C (−400°F), and life as we know it would not have arisen. This input of solar energy also helps recycle the carbon, oxygen, water, and other chemicals we and other organisms need to stay alive and healthy and to reproduce.

Broadly defined, **solar energy** includes perpetual *direct* energy from the sun and a number of *indirect* forms of energy produced by the direct input. Major indirect forms of solar energy include wind, falling and flowing water (hydropower), and biomass (solar energy converted to chemical energy in trees and other plants).

We have learned how to capture and use some of these direct and indirect forms of solar energy. *Passive* solar energy systems capture and store direct solar energy and use it to heat buildings and water without the use of mechanical devices. Examples are a well-insulated, airtight house with large insulating windows that face the sun and the use of rock, concrete, or water to store and release heat slowly.

Direct solar energy can also be captured by *active*

solar energy systems. For example, specially designed roof-mounted collectors concentrate direct solar energy; pumps transfer this heat to water, to the interior of a building, or to insulated storage tanks of stone or water. We have also learned how to make solar cells that convert solar energy directly to electricity in one simple, nonpolluting step. We have also developed wind turbines and hydroelectric power plants to convert indirect solar energy, in the form of wind and falling or flowing water, into electricity.

The sun's input of 99% of the energy used to heat the earth and make it livable is not sold in the marketplace. The remaining 1% of the energy we use on earth to supplement the massive solar input is commercial energy sold in the marketplace (Figure 3-4).

At one extreme, the United States, with 4.8% of the world's population, uses 25% of the world's commercial energy. At the other extreme, India, with about 16% of the world's people, uses only about 1.5% of the world's commercial energy. In 1990, the 251 million Americans used more energy for air conditioning alone than 1.1 billion Chinese used for all purposes.

The most important supplemental source of energy for LDCs is potentially renewable biomass—especially fuelwood—the main source of energy for heating and cooking for roughly half the world's population. One-fourth of the world's population in MDCs may soon face shortages of oil, but half the

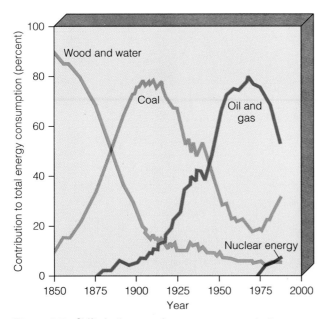

Figure 3-5 Shifts in the use of energy resources in the United States since 1850. Shifts from wood to coal and then from coal to oil and natural gas have each taken about 50 years. Because affordable oil is running out and the burning of fossil fuels is the major cause of projected global warming within your lifetime, most analysts believe we must make a new shift in energy resources over the next 50 years. Some believe we should use more nuclear energy, convert coal to a gas for use as a fuel, and install pollution control devices to sharply reduce air pollution emissions from burning coal and other remaining fossil fuels. Others believe we should depend more on renewable energy from the sun, wind, flowing water, and plant matter (biomass) and greatly reduce our present unnecessary waste of large amounts of energy. (Data from U.S. Department of Energy)

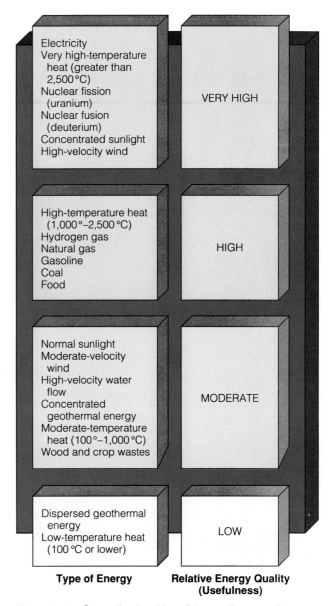

Figure 3-6 Generalized ranking of the quality or usefulness of different types of energy.

world's population in LDCs already face a fuelwood shortage.

In 1850 the United States and most other MDCs had a decentralized energy system based on locally available renewable resources, primarily wood. Today they have a centralized energy system based on nonrenewable fossil fuels (Figure 3-5), increasingly produced in one part of the world and transported to and used in another part. By 1988 about 83% of the commercial energy used in the United States was provided by burning oil, coal, and natural gas. This shift has fueled the rapid economic development of the MDCs, especially since the 1950s, but it has also caused most MDCs to become highly dependent on oil—a nonrenewable resource that is being rapidly depleted (see Case Study on p. 8).

To bring the energy use of LDCs up to levels currently used by MDCs by the year 2025 would require a fivefold increase in present global commercial energy use. Unless future energy use by all nations is based on increasing energy efficiency and greatly increased use of perpetual and renewable forms of energy, even a doubling of energy use based mostly on nonrenewable fossil fuels will seriously disrupt the earth's already damaged life-sustaining processes.

Energy Quality Energy varies in its quality, or ability to do useful work. **Energy quality** is a measure of energy usefulness (Figure 3-6). **High-quality energy** is organized or concentrated and has great ability to perform useful work. Examples of these useful forms of energy are electricity, coal, gasoline, concentrated sunlight, nuclei of uranium-235, and high-temperature heat.

By contrast, **low-quality energy** is disorganized or dilute and has little ability to do useful work. An example is the low-temperature heat in the air around you or in a river, lake, or ocean. For instance, the

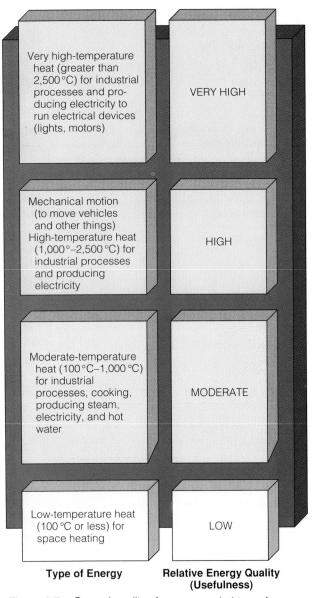

Type of Energy	Relative Energy Quality (Usefulness)
Very high-temperature heat (greater than 2,500 °C) for industrial processes and producing electricity to run electrical devices (lights, motors)	VERY HIGH
Mechanical motion (to move vehicles and other things) High-temperature heat (1,000°–2,500 °C) for industrial processes and producing electricity	HIGH
Moderate-temperature heat (100 °C–1,000 °C) for industrial processes, cooking, producing steam, electricity, and hot water	MODERATE
Low-temperature heat (100 °C or less) for space heating	LOW

Figure 3-7 General quality of energy needed to perform various energy tasks. To avoid unnecessary energy waste, it is best to match the quality of an energy source (Figure 3-6) to the quality of energy needed to perform a task—that is, not to use energy of a higher quality than necessary. This saves energy and usually saves money.

total amount of low-temperature heat stored in the Atlantic Ocean is greater than the amount of high-quality chemical energy stored in all the oil deposits in Saudi Arabia, but heat is so widely dispersed in the ocean that we can't do much with it. This dispersed heat, like that in the air around us, can't be used to move things and to heat things to high temperatures.

We use energy to accomplish certain tasks, each requiring a certain minimum energy quality (Figure 3-7). Electrical energy, which is very high-quality energy, is needed to run lights, electric motors, and electronic devices. We need high-quality mechanical energy to move a car. But we need only low-temperature heat (less than 100°C) to heat homes and other buildings. It makes sense to match the quality of an energy source (Figure 3-6) to the quality of energy needed to perform a particular task (Figure 3-7). This saves energy and usually saves money (see Guest Essay on p. 56).

Unfortunately, many forms of high-quality energy, such as high-temperature heat, electricity, gasoline, hydrogen gas (a useful fuel that can be produced by passing electricity through water), and concentrated sunlight, do not occur naturally. We must use other forms of high-quality energy such as fossil, wood, and nuclear fuels to produce, concentrate, and store them, or to upgrade their quality so they can be used to perform certain tasks.

3-3 PHYSICAL AND CHEMICAL CHANGES AND THE LAW OF CONSERVATION OF MATTER

Physical and Chemical Changes Elements and compounds can undergo physical and chemical changes; each change either gives off or requires energy, usually in the form of heat. A **physical change** is one that involves no change in chemical composition. For example, cutting a piece of aluminum foil into small pieces is a physical change. Each cut piece is still aluminum.

Changing a substance from one physical state to another is also a physical change. For example, when solid water, or ice, is melted or liquid water is boiled, none of the H_2O molecules involved are altered; instead they are organized in different spatial patterns.

In a **chemical change,** or **chemical reaction,** there is a change in the chemical composition of the elements or compounds involved. For example, when coal (which is mostly carbon, or C) burns completely, it combines with oxygen gas (O_2) from the atmosphere to form the gaseous compound carbon dioxide (CO_2). In this case energy is given off, making coal a useful fuel (C + O_2 → CO_2 + energy).

This reaction shows how the burning of coal or any carbon-containing compounds, such as those in wood, natural gas, oil, and gasoline, adds carbon dioxide gas to the atmosphere. The higher the concentration of carbon dioxide in the atmosphere, the warmer the average temperature near the earth's surface.

The Law of Conservation of Matter: There Is No Away The earth loses some gaseous molecules to space and it gains small amounts of matter from space, mostly in the form of stony or metallic bodies (meteorites). However, these losses and gains of matter are minute compared to the earth's total mass—somewhat like the world's beaches losing or adding a few grains of sand.

This means that *the earth has essentially all the matter it will ever have.* Fortunately, over billions of years natural processes have evolved for continuously recycling key chemicals back and forth between the nonliving environment (soil, air, water) and the living environment (plants, animals, decomposers).

You, like most people, probably talk about consuming or using up material resources, but the truth is that we don't consume any matter. We only use some of the earth's resources for a while. We take materials from the earth, carry them to another part of the globe, and process them into products. These products are used and then discarded, reused, or recycled.

In making and using products, we may change various elements and compounds from one physical or chemical form to another, but we neither create from nothing nor destroy to nothingness any measurable amount of matter. This fact, based on many thousands of measurements of matter undergoing physical and chemical changes, is known as the **law of conservation of matter.** In other words, *in all physical and chemical changes we can't create or destroy any of the atoms involved. All we can do is rearrange them, into different spatial patterns (physical changes) or different combinations (chemical changes).*

The law of conservation of matter means that there is no "away." *Everything we think we have thrown away is still here with us in one form or another.* We can collect dust and soot from the smokestacks of industrial plants, but these solid wastes must then go somewhere. We can collect garbage and remove solid grease and sludge from sewage, but these substances must be burned (perhaps causing air pollution), dumped into rivers, lakes, and oceans (perhaps causing water pollution), deposited on the land (perhaps causing soil and groundwater pollution), recycled, or reused.

Nobody wants a waste dump or incinerator near them. At the local level citizens who oppose dumps and incinerators in their communities are called NIMBYs—"Not In My Backyard"—by the industries they oppose. But the law of conservation of matter and the fact that hazardous chemicals we place in the ground or air don't stay put mean that we need to redefine our backyard. It extends beyond the boundaries of a particular piece of property, city, state, or country. It is everywhere. Once we understand this, we will shift from waste production and control to waste prevention and recycling. Then when it comes to hazardous substances, we will become NOPEs— "Nowhere on Planet Earth."

We can make the environment cleaner and convert some potentially harmful chemicals to less harmful or even harmless physical or chemical forms. But the law of conservation of matter means that we will always be faced with the problem of what to do with some quantity of wastes. However, by placing much greater emphasis on pollution prevention (input control) we can greatly reduce the amount of wastes we add to the environment (see Spotlight on p. 45).

NUCLEAR CHANGES

Natural Radioactivity In addition to physical and chemical changes, matter can undergo a third type of change known as a **nuclear change.** It occurs when nuclei of certain isotopes spontaneously change or are forced to change into one or more different isotopes. The three major types of nuclear change are natural radioactivity, nuclear fission, and nuclear fusion.

Natural radioactivity is a nuclear change in which unstable nuclei spontaneously shoot out "chunks" of mass (usually alpha or beta particles), energy (gamma rays), or both at a fixed rate. An isotope of an atom that spontaneously emits from its unstable nucleus fast-moving particles, high-energy radiation, or both is called a **radioactive isotope,** or **radioisotope.**

Radiation emitted by radioisotopes is called **ionizing radiation.** When it hits atoms, it has enough energy to dislodge one or more of their electrons to form positively charged ions that can react with and damage living tissue. The two most common types of particles emitted by radioactive isotopes are high-speed **alpha particles** (positively charged chunks of matter that consist of two protons and two neutrons) and **beta particles** (high-speed electrons). The most common form of ionizing energy released from radioisotopes is high-energy **gamma rays.** Figure 3-8 shows the relative penetrating power of alpha, beta, and gamma ionizing radiation. You are exposed to small amounts of harmful ionizing radiation from natural sources and from human sources.

Nuclear Fission: Splitting Nuclei **Nuclear fission** is a nuclear change in which nuclei of certain isotopes with large mass numbers (such as uranium-235) are split apart into lighter nuclei when struck by neutrons; this process releases more neutrons and energy (Figure 3-9). The two or three neutrons produced by each fission can be used to split many additional uranium-235 nuclei if enough nuclei are present to provide the **critical mass** needed for efficient capture of these neutrons.

These multiple fissions taking place within the critical mass represent a **chain reaction** that releases

Figure 3-8 The three major types of ionizing radiation emitted by radioactive isotopes vary considerably in their penetrating power.

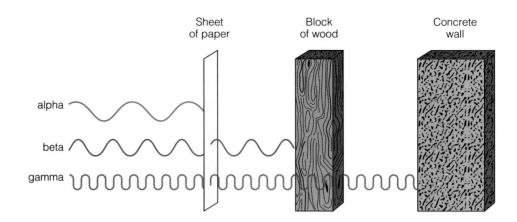

Figure 3-9 Fission of a uranium-235 nucleus by a slow-moving neutron.

Figure 3-10 A nuclear chain reaction initiated by one neutron triggering fission in a single uranium-235 nucleus. This shows only a few of the trillions of fissions caused when a single uranium-235 nucleus is split within a critical mass of other uranium-235 nuclei.

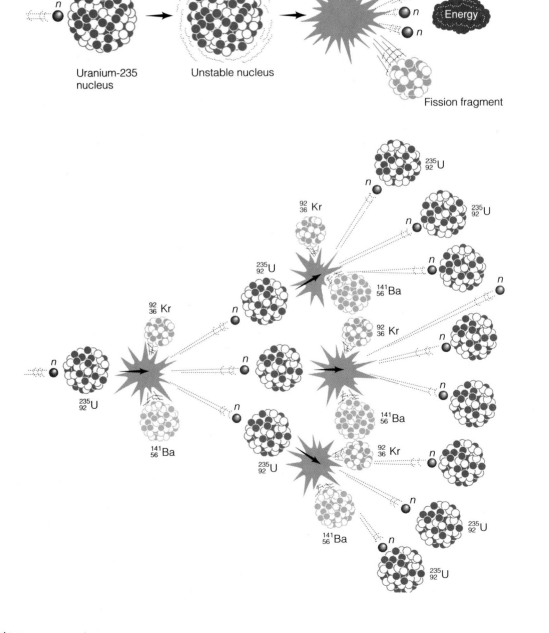

an enormous amount of energy (Figure 3-10). Living cells can be damaged by the ionizing radiation released by the radioactive lighter nuclei and the neutrons produced by nuclear fission.

In an atomic or nuclear fission bomb, a massive amount of energy is released in a fraction of a second in an uncontrolled nuclear fission chain reaction. This reaction is initiated by an explosive charge, which suddenly pushes a mass of fissionable fuel together from all sides, causing the fuel to reach the critical mass needed for a massive chain reaction.

In the nuclear reactor of a nuclear electric power plant, the rate at which the nuclear fission chain reaction takes place is controlled so that only one of each two or three neutrons released is used to split another nucleus. In conventional nuclear fission reactors, nuclei of uranium-235 are split apart and release energy. The heat released is used to produce high-pressure steam, which spins generators to produce electricity.

Nuclear Fusion: Forcing Nuclei to Combine **Nuclear fusion** is a nuclear change in which two nuclei of isotopes of light elements, such as hydrogen, are forced together at extremely high temperatures until they fuse to form a heavier nucleus, releasing energy in the process. Temperatures of 100 million°C to 1 billion°C are usually needed to force the positively charged nuclei (which strongly repel one another) to join together.

High-temperature fusion is much harder to initiate than fission, but once started, it releases far more energy per gram of fuel than fission. Fusion of

---

SPOTLIGHT Waste: An Outmoded and Dangerous Concept

For plants and nonhuman animals there is virtually no waste. The wastes or dead bodies of one form of life are food or nutrients for other forms of life. Sooner or later everything is recycled through natural processes.

Our challenge is to imitate these natural processes in our economic systems and lifestyles. *We need to recognize that most of what we call wastes are really wasted resources. They are potential resources that we are not recycling, reusing, or converting to useful raw materials or products.*

We should be teaching small children and adults to view trash cans and dumpsters as resource containers and trash as concentrated urban ore that needs to be separated into useful materials for recycling. Schools can be collection points for recycled materials as they were during World War II. Profits from school and university recycling centers run by students could be used to fund school activities. Students should also take field trips to recycling centers to see how the resources they collect are put back into use.

We can also make products that last longer and are easy to repair, recycle, and remanufacture. We should also recognize that each unit of energy that we waste unnecessarily harms the environment and sooner or later costs us money.

The basic problem is that we use economic systems to reward those who produce waste instead of those who try to use resources more efficiently. We give timber, mining, and energy companies tax write-offs and other subsidies to cut trees and to find and extract copper, oil, and coal from the earth's crust. But we give few if any such subsidies to companies and businesses that recycle copper or paper, use oil or coal more efficiently, or develop renewable alternatives to using nonrenewable fossil fuels.

This means that the market prices of nonfuel minerals, oil, gasoline, electricity, and water are much lower than their true costs to us and to society. Factories will produce more pollution than they should as long as their owners don't have to pay for each unit of pollution they produce. You and I will buy products that unnecessarily waste energy and other resources and produce more pollu- tion than necessary as long as the market prices we pay for them don't include the costs of the harmful effects from their extraction, production, and use.

In the end, we pay for pollution and resource waste as higher taxes, health bills, insurance, and other social costs. But these costs aren't attached directly to the cost of each type of product or service we use. Thus, we have no way of knowing the true short- and long-term harmful effects and costs of most of the products and services we use.

In a world with more and more people rapidly converting the world's resources to trash and waste heat, waste production is an outmoded and dangerous concept. It is not based on the way the natural processes that sustain life on earth work.

Using our economic and political systems to reward resource conservation instead of resource waste is the key to making the transition from a throwaway society to a sustainable-earth society.

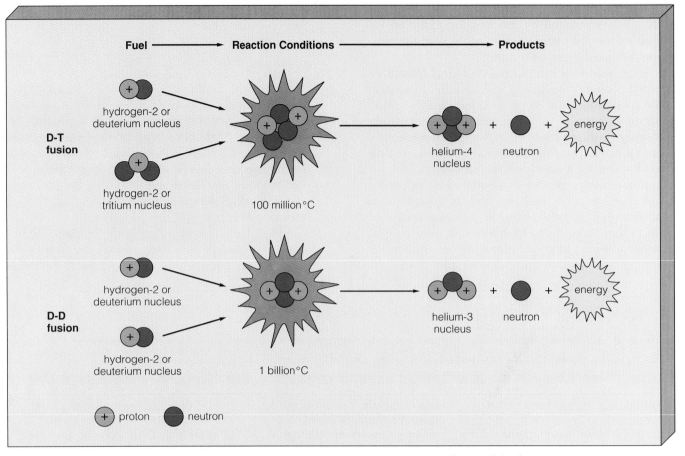

Figure 3-11 The deuterium-tritium (D-T) and deuterium-deuterium (D-D) nuclear fusion reactions, which take place at extremely high temperatures. Unconfirmed experiments in 1989 suggested that some type of nuclear fusion of deuterium nuclei may be possible at room temperature, but there is considerable controversy over these experiments.

hydrogen atoms to form helium atoms is the source of energy in the sun and other stars. There is some controversial evidence that it may be possible to carry out nuclear fusion near room temperature. Even if this is so, it is not clear whether the energy output will be large enough to yield high-temperature heat.

After World War II, the principle of *uncontrolled nuclear fusion* was used to develop extremely powerful hydrogen, or thermonuclear, bombs and missile warheads. These weapons use the D-T fusion reaction, in which a hydrogen-2, or deuterium (D), nucleus and a hydrogen-3, or tritium (T), nucleus are fused to form a larger, helium-4 nucleus, a neutron, and energy (Figure 3-11).

Scientists have also tried to develop *controlled nuclear fusion*, in which the D-T reaction is used to produce heat that can be converted into electricity. However, this process is still at the laboratory stage despite 41 years of research. If it ever becomes technologically and economically feasible, it is not projected to be a practical source of energy until 2050 or later, if ever.

3-5 THE FIRST AND SECOND LAWS OF ENERGY

First Law of Energy: You Can't Get Something for Nothing In studying millions of falling objects, physical and chemical changes, and changes of temperature in living and nonliving systems, scientists have observed and measured energy being changed from one form to another. However, they have never been able to detect any creation or destruction of energy in any chemical or physical change.

This information is summarized in the **law of conservation of energy,** also known as the **first law of energy or thermodynamics.** According to this scientific law, in any physical or chemical change, in any movement of matter from one place to another, and in any change in temperature, energy is neither created nor destroyed but merely transformed from one form to another. What this means is that it is impossible to either create or destroy energy. All we can do is transform energy from one form to another.

When a rabbit eats a garden plant, the plant vanishes from sight. But the matter and chemical

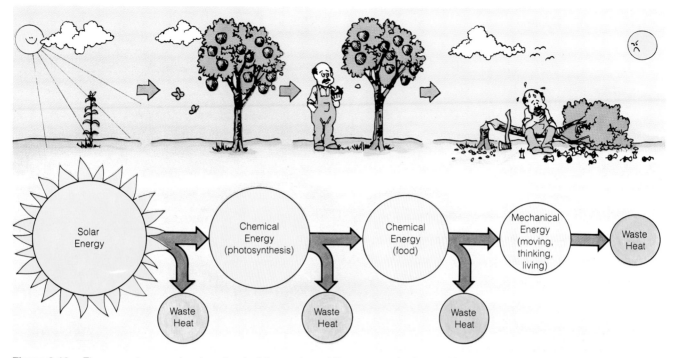

Figure 3-12 The second energy law in action in living systems. When energy is changed from one form to another, some of the initial input of high-quality energy is degraded, usually to low-quality heat, which is added to the environment.

energy the plant contained is converted to chemical energy and different forms of matter in the rabbit's body. Matter and energy have been changed to different forms, but no matter or energy has been created or destroyed.

This law means that we can never get more energy out of an energy transformation process than we put in: *Energy input always equals energy output: we can't get something for nothing; nor can we get nothing out of something in terms of energy quantity.*

Second Law of Energy: You Can't Break Even
Because the first law of energy states that energy can neither be created nor destroyed, you might think that there will always be enough energy. Yet if you fill a car's tank with gasoline and drive around, or if you use a flashlight battery until it is dead, you have lost something. If it isn't energy, what is it? The answer is energy quality (Figure 3-6).

Millions of measurements by scientists have shown that in any conversion of energy from one form to another, there is always a decrease in energy quality or the amount of useful energy. This summary of what we always find occurring in nature is known as the **second law of energy:** In any conversion of energy from one form to another, some of the initial energy input is always degraded to lower-quality, less-useful energy, usually low-temperature heat that flows into the environment. This low-quality energy

is so dispersed that it is unable to perform useful work.

When energy is changed from one form to another, useful energy is always converted to less useful, lower-quality energy. In other words, *we can't break even in terms of energy quality.* This law of degradation of energy quality is also known as the **second law of thermodynamics.** This law states that energy can be changed in only one direction—from usable to unusable, or from ordered to disordered. No one has ever found a violation of this fundamental scientific law.

Consider three examples of the second energy law in action. First, when a car is driven, only about 10% of the high-quality chemical energy available in its gasoline fuel is converted to mechanical energy to propel the vehicle and electrical energy to run its electrical systems. The remaining 90% is degraded to low-quality heat that is released into the environment and eventually lost into space. Second, when electrical energy flows through filament wires in an incandescent light bulb, it is changed into a mixture of about 5% useful radiant energy, or light, and 95% low-quality heat that flows into the environment. What we call a light bulb is really a heat bulb. A third example of the degradation of energy quality in living systems is illustrated in Figure 3-12.

The second energy law also means that *we can never recycle or reuse high-quality energy to perform useful*

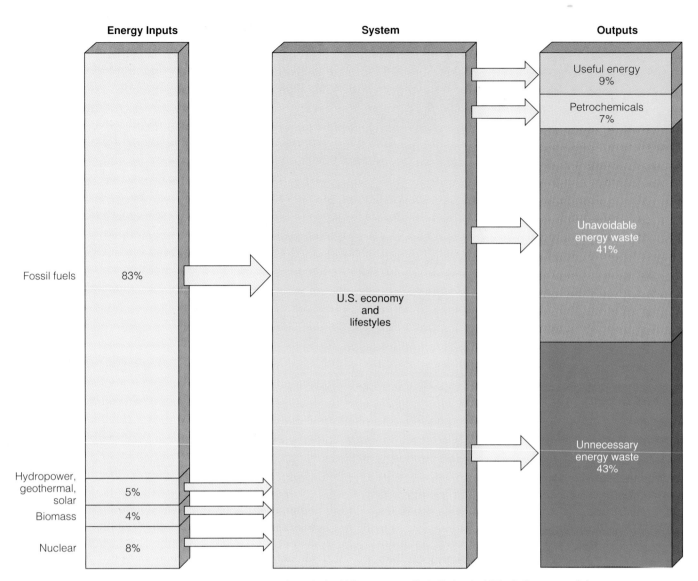

Energy Inputs	System	Outputs

Fossil fuels — 83%

Hydropower, geothermal, solar — 5%

Biomass — 4%

Nuclear — 8%

U.S. economy and lifestyles

Useful energy 9%

Petrochemicals 7%

Unavoidable energy waste 41%

Unnecessary energy waste 43%

Figure 3-13 Flow of commercial supplemental energy through the U.S. economy. Note that only 16% of all commercial energy used in the United States ends up performing useful tasks or is converted to petrochemicals. The rest is automatically wasted because of the second law of energy (41%) or is wasted unnecessarily (43%).

work. Once the concentrated, high-quality energy in a piece of food, a gallon of gasoline, a lump of coal, or a piece of uranium is released, it is degraded into dispersed, low-quality heat that flows into the environment. We can heat air or water at a low temperature and upgrade it to high-quality energy, but the second energy law tells us that it will take more high-quality energy to do this than we get.

Life and the Second Energy Law To form and preserve the highly ordered arrangement of molecules and the organized network of chemical changes in your body, you must continually get and use high-quality matter resources and energy resources from your surroundings. As you use these resources, you add disordered, low-quality heat and waste matter to your surroundings.

For example, your body continuously gives off heat equal to that of a 100-watt light bulb; this is the reason a closed room full of people gets warm. You also continuously give off molecules of carbon dioxide gas and water vapor that become dispersed in the atmosphere.

Planting, growing, processing, and cooking the foods you eat all require high-quality energy and matter resources that add low-quality heat and waste materials to the environment. In addition, enormous amounts of low-quality heat and waste matter are added to the environment when concentrated deposits of minerals and fuels are extracted from the earth, processed, and used or burned to heat and cool the buildings you use, to transport you, and to make roads, clothes, shelter, and other items you use.

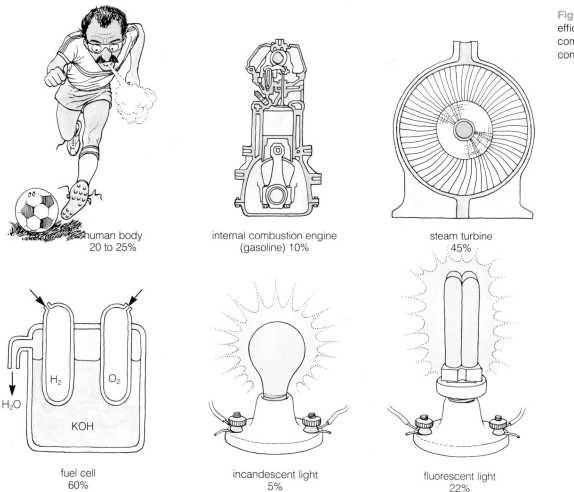

human body
20 to 25%

internal combustion engine
(gasoline) 10%

steam turbine
45%

fuel cell
60%

incandescent light
5%

fluorescent light
22%

3-6 ENERGY EFFICIENCY AND NET USEFUL ENERGY

Increasing Energy Efficiency The two energy laws are important tools in helping us decide how to reduce unnecessary energy waste by improving energy efficiency and in evaluating the net useful energy available from various energy resources. For example, only 16% of all the commercially produced energy that flows through the U.S. economy performs useful work or is used to make petrochemicals, which are used to produce plastics, medicine, and many other products (Figure 3-13). *This means that 84% of all energy used in the United States is wasted.* About 41% of this energy is wasted automatically because of the energy quality tax imposed by the second energy law, but 43% of the commercial energy used in the United States is unnecessarily wasted.

One way to cut much of this energy waste and save money is to increase **energy efficiency** (see Guest Essay on p. 56). This is the percentage of total energy input that does useful work and is not converted to low-quality, essentially useless heat in an energy conversion system. The energy conversion devices we use vary considerably in their energy efficiencies

(Figure 3-14). We can reduce waste by using the most efficient processes or devices available and by trying to make them more efficient.

We can save energy and money by buying the most energy-efficient home heating systems, water heaters, cars, air conditioners, refrigerators, and other household appliances available. The initial cost of the most energy-efficient models is usually higher, but in the long run they usually save money by having a lower **life-cycle cost:** the initial cost plus lifetime operating costs.

The net efficiency of the entire energy delivery process of a heating system, water heater, or car is determined by finding the efficiency of each energy conversion step in the process. These steps include extracting the fuel, purifying and upgrading it to a useful form, transporting it, and finally using it.

Figure 3-15 shows how net energy efficiencies are determined for heating a well-insulated home **(1)** passively with an input of direct solar energy through windows facing the sun and **(2)** with electricity produced at a nuclear power plant, transported by wire to the home, and converted to heat (electric resistance heating). This analysis shows that the process of

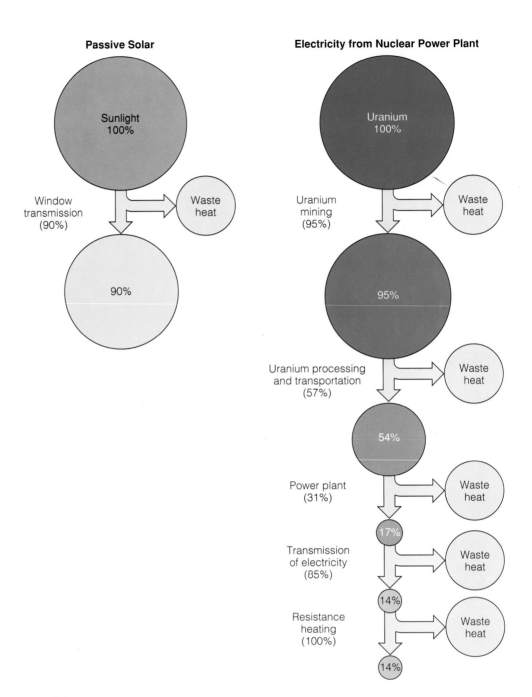

Figure 3-15 Comparison of net energy efficiency for two types of space heating. The cumulative net efficiency is obtained by multiplying the percentage shown inside the circle for each step by the energy efficiency for that step (shown in parentheses). Usually, the greater the number of steps in an energy conversion process, the lower its net energy efficiency.

converting the high-quality nuclear energy in nuclear fuel to high-quality heat at several thousand degrees, converting this heat to high-quality electricity, and then using the electricity to provide low-quality heat for warming a house to only about 20°C (68°F) is extremely wasteful of high-quality energy. Burning coal or any fossil fuel at a power plant to produce electricity for space heating is also inefficient. By contrast, it is much less wasteful to use a passive or active solar heating system to obtain low-quality heat from the environment and, if necessary, raise its temperature slightly to supply space heating.

Using high-quality electrical energy to provide low-quality heat for heating space or household water is like using a chain saw to cut butter or a sledgeham-

mer to kill a fly. A general rule of energy use is the *principle of matching energy quality to energy tasks* (see Guest Essay on p. 56). Don't use high-quality energy to do something that can be done with lower-quality energy (Figures 3-6 and 3-7).

Figure 3-16 lists the net energy efficiencies for a variety of space-heating systems. It shows that the most energy-efficient way to provide heating, especially in a cold climate, is to build a superinsulated house. Such a house is so heavily insulated and airtight that even in areas where winter temperatures fall to −40°C (−40°F), all of its space heating can usually be supplied by a combination of passive solar gain (about 59%), waste heat from appliances (33%), and body heat from occupants (8%).

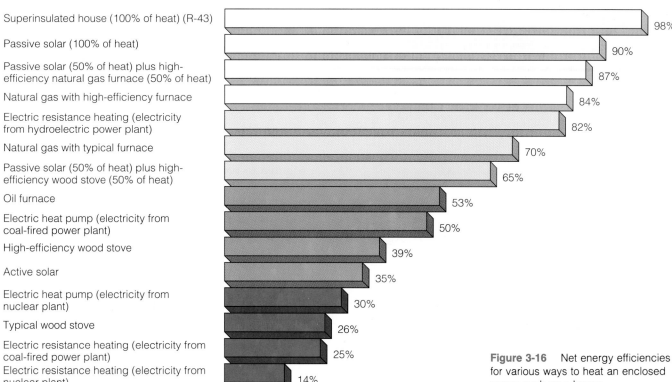

Net Energy Efficiency

Superinsulated house (100% of heat) (R-43)	98%
Passive solar (100% of heat)	90%
Passive solar (50% of heat) plus high-efficiency natural gas furnace (50% of heat)	87%
Natural gas with high-efficiency furnace	84%
Electric resistance heating (electricity from hydroelectric power plant)	82%
Natural gas with typical furnace	70%
Passive solar (50% of heat) plus high-efficiency wood stove (50% of heat)	65%
Oil furnace	53%
Electric heat pump (electricity from coal-fired power plant)	50%
High-efficiency wood stove	39%
Active solar	35%
Electric heat pump (electricity from nuclear plant)	30%
Typical wood stove	26%
Electric resistance heating (electricity from coal-fired power plant)	25%
Electric resistance heating (electricity from nuclear plant)	14%

Figure 3-16 Net energy efficiencies for various ways to heat an enclosed space such as a house.

Passive solar heating is the next most efficient and cheapest method of heating a house, followed by one of the new high-efficiency natural gas furnaces. The least efficient, most expensive way to heat a house is with electricity produced by nuclear power plants.

Using electricity to heat space or water by any method is extremely wasteful of energy and money because it involves using high-quality heat produced at a power plant to provide moderate-quality heat (Figure 3-7). For example, in 1988 the average price of obtaining 250,000 kilocalories (1 million British thermal units, or Btus) for heating space or water in the United States was $5.36 using natural gas, $5.87 using fuel oil, and $22.83 using electricity. If you like to throw away hard-earned dollars, then use electricity to heat your house and bath water.

Utility companies often run advertisements urging people to buy heat pumps, mostly because home heating and air conditioning systems run on electricity. A heat pump is an efficient way to heat a house as long as the outside temperature does not fall below −15°C (4.5°F), but when it does, these devices begin using electric resistance heating, the most expensive, energy-wasting way to heat any space. Heat pumps are useful for space heating in areas with fairly warm climates, but in such areas their main use is for air conditioning. The air conditioning units with most heat pumps are much less energy-efficient than many stand-alone units. Most heat pumps also require expensive repair every few years.

A similar analysis of net energy efficiency shows that the least efficient and most expensive way to heat water for washing and bathing is to use electricity produced by nuclear power plants. Indeed, we save money and waste less energy by not using high-quality electricity produced by any type of power plant to heat water for washing and bathing. The most efficient method is to use a tankless, instant water heater fired by natural gas or liquefied petroleum gas (LPG) (Figure 3-17). Such heaters fit under a sink or in a small closet and burn fuel only when the hot water faucet is turned on. They heat the water instantly as it flows through a small burner chamber and provide hot water only when and as long as it is needed. In contrast, conventional natural gas and electric resistance heaters keep a large tank of water hot all day and night and can run out after a long shower or two. Tankless heaters are widely used in many parts of Europe and are slowly beginning to appear in the United States. A well-insulated, conventional natural gas or LPG water heater is also efficient.

If engineers were asked to sit down and invent three devices that would waste enormous amounts of energy, they would probably come up with

■ the incandescent light bulb (which wastes 95% of its energy input)

Figure 3-17 Two LPG tankless instant hot water heaters that I use to provide backup hot water and space heating for my office (see Spotlight on p. 404). The unit on the right provides hot water for washing and bathing. Roof-mounted solar collectors (Figure 17-22, p. 405) preheat water stored in an insulated tank—a discarded conventional hot water heater wrapped in extra insulation. When I turn on a hot water faucet, the solar-heated water flows through the instant heater. If a sensor indicates that the water is below 49°C (120°F), the instant heater comes on to raise the water temperature to that level. About 50% to 60% of my space heat is provided passively by solar energy (Figure 17-22, p. 405). The rest is provided by a combination of active solar collectors and the tankless water heater shown on the left. Roof-mounted solar collectors store heat in a well-insulated tank—another discarded conventional hot water heater wrapped in extra insulation. When the thermostat calls for heat, this solar-heated water is pumped through the instant heater and then through a coil and back to the tank it came from. A fan blows air over the coil and transfers hot air through heating ducts, as with a conventional forced-air space heating system. Any heat not extracted from the hot water by the coil is returned to the insulated tank for reuse in this closed-loop system.

- a car or truck with an internal combustion engine (which wastes 90% of the energy in its fuel)

- a nuclear power plant with its electricity used to heat space or water for washing and bathing (which wastes 86% of the energy in its nuclear fuel) (Figure 3-15)

These devices were developed and widely used during a time when energy was cheap and plentiful. As this era draws to a close, we will have to replace or greatly improve the energy efficiency of these and other energy-conversion items in modern society (see Guest Essay on p. 56).

Using Waste Heat We cannot recycle high-quality energy, but we can slow the rate at which waste heat flows into the environment when high-quality energy is degraded. For instance, in cold weather an uninsulated, leaky house loses heat almost as fast as it is produced. By contrast, a well-insulated, airtight house can retain most of its heat for five to ten hours, and a well-designed, superinsulated house can retain most of its heat up to four days.

In some office buildings, waste heat from lights, computers, and other machines is collected and distributed to reduce heating bills during cold weather and exhausted to reduce cooling bills during hot weather. Waste heat from industrial plants and electrical power plants can be distributed through insulated pipes and used as a district heating system for nearby buildings, greenhouses, and fish ponds, as is done in some parts of Europe.

Waste heat from coal-fired and other industrial boilers can be used to produce steam to spin turbines and generate electricity at half the cost of buying it from a utility company. The electricity can be used by the plant or sold to the local power company for general use. This combined production of heat or steam and electricity from the same fuel source, known as **cogeneration,** is widely used in European industrial plants. If all large industrial boilers in the U.S. used cogeneration, there would be no need to build any electric power plants through the year 2020.

Net Useful Energy: It Takes Energy to Get Energy The usable amount of high-quality energy obtainable from a given quantity of an energy resource is its **net useful energy.** It is the total energy available from the resource over its lifetime minus the amount of energy used (the first energy law), automatically wasted (the second energy law), and unnecessarily wasted in finding, processing, concentrating, and transporting it to a user. For example, if nine units of fossil fuel energy are needed to supply ten units of nuclear, solar, or additional fossil fuel energy (perhaps from a deep well at sea), the net useful energy gain is only one unit of energy.

We can express this relationship as the ratio of useful energy produced to the useful energy used to produce it. In the example just given, the net energy ratio would be 10/9, or 1.1. The higher the ratio, the greater the net useful energy yield. When the ratio is less than 1, there is a net energy loss over the lifetime of the system. Figure 3-18 lists estimated net useful energy ratios for various alternatives to space heating, high-temperature heat for industrial processes, and gaseous and liquid fuels for vehicles.

Currently, oil has a relatively high net useful energy ratio because much of it comes from rich, accessible deposits such as those in Saudi Arabia and

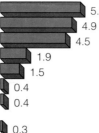

Space Heating

Passive solar — 5.8
Natural gas — 4.9
Oil — 4.5
Active solar — 1.9
Coal gasification — 1.5
Electric resistance heating (coal-fired plant) — 0.4
Electric resistance heating (natural-gas-fired plant) — 0.4
Electric resistance heating (nuclear plant) — 0.3

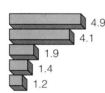

High-Temperature Industrial Heat

Surface-mined coal — 28.2
Underground-mined coal — 25.8
Natural gas — 4.9
Oil — 4.7
Coal gasification — 1.5
Direct solar (highly concentrated by mirrors, heliostats, or other devices) — 0.9

Transportation

Natural gas — 4.9
Gasoline (refined crude oil) — 4.1
Biofuel (ethyl alcohol) — 1.9
Coal liquefaction — 1.4
Oil shale — 1.2

Figure 3-18 Net useful energy ratios for various energy systems over their estimated lifetimes. (Data from Colorado Energy Research Institute, *Net Energy Analysis*, 1976, and Howard T. Odum and Elisabeth C. Odum, *Energy Basis for Man and Nature*, 3rd ed., McGraw-Hill, 1981)

other parts of the Middle East. When these sources are depleted, however, the net useful energy ratio of oil will decline and prices will rise. Then more money and high-quality fossil fuel will be needed to find, process, and deliver new oil from poorer deposits found deeper in the earth and in remote, hostile areas like Alaska, the Arctic, and the North Sea—far from where the energy is to be used.

Conventional nuclear fission energy has a low net energy ratio because large amounts of energy are required to extract and process uranium ore, to convert it to a usable nuclear fuel, and to build and operate power plants. Additional energy is needed to take nuclear plants apart after their 25 to 30 years of useful life and to store the resulting highly radioactive wastes for thousands of years.

3-7 MATTER AND ENERGY LAWS AND ENVIRONMENTAL AND RESOURCE PROBLEMS

Throwaway Societies Because of the law of conservation of matter and the second law of energy, resource use by each of us automatically adds some waste heat and waste matter to the environment. Your individual use of matter and energy resources

and your addition of waste heat and waste matter to the environment may seem small and insignificant, but you are only one of the 1.2 billion individuals in industrialized countries using large quantities of the earth's matter and energy resources at a rapid rate. Meanwhile, the 4.1 billion people in less developed countries hope to be able to use more of these resources.

Today's advanced industrialized countries are **throwaway societies,** sustaining ever-increasing economic growth by maximizing the rate at which matter and energy resources are used and wasted (Figure 3-19). The scientific laws of matter and energy tell us that if more and more people continue to use and waste resources at an increasing rate, sooner or later the capacity of the local, regional, and global environments to dilute and degrade waste matter and absorb waste heat will be exceeded.

Matter-Recycling Societies A stopgap solution to this problem is to convert from a throwaway society to a **matter-recycling society.** The goal of such a shift would be to allow economic growth to continue without depleting matter resources and without producing excessive pollution and environmental degradation, but as we have seen already, there is no free lunch.

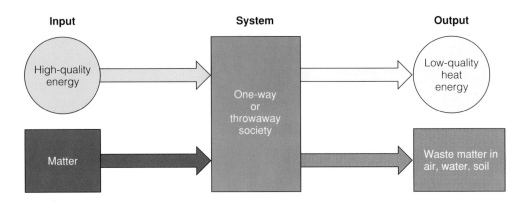

Figure 3-19 The one-way, or throwaway, society of most industrialized countries is based on maximizing the rates of energy flow and matter flow, rapidly converting the world's high-quality matter and energy resources to trash, pollution, and low-quality heat.

The two laws of energy tell us that *recycling matter resources always requires high-quality energy, which cannot be recycled.* Recycling of matter resources should be guided by four principles:

- High-quality energy cannot be recycled and should not be wasted.

- Recycling matter always requires high-quality energy, but efficient recycling usually takes less high-quality energy than producing new products by mining or harvesting and processing virgin materials.

- Don't dilute, disperse, or mix products containing matter resources that can be recycled.

- To promote matter recycling, give government tax breaks and other subsidies to manufacturers that use recycled materials; give few if any subsidies to industries that extract virgin resources. If funds are limited, eliminate all subsidies and let the resource extraction and recycling industries compete in the open marketplace.

In the long run, even a matter-recycling society based on indefinitely increasing economic growth must have an inexhaustible supply of affordable high-quality energy. The environment must also have an infinite capacity to absorb and disperse waste heat and to dilute and degrade waste matter.

Experts disagree on how much usable high-quality energy we have. However, supplies of coal, oil, natural gas, and uranium are clearly finite. And affordable supplies of oil, the most widely used supplementary energy resource, may be used up in several decades.

"Ah," you say, "but don't we have a virtually inexhaustible supply of solar energy flowing to the earth?" The problem is that the amount of solar energy reaching a particular small area of the earth's surface each minute or hour is low, and nonexistent at night.

With a proper collection and storage system, using passive and active systems to concentrate solar energy slightly to provide hot water and to heat a house to moderate temperatures makes good thermodynamic and economic sense. But to provide the high temperatures needed to melt metals or to produce electricity in a power plant, solar energy may not be cost effective. Why? Because its net useful energy ratio (Figure 3-18) is low. It takes a lot of energy to concentrate and raise its quality to a high level.

Suppose that affordable solar cells, nuclear fusion at room temperature, or some other breakthrough were to supply an essentially infinite supply of affordable useful energy. Would this solve our environmental problems? No.

Such a breakthrough would be important and useful. But the second energy law tells us that the faster we use more energy to transform more matter into products and to recycle these products, the faster large amounts of low-quality heat and waste matter are dumped into the environment. Thus, the more we attempt to "conquer" the earth, the more stress we put on the environment. Experts argue over how close we are to reaching overload limits, but the scientific laws of matter and energy indicate that such limits do exist.

Sustainable-Earth Societies The three scientific laws governing matter and energy changes indicate that the best long-term solution to our environmental and resource problems is to shift from a throwaway society based on maximizing matter and energy flow (and in the process wasting an unnecessarily large portion of the earth's resources) to a **sustainable-earth society** (Figure 3-20).

A sustainable-earth society would be modeled after what nature does to sustain us and other species. Such a society would go a step farther than a matter-recycling society. In addition to recycling and reusing much of the matter we now discard as trash, it would use energy more efficiently and not use high-quality

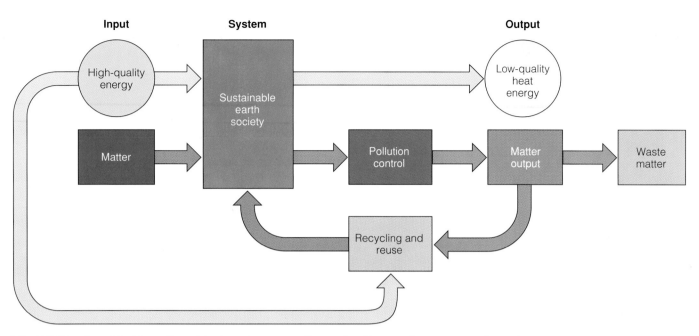

Figure 3-20 A sustainable-earth society, based on energy flow and matter recycling, reuses and recycles renewable matter resources, wastes less matter and energy, reduces unnecessary consumption, emphasizes pollution prevention and waste reduction, and controls population growth.

energy to do things that require moderate-quality energy (Figure 3-7).

Another goal of a sustainable-earth society would be to reduce use and waste of matter resources by making things that last longer and are easier to recycle, reuse, and repair. Our lifestyle motto should be: *Throwaway no, recycle yes, but reuse is best.* Just as important, a sustainable-earth society would cut down on the use of resources by regulating population growth.

Finally, a sustainable-earth society would use input approaches to reduce resource waste and to prevent pollution. The matter and energy laws show us why input or prevention approaches make more sense thermodynamically and economically than output approaches. For example, preventing a toxic chemical from reaching underground supplies of drinking water is much easier and cheaper than trying to remove the chemical once it has contaminated the groundwater.

Until we make this shift to input appproaches, we are merely treating the undesirable symptoms of unsustainable, throwaway societies instead of transforming them into sustainable-earth societies. If we add devices to car engines that reduce overall pollution by 50% and then more than double the number of cars or the number of kilometers driven, air pollution levels begin rising again. Air pollution also increases when we add devices to smokestacks of coal-burning power and industrial plants to cut air

pollution emissions in half and then more than double the use of coal.

Burning or burying wasted resources instead of not producing them or recycling and reusing them stimulates us to continue producing more waste, removing it from one part of the environment, and putting it in another part. Sooner or later even the best designed landfills leak wastes into water supplies and we also run out of affordable or politically acceptable land for landfills. Even the best designed waste incinerators release some toxic substances into the atmosphere and leave a highly toxic residue that must be disposed of—usually in landfills that ultimately leak toxic chemicals into underground water aquifers.

Because of the three basic scientific laws of matter and energy, we are all dependent on each other and on the other living and nonliving parts of nature for our survival. Everything is connected and we are all in it together. In the next chapter we will apply these laws to living systems and look at some biological principles that can teach us how to work with the rest of nature.

The second law of thermodynamics holds, I think, the supreme position among laws of nature. . . . If your theory is found to be against the second law of thermodynamics, I can give you no hope.

Arthur S. Eddington

Amory B. Lovins

Physicist and energy consultant Amory B. Lovins is one of the world's most influential and articulate experts on energy strategy. In 1989 he was the first recipient of the Delphi Prize for environmental work, and The Wall Street Journal *selected him as a "leader of tomorrow." He is director of research at the Rocky Mountain Institute in Old Snowmass, Colorado, which he and his wife Hunter cofounded in 1982. He has briefed five heads of state and served as a consultant to United Nations agencies, the U.S. Department of Energy, the Congressional Office of Technology Assessment, the U.S. Solar Energy Research Institute, and state and local governments. He is active in energy affairs in 20 countries and has published several hundred papers and a dozen books, including the widely discussed* Soft Energy Paths *(New York: Harper Colophon, 1979) and the nontechnical version of this work with coauthor L. Hunter Lovins,* Energy Unbound: Your Invitation to Energy Abundance *(San Francisco: Sierra Club Books, 1986).*

The answers you get depend on the questions you ask. But sometimes it seems so important to resolve a crisis that we forget to ask what problem we're trying to solve.

It is fashionable to suppose that we're running out of energy, and that the solution is obviously to get lots more of it. But asking how to get more energy begs the question of how much we need. That depends not on how much we used in the past but on what we want to do in the future and how much energy it will take to do those things.

How much energy it takes to make steel, run a sewing machine, or keep you comfortable in your house depends on how cleverly we use energy, and the more it costs, the smarter we seem to get. It is now cheaper, for example, to double the efficiency of most industrial electric motor drive systems than to get more electricity to run the old ones. (Just this one saving can more than replace the entire U.S. nuclear power program.) We know how to make lights five times as efficient as those presently in use and how to make household appliances that give us the same work as now, using one-fifth as much energy (saving money in the process).

Eight automakers have made good-sized, peppy, safe prototype cars averaging 30 to 52 kilometers per liter (70 to 120 miles per gallon). We know today how to make new buildings and many old ones so heat-tight (but still well ventilated) that they need essentially no energy to maintain comfort year-round, even in severe climates.

These energy-saving measures are uniformly cheaper than going out and getting more energy. Detailed studies in over a dozen countries have shown that supplying energy services in the cheapest way—by wringing more work from the energy we already have—would let us increase our standard of living while using several times less total energy (and electricity) than we do now. Those savings cost less than finding new domestic oil or operating existing power plants.

But the old view of the energy problem included a worse mistake than forgetting to ask how much energy we needed: It sought more energy, in any form, from any source, at any price—as if all kinds of energy were alike. This is like saying, "All kinds of food are alike; we're running short of potatoes and turnips and cheese, but that's OK, we can substitute sirloin steak and oysters Rockefeller."

Some of us have to be more discriminating than that. Just as there are different kinds of food, so there are many different forms of energy, whose different prices and qualities suit them to different uses (Figure 3-7, p. 42). There is, after all, no demand for energy as such; nobody wants raw kilowatt-hours or barrels of sticky black goo. People instead want energy services: comfort, light, mobility, ability to bake bread, ability to make cement. We ought therefore to start at that end of the energy problem: to ask, "What are the many different tasks we want energy for, and what is the amount, type, and source of energy that will do each task *in the cheapest way?*"

Electricity is a particularly special, high-quality, expensive form of energy. An average kilowatt-hour delivered in the United States in 1988 was priced at about eight cents, equivalent to buying the heat content of oil costing $128 per barrel—over seven times the average world price during 1988. The average cost of electricity from nuclear plants (including fuel and operating expenses) beginning operation in 1987 was 13.5 cents per kilowatt-hour, equivalent on a heat basis to buying oil at about $216 per barrel.

Such costly energy might be worthwhile if it were used only for the premium tasks that require it, such as lights, motors, electronics, and smelters. But those special uses, only 8% of all delivered U.S. energy needs, are already met twice over by today's power stations. Two-fifths of our electricity is already spilling over into uneconomic, low-grade uses such as water heating, space heating, and air conditioning. Yet no matter how efficiently we use electricity (even with heat pumps), we can never get our money's worth on these applications.

Thus, *supplying more electricity is irrelevant to the energy problem that we have.* Even though electricity accounts for almost all of the federal energy research and development budget and for at least half of national energy investment, it is the wrong kind of energy to meet our needs economically. Arguing about what kind of new power station to build—coal, nuclear, solar—is like shopping for the best buy in antique Chippendale chairs to burn in your stove or brandy to put in your car's gas tank. *It is the wrong question.*

Indeed, *any kind of new power station is so uneconomical that if you have just built one, you will save the country money by writing it off and never operating it.* Why? Because its additional electricity can be used only for low-temperature heating and cooling (the premium "electricity-specific" uses being already filled up) and is the most expensive way of supplying these services.

The real question is what is the cheapest way to do low-temperature heating and cooling. That means weatherstripping, insulation, heat exchangers, greenhouses, superwindows, window shades and overhangs, trees, and so on. These measures generally cost about half a penny per kilowatt-hour, whereas the running costs *alone* for a new nuclear plant will be nearly four cents per kilowatt-hour, so it is cheaper not to run it. In fact, under the crazy U.S tax laws, the extra saving from not having to pay the plant's future subsidies is probably so big that society can also recover the capital cost of having built the plant by shutting it down!

If we want more electricity, we should get it from the cheapest sources first. In approximate order of increasing price, these include:

1. Converting to efficient lighting equipment. This would save the U.S. electricity equal to the output of 120 large power plants plus $30 billion a year in fuel and maintenance costs.

2. Eliminating pure waste of electricity, such as lighting empty offices at headache level. Each kilowatt-hour saved can be resold without having to generate it anew.

3. Displacing with good architecture, and with passive and some active solar techniques, the electricity now used for water heating and space heating and cooling. Some U.S. utilities now give low- or zero-interest weatherization loans, which you need not start repaying for ten years or until you sell your house—because it saves the utility millions of dollars to get electricity that way compared with building new power plants. Most utilities also offer rebates for buying efficient appliances.

4. Making lights, motors, appliances, smelters, and the like cost-effectively efficient.

Just these four measures can quadruple U.S. electrical efficiency, making it possible to run today's economy, with no changes in lifestyles, using no thermal power plants, whether old or new, and whether fueled with oil, gas, coal, or uranium. We would need only the present hydroelectric capacity, readily available small-scale hydroelectric projects, and a modest amount of wind power. But if we still wanted more electricity, the next cheapest sources would include:

5. Industrial cogeneration, combined-heat-and-power plants, low-temperature heat engines run by industrial waste heat or by solar ponds, filling empty turbine bays and upgrading equipment in existing big dams, modern wind machines or small-scale hydroelectric turbines in good sites, steam-injected natural gas turbines, and perhaps recent developments in solar cells with waste heat recovery.

It is only after we had clearly exhausted all these cheaper opportunities that we would even consider:

6. Building a new central power station of any kind— the slowest and costliest known way to get more electricity (or to save oil).

To emphasize the importance of starting with energy end uses rather than energy sources, consider a sad little story from France, involving a "spaghetti chart" (or energy flowchart)—a device energy planners often use to show how energy flows from primary sources via conversion processes to final forms and uses. In the mid-1970s energy conservation planners in the French government started, wisely, on the right-hand side of the spaghetti chart. They found that their biggest single need for energy was to heat buildings, and that even with good heat pumps, electricity would be the most uneconomic way to do this. So they had a fight with their nationalized utility; they won; and electric heating was supposed to be discouraged or even phased out because it was so wasteful of money and fuel.

But meanwhile, down the street, the energy supply planners (who were far more numerous and influential in the French government) were starting on the left-hand side of the spaghetti chart. They said: "Look at all that nasty imported oil coming into our country! We must replace that oil. Oil is energy. . . . We need some other source of energy. Voila! Reactors can give us energy; we'll build nuclear reactors all over the country." But they paid little attention to what would happen to that extra energy, and no attention to relative prices.

Thus, the two sides of the French energy establishment went on with their respective solutions to two different, indeed contradictory, French energy problems: *more energy of any kind, versus the right kind to do each task cheapest.* It was only in 1979 that these conflicting perceptions collided. The supply side planners

(continued)

suddenly realized that the only way they would be able to *sell* all that nuclear electricity would be for electric heating, which they had just agreed not to do.

Every industrial country is in this embarrassing position (especially if we include in "heating" air conditioning, which just means heating the outdoors instead of the indoors). Which end of the spaghetti chart we start on, or *what we think the energy problem is*, is not an academic abstraction: It *determines what we buy*. It is the most fundamental source of disagreement about energy policy.

People starting on the left side of the spaghetti chart think the problem boils down to whether to build coal or nuclear power stations (or both). People starting on the right realize that *no* kind of new power station can be an economic way to meet the needs for low- and high-temperature heat and for vehicular liquid fuels that are 92% of our energy problem.

So if we want to provide our energy services at a price we can afford, let's get straight what question our technologies are supposed to provide the answer to. Before we argue about the meatballs, let's untangle the strands of spaghetti, see where they're supposed to lead, and find out what we really need the energy for!

Guest Essay Discussion

1. List the energy services you would like to have, and note which of these must be furnished by electricity.

2. The author argues that building more nuclear, coal, or other electrical power plants to supply electricity for the United States is unnecessary and wasteful. Summarize the reasons for this conclusion and give your reasons for agreeing or disagreeing with this viewpoint.

3. Do you agree or disagree that increasing the supply of energy, instead of concentrating on improving energy efficiency, is the wrong answer to U.S. energy problems? Explain.

DISCUSSION TOPICS

1. Explain why we don't really consume anything and why we can never really throw matter away.

2. A tree grows and increases its mass. Explain why this isn't a violation of the law of conservation of matter.

3. If there is no "away," why isn't the world filled with waste matter?

4. Use the second energy law to explain why a barrel of oil can be used only once as a fuel.

5. Explain why most energy analysts urge that improving energy efficiency forms the basis of any individual, corporate, or national energy plan. Is it an important part of your personal energy plan or lifestyle (see Individuals Matter inside the back cover)? Why or why not?

6. Explain why using electricity to heat a house and to supply household hot water is expensive and wasteful of energy. What energy tasks can be done best by electricity?

7. You are about to build a house. What energy supply (oil, gas, coal, or other) would you use for space heating, cooking food, refrigerating food, and heating water? Consider the long-term economic and environmental impact. Would you decide differently if you planned to live in the house for only 5 years instead of 25 years? If so, how?

8. a. Use the law of conservation of matter to explain why a matter-recycling society will sooner or later be necessary.

   b. Use the first and second laws of energy to explain why in the long run a sustainable-earth society, not just a matter-recycling society, will be necessary.

FURTHER READINGS

Carrying Capacity, Inc. 1987. *Beyond Oil*. New York: Ballinger.

Christensen, John W. 1990. *Global Science: Energy, Resources, and Environment*. 3rd ed. Dubuque, Iowa: Kendall/Hunt.

Colorado Energy Research Institute. 1976. *Net Energy Analysis: An Energy Balance Study of Fossil Fuel Resources*. Golden, Colo.: Colorado Energy Research Institute.

Fowler, John M. 1984. *Energy and the Environment*. 2nd ed. New York: McGraw-Hill.

Lovins, Amory B. 1977. *Soft Energy Paths*. Cambridge, Mass.: Ballinger.

Lovins, Amory B., and L. Hunter Lovins. 1986. *Energy Unbound: Your Invitation to Energy Abundance*. San Francisco: Sierra Club Books.

Nash, Hugh, ed. 1979. *The Energy Controversy: Soft Path Questions and Answers*. San Francisco: Friends of the Earth.

Odum, Howard T., and Elisabeth C. Odum. 1981. *Energy Basis for Man and Nature*. 3rd ed. New York: McGraw-Hill.

Rifkin, Jeremy. 1989. *Entropy: Into the Greenhouse World: A New World View*. New York: Bantam Books.

CHAPTER FOUR

ECOSYSTEMS:

WHAT ARE THEY

AND

HOW DO THEY WORK?

General Questions and Issues

1. What two major natural processes keep us and other organisms alive?

2. What is an ecosystem, and what are its major living and nonliving components?

3. What happens to energy in an ecosystem?

4. What happens to matter in an ecosystem?

5. What roles do different organisms play in an ecosystem, and how do organisms interact?

If we love our children, we must love the earth with tender care and pass it on, diverse and beautiful, so that on a warm spring day 10,000 years hence they can feel peace in a sea of grass, can watch a bee visit a flower, can hear a sandpiper call in the sky, and can find joy in being alive.

HUGH H. ILTIS

WHAT ORGANISMS LIVE in a forest or a pond? How do they get enough matter and energy resources to stay alive? How do these organisms interact with one another and with their physical and chemical environment? What changes will this forest or pond undergo through time?

Ecology is the science that attempts to answer such questions about how nature works. In 1869 German biologist Ernst Haeckel coined the term *ecology* from two Greek words: *oikos*, meaning "house" or "place to live," and *logos*, meaning "study of."

Ecology is the study of living things in their home or **environment**: all the external conditions and factors, living and nonliving, that affect an organism (Figure 4-1). In other words, *ecology* is the study of the interactions between organisms and their living (biotic) and nonliving (abiotic) environment. The key word is *interactions*. Scientists usually carry out this study by examining different **ecosystems**: forests, deserts, grasslands, ponds, lakes, oceans, or any set of organisms interacting with one another and with their nonliving environment.

This chapter will consider the major nonliving and living components of ecosystems and how they interact. The next chapter will consider major types of ecosystems and the changes they can undergo because of natural events and human activities.

 THE EARTH'S LIFE-SUPPORT SYSTEMS: AN OVERVIEW

The Biosphere and the Ecosphere The earth has several major parts that play a role in sustaining life (Figure 4-2). You are part of what ecologists call the **biosphere**—the living and dead organisms found near the earth's surface in parts of

Figure 4-1 Ecology is a study of how organisms interact with other living things and nonliving things such as sunlight, air, water, and soil. This arctic fox, with its winter coat that helps hide it in the snow, and other animals depend on sunlight, water, plants, and decomposers (mostly bacteria and fungi) for their survival.

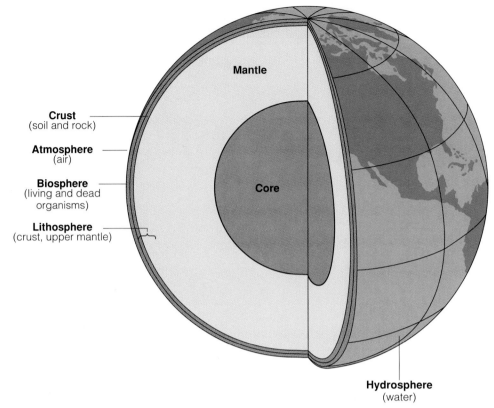

Figure 4-2   Our life-support system: the general structure of the earth.

Mantle

Crust
(soil and rock)

Atmosphere
(air)

Biosphere
(living and dead organisms)

Lithosphere
(crust, upper mantle)

Core

Hydrosphere
(water)

the atmosphere, hydrosphere, and lithosphere. Virtually all life on earth exists in a thin film of air, water, and rock in a zone extending from about 61 meters (200 feet) below the ocean's surface to about 6,000 meters (20,000 feet) above sea level.

The living organisms that make up the biosphere interact with one another, with energy from the sun, and with various chemicals in the atmosphere, hydrosphere, and lithosphere (Figure 4-2). This collection of living and dead organisms (the biosphere) interacting with one another and their nonliving environment (energy and chemicals) throughout the world is called the **ecosphere.** If the earth were an apple, the ecosphere would be no thicker than the apple's skin. *The goal of ecology is to learn how the ecosphere works.*

Energy Flow and Matter Recycling   What keeps you, me, and most other organisms alive on this tiny planet? The answer is that life on earth depends largely on two fundamental processes (Figure 4-3):

■ the *one-way flow of high-quality energy* from the sun, through materials and living things on or near the earth's surface, then into the atmosphere, and eventually into space as low-quality heat

■ the *recycling* of chemicals required by living organisms through parts of the ecosphere

The Sun: Source of Energy for Life   The source of the energy that sustains life on earth is the sun. It lights and warms the earth and supplies the energy used by green plants to synthesize the compounds that keep them alive and serve as food for almost all other organisms. Solar energy also powers the recycling of key chemicals and drives the climate and weather systems that distribute heat and fresh water over the earth's surface.

The sun is a gigantic fireball composed of hydrogen (72%) and helium (28%) gases. Temperatures and pressures in its inner core are so high that the hydrogen nuclei found there are compressed and fused to form helium gas. This nuclear fusion reaction (Figure 3-11, p. 46) taking place at the center of the sun continually releases massive amounts of energy.

This energy is radiated into space as a spectrum of ultraviolet light, visible light, infrared radiation, and other forms of radiant energy (Figure 4-4). These forms of energy travel outward in all directions through space and make the 150-million-kilometer (93-million-mile) trip to the earth in about eight minutes.

Figure 4-5 shows what happens to the solar radiant energy reaching the earth. About 34% is immediately reflected back to space by clouds, chemicals, and dust in the atmosphere and by the earth's surface. Most of the remaining 66% warms the atmosphere and land, evaporates water and cycles it

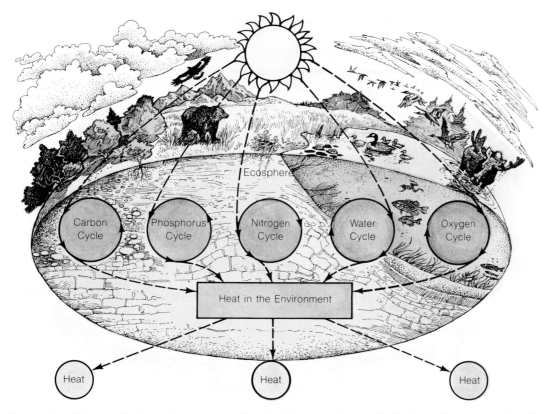

Figure 4-3 Life on earth depends on the recycling of critical chemicals (solid lines) and the one-way flow of energy through the ecosphere (dashed lines). This greatly simplified overview shows only a few of the many chemicals that are recycled.

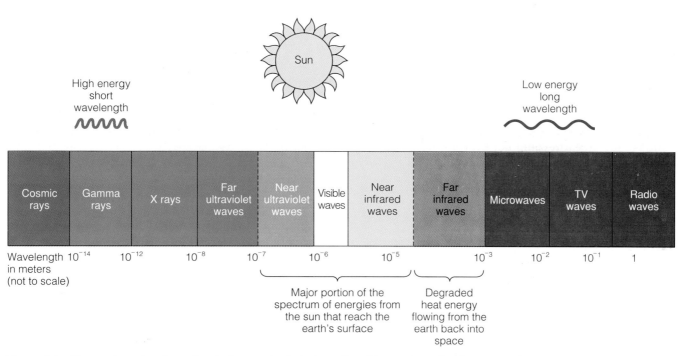

Figure 4-4 The sun is a gigantic nuclear fusion reactor that gives off a wide spectrum of radiant energy into space. When this energy reaches the earth, much of it is either reflected or absorbed by chemicals in the atmosphere, which prevents most of the harmful, high-energy cosmic rays, gamma rays, X rays, and far-ultraviolet ionizing radiation from reaching the earth's surface.

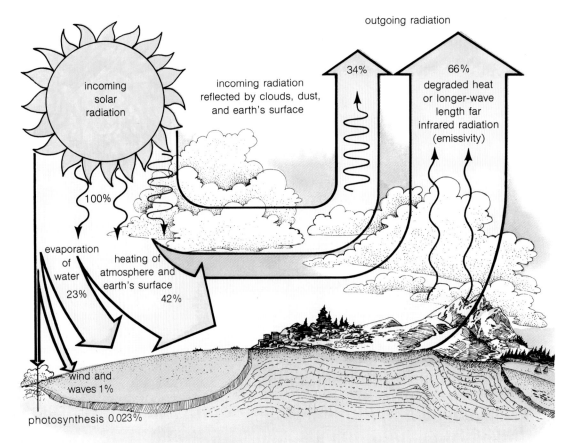

Figure 4-5 The flow of energy to and from the earth.

outgoing radiation

incoming
solar
radiation

incoming radiation
reflected by clouds, dust,
and earth's surface

34%

66%
degraded heat
or longer-wave
length far
infrared radiation
(emissivity)

100%

evaporation
of
water
23%

heating of
atmosphere and
earth's surface
42%

wind and
waves 1%

photosynthesis 0.023%

through the ecosphere, and generates winds. A tiny fraction (0.023%) is captured by green plants and used in the process of photosynthesis to make organic compounds that organisms need to survive.

Most of the radiant energy reaching the earth's surface is in the form of light and near-infrared radiation (Figure 4-4). Most of the harmful forms of ionizing radiation emitted by the sun, especially ultraviolet radiation, are absorbed by molecules of ozone (O_3) in the upper atmosphere (stratosphere) and water vapor in the lower atmosphere (troposphere). Without this screening effect, most present forms of life on earth could not exist.

Most of the incoming solar radiation not reflected away is degraded into low-quality heat (far-infrared radiation) in accordance with the second law of energy, and flows into space. The rate at which energy returns to space as low-quality heat is affected by the presence of molecules such as water, carbon dioxide, methane, nitrous oxide, and ozone and by some forms of solid particulate matter in the lower atmosphere. These substances, acting as gatekeepers, allow some high-quality forms of radiant energy from the sun to pass through the atmosphere. They also absorb and reradiate some of the resulting low-quality heat back toward the earth's surface.

Scientists are becoming increasingly concerned that our activities are adding molecules to the lower atmosphere that can slow the rate at which heat returns to space, causing global warming that will disrupt food growing patterns and wildlife habitats and raise average sea levels (as discussed in more detail in Chapter 10).

Biogeochemical Cycles Any element or compound an organism needs to live, grow, and reproduce is called a **nutrient.** Nutrients include organic compounds, such as sugars and proteins, and inorganic materials, such as water, carbon dioxide, oxygen gas, nitrate ions, phosphate ions, and ions of elements such as iron and copper.

Most of the earth's chemicals do not occur naturally in forms useful to the organisms that make up the biosphere. Fortunately, elements and compounds required as nutrients for life on earth are continuously cycled in complex paths through the living and nonliving parts of the ecosphere and converted to useful forms by a combination of biological, geological, and chemical processes.

This recycling of nutrients from the nonliving environment (reservoirs in the atmosphere, hydrosphere, and earth's crust) to living organisms, and back to the nonliving environment, takes place in **biogeochemical cycles** (*bio* meaning "life," *geo* for "earth," and *chemical* for the changing of matter from one form to another). These cycles, driven directly or indirectly by incoming energy from the sun, include the carbon, oxygen, nitrogen, phosphorus, sulfur, and hydrologic (water) cycles (Figure 4-3).

Thus, a chemical may be part of an organism at one moment and part of its nonliving environment at another moment. For example, one of the oxygen molecules you just inhaled may be one inhaled previously by you, your grandmother, King Tut thousands of years ago, or a dinosaur millions of years ago. Similarly, some of the carbon atoms in the skin covering your right hand may once have been part of a leaf, a dinosaur hide, or a limestone rock.

4-2 ECOSYSTEMS: TYPES AND COMPONENTS

The Realm of Ecology Ecology is primarily concerned with interactions among five of the levels of organization of matter shown in Figure 3-1 (page 37): organisms, populations, communities, ecosystems, and the ecosphere. An **organism** is any form of life. Biologists classify the earth's organisms in anywhere from 3 to 20 categories. To understand the key ideas in this introductory book, we will broadly classify organisms as *plants*, *animals*, or *decomposers*.*

The earth contains a variety of plants. They range from microscopic, one-celled, floating and drifting plants known as phytoplankton (mostly diatoms) to the largest of all living things, the giant sequoia trees of western North America.

Green plants use sunlight to produce food for themselves and for other plants, animals, and decomposers that cannot produce their own food. Some green plants are **evergreens** that retain some of their leaves or needles throughout the year. Examples are ferns and cone-bearing trees (conifers) such as firs, spruces, and pines. **Deciduous plants** such as oak and maple lose their leaves and become dormant during winter. **Succulent plants** such as desert cacti (Figure 4-6) survive by having no leaves, thus reducing the loss of scarce water; they store water and use sunlight to produce the food they need in the thick fleshy tissue of their green stems and branches.

Some plants are **annuals** that grow, set seed, and then die in a single year. Wheat, rice, corn, and most other plants we depend on for food are annuals that must be replanted each year. Other plants are **perennials** that grow from the same root stock each

---

* Your instructor may wish to expand this to the five kingdoms—bacteria, protists (such as diatoms and ameba), fungi, plants, and animals—used in most introductory biology courses. But the material in this text can be understood without going into this level of detail.

Figure 4-6 These saguaro cacti in Arizona are succulent green plants that store water and produce food in the fleshy tissue of their stems and branches. They reduce water loss in the hot desert climate by having no leaves.

François Gohier/Ardea London

year. Examples are grasses and daffodils that spring up each year on their own.

There are also a variety of animals on earth. They range in size from tiny floating and moving zooplankton (which feed on phytoplankton) to the 4-meter- (14-foot-) high male African elephant and the 30-meter- (100-foot-) long blue whale. Animals may be covered with skin, scales, plates, feathers, or hair. Some called **vertebrates** have backbones (Figure 4-7) and others called **invertebrates** have no backbones (Figure 4-8). Some are cold blooded (invertebrates, fish, amphibians, and reptiles) and others have warm blood (birds and mammals). Some animals feed on plants, some feed on other animals, and some feed on both.

Decomposers are microscopic bacteria and larger fungi (mostly molds and mushrooms) (Figure 4-9). They get their food by breaking down the complex organic compounds in animal and other wastes and the tissues of dead plants and animals into simpler compounds. These compounds are returned to water in the soil or to bodies of water as nutrients for plants

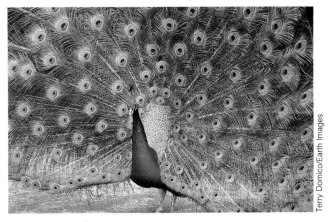

Figure 4-7 This peacock with its courtship display is a vertebrate because it has a backbone.

Figure 4-8 This cobalt blue sea star has no backbone and is classified as an invertebrate. Other invertebrates are insects, crabs, jellyfish, sponges, and mollusks.

Figure 4-9 Two types of decomposers are shelf fungi and *Boletus luridus* mushrooms.

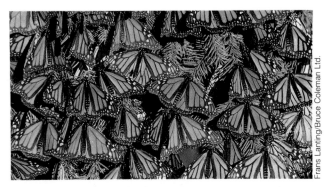

Figure 4-10 Population of monarch butterflies hibernating during winter in Michoacan, Mexico.

to start the cycle of life from plants to animals to decomposers again. Without decomposers, plants and animals couldn't survive and we would be knee deep in wastes.

A **population** is a group of organisms of the same kind living in a particular location (Figure 4-10). Examples of populations are all the sunfish in a pond, gray squirrels in a forest, white oak trees in a forest, people in a country, or people in the world. Populations are dynamic groups of organisms that adapt to changes in environmental conditions by changing their size, distribution of various age groups (age structure), and genetic makeup.

Figure 4-11 Two of the earth's incredible diversity of species. On the left is the world's largest flower, called the flesh flower, growing in a tropical forest in Sumatra. It is 0.9 meter (3 feet) in diameter. On the right is a cottontop tamarin, another resident of a tropical forest.

For organisms that reproduce sexually, a **species** is one or more populations whose members actually or potentially interbreed under natural conditions (Figure 4-11). Worldwide, it's estimated that 5 to 30 million different species exist. So far, about 1.6 million of the earth's species have been described and named, and about 10,000 new species are added to the list each year.

Each organism and population has a **habitat**: the place or type of place where it naturally lives. When several populations of different species live together and interact with one another in a particular place, they make up what is called a **community, or biological community**. Examples are all the plants, animals, and decomposers found in a forest, a pond, a desert, a dead log, or an aquarium.

An **ecosystem** is the combination of a community and the chemical and physical factors making up its nonliving environment. It is an ever-changing (dynamic) network of biological, chemical, and physical interactions that sustain a community and allow it to respond to changes in environmental conditions. All the earth's ecosystems together make up the **ecosphere.**

Major Terrestrial and Aquatic Ecosystems Ecosystems can be classified into general types that contain similar types of organisms. The ecosphere's major land ecosystems are called **terrestrial ecosystems**. These different types of habitats are described primarily by their types of vegetation as various types of forests, grasslands, or deserts. The differences

among these land ecosystems in various parts of the world are caused mostly by differences in average temperature and average precipitation (Figure 4-12).

Major ecosystems in the hydrosphere are called **aquatic ecosystems**. Examples include ponds, lakes, rivers, open ocean, coral reefs, estuaries (mouths of river segments or ocean inlets where salt water and fresh water mix), and coastal and inland wetlands (such as swamps, marshes, and prairie potholes that are covered with water all or part of the time). The major differences between these ecosystems are the result of differences in the amount of various nutrients dissolved in the water (salinity), differences in the depth of sunlight penetration, and differences in average water temperature. Major terrestrial and aquatic ecosystems are discussed in more detail in Chapter 5.

Abiotic Components of Ecosystems Ecosystems consist of various nonliving and living components. Figures 4-13 and 4-14 are greatly simplified diagrams showing a few of the components of ecosystems in a freshwater pond and in a field.

The nonliving, or **abiotic,** components of an ecosystem include various physical and chemical factors. Major physical factors affecting ecosystems are

- sunlight and shade

- average temperature and temperature range

- average precipitation and its distribution throughout each year

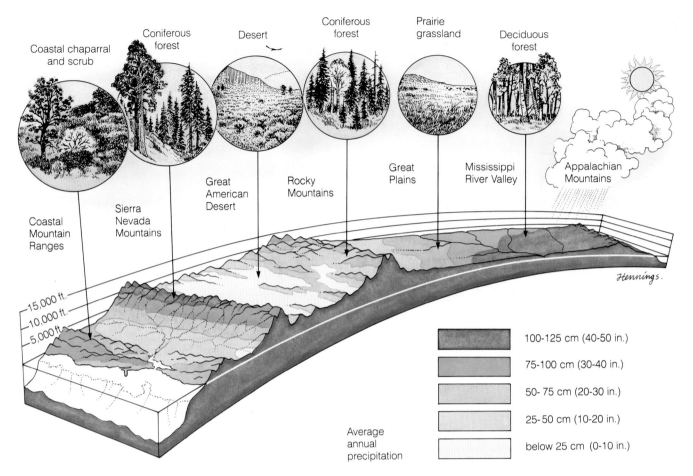

Figure 4-12 Gradual transition from one major land ecosystem or biome to another along the 39th parallel crossing the United States. The major factors causing these transitions are changes in average temperature and precipitation.

■ wind

■ latitude (distance from the equator)

■ altitude (distance above sea level)

■ nature of soil (for terrestrial ecosystems)

■ fire (for terrestrial ecosystems)

■ water currents (for aquatic ecosystems)

■ amount of suspended solid material (for aquatic ecosystems)

Major chemical factors affecting ecosystems are

■ level of water and air in soil

■ level of plant nutrients dissolved in soil water in terrestrial ecosystems and in water in aquatic ecosystems

■ level of natural or artificial toxic substances dissolved in soil water in terrestrial ecosystems and in water in aquatic ecosystems

■ salinity of water for aquatic ecosystems

■ level of dissolved oxygen in aquatic ecosystems

Biotic Components of Ecosystems The major types of organisms that make up the living, or **biotic,** components of an ecosystem are usually classified as *producers, consumers,* and *decomposers.* This classification is based on organisms' general nutritional habits.

Producers—sometimes called **autotrophs** (self-feeders)—are organisms that can manufacture the organic compounds they use as sources of energy and nutrients. Most producers are green plants that make the organic nutrients they require through **photosynthesis.** This complicated process begins when sunlight is absorbed by pigments such as chlorophyll, which gives the plants their green color. The plants use this energy to combine carbon dioxide (which they get from the atmosphere or from water) with water (which they get from the soil or aquatic surroundings) to make carbohydrates—sugars (such as glucose), starches, and celluloses. Oxygen gas is given off as a by-product of photosynthesis. Photosynthesis can be summarized as follows:

$$\frac{\text{carbon}}{\text{dioxide}} + \text{water} + \textbf{solar energy} \rightarrow \text{glucose} + \text{oxygen}$$

$CO_2 + H_2O + \text{energy} \rightarrow C_6H_{12}O_6 + H_2O$

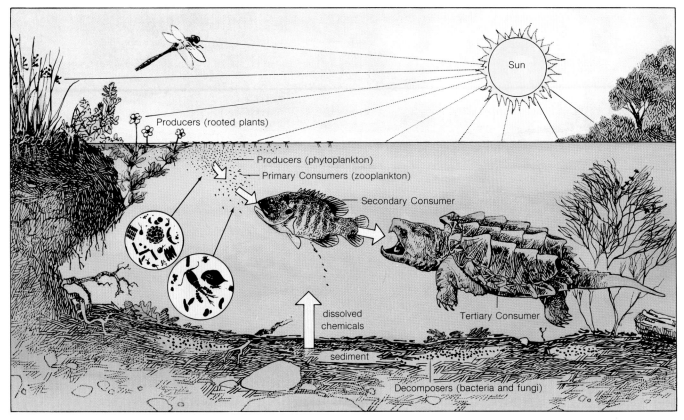

Figure 4-13 The major components of a freshwater pond ecosystem.

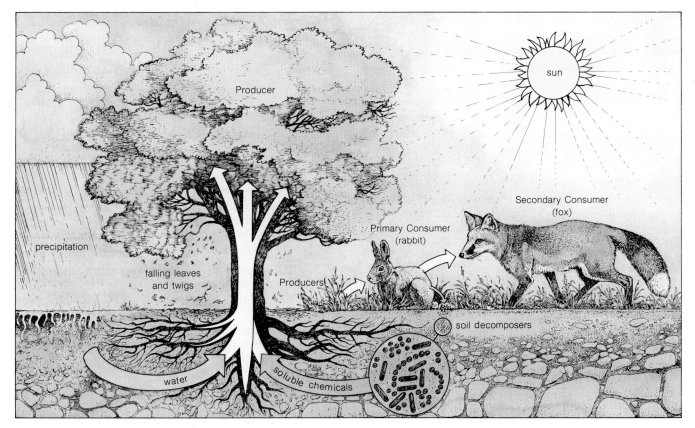

Figure 4-14 The major components of an ecosystem in a field.

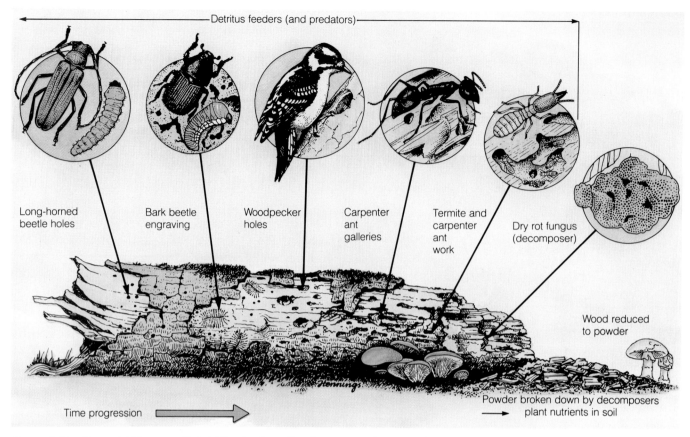

Long-horned
beetle holes

Bark beetle
engraving

Woodpecker
holes

Carpenter
ant
galleries

Termite and
carpenter
ant
work

Dry rot fungus
(decomposer)

Wood reduced
to powder

Time progression

Powder broken down by decomposers
plant nutrients in soil

Figure 4-15 Some detritivores, called detritus feeders, directly consume dead organic matter in a fallen tree. Other detritivores, called decomposers, break down complex organic chemicals in the dead wood into simpler nutrient chemicals that are returned to soil water for reuse by plants.

In essence this complex process converts radiant energy from the sun into chemical energy stored in the chemical bonds that hold glucose and other carbohydrates together. This stored chemical energy produced by photosynthesis is the direct or indirect source of food for most organisms. Most of the oxygen in the atmosphere is also a product of photosynthesis. An estimated 59% of the earth's photosynthesis takes place on land and the remaining 41% in the oceans and other aquatic ecosystems.

Some producer organisms, mostly specialized bacteria, can extract inorganic compounds from their environment and convert them to organic nutrients without the presence of sunlight. This process is called **chemosynthesis**. For example, hydrothermal vents in some parts of the ocean floor spew forth large amounts of superheated ocean water and rotten-egg-smelling hydrogen sulfide gas. In this pitch-dark, hot environment, specialized producer bacteria carry out chemosynthesis to convert inorganic hydrogen sulfide to nutrients they require.

Only producers can make their own food. In addition, they provide food directly or indirectly for animals and decomposers. You, I, and most other animals get nutrients either by eating plants or by eating animals that feed on plants; all flesh is grass, so to speak.

Organisms that get the nutrients and energy they require by feeding either directly or indirectly on producers are called **consumers,** or **heterotrophs** (other-feeders). Some consumers feed on living plants and animals, and others feed on small fragments of dead plant and animal matter, called **detritus**.

Depending on their food sources, consumers that feed on living organisms fall into three major classes:

- **Herbivores** (plant eaters)—*primary consumers,* which feed directly and only on all or part of living plants. Some birds eat seeds, buds, and foliage. Deer and rabbits eat twigs and leaves. Gophers attack plant roots. Grasshoppers and many other kinds of insects eat all parts of plants. In aquatic ecosystems, zooplankton feed on phytoplankton (Figure 4-13).

- **Carnivores** (flesh eaters)—*secondary consumers,* which feed only on plant-eating animals (herbivores), and *tertiary* or *higher-level consumers,* which feed only on animal-eating animals. Spiders and birds that eat plant-eating insects, and tuna, which feed on herrings and anchovies, are secondary consumers. Hawks, which eat snakes and weasels, and sharks, which eat other fish, are tertiary or higher-level consumers.

- **Omnivores** ("everything eaters")—which can eat both plants and animals. Examples are pigs, rats, foxes, cockroaches, and humans.

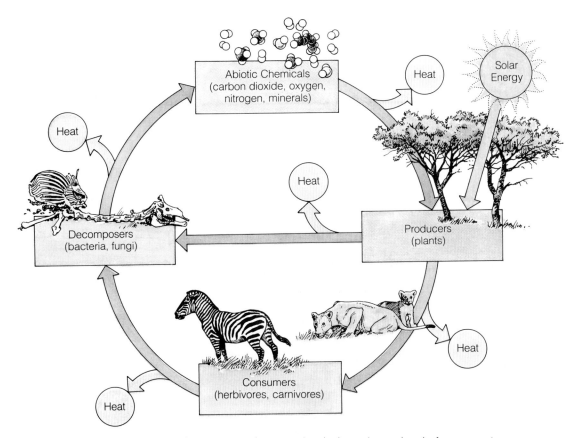

Figure 4-16 The major structural components (energy, chemicals, and organisms) of an ecosystem are connected through the functions of energy flow and matter recycling. There is a one-way flow of energy from the sun through producer organisms (mostly plants), through consumers (mostly animals), through decomposers (bacteria and fungi), and back into the environment as low-quality heat. Because of the second energy law, the quality of this energy is degraded as it flows through the ecosystem. Nutrients are transferred from one organism to another and modified as needed. Decomposers break down the complex organic chemicals in dead organisms and in their wastes to simpler inorganic chemicals for use by producers to begin the cycle again.

Consumer organisms that feed on **detritus**, or dead organic plant and animal matter, are known as **detritivores** (Figure 4-15). There are two major classes of detritivores: detritus feeders and decomposers. **Detritus feeders** ingest fragments of dead organisms and their cast-off parts and organic wastes. Examples are crabs, earthworms, and clams.

detritus feeder + decomposer

Much of the detritus in ecosystems—especially dead wood and leaves—undergoes decay, rot, or decomposition, in which its complex organic molecules are broken into simpler inorganic compounds containing nutrient elements. This decomposition process is brought about by the feeding activity of the other type of detritus consumer, **decomposers**. They digest dead tissue or wastes and absorb their soluble nutrients. Decomposers consist of two classes of organisms: microscopic, single-celled *bacteria* and *fungi* (mostly molds and mushrooms) (Figures 4-9 and 4-15). Bacteria and fungi decomposers in turn are an important source of food for organisms such as worms and insects living in the soil and water.

bacteria + fungi

The chemical energy stored in glucose and other carbohydrates is used by producers, consumers, and decomposers to drive their life processes. This is part of the one-way flow of energy through organisms and ecosystems governed by the second law of energy. Aerobic (oxygen-consuming) organisms change part of the glucose and other, more complex organic compounds they synthesize (producers), eat (consumers), or decompose (decomposers) back into carbon dioxide and water by the process of **cellular aerobic respiration**:

$$\text{glucose} + \text{oxygen} \rightarrow \text{carbon dioxide} + \text{water} + \textbf{energy}$$

Respiration is the slow "burning process" in which oxygen is used to release the energy stored in the chemical bonds of carbohydrates and other organic nutrient compounds. Although the detailed steps in the complex processes of photosynthesis and respiration differ, respiration is, in essence, the opposite of photosynthesis.

The survival of any individual organism depends on *matter flow and energy flow* through its body. However, the community of organisms in an ecosystem survives primarily by a combination of *matter recycling and a one-way flow of energy* (Figure 4-16).

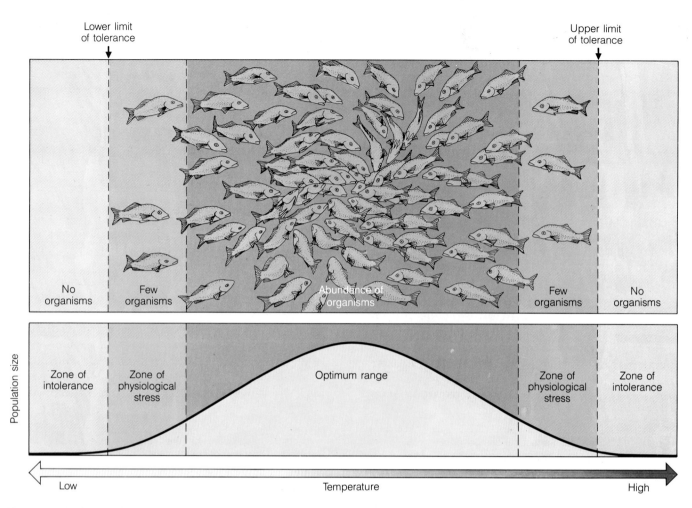

Lower limit
of tolerance

Upper limit
of tolerance

No
organisms

Few
organisms

Abundance of
organisms

Few
organisms

No
organisms

Population size

Zone of
intolerance

Zone of
physiological
stress

Optimum range

Zone of
physiological
stress

Zone of
intolerance

Low

Temperature

High

Figure 4-17   Range of tolerance for a population of organisms of the same species to an abiotic environmental factor—in this case, temperature.

Figure 4-16 shows that decomposers are responsible for completing the cycle of matter. They carry out waste disposal in nature by breaking down organic compounds in organic wastes and dead organisms into inorganic nutrients for producers so that the cycle of life can begin again. Without decomposers, the entire world would soon be knee-deep in plant litter, dead animal bodies, animal wastes, and garbage. Figure 4-16 also shows that an ecosystem can exist without consumers because chemicals can be cycled directly from producers to decomposers and back to producers. This means that we are an unnecessary part of the ecosphere.

Tolerance Ranges of Species to Abiotic Factors
The reason that organisms don't spread everywhere is that each species and each individual organism of a species has a particular **range of tolerance** to variations in chemical and physical factors in its environment, such as temperature (Figure 4-17). A plant, such as the magnolia tree, that is killed by freezing temperatures will not be found in the far north. And

you won't find plants that need lots of water in a desert.

The tolerance range includes an optimum range of values within which populations of a species thrive and operate most efficiently (Figure 4-17). This range also includes values slightly above or below the optimum level of each abiotic factor—values that usually support a smaller population size. When values exceed the upper or lower limits of tolerance, few if any organisms of a particular species survive.

These observations are summarized in the **law of tolerance**: *The existence, abundance, and distribution of a species in an ecosystem are determined by whether the levels of one or more physical or chemical factors fall above or below the levels tolerated by the species.*

Individual organisms within a large population of a species may have slightly different tolerance ranges because of small differences in their genetic makeup, health, and age. For example, it may take a little more heat or a little more of a poisonous chemical to kill one frog or one person than another. That is why the tolerance curve shown in Figure 4-17 represents the response of a population composed of many individuals of the same species to changes in an environmental factor such as temperature.

Many types of organisms can change their tolerance to physical factors such as temperature if exposed to gradually changing conditions. For example, you can tolerate a higher water temperature by getting into a tub of fairly hot water and then slowly adding hotter and hotter water.

This adaptation to slowly changing new conditions, or acclimation, is a useful protective device, but it can also be dangerous. With each change, the organism comes closer to its limit of tolerance. Suddenly, without any warning signals, the next small change triggers a **threshold effect**—a harmful or even fatal reaction as the tolerance limit is exceeded—much like adding the single straw that breaks an already overloaded camel's back.

The threshold effect partly explains why many environmental problems seem to arise suddenly even though they have been building for a long time. For example, trees in certain forests begin dying in large numbers after prolonged exposure to numerous air pollutants. We usually notice the problem only when entire forests die, as is happening in parts of Europe and North America. By then we are 10 to 20 years too late to prevent the damage. The threshold effect also explains why we must emphasize input approaches to prevent pollution thresholds from being exceeded.

Limiting Factors in Ecosystems Another ecological principle related to the law of tolerance is the **limiting factor principle**: *Too much or too little of any abiotic factor can limit or prevent growth of a population of a species in an ecosystem even if all other factors are at or near the optimum range of tolerance for the species.* A single factor found to be limiting the population growth of a species in an ecosystem is called the **limiting factor**.

Examples of limiting factors in terrestrial ecosystems are temperature, water, light, and soil nutrients. For example, suppose a farmer plants corn in a field where the soil has too little phosphorus. Even if the corn's needs for water, nitrogen, potassium, and other nutrients are met, the corn will stop growing when it has used up the available phosphorus. In this case, availability of phosphorus is the limiting factor that determines how much corn will grow in the field. Growth can also be limited by the presence of too much of a particular abiotic factor. For example, plants can be killed by too much water or by too much fertilizer.

In aquatic ecosystems, **salinity** (the amounts of various salts dissolved in a given volume of water) is a limiting factor. It determines the species found in marine ecosystems, such as oceans, and freshwater ecosystems, such as rivers and lakes. Three major limiting factors determining the numbers and types of organisms at various layers in aquatic ecosystems are temperature, sunlight, and **dissolved oxygen content** (the amount of oxygen gas dissolved in a given volume of water at a particular temperature).

4-3 ENERGY FLOW IN ECOSYSTEMS

Food Chains and Food Webs *There is no waste in functioning natural ecosystems.* All organisms, dead or alive, are potential sources of food for other organisms. A caterpillar eats a leaf; a robin eats the caterpillar; a hawk eats the robin. When the plant, caterpillar, robin, and hawk die, they are in turn consumed by decomposers.

A series of organisms, each eating or decomposing the preceding one, is called a **food chain** (Figure 4-18). Food chains are channels for the one-way flow of a tiny part of the sun's high-quality energy captured by photosynthesis, through the living components of ecosystems, and into the environment as low-quality heat. Food chains are also pathways for the recycling of nutrients from producers, consumers (herbivores, carnivores, and omnivores), and decomposers back to producers (Figure 4-16).

All organisms that share the same general types of food in a food chain are said to be at the same **trophic level** (from the Greek *trophos*, "feeder"). As shown in Figure 4-18, all producers belong to the first trophic level; all primary consumers, whether feeding on living or dead producers, belong to the second trophic level; and so on.

The food chain concept is useful for tracing chemical recycling and energy flow in an ecosystem, but simple food chains like the one shown in Figure 4-18 rarely exist by themselves in nature. Few herbivores or primary consumers feed on just one kind of plant, nor are they eaten by only one type of carnivore or secondary consumer. In addition, omnivores eat several different kinds of plants and animals at several trophic levels.

This means that the organisms in a natural ecosystem are involved in a complex network of many interconnected food chains, called a **food web**. A simplified food web in the Antarctic is diagrammed in Figure 4-19, which shows that trophic levels can be assigned in food webs just as in food chains.

Energy Flow Pyramids At each transfer from one trophic level to another in a food chain or web, work is done, low-quality heat is given off to the environment, and the availability of high-quality energy to organisms at the next trophic level is reduced. This reduction in high-quality energy available at each trophic level is the result of the inevitable energy quality tax imposed by the second law of energy.

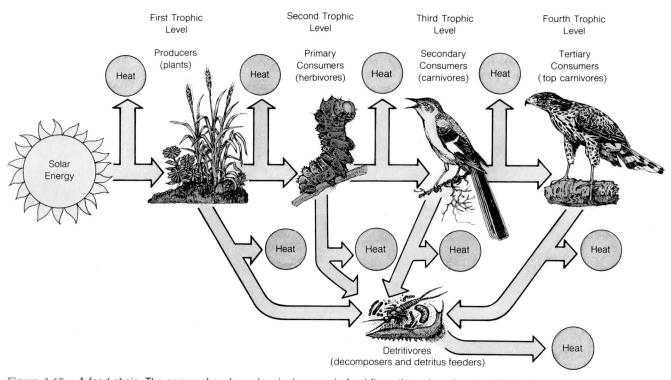

First Trophic Level

Second Trophic Level

Third Trophic Level

Fourth Trophic Level

Producers (plants)

Primary Consumers (herbivores)

Secondary Consumers (carnivores)

Tertiary Consumers (top carnivores)

Solar Energy

Heat

Detritivores
(decomposers and detritus feeders)

Figure 4-18 A food chain. The arrows show how chemical energy in food flows through various trophic levels, with most of the high-quality chemical energy being degraded to low-quality heat in accordance with the second law of energy.

The percentage of available high-quality energy transferred from one trophic level to another varies from 2% to 30%, depending on the types of species involved and the ecosystem in which the transfer takes place. In the wild, ecologists estimate that an average of about 10% of the high-quality chemical energy available at one trophic level is transferred and stored in usable form as chemical energy in the bodies of the organisms at the next level. The rest of the energy is used to keep the organisms alive, and most is eventually degraded and lost to the environment as low-quality heat in compliance with the second law of energy.

Figure 4-20 illustrates this loss of usable high-quality energy at each step in a simple food chain. The **pyramids of energy flow and energy loss** in this diagram show that the greater the number of trophic levels or steps in a food chain or web, the greater the cumulative loss of usable high-quality energy.

The energy-flow pyramid explains why a larger population of people can be supported if people shorten the food chain by eating grains directly (for example, rice → human) rather than eating animals that feed on grains (grain → steer → human).

Net Primary Productivity of Plants The rate at which the plants in an ecosystem produce usable chemical energy or biomass is called **net primary productivity.** It is equal to the rate at which plants

use photosynthesis to store chemical energy in biomass minus the rate at which they use some of this chemical energy in aerobic cellular respiration to live, grow, and reproduce:

| **net primary productivity** | = rate at which plants produce chemical energy through photosynthesis | − rate at which plants use chemical energy through aerobic cellular respiration |

Net primary productivity is usually reported as the amount of energy produced by the plant material in a specified area of land over a given time.

Net primary productivity can be thought of as the basic food source or "income" of animals. Ecologists have estimated the average annual net primary production per square meter for the major terrestrial and aquatic ecosystems. Figure 4-21 shows that ecosystems with the highest average net primary productivities are estuaries, swamps, marshes, and tropical rain forests; the lowest are tundra (Arctic grasslands), open ocean, and desert.

You might conclude that we should clear tropical forests to grow crops and that we should harvest plants growing in estuaries, swamps, and marshes to help feed the growing human population. That conclusion is incorrect. One reason is that the plants—mostly grasses—in estuaries, swamps, and marshes cannot be eaten by people. But they are extremely

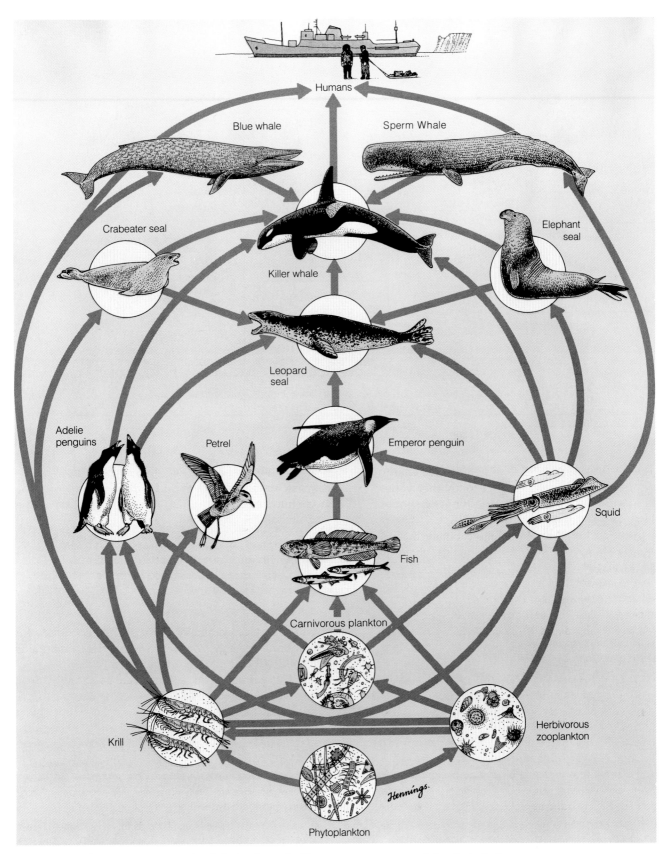

Humans

Blue whale

Sperm Whale

Crabeater seal

Killer whale

Elephant seal

Leopard seal

Adelie penguins

Petrel

Emperor penguin

Squid

Fish

Carnivorous plankton

Krill

Herbivorous zooplankton

Phytoplankton

Hennings.

Figure 4-19 Greatly simplified food web in the Antarctic. There are many more participants, including an array of decomposer organisms.

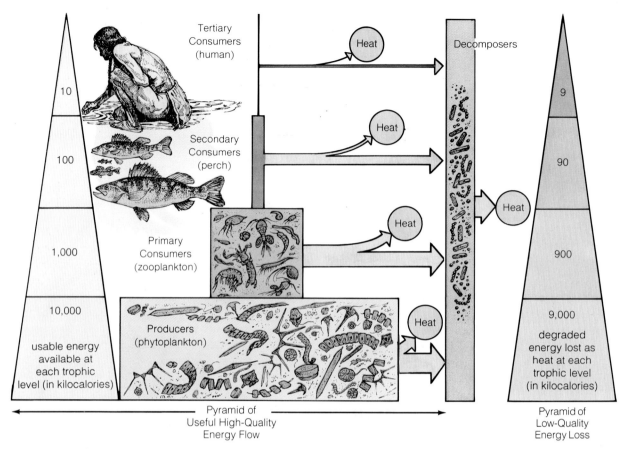

Figure 4-20 Generalized pyramids of energy flow and energy loss, showing the decrease in usable high-quality energy available at each succeeding trophic level in a food chain or web.

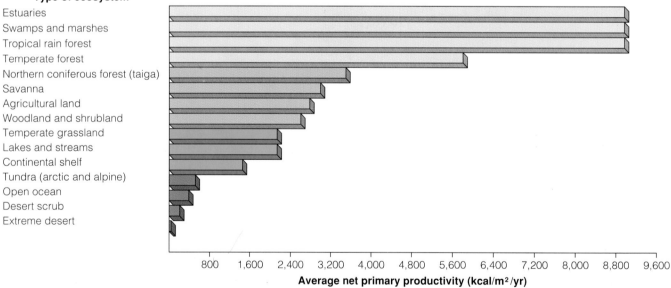

Figure 4-21 Estimated annual average net productivity of plants per square meter in major types of ecosystems.

important as food sources and spawning areas for fish, shrimp, and other forms of aquatic life that provide us with protein. So we should protect, not harvest, these plants.

In tropical forests most of the nutrients are stored in the trees and other vegetation rather than in the soil. When the trees are cleared, the low levels of nutrients in the exposed soil are rapidly depleted by frequent rains and by growing crops. Thus, food crops can be grown only for a short time without massive, expensive inputs of commercial fertilizers. So we should protect, not cut down, these forests.

We are already consuming, diverting, or wasting about 40% of the earth's potential net primary productivity on land. What will happen if we double the human population within your lifetime?

CO₂ 0.04% atmosphere (handwritten annotation)

4-4 MATTER RECYCLING IN ECOSYSTEMS

Biogeochemical Cycles Nutrients, chemicals essential for life, are recycled in the ecosphere (Figure 4-3) and in mature ecosystems in biogeochemical cycles (Figure 4-22). In these cycles the nutrients move from the environment, to organisms, then back to the environment.

There are two basic types of biogeochemical cycles: gaseous and sedimentary. *Gaseous cycles* primarily move nutrients back and forth between reservoirs in the atmosphere and hydrosphere and living organisms. Most of these cycles recycle elements rapidly, often within hours or days. The major gaseous cycles are the carbon, oxygen, hydrogen, and nitrogen cycles. *Carbon, O₂, H₂, N₂* (handwritten annotation)

Sedimentary cycles move nutrients mostly back and forth between reservoirs in the earth's crust (soil and rocks), hydrosphere (water), and living organisms. Elements in these cycles are usually recycled much more slowly than those in gaseous cycles because the elements are tied up in sedimentary rocks for long

periods of time, often thousands to millions of years. Phosphorus and sulfur are 2 of the 36 or so elements recycled in this manner.

Carbon Cycle Carbon is the basic building block of the carbohydrates, fats, proteins, nucleic acids such as DNA and RNA, and other organic compounds necessary for life. Most land plants get their carbon by absorbing carbon dioxide gas, which makes up about 0.04% of the gaseous atmosphere, through pores in their leaves. Phytoplankton, the microscopic plants that float in aquatic ecosystems, get their carbon from atmospheric carbon dioxide that has dissolved in water.

These producer plants then carry out photosynthesis, which converts the carbon in carbon dioxide to carbon in complex organic compounds such as glucose. Then the cells in oxygen-consuming plants, animals, and decomposers carry out aerobic cellular respiration, which breaks down glucose and other complex organic compounds and converts the carbon back to carbon dioxide for reuse by producers.

This linkage between photosynthesis and aerobic respiration circulates carbon in the ecosphere and is a major part of the carbon cycle (Figure 4-23). Oxygen and hydrogen, the other elements in glucose and other carbohydrates, cycle almost in step with carbon.

Figure 4-24 shows some other parts of the carbon cycle. It shows that some of the earth's carbon is tied up for long periods in fossil fuels—coal, petroleum, natural gas, peat, oil shale, tar sands, and lignite—formed over millions of years in the lithosphere as

Phosphorus, Sulfur (handwritten annotation)

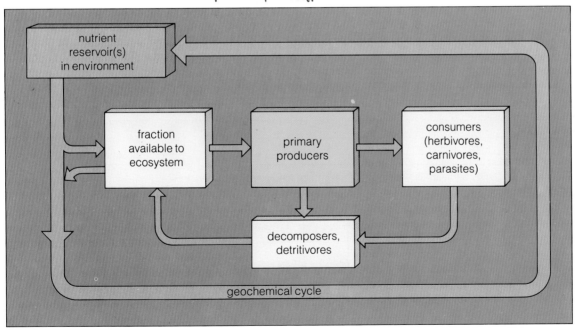

Figure 4-22 Generalized model of nutrient cycling in a mature ecosystem. Nutrients move from the environment, through organisms, and back to the environment in biogeochemical cycles. (Used by permission from Cecie Starr and Ralph Taggart, *Biology*, 5th ed., Wadsworth, 1989)

the organic compounds in dead plants and animals are buried and subjected to high pressures and temperatures. The carbon in these mineral deposits remains locked up until it is released to the atmosphere as carbon dioxide when fossil fuels are extracted and burned.

In aquatic ecosystems, carbon and oxygen combine with calcium to form insoluble calcium carbonate in shells and rocks. When shelled organisms die, they sink and their shells are buried in bottom sediments (Figure 4-24). Carbon in these sediment deposits reenters the cycle as carbon dioxide very slowly. The melting of rocks in long-term geological processes and volcanic eruptions also releases carbon dioxide into the air or water.

We have intervened in the carbon cycle mainly in two ways, especially since 1950, as world population and resource use have increased rapidly:

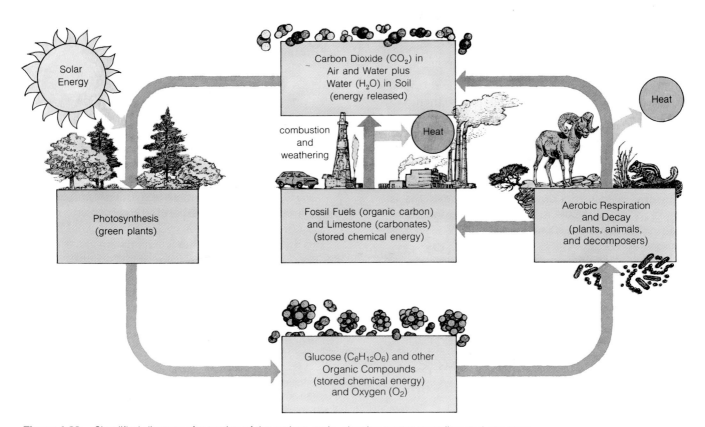

Figure 4-23 Simplified diagram of a portion of the carbon cycle, showing matter recycling and one-way energy flow through the processes of photosynthesis and aerobic respiration. This cyclical movement of matter through ecosystems and the ecosphere is also an important part of the oxygen and hydrogen cycles.

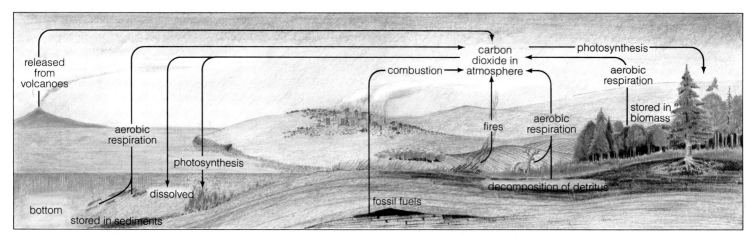

Figure 4-24 Simplified diagram of other parts of the carbon cycle. (Used by permission from Cecie Starr and Ralph Taggart, *Biology*, 5th ed., Wadsworth, 1989)

- Removal of forests and other vegetation without sufficient replanting, which leaves less vegetation to absorb CO_2. Also, carbon dioxide is added to the atmosphere when unharvested wood and plant debris decay and organic matter and roots in the exposed soil react with oxygen in the atmosphere.

- Burning carbon-containing fossil fuels and wood (Figure 4-24). This produces carbon dioxide that flows into the atmosphere. Scientists project that this carbon dioxide, along with other chemicals we are adding to the atmosphere, will warm the earth's atmosphere in coming decades, disrupt global food production and wildlife habitats, and raise average sea levels.

Nitrogen Cycle Organisms require nitrogen in various chemical forms to make proteins and genetically important nucleic acids such as DNA. Most green plants need nitrogen in the form of nitrate ions (NO_3^-) and ammonium ions (NH_4^+). The nitrogen gas (N_2) that makes up about 78% of the volume of the earth's atmosphere is useless to such plants, people, and most other organisms. Fortunately, nitrogen gas is converted into water-soluble ionic compounds containing nitrate ions and ammonium ions, which are taken up by plant roots as part of the **nitrogen cycle**. This gaseous cycle is shown in simplified form in Figure 4-25.

The conversion of atmospheric nitrogen gas into other chemical forms useful to plants is called **nitrogen fixation**. It is carried out mostly by cyanobacteria (once known as blue-green algae) and other kinds of bacteria in soil and water and by rhizobium bacteria living in small swellings called nodules on the roots of alfalfa, clover, peas, beans, and other legume plants (Figure 4-26). Also playing a role in nitrogen fixation, lightning converts nitrogen gas and oxygen gas in the atmosphere to nitric oxide and nitrogen dioxide gas. These gases react with water vapor in the atmosphere and are converted to nitrate ions that return to the earth as nitric acid dissolved in precipitation and as particles of solid nitrate compounds (salts).

Plants convert inorganic nitrate ions and ammonium ions obtained from soil water into proteins, DNA, and other large, nitrogen-containing organic compounds they require. Animals get most of their nitrogen-containing nutrients by eating plants or other animals that have eaten plants.

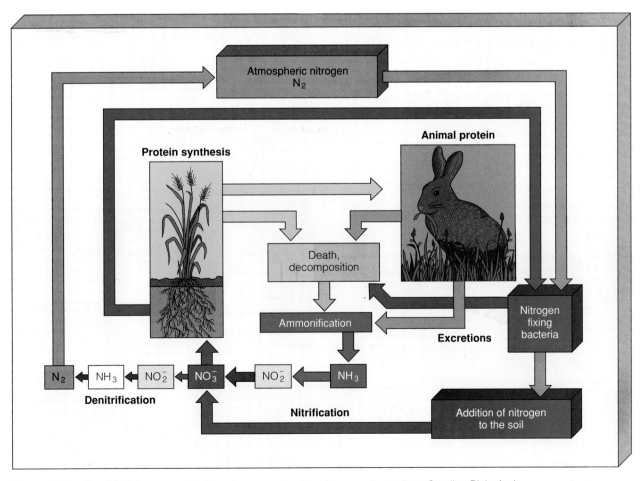

Figure 4-25 Simplified diagram of the the nitrogen cycle. (Used by permission from Carolina Biological Supply Company)

Specialized decomposer bacteria convert the nitrogen-containing organic compounds found in detritus (wastes and dead bodies of organisms) into inorganic compounds such as ammonia gas (NH_3) and water-soluble salts containing ammonium ions (NH_4^+). Other specialized groups of bacteria then convert these inorganic forms of nitrogen back into nitrate ions in the soil and into nitrogen gas, which is released to the atmosphere to begin the cycle again.

We intervene in the nitrogen cycle in several ways:

- Emission of large quantities of nitric oxide (NO) into the atmosphere when wood or any fossil fuel is burned. It is formed when oxygen and nitrogen molecules in the air combine at the high temperatures involved when fuels are burned in air. The nitric oxide then combines with oxygen gas in the atmosphere to form nitrogen dioxide (NO_2) gas, which can react with water vapor in the atmosphere to form nitric acid (HNO_3). This acid is a component of acid deposition, which is damaging trees and killing fish in parts of the world.

- Emission of the earth-warming gas nitrous oxide (N_2O) into the atmosphere by the action of certain bacteria on fertilizers and livestock wastes.

- Mining mineral deposits of compounds containing nitrate and ammonium ions for use as commercial fertilizers.

- Depleting nitrate ions and ammonium ions from soil by harvesting nitrogen-rich crops.

- Adding excess nitrate ions and ammonium ions to aquatic ecosystems in runoff of animal wastes from livestock feedlots, runoff of commercial nitrate fertilizers from cropland, and discharge of untreated and treated municipal sewage. This excess supply of plant nutrients stimulates rapid growth of algae and other

aquatic plants. The breakdown of dead algae by aerobic decomposers depletes the water of dissolved oxygen gas, causing massive fish kills.

Phosphorus Cycle Phosphorus, mainly in the form of certain types of phosphate ions (PO_4^{3-} and HPO_4^{2-}), is an essential nutrient of both plants and animals. It is a part of DNA molecules, which carry genetic information; ATP and ADP molecules, which store chemical energy for use by organisms in cellular respiration; certain fats in the membranes that encase plant and animal cells; and bones and teeth in animals.

Various forms of phosphorus are cycled mostly through the water, soil, and living organisms by the **phosphorus cycle,** shown in simplified form in Figure 4-27. In this cycle, phosphorus moves slowly from phosphate deposits on land and shallow ocean sediments to living organisms and back to the land and ocean. Bacteria are less important in the phosphorus cycle than in the nitrogen cycle.

Phosphorus released by the slow breakdown, or weathering, of phosphate rock deposits is dissolved in soil water and taken up by plant roots. However,

Figure 4-26 Plants in the the legume family have root nodules where rhizobium bacteria fix nitrogen by converting gaseous nitrogen (N_2) in the atmosphere to ammonia (NH_3), which in soil water forms ammonium ions (NH_4^+) that are taken up by the roots of plants.

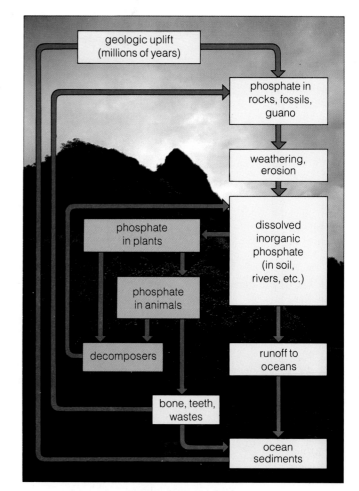

Figure 4-27 Simplified diagram of the phosphorus cycle. (Used by permission from Cecie Starr and Ralph Taggart, *Biology*, 5th ed., Wadsworth, 1989)

most soils contain only small amounts of phosphorus because phosphate compounds are fairly insoluble in water and are found only in certain kinds of rocks. Thus, phosphorus is the limiting factor for plant growth in many soils and aquatic ecosystems.

Animals get their phosphorus by eating plants or by eating animals that have eaten plants. Animal wastes and the decay products of dead animals and plants return much of this phosphorus to the soil, to rivers, and eventually to the ocean bottom as insoluble deposits of phosphate rock.

Some phosphate is returned to the land as guano—the phosphate-rich manure produced by fish-eating birds such as pelicans, gannets, and cormorants. But this return is small compared to the much larger amounts of phosphate eroded from the land to the oceans each year by natural processes and human activities.

Over millions of years geologic processes may push up and expose the seafloor, forming islands and other land surfaces. Weathering then slowly releases phosphorus from the exposed rocks and allows the cycle to begin again.

We intervene in the phosphorus cycle chiefly in two ways:

■ Mining large quantities of phosphate rock (Figure 4-28) to produce commercial inorganic fertilizers and detergent compounds.

■ Adding excess phosphate ions to aquatic ecosystems in runoff of animal wastes from livestock feedlots, runoff of commercial phosphate

fertilizers from cropland, and discharge of untreated and treated municipal sewage. As with nitrate and ammonium ions, an excessive supply of this plant nutrient causes explosive growth of algae and other aquatic plants that disrupt life in aquatic ecosystems.

Hydrologic Cycle The **hydrologic cycle** or **water cycle,** which collects, purifies, and distributes the earth's fixed supply of water, is shown simplified in Figure 4-29. Solar energy and gravity continuously

Figure 4-28 Surface mining of phosphate rock in Illinois. Because phosphorus is recycled so slowly to the land, it is the major plant nutrient likely to be in short suppy for use as a commercial fertilizer to grow more food for the world's rapidly increasing population.

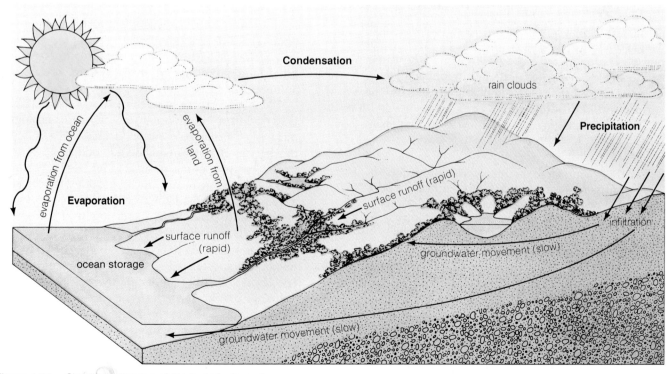

Figure 4-29   Simplified diagram of the hydrologic cycle.

move water among the ocean, air, land, and living organisms. The main processes in this water recycling and purifying cycle are *evaporation* (conversion of water to water vapor), *condensation* (conversion of water vapor to droplets of liquid water), *transpiration* (the process in which water is absorbed by the root systems of plants, passes through their living structure, then evaporates into the atmosphere as water vapor), *precipitation* (dew, rain, sleet, hail, snow), and *runoff* back to the sea to begin the cycle again.

Incoming solar energy evaporates water from oceans, rivers, lakes, soil, and vegetation into the atmosphere. Winds and air masses transport this water vapor over various parts of the earth's surface. Decreases in temperature in parts of the atmosphere cause the water vapor to condense and form tiny droplets of water in the form of clouds or fog. Eventually these droplets combine and become heavy enough to fall to the land and into bodies of water as precipitation. About half of the earth's precipitation falls on tropical forests.

Some of the fresh water returning to the earth's surface as precipitation becomes locked in glaciers. Much of it collects in puddles and ditches and runs off into nearby lakes, streams, and rivers, which carry water back to the oceans, completing the cycle. This runoff of fresh water from the land also causes soil erosion, which moves various chemicals through portions of other biogeochemical cycles.

A large portion of the water returning to the land seeps or infiltrates into surface soil layers and some percolates downward deep into the ground. There it is stored as groundwater in *aquifers*—spaces in and between rock formations. Underground springs and streams eventually return this water to the surface and to rivers, lakes, and surface streams, from which it evaporates or returns to the ocean. Fresh water percolates downward through the soil to replenish aquifers. However, the underground circulation of water is extremely slow compared to that on the surface and in the atmosphere.

We intervene in the water cycle in two main ways:

- Withdrawing large quantities of fresh water from rivers, lakes, and aquifers. In heavily populated or heavily irrigated areas, withdrawals have led to groundwater depletion or intrusion of ocean salt water into underground water supplies.

- Clearing vegetation from land for agriculture, mining, roads, parking lots, construction, and other activities. This reduces seepage that recharges groundwater supplies, increases the risk of flooding, and increases the rate of surface runoff, which increases soil erosion.

4-5 ROLES AND INTERACTIONS OF SPECIES IN ECOSYSTEMS

Types of Species Found in Ecosystems

If you observe various ecosystems you will find that they can have four types of species:

- **Native species** that normally live and thrive in a particular ecosystem.

- **Immigrant** or **alien species** that migrate into an ecosystem or that are deliberately or accidentally introduced into an ecosystem by humans. Some of these species are beneficial while others can take over and eliminate many native species.

- **Immigrant** or **alien species** that migrate into an that an ecosystem is being degraded. For example, the present decline of migratory, insect-eating songbirds in North America indicates a loss of habitat in their summer homes in North America and in their winter homes in the rapidly disappearing tropical forests in Latin America and the Caribbean Islands. Some indicator species, such as the American bald eagle and the brown pelican, feed at high trophic levels in food chains and webs. This makes them vulnerable to high levels of fat-soluble chemicals such as DDT, whose concentrations are increased in the tissues of organisms at each successive trophic level.

- **Keystone species** that play roles affecting many other organisms in an ecosystem. The loss of a keystone species can lead to sharp population drops and extinctions of other species that depend on it for certain services. An example is the alligator (see Case Study on p. 81).

Ecological Niche An **ecological niche** (pronounced "nitch") is a description of all the physical, chemical, and biological factors that a species needs to live, grow, and reproduce in an ecosystem. It describes an organism's role in an ecosystem.

A common analogy is that an organism's habitat is its "address" in an ecosystem, while its ecological niche is its "occupation" and "lifestyle." For example, the habitat of a robin includes woodlands, forests, parks, pasture lands, meadows, orchards, gardens, and yards. Its ecological niche includes factors such as nesting and roosting in trees, eating insects, earthworms, and fruit, and dispersing fruit and berry seeds in its droppings.

Information about ecological niches helps people manage domesticated and wild plant and animal species as sources of food and other resources. It also helps us predict the effects of either adding or removing a species to or from an ecosystem. But determining the interacting factors that make up an organism's ecological niche is very difficult.

Ecological Importance of the American Alligator

People tend to divide plants and animals into "good" and "bad" species and to assume that we have a duty to wipe out the villains or use them up to satisfy our needs and wants. One species that we drove to near extinction in many of its marsh and swamp habitats is the American alligator (Figure 4-30).

Alligators have no natural predators, except people. Hunters once killed large numbers of these animals for their exotic meat and supple belly skin, which was used to make shoes, belts, and other items. Between 1950 and 1960, hunters wiped out 90% of the alligators in Louisiana. The alligator population in the Florida Everglades also was threatened.

Many people might say, "So what?" But they are overlooking the key role the alligator plays in subtropical, wetland ecosystems such as the Everglades. Alligators dig deep depressions, or "gator holes," which collect fresh water during dry spells. These holes are refuges for aquatic life and supply fresh water and food for birds and other animals.

Large alligator nesting mounds also serve as nest sites for birds such as herons and egrets. As alligators move from gator holes to nesting mounds, they help keep waterways open. They also eat large numbers of gar, a fish that preys on other fish. This means that alligators help maintain populations of game fish such as bass and bream.

Figure 4-30 The American alligator is a keystone species in its marsh and swamp habitats in the southeastern United States. In 1967 the species was classified as an endangered species in the United States. This protection allowed the population of the species to recover to the point that its status has been changed from endangered to threatened.

In 1967 the U.S. government placed the American alligator on the endangered species list. Averaging about 40 eggs per nest and protected from hunters, by 1975 the alligator population had made a strong comeback in many areas—too strong, according to some people who found alligators in their backyards and swimming pools.

The problem is that both human and alligator populations are increasing rapidly, and people are taking over the natural habitats of the alligator. A gator's main diet is snails, apples, sick fish, ducks, raccoons, and turtles. But if a pet or a person falls into or swims in a canal, pond, or other area where a gator lives, the reptile's jaws will shut by reflex on just about anything it encounters.

In 1977 the U.S. Fish and Wildlife Service reclassified the American alligator from endangered to threatened in Florida, Louisiana, and Texas, where 90% of the animals live. In 1987 this reclassification was extended to seven other states.

As a threatened species, alligators are still protected from excessive harvesting by hunters. However, limited hunting is allowed in some areas to keep the population from growing too large. Florida, with at least 1 million alligators, permits 7,000 kills a year. Another 1 million live in southern states extending from Texas to South Carolina. The comeback of the American alligator is an important success story in wildlife conservation.

Specialist and Generalist Niches The niche of an organism can be classified as specialized or generalized. Most species of plants and animals can tolerate only a narrow range of climatic and other environmental conditions and feed on a limited number of different plants or animals. Such species have a *specialized niche*, which limits them to specific habitats in the ecosphere.

The giant panda, for example, has a highly specialized niche because it obtains 99% of its food by consuming bamboo plants. The destruction of several species of bamboo in parts of China, where the panda is found, has led to the animal's near extinction.

In a tropical rain forest, an incredibly diverse array of plant and animal life survives by occupying a variety of specialized ecological niches in distinct layers of the forest's vegetation (Figure 4-31). The widespread clearing of such forests is dooming millions of specialized species to extinction.

Figure 4-31 Stratification of specialized plant and animal niches in a tropical rain forest. Species occupy specialized niches in the various layers of vegetation.

Species with a *generalist niche* are very adaptable. They can live in many different places, eat a wide variety of foods, and tolerate a wide range of environmental conditions. This explains why they are usually in less danger of extinction than species with a specialized niche. Examples of generalist species include flies, cockroaches, mice, rats, and human beings.

The ecological niches of species include how they interact with other species in an ecosystem. The major types of species interaction are interspecific competition, predation, parasitism, mutualism, and commensalism.

Competition Between Species As long as commonly used resources are abundant, different species can share them. But when two or more species in the same ecosystem attempt to use the same scarce resources, they are said to be engaging in **interspecific competition**. At least part of their ecological niches overlap. The scarce resource may be food, water,

carbon dioxide, sunlight, soil nutrients, space, shelter, or anything required for survival.

One competing species may gain an advantage over the other species by producing more young, getting more food or solar energy, or defending itself better. It may also have an advantage by being able to tolerate a wider range of temperature, light, water salinity, or concentrations of certain poisons.

Other species gain greater access to a scarce resource by preventing competing species from using the resource. For example, hummingbirds chase bees and other hummingbird species away from clumps of flowers.

Populations of some animal species avoid or reduce competition with more dominant species by moving to another area, switching to a less accessible or less readily digestible food source, or hunting for the same food source at different times of the day or in different places. For example, hawks and owls feed on similar prey, but hawks hunt during the day and owls hunt at night. Where lions and leopards occur

Sharks have lived in the oceans for over 400 million years, long before dinosaurs appeared (Figure 4-32). During their long history sharks have evolved into more than 350 species whose size, behavior, and other characteristics differ widely.

The smallest species, the cigar shark, is only a foot long at maturity. Another species, called "cookie cutters," is also about a foot long. Cookie cutters survive by taking bites out of the sides of porpoises, whales, and large fish such as bluefin tuna. At the other end of the scale is the whale shark, the world's biggest fish. It can grow to 18 meters (60 feet).

Sharks have extremely sensitive sense organs. They can detect the scent of decaying fish or blood even when it is diluted to only one part per million parts of seawater. They have superb hearing and better night vision than we do. They also sense the electrical impulses radiated by the muscles of animals, making it difficult for their prey to escape detection. They are powerful and rapid swimmers. Because their bodies are denser than seawater, most types of shark must always keep moving in order not to sink.

Sharks are the key predators in the world's oceans, helping control the numbers of many other ocean predators. Without sharks the oceans would be overcrowded with dead and dying fish and depleted of many healthy ones that we rely on for food. Eliminating sharks would upset the balance of ocean ecosystems.

Yet this is precisely what we are in danger of doing. Every year we catch over 100 million sharks, mostly for food and for their fins, which are sent off to Asia for shark-fin soup. Since 1986, the demand for sharks has increased dramatically. Others are killed for sport and out of fear. Some shark species, such as edible thresher and mako sharks, are being commercially exploited and could face extinction. Sharks are vulnerable to overfishing because it takes most species about 12 years to begin reproducing and they produce only a few offspring.

Influenced by movies and popular novels, most people see sharks as people-eating monsters. This is far from the truth. Each year a few types of shark—great white, bull, tiger, and oceanic whitetip—injure about 100 people and kill perhaps 25. Most attacks are by great white sharks, which often feed on sea lions and other marine mammals and sometimes mistake human swimmers for their normal prey. Nevertheless, with hundreds of millions of people swimming in the ocean each year, the chances of being killed by a shark are minute—about 1 in 5 million. You are thousands of times more likely to get killed when you drive or ride in a car.

Furthermore, sharks help save lives. In addition to providing people with food, they are helping us learn how to fight cancer, bacteria, and viruses. Sharks are very healthy and have aging processes similar to ours. Their highly effective immune system allows wounds to heal quickly without becoming infected. A chemical extracted from shark cartilage is being used as an artificial skin for burn victims. Sharks are among the few animals in the world that almost never get cancer and eye cataracts. Understanding why can help us improve human health.

Sharks are needed in the world's ocean ecosystems. Although they don't need us, we need them.

Figure 4-32 This blue shark and other types of sharks are key predators in ocean ecosystems.

Howard Hall/Earth Images

together, lions take mostly larger animals as prey and leopards take smaller ones.

Predation and Parasitism The most obvious form of species interaction in food chains and webs is **predation**: An individual organism of one species, known as the **predator,** feeds on parts or all of an organism of another species, the **prey,** but does not live on or in the prey. Together, the two kinds of organisms involved, such as lions and zebras, are said to have a **predator-prey relationship**. Sharks are one of the most important predators in the world's oceans (see Case Study above).

Prey species have various protective mechanisms. Otherwise, they would easily be captured and eaten. Some can run, swim, or fly fast. Others have thick

Figure 4-33 Parasitism. Sea lampreys are parasites that use their suckerlike mouths to attach themselves to the sides of fish on which they prey. Then they bore a hole in the fish with their teeth and feed on its blood.

Figure 4-34 Mutualism. These oxpeckers are feeding on the ticks that infest this endangered black rhinoceros in Kenya. The rhino benefits from regular removal of these parasites. Only about 3,500 black rhinos are left in Africa. This and other species of rhinoceros face extinction because they are illegally killed for their horns, which can sell for as much as $44,000 a kilogram ($20,000 a pound), and because of a loss of habitat. Private and government conservationists are trying to protect rhinos from further poaching by creating fenced or heavily guarded sanctuaries and private ranches, relocating some animals to protected areas, and building up captive breeding populations for all species. These efforts have led to a slow increase in numbers in protected areas and in captivity. But it is an expensive and dangerous uphill fight. Wildlife officials have even resorted to cutting off and burning the horns of surviving rhinos to protect them from being killed.

skins or shells. Still others have camouflage colorings or the ability to change color so that they can hide by blending into their environment. Others give off chemicals that smell or taste bad to their predators, or poison them. Some live in large groups (schools of fish, herds of antelope).

Predators have several methods of capturing their prey. A carnivore generally has to chase and catch its food; a herbivore doesn't. One way a carnivore can get enough food is to run fast, as the cheetah does. Another way is to cooperate by hunting in packs, as spotted hyenas, African lions, wolves, and Cape hunting dogs do. Under natural conditions such species are often more abundant than those, such as leopards, tigers, and cougars, that do not cooperate in hunting prey.

A third way many predators get enough to eat is to look for prey individuals that appear to be sick, crippled, or in some way disabled. This natural weeding out of diseased and weak individuals also benefits the prey species by preventing the spread of disease and leaving stronger and healthier individuals for breeding. A fourth way to capture prey is to invent weapons and traps and learn how to domesticate animals, as humans have done.

Another type of species interaction is parasitism. **Parasites** feed off another organism, called their **host,** but unlike predators, they live on or in the host for a good part of their life cycle. The parasite draws nourishment from and gradually weakens its host. This may or may not kill the host.

Both plants and animals can be attacked by parasites, and both plants and animals may be parasites. Tapeworms, disease-causing bacteria, and other parasites live inside their hosts. Lice, ticks, mistletoe plants, and lampreys (Figure 4-33) attach themselves to the outside of their hosts.

Some parasites can move from one host to another, as dog fleas do. Others may spend their adult lives attached to a single host. Examples are mistletoe, which feeds on oak tree branches, and tapeworms, which feed in the intestines of humans and other animals. Fleas and ticks may have internal parasites even smaller than they are.

Mutualism and Commensalism In some cases two different types of organisms interact directly in ways that benefit each species. Such a mutually beneficial interaction between species is called **mutualism**. The honeybee and certain flowers have a mutualistic relationship. The honeybee feeds on the flower's nectar and in the process picks up pollen and pollinates female flowers when it feeds on them. Another example is the relationship between rhinos and oxpeckers (Figure 4-34). Another important case is the rhizobium bacteria that live in nodules in the roots of legumes (members of the pea family) (Figure 4-26). The bacteria capture atmospheric nitrogen and convert it to a form usable by the plants, and the legume provides the bacteria with sugar. This mutualistic relationship is a vital part of the nitrogen cycle (Figure 4-25).

In another type of species interaction, called **commensalism,** one type of organism benefits, while the other is neither helped nor harmed to any great degree. For example, in the open sea certain types of barnacles live on the jawbones of whales. The barnacles benefit by having a safe place to live and a steady food supply. The whale apparently gets no benefit from this relationship, but it suffers no harm from it either.

This chapter has shown that the essential feature of the living and nonliving parts of individual terrestrial and aquatic ecosystems and of the global ecosystem, or ecosphere, is interdependence and connectedness. Without the services performed by diverse communities of plants, animals, and decomposers, we would be starving, gasping for breath, and drowning in our own wastes. The next chapter shows how this interdependence is the key to understanding the earth's major types of ecosystems and how ecosystems change in response to natural and human stresses.

We sang the songs that carried in their melodies all the sounds of nature—the running waters, the sighing of winds, and the calls of the animals. Teach these to your children that they may come to love nature as we love it.

GRAND COUNCIL FIRE OF AMERICAN INDIANS

DISCUSSION TOPICS

1. a. A bumper sticker asks, "Have you thanked a green plant today?" Give two reasons for appreciating a green plant.
    b. Trace the sources of the materials that make up the sticker and see whether the sticker itself is a sound application of the slogan.
    c. Explain how decomposers help keep you alive.

2. a. How would you set up a self-sustaining aquarium for tropical fish?
    b. Suppose you have a balanced aquarium sealed with a clear glass top. Can life continue in the aquarium indefinitely as long as the sun shines regularly on it?
    c. A friend cleans out your aquarium and removes all the soil and plants, leaving only the fish and water. What will happen?

3. Using the second law of energy, explain why there is such a sharp decrease in high-quality energy as energy flows through a food chain or web. Doesn't an energy loss at each step violate the first law of energy? Explain.

4. Using the second law of energy, explain why many poor people in less developed countries exist mostly on a vegetarian diet.

5. Using the second law of energy, explain why on a per weight basis steak costs more than corn.

6. Why are there fewer lions than mice in an African ecosystem supporting both types of animals?

FURTHER READINGS

Colinvaux, Paul A. 1978. *Why Big Fierce Animals Are Rare.* Princeton, N.J.: Princeton University Press.

Ehrlich, Paul R. 1986. *The Machinery of Life: The Living World Around Us and How It Works.* New York: Simon & Schuster.

Ehrlich, Paul R., Anne H. Ehrlich, and John P. Holdren. 1977. *Ecoscience: Population, Resources and Environment.* San Francisco: W. H. Freeman.

Kormondy, Edward J. 1984. *Concepts of Ecology.* 3rd ed. Englewood Cliffs, N.J.: Prentice-Hall.

Odum, Eugene P. 1989. *Ecology and Our Endangered Life-Support Systems.* Sunderland, Mass.: Sinauer.

Ramadé, Francois. 1984. *Ecology of Natural Resources.* New York: John Wiley.

Rickleffs, Robert E. 1989. *Ecology.* 3rd ed. Salt Lake City, Utah: W. H. Freeman.

Smith, Robert L. 1990. *Elements of Ecology.* 4th ed. New York: Harper & Row.

Watt, Kenneth E. F. 1982. *Understanding the Environment.* Newton, Mass.: Allyn & Bacon.

CHAPTER FIVE

ECOSYSTEMS: WHAT ARE THE MAJOR TYPES AND WHAT CAN HAPPEN TO THEM?

General Questions and Issues

1. What are the major types of terrestrial ecosystems, and how does climate influence the type found in a given area?

2. What are the major types of aquatic ecosystems, and what major factors influence the kinds of life they contain?

3. What are the major effects of environmental stress on living systems?

4. How can populations of plant and animal species adapt to natural and human-induced stresses to preserve their stability and sustainability?

5. How can communities and ecosystems change and adapt to small- and large-scale natural and human-induced stresses?

6. What major impacts do human activities have on populations, communities, and ecosystems?

When we try to pick out anything by itself, we find it hitched to everything else in the universe.

JOHN MUIR

HE ECOSPHERE, in which all organisms are found, contains an astonishing variety of *terrestrial ecosystems,* such as deserts, grasslands, and forests, and *aquatic ecosystems,* such as lakes, reservoirs, ponds, wetlands, and oceans. These ecosystems are the life-support systems for us and other species (Figure 5-1). Each ecosystem has a characteristic plant and animal community adapted to certain environmental conditions, especially climate. Each varies in its average productivity of plant life that supports most other forms of life (Figure 4-21, p. 74, and Figure 5-1).

The world's ecosystems and biological communities are dynamic, not static. Their populations of various species are always changing and adapting in response to major and minor changes in environmental conditions. Some of these changes occur because of natural events such as climate change, floods, and volcanic eruptions. Others are caused by land clearing, emissions of various pollutants, and other human activities. However, there are limits to how rapidly such adaptations to stress can occur. Understanding how the world's major ecosystems work and change can help us in sustaining them rather than degrading and destroying them.

5-1 MAJOR TYPES OF TERRESTRIAL ECOSYSTEMS: DESERTS, GRASSLANDS, AND FORESTS

Effects of Precipitation and Temperature on Plant Types Figure 5-2 shows the distribution of the world's major land or terrestrial ecosystems. Why is one area of the earth's land surface a desert, another a grassland, and another a forest? Why are there different types of deserts, grasslands, and forests? What determines the types of life found in these ecosystems?

The general answer to these questions is differences in **climate:** the average of the day-to-day weather conditions in a region, usually over a 30-year period or longer. Many factors affect the climate of an area, but the two most important are its *average temperature* and *average precipitation.* Figure 5-3 shows the global distribution of the major types of climate based on these two factors. By comparing Figures 5-2 and 5-3 you can see the world's major terrestrial ecosystems vary with climate.

With respect to plants, *precipitation generally is the limiting factor that determines whether most of the world's terrestrial ecosystems are desert, grassland, or forest.* A **desert** is an area where evaporation exceeds precipitation and the average amount of precipitation is less than 25 centimeters (10 inches) a year. Such areas have little vegetation or have widely spaced, mostly low vegetation.

Gene Carl Feldman, Compton J. Tucker—NASA/Goddard Space Flight Center

Figure 5-1 The biosphere. Three years of satellite data were combined to produce this picture of the earth's biological productivity. Rain forests and other highly productive areas appear as dark green, deserts as yellow. The concentration of phytoplankton, a primary indicator of ocean productivity, is represented by a scale that runs from red (highest) to orange, yellow, green, and blue (lowest).

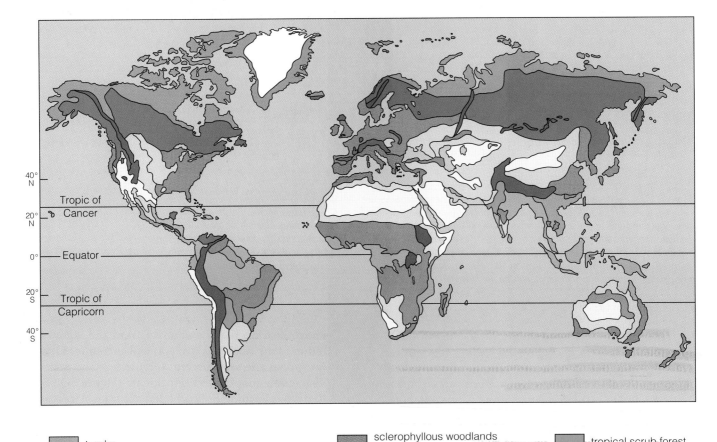

▰ tundra	▰ sclerophyllous woodlands and shrublands (chaparral)	▰ tropical scrub forest
▰ boreal forest (taiga), evergreen coniferous forest (e.g., montane coniferous forest)	☐ desert	▰ tropical savanna, thorn forest
▰ temperate deciduous forest	▰ tropical rain forest, tropical evergreen forest	☐ semidesert, arid grassland
☐ temperate grassland	▰ tropical deciduous forest	▰ mountains (complex zonation)

Figure 5-2 The world's major terrestrial ecosystems.

BIOME
MARINE ECOSYSTEM
ZONES

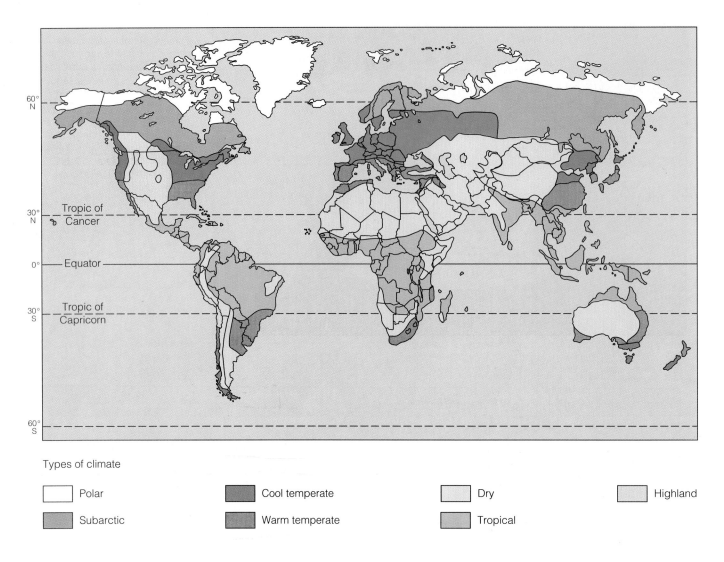

Types of climate

☐ Polar   ■ Cool temperate   ☐ Dry   ▨ Highland

▨ Subarctic   ■ Warm temperate   ☐ Tropical

Figure 5-3 Generalized map of global climates. These climates are dictated generally by two variables: the temperature with its seasonal variations and the quantity and distribution of precipitation over each year. The average temperature of an area depends upon the strength of the sun's rays reaching it and thus upon latitude or distance from the equator, which receives the largest solar input. But this is modified by global air circulation, winds, ocean currents, and topographical features that distribute humidity and heat over the face of the earth, as discussed in more detail in Section 10-1. Because of these factors, the limits of the earth's climate zones do not exactly follow the parallels of latitude we draw around the earth's surface.

Grasslands are regions where the average annual precipitation is great enough to allow grass to prosper, yet so erratic that periodic drought and fire prevent large stands of trees from growing. Undisturbed areas with moderate to high average annual precipitation tend to be covered with **forest,** containing various species of trees and smaller forms of vegetation.

Average precipitation and average temperature, along with soil type, are the major factors determining the particular type of desert, grassland, or forest in a particular area. Acting together, these factors lead to tropical, temperate, and polar deserts, grasslands, and forests (Figure 5-4).

Climate and vegetation both vary with **latitude** (distance from the equator) and **altitude** (height

above sea level). If you travel from the equator to either of the earth's poles, you will encounter increasingly cold and wet climates (Figure 5-3) and zones of vegetation adapted to each climate (Figure 5-5). Similarly, as elevation or height above sea level increases, the climate becomes wetter and colder. If you climb a tall mountain from its base to its summit, you will find changes in plant life similar to those you would find in travelling from the equator to one of the earth's poles (Figure 5-5).

Major Types of Deserts Three major types of deserts occur because of combinations of low average precipitation with different average temperatures: tropical, temperate, and cold (Figure 5-4). *Tropical deserts,* such

Tropical – Sahara
Temperate – Mojave
Cold – Gobi

Handwritten annotations on figure: boreal/taigas · 4 seasons · ARCTIC · HERE · TALL & SHORT GRASS PRAIRIES · Gobi · Mojave · Sahara · Serengeti

Figure 5-4 Average precipitation and average temperature act together over a period of 30 years or more as limiting factors that determine the type of desert, grassland, or forest ecosystem found in a particular area.

as the southern Sahara, make up about one-fifth of the world's desert area (Figure 5-6). They typically have few plants and a hard, windblown surface strewn with rocks and some sand. Temperatures are hot year-round. In *temperate deserts*, such as the Mojave in southern California, daytime temperatures are hot in summer and cool in winter (Figure 5-7). In *cold deserts*, such as the Gobi lying south of Siberia, winters are cold and summers are warm or hot.

Handwritten annotation: Temperate hot summer cool winter

The atmosphere above deserts is a poor insulator because it contains little water vapor. Because the ground radiates heat rapidly after the sun goes down, nights in deserts are often cold.

Plants and animals in all deserts are adapted to capture and conserve scarce water. Plants in temperate and cold deserts usually are widely spaced, minimizing competition for water. Other ways plants get water is through roots reaching deep into the soil to tap groundwater, or wax-coated leaves to reduce the amount of water lost by evaporation; flesh-stemmed cacti store water in their succulent tissues (Figure 4-6, p. 63).

Most desert animals escape the daytime heat by staying underground in burrows during the day and being active at night. Desert animals also have special adaptations to help them conserve water (Figure 5-8). Insects and reptiles have thick outer coverings to minimize water loss through evaporation. Some desert animals become dormant during periods of extreme heat or drought.

The slow growth rate of plants, low species diversity, and shortages of water make deserts quite fragile. For example, vegetation destroyed by human activities such as livestock grazing and driving of motorcycles and other off-road driving may take decades to grow back. Vehicles can also collapse some underground burrows that are habitats for many desert animals.

Major Types of Grasslands *Tropical grasslands*, or *savannas*, are found in areas with high average temperatures, two prolonged dry seasons during winter and summer, and abundant rain the rest of the year. They occur in a wide belt on either side of the equator

Handwritten annotation: 2 dry seasons - winter + summer

Handwritten annotation: TROPICAL GRASSLAND - SAVANNA

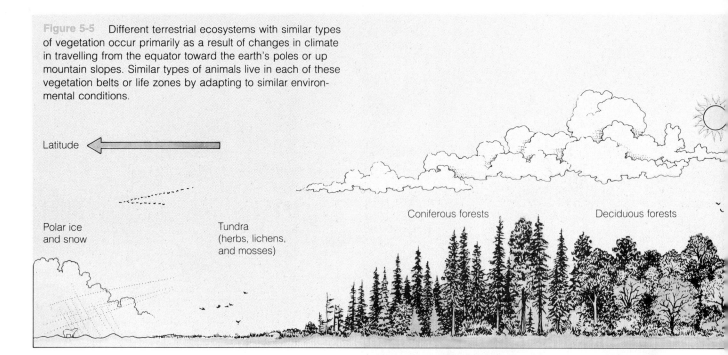

Figure 5-5   Different terrestrial ecosystems with similar types of vegetation occur primarily as a result of changes in climate in travelling from the equator toward the earth's poles or up mountain slopes. Similar types of animals live in each of these vegetation belts or life zones by adapting to similar environmental conditions.

Latitude

Polar ice and snow

Tundra (herbs, lichens, and mosses)

Coniferous forests

Deciduous forests

Gert Behrens/Ardea London

Figure 5-6   Tropical desert in southwest Africa. Each of the world's major ecosystems contains a diverse community of plants, animals, and decomposers that sustain the ecosystem and adapt to changing environmental conditions.

Richard Frear/National Park Service

Figure 5-7   Temperate desert in Arizona. The vegetation includes creosote bushes, ocotillo, saguaro cacti, and prickly pear cacti. After a brief and infrequent rain the ground is covered with a variety of wildflowers. Most animals escape the hot days by living underground and coming out at night.

Tropical grassland

SERENGETI

beyond the borders of tropical rain forests (Figure 5-2).

Land covered with native grasses has important ecological roles. Such grasslands act like a giant sponge, soaking up scarce rainfall so that it does not run off the land in one big rush. This helps provide water for springs and streams year-round. Grasses also shield the life-nurturing topsoil from the erosive force of wind and rain. The grasses, shrubs, small trees, and soil on grasslands provide food and habitats for many types of wildlife.

Some tropical grasslands, such as Africa's Serengeti Plain, consist mostly of open plains covered with low or high grasses (Figure 5-9). Others contain grasses along with varying numbers of widely spaced, small, mostly deciduous trees and shrubs, which shed their leaves during the dry season to avoid excessive water loss.

Temperate grasslands are found in the large interior areas of most continents (Figure 5-2). Types of temperate grasslands are the *tall-grass prairies* (Figure 5-10) and *short-grass prairies* (Figure 5-11) of the midwestern and western United States and Canada, the *pampas* of South America, the *veld* of southern Africa, and the *steppes* that stretch from central Europe into Siberia. In these ecosystems winds blow almost con-

Temperate Grasslands: *steppes*
Tall-grass prairies
short-grass "
pampas
veld

Altitude

Mountain
ice and snow

Tundra (moss,
lichen, herbs)

Coniferous
forests

Deciduous
forests

Tropical forests

Tropical
forests

Figure 5-9 African elephant in a tropical grassland (savanna), Serengeti Plain in Tanzania.

Figure 5-8 This nocturnal kangaroo rat in a California desert is a master of water conservation. Instead of drinking water it gets the water it needs from its food and cellular respiration. It also conserves water by excreting dry feces and thick, nearly solid urine.

tinuously and evaporation is rapid. As long as it is not plowed up, the soil is held in place by the thick network of grass roots. However, because of their fertile soils the world's temperate grasslands are major areas for growing crops (Figure 5-12).

Polar grassland, or *Arctic tundra,* is found in areas just below the Arctic ice region (Figure 5-2). During most of the year this ecosystem is bitterly cold, with icy galelike winds, and is covered with ice and snow. Winters are long and dark, and average annual precipitation is low and occurs mostly as snow. The Arctic tundra is carpeted with a thick, spongy mat of low-growing plants such as lichens (growths of algae

and fungi), sedges (grasslike plants often growing in dense tufts in marshy places), mosses, grasses, and low shrubs (Figure 5-13). Because of the cold temperatures, decomposition is slow.

One effect of the tundra's extreme cold is **permafrost**—water frozen year-round in thick, underground layers of soil. During the six-to-eight-week summer, when sunlight persists almost around the clock, the surface layer of soil thaws. But the layer of permanent ice a few feet below the surface prevents the water from seeping deep into the ground. As a result, during summer the tundra turns into a soggy landscape dotted with shallow lakes, marshes, bogs, and ponds. Hordes of mosquitoes, deerflies, blackflies, and other insects thrive in the shallow surface pools. They serve as food for large colonies of migratory birds, especially waterfowl, which migrate from the south to nest and breed in the bogs and ponds.

The low rate of decomposition, shallow soil, and slow growth rate of plants make the Arctic tundra perhaps the earth's most fragile ecosystem (Figure 5-14). Vegetation destroyed by this and other human activities can take decades to grow back. Buildings, roads, pipelines, and railroads must be built over bedrock, on insulating layers of gravel, or on deep-seated pilings. Otherwise, the structures melt the upper layer of permafrost and tilt or crack as the land beneath them shifts and settles.

Major Types of Forests *Tropical rain forests,* found in certain areas near the equator (Figures 5-1 and 5-2), have a warm but not hot annual mean temperature that varies little daily or seasonally. They have high humidity and heavy rainfall almost daily. These eco-

Figure 5-10 A patch of tall-grass prairie temperate grassland in Mason County, Illinois, in early September. Only about 1% of the original tall-grass prairies that once thrived in the midwestern United States and Canada remain. Because of their highly fertile soil, most have been cleared for crops such as corn, wheat, and soybeans, and for hog farming.

John Schwegman

Figure 5-11 Sheep grazing on a temperate grassland (shortgrass prairie) in Idaho.

Soil Conservation Service

systems are dominated by evergreen trees, which keep most of their leaves or needles throughout the year (Figure 5-15).

The almost unchanging climate in rain forests means that water and temperature are not limiting factors as they are in other terrestrial ecosystems. In this ecosystem, nutrients from the often nutrient-poor soils are the major limiting factors.

A mature rain forest has a greater diversity of plant and animal species per unit of area than any other terrestrial ecosystem. These diverse forms of plant and animal life occupy a variety of ecological niches in distinct layers (Figure 4-31, p. 82).

After hundreds of thousands of years of evolution in a fairly constant climate, species in these ecosystems have evolved numerous mechanisms to defend against their predators and to increase their populations. You can find caterpillars masquerading as snakes, trees that rely on rivers to distribute their seeds when the rivers flood, and plants that give off the smell of rotting meat to attract flies as pollinators.

Rain forests are deceptively fragile. Left to themselves, they can sustain themselves indefinitely. But clear large areas and you end up with patches of grassland and eventually desert.

The reason for this fragility is that most of the nutrients in this ecosystem are in the vegetation, not in the upper layers of soil as in most other terrestrial ecosystems. Once the vegetation is removed, the soils rapidly lose the few nutrients they have and cannot grow crops for more than a few years without large-scale use of commercial fertilizers. Furthermore, when vegetation is cleared, the heavy rainfall washes away most of the thin layer of topsoil. This means that

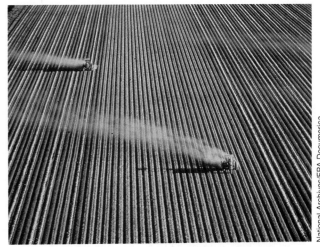

Figure 5-12 Replacement of a temperate grassland with a monocultured cropland near Blythe, California. When the tangled network of natural grasses is removed, the vital topsoil is subject to severe wind erosion unless it is kept covered with some type of vegetation. If global warming accelerates in your lifetime, many of these grasslands are expected to become too hot and dry for farming, thus threatening the world's food supply.

National Archives/EPA Documerica

regenerating a mature rain forest on large cleared areas is almost impossible.

These unique and valuable forests are being cleared at a rapid rate to harvest timber, to mine minerals, and to plant crops and graze livestock on unsustainable soils. Some ecologists project that if the clearing of tropical rain forests continues at the present rate, within 50 years only a few of these diverse ecosystems will remain. Also gone will be hundreds of thousands of animal and plant species with highly specialized niches in these forests.

Temperate deciduous forests grow in areas with

Figure 5-13 Polar grassland (Arctic tundra) in Alaska in summer.

Kenneth W. Fink/Ardea London

Figure 5-14 Arctic tundra is a fragile biome as shown by this degradation of tundra on Victoria Island, Northwest Territories, Canada. Vehicles have broken the thin layer of vegetation and soil that covers the permafrost (ice) and shields it from the sun. The permafrost melts and the tire tracks turn into thin ribbons of water and ice.

Brian Rogers/Biofotos

Figure 5-15 Tropical rain forest in Monteverde Cloud Forest Reserve in Costa Rica. These storehouses of the earth's biodiversity are habitats for 50% to 80% of the earth's species.

moderate average temperatures that change significantly during four distinct seasons (Figure 5-2). These ecosystems are dominated by a few species of broadleaf deciduous trees such as oak, hickory, maple, poplar, sycamore, and beech. These plants survive during winter by dropping their leaves and going into an inactive state (Figure 5-16).

Temperate forests have nutrient-rich soil (helped by the annual fall of leaves) and valuable timber. All but about 0.1% of the original stands of temperate forests in North America have been cleared for farms, orchards, timber, and urban development. Some have

been converted to intensely managed *tree farms*, where a single species is grown for timber or pulpwood (Figure 5-17).

Northern coniferous forests, also called *boreal forests* and *taigas*, are found in regions with a subarctic climate (Figure 5-3). Winters are long and dry with light snowfall and short days, and temperatures range from cool to extremely cold. Summers are short with mild to warm temperatures, but the sun typically shines for 19 hours each day.

These coniferous forests form an almost unbroken belt just south of the Arctic tundra across North

Figure 5-16 Temperate deciduous forest in Rhode Island during winter, spring, summer, and fall.

Paul W. Johnson/Biological Photo Service

Paul W. Johnson/Biological Photo Service

Paul W. Johnson/Biological Photo Service

Paul W. Johnson/Biological Photo Service

Figure 5-17 Tree farm of southern pine in North Carolina. Converting a diverse deciduous forest to a single species (monoculture) tree farm increases the production of wood for timber or pulpwood. But it represents a loss of biological diversity, and such monocultures are vulnerable to attacks by pests and disease.

Figure 5-18 Northern coniferous forest (taiga or boreal forest) in Washington.

America, Asia, and Europe (Figure 5-2). They are dominated by a few species of coniferous (cone-bearing) evergreen trees such as spruce, fir, cedar, and pine (Figure 5-18). The needle-shaped, waxy-coated leaves of these trees conserve heat and water during the long, cold, dry winters. Plant diversity is low in these forests because few species can survive the winters, when soil moisture is frozen. Parts of the forests are dotted with wet bogs, or *muskegs*.

Beneath the dense stands of trees, a carpet of fallen needles and leaf litter covers the nutrient-poor soil, making the soil acidic and preventing most other plants from growing on the forest floor. During the brief summer the soil becomes waterlogged.

5-2 AQUATIC ECOSYSTEMS

Limiting Factors of Aquatic Ecosystems Five main factors affect the types and numbers of organisms in aquatic ecosystems. One is **salinity**: the concentration of dissolved salts, especially sodium chloride, in a body of water. The other factors are the depth to which sunlight penetrates, the amount of dissolved oxygen, the availability of plant nutrients, and the water temperature.

Salinity levels are used to divide aquatic ecosystems into two major classes. Those with high to very high salinity levels are marine or saltwater ecosys-

tems. They include oceans, estuaries (where fresh water from rivers and streams mixes with seawater), coastal wetlands, and coral reefs. Freshwater ecosystems have low salinity. Examples are inland bodies of standing water (lakes, reservoirs, ponds, and inland wetlands) and flowing water (the different segments of streams and rivers).

Marine Aquatic Ecosystems: Why Are the Oceans Important? As landlubbers, we tend to think of the earth in terms of land. It is more accurately described, however, as the "water planet" because 71% of its surface is covered by water (Figure 5-19). The oceans make up 97% of that water, and they play key roles in the survival of life on earth.

The oceans are the ultimate receptacle of terrestrial water flowing from rivers. Because of their size and currents, the oceans mix and dilute many human-produced wastes to less harmful or harmless levels. They play a major role in regulating the climate of the earth by distributing solar heat through ocean currents and evaporation as part of the hydrologic cycle (Figure 4-29, p. 79). They also participate in other major biogeochemical cycles.

In addition, by serving as a gigantic reservoir for carbon dioxide, the oceans help regulate the temperature of the atmosphere. Oceans provide habitats for about 250,000 species of marine plants and animals, which are food for many organisms, including human beings. They also serve as a source of iron, sand, gravel, phosphates, lime, magnesium, oil, natural gas, and many other valuable resources.

Figure 5-19 The ocean planet. About 71% of the earth's surface is covered with water. About 97% of this water is in the interconnected oceans that cover 90% of the planet's mostly ocean hemisphere (left) and 50% of its land-ocean hemisphere (right).

Ocean hemisphere   Land-ocean hemisphere

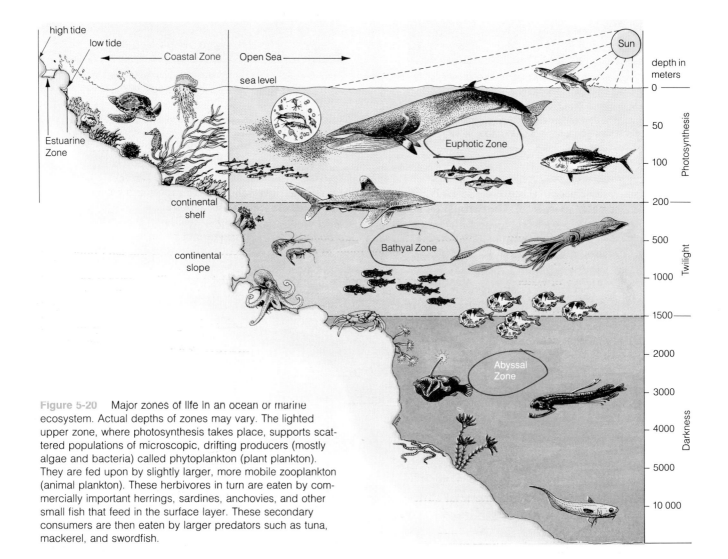

Figure 5-20 Major zones of life in an ocean or marine ecosystem. Actual depths of zones may vary. The lighted upper zone, where photosynthesis takes place, supports scattered populations of microscopic, drifting producers (mostly algae and bacteria) called phytoplankton (plant plankton). They are fed upon by slightly larger, more mobile zooplankton (animal plankton). These herbivores in turn are eaten by commercially important herrings, sardines, anchovies, and other small fish that feed in the surface layer. These secondary consumers are then eaten by larger predators such as tuna, mackerel, and swordfish.

Major Ocean Zones Each of the world's oceans can be divided into two major zones: coastal and open sea (Figure 5-20). The **coastal zone** is the relatively warm, nutrient-rich, shallow water that extends from the high-tide mark on land to the edge of a shelflike extension of the continental land mass known as the *continental shelf*. The coastal zone, representing less than 10% of the total ocean area, contains 90% of all ocean plant and animal life and is the site of most major commercial marine fisheries. This thin zone is the source of most of the oceans' biological productivity which supports oceanic animal life (Figure 5-1).

10% water 90% animal life

Many people view estuaries and coastal wetlands (Figures 5-20 and 5-21) as desolate, mosquito-infested, worthless lands. They believe these ecosystems should be drained, dredged, filled in, built on, or used as depositories for human-generated pollutants and waste materials. Essential water flows that sustain these ecosystems are diverted for human use.

In fact, these highly productive areas (Figure 4-21, p. 74) supply food and serve as spawning and nursery grounds for many species of marine fish and shellfish. In the United States, estuaries and coastal wetlands are spawning grounds for more than 70% of the country's commercial fish and shellfish, generating $5.5 billion a year and employing 330,000 people. They are also breeding grounds and habitats for waterfowl and other wildlife, including many endangered species. Each year millions of people visit coastal zones for whale or bird watching, waterfowl hunting, and other recreational activities.

Coastal areas also dilute and filter out large amounts of waterborne pollutants, helping protect the quality of waters used for swimming, fishing, and wildlife habitats. It is estimated that one acre of tidal estuary substitutes for a $75,000 waste treatment plant and has a total land value of $83,000 when its production of fish for food and recreation is included. By comparison, one acre of prime farmland in Kansas has a top value of $1,200 and an annual production value of $600.

Estuaries and coastal wetlands help protect coastal areas. They absorb damaging waves caused by violent storms and hurricanes and serve as giant sponges to absorb floodwaters.

Clearly, estuaries and coastal wetlands are vital natural ecosystems for us and other species. But they are also among our most intensely populated, used, and stressed ecosystems. Nearly 55% of the estuaries and coastal wetlands in the United States have been destroyed or damaged, primarily by dredging and filling and contamination by wastes from upstream sources and highly developed coastal areas. California has lost 90% of its wetlands.

These ecosystems are particularly vulnerable to toxic contamination because they trap pollutants, concentrating them to very high levels. Among America's most polluted estuaries are Massachusetts Bay near Boston, Long Island Sound in New York, Chesapeake Bay in Maryland, California's San Francisco Bay, and Washington's Puget Sound. Flounder in Boston Harbor have the highest rate of cancers and lesions of any on the East Coast. English sole from the Puget Sound are often riddled with liver cancer. On any given day one-third of U.S. shellfish beds are closed to commercial or sport fishing because of contamination.

Large areas of coastal wetlands are still being lost or severely degraded, especially in the southeastern United States, where 83% of the remaining wetlands in the lower 48 states are found. In 1988, about 53% of the U.S. population lived within 80 kilometers (51 miles) of coastal waters. In our desire to live near the coast, we are destroying the values that make coastal areas so enjoyable.

Wetlands fare no better in LDCs. In India and Bangladesh, for example, firewood gatherers have reduced vital mangrove swamps by more than 90%.

Protecting the remaining estuaries and coastal wetlands in the United States and throughout the world from destruction and degradation must be an urgent environmental priority. If this isn't done, people living near and visiting coastal areas will experience an ugly world of eroding, littered beaches, algae and sewage-laden surf, and degraded wetlands devoid of many of their once-familiar varieties of fish and birds.

The sharp increase in water depth at the edge of the continental shelf marks the separation of the coastal zone from the **open sea** (Figure 5-20). This marine zone contains about 90% of the total surface area of the ocean but only about 10% of its plant and animal life. The open sea is divided into three zones based primarily on the ability of sunlight to penetrate to various depths (Figure 5-20).

The Coastal Zone: A Closer Look   The coastal zone includes a number of different habitats. **Estuaries** are coastal areas where fresh water from rivers, streams, and land runoff mixes with salty seawater. Land that is flooded all or part of the year with fresh or salt water is called a **wetland**. Wetlands extending inland from estuaries and covered all or part of the year with salt water are known as **coastal wetlands**. In temperate areas, coastal wetlands usually consist of a mix of bays, lagoons, and salt marshes, where grasses are the dominant vegetation (Figure 5-21). In tropical areas, we find swamps dominated by mangrove trees. These nutrient-rich areas are among the world's most productive (Figure 4-21, p. 74) and important ecosystems (see Spotlight above).

Some coasts have steep *rocky shores* pounded by waves (Figure 5-22). Many organisms live in the numerous intertidal pools in the rocks. Other coasts have gently sloping *barrier beaches* at the water's edge,

LAND - fresh

Figure 5-21 Salt marsh (coastal wetland) and estuary on Cape Cod, off the coast of Massachusetts. These and other coastal wetlands trap plant nutrients from rivers and nearby land and thus have very high biological productivity. They also help remove and degrade some of the pollutants deposited by rivers and land runoff.

E. R. Degginger

Figure 5-22 Rocky shore beach in Acadia National Park, Maine.

Richard Frear/National Park Service

Figure 5-23 Developed barrier beach, Myrtle Beach, South Carolina. Note that the protective dunes have been eliminated.

Myrtle Beach Area Chamber of Commerce

which are prime sites for human developments (Figure 5-23). If not destroyed by human activities, one or more rows of sand dunes on such beaches serve as the land's first line of defense against the ravages of the sea (Figure 5-24). When coastal developers remove the dunes or build behind the first set of dunes, minor hurricanes and sea storms can flood and even sweep away houses and other buildings. Coastal dwellers mistakenly call these human-assisted disasters natural disasters.

Strings of thin *barrier islands* in some coastal areas (such as portions of North America's Atlantic and Gulf coasts) help protect the mainland, estuaries, lagoons, and coastal wetlands by dispersing the en-

ergy of approaching storm waves. People build cottages and other structures on these slender ribbons of sand with water on all sides (Figure 5-25), but sooner or later most of these structures are damaged or destroyed by flooding, major storms and hurricanes, and gradual movement of the islands toward the mainland by currents and winds.

The coastal zones of warm tropical and subtropical oceans often contain *coral reefs* (Figure 5-26). Because of their complexity and diversity, coral reefs support at least one-third of all marine fish species as well as numerous other marine organisms.

Along some steep, western coasts of continents, almost constant trade winds blow offshore and push surface water away from the shore. This outwardly moving surface water is replaced by an **upwelling** of cold, nutrient-rich bottom water. An upwelling brings plant nutrients from the deeper parts of the ocean to the surface and supports large populations of plankton, fish, and fish-eating seabirds. Although they make up only about 0.1% of the world's total ocean area, upwellings are highly productive (Figure 5-1). However, periodic changes in climate and ocean currents coupled with overfishing can reduce their high productivity and cause sharp drops in the annual catch of some important marine fish species, such as anchovies, used mostly as livestock feed.

Freshwater Lakes and Reservoirs **Lakes** are large natural bodies of standing fresh water formed when precipitation, land runoff, or flowing groundwater fills depressions in the earth. They normally consist of four distinct zones (Figure 5-27), which provide a variety of habitats and ecological niches for different species.

A lake with a small supply of plant nutrients (mostly nitrates and phosphates) is called an **oligo-trophic lake** (Figure 5-28). Such a lake is usually deep

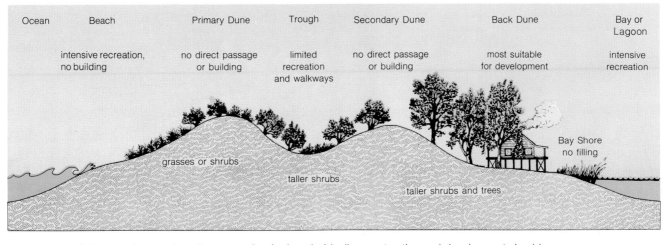

Ocean | Beach | Primary Dune | Trough | Secondary Dune | Back Dune | Bay or Lagoon

intensive recreation, no building | no direct passage or building | limited recreation and walkways | no direct passage or building | most suitable for development | intensive recreation

grasses or shrubs

taller shrubs

taller shrubs and trees

Bay Shore no filling

Figure 5-24 Primary and secondary dunes on a barrier beach. Ideally, construction and development should be allowed only behind the second strip of dunes, with walkways to the beach built over the dunes to keep them intact. This helps protect structures from being damaged and washed away by wind, high tides, beach erosion, and flooding from storm surges. Protection of barrier beaches is rare, however, because the short-term economic value of limited ocean-front land is considered to be much higher than its long-term ecological and economic values.

Nick Demetrakas

Figure 5-25 Developed barrier island, Ocean City, Maryland. Developing and living on a barrier island is a risky business. There is no effective protection against flooding and damage from hurricanes and severe storms, as residents on barrier islands in South Carolina learned when a devastating hurricane hit in 1989. If global warming raises average sea levels as projected, sometime in the next century most of these valuable pieces of real estate will be underwater.

Peter Scoones/Planet Earth Pictures

Figure 5-26 Coral reef in the Red Sea. Many of the world's diverse tropical reefs are being degraded by pollution and changes in water temperature.

and has crystal-clear water with cool to cold temperatures. It has small populations of phytoplankton and fish such as smallmouth bass and lake trout.

A lake with a large or excessive supply of plant nutrients is called a **eutrophic lake** (Figure 5-28). This type of lake is usually shallow and has cloudy, warm water. It has large populations of phytoplankton (especially algae) and zooplankton, and diverse populations of fish, particularly bullhead, catfish, and carp. In warm summer months, the bottom layer of a eutrophic lake is often depleted of dissolved oxygen.

Many lakes fall somewhere between the two extremes of nutrient enrichment and are called **mesotrophic lakes**.

Eutrophication is the physical, chemical, and biological changes that take place after a lake receives

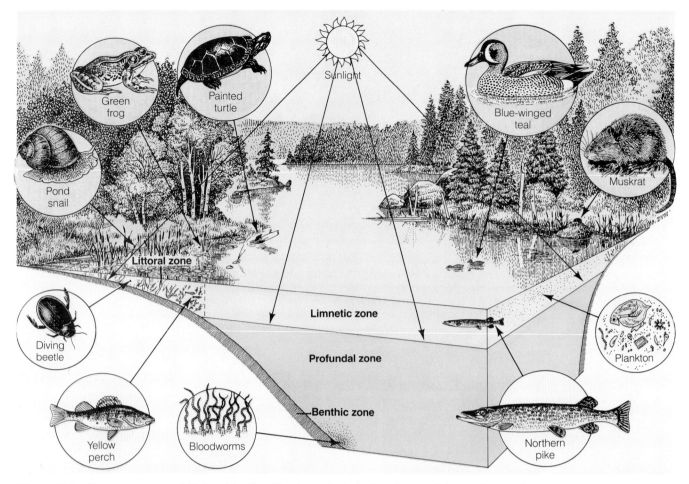

Figure 5-27 Four major zones of life in a lake. The *littoral zone* includes the shore and the shallow, nutrient-rich waters near the shore, in which sunlight penetrates to the lake bottom. It contains a variety of free-floating producers, rooted aquatic plants, and other forms of aquatic life such as frogs, snails, and snakes. The *limnetic zone* is the open-water surface layer that gets enough sunlight for photosynthesis. It contains varying amounts of floating phytoplankton, plant-eating zooplankton, and fish, depending on the supply of plant nutrients. The *profundal zone* is the deep, open water where it is too dark for photosynthesis. It is inhabited by fish adapted to its cooler, darker water. The *benthic zone* at the bottom of a lake is inhabited mostly by large numbers of decomposers, detritus-feeding clams, and wormlike insect larvae. These detritivores feed on dead plant debris, animal remains, and animal wastes that descend from above.

inputs of plant nutrients from the surrounding land basin from natural erosion and runoff over a long period of time. Near urban or agricultural centers the natural input of plant nutrients to a lake can be greatly accelerated by human activities. This **accelerated,** or **cultural, eutrophication** is caused mostly by nitrate- and phosphate-containing effluents from sewage treatment plants, runoff of fertilizers and animal wastes, and accelerated soil erosion.

Reservoirs are normally large, deep, human-created bodies of standing fresh water. They are often built behind dams to collect water running down from mountains in streams and rivers (Figure 5-29). Reservoirs are built primarily to store and release water in a controlled manner. The released water may be used for hydroelectric power production at the dam site. Water can also be released over the dam and diverted into irrigation canals to grow crops

on dry land. It can be stored and released slowly to prevent flooding, and it can be carried by aqueduct to towns and cities for use by homes, businesses, and industries. Reservoirs are also used for recreation such as swimming, fishing, and boating.

Freshwater Streams Precipitation that doesn't infiltrate into the ground or evaporate into the atmosphere remains on the earth's surface as **surface water**. This water becomes **runoff,** which flows into streams and eventually downhill to the oceans for reuse in the hydrologic cycle (Figure 4-29, p. 79). The entire land area that delivers the water, sediment, and dissolved substances via streams to a major river, and ultimately to the sea, is called a **watershed,** or **drainage basin**.

The downward flow of water from mountain highlands to the sea takes place in three phases

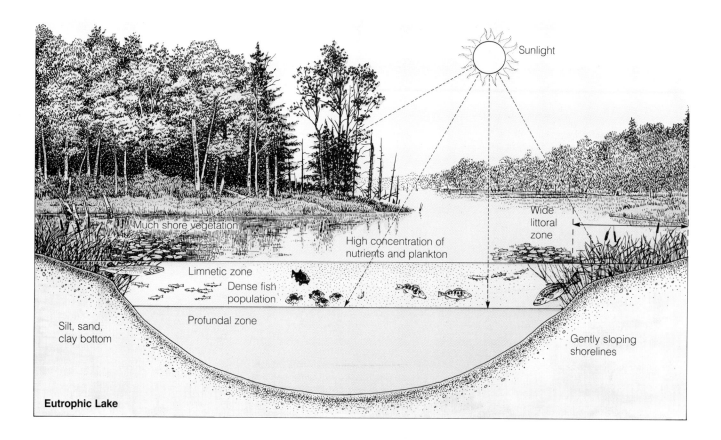

Much shore vegetation

High concentration of
nutrients and plankton

Wide
littoral
zone

Limnetic zone

Dense fish
population

Profundal zone

Silt, sand,
clay bottom

Gently sloping
shorelines

Sunlight

Eutrophic Lake

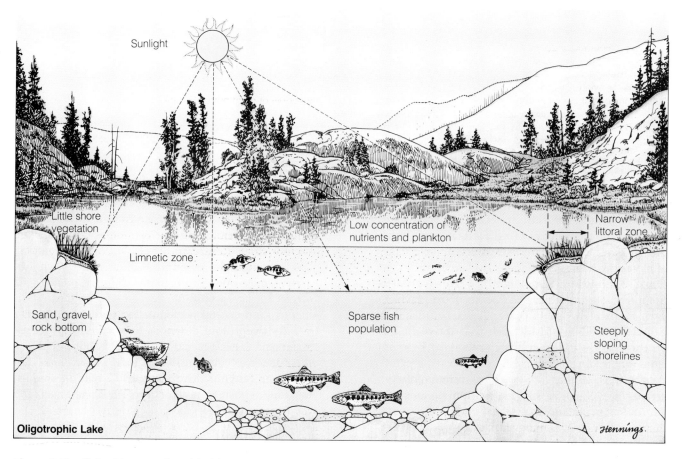

Sunlight

Little shore
vegetation

Low concentration of
nutrients and plankton

Narrow
littoral
zone

Limnetic zone

Sand, gravel,
rock bottom

Sparse fish
population

Steeply
sloping
shorelines

Oligotrophic Lake

Hennings.

Figure 5-28 Eutrophic, or nutrient-rich, lake, and oligotrophic, or nutrient-poor, lake. Mesotrophic lakes fall between these two extremes of nutrient enrichment.

Figure 5-29 Reservoir formed behind Shasta Dam on the Sacramento River north of Redding, California. This reservoir and dam are used to produce electricity (hydropower) and to help control flooding in areas below the dam by storing and releasing water slowly. Water flowing through the base of the dam drives turbines that produce electricity. About 13.5% of the electrical power used in the United States is produced by large hydroelectric power plants such as this one.

U. S. Department of Interior/Bureau of Reclamation

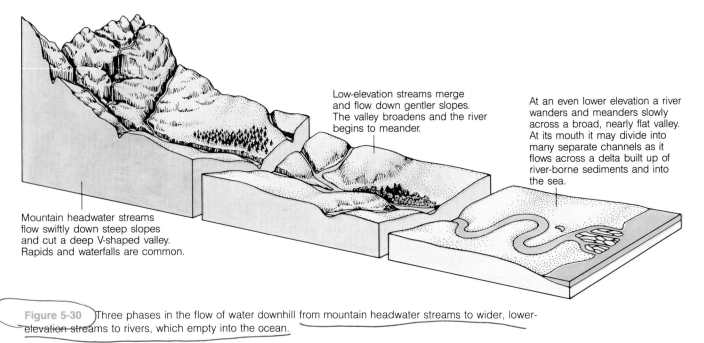

Low-elevation streams merge and flow down gentler slopes. The valley broadens and the river begins to meander.

At an even lower elevation a river wanders and meanders slowly across a broad, nearly flat valley. At its mouth it may divide into many separate channels as it flows across a delta built up of river-borne sediments and into the sea.

Mountain headwater streams flow swiftly down steep slopes and cut a deep V-shaped valley. Rapids and waterfalls are common.

Figure 5-30 Three phases in the flow of water downhill from mountain headwater streams to wider, lower-elevation streams to rivers, which empty into the ocean.

(Figure 5-30). Because of the differences in environmental conditions in each phase, a river consists of a series of different ecosystems. First, narrow headwater or mountain highland streams with cold, clear water rush down steep slopes. As this turbulent water flows and tumbles downward over waterfalls and rapids, it dissolves large amounts of oxygen from the air. Most plants in this flow survive because they are attached to rocks. Most fish that thrive in this environment are cold-water fish, such as trout, which require a high level of dissolved oxygen.

In the second phase, the headwater streams merge to form wider, deeper, lower-elevation streams that flow down gentler slopes and meander through wider valleys. Here the water is warmer and usually less turbulent. It can support a variety of cold-water and warm-water fish species with slightly lower oxygen requirements.

Gradually these streams coalesce into wider and deeper rivers that meander across broad, flat valleys. The main channels of these rivers support a distinctive variety of fish, whereas their backwaters support species similar to those found in lakes. At its mouth a river may divide into many channels as it flows across a *delta*—a built-up deposit of riverborne sediments—and coastal wetlands and estuaries.

As rivers flow downhill they become powerful shapers of land. Over millions and billions of years, the friction of moving water levels mountains and cuts deep canyons. Rivers also make hills and mountains by depositing sediment in low-lying areas.

Inland Wetlands Lands covered with fresh water all or part of the year and located away from coastal areas are called **inland wetlands**. They include inland bogs, marshes, swamps, mud flats, river-overflow

Away coastal areas
Inland wetlands

Inland wetlands provide habitats for a variety of fish, waterfowl, and other wildlife, including many rare and endangered species. Wetlands near rivers help regulate stream flow by storing water during periods of heavy rainfall and releasing it slowly. Thus, they reduce the frequency, level, and velocity of floods and riverbank erosion. Wetlands also improve water quality by trapping stream sediments and absorbing, diluting, and degrading many toxic pollutants.

Wetlands are also great places to visit. They are peaceful oases in an increasingly hectic world.

By holding water, many wetlands allow increased ground infiltration, thus helping recharge groundwater supplies. Wetlands are used for recreation, especially waterfowl hunting, and to grow crops such as blueberries, cranberries, and especially rice, which is a fundamental part of the diet for half the world's people. Wetlands also play significant roles in the global cycles of carbon, nitrogen, and sulfur.

Because people are unaware of the ecological importance of inland wetlands, wetlands often are dredged or filled in and used as croplands, garbage dumps, and sites for urban and industrial development. They are viewed as wastelands.

Since the Europeans first arrived, about 56% of the coastal and inland wetland acreage in the lower 48 states has been destroyed—enough to cover an area four times the size of Ohio. About 80% of this loss was due to draining and clearing of inland wetlands for agriculture. Most of the rest was used for real estate development and highways. Conversion of meandering rivers into straight channels for commercial boat traffic has severely damaged or eliminated floodplain wetlands along many major rivers. Wetland destruction has greatly reduced the habitat of birds and other wildlife that live on or near these ecosystems, threatening some species with extinction. It is estimated that each year the United States is still losing 121,000 to 182,000 hectares (300,000 to 450,000 acres) of coastal and inland wetlands.

About 95% of remaining U.S. wetlands are inland wetlands, usually on or adjacent to agricultural property. Attempts have been made to slow the rate of inland wetland loss. The Farm Act of 1985 has a "swampbuster" provision that withholds agricultural subsidies from landowners who convert wetlands to croplands.

The Emergency Wetlands Resource Act raises about $20 million a year for federal wetland acquisition. A federal permit is now required to fill wetlands or deposit polluting material in them. The North American Waterfowl Management Plan should help the Canadian and American governments and conservation organizations acquire and restore massive wetland areas in the next 15 years.

A major problem is that only about 8% of inland wetlands is under federal protection, and federal, state, and local protection of wetlands are weak. Existing laws are poorly enforced and subsidies and tax incentives are still given to developers who drain, dredge, fill, and destroy wetlands. Despite the swampbuster provisions of the 1985 U.S. Farm Act, wetland drainage in 1987 was the highest in 20 years.

The United States urgently needs a better system for protecting and managing its wetlands, both coastal and inland. The immediate goal of such a program should be to prevent further loss of the country's wetlands. The long-term goal should be to restore the quantity and quality of the country's wetlands.

---

lands, prairie potholes, and the wet tundra during summer. These important ecosystems are being rapidly destroyed and degraded (see Spotlight above).

5-3   RESPONSES OF LIVING SYSTEMS TO ENVIRONMENTAL STRESS

Stability of Living Systems   The organisms that make up the populations of various biological communities and ecosystems have some ability to withstand or recover from externally imposed changes or stresses—provided these external stresses are not too severe. In other words, organisms have some degree of *stability*. This stability is maintained only by constant dynamic change.

It is useful to distinguish between three aspects of stability in living systems. **Inertia,** or **persistence,** is the ability of a living system to resist being disturbed or altered. **Constancy** is the ability of a living system, such as a population, to maintain a certain size. **Resilience** is the ability of a living system to restore itself to an original condition after being exposed to an outside disturbance that is not too drastic.

Nature is remarkably resilient. For example, human societies have survived natural disasters, plagues, and devastating wars. Changes in the genetic makeup of rapidly producing insect populations enable many individual insects to survive massive doses of deadly pesticides and ionizing radiation. Plants recolonize areas devastated by volcanoes, retreating glaciers, mining, bombing, and farming, although

Table 5-1 Changes Affecting Ecosystems

Natural Changes

Catastrophic	Drought
	Flood
	Fire
	Volcanic eruption
	Earthquake
	Hurricane
	Disease
Gradual	Changes in climate
	Immigration and emigration of species
	Adaptation and evolution of species as a response to environmental stress
	Changes in plant and animal life (ecological succession)

Human-Caused Changes

Catastrophic	Deforestation
	Overgrazing of grasslands
	Plowing of grasslands
	Soil erosion
	Using pesticides
	Excessive or inappropriate use of fire
	Release of toxic substances into the air, water, or soil
	Urbanization
	Mining
Gradual	Salt buildup in soil from irrigation (salinization)
	Waterlogging of soil from irrigation
	Compaction of soil from agricultural equipment
	Pollution of surface waters (rivers, lakes, reservoirs, wetlands, oceans)
	Depletion and pollution of underground aquifers
	Air pollution (can also be catastrophic)
	Loss and degradation of wildlife habitat (can also be catastrophic)
	Killing of undesirable predator and pest species
	Introduction of alien species
	Release of toxic substances into the air, water, and soil
	Overhunting
	Overfishing
	Excessive tourism

Table 5-2 Some Effects of Enviromental Stress

Organism Level

Physiological and biochemical changes
Psychological disorders
Behavioral changes
Fewer or no offspring
Genetic defects in offspring (mutagenic effects)
Birth defects (teratogenic effects)
Cancers (carcinogenic effects)
Death

Population Level

Population increase or decrease
Change in age structure (old, young, and weak may die)
Survival of strains genetically resistant to stress
Loss of genetic diversity and adaptability
Extinction

Community-Ecosystem Level

Disruption of energy flow
   Decrease or increase in solar energy input
   Changes in heat output
   Changes in trophic structure in food chains and food webs

Disruption of chemical cycles
   Depletion of essential nutrients
   Excessive addition of nutrients

Simplification
   Reduction in species diversity
   Reduction or elimination of habitats and filled ecological niches
   Less complex food webs
   Possibility of lowered stability
   Possibility of ecosystem collapse

such natural restoration usually takes a long time on a human time scale.

Types and Effects of Environmental Stress Ecosystems are affected by a number of natural and human-caused changes summarized in Table 5-1. Some of these changes are gradual and some are sudden or catastrophic.

Table 5-2 summarizes what can happen to organisms, populations, communities, and ecosystems as a result of environmental stress. The stresses that can cause the changes shown in Table 5-2 may result from natural hazards (such as earthquakes, volcanic eruptions, hurricanes, drought, floods, and fires) or from human activities (industrialization, warfare, transportation, and agriculture).

A population of a species that is well adapted to its environment has four major ways to deal with an environmental stress:

- decrease its birth rate or suffer an increase in its death rate

- migrate to another area with a similar but less stressful environment (slow or very difficult for most plants)

- adapt to changed environmental conditions through natural selection (very slow for species that reproduce slowly and have few offspring)

- become extinct

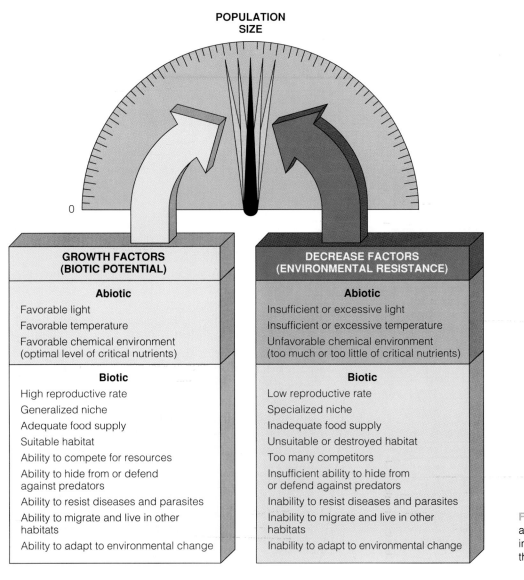

POPULATION
SIZE

GROWTH FACTORS (BIOTIC POTENTIAL)	DECREASE FACTORS (ENVIRONMENTAL RESISTANCE)
Abiotic	**Abiotic**
Favorable light	Insufficient or excessive light
Favorable temperature	Insufficient or excessive temperature
Favorable chemical environment (optimal level of critical nutrients)	Unfavorable chemical environment (too much or too little of critical nutrients)
Biotic	**Biotic**
High reproductive rate	Low reproductive rate
Generalized niche	Specialized niche
Adequate food supply	Inadequate food supply
Suitable habitat	Unsuitable or destroyed habitat
Ability to compete for resources	Too many competitors
Ability to hide from or defend against predators	Insufficient ability to hide from or defend against predators
Ability to resist diseases and parasites	Inability to resist diseases and parasites
Ability to migrate and live in other habitats	Inability to migrate and live in other habitats
Ability to adapt to environmental change	Inability to adapt to environmental change

Figure 5-31 Population size is a balance between factors that increase numbers and factors that decrease numbers.

5-4 POPULATION RESPONSES TO STRESS

Changes in Population Size, Structure, and Distribution Changes in the size, structure, and distribution of a population in response to changes in environmental conditions, such as an excess or a shortage of food or other critical nutrients, are called **population dynamics.** A change in the birth rate or death rate is the major way that populations of most species respond to changes in resource availability or other environmental changes (Figure 5-31).

Members of some animal species can avoid or reduce the effects of an environmental stress by leaving one area (emigration) and migrating to an area (immigration) with more favorable environmental conditions and resource supplies. Plants can migrate to other areas but this often takes decades or centuries. The structure of the population in terms of the numbers of individuals of different ages and

sex may change. Old, very young, and weak members may die when exposed to an environmental stress. The remaining population is then better equipped to survive such stresses as a more severe climate, an increase in predators, or an increase in disease organisms.

J and S Curves: Idealized Model of Population Dynamics With unlimited resources and ideal environmental conditions, a species can produce offspring at its maximum rate. Such growth starts off slowly and then increases rapidly to produce an exponential, or J-shaped, curve of population growth (Figure 5-32). Almost any kind of organism would be capable of increasing its population to crowd out the entire world if given ample food, water, space, and protection from its enemies.

But this doesn't happen because environmental conditions are less than ideal and resources are

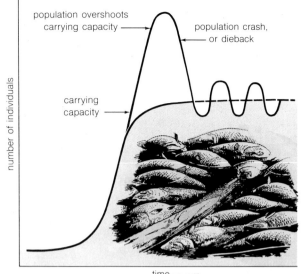

Figure 5-32 The J-shaped curve of population growth of a species is converted to an S-shaped curve when the population growth is limited by one or more environmental factors. A population crash can occur when a rapidly expanding population overshoots the carrying capacity of its environment or a change in environmental conditions lowers that carrying capacity.

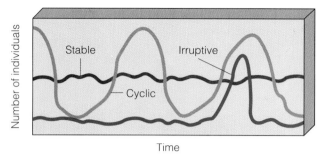

Figure 5-33 Basic types of population curves for living species.

normally limited. Factors such as predation, competition within and between species, food shortages, disease, adverse climatic conditions, and lack of suitable habitat usually act to keep the growth rate of a population below its maximum rate. The maximum species population size that a natural ecosystem can support indefinitely under a given set of environmental conditions is called that ecosystem's **carrying capacity** for that species.

Without unlimited resources the population size of a species is limited and its death rate begins to rise as it reaches or temporarily exceeds its ecosystem's carrying capacity. When this happens, the J-shaped curve of population growth bends away from its steep incline and eventually levels off to form an S-shaped curve (Figure 5-32). Then the population size typically fluctuates slightly above and below the carrying capacity of the environment.

This transition can be fairly smooth, or it can be a sharp drop in population size known as a population crash (Figure 5-32). A crash occurs when a reproducing population of a species overshoots the carrying capacity of its environment or when a change in conditions suddenly lowers the carrying capacity. Large numbers of individuals then die if they cannot migrate to other areas.

Crashes have occurred in the human populations of various countries throughout history. Ireland, for example, experienced a population crash after a fungus infection ruined the potato crop in 1845. De-

pendent on the potato for a major portion of their diet, by 1900 half of Ireland's 8 million people had died of starvation or emigrated to other countries.

In spite of such local and regional disasters, the overall human population on earth has continued to grow. Human beings have made technological, social, and other cultural changes that have extended the earth's carrying capacity for their species (Figure 2-2, p. 22). The human species has been able to alter its ecological niche by increasing food production, controlling disease, and using large amounts of energy and matter resources to make normally uninhabitable areas of the earth inhabitable.

The idealized J-shaped to S-shaped mathematical model of population dynamics shown in Figure 5-32 can be observed in laboratory experiments. In nature, however, we find that species have three general types of population growth curves: relatively stable, irruptive, and cyclic (Figure 5-33).

Natural Selection and Biological Evolution A population of a particular species can undergo changes in its genetic composition, or gene pool, that enable it to better adapt to changes in environmental conditions. All individuals of a population do not have exactly the same genes. Each has a unique combination of such traits as size, shape, color, and ability to withstand temperature extremes and exposure to certain toxic substances. Individuals with a protective genetic composition generally produce more offspring than those that don't have these traits and pass these traits on to their offspring, a process known as **differential reproduction**.

The process by which some genes and gene combinations in a population are reproduced more than others is called **natural selection**. Charles Darwin, who proposed this idea in 1858, described natural selection as "survival of the fittest." This phrase has often been misinterpreted to mean survival of the strongest, biggest, or most aggressive. Instead, *fittest* means that individuals with the genetic traits best adapted to their environment on the average produce

J. A. Bishop and L. M. Cook

Figure 5-34 Two varieties of peppered moths found in England in the 1800s. In the mid-1800s before the Industrial Revolution, the speckled light-gray form of this moth was prevalent. This moth, which is active at night, rests on light-gray speckled lichens on tree trunks during the day. Their color helped camouflage them from their bird predators. A dark-gray form also existed but was quite rare. However, as the Industrial Revolution proceeded during the last half of the 1800s, the dark form of this moth sharply increased in frequency, especially near industrial cities. The soot and other pollutants from factory smokestacks began killing lichens and darkening tree trunks. In this new environment the dark form blended in with the blackened trees and the light form was highly visible to its bird predators. Through natural selection the dark form began to survive and reproduce at a greater rate than its light-colored kin.

the most offspring in a given generation and tend to replace the less successful ones (Figure 5-34).

The resulting change in the genetic composition of a population exposed to new environmental conditions because of differential reproduction and natural selection is called **biological evolution** or simply **evolution**.

Species differ widely in how rapidly they can undergo evolution through natural selection. Those that can quickly produce a large number of tiny offspring with short average life spans (weeds, insects, rodents, bacteria) can adapt to a change in environmental conditions through natural selection in a short time.

For example, when a chemical is used as a pesticide to reduce the insect population, a small number of resistant individuals survive. They can then rapidly breed new populations with a larger number of individuals genetically resistant to the toxic effects of the chemical. Thus, in the long run our present chemical approach to pest control usually increases, not decreases, the populations of species we consider pests. As one observer put it, "I hope that when the insects take over the world, they will remember that we always took them along on our picnics."

Other species, such as elephants, horses, tigers, white sharks, and humans, have long generation times and a small number in each litter. This means that they cannot reproduce a large number of offspring rapidly. For such species, adaptation to an environmental stress by natural selection typically takes thousands to millions of years.

Often, the plants, animals, and decomposers in various ecosystems have coevolved over a long period of time. For example, many plants, having coevolved with their consumers for ages, have developed a variety of poisons, odors, and flavors to protect themselves. Such coevolution also has led animals to develop different ways of defending themselves against predators and caused predators to evolve ways of overcoming the defenses of their prey. Coevolution also has led to mutualism (Figure 4-34, p. 84), commensalism, and other relationships between species.

Speciation and Extinction The earth's present inventory of 5 to 30 million species is believed to be the result of a combination of two processes taking place over billions of years. One is **speciation**: A new species, adapted to changes in environmental conditions, originates as a result of natural selection. The other is **extinction**: A species ceases to exist because it cannot genetically adapt and successfully reproduce under new environmental conditions. Most of the evidence we have about extinction and speciation comes from the discovery of fossils.

Biologists estimate that 98% to 99% of all the species that have ever lived are now extinct. But the fact that we have 5 to 30 million species on earth today means that speciation, on average, has kept ahead of extinction.

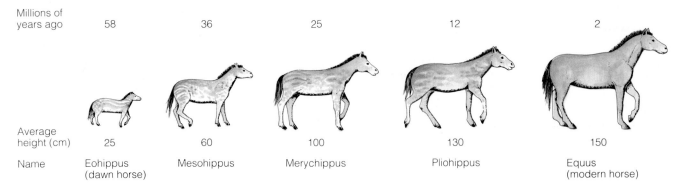

Millions of years ago	58	36	25	12	2
Average height (cm)	25	60	100	130	150
Name	Eohippus (dawn horse)	Mesohippus	Merychippus	Pliohippus	Equus (modern horse)

Figure 5-35   Speciation of the horse through natural selection along the same genetic line into different equine species.

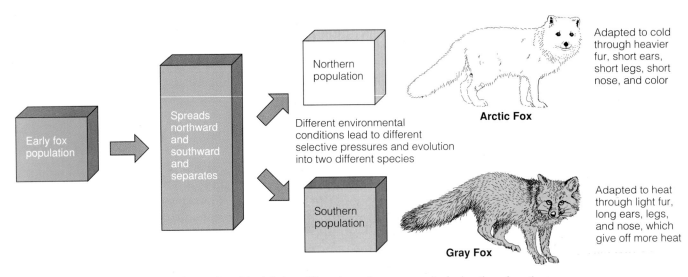

Figure 5-36   Speciation of an early species of fox into two different species as a result of migration of portions of the original fox population into areas with different climates. See Figure 4-1, p. 59, for a photograph of an Arctic fox with its winter coat.

Speciation can occur from changes within a single genetic line over a long period of time, often because of long-term changes in climate (Figure 5-35). It can also take place by the gradual splitting of lines of descent into two or more new species in response to new environmental conditions. This type of speciation is believed to occur when a population of a particular species becomes distributed over areas with different climates, rivers, food sources, soils, and other environmental conditions for long periods—typically for 1,000 to 100,000 generations.

Populations of some animal species also split up when part of the group migrates in search of food (Figure 5-36). If the isolated populations remain separated under different conditions long enough, members of the two populations will be unable to interbreed if they occupy the same area again. Thus, one species has become two different species. In some rapidly producing organisms, speciation may take place in thousands or even hundreds of years. In

most cases, however, it takes from tens of thousands to millions of years.

Over billions of years the combination of speciation and extinction operating through natural selection has produced the planet's most valuable resource: *biological diversity*, or *biodiversity* (see Spotlight on p. 7). One of our key goals should be not to reduce the *genetic*, *species*, and *ecological diversities* that make up this part of the earth's natural capital.

If we compress the development of different forms of life on earth through biological evolution to a 24-hour time scale, then our species, *homo sapiens sapiens*, appeared less than 1 second before midnight. Agriculture began only a quarter second before midnight, and the Industrial Revolution has been around for only seven-thousandths of a second. But despite our brief time on earth, we are now in the process of hastening the extinction of more of the earth's species in a shorter time than at any other time in the earth's almost 5-billion-year history.

5-5 COMMUNITY-ECOSYSTEM RESPONSES TO STRESS

Responses to Slight and Moderate Stress
Communities and ecosystems are so complex and variable that ecologists have little understanding of how they maintain their inertia and resilience. One major problem is the difficulty of conducting controlled experiments. Identifying and observing even a tiny fraction of the interacting variables in simple communities and ecosystems are virtually impossible.

Greatly simplified ecosystems can be set up and observed under laboratory conditions. But extrapolating the results of such experiments to much more complex natural communities and ecosystems is difficult, if not impossible.

At one time it was believed that the higher the species diversity in an ecosystem, the greater its stability. According to this idea, an ecosystem with a diversity of species has more ways to respond to most environmental stresses because it does not "have all its eggs in one basket." However, research indicates exceptions to this intuitively appealing idea. Part of the problem is that there are many different ways to define stability and diversity.

Does an ecosystem need both high inertia and high resilience to be considered stable? Evidence indicates that some ecosystems have one of these properties but not the other. For example, California redwood forests and tropical rain forests have high species diversity and high inertia. This means that they are hard to alter significantly or destroy. However, once large tracts of these diverse ecosystems are cleared or severely degraded, they have such low resilience that restoring them is nearly impossible.

On the other hand, grasslands, with much lower species diversity, burn easily and thus have low inertia. However, because most of their plant matter consists of roots beneath the ground surface, these ecosystems have high resilience, which allows them to recover quickly. A grassland can be destroyed only if its roots are plowed up and wheat or some other crop is planted in its soil (Figure 5-12, p. 92).

Clearly, we have a long way to go in understanding how the factors involved in natural communities and ecosystems interact. But considerable evidence indicates that simplifying a natural community or ecosystem by the intentional or accidental removal of a species often has unpredictable short- and long-term harmful effects (see Spotlight above).

Responses to Large-Scale Stress: Ecological Succession The organisms in most communities and ecosystems can adapt not only to small and moderate changes in environmental conditions but also to quite severe changes. Sometimes, for example, little vegetation and soil are left as a result of a natural envi-

---

SPOTLIGHT   Ecosystem Interference Is Full of Suprises!

Malaria once infected 9 out of 10 people in North Borneo, now known as Brunei. In 1955 the World Health Organization (WHO) began spraying dieldrin (a pesticide similar to DDT) to kill malaria-carrying mosquitoes. The program was so successful that the dreaded disease was almost eliminated from the island.

But other unexpected things happened. The dieldrin killed other insects, including flies and cockroaches, living in houses. The islanders applauded. But then small lizards that also lived in the houses died after gorging themselves on dead insects. Then cats began dying after feeding on the dead lizards. Without cats, rats flourished and overran the villages. Now people were threatened by sylvatic plague carried by the fleas on the rats. The situation was brought under control when WHO parachuted healthy cats onto the island.

Then roofs began to fall in. The dieldrin had killed wasps and other insects that fed on a type of caterpillar that either avoided or was not affected by the insecticide. With most of its predators eliminated, the caterpillar population exploded. The larvae munched their way through one of their favorite foods, the leaves used in thatching roofs.

In the end, the Borneo episode was a success story; both malaria and the unexpected effects of the spraying program were brought under control. But it shows the unpredictable results of interfering in an ecosystem.

---

ronmental change (retreating glaciers, fires, floods, volcanic eruptions, earthquakes) or a human-induced change (fires, land clearing, surface mining, flooding to create a pond or reservoir, pollution).

After such a large-scale disturbance, life usually begins to recolonize a site in a series of stages. First, a few hardy **pioneer species** invade the environment and start creating soil or, in aquatic ecosystems, sediment. Eventually the species that make up this pioneer community change the soil or bottom sediments and other conditions so much that the area is less suitable for them and more suitable for a new group of plants and animals with different ecological niche requirements. This process, in which communities of plant, animal, and decomposer species change into at least partially different and usually more complex communities, is called **ecological succession** or **biotic development**.

canopy

lower
canopy trees

tall shrub
understory

| annual weeds | perennial weeds and grasses | shrubs | young pine forest | mature oak forest |

Time ——→

Figure 5-37 Secondary ecological succession of plant communities on an abandoned farm field in North Carolina over about 150 years. Succession of animal communities is not shown.

If not severely disrupted, ecological succession often continues until the community becomes more self-sustaining than the preceding ones. When this happens, what ecologists call a **mature community** occupies the site. Depending primarily on the climate, such communities may be various types of mature grasslands, forests, or deserts (Figure 5-4, p. 89).

Ecologists recognize two types of ecological succession: primary and secondary. Which type takes place depends on the conditions at a particular site at the beginning of the process. **Primary succession** is the sequential development of biotic communities in a bare area. Examples of such areas include the rock or mud exposed by a retreating glacier or mudslide, cooled volcanic lava, a new sandbar deposited by a shift in ocean currents, and surface-mined areas from which all topsoil has been removed. On such barren surfaces, primary succession from bare rock to a mature forest may take hundreds to thousands of years.

The more common type of succession is **secondary succession**. This is the sequential development of communities in an area where the natural vegetation has been removed or destroyed, but the soil or bottom sediment has not been destroyed. Examples of areas that can undergo secondary succession include abandoned farmlands, burned or cut forests, land stripped of vegetation for surface mining, heavily

polluted streams, and land that has been flooded naturally or artificially to produce a reservoir or pond. Because some soil or sediment is present, new vegetation can usually sprout within only a few weeks.

In the central (Piedmont) region of North Carolina, European settlers cleared away the native oak and hickory climax forests and planted the land in crops. Figure 5-37 shows how a patch of this abandoned farmland, covered with a thick layer of soil, has undergone secondary succession over a period of about 150 years until the area is again covered with a mature oak and hickory forest.

Lakes, reservoirs, and ponds also undergo secondary succession. They gradually fill up with bottom sediments and eventually become terrestrial ecosystems.

Comparison of Immature and Mature Ecosystems
Immature ecosystems and mature ecosystems have strikingly different characteristics, as summarized in Table 5-3. Immature communities at the early stages of biotic development have only a few species (low species diversity) and fairly simple food webs, made up mostly of producers fed upon by herbivores and relatively few decomposers.

Most of the plants in a pioneer community are small annuals that grow close to the ground. They use most of their energy to produce large numbers

Table 5-3 Ecosystem Characteristics at Immature and Mature Stages of Ecological Succession

Characteristic	Immature Ecosystem	Mature Ecosystem
Ecosystem Structure		
Plant size	Small	Large
Species diversity	Low	High
Trophic structure	Mostly producers, few decomposers	Mixture of producers, consumers, and decomposers
Ecological niches	Few, mostly generalized	Many, mostly specialized
Community organization (number of interconnecting links)	Low	High
Ecosystem Function		
Food chains and webs	Simple, mostly plant → herbivore with few decomposers	Complex, dominated by decomposers
Efficiency of nutrient recycling	Low	High
Efficiency of energy use	Low	High

of small seeds for reproduction rather than to develop large root, stem, and leaf systems. They receive some matter resources from other ecosystems, for example, as runoff because they are too simple to hold and recycle many of the nutrients they receive.

In contrast, the community in a mature ecosystem has high species diversity, relatively stable populations, and complex food webs dominated by decomposers that feed on the large amount of dead vegetation and animal wastes. Most plants in mature ecosystems are larger herbs and trees that produce a small number of large seeds. They use most of their energy and matter resources to maintain their large root, trunk, and leaf systems rather than to produce large numbers of new plants. They also have the complexity necessary to entrap, hold, and recycle most of the nutrients they need.

5-6 HUMAN IMPACTS ON ECOSYSTEMS

Human Beings and Ecosystems In modifying ecosystems for our use, we simplify them. For example, we bulldoze and plow grasslands and forests. Then we replace the thousands of interrelated plant and animal species in these ecosystems with greatly simplified, single-crop ecosystems (Figures 5-12 and 5-17), or monocultures, or with structures such as buildings, highways, and parking lots.

Modern agriculture is based on the practice of deliberately keeping ecosystems in early stages of succession, in which the biomass productivity of one or a few plant species (such as corn or wheat) is high

(Figure 5-12). But such simplified ecosystems are highly vulnerable.

A major problem is the continual invasion of crop fields by unwanted pioneer species, which we call *weeds* if they are plants, *pests* if they are insects or other animals, and *diseases* if they are harmful microorganisms such as bacteria, fungi, and viruses. Weeds, pests, or diseases can wipe out an entire monoculture food or tree plantation crop unless it is artificially protected with pesticides such as insecticides (insect-killing chemicals) and herbicides (plant-killing chemicals) or by some form of biological control.

When quickly breeding species develop genetic resistance to these chemicals, farmers must use ever-stronger doses or switch to a new product. But this increases the rate of natural selection of the pests to the point that the effectiveness of each chemical is eventually lost. This illustrates biologist Garrett Hardin's **first law of ecology**: *We can never do merely one thing. Any intrusion into nature has numerous effects, many of which are unpredictable.*

Cultivation is not the only way people simplify ecosystems. Ranchers, who don't want bison or prairie dogs competing with sheep for grass, eradicate these species as well as wolves, coyotes, eagles, and other predators that occasionally kill sheep. Far too often, ranchers allow livestock to overgraze grasslands until excessive soil erosion converts these ecosystems to simpler and less productive deserts.

The cutting of vast areas of diverse tropical rain forests is causing the irreversible loss of many plant and animal species. People also tend to overfish and overhunt some species to extinction or near extinction, another way of simplifying ecosystems. The burning

Table 5-4 Comparison of a Natural Ecosystem and a Simplified Human System	
Natural Ecosystem (marsh, grassland, forest)	**Simplified Human System (cornfield, factory, house)**
Captures, converts, and stores energy from the sun	Consumes energy from fossil or nuclear fuels
Produces oxygen and consumes carbon dioxide	Consumes oxygen and produces carbon dioxide from the burning of fossil fuels
Creates fertile soil	Depletes or covers fertile soil
Stores, purifies, and releases water gradually	Often uses and contaminates water and releases it rapidly
Provides wildlife habitats	Destroys some wildlife habitats
Filters and detoxifies pollutants and waste products free of charge	Produces pollutants and waste, which must be cleaned up at our expense
Usually capable of self-maintenance and self-renewal	Requires continual maintenance and renewal at great cost

of fossil fuels in industrial plants, homes, and vehicles creates atmospheric pollutants that fall to the earth as acidic compounds. These chemicals simplify forest ecosystems by killing trees and aquatic ecosystems by killing fish.

It is becoming increasingly clear that the price we pay for simplifying, maintaining, and protecting such stripped-down ecosystems is high: It includes time, money, increased use of matter and energy resources, loss of genetic diversity, and loss of natural landscape (Table 5-4). There is also the danger that as the human population grows, we will convert too many of the world's wild, mature ecosystems to simple, young, productive, but highly vulnerable forms. The challenge is to maintain a balance between simplified, human ecosystems and the neighboring, more complex, natural ecosystems our simplified systems and other forms of life depend on.

Some Environmental Lessons It should be clear from the brief discussion of principles in this and the preceding chapter that living systems have six major features: *interdependence, diversity, resilience, adaptability, unpredictability,* and *limits* (see Spotlight above).

SPOTLIGHT Nature's Secrets for Sustainable Living

Nature sustains itself by several processes:

- Relying on solar energy by using plants to capture solar energy and convert it to chemical energy used to keep plants and plant-eating animals alive.

- Using biological, chemical, and geological processes to recycle vital nutrients.

- Relying on renewable resources by having soil, water, air, plants, and animals that are renewed through natural processes.

- Biodiversity—evolving a variety of species (species diversity), genetic variety within species (genetic diversity), and ecosystems (ecological diversity) in response to environmental changes over billions of years and as a mechanism for responding to future changes.

- Adaptation in which natural populations can change their genetic makeup in response to changes in environmental conditions.

- Population control in which the birth rates, death rates, age distribution, and migration patterns of natural populations respond to changes in environmental conditions.

- Resource conservation. There is little waste in nature. Organisms generally use only what they need to survive, stay healthy, and reproduce.

These are the secrets of nature that we must understand and mimic as we alter nature. Understanding these secrets does not mean that we should stop growing food, building cities, and making other changes that affect the earth's biological communities. But we do need to recognize that such human-induced changes have far-reaching and unpredictable consequences. We need wisdom, care, and restraint as we alter the ecosphere (Section 2-3).

In addition to the first law of ecology, our actions should take into account the **second law of ecology** or principle of interrelatedness: *Everything is connected to and intermingled with everything else; we are all in it together.* Another cardinal rule is the **third law of ecology**: *Any substance that we produce should not interfere with any of the earth's natural biogeochemical cycles.*

We also have an obligation to work with nature to repair many of the wounds we have already inflicted. This is the goal of the emerging science and

Edward J. Kormondy

Edward J. Kormondy is chancellor and professor of biology at the University of Hawaii-Hilo/West Oahu College. He has taught at the California State University at Los Angeles, the University of Southern Maine, the University of Michigan, Oberlin College, and Evergreen State College. Among his many research articles and books are Concepts of Ecology *and* Readings in Ecology *(both published by Prentice-Hall). He has been a major force in biological education and for several years was director of the Commission on Undergraduate Education in the Biological Sciences.*

Energy flows—but downhill only in terms of its quality; chemical nutrients circulate—but some stagnate; populations stabilize—but some go wild; communities age—but some age faster. These dynamic and relentless processes are as characteristic of ecosystems as are thermonuclear fusion reactions in the sun.

Thinking one can escape the operation of these and other laws of nature is like thinking one can stop the earth from revolving or make rain fall up. Yet we have consciously peopled the earth only for hundreds of millions to endure starvation and malnutrition, deliberately dumped wastes only to ensure contamination, purposefully simplified agricultural systems only to cause widespread crop losses from pest invasions. Such actions suggest that we believe that energy and food automatically increase as people multiply, that things stay where they are put, that simplification of ecosystems aids in their productivity. Such actions indicate that we have ignored basic, inexorable, and unbreakable physical laws of ecosystems. We have pro-

posed, but nature has disposed, often in unexpected ways counter to our intent.

We proposed more people, more mouths to be fed, more space to be occupied. Nature disposed by placing an upper limit on the rate at which the earth's plants can produce organic nutrients for themselves and for the people and other animals that feed on them. It also disposed by using and degrading energy quality at and between all trophic levels in the biosphere's intricate food webs and by imposing an upper limit on the total space that is available and can be occupied by humans and other species.

Ultimately, the only way there can be more and more people is for each person to have less and less food and fuel energy and less and less physical space. Absolute limits to growth are imposed both by thermodynamics and space. We may argue about what these limits are and when they will be reached, but there are limits and, if present trends continue, they will be reached. The more timely question then becomes a qualitative one. What quality of life will we have within these limits? What kind of life do you want? What quality of life will future generations have?

We proposed exploitative use of resources and indiscriminate disposal of human and technological wastes. Nature disposed, and like a boomerang, the consequences of our acts came back to hit us. On the one hand, finite oil, coal, and mineral resource supplies are significantly depleted—some nearing exhaustion. On the other hand, air, water, and land are contaminated, some beyond restoring.

Nature's laws limit each resource; some limits are more confining than others, some more critical than others. The earth is finite, and its resources are therefore finite. Yet another of nature's laws is that fundamental resources—elements and compounds—circulate, some fully and some partially. They don't stay where they are put. They move from the land to the water and the air, just as they move from the air and water to the land. Must not our proposals for using resources and discharging wastes be mindful of ultimate limits and the earth's chemical recycling

(continued)

art of *rehabilitation and restoration ecology.* But our primary goal should be to reduce the destruction and degradation of wild natural systems instead of having to depend on expensive, incomplete, and time consuming ecosystem rehabilitation and restoration. We must tune our senses to how nature really works and sustains itself, sensing in nature fundamental rhythms we can trust and cooperate with even though we will never fully understand them.

What has gone wrong, probably, is that we have failed to see ourselves as part of a large and indivisible whole. For too long we have based our lives on a primitive feeling that our "God-given" role was to have "dominion over the fish of the sea and over the fowl of the air and over every living thing that moveth upon the earth." We have failed to understand that the earth does not belong to us, but we to the earth.
ROLF EDBERG

processes? What about your own patterns of resource use and waste disposal?

We proposed simplification of our agricultural systems to ease the admittedly heavy burden of cultivation and harvest. Nature has disposed otherwise, however. Simple ecosystems such as a cornfield are youthful ones, and like our own youth, are volatile, unpredictable, and unstable. Young ecosystems do not conserve nutrients, and agricultural systems in such a stage must have their nutrients replaced artificially and expensively by adding commercial inorganic fertilizers. Young agricultural systems essentially lack resistance to pests and disease and have to be protected artificially and expensively by pesticides and other chemicals. These systems are also more subject to the whims of climate and often have to be expensively irrigated. Must not our proposals for managing agricultural systems be mindful of nature's managerial strategy of providing biological diversity to help sustain most complex ecosystems? What of your own manicured lawn?

The take-home lesson is a rather straightforward one: We cannot propose without recognizing how nature disposes of our attempts to manage the earth's resources for human use. We are shackled by basic ecological laws of energy flow, chemical recycling, population growth, and community aging processes. We have plenty of freedom within these laws, but like it or not we are bounded by them. You are bounded by them. What do you propose to do? And what might nature dispose in return?

Guest Essay Discussion

1. List the patterns of your life that are in harmony with the laws of energy flow and chemical recycling and those that are not.

2. Can you think of other examples of "we propose" and "nature disposes"?

3. Set up a chart with examples of "we propose" and "nature disposes," but add a third column titled "we repropose," based on using ecological principles to work with nature.

DISCUSSION TOPICS

1. List a probable limiting factor for each of the following ecosystems:
   a. a desert
   b. the surface layer of the open sea
   c. the Arctic tundra
   d. the floor of a tropical rain forest
   e. the bottom of a deep lake

2. If possible, visit a nearby lake. Would you classify it as oligotrophic, mesotrophic, or eutrophic? What are the major factors contributing to its nutrient enrichment? Which of these are related to human activities?

3. Since the deep oceans are vast, self-sustaining ecosystems located far away from human habitats, why not use them as a depository for essentially all of our radioactive and other hazardous wastes? Give your reasons for agreeing or disagreeing with this proposal.

4. Why are coastal and inland wetlands considered to be some of the planet's most important ecosystems? Why have so many of these vital ecosystems been destroyed by human activities? What factors in your lifestyle contribute to the destruction and degradation of wetlands?

5. Someone tells you not to worry about air pollution because the human species through natural selection will develop lungs that can detoxify pollutants. How would you reply?

6. Explain why a simplified ecosystem such as a cornfield is much more vulnerable to harm from insects, plant diseases, and fungi than a more complex, natural ecosystem such as a grassland. Why are natural ecosystems less vulnerable?

FURTHER READINGS

Attenborough, David. 1984. *The Living Planet.* Boston: Little, Brown.

Attenborough, David, et al. 1989. *The Atlas of the Living World.* Boston: Houghton Mifflin.

Brown, J. H., and A. C. Gibson. 1983. *Biogeography.* St. Louis: C. V. Mosby.

Clapham, W. B., Jr. 1984. *Natural Ecosystems.* 2d ed. New York: Macmillan.

Hardin, Garrett. 1985. "Human Ecology: The Subversive, Conservative Science." *American Zoologist,* vol. 25, 469-476.

Lovelock, James E. 1988. *The Ages of Gaia: A Biography of Our Living Earth.* New York: Norton.

Maltby, Edward. 1986. *Waterlogged Wealth.* Washington, D.C.: Earthscan.

Odum, Eugene P. 1969. "The Strategy of Ecosystem Development." *Science,* vol. 164, 262–270.

Pilkey, Orin H., Jr., and William J. Neal, eds. 1987. *Living with the Shore.* Durham, N.C.: Duke University Press.

Teal, J., and M. Teal. 1969. *Life and Death of a Salt Marsh.* New York: Ballantine.

Tudge, Colin. 1988. *The Environment of Life.* New York: Oxford University Press.

Wallace, David. 1987. *Life in the Balance.* New York: Harcourt Brace Jovanovich.

Wilson, E. O. 1984. *Biophilia.* Cambridge, Mass.: Harvard University Press.

Wilson, E. O., ed. 1988. *Biodiversity.* Washington, D.C.: National Academy Press.

CHAPTER SIX

HUMAN POPULATION DYNAMICS: GROWTH, URBANIZATION, AND REGULATION

General Questions and Issues

1. How is population size affected by birth rates, death rates, migration rates, and fertility rates?

2. How is population size affected by the percentage of males and females at each age level?

3. How is the world's population distributed between rural and urban areas, and how do transportation systems affect population distribution and urban growth?

4. What methods can be used to regulate the size and the rate of change of the human population?

5. What success have the world's two most populous countries, China and India, had in trying to control the rate of growth of their populations?

The population of most less developed countries is doubling every twenty to thirty years. Trying to develop into a modern industrial state under these conditions is like trying to work out the choreography for a new ballet in a crowded subway car.

GARRETT HARDIN

 OUNT SLOWLY TO SIXTY. During the minute it took you to do this, there were 181 more people in the world to feed, clothe, educate, and house. By this time tomorrow there will be 260,000 more people on our only home (Figure 6-1). We are now adding more people each day than at any other time in human history.

The reason that the world's population continues to grow rapidly by 1 billion people every ten years is simple. There are 2.8 births for each death, with 1.6 births for each death in MDCs and 3.2 births for each death in LDCs. We have brought average death rates and birth rates down, but death rates have fallen more sharply than birth rates. If this continues, one of two things will probably happen during your lifetime: the number of people on earth will double before population growth comes to a halt or the world will experience an unprecedented population crash with hundreds of millions, perhaps billions, of people dying prematurely.

6-1 MAJOR FACTORS AFFECTING HUMAN POPULATION SIZE

Birth Rates and Death Rates In the world as a whole or in a country experiencing little if any migration, annual changes in population size are determined by the difference between the number of people born and the number that die each year. Demographers, or population specialists, normally use the annual crude birth rate and crude death rate, not total live births and deaths, to describe population

Figure 6-1 City pool in Tokyo, Japan. Japan has developed numerous social mechanisms for coping with its high population density and, like most MDCs, has slowed its population growth. In 1990 Japan's population of 124 million was growing exponentially at 0.4%. Japan has about half the population of the United States, but a land area only equal to the size of California. However, living in a highly industrialized country means that each Japanese citizen has an enormous impact on the earth's environment (Figure 1-11, p. 14). Because Japan must import most of its timber, mineral, and energy resources, its population has a severe impact on tropical forests, oceans, and other ecosystems worldwide.

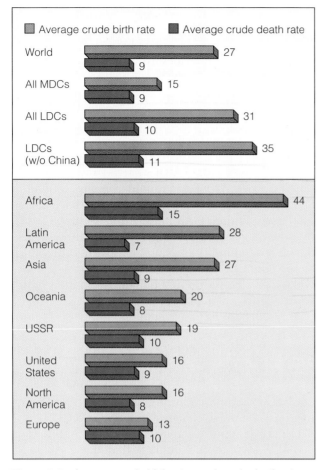

Figure 6-2 Average crude birth rates and crude death rates of various groups of countries in 1990. (Data from Population Reference Bureau)

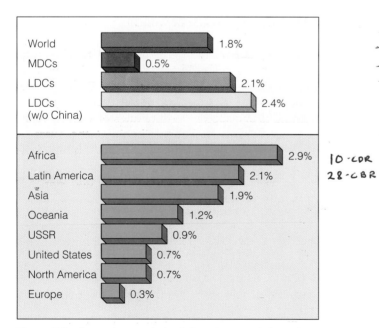

Figure 6-3 Average annual population change rate in various groups of countries in 1990. (Data from Population Reference Bureau)

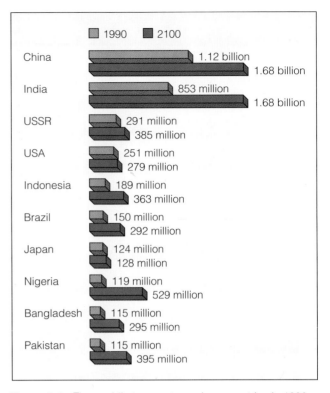

Figure 6-4 The world's ten most populous countries in 1990, with projections of their population size in 2100. (Data from World Bank)

change. The **crude birth rate** is the annual number of live births per 1,000 persons in a population at the midpoint of a given year (July 1). The **crude death rate** is the annual number of deaths per 1,000 persons in a population at the midpoint of a given year. Figure 6-2 shows the crude birth rates and death rates for the world and various groups of countries in 1990.

The annual rate at which the size of a population changes is called the **annual rate of natural population change**. It is usually expressed as a percentage representing the difference between the crude birth rate and the crude death rate divided by 10. It indicates how fast the population size of the world or other region (assuming no migration) is growing or decreasing. For example, in 1990 the crude birth rate for the world was 27, and the death rate 9. Thus, the world's population grew exponentially in 1990 at a rate of 1.8%: (27 − 9)/10 = 18/10 = 1.8%.

The annual rate at which the world's population was growing decreased from a high of 2% in the mid-1960s to 1.8% in 1990, but the base of the population undergoing this exponential growth has increased by more than 2 billion. This slowdown in the annual exponential growth rate is good news. But it is like learning that a truck heading straight at you has slowed down from 161 kilometers (100 miles) per hour to 145 kilometers (90 miles) per hour and its weight has increased by two-thirds.

When the crude death rate equals the crude birth rate, population size remains stable (assuming no

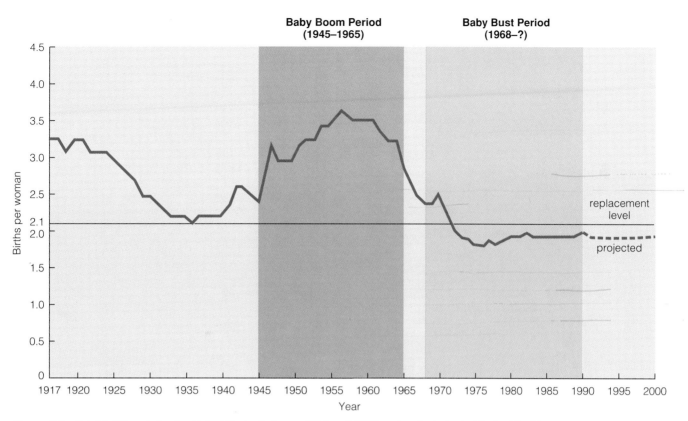

Figure 6-5 Total fertility rate for the United States between 1917 and 1990 and projected rate (dashed line) to 2000. (Data from Population Reference Bureau and U.S. Census Bureau)

migration). This is known as **zero population growth (ZPG)**. When the crude death rate is higher than the crude birth rate, population size decreases.

Figure 6-3 gives the annual population change rates in major parts of the world. In 1990 population change rates ranged from a growth rate of 4.3% in Gaza in western Asia to a decline rate of −0.15% in Hungary.

The impact of exponential population growth on population size is much greater in countries with a large existing population base. In sheer numbers, China and India dwarf all other countries, making up 37% of the world's population (Figure 6-4).

Rapid population growth in LDCs creates pressures on their physical and financial resources, making it difficult to raise living standards. It has also helped widen the differences in average income among rich and poor countries (see Section 7-3). And in both LDCs and MDCs population growth contributes to resource depletion and degradation (Table 1-1, p. 10, and Figure 1-11, p. 14).

Factors Affecting Birth Rates and Fertility Rates

In addition to the crude birth rate, two types of fertility rates affect a country's population size and growth rate. **Replacement-level fertility** is the number of children a couple must have to replace themselves. You might think that two parents need have only two children to replace themselves. The actual

average replacement-level fertility rate, however, is slightly higher, primarily because some female children die before reaching their reproductive years. In MDCs, the average replacement-level fertility is 2.1 children per couple or woman. In LDCs with high infant mortality rates (deaths of children under the age of one), the replacement level may be as high as 2.5 children per couple or woman.

The most useful measure of fertility for projecting future population change is the **total fertility rate (TFR)**. This is an estimate of the number of live children the average woman will bear if she passes through all her childbearing years (ages 15–44) conforming to age-specific fertility rates of a given year. In simpler terms, it is an estimate of the average number of children women will have during their childbearing years.

In 1990 the average total fertility rate was 3.5 children per woman for the world as a whole, 2.0 in MDCs, and 4.0 in LDCs (4.6 if China is excluded). These rates ranged from a low of 1.3 in Italy to a high of 8.3 in Rwanda in eastern Africa. Population experts expect TFRs in MDCs to remain around 1.9 and those in LDCs to drop to around 2.3 by 2025. This is good news, but it will still lead to a projected population of around 9 billion by then.

Since 1972 the United States has had a total fertility rate of 1.8 (2.0 in 1990)—below the replacement level (Figure 6-5). But a total fertility rate below

Replacement-level fertility

CASE STUDY   U.S. Population Stabilization

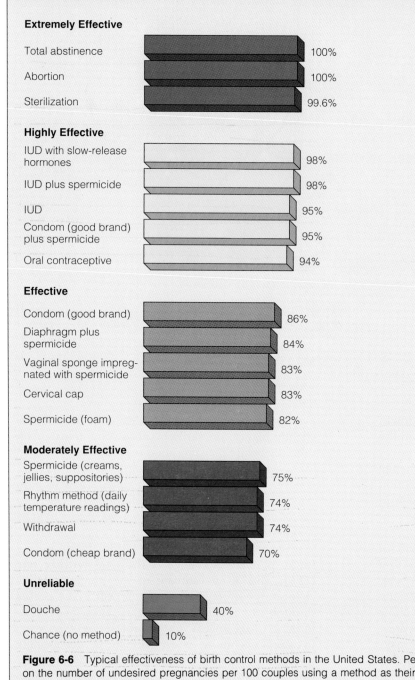

Extremely Effective

Total abstinence	100%
Abortion	100%
Sterilization	99.6%

Highly Effective

IUD with slow-release hormones	98%
IUD plus spermicide	98%
IUD	95%
Condom (good brand) plus spermicide	95%
Oral contraceptive	94%

Effective

Condom (good brand)	86%
Diaphragm plus spermicide	84%
Vaginal sponge impregnated with spermicide	83%
Cervical cap	83%
Spermicide (foam)	82%

Moderately Effective

Spermicide (creams, jellies, suppositories)	75%
Rhythm method (daily temperature readings)	74%
Withdrawal	74%
Condom (cheap brand)	70%

Unreliable

| Douche | 40% |
| Chance (no method) | 10% |

Figure 6-6   Typical effectiveness of birth control methods in the United States. Percent effectiveness is based on the number of undesired pregnancies per 100 couples using a method as their sole form of birth control for a year. A 94% effectiveness for oral contraceptives means that for every 100 women using the pill regularly for one year, six are likely to get pregnant. (Data from Allan Guttmacher Institute)

The population of the United States has grown from 4 million in 1790 to 251 million in 1990—a sixty-two-fold increase. The total fertility rate in the United States has oscillated wildly (Figure 6-5). At the peak of the post-World War II baby boom (1945–65) in 1957, the TFR reached 3.7 children per woman. Since then it has generally declined and has been at or below replacement level since 1972.

Various factors contributed to this decline:

- Widespread use of effective birth control methods (Figure 6-6).

- Availability of legal abortions.

- Social attitudes favoring smaller families.

- Greater social acceptance of childless couples.

- Rising costs of raising a family. It will cost about $150,000 to raise a child born in 1988 to age 18. Four years of college cost another $160,000 at a private college and $77,000 at an in-state publicly supported college.

- Increases in the average marriage age between 1958 and 1989 from 20.1 to 23.9 for women and from 22.8 to 26.2 for men.

- An increasing number of women working outside the

replacement level doesn't necessarily mean that a country's population has stabilized or is declining (see Case Study above).

A number of socioeconomic and cultural factors affect a country's average birth rate and total fertility rate. The following are the most significant factors:

- *Average levels of education and affluence.* Birth and fertility rates are usually lower in MDCs, where both of these factors are high.

- *Importance of children as a part of the family labor force.* Birth and fertility rates tend to be high in LDCs (especially in rural areas). They are lower

Labor force
education
affluence

Urbanization
high costs of educating

Labor force
education
affluence

Urbanization
high costs of educating

118   PART TWO   Basic Concepts

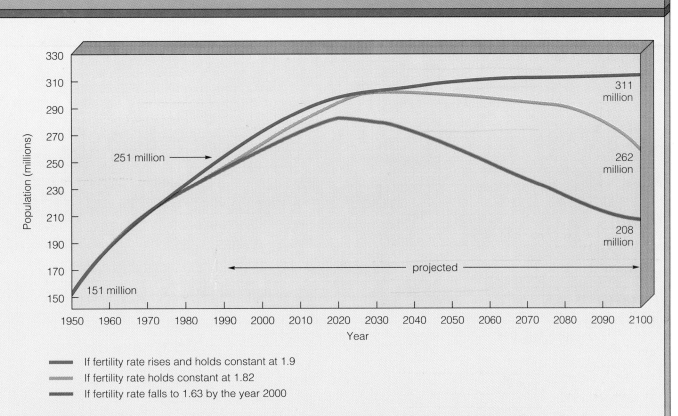

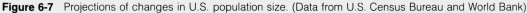

If fertility rate rises and holds constant at 1.9

If fertility rate holds constant at 1.82

If fertility rate falls to 1.63 by the year 2000

Figure 6-7 Projections of changes in U.S. population size. (Data from U.S. Census Bureau and World Bank)

home. By 1990 more than 70% of American women of child-bearing age worked outside the home and had a childbearing rate one-third the rate of those not in the paid labor force.

The decline in the total fertility rate since 1969 has led to a decline in the annual rate of population growth in the United States. The United States has not reached zero population growth (ZPG) in spite of the dramatic drop in the average total fertility rate to below the replacement level. The major reasons for this are

- the large number of women still moving through their childbearing years
- high levels of legal and illegal immigration
- an increase in the number of unmarried young women (including teenagers) having children

In 1990 the U.S. population grew by 0.7%. This added 2.8 million people: 1.9 million more births than deaths, 0.7 million legal immigrants, and an estimated 0.2 million illegal immigrants.

Given the erratic history of the

U.S. total fertility rate (Figure 6-5), no one knows whether or how long it will remain below replacement level. The Census Bureau and the World Bank have made various projections of future U.S. population growth, assuming different average total fertility rates, life expectancies, and net legal immigration rates (Figure 6-7).

All Americans will play a role in determining which of these and other demographic possibilities becomes a reality. What role do you intend to play in determining your country's future demographic history?

in countries where a compulsory mass education system removes children from the family labor force during most of the year.

- *Urbanization.* People living in urban areas tend to have fewer children than those living in rural areas, where children are needed to help with

growing food, collecting firewood and water, and other survival tasks.

- *High costs of raising and educating children.* Rates tend to be low in MDCs, where raising children is more costly since they don't enter the labor force until their late teens or early twenties.

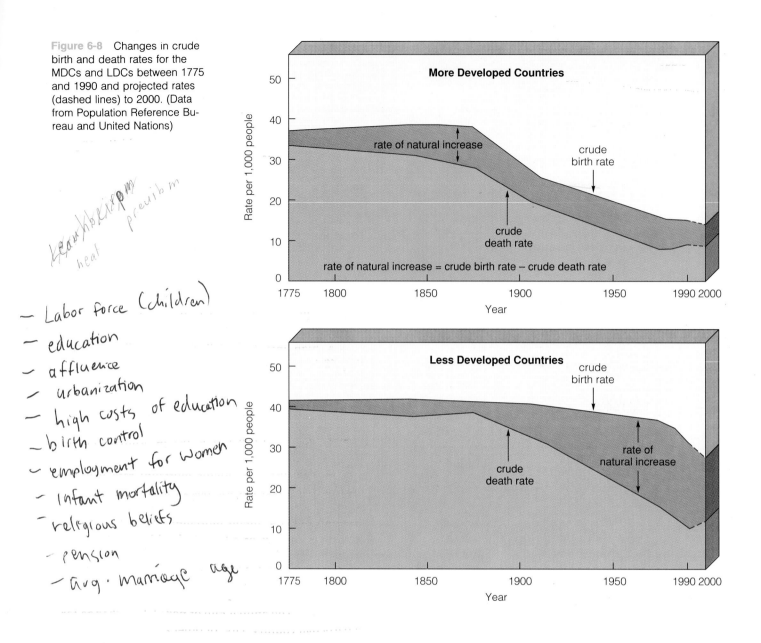

Figure 6-8 Changes in crude birth and death rates for the MDCs and LDCs between 1775 and 1990 and projected rates (dashed lines) to 2000. (Data from Population Reference Bureau and United Nations)

(handwritten margin notes)
- Labor force (children)
- education
- affluence
- urbanization
- high costs of education
- birth control
- employment for women
- Infant mortality
- religious beliefs
- pension
- avg. marriage age

- *Educational and employment opportunities for women.* Rates tend to be high when women have little or no access to education and to paid employment outside the home.

- *Infant mortality rates.* In areas with high infant mortality rates people tend to have more children to replace those who have died.

- *Average marriage age* (or, more precisely, the average age at which women give birth to their first child). People have fewer children when the average marriage age of women is 25 or higher. This reduces the typical childbearing years (ages 15–44) by 10 or more years and cuts the prime reproductive period (ages 20–29), when most women have children, by half or more.

- *Availability of private and public pension systems.* Pensions eliminate the need for parents to have many children to support them in old age.

- *Availability of reliable methods of birth control* (Figure 6-6). Widespread availability tends to reduce birth and fertility rates. However, this factor can be counteracted by religious beliefs that prohibit or discourage the use of abortion or certain forms of contraception.

- *Religious beliefs, tradition, and cultural norms that influence the number of children couples want to have.*

Figure 1-2, p. 4, shows three projections of world population growth through the next century. Each projection is based on a different set of assumptions about fertility rates, death rates, and international migration patterns. No one really knows whether any of these projections will prove accurate. Demographic projections are not predictions of what will necessarily take place. Instead, they represent possibilities based on present trends and certain assumptions about people's future reproductive behavior. Despite their

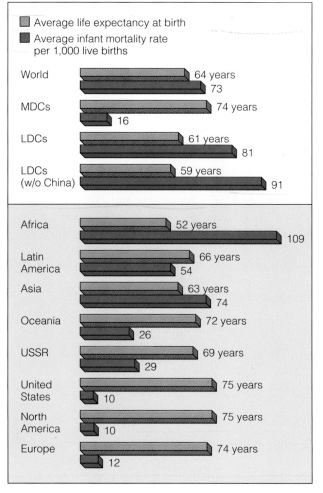

Average life expectancy at birth
Average infant mortality rate per 1,000 live births

	Life expectancy	Infant mortality
World	64 years	73
MDCs	74 years	16
LDCs	61 years	81
LDCs (w/o China)	59 years	91
Africa	52 years	109
Latin America	66 years	54
Asia	63 years	74
Oceania	72 years	26
USSR	69 years	29
United States	75 years	10
North America	75 years	10
Europe	74 years	12

Figure 6-9 Life expectancy at birth and average infant mortality rate for various groups of countries in 1990. (Data from Population Reference Bureau)

uncertainty, demographic projections help us focus our energies on converting the most desirable possibilities into reality.

Factors Affecting Death Rates The rapid growth of the world's population over the past 100 years was not caused by a rise in crude birth rates. Rather, it is due largely to a decline in crude death rates, especially in the LDCs (Figure 6-8).

The major interrelated reasons for this general drop in death rates are

- better nutrition because of increased food production and better distribution

- reduced incidence and spread of infectious diseases because of improved personal hygiene and improved sanitation and water supplies

- improvements in medical and public health technology, including antibiotics, immunization, and insecticides

Two useful indicators of overall health in a country or region are **life expectancy**—the average number of years a newborn infant can be expected to live—and the **infant mortality rate**—the number of babies out of every 1,000 born that die before their first birthday (Figure 6-9).

In 1990, average life expectancy at birth ranged from a low of 41 years in Sierra Leone in western Africa, in Afghanistan in southern Asia, and in Ethiopia in eastern Africa to a high of 79 years in Japan. In the world's 41 poorest countries, mainly in Asia and Africa, average life expectancy is only 47 years.

Between 1900 and 1990, average life expectancy at birth increased sharply in the United States from 42 to 75 (78.3 for females and 71.3 for males). But in 1990, people in 25 countries and colonies had an average life expectancy at birth one to three years higher than people in the United States.

A high infant mortality rate usually indicates insufficient food (undernutrition), poor nutrition (malnutrition), and a high incidence of infectious disease (usually from contaminated drinking water). In 1990 infant mortality rates ranged from a low of 4.8 deaths per 1,000 live births in Japan to a high of 182 deaths per 1,000 live births in Afghanistan in southern Asia.

In 1990 the U.S. infant mortality rate was 9.7 deaths per 1,000 (18 for blacks and 9 for whites), higher than that in 23 other countries. Two factors keeping the U.S. infant mortality rate higher than it could be are lack of adequate health care for poor women during pregnancy and for their babies after birth and the high birth rate for teenage women in the United States (see Case Study on p. 122). *health care + teenage births*

Migration The annual rate of population change for a particular country, city, or other area is also affected by movement of people into (immigration) and out of (emigration) that area:

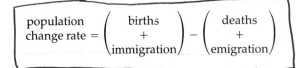

$$\text{population change rate} = \left(\begin{array}{c} \text{births} \\ + \\ \text{immigration} \end{array}\right) - \left(\begin{array}{c} \text{deaths} \\ + \\ \text{emigration} \end{array}\right)$$

Most countries control their rates of population growth to some extent by restricting immigration. Only a few countries annually accept a large number of immigrants or refugees. This means that population change for most countries is determined mainly by the differences between their birth rates and death rates.

However, migration within countries, especially from rural to urban areas, plays an important role in the population dynamics of cities, towns, and rural areas. This migration affects the way population is distributed within countries, as discussed in Section 6-3.

The United States has the highest teenage pregnancy rate of any industrialized country: ten times higher than Japan's and three times higher than that of most European countries. Every year in the United States about 900,000 teenage women—one in every ten—become pregnant. About 547,000 of these young women give birth. The remaining 330,000 have abortions, accounting for almost one of every four abortions performed in the United States.

Almost half of the babies born to teenage mothers in the United States depend on welfare to survive. Each of these babies costs taxpayers an average of $15,600 over the 20 years following its birth.

• Babies born to teenagers are more likely to have a low birth weight—the most important factor in infant deaths—thus increasing the country's infant mortality rate. The United States has developed the medical technology for saving low-weight babies, but babies of poor teenage women with no health insurance often do not have access to these expensive, life-saving procedures.

Why are teenage pregnancy rates in the United States so high? UN studies show that U.S. teenagers aren't more sexually active than those in other MDCs, but they are less likely to take precautions to prevent pregnancy. They aren't getting effective sex education at home or at school.

In Sweden, which has a much lower teenage pregnancy rate than the United States, every child receives a thorough grounding in basic reproductive biology by age 7. By age 12, each child has been told about the various types of contraceptives.

Polls show that 75% of Americans favor sex education in the schools, including information about birth control, and school clinics that dispense contraceptives with parental consent. Yet effective sex education is not widely available and is not given at an early enough age. Why? Mostly because of strong opposition to sex education by small but highly vocal religious groups who fear that such education will increase sexual activity and abortions. What do you think should be done about teenage pregnancy?

6-2 POPULATION AGE STRUCTURE

Age Structure Diagrams Why will world population probably keep growing for at least 60 years after the average world total fertility rate has reached or dropped below the replacement-level fertility of 2.1? The answer to this question lies in an understanding of the **age structure,** or age distribution, of a population: the percentage of the population, or the number of people of each sex, at each age level in a population.

Demographers make a population age structure diagram by plotting the percentages or numbers of males and females in the total population in three age categories: *preproductive* (ages 0–14), *reproductive* (ages 15–44), and *postproductive* (ages 45–85 +). Figure 6-10 shows the age structure diagrams for countries with rapid, slow, and zero growth rates.

Mexico and most LDCs with rapidly growing populations have pyramid-shaped age structure diagrams. This indicates a high ratio of children under age 15 to adults over age 65. In contrast, the diagrams for the United States, Sweden, and most MDCs undergoing slow or no population growth have a narrower base. This shows that such countries have a much smaller percentage of population under age 15 and a larger percentage above 65 than countries experiencing rapid population growth.

MDCs, such as Sweden and Denmark, that have achieved or nearly achieved zero population growth have roughly equal numbers of people at each age level (Figure 6-10). Hungary, West Germany, and Monaco, which are experiencing a slow population decline, have roughly equal numbers of people at most age levels but lower numbers under age five.

Age Structure and Population Growth Momentum Any country with a large number of people below age 15 has a powerful built-in momentum to increase its population size unless death rates rise sharply. The number of births rises even if women have only one or two children. This happens because the number of women who can have children increases greatly as females reach their reproductive years.

In 1990 one of every three persons on this planet was under 15 years old. In LDCs the number is even higher—36% compared to 22% in MDCs. Figure 6-11 shows the massive momentum for population growth in LDCs because of their large numbers of people under age 15 who will be moving into their reproductive years. This explains why population experts project that the world's population will not level off until around the middle of the next century and perhaps not until the end of the next century. Within your lifetime world population is expected to stop growing. But the key question is whether it will peak at 7 billion, 10 billion, or 14 billion (Figure 1-2, p. 4).

The powerful force for continued population growth, mostly in LDCs, will be slowed only by a massive program to reduce birth rates or a catastrophic rise in death rates. Greatly increasing efforts to reduce birth rates now could bring the projected peak population down to 7 billion instead 10 billion, or even 14 billion (Figure 1-2, p. 4). Many analysts consider this one of our most important challenges.

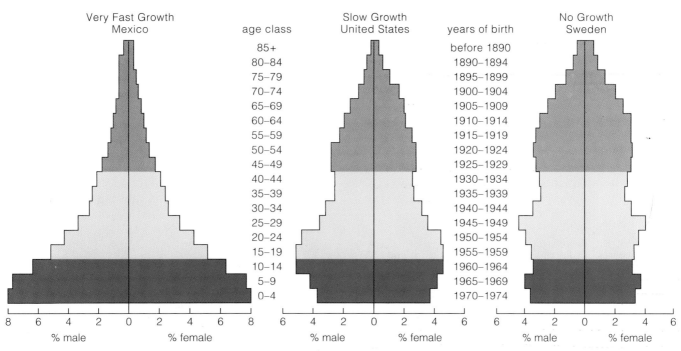

Figure 6-10 Population age structure diagrams for countries with rapid, slow, and zero population growth rates. Bottom portions represent preproductive years (0–14), middle portions represent reproductive years (15–44), and top portions represent postproductive years (45–85+). (Data from Population Reference Bureau)

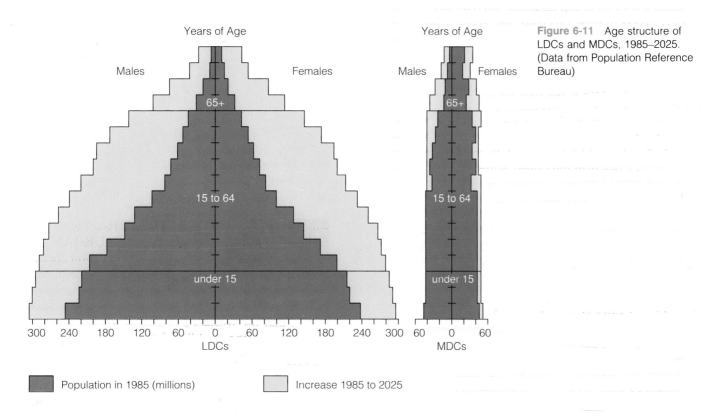

Figure 6-11 Age structure of LDCs and MDCs, 1985–2025. (Data from Population Reference Bureau)

Population in 1985 (millions)   Increase 1985 to 2025

Making Projections from Age Structure Diagrams

A baby boom took place in the United States between 1945 and 1965. This 75-million-person bulge will move upward through the country's age structure during the 80-year period between 1945 and 2025 as baby boomers move through their youth, young adulthood, middle-age, and old age (Figure 6-12). In 1970 the median age of the U.S. population was about 29. By 1990 it was almost 33 and is projected to reach 36 by 2000 and 39 by 2010.

Today baby boomers make up nearly half of all adult Americans. In sheer numbers they dominate the population's demand for goods and services. Companies not providing products and services for

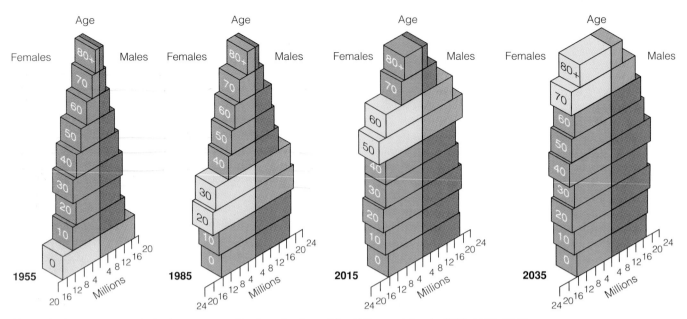

Figure 6-12   Tracking the baby-boom generation. Age structure of the U.S. population in 1955, 1985, 2015, and 2035. (Data from the Population Reference Bureau and U.S. Census Bureau)

this aging bulge in the population can go bankrupt. Young people making career choices should also consider these demographic facts of life. Baby boomers, who made up an estimated 60% of registered ~voters~ [votes] voters in 1990, will play an increasingly important role in deciding who gets elected and what laws are passed between now and 2030.

During their working years baby boomers will create a large surplus of money in the Social Security trust fund. However, even if elected officials resist the temptation to dip into these funds for balancing the budget or other purposes, the large number of retired baby boomers will quickly use up this surplus. Unless changes in funding are made, the Social Security fund will be depleted by 2048. Elderly baby boomers will also put severe strains on health care services.

The economic burden of helping support so many retired baby boomers will be on the baby bust generation, the much smaller group of people born between 1968 and 1990, when average fertility rates fell sharply (Figure 6-5). Retired baby boomers may use their political clout to force members of the baby bust generation to pay greatly increased income, health care, and Social Security taxes.

In many respects, the baby bust generation should have an easier time than the baby boom generation. Fewer people will be competing for education, jobs, and services. Labor shortages should also drive up their wages. But three out of four new jobs available between now and 2010 will require education or technical training beyond high school. People without such training may face economic hard times.

Although they will probably have no trouble getting entry-level jobs, the baby bust group may find it hard to get job promotions as they reach middle age because most upper-level positions will be occupied by the much larger baby boom group. Many baby boomers may delay retirement because of improved health and the need to build up adequate retirement funds. From these few projections, we see that any baby boom bulge or baby bust indentation in the age structure of a population creates a number of social and economic changes that ripple through a society for decades.

urban ¦ pop 7 2500

6-3   POPULATION DISTRIBUTION: URBANIZATION AND URBAN GROWTH

The World Situation   Economic, environmental, and social conditions are affected not only by population growth and age structure but also by how population is distributed geographically in rural or **urban areas,** often defined as a village, town, or city with a population of more than 2,500 people.

It is important to distinguish between urbanization and urban growth. A country's **urbanization** is the percentage of its population living in an urban area. **Urban growth** is the rate of growth of urban populations. Three things lead to both urbanization and urban growth: rapid population growth, economic development, and rural poverty.

Several trends are important in understanding the problems and challenges of urbanization and urban growth:

1) Rapid population growth
2) economic development
3) Rural poverty

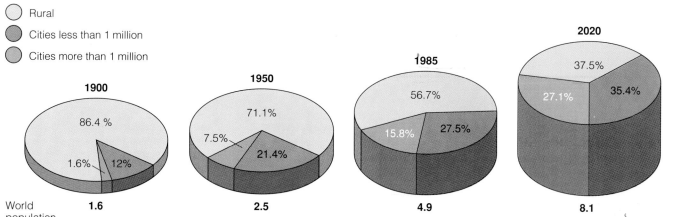

○ Rural

● Cities less than 1 million

● Cities more than 1 million

1900
86.4 %
1.6% 12%

1950
71.1%
7.5%
21.4%

1985
56.7%
15.8%
27.5%

2020
37.5%
27.1%
35.4%

World population in billions

1.6   2.5   4.9   8.1

Figure 6-13 Patterns of world urbanization from 1900 to 1985, with projections to 2020. (Data from United Nations and Population Reference Bureau)

[handwritten: no. people in 14% to 43% urban areas]

[handwritten: 4) Gov't policies / 3) No jobs needed / 1) Urban job opportunities / 2) Rural poverty.]

- Between 1900 and 1985 the percentage of the world's population living in urban areas increased from 14% to 43% (Figure 6-13).

- The number of large cities is increasing rapidly. *[handwritten: no. of large cities up.]* In 1940 only one person in 100 lived in a city with a million or more inhabitants. Today one of every ten persons lives in such cities and many of these live in megacities with 10 million or more people. The United Nations projects that by 2000 there will be 26 megacities, more than two-thirds of them in LDCs. At that time the world's two largest cities will be Mexico City with a projected population of 26.3 million and São Paulo, Brazil, with 24 million people.

- LDCs, with 32% urbanization, are simultaneously experiencing high rates of natural population increase and rapidly increasing urban growth. *[handwritten: LDCs ↑]*

- In MDCs, with 73% urbanization, urban growth is slowing. *[handwritten: MDs ↓]*

- The distribution of people living in absolute poverty is shifting from rural to urban areas at an increasing rate. *[handwritten: poverty]*

If these trends continue, by 2020 almost two-thirds of the world's population will be living in urban areas (Figure 6-13). Accommodating the 2.9 billion more people in urban areas by 2020 will be a monumental task. Most of these people will be struggling to survive in the urban areas of LDCs that can't provide adequate services, shelter, or jobs for much of their present urban populations.

The Worsening Urban Crisis in LDCs In LDCs more than 20 million rural people migrate to cities each year. Several factors are responsible for this population shift (see Case Study on p. 126). Part is caused by the pull of urban job opportunities, but much of this migration is caused by rural poverty, which pushes people into cities. Modern mechanized agriculture decreases the need for farm labor and allows large landowners to buy out small-scale, subsistence farmers who cannot afford to modernize. Without jobs or land these people are forced to move to cities. Urban growth in LDCs is also caused by government policies that distribute most income and social services to urban dwellers at the expense of rural dwellers.

For most of the rural poor moving to urban areas in LDCs, as well as the urban poor in MDCs, the city becomes a poverty trap, not an oasis of economic opportunity and cultural diversity (see Case Study on p. 129). In most LDCs, between one-fourth and one-half of those needing a job cannot find enough work to meet their basic needs. Those few fortunate enough to get a job must work long hours for low wages. To survive they often have to take jobs that expose them to dust, hazardous chemicals, excessive noise, and dangerous machinery.

Many of the urban poor in LDCs are forced to live on the streets (Figure 1-4, p. 5). Others crowd into mushrooming slums and shantytowns, their shelters made from corrugated steel, plastic sheets, tin cans, and packing boxes, which ring the outskirts of most cities in these countries. In most large cities in LDCs, shantytown populations double every five to seven years—four to five times the population growth rate of the entire city.

Probably between a third and a half of the residents of cities in LDCs live on public or private lands illegally. The people in these settlements live in constant fear of eviction or of having their makeshift shelters destroyed by bulldozers. Many shantytowns are also located on land subject to landslides, floods, or tidal waves or in the most polluted districts of cities because this is the only land no one else wants.

Brazil, the largest country in South America, ranks fifth in land area in the world and eighth in population (Figure 6-14). In 1990 its population of about 150 million was expanding by 1.9%, adding 2.9 million people a year. It is encouraging that between 1965 and 1990 Brazil's total fertility rate fell from 6.8 to 3.3. But like most LDCs, a large share of Brazil's population (36%) is under age 15. This explains why the country's population is projected to reach 234 million by 2020.

Brazil's gross national product averages almost $2,300 per person, making it a middle income developing country. This average is deceiving, however, because most of the country's wealth is concentrated in the hands of a small fraction of the population. Most other people are poor, and they must survive on an average income of only several hundred dollars a year.

Some 81% of the country's farmland is owned by only 4.5% of the country's landowners. About 70% of the country's rural families are landless.

Brazil's economic status as a middle income country is also deceiving. It has fueled much of its recent economic growth by borrowing abroad. In 1989 its foreign debt was over $115 billion, the largest of any LDC. Of every $100 Brazil earns from exports of minerals, beef, soybeans, and manufactured products, $41 is paid out in interest to its creditors.

This heavy debt burden pushes the country to expand its exports of timber, minerals, and other items and in the process deplete and degrade its natural resource base. Brazil is like a family giving the appearance of affluence by living off credit cards and having to use much of their capital and income to finance its debt.

Brazil is divided geographically into a largely impoverished tropical north and a temperate south, where most industry and wealth are concentrated. The Amazon basin, which covers about one-third of the country's territory, remains largely unsettled, but this is changing as landless poor migrate there hoping to grow enough food to survive and its forests are cut down for grazing livestock, timber, and mining, or flooded to create massive reservoirs for hydroelectric dams.

Brazil is much more urbanized than most LDCs. Between 1950 and 1990, the urban population in Brazil increased from 34% to 74%, compared to an average of 32% for all LDCs. Attracted by the prospect of jobs, many of the rural poor in the north and northeast have flooded into Rio de Janeiro and São Paulo in the south. These modern cosmopolitan centers are surrounded by sprawling urban slums and massive rural poverty (see Figure 7-1, p. 148).

Overpopulation is one factor increasing the ranks of the unemployed and underemployed in rural and urban Brazil. Many poor people have been displaced from the land by the conversion of small farms in the south to industrialized agriculture for export crops. Deforestation and desertification in the extremely poor northeast have also helped push environmental refugees to southern cities and to development projects in the tropical forests of the Amazon basin (Figure 6-14).

As a safety valve for its exploding population, the Brazilian gov-

Fires are common because most residents use kerosene stoves, fuelwood, or charcoal for heating and cooking.

Most cities refuse to provide shantytowns and slums with adequate drinking water, sanitation, food, health care, housing, schools, and jobs. Not only lacking the money, officials also fear that improving services will attract even more of the rural poor.

Despite joblessness, squalor, overcrowding, and rampant disease, shantytown residents cling to life with resourcefulness, tenacity, and hope. Most urban migrants do have more opportunities and are better off than the rural poor they left behind. With better access to family planning programs, they tend to have fewer children; also the children in most cities have better access to schools than do rural dwellers. Even so, nearly half of all school-age children in urban areas of LDCs drop out of school before they finish the fourth grade to work or take care of younger children.

The U.S. Situation In 1800 only 5% of Americans lived in cities. Since then three major internal population shifts have taken place in the United States. They are migration from rural to urban areas, from central cities to suburbs and smaller cities, and from the North and East to the South and West.

The major shift was from rural to urban areas as the country industrialized and needed fewer farmers to produce sufficient food. Currently about 77% of Americans live in the nation's 283 metropolitan areas—defined as cities and towns with at least 50,000 people that are linked socially or economically (Figure 6-15). Nearly half (48%) of the American people live in metropolitan areas of 1 million or more.

Since 1970 many people have moved from large central cities to suburbs, and to smaller cities and rural areas, primarily because of the large numbers of new jobs in such areas. Today about 41% of the country's urban dwellers live in central cities and 59% live in suburbs.

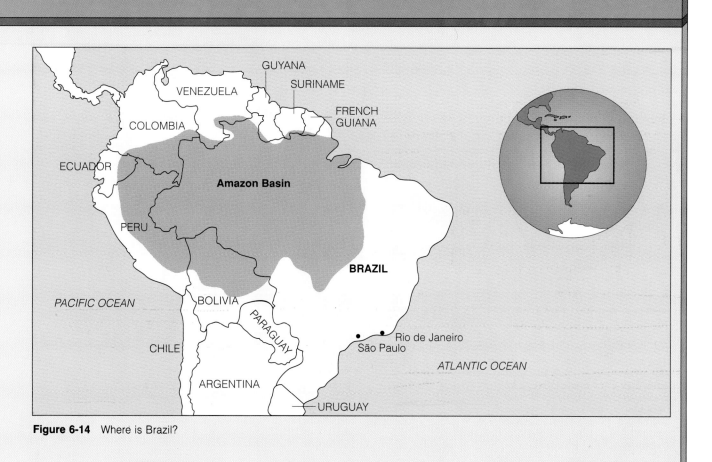

Figure 6-14 Where is Brazil?

ernment has encouraged migration to the Amazon basin, with economic aid from international lending agencies such as the World Bank. This policy is supported by wealthy Brazilians, who want to diffuse pressures for more equitable land distribution, and by ranchers, who receive government subsidies to establish cattle ranches by clearing tropical forests.

Since 1980 about 80% of the population increase in the United States has occurred in the South and West, particularly near the coasts. Most of this increase is due to migration from the North and East. This shift is projected to continue, with the South having the largest population increase between 1987 and 2010, followed by the West.

Many cities in the United States and other industrialized countries are facing problems of deteriorating services and infrastructure, environmental degradation, inner-city decay, and neighborhood collapse. Many of the unemployed, racial and ethnic minorities, and the elderly are trapped in a downward spiral of poverty and degradation.

Municipal governments in many cities are paralyzed by a combination of deteriorating roads, sewers, and other services, rising costs, declining tax bases, and decreased aid from state and federal sources. Unlike most LDCs, most MDCs have the means and resources to improve urban life for most of their inhabitants. Whether they do this depends mostly on public pressure to force elected officials to adopt sustainable forms of urban economic development.

Population Distribution, Transportation Systems, and Energy Efficiency If suitable rural land is not available for conversion to urban land, a city grows upward, not outward; it occupies a relatively small area and develops a high population density. Most people living in such compact cities walk, ride bicycles, or use energy-efficient mass transit. Most residents live in multistory apartment buildings with shared walls, reducing heating and cooling costs. Because of the lack of land, many European cities and Asian cities such as Hong Kong, Tokyo, and Singapore are compact and tend to be more energy efficient than the dispersed cities of the United States.

A combination of cheap gasoline, a large supply of rural land suitable for urban development, and a network of highways usually results in a dispersed

Figure 6-15 Major urban regions in the United States. (Data from U.S. Census Bureau)

1 **Bosman** (Boston–Manchester)
2 **Mega York** (New York–Philadelphia)
3 **Philwil** (Philadelphia–Wilmington)
4 **Washbalt** (Washington–Baltimore)
5 **Pittyoung** (Pittsburgh–Youngstown)
6 **Miamilaud** (Miami–Fort Lauderdale)
7 **Hougalbeau** (Houston–Galveston–Beaumont)
8 **Dalworth** (Dallas–Fort Worth)

9 **Denboul** (Denver–Boulder)
10 **San Angeles** (Los Angeles–San Diego)
11 **San Franjose** (San Francisco–San Jose)
12 **Seatac** (Seattle–Tacoma)
13 **Chimilgar** (Chicago–Milwaukee–Gary)
14 **Detanntol** (Detroit–Ann Arbor–Toledo)
15 **Clevak** (Cleveland–Akron)

city with a low population density. Most people living in such a city rely on cars with low energy efficiencies for transportation within the central city, to and from the central city, and within and between suburbs (see Pro/Con on p. 130). Most people live in single-family houses, whose unshared walls lose and gain heat rapidly unless they are well insulated.

Resource and Environmental Problems of Urban Areas Cities survive only by importing large quantities of food, water, energy, minerals, and other resources from near and distant farmlands, forests, mines, and watersheds. Instead of being recycled, most of the solid, liquid, and gaseous wastes of cities are discharged into or eventually end up in air, water, and land outside their boundaries (Figure 6-19).

As urban areas grow, their resource input needs and pollution outputs place increasing stress on distant aquifers, wetlands, estuaries, forests, croplands, rangelands, wilderness, and other ecosystems. In the words of Theodore Roszak:

The supercity . . . stretches out tentacles of influence that reach thousands of miles beyond its already sprawling parameters. It sucks every hinterland and wilderness into its technological metabolism. It forces rural populations off the land and replaces them with vast agroindustrial combines. Its investments and technicians bring the roar of the bulldozer and oil derrick into the most uncharted quarters. It runs its conduits of transport and communication, its lines of supply and distribution through the wildest landscapes. It flushes its waste into every nearby river, lake, and ocean or trucks them away into desert areas. The world becomes its garbage can.

Major urban resource and environmental problems are:

- Scarcity of trees, shrubs, and other natural vegetation that absorb air pollutants, give off oxygen, help cool the air as water evaporates from their leaves, muffle noise, provide wildlife habitats, and provide aesthetic pleasure. As one observer remarked, "Most cities are places where they cut down the trees and then name the streets after them."

aesthetic

In 1990 Mexico's population was 89 million and was growing exponentially at 2.4% a year. Between 1965 and 1990 Mexico's average total fertility rate dropped from 6.7 to 3.8. But because 42% of Mexico's population is under age 15 (Figure 6-10, p. 123), its population is projected to reach 142 million by 2020.

In 1969 the population of Mexico City, the capital of Mexico (Figure 6-16), was 9 million. In 1990 it had 21 million residents—almost one of every four Mexicans—making it the world's most populous urban area. Every day another 1,500 poverty-stricken rural peasants pour into the city, hoping to find a better life.

Today the city suffers from severe air pollution, massive unemployment (close to 50%), and a soaring crime rate. One-third of

the city's people live in crowded slums, or barrios, without running water or electricity. With at least 5 million people living without sewer facilities, tons of human

waste are left in gutters and vacant lots every day. When the winds pick up dried excrement a "fecal snow" often falls on parts (continued)

United Nations

Figure 6-17   The air in Mexico City is among the dirtiest in the world. This is due to a combination of topography, a large population, industrialization, large numbers of motor vehicles, and too little emphasis on reducing rural-to-urban migration and preventing and controlling pollution. Breathing the city's air has been compared to smoking two packs of cigarettes a day.

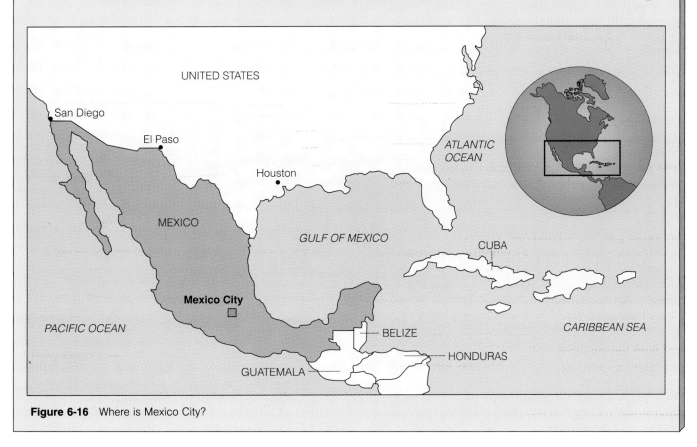

Figure 6-16   Where is Mexico City?

of the city. About half of the city's garbage is left in the open to rot, attracting armies of rats and swarms of flies.

Over 3 million cars, 7,000 diesel buses, and 130,000 factories spew pollutants into the atmosphere. Air pollution is intensified because the city lies in a basin surrounded by mountains and has frequent thermal inversions that trap pollutants near ground level (Figure 6-17).

Breathing the city's air is like smoking two packs of cigarettes a day from birth. The city's air and water pollution cause an estimated 100,000 premature deaths a year. According to a World Health Organization study, 7 out of 10 babies born in Mexico City have unsafe levels of lead in their blood—threatening an entire generation of children with intellectual stunting. These problems, already at crisis levels, will become even worse if this urban area, as projected, grows to 26.3 million people by the end of this century.

The Mexican government is in-dustrializing other parts of the country in an attempt to stop or at least slow migration to Mexico City. Also Mexico's massive $100 billion foreign debt severely strains its ability to mount a massive effort to reduce pollution. By 1993, all new cars in Mexico must have catalytic converters (output devices for reducing pollution emissions). But new cars represent only about 5% of the cars on the road each year. What do you think should be done?

PRO/CON Advantages and Disadvantages of the Automobile

In the United States the car is now used for about 98% of all urban transportation, 85% of all travel between cities, and 84% of all travel to and from work (Figure 6-18). In widely dispersed American cities like Los Angeles, Detroit, Denver, Phoenix, and Houston, 88% to 94% of the people drive to and from work. Instead of using car or van pools, most people commute alone.

The number of cars and trucks in the United States is growing twice as fast as the number of people. With only 4.8% of the world's people, the United States has one-third of the world's 400 million cars. By contrast, China and India, with 37% of the world's people, have only 0.5% of its cars.

In America's dispersed cities cars are not just a necessity, they are a way of life. The average distance each U.S. motorist travels has doubled since 1960. No wonder British author J. B. Priestly remarked, "In America, the cars have become the people."

The automobile has many advantages. Above all, it offers peo-ple freedom to go where they want to go, when they want to go there. To most people cars are also personal fantasy machines that serve as symbols of power, speed, excitement, sexiness, spontaneity, and adventure.

In addition, much of the world's economy is built around producing motor vehicles and supplying roads, services, and repairs for these vehicles. Half of $\frac{1}{2}$ the world's paychecks and resource use are auto-related.

In the United States one of every six dollars spent and one of every six nonfarm jobs are connected to the automobile or related industries such as oil, steel, rubber, plastics, automobile services, and highway construction. This industrial complex accounts for 20% of the annual GNP and provides about 18% of all federal taxes.

In spite of their advantages, motor vehicles have many harmful effects on human lives and on air, water, and land resources. Since 1885 when Karl Benz built the first automobile, about 17 million people have been killed by motor ve-hicles—130 times the number killed at Hiroshima. Though we tend to deny it, riding in cars is one of the most dangerous things we do in our daily lives.

Worldwide, cars and trucks kill an average of 320,000 people, maim 500,000, and injure 10 million a year. Half of the world's people will be involved in an auto accident at some time during their lives.

Each year in the United States motor vehicle accidents kill around 48,000 people and seriously injure at least 300,000. Since the automobile was introduced, almost 2 million Americans have been killed on the highways—about twice the number of Americans killed in all U.S. wars. These accidents cost society about $60 billion annually in lost income and in insurance, administrative, and legal expenses.

By providing almost unlimited mobility, automobiles and highways have been a major factor in urban sprawl in the United States and other countries with large livable land areas. This dispersal of cities has made it increasingly dif-

PACT

(handwritten margin notes: 1) aesthetics 2) temp, precipitations fog, higher 3) no water 4) water pollution, flooding 5) Noise 6) Loss land, soil, wildlife habitats)

- Alteration of local and sometimes regional climate. Average temperatures, precipitation, fog, and cloudiness are generally higher in cities than in suburbs and nearby rural areas. The enormous amount of heat generated by cars, factories, furnaces, air conditioners, and people in cities creates a microclimatic effect known as an **urban heat island** (Figure 6-20). This dome of heat also traps pollutants, especially tiny solid particles (suspended particulate matter), creating a **dust dome** above urban areas.

- Lack of water, requiring expensive reservoirs, aqueducts, and deep wells (Chapter 11).

- Rapid runoff of water from asphalt and concrete. This can overload sewers and storm drains, contributing to water pollution and flooding in cities and downstream areas.

- Production of large quantities of air pollution (Chapter 9), water pollution (Chapter 11), garbage, and other solid waste (Chapter 19).

- Excessive noise (Table 6-1).

- Loss of rural land, fertile soil, and wildlife habitats as cities expand.

Making Urban Areas More Livable and Sustainable We must maintain and improve life throughout cities where half of the world's people live and where two out of three people are expected to live by 2020. The world's population is already so large that dispersing much of its urban population would destroy the countryside and remaining wilderness areas and accelerate the already serious wildlife extinction epidemic. City dwellers generally consume more re-

---

ficult for subways, trolleys, and buses to be economically feasible alternatives to the private car.

Worldwide, at least a third of the average city's land is devoted to roads and parking. Roads and parking space take up two-thirds of the total land area of Los Angeles, more than half of Dallas, and more than one-third of New York City and Washington, D.C. This prompted urban expert Lewis Mumford to suggest that the U.S. national flower should be the concrete cloverleaf.

Instead of reducing automobile congestion, the construction of roads encourages more automobiles and travel, causing even more congestion. As economist Robert Samuelson put it, "Cars expand to fill available concrete." In 1975 some 40% of U.S. rush hour traffic was rated as congested. By 1990 this figure had risen to 77%.

If present trends continue, U.S. motorists will spend an average of two years of their lifetimes in traffic jams. Companies are losing billions of dollars because many of their employees can't get to work

on time or arrive at work tired and irritated.

In 1907 the average speed of horse-drawn vehicles through the borough of Manhattan was 18.5 kilometers (11.5 miles) per hour. Today, cars and trucks with the potential power of 100 to 300 horses creep along Manhattan streets at an average speed of 8.4 kilometers (5.2 miles) per hour. In London average auto speeds are about 13 kilometers (8 miles) per hour, and even lower in Tokyo, where everyday traffic is

called *tsukin jigoku*, or commuting hell.

In the United States motor vehicles account for 63% of the country's oil consumption (up from 50% in 1973) and produce at least 50% of the country's air pollution, even though U.S. emission standards are as strict as any in the world. Motor vehicle use is also responsible for water pollution from oil spills, gasoline spills, and the dumping of used engine oil. What do you think should be done?

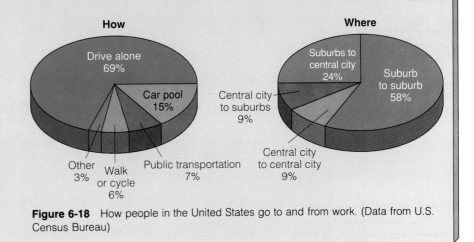

Figure 6-18 How people in the United States go to and from work. (Data from U.S. Census Bureau)

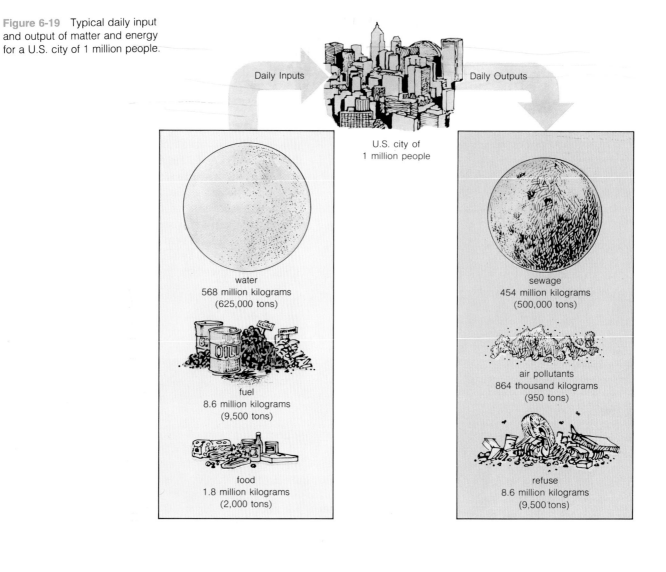

Figure 6-19 Typical daily input and output of matter and energy for a U.S. city of 1 million people.

Daily Inputs

Daily Outputs

U.S. city of 1 million people

water
568 million kilograms
(625,000 tons)

fuel
8.6 million kilograms
(9,500 tons)

food
1.8 million kilograms
(2,000 tons)

sewage
454 million kilograms
(500,000 tons)

air pollutants
864 thousand kilograms
(950 tons)

refuse
8.6 million kilograms
(9,500 tons)

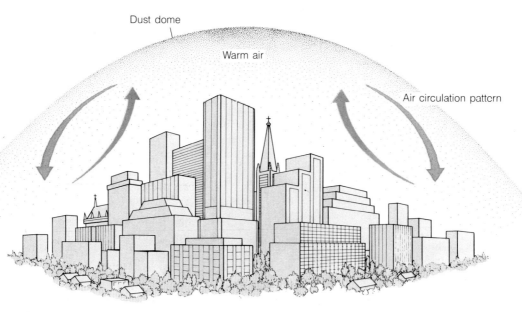

Dust dome

Warm air

Air circulation pattern

Figure 6-20 An urban heat island causes air circulation patterns that create a dust dome over the city. Winds elongate the dome toward downwind areas. A strong cold front can blow the dome away and lower urban pollution levels.

| Table 6-1 | Effects of Common Sounds |

Example	Sound Pressure (dbA)	Effect from Prolonged Exposure
Jet takeoff (25 meters away*)	150	Eardrum rupture
Aircraft carrier deck	140	
Armored personnel carrier, jet takeoff (100 meters away), earphones at loud level	130	
Thunderclap, textile loom, live rock music, jet take-off (161 meters away), siren (close range), chain saw	120	Human pain threshold
Steel mill, riveting, automobile horn at 1 meter, "jam box" stereo held close to ear	110	
Jet takeoff (305 meters away), subway, outboard motor, power lawn mower, motorcycle at 8 meters, farm tractor, printing plant, jackhammer, garbage truck	100	
Busy urban street, diesel truck, food blender, cotton spinning machine	90	Hearing damage (8 hours), speech interference
Garbage disposal, clothes washer, average factory, freight train at 15 meters, dishwasher, blender	80	Possible hearing damage
Freeway traffic at 15 meters, vacuum cleaner, noisy office or party, TV audio	70	Annoying
Conversation in restaurant, average office, background music, chirping bird	60	Intrusive
Quiet suburb (daytime), conversation in living room	50	Quiet
Library, soft background music	40	
Quiet rural area (nighttime)	30	
Whisper, rustling leaves	20	Very quiet
Breathing	10	
	0	Threshold of hearing

* To convert meters to feet, multiply by 3.3.

[handwritten: 4) Flowers, wildflows vegetation]

sources per person than people living in rural areas. But urban residents take up less space, tend to be better educated, and have lower birth rates than rural dwellers.

An important goal in coming decades should be to make existing and new urban areas more self-sufficient, sustainable, and enjoyable places to live (see Case Study on p. 134). Ways to do this include:

Land Use and Maintenance

- Establishing ecological land-use planning and control to reduce or eliminate harmful development on wildlands, agricultural lands, and sites of cultural or historic importance. *[handwritten: 1) Land-use control harmful]*

- Repairing and revitalizing existing cities. *[handwritten: 2) Repair]*

- Stimulating sustainable growth of medium-size cities and new towns to take pressure off over-populated and overstressed large urban areas. *[handwritten: 3) Growth of medium sized cities]*

- Planting lawns and public areas with wildflowers and natural groundcover vegetation instead of lawns that must be drenched with water, fertilizer, and pesticides.

- Restoring shorefronts, marshes, springs, creeks, and rivers.

- Establishing greenbelts of undeveloped forest-land and open space within and around urban areas; preserving nearby wetlands and agricultural land.

- Planting large numbers of trees in greenbelts, on unused lots, and along streets to reduce air pollution and noise and to provide recreational areas and wildlife habitats.

- Legalizing shantytowns and providing residents with support and low-cost loans to develop housing, water, sanitation, and utility systems, and to plant community gardens and trees for fruit, shade, and fuel.

Davis, California (Figure 6-21), has ample sunshine, a flat terrain, and about 38,000 people. Its citizens and elected officials have committed themselves to making it an ecologically sustainable city.

The city's building codes encourage the use of solar energy to provide space heating and hot water and require all new homes to meet high standards of energy efficiency. When any existing home is sold, it must be inspected and the buyer must bring it up to the energy conservation standards for new homes. The community also has a master plan for planting deciduous trees, which provide shade and reduce heat gain in the summer and allow solar gain during winter.

The city has adopted several policies that discourage the use of automobiles and encourage the use of bicycles. Some streets are closed to automobiles, and people are encouraged to work at home.

A number of bicycle paths and lanes have been built and some city employees are given bikes (Figure 6-22). Any new housing tract must have a separate bicycle lane. As a result, 28,000 bikes account for 40% of all in-city transportation and much less land is needed for parking spaces. This heavy dependence on the bicycle is made possible by the city's warm climate and flat terrain.

Davis also limits the type and rate of growth and development and maintains a mix of homes for people with low, medium, and high incomes. Development of the fertile farmland surrounding the city for residential or commercial use is restricted. What things are being done to make the area where you live more sustainable?

Figure 6-22   Bicyclists in Davis, California.

Figure 6-21   Where is Davis, California?

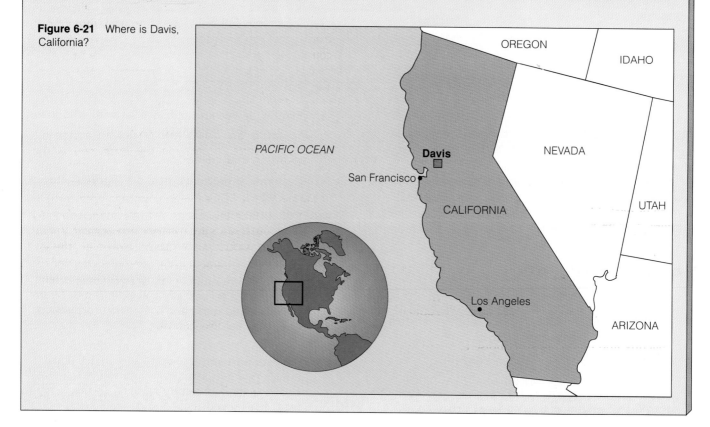

Energy Efficiency

■ Discouraging excessive dependence on motor vehicles within urban areas by providing efficient bus and trolley service and bike lanes, charging car commuters fees to enter cities and to park their vehicles, and establishing express lanes for buses and carpool vans and cars.

use bikes

■ Getting more energy from locally available resources. Many cities can increase the energy they get from perpetual and renewable energy resources by relying more on firewood (with adequate reforestation and pollution control for wood stoves), solar energy, small-scale hydroelectric plants, farms of wind turbines, geothermal deposits, methane gas from landfills, and cogeneration of electricity and heat at industrial plants (Chapter 17).

firewood
solar
geothermal

■ Bringing home, workplace, and services closer together to cut down on energy use, transportation, and the land devoted to accommodating cars.

bring home + work closer

■ Enacting building codes that require new and existing buildings to be energy efficient and responsive to climate (Section 17-1).

Water

■ Encouraging water conservation by installing water meters in all buildings and raising the price of water to reflect its true cost.

price H2O ↑

■ Establishing small neighborhood water-recycling plants.

Water recycle

■ Enacting building codes that require water conservation in new and existing buildings and businesses.

Conservation

Food

■ Growing food in abandoned lots, community garden plots, small fruit tree orchards, rooftop gardens and greenhouses, apartment window boxes, and solar-heated fish ponds and tanks.

■ Giving tax breaks to grocery stores that carry a full line of food tested and certified to have been produced by organic methods without the use of pesticides, commercial inorganic fertilizers, growth hormones, antibiotics, or other synthetic chemicals.

discourage inorganically grown food

Pollution and Wastes

■ Discouraging industries that produce large quantities of pollution and use large amounts of water or energy.

discourage Pollution

■ Enacting and enforcing strict noise control laws to reduce stress from rising levels of urban noise.

Noise Control, stress

■ Giving economic incentives to businesses that recycle and reuse resources and that emphasize

incentives for those that save

pollution prevention (input approaches) instead of relying mostly on output methods for controlling pollution.

■ Establishing urban composting centers to convert yard and food wastes to soil conditioner for use in parks and other public lands.

■ Recycling food wastes, effluents, and sludge from sewage treatment plants as fertilizer for parks, roadsides, flower gardens, and recreation areas.

■ Recycling or reusing solid waste and some types of hazardous waste.

Economic Development and Population Regulation

■ Reducing the flow of people from rural to urban areas by increasing investments and social services in rural areas and by giving equal food, energy, and other subsidies to rural and urban dwellers.

■ Reducing national population growth rates.

6-4 METHODS FOR REGULATING POPULATION CHANGE

Controlling Births, Deaths, and Migration A government can influence the size and rate of growth or decline of its population by encouraging a change in any of the three basic demographic variables: births, deaths, and migration. The governments of most countries achieve some degree of population regulation by allowing little immigration from other countries. Some governments also encourage emigration to other countries to reduce population pressures. Only a few countries, chiefly Canada, Australia, and the United States (see Case Study on p. 136), allow large annual increases in their population from immigration.

Increasing the death rate is not an acceptable alternative. Thus, decreasing the birth rate is the focus of most efforts to slow population growth. Today about 93% of the world's population and 91% of the people in LDCs live in countries with fertility reduction programs. Three general approaches to decreasing birth rates are *economic development, family planning,* and *socioeconomic change.* However, the effectiveness and funding of these programs vary widely from country to country. Few governments spend more than 1% of the national budget on them. There is also controversy over whether population growth is good or bad (see Pro/Con on p. 138).

Economic Development and the Demographic Transition Demographers examined the birth and death rates of western European countries that industrialized during the nineteenth century. On the

The United States, founded by immigrants and their children, has admitted more immigrants and refugees than any other country in the world. Between 1820 and 1989 the United States admitted almost twice as many immigrants as all other countries combined.

The number of legal immigrants entering the United States since 1820 has varied during different periods as a result of changes in immigration laws and economic growth (Figure 6-23). Between 1820 and 1960 most legal immigrants came from Europe. Since then most have come from Asia and Latin America.

Between 1960 and 1990 the number of legal immigrants admitted per year more than doubled from 250,000 to 697,000. Each year another 200,000 to 500,000 enter the country illegally, most from Mexico and other Latin American countries. This means that in 1990 legal and illegal immigrants increased the U.S. population by 800,000 to 1.1 million people, accounting for almost 40% of the country's population growth. If rates of birth, death, and immigration continue at present levels, soon immigration will be the major factor increasing the population of the United States.

Some analysts have called for an annual ceiling of no more than 450,000 for all categories of legal immigration, including refugees, to reduce the intensity of some of the country's social, economic, and environmental problems and reach zero population growth sooner. Other analysts favor increasing the number of immigrants and changing immigration law to give preference to those with valuable professional skills, high levels of education, and a knowledge of English. This could help offset the projected decrease in skilled younger workers as U.S. population growth slows (Figure 6-7) and the U.S. median age increases.

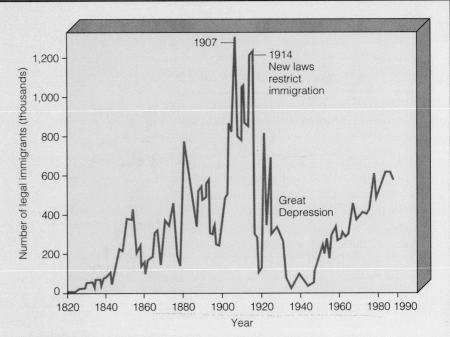

Figure 6-23 Legal immigration to the United States: 1820–1990. (Data from U.S. Immigration and Naturalization Service)

But some people oppose a policy of immigration based primarily on skills. They argue that it amounts to a brain drain of educated and talented people from LDCs that need these important human resources. It would also diminish the historic role of the United States in serving as a place of opportunity for the world's poor and oppressed.

In 1986 Congress passed a new immigration law designed to help control illegal immigration. This law included an amnesty program for some illegal immigrants and allowed them to become temporary residents and apply for citizenship after 6.5 years. By the May 4, 1988, deadline about 2.1 million of the estimated 5 million illegal aliens had signed up for the amnesty program.

The 1986 law also prohibits the hiring of illegal immigrants. Employers must examine the identity documents of all new employees. Employers who knowingly hire illegal aliens are subject to fines of $250 to $10,000 per violation, and repeat offenders can be sentenced to prison for up to six months. The bill also authorized funds to beef up the border patrol staff by 50%, to increase efforts to detect employers violating the new law, and to deport illegal aliens.

Critics charge that illegal immigrants can get around the law with readily available fake documents. Employers are not responsible for verifying the authenticity of documents or for keeping copies. Also, the Immigration and Naturalization Service does not have enough money or staff to check most employers, prosecute repeat violators, or effectively patrol more than a fraction of the 3,140-kilometer (1,950-mile) U.S.–Mexico border.

With nearly 60% of Mexico's labor force unemployed or underemployed, many Mexicans and immigrants from other Latin American countries think being caught and sent back is a minor risk compared to remaining in poverty. What, if anything, do you think should be done about legal and illegal immigration into the United States?

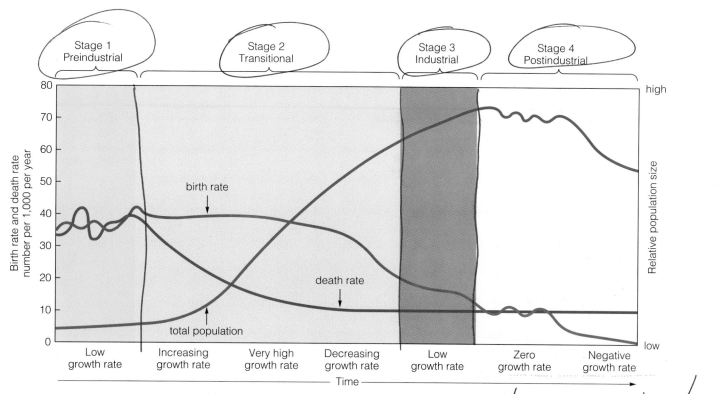

Stage 1
Preindustrial

Stage 2
Transitional

Stage 3
Industrial

Stage 4
Postindustrial

Figure 6-24   Generalized model of the demographic transition.

(handwritten) HUNGARY + W. Ga~

basis of these data they developed a hypothesis of population change known as the **demographic transition**. Its basic idea is that as countries become industrialized, they have declines in death rates followed by declines in birth rates.

This transition takes place in four distinct phases (Figure 6-24). In the *preindustrial stage* harsh living conditions lead to a high birth rate (to compensate for high infant mortality) and a high death rate, and the population grows slowly, if at all. The *transitional stage* begins shortly after industrialization begins. In this phase the death rate drops, mostly because of increased food production and improved sanitation and health. But the birth rate remains high and the population grows rapidly (typically 2.5% to 3% a year).

(handwritten: 2.5% 3%)

In the *industrial stage* industrialization is widespread. The birth rate drops and eventually approaches the death rate. The major reason for this is that couples in cities realize that children are expensive to raise and that having too many children hinders them from taking advantage of job opportunities in an expanding economy. Population growth continues but at a slower and perhaps fluctuating rate, depending on economic conditions. Most MDCs are now in this third phase.

A fourth phase, the *postindustrial stage*, takes place when the birth rate declines even further to equal the death rate, thus reaching zero population growth. Then the birth rate falls below the death rate and total population size slowly decreases. By 1990, eighteen European countries had reached or were close

to ZPG. Two of these countries—Hungary and West Germany—were experiencing population declines.

Can Most of Today's LDCs Make the Demographic Transition? In most LDCs today, death rates have fallen much more than birth rates (Figure 6-8). In other words, these LDCs are still in the transitional phase, halfway up the economic ladder, with high population growth rates. Some economists believe that LDCs will make the demographic transition over the next few decades without increased family planning efforts. But many population analysts fear that the rate of economic growth in many LDCs will never exceed their high rates of population growth. Without rapid and sustained economic growth, LDCs could become stuck in the transitional stage of the demographic transition.

Furthermore, some of the conditions that allowed today's MDCs to develop are not available to today's LDCs. Even with large and growing populations, many LDCs do not have enough skilled workers to produce the high-technology products needed to compete in today's economic environment. Most low- and middle-income LDCs also lack the capital and resources needed for rapid economic development. And the amount of money being given or lent to LDCs—struggling under tremendous debt burdens—has been decreasing since 1980.

LDCs also face stiff competition from MDCs and recently industrialized LDCs in selling the products on which their economic growth depends. Also, many LDCs with the fastest rates of population growth do

Some analysts believe that economic growth and technological advances will enable us to provide food and other necessities for 10 billion people and perhaps as many as 30 billion without seriously disrupting the earth's life support systems. Mostly because of technological advances, today the world supports 5.3 billion people whose average life span is longer than ever before. Things are getting better, not worse, for many of the world's people.

Proponents of population growth argue that people are the world's most valuable resource for finding solutions to our problems. They are not drains on the wealth and resources of a country. More people leads to increased economic productivity by creating and applying new knowledge. If a person produces more than he or she consumes after society invests in his or her childhood and education, then that person is a net gain to that society. These analysts are confident that the nature of the physical world combined with human ingenuity permits continued improvement in humanity's lot in the long run, indefinitely.

Some believe that a ''birth dearth'' eventually leading to a population decline in MDCs will decrease their economic growth and power. They argue that without more babies, MDCs with declining populations will face a shortage of workers, taxpayers, scientists, engineers, consumers, and soldiers needed to maintain healthy economic growth, national security, and global power and influence. They also contend the aging societies in MDCs will be less innovative and dynamic. These analysts urge the governments of the United States and other MDCs to prevent this by giving tax breaks and other economic incentives to couples who have more than two children.

To these analysts the primary cause of poverty and despair for one out of five people on earth is not population growth. Instead it is a lack of free and productive economic systems in LDCs.

Others opposed to population regulation feel that everyone should have the freedom to have as many children as they want. To some, population regulation is a violation of their deep religious beliefs. To others, it is an intrusion into their personal privacy and freedom. To minorities, population regulation is sometimes seen as a form of genocide to keep their numbers and power from rising.

Proponents of population regulation point to the fact that we are not providing adequate basic necessities for one out of five people on earth today who don't have the opportunity to be a net economic gain for their country. They see people overpopulation in LDCs and consumption overpopulation in MDCs (Figure 1-11, p. 14) as threats to the earth's life support systems for us and other species (Table 1-1, p. 10).

These analysts recognize that population growth is not the only cause of our environmental and resource problems. But they be-

---

not have enough natural resources to support economic development similar to that in Europe and North America.

Family Planning Recent evidence suggests that improved and expanded family planning programs may bring about a more rapid decline in the birth rate and at a lower cost than economic development alone. **Family planning** programs provide educational and clinical services that help couples choose how many children to have and when to have them.

Such programs vary from culture to culture. But most provide information on birth spacing, birth control, breastfeeding, and prenatal care and distribute contraceptives. In some cases they also perform abortions and sterilizations, often without charge or at low rates. For religious and cultural reasons most of these efforts focus on married couples.

Family planning saves a government money by reducing the need for various social services. It also has health benefits. In LDCs about 1 million women die from pregnancy-related causes. Half of these deaths could be prevented by effective family planning and health care programs. Family planning programs also help control the spread of AIDS and other sexually transmitted diseases.

Family planning has been a major factor in reducing birth and fertility rates in highly populous China and Indonesia (see Case Study on p. 140), in Brazil (see Case Study on p. 126), and several other LDCs with moderate to small populations. These successful programs have been based on committed leadership, local implementation, and wide availability of contraceptive services.

Family planning has had moderate to poor results in more populous LDCs such as India, Egypt, Bangladesh, Pakistan, and Nigeria. Results have also been poor in 79 less populous LDCs—especially in Africa and Latin America—where population growth rates are usually very high.

Family planning could be provided in LDCs to all couples who want it for about $8 billion a year—less than four days of world military spending. Currently only about $3.2 billion is being spent. If MDCs provided half of the $8 billion, each person in the MDCs would spend only $3.33 a year to help reduce world population by 2.7 billion.

CLOUDS

lieve that adding several hundred million more people in MDCs and several billion more in LDCs will intensify many environmental and social problems by increasing resource use and waste, environmental degradation, and pollution. To proponents of population regulation it is unethical for us not to encourage a sharp drop in birth rates and unsustainable forms of resource use to prevent a sharp rise in death rates and human misery and a decrease in the earth's biodiversity in the future (see Guest Essay on p. 145).

Despite promises about sharing the world's wealth, since 1960 the gap between the rich and the poor has been getting larger (see Section 7-3). Proponents of population regulation believe this is caused by a combination of population growth and unwillingness of the wealthy to share the world's wealth and resources more fairly. They call for MDCs to use their economic systems to reward population regulation and sustainable forms of economic growth instead of continuing their unsustainable forms of economic growth and encouraging LDCs to follow this eventually unsustainable and disastrous path for the planet.

Those favoring population regulation point out that technological innovation—not sheer numbers of people—is the key to military and economic power in today's world. Otherwise, England, West Germany, Japan, and Taiwan, with fairly small populations, should have little global economic and military power and China and India should rule the world. Also, as world military tensions ease, people are becoming aware that environmental security is now a key to economic and national security.

History does not show that an older society is necessarily more conservative and less innovative than one dominated by younger people. A society with a higher average age tends to have a larger pool of collective wisdom based on experience. Indeed, the most conservative and least innovative societies in the world today are LDCs with a large portion of their populations under age 29.

Instead of encouraging births, these analysts believe that the United States and other MDCs should establish an official goal of stabilizing their populations by 2025. This would help reduce the severe environmental impact of these MDCs on the biosphere. It would also set a good example for LDCs to reduce their population growth more rapidly and adopt sustainable forms of economic growth.

Proponents of population regulation believe that we should have the freedom to produce as many children as we want only as long as this does not reduce the quality of other people's lives now and in the future by impairing the ability of the earth to sustain life. Limiting individual freedom to protect the freedom of other individuals is the basis of most laws in modern societies. What do you think?

But even the present inadequate level of expenditure for family planning is decreasing. The United States has sharply curtailed its funding of international family planning agencies since 1985, mostly as a result of political pressure by pro-life activists. The Reagan administration slashed aid to international family planning and the Bush administration has not restored it. In 1989 President Bush vetoed a foreign aid bill because it contained a tiny $15 million targeted for the UN Population Fund. Unless the United States reverses this policy, critics believe that it sends a message to the world that the United States considers mass starvation preferable to helping people prevent unwanted pregnancies.

Some analysts believe that family planning and effective sex education must be expanded to include teenagers who make up almost half of the world's population. In most LDCs and MDCs pregnancy rates among unmarried teenagers have been increasing and threaten to overcome family planning efforts focused mostly on married couples (see Case Study on p. 122). Increased teenage pregnancy also contributes to higher legal and illegal abortion rates and the spread of sexually transmitted diseases, including AIDS. But teaching effective sex education in schools and extending family planning to teenagers are highly controversial issues that often go against religious and cultural taboos.

Economic Rewards and Penalties Some population experts argue that family planning, even coupled with economic development, cannot lower birth and fertility rates fast enough to avoid a sharp rise in death rates in many LDCs. Why? Because most couples in LDCs want 3 or 4 children, well above the 2.1 fertility rate needed to bring about eventual population stabilization.

These experts call for increased emphasis on bringing about socioeconomic change to help regulate population size. Governments can discourage births by using economic rewards and penalties. And increased rights, education, and work opportunities for women would reduce fertility rates.

About 20 countries offer small payments to individuals who agree to use contraceptives or to be sterilized. They also pay doctors and family planning workers for each sterilization they perform and each IUD they insert. For example, in India a person

Indonesia is the world's fifth most populous country with a population of 189 million in 1990 (Figure 6-25). It also has one of the world's most successful family planning programs. Between 1970 and 1990 its birth rate dropped from 44 to 27 per 1,000 and the number of couples using contraceptives jumped from around zero to 48%.

Even with this massive effort, the country's population is growing exponentially at 1.8% a year, adding 3.4 million people in 1990. Mostly because 38% of its population is under age 15, Indonesia's population is projected to reach almost 288 million by 2020.

Indonesia, like most LDCs and MDCs, is faced with increasing pregnancy among unmarried teenagers. In 1987 the government began family planning education for out-of-school teenagers. Sometime in the 1990s government officials expect to have a required population education course in grades 4 through 12, where students will learn about the economic and environmental implications of population growth.

However, the government has held back from including sex education in the public school curriculum. In a country where 90% of the population is Muslim, the government is fearful that doing this would provoke political instability by alienating powerful Muslim religious leaders.

Figure 6-25 Where is Indonesia?

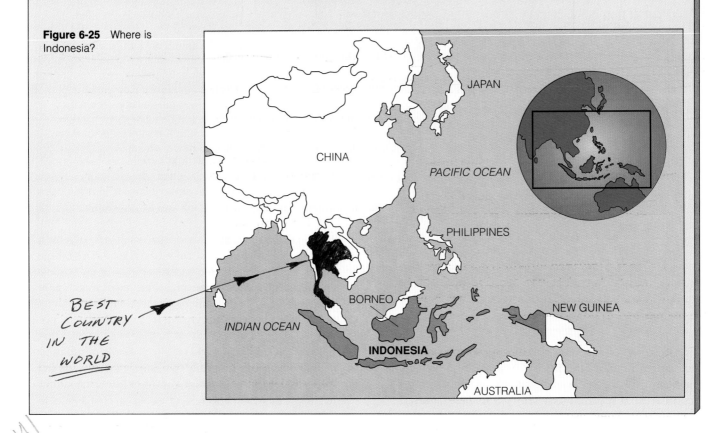

BEST COUNTRY IN THE WORLD

receives about $15 for being sterilized, the equivalent of about two weeks' pay for an agricultural worker.

Such payments, however, are most likely to attract people who already have all the children they want. In some cases the poor feel they have to accept them in order to survive.

Some countries, such as China, penalize couples who have more than a certain number of children—usually one or two. Penalties may be extra taxes and other costs or not allowing income tax deductions for a couple's third child (as in Singapore, Hong Kong, Ghana, and Malaysia). Families who have more children than the desired limit may also suffer reduced free health care, decreased food allotments, and loss of job choice.

Like economic rewards, economic penalties can be psychologically coercive for the poor. Programs that withhold food or increase the cost of raising children punish innocent children for actions by their parents.

Experience has shown that economic rewards and penalties designed to reduce fertility work best if they

- nudge rather than push people to have fewer children

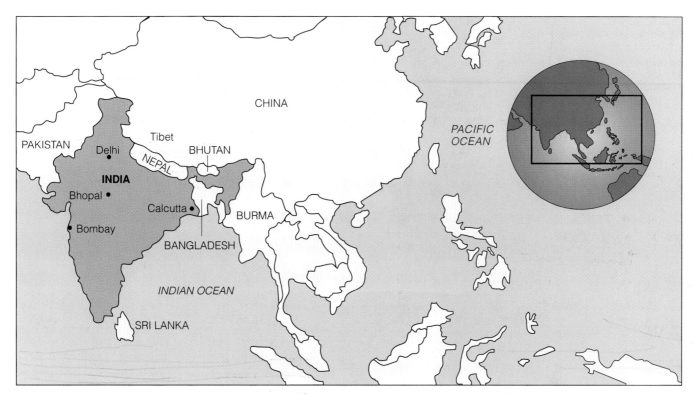

Figure 6-26   Where is India?

■ reinforce existing customs and trends toward smaller families

■ do not penalize people who produced large families before the programs were established

■ increase a poor family's income or land

However, once population growth is out of control a country may be forced to use coercive methods to prevent mass starvation and hardship. This is what China has had to do (Section 6-5).

Changes in Women's Roles   Another socioeconomic method of population regulation is to improve the condition of women. Today women do almost all of the world's domestic work and child care, mostly without pay. They also do more than half the work associated with growing food, gathering fuelwood, and hauling water. Women also provide more health care with little or no pay than all the world's organized health services put together. As one Brazilian woman put it, "For poor women the only holiday is when you are asleep."

Despite their vital economic and social contributions, most women in LDCs don't have a legal right to own land or to borrow money to increase agricultural productivity. Although women work two-thirds of all hours worked in the world, they get only one-tenth of the world's income and own a mere 1% of the world's land. Many are abused or beaten by their husbands who in effect own them as slaves.

At the same time, women make up about 60% of the world's almost 900 million adults who can neither read nor write. Women also suffer the most malnutrition, because men and children are usually fed first where food supplies are limited.

Numerous studies have shown that increased education is a strong factor leading women to have fewer children. Educated women are more likely than uneducated women to be employed outside the home rather than to stay home and raise children. They marry later, thus reducing their prime reproductive years, and lose fewer infants to death, a major factor in reducing fertility rates.

Giving more of the world's women the opportunity to become educated and to express their lives in meaningful, paid work and social roles outside the home will require some major social changes. But making these changes will be difficult because of the long-standing political and economic domination of society by men throughout the world.

6-5   CASE STUDIES: POPULATION REGULATION IN INDIA AND CHINA

India   India (Figure 6-26) started the world's first national family planning program in 1952, when its population was nearly 400 million. In 1990, after 38 years of population control effort, India was the world's second most populous country, with a population of 853 million.

Figure 6-27 A camp for starving refugees in New Delhi, India. There is fear that hunger and malnutrition will increase in India, as more of its rapidly growing, mostly poverty-stricken population will be unable to grow or buy enough food.

R. Koch/Contrasto/Picture Group

In 1952 it was adding 5 million people to its population each year. In 1990 it added 18 million. India is expected to become the world's most populous country around 2015. Its population is projected to more than double to 1.6 billion before leveling off early in the twenty-second century.

In 1990 India's average per capita income was about $330 a year. At least one-third of its population had an annual income per person of less than $100 a year and 11 out of every 100 babies born died before their first birthday. To add to the problem, nearly half of India's labor force is unemployed or can find only occasional work. India produces enough food to give its population an adequate survival diet, but widespread poverty means that many people don't have enough land to grow the food they need or enough money to buy sufficient food (Figure 6-27). As India's population continues to grow rapidly, some analysts fear that hunger and malnutrition will increase in India.

Without its long-standing family planning program, India's numbers would be growing even faster. But the results of the program have been disappointing. Factors contributing to this failure have been poor planning, bureaucratic inefficiency, the low status of women (despite constitutional guarantees of equality), extreme poverty, and too little administrative and financial support.

But the roots of the problem are deeper. About 3 out of every 4 people in India live in 560,000 rural villages, where crude birth rates are still close to 40 births per 1,000 people. The overwhelming economic and administrative task of delivering contraceptive services and education to the mostly rural population is complicated by an illiteracy rate of about 71%, with 80% to 90% of the illiterate people being rural women.

For years the government has provided information about the advantages of small families. Yet Indian women still have an average of 4.2 children because most couples believe they need many children as a source of cheap labor and old-age survival insurance. This belief is reinforced by the fact that almost one-third of all Indian children die before age five.

In 1978 the government took a new approach, raising the legal minimum age for marriage from 18 to 21 for men and from 15 to 18 for women. The 1981 census, however, showed that there was no drop in the population growth rate between 1971 and 1981. Since then the government has increased family planning efforts and funding, with the goal of achieving replacement-level fertility by 2000. Whether such efforts will succeed remains to be seen.

China Between 1958 and 1962 an estimated 30 million people died from famine in China (Figure 6-28). Since 1970, however, China has made impressive efforts to feed its people and bring its population growth under control. In 1990, however, the number of homeless and hungry people in China began rising.

Today China has enough grain both to export and to feed its population of 1.1 billion. Between 1972 and 1985, China achieved a remarkable drop in its crude birth rate, from 32 to 18 per 1,000 population, and its total fertility rate dropped from 5.7 to 2.1 children per woman.

To accomplish a sharp drop in fertility, China has established the most extensive and strictest population control program in the world. The following are its major features:

- strongly encouraging couples to postpone marriage

- expanding educational opportunities

- providing married couples with easy access to free sterilization, contraceptives, and abortion

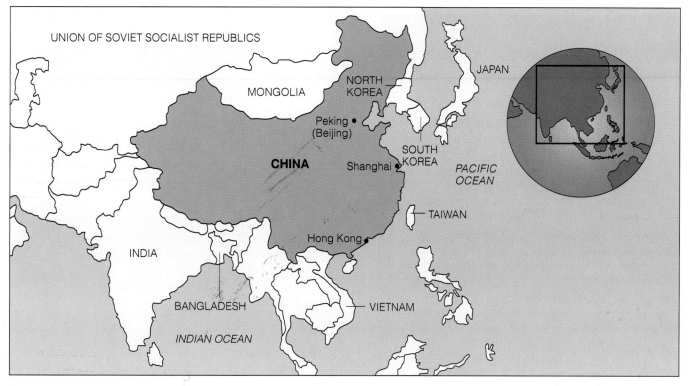

Figure 6-28 Where is China?

Figure 6-29 Poster encouraging couples in China to have no more than one child. Couples who do this are given economic rewards and those who do not suffer economic penalties.

Possible method
How feed poor poverty?
What do you do about jobs?

What is one method, U can use to feed all the people in poverty? U.S. can help.

- urging couples to have no more than one child (Figure 6-29)

- giving couples who sign pledges to have no more than one child economic rewards such as salary bonuses, extra food, larger pensions, better housing, free medical care and school tuition for their child, and preferential treatment in employment when the child grows up

- requiring those who break the pledge to return all benefits

- exerting pressure on women pregnant with a third child to have abortions

- requiring one of the parents in a two-child family to be sterilized

- using mobile units and paramedics to bring sterilization, family planning, health care, and education to rural areas

- training local people to carry on the family planning program

- expecting all leaders to set an example with their own family size

By 1987, however, China's birth rate had risen slightly, to 21, and its total fertility rate had increased

this program, however, could be used in many LDCs. Especially useful is the practice of localizing the program, rather than asking the people to go to distant centers. Perhaps the best lesson that other countries can learn from China's experience is not to wait to curb population growth until the choice is between mass starvation and coercive measures. The lesson of China shows how rapid population growth can help limit individual freedom.

Population programs aren't simply a matter of promoting smaller families. They also mean guaranteeing that our children are given the fullest opportunities to be educated, to get good health care, and to have access to the jobs and careers they eventually want. It is really a matter of increasing the value of every birth, of expanding the potential of every child to the fullest, and of improving the life of a community.

PRANAY GUPTE

DISCUSSION TOPICS

1. Explain the difference between achieving replacement-level fertility and achieving zero population growth (ZPG).

2. How many children do you plan to have? Why?

3. List the advantages and disadvantages of living in
   a. the downtown area of a large city
   b. suburbia
   c. a small town in a rural area
   d. a small town near a large city
   e. a rural area
   Which would you prefer to live in? Why? Which will you probably end up living in? Why?

4. What conditions, if any, would encourage you to rely less on the automobile? Would you regularly travel to school or work in a car pool, on a bicycle or motor scooter, on foot, or by mass transit? Explain.

5. Should world population growth be controlled? Explain.

6. Debate the following resolution: The United States has a serious consumption overpopulation problem and should adopt an official policy to stabilize its population and reduce unnecessary resource waste and consumption as rapidly as possible.

7. Do you believe that the demographic transition hypothesis applies to most of today's LDCs? Explain.

8. a. Should the number of legal immigrants and refugees allowed into the United States each year be sharply reduced? Explain.
   b. Should illegal immigration into the United States be sharply decreased? Explain. If so, how would you go about achieving this?

9. Why has China been more successful than India in reducing its rate of population growth? Do you agree with China's present population control policies? Explain. What alternatives, if any, would you suggest?

to 2.3. These rates have remained at these levels. The major reasons for the increases were the large number of women moving into their childbearing years, some relaxation of the government's stringent policies, and a strong preference for male children. Most couples who have a female child are eager to try again for a son, who by custom helps support his parents as they grow old.

China's leaders have a goal of reaching ZPG by 2000 with a population at 1.2 billion, followed by a slow decline to 0.6 to 1.0 billion by 2100. Achieving this goal will be very difficult, because 27% of the Chinese people are under age 15. As a result, the United Nations projects that the population of China may be around 1.5 billion by 2020.

Most countries cannot or do not want to use the coercive elements of China's program. Other parts of

Garrett Hardin

As professor of human ecology at the University of California at Santa Barbara for many years, Garrett Hardin made important contributions to the joining of ethics and biology. He has raised hard ethical questions, sometimes taken unpopular stands, and forced people to think deeply about environmental problems and their possible solutions. He is best known for his 1968 essay, "The Tragedy of the Commons," which has had significant impacts on economics, political science, and the management of potentially renewable resources. His many books include Promethean Ethics *and* Filters Against Folly: How to Survive Despite Economists, Ecologists, and the Merely Eloquent.

A competent physicist has placed the human carrying capacity of the globe at 50 billion—about 10 times the present world population. Before you are tempted to urge women to have more babies, consider what Robert Malthus said nearly 200 years ago: "There should be no more people in a country than could enjoy daily a glass of wine and piece of beef for dinner."

A diet of grain or bread is symbolic of minimum living standards; wine and beef are symbolic of all forms of higher living standards that make greater demands on the environment. When land used for the direct production of plants for human consumption is converted to growing crops for wine or corn for cattle, fewer calories get to the human population. Since carrying capacity is defined as the *maximum* number of animals (humans) an area can support, using part of the area to support such cultural luxuries as wine and beef reduces the carrying capacity. This reduced capacity is called the *cultural carrying capacity*. Cultural carrying capacity is always less than simple carrying capacity.

Energy is the common coin in which all competing demands on the environment can be measured. Energy saved by giving up a luxury can be used to produce more bread and support more people. We could increase the simple carrying capacity of the earth by giving up any (or all) of the following "luxuries": street lighting; vacations; most private cars; air conditioning; and artistic performances of all sorts—drama, dancing, music, and lectures. Since the heating of buildings is not as efficient as multiple layers of clothing, space heating would be forbidden.

Is that all? By no means: to come closer to home, look at this book. The production and distribution of such an expensive treatise consume a great deal of energy. In fact, the energy bill for the whole of higher education is very high (which is one reason tuition costs so much). By giving up all education beyond the eighth grade, we could free enough energy to sustain millions more human lives.

At this point a skeptic might well ask: "Does God give a prize for the maximum population?" From this brief analysis we can see that there are two choices. We can maximize the number of human beings living at the lowest possible level of comfort, or we can try to optimize the quality of life for a much smaller population.

What is the carrying capacity of the earth? is a scientific question. Scientifically, it may be possible to support 50 billion people at a "bread" level. But is this what we want? What is the cultural carrying capacity? requires that we debate questions of value, about which opinions differ.

An even greater difficulty must be faced. So far we have been treating the capacity question as a *global* question, as if there were a global sovereignty to enforce a solution on all people. But there is no global sovereignty ("one world"), nor is there any prospect of one in the foreseeable future. We must make do with nearly 200 different national sovereignties. That means, as concerns the capacity problem, we must ask how nations are to coexist in a finite global environment if different sovereignties adopt different standards of living.

Consider a redwood forest. It produces no human food. Protected in a park, the trees do not even produce lumber for houses. Since people have to travel long distances to visit it, the forest is a net loss in the national energy budget. But those who are fortunate enough to wander quietly through the cathedral-like aisles of soaring trees report that the forest does something precious for the human spirit.

Now comes an appeal from a distant land where millions are starving because their population has overshot the carrying capacity. We are asked to save lives by sending food. So long as we have surpluses we may safely indulge in the pleasures of philanthropy. But the typical population in such poor countries increases by 2.1% a year—*or more*; that is, the country's population doubles every 33 years—*or less*. After we have run out of our surpluses, then what?

A spokesperson for the needy makes a proposal: "If you would only cut down your redwood forests, you could use the lumber to build houses and then grow potatoes on the land, shipping the food to us. Since we are all passengers together on Spaceship Earth, are

(continued)

you not duty bound to do so? Which is more precious, trees or human beings?"

The last question may sound ethically compelling, but let's look at the consequences of assigning a preemptive and supreme value to human lives. There are at least 2 billion people in the world who are poorer than the 32 million legally "poor" in America, and they are increasing by about 40 million per year. Unless this increase is brought to a halt, sharing food and energy on the basis of need would require the sacrifice of one amenity after another in rich countries. The final result of sharing would be complete poverty everywhere on the face of the earth to maintain the earth's simple carrying capacity. Is that the best humanity can do?

To date, there has been overwhelmingly negative reaction to all proposals to make international philanthropy conditional upon the stopping of population growth by the poor, overpopulated recipient nations. Foreign aid is governed by two apparently inflexible assumptions:

- The right to produce children is a universal, irrevocable right of every nation, no matter how hard it presses against the carrying capacity of its territory.

- When lives are in danger, the moral obligation of rich countries to save human lives is absolute and undeniable.

Considered separately each of these two well-meaning doctrines might be defended; together they constitute a fatal recipe. If humanity gives maximum carrying capacity questions precedence over problems of cultural carrying capacity, the result will be universal poverty and environmental ruin. The moral is a simple ecological commandment: *Thou shalt not transgress the carrying capacity.*

Or do you see an escape from this harsh dilemma?

Guest Essay Discussion

1. What items would you include as essential in maintaining your own quality of life? Do you feel that everyone in the world should have or should strive for that quality of life? Explain.

2. What population size do you believe would allow the world's people to have the quality of life you described in the previous question? What do you believe is the cultural carrying capacity of the United States? Should the United States have a national policy to establish this population size as soon as possible? Explain.

3. Do you agree with the two principles the author of this essay says are the basis of foreign aid to needy countries? If not, what changes would you make in the requirements for receiving such aid?

FURTHER READINGS

Berg, Peter. 1989. *A Green City Program*. San Francisco: Planet Drum Foundation.

Bouvier, Leon F. 1984. "Planet Earth 1984–2034: A Demographic Vision." *Population Bulletin*, vol. 39, no. 1, 1–39.

Brown, Lester R., and Jodi Jacobson. 1986. *Our Demographically Divided World*. Washington, D.C.: Worldwatch Institute.

Brown, Lester R.,and Jodi Jacobson. 1987. *The Future of Urbanization: Facing the Ecological and Economic Restraints*. Washington, D.C.: Worldwatch Institute.

Croll, Elisabeth, et al. 1985. *China's One-Child Family Policy*. New York: St. Martin's Press.

Dantzig, George B., and Thomas L. Saaty. 1973. *Compact City: A Plan for a Liveable Environment*. San Francisco: W. H. Freeman.

Ehrlich, Paul R., and Anne H. Ehrlich. 1990. *The Population Explosion*. New York: Doubleday.

Formos, Werner. 1987. *Gaining People, Losing Ground: A Blueprint for Stabilizing World Population*. Washington, D.C.: Population Institute.

Grant, James P. 1989. *The State of the World's Children 1989*. New York: Oxford University Press.

Gupte, Pranay. 1984. *The Crowded Earth: People and the Politics of Population*. New York: W. W. Norton.

Haupt, Arthur, and Thomas T. Kane. 1985. *The Population Handbook: International*. 2nd ed. Washington, D.C.: Population Reference Bureau.

Jacobs, Jane. 1984. *Cities and the Wealth of Nations*. New York: Random House.

Lowe, Marcia D. 1989. *The Bicycle: Vehicle for a Small Planet*. Washington, D.C.: Worldwatch Institute.

McHarg, Ian L. 1969. *Design with Nature*. Garden City, N.Y.: Natural History Press.

Population Reference Bureau. Annual. *World Population Data Sheet*. Washington, D.C.: Population Reference Bureau.

Renner, Michael. 1988. *Rethinking the Role of the Automobile*. Washington, D.C.: Worldwatch Institute.

Ryn, Sin van der, and Peter Calthorpe. 1986. *Sustainable Communities*. San Francisco: Sierra Club.

Simon, Julian L. 1981. *The Ultimate Resource*. Princeton, N.J.: Princeton University Press.

Todd, Nancy Jack, and John Todd. 1984. *Bioshelters, Ocean Arks, City Farming: Ecology As the Basis of Design*. San Francisco: Sierra Club Books.

Wattenberg, Ben J. 1987. *The Birth Dearth*. New York: Pharos Books.

Weber, Susan, ed. 1988. *USA by Numbers: A Statistical Portrait of the United States*. Washington, D.C.: Zero Population Growth.

CHAPTER SEVEN

ENVIRONMENTAL ECONOMICS AND POLITICS

General Questions and Issues

1. What is economic growth? How can it and economic systems be redirected and managed to sustain the earth's life-support systems?

2. How can economic systems be used to regulate resource use and reduce environmental degradation and pollution?

3. What are the main causes of poverty and the human and environmental degradation it leads to? What can be done to help people escape from the global poverty trap?

4. How do political decisions affect resource use and environmental quality?

5. How can we bring about change?

As important as technology, politics, law, and ethics are to the pollution question, all such approaches are bound to have disappointing results, for they ignore the primary fact that pollution is primarily an economic problem, which must be understood in economic terms.

LARRY E. RUFF

 IFE FORCES US TO MAKE trade-offs or choices to get as many of our needs and wants as possible and to have as much control over our lives as possible. We do this individually and in groups by making *economic and political decisions.* But often the poor, who don't own enough land or have enough work income to meet their basic needs, are pushed into a downward spiral of poverty by economic and political decisions beyond their control (Figure 7-1).

The basic economic problem is that we cannot use the world's limited resources to produce enough material goods and services to satisfy everyone's wants. Therefore, individuals, businesses, and societies must make **economic decisions** about what goods and services to produce, how to produce them, how much to produce, how to distribute them, and what to buy and sell. Because producing and using anything require resources and have some harmful impact on the environment, economic decisions affect resource use and the quality of the environment.

Politics is the process by which individuals and groups try to influence or control the policies and actions of governments of the local, state, national, or international community. Politics is concerned with the distribution of resources and benefits—who gets what, when, and how. Thus, it plays a major role in regulating the world's economic systems and influencing economic decisions. Political decisions can also help prevent the degradation of commonly owned or shared resources such as air, water, wildlife, and public land.

7-1 ECONOMIC GROWTH, PRODUCTIVITY, AND EXTERNAL COSTS

Economic Goods, Needs, and Wants An **economic good** is any material item or service that gives people satisfaction and whose present or ultimate supply is limited. The types and amounts of certain economic goods—food, clothing, water, oxygen, shelter, health care—that you must have to survive and to stay healthy are your **economic needs.** Anything beyond these is an **economic want.** What you believe you need and want is influenced by the customs and conventions of the society you live in and your level of affluence.

Economic Resources The things used in an economy to produce material goods and services are called **economic resources** or **factors of production.** They are usually divided into three groups:

Figure 7-1 Extreme poverty forces hundreds of millions of people to live in slums such as this one in Rio de Janeiro, Brazil, where adequate water supplies, sewage disposal, and other services don't exist. These people and the much larger number of the desperately poor trying to survive in rural areas are pushed deeper into poverty by local, national, and global economic and political forces beyond their control. Yet, these people that the world's economic systems are throwing away are the world's experts on recycling, reuse, and living sustainably on the land. Otherwise they die.

1. **Natural resources**—resources produced by the earth's natural processes. These include the actual area of the earth's solid surface, nutrients and minerals in the soil and deeper layers of the earth's crust, wild and domesticated plants and animals (biodiversity), water, air, and nature's waste disposal and recycling services.

2. **Capital or intermediate goods**—manufactured items made from natural resources and used as inputs to produce and distribute economic goods and services bought by consumers. These include tools, machinery, equipment, factory buildings, and transportation and distribution facilities.

3. **Labor**—the physical and mental efforts of workers, managers, and investors in producing and distributing economic goods and services.

Virtually everything we have or will have comes ultimately from the sun and the earth. However, our economic systems often refuse to treat the earth's natural resources as capital that supports our economic activities. Instead, we spend this natural capital for short-term economic gain and don't keep track of natural resource depletion and degradation in the GNP indicators used to measure economic growth. Then we are surprised when nature begins bouncing our checks because we have squandered our one-time inheritance instead of living sustainably on the renewable income it provides.

Gross National Product and Economic Growth The **gross national product (GNP)** is the market value in current dollars of all goods and services produced by an economy for final use during a year. To get a better idea of how much economic output is actually growing or declining economists use the **real GNP:** the gross national product adjusted for *inflation*—any increase in the average price level of final goods and services.

All market and centrally planned mixed economies in the world today seek to increase their **economic growth:** an increase in the capacity of the economy to provide goods and services for final use. It is almost always identified with increases in real GNP.

Economic growth creates a larger economic pie. But most people care little about how much bigger the pie is if they are not getting a bigger slice. To show how the average person's slice of the economic pie is changing, economists often calculate the **average per capita real GNP:** the real GNP divided by the total population. If population expands faster than economic growth, the average per capita (per person) GNP falls. The pie has grown but the average slice per person has shrunk. Average per capita GNP can be a useful measure, but average slice size may hide the fact that the wealthy few have enormous slices and the many poor have only a few crumbs.

GNP, Quality of Life, and Environmental Degradation Since 1942 most governments have used real GNP and average per capita real GNP as measures of their society's well-being. But these indicators do not and were never intended to measure social welfare or quality of life. Instead, they measure the speed at which an economy is running.

A serious problem with these economic indicators is that they include the values of both beneficial and harmful goods and services. For example, producing more cigarettes raises real GNP. But it also causes more cancer and heart disease, which also increase real GNP by increasing health and insurance costs.

GNP indicators also don't tell us how resources and income are distributed among the people in a country—how many people have large slices and how many have only a few crumbs of the economic pie. Also depletion and degradation of natural resources upon which all economies ultimately depend are not subtracted from GNP. This means that a country can exhaust its mineral resources, erode its soils, pollute its aquifers, cut down its forests, and hunt its wildlife and fisheries to extinction and none of this would show up as a loss in its GNP while this was being done.

This promotes the idea that economic growth can be sustained indefinitely by depleting the natural resource base upon which it is built. Using GNP as an indicator of progress sends a false and dangerous message to economists, politicians, and consumers. The result is a throwaway economy built upon illusory gains in income and permanent losses in wealth.

Social and Environmental Indicators Economists William Nordhaus and James Tobin have developed an indicator called **net economic welfare (NEW)** to estimate the annual change in quality of life in a country. They calculate the NEW by putting a price tag on pollution and other "negative" goods and services included in GNP—those that do not improve the quality of life. The costs of these negative factors are then subtracted from GNP to give NEW. The net economic welfare can then be divided by a country's population to estimate the **average per capita net economic welfare**. These indicators can then be adjusted for inflation. Applying this indicator to the United States shows that since 1940 average real NEW per person has risen at about half the rate of the average real GNP per person. The net economic welfare indicator was developed in 1972, but is still not widely used. One reason is that putting a price tag on the "bads" is not easy and is controversial. Another reason is that some politicians prefer using the real GNP per person because it can make people think they are better off than they are.

Economist Kenneth Boulding has suggested an indicator that would measure the value of goods and services based on sustainable use of perpetual and renewable resources and on increased recycling and reuse of nonrenewable resources. A second indicator would represent the value of goods and services based on the throwaway use of nonrenewable resources with little or no recycling or reuse. Progress toward a sustainable-earth economy would be indicated by an increase in the first indicator over the value of the second indicator. The result would be a *gross sustainable product (GSP)* indicator. Economist Robert Repetto at the World Resources Institute has also developed national income indicators that include resource depletion and environmental degradation.

These social and environmental indicators, like all indicators, are not perfect. But without such indicators we know too little about what is happening to people, the environment, and the planet's natural resource base, what needs to be done, and what types of policies work. We have blindfolded ourselves from what we are doing at a time when we have immense power to harm ourselves and other living things. We have refused to recognize that the earth's ecological systems and economic systems are interlocked. When we, elected officials, or business leaders put short-term economic gain above long-term ecological sustainability, we are living as if the earth doesn't matter.

In 1989 the U.S. Congress passed laws requiring the federal government each year to calculate a gross sustainable productivity for the United States in addition to the conventional GNP figures. This is a hopeful beginning.

Internal and External Costs The price you pay for a car reflects the costs of building and operating the factory, raw materials, labor, marketing, shipping, and company and dealer profits. After you buy the car, you also have to pay for gasoline, maintenance, and repair. All these direct costs, paid for by the seller and buyer of an economic good, are called **internal costs.**

Making, distributing, and using any economic good also involve what economists call **externalities**. These are social benefits ("goods") and social costs ("bads") outside the market process. They are not included in the market price of an economic good or service. For example, if a car dealer builds an aesthetically pleasing sales building, this is an **external benefit** to other people who enjoy the sight at no cost to them.

On the other hand, when a car factory and the cars sold emit pollutants into the environment, their harmful effects are an **external cost** passed on to society and in some cases future generations. Pollution from making cars and driving them and accidents caused by unsafe cars harm people and kill some of them unnecessarily. This means that car insurance, health insurance, and medical bills go up for everyone. Air pollution from cars also kills or weakens some types of trees, raising the price of lumber, paper, and this textbook. Taxes may also go up because the public may demand that the government spend a lot of money to regulate the land, air, and water pollution and degradation caused by producing and using cars and by mining and processing the raw materials used to make them.

Because these harmful costs are external and hence aren't included in the market price, you don't connect them with the car or type of car you are driving. But as a consumer and taxpayer, you pay these hidden costs sooner or later.

If you use a car, you can pass other external costs on to society. You increase these costs when you throw trash out of a car, drive a car that gets poor gas mileage and thus adds more air pollution per kilometer than a more efficient car, dismantle or don't maintain a car's air pollution control devices, drive with a noisy muffler or faulty brakes, and don't keep your motor tuned. You don't pay directly for these harmful activities, but you and others pay indirectly in the form of higher taxes, higher health costs, higher health insurance, and higher cleaning and maintenance bills.

Internalizing External Costs As long as people are rewarded for polluting, depleting, degrading, and wasting resources, few people are going to volunteer to change; doing so would be committing economic suicide. Suppose you own a company and believe that it is wrong to pollute the environment any more than can be handled by the earth's natural processes. If you voluntarily install expensive pollution controls and your competitors don't, your product will cost more. Your profits would decline and sooner or later you would probably go bankrupt and your employees would lose their jobs.

However, there is a way out of this trap if you and your workforce are smart and creative enough to invest time and money into input approaches rather than more expensive and temporary output approaches to pollution control. This reduces production costs, makes you more competitive with those using antiquated and more costly output pollution control, can increase your profits, and helps sustain the earth.

A general way to deal with the problem of external costs is for the government to add taxes, pass laws, or use other devices to force producers to include all or most of this expense in the market price of all economic goods. Then the market price of an economic good would be its **true cost**: its internal costs plus its short- and long-term external costs. This is what economists call *internalizing the external costs.*

What would happen if we internalized enough of the external costs of pollution and resource waste to help prevent pollution and to use resources more efficiently? Economic growth would be redirected. We would increase the beneficial parts of the GNP, decrease the harmful parts, increase production of beneficial goods, raise the net economic welfare, and help sustain the earth.

On the other hand, some things you like would not be available anymore because they would cost producers so much to make that few people could afford to buy them. You would pay more for most things because their market prices would be closer to their true costs. But everything would be "up front."

External costs would no longer be hidden. You would have the information you need to make informed economic decisions about the effects of your lifestyle on the planet's life support systems, which affect you, other people, and other living things.

Moreover, real market prices wouldn't always be higher. Some things could even get cheaper. Internalizing external costs stimulates producers to find ways to cut costs by increasing productivity and inventing new ways do produce things. Doing so helps them compete with producers in countries where external costs are not internalized.

Internalizing external costs would make pollution prevention more profitable than pollution control. Waste reduction, recycling, and reuse would be more profitable than waste management based on producing more and more throwaway products and wastes to support eventually unsustainable economic growth and then burying or burning these wasted resources.

Jobs would be lost in the waste production, pollution control, and waste management businesses, but a larger number of jobs would be created in the pollution prevention, waste reduction, recycling, and reuse businesses. This shift would also unleash the creativity of investors, business leaders, workers, scientists, and engineers for developing and improving these input approaches. Now these vital human resources are being used mostly to deplete resources at an increasing rate and then shuffle the wastes produced from one part of the environment to another by using temporary output approaches. This is built upon the illusion of infinite resources and an infinite "away" in which to burn, dump, or bury the wasted resources we produce.

Internalizing external costs makes so much sense you might be wondering why it's not more widely done. One reason is that many producers of harmful and wasteful goods fear they would have to charge so much that they couldn't stay in business or would have to give up government subsidies that have helped hide the external costs. They would have to change the things they produce and the ways they produce things. Their philosophy is: "If it isn't broken and we are making money, then why fix it?"

Environmentalists believe that our throwaway economic system is broken and we don't have much time to fix it. It and other economic systems have built-in ways to encourage and discourage various types of economic activities. The problem is that we are rewarding waste and discouraging thrift in the use of the earth's natural capital.

Another problem is that it's not easy to put a price tag on all the harmful effects of making and using an economic good. People disagree on the values they attach to various costs and benefits. But making difficult choices about resource use is what economics and politics are all about.

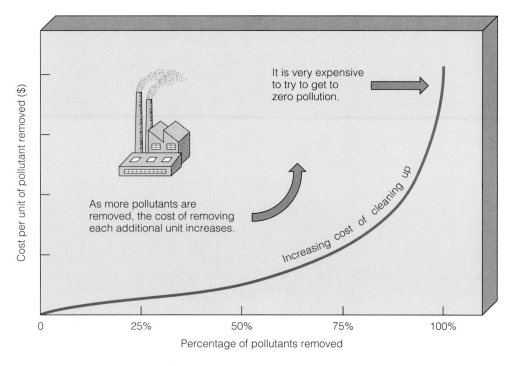

Figure 7-2 The cost of removing each additional unit of pollution rises exponentially. This explains why it is better to use an input approach that prevents a pollutant from reaching the environment or that keeps its concentration very low.

Cost per unit of pollutant removed ($)

It is very expensive to try to get to zero pollution.

As more pollutants are removed, the cost of removing each additional unit increases.

Increasing cost of cleaning up

0   25%   50%   75%   100%

Percentage of pollutants removed

7-2 ECONOMIC APPROACHES TO IMPROVING ENVIRONMENTAL QUALITY AND CONSERVING RESOURCES

How Far Should We Go? Shouldn't our goal always, be zero pollution? For most pollutants economists say the answer is no. First, because everything we do produces some potential pollutants, and nature can handle some of our wastes. The trick is not to destroy, degrade, or overload these natural processes. Exceptions are toxic products that cannot be degraded by natural processes or that break down very slowly in the environment. They should neither be produced nor used except in small amounts with special permits.

Second, we can't afford to have zero pollution for any but the most harmful substances. Removing a small percentage of the pollutants in air, water, or soil is not too costly, but when we remove more, the price per unit multiplies. The cost of removing pollutants follows a J curve of exponential growth (Figure 7-2).

If we go too far in cleaning up, the costs of pollution control will be greater than its harmful effects. This may cause some businesses to go bankrupt. You and others may lose jobs, homes, and savings. If we don't go far enough, however, the harmful external effects will cost us more than reducing the pollution to a lower level would cost. Then you and others may get sick or even die. Getting the right balance is crucial.

How do we do this? Theoretically, we begin by plotting a curve of the estimated social costs of cleaning up pollution and a curve of the estimated social costs of pollution. Adding the two curves together, we get a third curve showing the total costs. The lowest point on this third curve is the optimal level of pollution (Figure 7-3).

On a graph this looks neat and simple. But environmentalists and business leaders often disagree in their estimates of the social costs of pollution. Furthermore, the optimal level of pollution is not the same in different areas. Areas with lots of people and industry have lower optimal pollution levels. Soils and lakes in some areas are more sensitive to acids and other pollutants than those in other places. But some believe we should go much further than this (see Spotlight p. 153).

Improving Environmental Quality and Reducing Resource Waste Preventing pollution and reducing unnecessary resource use and waste require government intervention in the marketplace. There are four ways local, state, and federal governments can intervene:

- *Make harmful actions illegal.* Pass and enforce laws that set pollution standards, regulate harmful activities, ban the use of highly toxic or earth degrading chemicals, and require that certain resources be conserved.

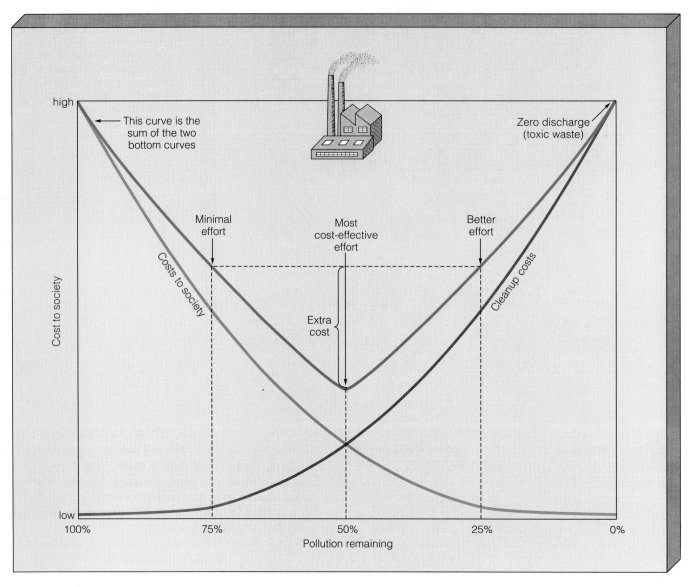

Figure 7-3 Cost-effective analysis for finding the optimal level of pollution. Note that the extra costs to society for the minimal effort (left) and the better effort (right) are the same.

- *Penalize harmful actions.* Levy taxes on each unit of pollution discharged into the air or water and each unit of unnecessary resource waste; require polluters to carry high levels of liability insurance; and have pollution control laws carry heavy automatic fines and automatic jail sentences. Taxes, fees, and fines must be high enough to force pollution prevention and waste reduction. Otherwise, some states or communities could become hooked on the tax dollars from pollution production and waste production and become major dumping grounds.

- *Market pollution rights and resource use rights.* Sell rights that allow pollution up to the estimated optimal level; sell the right to harvest or extract a sustainable amount of resources from public lands or common property resources.

- *Reward beneficial actions.* Use tax dollars to pay subsidies to businesses and individuals that install pollution control equipment and reduce unnecessary resource use and waste by recycling and reusing resources and by inventing more efficient processes and devices.

The first three are *polluter-or-resource-waster-pays* approaches that internalize some or most external costs of pollution and resource waste. Because the internalized costs would be passed on to consumers, these measures make each of us pay directly for the unnecessary pollution added to the environment and the resources wasted in the production of the economic goods we choose to buy.

Most economists prefer the second and third methods because they use the marketplace to control pollution and resource waste and do a better job of internalizing the external costs. Most environmentalists favor a combination of the first three methods.

The first three approaches share several disadvantages. Because pollution costs are internalized, the initial cost of products may be higher unless new, more cost-effective and productive technologies are developed. This can put a country's products at a disadvantage in the international marketplace. Higher initial costs also mean that the poor are penalized unless they are given tax relief or other subsidies from public funds. Also, fines and other punishments must be severe enough and enforced quickly enough to deter violations.

The fourth approach is a *taxpayer-pays* approach that does little to internalize external costs. It leads to higher-than-optimal levels of pollution and resource waste. It is not surprising that polluting industries and resource wasters usually prefer this approach which has taxpayers pay them not to pollute or waste resources.

All four approaches are limited by incomplete and disputed information about the short- and long-term effects of pollutants and require greatly increased environmental monitoring to determine how well they work. Widespread monitoring is also necessary to catch violators of antiwaste and antipollution laws. Lack of information may cause us to make mistakes in our attempts to reduce pollution and resource waste, but not dealing with these problems will be much more harmful and costly in the long run.

Cost-Benefit Analysis One method used to help make economic decisions is **cost-benefit analysis**. It involves comparing the estimated short-term and long-term costs (losses) and benefits (gains) of an economic decision. If the estimated benefits exceed the estimated costs, the decision to produce or buy an economic good or provide a public good is considered worthwhile. You intuitively make such evaluations when you decide to buy a particular economic good or service.

More formal cost-benefit analysis is often used in evaluating whether to build a large hydroelectric dam, to clean up a polluted river, or to reduce air pollution emissions to an optimal level (Figure 7-3). But use of cost-benefit analysis is controversial (see Pro/Con on p. 154).

A Sustainable-Earth Economy Conservationists and a few economists, including Herman Daly, Kenneth Boulding, and Nicholas Georgescu-Roegen, have proposed that the world's countries make a transition to a **sustainable-earth economy** (see Spotlight on p. 156). In its present form, the industrial economy is not a sustainable economy because it is based on

---

SPOTLIGHT   **The Case for Zero Discharge and for Assuming Chemicals Are Guilty Until Proven Innocent**

Some believe that we should set zero or close to zero discharge goals for all hazardous substances. They argue that setting optimal or politically acceptable levels of pollutants based on using output approaches is a legalized way of justifying the killing or harming of an "acceptable" number of people. To them this approach is ethically unacceptable.

Instead of spending a lot of money and time using output approaches to clean up chemicals we release into the environment, they say don't produce or release harmful chemicals into the environment in the first place. Use the economic system to reward those who do this and punish those who don't. Then most of the expensive time- and talent-consuming apparatus of setting standards and arguing over optimal levels would no longer be needed and human talents would be focused on sustaining rather than degrading the earth's natural capital.

These environmentalists also call for reversing the present legal principle that a chemical is assumed innocent until it is proved to have caused harm by the people it has harmed. They point out that by then it is too late and the people who have been harmed usually don't have the money and other resources needed to establish this in the courts.

Instead, these environmentalists believe that a chemical should be assumed to be guilty until proven otherwise by the people proposing to make or use it. Why should chemicals be given the same legal rights as people? they ask.

Critics of this proposal and the idea of zero discharge argue that these changes would bring the production of most things to a halt, wreck the economy, and put large numbers of people out of work. But proponents counter that not doing this will eventually wreck the environment and thus the economy, eventually put more people out of work, kill large numbers of workers and citizens, and reduce the earth's biodiversity. What do you think?

---

depleting and degrading the earth's natural capital and calling it "growth." We are running up a massive earth debt—the ultimate debt—that can't be ignored any longer. They call for us to move from an earth-plundering economy to a more sustainable ecological economy.

Cost-benefit analysis is a useful way to gather and analyze data on a proposed project or course of action. The main argument for its use is to improve economic efficiency by finding the cheapest way to do something. But unless decision makers and citizens are also aware of its severe limitations, cost-benefit analysis can be used as a device for justifying something that should not be done or that could be done in a less harmful and cheaper way.

One problem is that putting a price tag on future benefits and costs is difficult. Because the future is unknown, all we can do is make educated guesses based on various assumptions about what the future value of a resource might be.

Business people give more weight than environmentalists do to immediate profits and values over possible future profits and values. They worry that inflation will make the value of their earnings less in the future than now. They also fear that innovation or changed consumer preferences will make a product or service obsolete.

Many business leaders and economists also assume that economic growth through technological progress will automatically raise average living standards in the future. So why should the current generation pay higher prices and taxes to benefit future generations who will be better off anyway? Environmentalists believe that this is not a reasonable assumption as long as our economic systems are based upon depleting the natural capital that supports them.

Another problem is determining who gets the benefits and who is harmed by the costs. For example, suppose a cost-benefit analysis concludes that it is too expensive to meet certain safety and environmental standards in a manufacturing plant. The owners of the company benefit by not having to spend money on making the plant less hazardous. Consumers may also benefit from lower prices. But the workers are harmed by having to work under hazardous and unhealthful conditions or they may lose their jobs if the plant shuts down because its owners can't or won't spend the

money to meet stricter safety and environmental standards.

In the United States, for example, an estimated 100,000 Americans die each year from exposure to hazardous chemicals and other safety hazards at work. Another 400,000 are seriously injured from such exposure. Is this a necessary or unnecessary (and unethical) cost of doing business?

The most serious limitation of cost-benefit analysis is that many things we value cannot be reduced to dollars and cents. Some of the costs of air pollution, such as extra laundry bills, house repainting, and ruined crops, are fairly easy to estimate. But how do we put meaningful price tags on human life, clean air and water, beautiful scenery, a wilderness area, whooping cranes, and the ability of natural systems to degrade and recycle some of our wastes and replenish timber, fertile soil, and other vital potentially renewable resources?

The dollar values we assign to such items will vary widely because of different assumptions and value judgments, leading to a wide range of projected costs and

---

The natural processes that sustain the earth are the best model for any human economy in their view. The earth's processes use, conserve, and recycle resources with virtually no waste and with far greater efficiency and productivity than any economy we have invented so far. This means that our present linear economic systems (Figure 3-19, p. 54) must become circular (Figure 3-20, p. 55).

According to the Worldwatch Institute, making the transition to a sustainable-earth economy will cost about $150 billion a year—a small price to pay for economic and environmental security. This involves spending about one-sixth as much on environmental security as the world now spends each year on military security.

Making the transition to a sustainable-earth economy will require government intervention into the marketplace using policies and laws that integrate economics and ecology. Consumers and investors will also have to exercise the enormous power they

have over corporate behavior, what products are produced, and how they are produced (see Individuals Matter on p. 158).

7-3 POVERTY: A HUMAN AND ENVIRONMENTAL TRAGEDY

The Widening Gap Between the Rich and the Poor Since 1900 the annual value of goods and services produced worldwide has increased twentyfold, the products of industry fiftyfold, and the use of energy thirtyfold. Average income per person has doubled worldwide since 1950, but most of the fruits of this economic growth have gone to wealthy people, mostly in MDCs. Since 1950 the gap between the rich and the poor has grown (Figure 7-4). These curves began diverging even more in the 1980s, a period

benefits. For example, values assigned to a human life in various cost-benefit studies vary from nothing to about $7 million. If you were asked to put a price tag on your life, you might say it is priceless, or you might contend that making such an estimate would be impossible or even immoral.

Although you may not want others to place a low monetary value on your life, you do so if you choose to smoke cigarettes, not eat properly, drive without a seat belt, drive while impaired by alcohol or some other drug, or refuse to pay more for a safer car. In each case you decide that the benefits—pleasure, convenience, or a lower purchase price—outweigh the potential costs—poorer health, injury, or death. In these cases, however, you—not bureaucrats—are making these decisions.

Critics of cost-benefit analysis argue that because estimates of many costs and benefits are so uncertain, they can easily be weighted to achieve the desired outcome by proponents or opponents of a proposed project or action. The experts making or evaluating such analyses have to be

paid by somebody, so they often represent the point of view of that somebody.

The difficulty in making cost-benefit analyses does not mean that they should not be made or that they are not useful. But they can be useful as long as decision makers and the public are aware that they give only rough estimates and guidelines for resource use and management based on assumptions, and that they can easily be distorted.

And we should understand that they require assigning market price values to things that cannot be valued in this way. The industrial economy is based on reducing everything to its market price. With this approach it is not surprising that resources are destined to be wasted.

To environmentalists the way out of this dilemma is to recognize that the value of air, water, soil, minerals, and biological diversity that support us and other species is infinite and cannot be assigned a meaningful price; therefore it must be protected at all costs. Otherwise we eventually commit economic and environmental sui-

cide by destroying the earth capital that supports us.

But to growth-oriented business leaders assigning infinite values to the earth's natural resources is a form of economic suicide that will price their goods and services out of the marketplace and put many people out of work. Environmentalists agree that this will happen until business leaders and investors get out of earth-degrading businesses and create jobs and make their profits in new earth-sustaining businesses.

Money can be made by working against or working with the earth. Those with vision and creativity are busy creating the earth-sustaining businesses of the future. Those scared of change and risk taking will be left behind and remain part of the problem instead of part of the solution until they are forced to change or go out of business. What do you think?

sometimes called the *decade of greed*. By 1990 people in a majority of the LDCs had a lower average per capita income (adjusted for inflation) than they did in 1980.

For decades economists have talked of wealth produced by economic growth "trickling down" to the poor; but Figure 7-4 shows that little has trickled down. The rich have gotten much richer while the poor have stayed poor and some have gotten even poorer. This poverty epidemic has created a growing chasm of misery between the haves and have-nots.

Actual conditions for the poor are much worse than that suggested by average national income figures, especially in Latin America. Two-thirds of the people in most countries earn less than their country's average income. Brazil, with the world's eighth largest economy, has one of the widest gaps between its rich and poor found in the world (see Case Study on p. 126).

Today, perhaps one in five people on earth enjoys a high level of affluence. The next three get by, while the fifth is desperately poor and must constantly struggle to survive (see Spotlight on p. 160).

Poverty is also found in MDCs. One-fifth of Soviet citizens reportedly live below the country's official poverty line. Since 1980 there has been a widening gap between the rich and the poor in the United States. In 1988, some 32 million Americans were below the official poverty line. This burden of poverty falls most heavily on minorities, female-headed households, and the young. One out of five American children is growing up in poverty.

Most poverty in MDCs is not as severe as that for the 1.2 billion desperately poor in LDCs. But it still represents an unnecessary degradation of human life.

The Global Poverty Trap The world's poor are caught in a poverty trap by local, national, and global

A sustainable-earth economy discourages certain types of economic growth and encourages other types to prevent overloading and degradation of the earth's life-support systems now and in the future.

Discourages:

- Throwaway and nondegradable products, oil and coal use, nuclear energy, deforestation, overgrazing, groundwater depletion, soil erosion, resource waste, and output pollution control.
- Creation and satisfaction of wants that cause high levels of pollution, environmental degradation, and resource waste.

Does this by:

- Using fees and marketable permits to internalize the external costs of goods and services.
- Removing government subsidies from highly pollution-producing, resource-depleting, and resource-wasting economic activities.
- Discouraging policies and practices that support current living standards by depleting the earth's natural resource capital for us, future generations, and other species.
- Requiring an environmental audit for all products from "cradle to grave."
- Giving sustainable-earth products and services a green seal of approval after a full environmental audit by a nonprofit, independent testing firm whose board of directors is controlled by environmentalists.

Encourages:

- Recycling, reuse, solar energy, energy conservation, education, prevention of health problems, ecological restoration, input pollution control (pollution prevention), appropriate technology, waste reduction, and long-lasting, reusable, easily repaired products.

- Consumption of goods and services that satisfy essential needs, not artificially created wants.
- Growth in productivity—not mere production—of beneficial goods and services (do more with less).
- Sustainable development—economic growth that relies on sustainable use of renewable resources and that does not deplete or degrade the earth's natural capital for current and future generations.
- Use of renewable resources at a sustainable rate.
- Use of locally available matter and energy resources.
- Decentralization of some production facilities to reduce transportation costs, make better use of locally available resources, enhance national security by spreading out targets, and increase employment.
- Preservation of biological diversity at local, national, and global levels by setting aside and controlling the use of forests, wetlands, grasslands, soil, wildlife, and representative aquatic ecosystems.
- Self-sufficiency of families, urban areas, rural areas, and countries.
- Regulation of human population growth.
- Global economic and political cooperation to promote peace and sustain the earth's life-support systems for everyone now and in the future.
- A fairer distribution of the world's resources and wealth with primary emphasis on meeting the basic needs of the poor and helping them sustain themselves.
- A broadened definition of national security to include resource, environmental, and economic security and to consider demographic issues. In 1990 the U.S. federal govern-

ment spent $303 billion on military security and only $14 billion on environmental security.

Does this by:

- Integrating economics and ecology in decision making (the most important condition).
- Using government subsidies to encourage pollution prevention, resource conservation, and waste reduction and selling marketable permits for resource extraction.
- Educating people to understand and value the earth's life-sustaining processes for present and future generations and for all species.
- Increasing aid from rich countries to poor countries that helps LDCs become more self-reliant rather than more dependent on MDCs. Since 1982 the traditional flow of capital from MDCs to LDCs has been reversed, with more than $50 billion annually transferred to MDCs from LDCs, mostly to pay interest on their massive debt. LDCs must deplete their natural capital just to pay the interest on their debt.
- Eliminating much of the trillion dollar debt that LDCs owe to MDCs and international lending agencies through a combination of debt forgiveness and debt-for-nature swaps in which debts are forgiven in exchange for countries agreeing to set aside and use areas as wilderness or for carefully monitored sustainable development.

Determines progress with indicators that measure:

- Changes in the quality of life.
- Sustainable use of renewable resources.
- Recycling and reuse of nonrenewable resources.
- Pollution prevention and waste reduction.
- Improvements in energy efficiency.

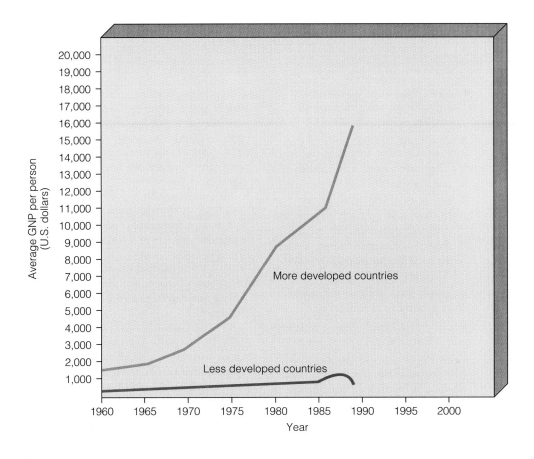

Figure 7-4 The gap in average GNP per person in MDCs and LDCs has been widening since 1960 and has accelerated in the 1980s. When adjusted for inflation, this gap is even wider than shown here. (Data from the United Nations)

forces beyond their control. Dismantling this trap is a vital step in sustaining the earth. At the local level, there are four parts of the poverty trap:

■ Lack of access to enough land and income to meet basic needs.

■ Physical weakness and poor health caused by not having enough land to grow or income to buy enough food for good health. This decreases the ability of the poor to work and plunges them deeper into poverty (Figure 7-5).

■ Rapid population growth, which produces more workers than can be employed and forces wages down as the poor compete with each other for scarce work. Some people in affluent countries consider poor people ignorant for having so many children, but poor parents need children as a form of economic security to help grow food, work, or beg in the streets. The two or three of their children who typically survive to adulthood are also a form of social security to help their parents survive in old age (typically their forties). This strength in numbers may lower the family's chances of escaping poverty but it reduces their risk of starvation.

■ Powerlessness that can subject the poor to being tricked into signing away the little land or livestock they own, paying such high interest rates on loans that they lose their land and livestock, and having to pay bribes to get work.

These local parts of the poverty trap are reinforced by government policies at the national level. They include national budgets that favor urban and industrial over rural development and military over social expenditures.

Additional layers of the global poverty trap are added at the international level. They include:

■ The $1.2 trillion debt LDCs owe to MDC banks and governments. UNICEF blames the death of 500,000 children a year on the debt burden of the LDCs.

■ Sharp declines in the income of LDCs that depend on exports of cash crops such as coffee, sugar, and cotton and raw materials such as iron ore and copper because of drops in the market prices of these commodities since 1980. This has caught most LDCs in a vise between rising debt and falling earnings and pushes them deeper into the jaws of the global poverty trap.

■ Rising trade barriers in rich countries that each year cost LDCs $50 to $100 billion in lost sales and depressed prices.

■ Decreased investment in LDCs by MDCs because of the economic turmoil and uncertainty in these poor countries.

■ Loss of investment capital because wealthy elites in LDCs have invested or deposited much

Individual consumers are the catalyst for producing green or earth-sustaining products and services. For example, U.S. consumers who collectively spend $3.5 trillion per year on goods and services have the world's greatest impact on environmental pollution and degradation. By choosing what to buy and what companies to invest in, American and other consumers can also force companies and elected leaders to become more environmentally responsible.

A 1989 nationwide poll found that 89% of Americans are concerned about the impact on the environment of the products they purchase. The problem is that major stores and chains don't have green-product sections. There has been too little information evaluating what products are truly green, partially green, and not green but labeled as such with advertising buzzwords such as biodegradable, recyclable, and environmentally safe.

For example, biodegradable plastics are an expensive sham. They take many decades to biodegrade in today's oxygen-deficient, packed landfills. Their use also encourages the continuing production of throwaway plastics made from chemicals derived from oil. Instead, the emphasis should be on recycling plastics. Better yet we should not be using plastics for many products that could be produced in reusable bottles or other containers or that could have much less packaging of any type.

This lack of consumer information about green products is beginning to change. In 1990 *The Green Consumer* was published to provide such information (see Further Readings). Consumers are urged to avoid products that

- endanger the health of the consumer or others
- cause significant damage to the environment during manufacture, use, or disposal

- consume a large amount of energy during manufacture, use, or disposal
- cause unnecessary waste because of overpackaging, throwaway packaging, and unduly short useful life
- use materials derived from threatened species (Chapter 16) or from unsustainable use of threatened environments such as ancient and tropical forests (Chapter 15)
- involve the unnecessary use of, or cruelty to, animals
- adversely affect other countries, particularly LDCs, and the poor

Learning about and buying truly and partially green products is important in helping sustain the earth. Following are guidelines for green consuming:

1. Begin by asking yourself if you really need this product.

2. When possible buy products that are durable, reusable, and used rather than new.

3. When this is not possible, buy products that are made from recycled materials or renewable resources and that are recyclable. Remember that just because something is recyclable does not mean that it will be recycled unless you make sure that it gets to a recycling center and unless you buy recycled products to create a demand for such products.

4. Buy products with the least packaging.

5. Boycott harmful products.*

6. Help elect people to local, state, and national offices who make sustaining the earth their major priority.

Buying green makes us feel good, but it should not divert us from the urgent need to change political and economic structures so that they sustain rather than

degrade the earth. Green consuming is still consuming, much of it unnecessary and harmful. Buying green, like output pollution control, puts a dent in our pollution problems, but it does not prevent them.

If you invest in money market funds or in stocks, invest in green funds and companies. Also, try to work for or start green companies. Each year the Council on Economic Priorities publishes a small book rating companies on their social and environmental responsibility.†

The Coalition for Environmentally Responsible Economies has drawn up ten Valdez Principles, criteria that can be used to evaluate socially and environmentally responsible companies. These principles cover pollution release and prevention to protect the ecosphere, sustainable use of natural resources, reduction and disposal of waste, energy efficiency and conservation, risk reduction to employees and surrounding communities, marketing of safer products and services, damage compensation, disclosure of potential hazards, the inclusion of environmental representatives on corporate boards, and annual corporate environmental audits by independent environmental auditors to evaluate compliance with these principles. All environmental and conservation groups (see Appendix 1) should also agree to abide by the Valdez Principles and open their books to an annual environmental audit, with the results to be made public.

On Earth Day 1990 millions of people signed the Earth Day Pledge, promising to honor the environment when they vote, purchase, consume, and invest. Honoring this pledge is a way of exercising the most important economic and political power we have to help sustain the earth.

---

* For information on boycotted products subscribe to *National Boycott News*, 6506 28th Ave., NE, Seattle, WA 98115 ($10 a year).

† For information on green investing contact The Social Investment Forum, 711 Atlantic Ave., Boston, MA 02111.

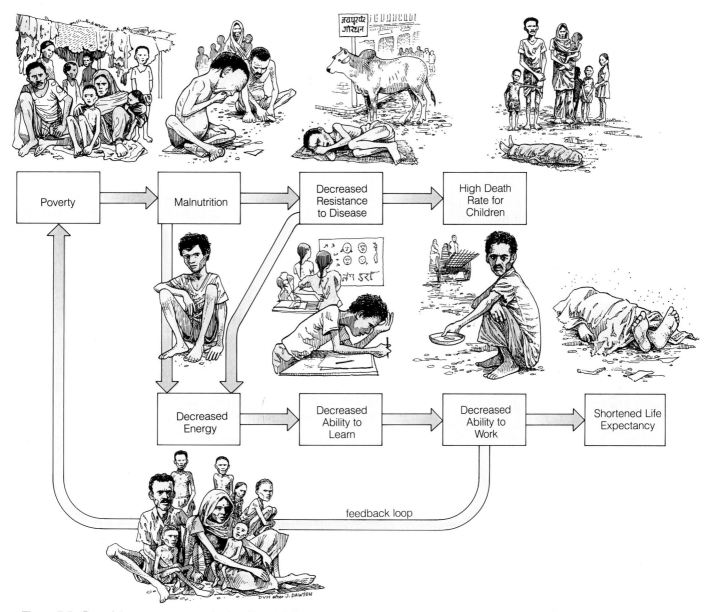

Figure 7-5 Part of the poverty trap at the local level. Interactions among poverty, malnutrition, and disease form a tragic cycle that can perpetuate such conditions in succeeding generations of families.

of their money abroad, where it is safe from taxation and political and economic disruptions.

Poverty and Environmental Degradation Most of the world's air pollution, water pollution, ozone depletion, and global warming are caused by high rates of resource use and waste production in MDCs (Figure 1-11, p. 14, and Figure 3-19, p. 54). But poverty is the major cause of environmental degradation in LDCs.

For the rural poor sustaining soil fertility, forest productivity, and wildlife populations is not just an idea, it is what keeps them alive. The poor are also the world's greatest recyclers and reusers. They can't afford to waste anything. The world's poor have more to teach us about sustaining the earth than anyone. They are the earth's frontline sustainable-earth citizens.

But when the rural poor are faced with starvation they are driven to knowingly overexploit their vital resource base. The result is increased deforestation, soil erosion and flooding, spreading deserts, and loss of biodiversity.

This environmental decline further decreases the ability of the poor to grow enough food or get enough fuel and pushes them deeper into poverty. They become locked into a downward spiral of increasing poverty, desperation and misery, and environmental degradation. This is a tragedy for the poor, the rich, and the earth.

In effect, LDCs are being coereced into depleting their resources to help support the wasteful, earth-degrading lifestyles of people in MDCs and the rich in their own countries. The natural resources of most debt-ridden poor countries are not being used for

SPOTLIGHT The World's Desperately Poor People

One of every five people on earth is desperately poor—too poor to grow or buy enough food to maintain good health or perform a job. About 80% of these people live in rural areas, except in Latin America where nearly half live in cities. Two-thirds of those living in absolute poverty are children under age 15.

Each year at least 20 million (and probably 40 million) of the world's 1.2 billion desperately poor people die unnecessarily from preventable malnutrition (lack of enough protein and other nutrients needed for good health) and diseases. Half of those who die are children under the age of five (Figure 1-3, p. 5). Most of these children die from diarrhea and measles, deadly diseases for people weakened by malnutrition.

During your lunch hour, at least 2,300 (probably 4,600) people died prematurely from starvation, malnutrition, and/or poverty-related diseases. By the time you eat lunch tomorrow, at least 55,000 (probably 110,000) more will have died. This death toll is equivalent to 137 to 275 jumbo jet planes, each carrying 400 passengers, crashing every day with no survivors.

Yet this tragic news is rarely covered by the media. Why? Because it happens every day. Because it happens most in rural areas and urban slums (Figure 7-1) in LDCs away from the glare of TV cameras and reporters.

In 1990 at least 14 million children died from simple diseases that could have been prevented at a cost of only five dollars per child. This preventable death toll of children in just one year equals about one-tenth of the number of people killed in all wars fought during the past 200 years. Such unnecessary deaths will continue until we expand the concept of national security to include economic and environmental security and greatly increase funding for these vital elements of our individual and collective security.

What Can Be Done? The root cause of the global poverty-environmental degradation trap is lack of access by one out of five people to enough land or income to meet their basic needs. Thus, the solution to this crisis must begin at the bottom.

Virtually all forms of aid from MDCs and from the governments of LDCs must be directed to the bottom fifth of humanity. This is based on Mahatma Ghandi's concept of *antyodaya*: Putting the poor and their environment first, not last.

Instead of asking experts and consultants what to do, we must ask the poor. They must be actively involved as advisers, leaders, and participants in what they need and in the design and running of programs that enable them to help themselves. They know far more poverty, survival, and environmental sustainability than any bureaucrats or experts. How many of the world's experts could survive by growing food on a steeply sloping plot or raising and keeping a family of six alive on eighty cents a day?

The role of MDCs and the governments of LDCs is to give the poor enough land and job income to meet their basic needs, putting them in charge, getting out of their way, spotlighting what works, and transferring this information to others. The layers of the poverty trap at the national level must be dismantled by drastic and difficult changes in government policies. They include:

- Shifting more of the national budget to the rural and urban poor.

- Seeing that the present trickle of aid to the poor becomes a healthy flow and that this flow is not diverted by the greedy before it reaches the needy.

- Giving villages, villagers, and the urban poor title to common lands and to crops and trees they plant on common lands.

- Redistributing some of the land owned by the wealthy to the poor, as has been done in South Korea and China. Somehow the rich must be made to see that their long-term survival and economic well-being depend on helping the poor sustain themselves so that an entire country's economic and environmental future is not sacrificed for short-term greed.

- Allocating much more money for education, health care, family planning, clean drinking water, and sanitation for the poor in rural villages and in urban slums with these programs planned and run by local residents.

- Greatly increasing the rights of poor women who grow and cook most of the food, collect most of the firewood, haul most of the water, and provide most of the health care for the poor with no pay and few human rights.

The local poor and the governments of LDCs cannot escape the widening jaws of the poverty–

their own economic development or to raise living standards, but to pay the $178 billion annual interest on their debt to industrialized country creditors. If these LDCs don't sell off their resources at bargain basement prices they face increased starvation and economic decline. But by selling off and degrading their resource base, these LDCs face an even bleaker economic and environmental future. Without major shifts in policies by both MDCs and LDCs, perhaps 3 to 5 billion people—half of humanity—could be living in absolute poverty some time between 2050 and 2075.

environmental degradation trap unless MDCs and the rich in LDCs dismantle their layers of this trap. This begins by recognizing that the fate of the rich, the poor, and other forms of life on our planetary home are intertwined. Ways to dismantle these layers include:

- Forgiving much of the present debt owed by LDCs to MDCs and recognizing that this is a vital investment in global environmental and economic security for the rich and the poor. Much of this debt can be forgiven in exchange for agreements by the governments of LDCs to increase expenditures for rural development, family planning, health care, and education and better land redistribution, protection of remaining wilderness areas, and sustainable use of other lands and renewable resources.

- Increasing the nonmilitary aid given by MDCs to LDCs to 5% of the annual GNP of the MDCs. Currently, the United States contributes less than 0.25% of its GNP as nonmilitary aid to LDCs. This aid should be given directly to the poor to help them sustain themselves. National and international lending agencies should not lend money to projects unless a favorable environmental impact assessment has been made.

- Lifting trade barriers that hinder the export of commodities from LDCs to MDCs. Businesses in MDCs now being protected from cheaper foreign imports will oppose this. But it is time for protected businesses to innovate and become more competitive instead of resisting change in the name of protecting short-term profits and keeping prices for consumers higher than they need be. They should practice the basic principle of free enterprise: If you can't compete, you shouldn't be in business.

- Having governments throughout the world cooperate in tracking the flight and concealment of capital from LDCs to MDCs and requiring the owners of this capital to pay taxes on this capital and any income it generates to their national treasuries to help finance the economic recovery of their homeland.

- Recognizing that the greatest threat to the global environment for the rich and the poor and other species are the throwaway economic systems (Figure 3-19, p. 54) in MDCs that are fueled by depleting and degrading the earth's natural capital and replacing them with sustainable-earth economic systems (see Spotlight on p. 156).

- Aiding LDCs in developing new, diversified sustainable-earth economies (see Spotlight on p. 156) instead of following the throwaway economic development model of the MDCs that must now be modified and replaced because it threatens the life-support systems for everyone.

7-4 POLITICS AND ENVIRONMENTAL AND RESOURCE POLICY

Influencing Public Policy Politics is concerned with the distribution of resources and benefits—who gets what, when, and how. Because there is always competition for scarce resources to satisfy growing wants, decision makers in democratic governments must deal with an array of conflicting groups. Each special-interest group is asking for resources or money or relief from taxes to help purchase or control more of certain resources. Interest groups that are highly organized and well funded usually have the most influence.

Decisions about environmental and resource use policies are influenced by a mixture of governmental and nongovernmental organizations operating at the global, regional, national, state, and local levels. Individuals and organized groups influence and change government policies in democracies mainly by

- voting

- contributing money and time to candidates running for office

- lobbying and writing elected representatives to pass certain laws, establish certain policies, and fund various programs

- providing financial support and free time to organizations that lobby elected officials and file lawsuits to bring about changes such as better environmental protection and resource conservation

- using the formal education system and the media to influence public opinion

- filing lawsuits asking the courts to overturn, enforce, or interpret the meaning of existing laws

- carrying out grassroots activities such as marches, mass meetings, sit-ins, hugging trees to prevent them from being cut (see Case Study on p. 32), protesting the location of waste landfills and incinerators, organizing product boycotts, and using consumer buying power (see Individuals Matter on p. 158)

The Nature of Democratic Political Systems Political systems in democracies are designed to bring about gradual or incremental change, not revolutionary change. Rapid change is difficult because of distribution of power among different branches of government, conflicts among interest groups, conflicting information from experts, and lack of money (see Case Study on p. 162).

Because tax income is limited, developing and adopting a budget is the most important thing decision makers do. This involves answering two key questions: What resource use and distribution problems

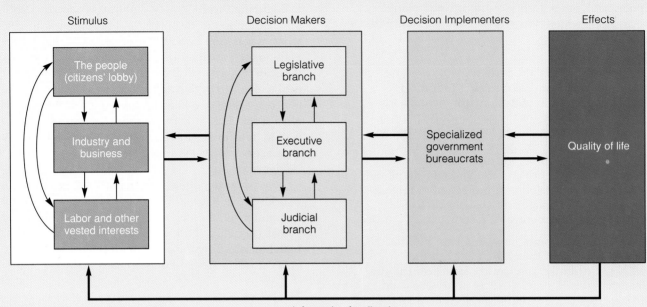

Figure 7-6 Crude model of the U.S. political system.

The writers of the U.S. Constitution wanted to develop a political system strong enough to provide security and order and to protect liberty and property but without giving too much power to the federal government. This was done by dividing political power between the federal and state governments and within the three branches of the federal government—the legislative, executive, and judicial (Figure 7-6).

These branches are connected and controlled by a series of checks and balances to prevent one branch from gaining too much power. Once federal laws are passed, they are supposed to be implemented and enforced by various bureaucratic agencies in the executive branch of the federal government and by delegation of certain responsibilities to state governments.

Actions that affect environmental quality and resource use are controlled by an elaborate network of laws and regulations at the federal, state, and local levels. Figure 7-7 summarizes the major forces involved in environmental policy making at the federal level.

The end result is usually a compromise that satisfies no one but muddles through, mostly by making short-term incremental changes. Even if a tough environmental law is passed, the next hurdle is to see that Congress appropriates enough funds to see that the law is adequately enforced. Often environmental laws contain glowing rhetoric about goals but only vague, unrealistic, or indirect guidance about how these goals are to be achieved.

The details of implementation are left up to a regulatory agency such as the EPA and the courts. Fragmentation of management responsibility among many different federal and state agencies often leads to contradictory policies, duplicated efforts, and wasted funds, while prohibiting an effective integrated approach to interrelated problems.

The government established by the U.S. Constitution was not designed for efficiency. Instead it was designed for consensus and accommodation to promote survival and adaptation through gradual change. By staying as close to the middle of the road as

possible, the government attempts to steer its way through crises. Ralph Waldo Emerson once said, "Democracy is a raft which will never sink, but then your feet are always in the water."

Despite serious shortcomings in the U.S. political system, environmental and resource conservation laws and agencies have improved the quality of the environment and reduced some forms of resource degradation. On a per capita basis, few other countries spend as much as the United States does to protect the environment.

Some analysts believe that the U.S. political-economic system is working reasonably well and no fundamental changes need to be made. "If it's not broken, don't fix it," they say.

Others think that the system must undergo changes that will improve its ability to deal with, anticipate, or prevent the growing number of regional, national, and global environmental and resource problems we face today. They say, "Fix the system better and use it to reward pollution prevention, waste reduction, and resource

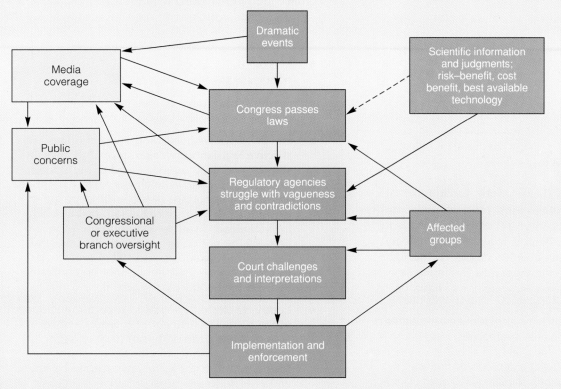

Figure 7-7 Major forces involved in making environmental policy at the federal level in the United States.

conservation instead of concentrating mostly on pollution control, waste management, and resource waste.''

We face an interlocking set of environmental, economic, and social problems that must be dealt with using integrated, comprehensive approaches. Yet most of our political institutions are compartmentalized and fragmented into specialized, narrowly focused cells that bear little relationship to the real world.

We must integrate economic and environmental policies and reorganize local, national, and global institutions to reflect the way the earth works. Otherwise many analysts believe that we will continue our earth-degrading ways until we impoverish the planet for ourselves and many other species.

This means we must recognize four facts of life:

- The earth can easily get along without our species, which has been around for only an eye blink of the planet's existence.

- We cannot get along without the earth.

- We have done a lot of damage to the earth during our brief existence, especially during the past 200 years, but nature always bats last and has a bat that makes our bat look like a toothpick.

- We must shift from economic and political policies that rely largely on increasingly expensive and ineffective after-the-fact repair of environmental damage to those that anticipate and prevent such damage.

Writer Kurt Vonnegut has suggested a prescription for choosing leaders.

I hope you have stopped choosing abysmally ignorant optimists for positions of leadership. . . . The sort of leaders we need now are not those who promise ultimate victory over Nature, but those with the courage and intelligence to present what appear to be Nature's stern, but reasonable surrender terms:

1. Reduce and stabilize your population.

2. Stop poisoning the air, the water, and the topsoil.

3. Stop preparing for war and start dealing with your real problems.

4. Teach your kids, and yourselves too, while you're at it, how to inhabit a small planet without killing it.

5. Stop thinking science can fix anything, if you give it a trillion dollars.

6. Stop thinking your grandchildren will be OK no matter how wasteful or destructive you may be, since they can go to a nice new planet on a spaceship. That is really mean and stupid.

will be addressed? How much of limited tax income will be used to address each problem? Someone once said that the way to understand human history is to study budgets.

Most political decisions are made by bargaining, accommodation, and compromise between leaders of competing elites or power groups within a society. Most politicians who remain in power become good at finding compromises and making trade-offs that give a little to each side. They play an important role in holding society together, preventing chaos and disorder, and making incremental changes, but these same processes hinder major changes and dealing with long-range problems.

U.S. Environmental Legislation Environmentalists, with backing from many other citizens and members of Congress, have pressured Congress to enact a number of important federal environmental and resource protection laws, as discussed throughout this text and listed in Appendix 2. Similar laws, and in some cases even stronger laws, have been passed by most states.

These laws attempt to provide environmental protection using five major approaches:

1. Setting pollution level standards or limiting emissions or effluents for various classes of pollutants (Federal Water Pollution Control Act and the Clean Air Act).

2. Screening new substances before they are widely used in order to determine their safety (Toxic Substances Control Act of 1976).

3. Requiring a comprehensive evaluation of the environmental impact of an activity before it is undertaken (National Environmental Policy Act).

4. Setting aside or protecting various ecosystems, resources, or species from harm (Wilderness Act and Endangered Species Act).

5. Encouraging resource conservation (Resource Conservation and Recovery Act and National Energy Act).

Most current environmental laws legalize certain levels of pollution and waste and then move these wasted resources from one part of the environment to another in a futile search for an infinite away. Instead, we should be passing and strictly enforcing a new set of laws that emphasize pollution prevention, resource reduction, and integrated pollution management that considers the effects of pollutants on all parts of the environment.

Strategies of Polluters and Resource Depleters It is natural that producers of pollution and resource degradation resist having to internalize their external costs. It costs them money, can reduce their profits, and may make it more difficult for them to compete in the international marketplace with companies in other countries that don't require such internalization.

Companies use several basic strategies to avoid external cost internalization:

- Making donations to the election campaigns of politicians favoring their positions.

- Establishing groups of well-paid lobbyists and lawyers in national and state capitals.

- Having hired lobbyists and lawyers oppose restrictive legislation, weaken proposed laws and standards, inject loopholes and opportunities for delays and legal challenges, divert attention away from important issues such as pollution prevention and waste reduction, make penalties for violations trivial compared to the profits to be made by not complying with the law, and make laws so complex and full of unnecessary technical jargon that they cannot be understood by even well-educated laypersons.

- Lobbying elected officials to reduce the budgets of the EPA, Department of Interior, and other agencies so that they do not have enough money or personnel to effectively monitor, implement, and enforce the laws passed by federal or state legislatures.

- Pressuring elected officials to appoint agency heads and middle- and upper level managers who support the position of industries threatened with environmental regulation—the "put the fox in the henhouse approach."

- Hindering implementation and enforcement of environmental and resource conservation laws by offering jobs to underpaid, unappreciated, upper- and middle-level managers in regulatory agencies. If regulators view the industries they are supposed to control as their potential employers at much higher pay, they are more likely to compromise or delay implementing or enforcing environmental regulations.

- Making donations or giving research grants to environmental and resource conservation organizations with the goal of diluting or influencing how far they go and withdrawing support if they go too far.

- Influencing media by directly or implicitly threatening to withdraw vital advertising income if these organizations probe too deeply. If necessary, they buy up media businesses.

- Mounting well-funded advertising and political campaigns encouraging people to support their positions, scaring people by saying that certain laws and environmental regulations will put them out of work, and opposing tougher environmental and resource laws.

- Setting up and highly publicizing showcase environmental and resource conservation projects while continuing to do most of their business as usual.

- Adopting the latest environmental slogans such as sustainable development, pollution prevention, recycling, reuse, resource reduction, and biodegradable products, and using them to give the appearance of change while continuing to do business as usual.

- Deciding what is most profitable to manufacture, using advertising to create a demand for mostly throwaway products, and telling people that they are the problem because businesses are merely responding to consumer demands for a throwaway society.

Not all corporations follow this model, but too many do. According to a study by Amitai Etzioni of the Harvard Business School, two-thirds of the Fortune 500 companies have been charged with serious crimes, from price-fixing to illegal dumping of hazardous wastes. Some businesses, recognizing the growing political and economic power of the national and global environmental movement, are changing their ways. They realize that producing green products that help sustain the earth is a major source of future economic growth. Corporations that help sustain the earth also help sustain themselves and in the process can still make a healthy profit.

In a rapidly changing world and economic climate, the companies that will thrive are those that encourage continuous change and innovation. Instead of digging in their heels and adopting the philosophy that "if it isn't broken don't fix it," they say "if it isn't broken, break it and fix it better." These are the companies of the future. Invest in or work for this type of earth-sustaining company or start one of your own. You'll make money and you'll feel much better about yourself.

BRINGING ABOUT CHANGE

The Role of Individuals A major theme of this book is that individuals matter. History shows that significant change comes from the bottom up, not the top down. Leaders with vision can lead only when they have the support of the people. Leaders without vision or courage must be pushed by the people. The earth is too vital and under too much stress to be left in the hands of politicians alone.

Without the grass-roots political actions of millions of individual citizens and organized groups, the air you breathe and the water you drink today would be much more polluted. The main reason the U.S. Congress passed many environmental and resource conservation laws in the 1970s (see inside back cover) was that 20 million citizens took to the streets on the first Earth Day on April 22, 1970, and demanded action. Politicians got the message and those that already understood the message were able to lead. But the 1970s were only a warmup for the real work we need to do.

There are three types of environmental leadership:

- *Leading by working within the system*—bringing about environmental improvement by using existing economic and political systems, often in new creative ways.

- *Leading by example*—using your own life and lifestyle to show others that change is possible and beneficial.

- *Leading by challenging the system*—raising public awareness and building political support for far-reaching changes by challenging existing political and economic systems.

All three types of leadership are needed to sustain the earth. Many lawyers, lobbyists, and technical experts are playing important roles in sustaining the earth by working within the system (see Spotlight on p. 166). They are supported, pushed, and challenged by grass-roots activists who are leading by example and by challenging the system. Find the type of leadership you are most comfortable with and become such a leader or work with such leaders.

The Grass-Roots Action Level of the Environmental Movement The base of the environmental movement in the United States and in other countries consists of thousands of grass-roots groups of citizens who have organized to protect themselves from pollution and environmental degradation at the local level. In the United States alone there are nearly 6,000 such groups. Their motto is *think globally and act locally.*

Practicing green global politics means working from bottom up to protect the earth. Caring for the planet means adopting a worldview and personal lifestyle that walk gently on the earth and then learning how to care for, protect, and heal each of the planet's millions of human and natural neighborhoods (see Case Study on p. 32).

Unlike environmental organizations at the national and state levels, most grass-roots organizations are unwilling to compromise or negotiate. Instead of dealing with environmental goals and abstractions, they are fighting for immediate threats to their lives and the lives of their children and grandchildren and to the value of any property they own. They are inspired by the words of ecoactivist Edward Abbey: "At some point we must draw a line across the ground of our home and our being, drive a spear

The Natural Resources Defense Council (NRDC) was founded in 1970. Since then its teams of scientists, lawyers, and resource specialists have been working on critical environmental and resource problems. These efforts are supported by membership fees and contributions from about 105,000 individuals.

The following are some of the many accomplishments of the NRDC:

- 1973—compelled the EPA to establish regulations restricting lead additives in gasoline

- 1975—forced the Nuclear Regulatory Commission to adopt tougher regulations for storage and disposal of radioactive wastes from uranium mining and processing

- 1976—led the successful fight to ban the use of chlorofluorocarbons (CFCs) in aerosol products

- 1983—filed a lawsuit that forced the National Steel Company to comply with air pollution control laws and pay $2.5 million in back penalties; spearheaded a successful campaign to protect 40 million acres of fragile coastal areas in Florida, California, and Massachusetts from an offshore oil leasing program pushed by the Reagan administration

- 1984—filed a lawsuit compelling oil refineries to tighten pollution control and reduce toxic discharges; won a Supreme

Court case giving the public the right to obtain chemical industry data on the health effects of pesticides

- 1985—led a coalition of citizen groups in successful negotiations with the chemical industry to strengthen safety provisions of the federal pesticide law; won an appeal against the U.S. Forest Service's 50-year management plans that would have increased environmentally damaging logging in four Colorado national forests

- 1986—launched a history-making agreement with the Soviet Academy of Sciences that will allow scientists to monitor nuclear test sites in both countries; led negotiations with the oil industry to protect environmentally sensitive areas in Alaska's Bering Sea

- 1987—won an environmental penalty of $1.5 million against the Bethlehem Steel Company for polluting the Chesapeake Bay; played a key role in promoting the International Ozone Treaty designed to cut worldwide use of CFCs at least 35% by the end of this century; years of lobbying led to new appliance energy-efficiency standards that will save energy equal to that of 40 large coal-fired or nuclear plants and 1.5 billion barrels of oil and natural gas; won a lawsuit on the disposal of radioactive waste from government and commercial nuclear facilities

- 1988 and 1989—published a research study projecting increased cancer rates among preschool children from exposure to pesticides in their food; pressured Congress to pass tough legislation to protect the ozone layer; filed a lawsuit to force the EPA to phase out all production and use of CFCs in the United States; mounted a consumer education and action program against CFC products; petitioned the President to declare 1989 the "International Year of the Climate" and convene an international summit meeting of world leaders on global warming; lobbied Congress for legislation that would cut U.S. fossil fuel use by 50% by 2015, increase the use of renewable energy resources (Chapter 17), sharply reduce sulfur dioxide emissions, and set new deadlines for cities to control smog (Chapter 9); worked with the Soviet Academy of Sciences on demonstration projects to combat global warming; filed suit against President Bush's Department of Interior for failure to comply with the nation's environmental laws in recommending that oil and natural gas be developed in Alaska's Arctic National Wildlife Refuge (see Pro/Con on p. 367); filed suit against the EPA to enforce existing water pollution laws

into the land, and say to the bulldozers, earthmovers, and corporations, 'this far and no further.' " Most of these groups use legal, nonviolent tactics but some go further and practice what they call aggressive nonviolence (see Pro/Con on p. 167).

Environmental democracy in action is a major reason the nuclear power industry is virtually dead. Public opposition to waste incinerators and landfills is a key contributor to the upsurge in recycling.

The necessary political action and intervention into the marketplace that determines what we produce and how it is produced won't happen unless enough people adopt a sustainable-earth worldview (Section 2-3), live their lives and base their consumption patterns on this worldview, and carry out political actions to bring about changes in the ways we think and act. This begins with the realization that the most important things that sustain us and other forms of

The world's largest environmental group is Greenpeace. Between 1980 and 1989 membership in this activist organization increased from 240,000 to 1.5 million in the United States and worldwide the group has 2.5 million members. Greenpeace members have risked their lives by placing themselves in small boats between whales and the harpoon guns of Icelandic and Soviet whaling ships. Its members have dangled from a New York bridge to stop traffic and protest a garbage barge heading to sea; protested the dumping of toxic wastes into rivers by industries and sewage treatment plants (Figure 7-8); skydived from the smokestacks of coal-burning power plants to protest acid rain; dumped toxic sludge on the steps of East Berlin's courthouse to protest water pollution; sneaked into plants to document illegal pollution and dumping; led countless demonstrations; and helped organize local activist organizations.

Two more radical environmental groups are Earth First!, led by Dave Foreman, and the Sea Shepherd Conservation Society, headed by Paul Watson. They use bold and aggressive tactics because they believe that the earth can't wait for the beneficial, but much too slow, pace of change accomplished by working only within the system and by using only legal means.

Each of these groups expanded from a handful in 1980 to about 15,000 members by 1990. They practice civil disobedience and aggressive nonviolence. This means absolute nonviolence against humans and other living things and *strategic* violence against nonliving objects such as bulldozers, power lines, and whaling ships. Their goals are to prevent environmental destruction, increase citizen awareness, and raise the costs of business for loggers, whalers, and others practicing planet wrecking.

Figure 7-8   Greenpeace activists protesting discharge of toxic wastes from the "Pig's Eye" sewage treatment plant in St. Paul, Minnesota. This plant is the largest discharger of toxic chemicals into the Mississippi River north of St. Louis.

Figure 7-9   Earth First! activists blocking a logging truck in the Siskiyou National Forest in Oregon.

These environmental shock troops use tactics known as monkeywrenching or environmental sabotage or "ecotage." Examples include chaining themselves to the tops of trees to keep loggers from cutting them down, driving spikes in trees to prevent them from being cut into lumber and labelling these trees (the spikes don't hurt the trees but shatter saw-blades that could hurt loggers or mill workers if the trees are not labelled to prevent them from being cut down), blocking bulldozers with their bodies (Figure 7-9), blocking or sinking illegal whaling ships, taking photos and videos of illegal or brutal commercial fishing and hunting activities, pulling up survey stakes, felling high-voltage towers, dying the fur of harp seals to prevent them from being killed for their furs, and sabotaging bulldozers, road

(continued)

graders, power shovels, and back-hoes.

Environmentalists disagree over the use of such tactics. A few applaud and join or financially support such groups, contending that it is important to protect forests, wetlands, wildlife, and people from being ravaged before they can be protected by the slow place of political change. Some also fear that many of the mainstream environmental organizations are being co-opted by the system into too much compromise of the ideals they once held. Leaders of mainstream environmental groups need to have more militant groups nipping at their heels to make them take stronger positions.

In May 1989 Dave Foreman was arrested by FBI agents and charged with conspiracy for allegedly helping finance the destruction of an electric power tower near Phoenix, Arizona. An FBI informant had apparently infiltrated the group. Foreman charged that the government was attempting to intimidate and destroy the group just as it did in the 1960s when civil rights and anti-war groups opposed what they considered unjust government policies.

Members of more militant environmental groups point to the long history of civil disobedience against laws believed to be unjust used in the American revolution, the fight to allow women to vote, the civil rights movement, the antiwar movement, and now the environmental movement. Benjamin White, Jr., Atlantic Director of the Sea Shepherd Society, summarizes why we must all become environmental activists:

We must begin by declaring a state of planetary emergency. . . . We must stop compromising our basic right to clean air, water, soil, and bloodstreams, and a future with wild animals and wilderness. . . . We must also be willing to take risks. If your family were threatened, would you put your life on the line? Would you go to jail if necessary? Your family is threatened. It's time to take direct action.

An increasing number of ordinary citizens are directly or indirectly supporting more militant grass-roots environmental groups because they fear for their children's environmental future. They are fed up with presidents and other politicians who make nice speeches about protecting the environment, support a few symbolic projects, and behind the scenes allow the continuing rape of the earth in the name of short-term economic growth.

Other individuals are taking direct action. In 1988 a Kansas woman who lived near Wichita's Vulcan Chemical plant and whose family had been beset by health problems handcuffed herself to a chair outside the governor's office until he saw her. In 1989 protestors of Conoco, Inc., a refinery in Ponca City, Oklahoma, set up a tent city on the grounds of the state capitol. In 1990 Conoco offered the families who lived near the refinery up to $27 million to relocate. In 1989 a social studies class at a New Jersey high school persuaded the school board to switch from Styrofoam lunch trays to old-fashioned washable dishes. Since then these students have protested McDonald's recycling practices and are raising money to buy and protect 121 hectares (300 acres) of rain forest in Belize.

Other environmentalists don't support radical efforts and fear that any illegal or violent actions could cause a public backlash against other environmental efforts and groups. What do you think?

life cannot be assigned a dollar value and that economic and military security are impossible without environmental security.

There is something fundamentally wrong in treating the earth as if it were a business in liquidation.

HERMAN E. DALY

DISCUSSION TOPICS

1. Some economists argue that only through unlimited economic growth will we have enough money to eliminate poverty and protect the environment. Explain why you agree or disagree with this view. If you disagree, how should we deal with these problems? For example, do you agree or disagree with the proposals for dismantling the global poverty trap listed on pp. 160–161? Explain.

2. Do you believe that cost-benefit analysis should be used to make all decisions about how limited federal, state, and local government funds are to be used? Explain. If not, what decisions should not be made in this way?

3. Do you favor internalizing the external costs of pollution and unnecessary resource waste? Explain. How might it affect your lifestyle? The lifestyle of the poor? Wildlife?

4. Do you favor making a shift to a sustainable-earth economy? Explain. How might this affect your lifestyle? The lifestyle of the poor? Wildlife?

5. a. Do you believe that we should establish optimal levels or zero discharge levels for most of the

chemicals we release into the environment? Explain. What effects would adopting zero discharge levels have on your life and lifestyle?

b. Do you believe that all chemicals that we release or propose to release into the environment should be assumed to be guilty of causing harm until proven otherwise? Explain. What effects would adopting this legal principle have on your life and lifestyle?

6. What do you believe are the major strengths and weaknesses of the form of government in the United States (or in the country where you live) for protecting the environment and sustaining the earth? What major changes, if any, would you make in this system?

7. To what degree do you honor the Earth Day 1990 Pledge to honor the environment when you vote, purchase, consume, and invest?

FURTHER READINGS

Berry, Wendell. 1987. *Home Economics*. San Francisco: North Point Press.

Branch, Melville C. 1990. *Planning: Universal Process*. New York: Praeger.

CEIP Fund. 1989. *The Complete Guide to Environmental Careers*. Washington, D.C.: CEIP Fund.

Collard, David, et al., eds. 1988. *Economics, Growth, and Sustainable Environments*. New York: St. Martin's Press.

Costner, Pat, and Dave Rapaport. 1990. "What Works: An Oral History of Five Greenpeace Campaigns." *Greenpeace*, January/February, 9–13.

Dahlberg, Kenneth A., et al. 1985. *Environment and the Global Arena*. Durham, N.C.: Duke University Press.

Daly, Herman E., ed. 1980. *Economics, Ecology, and Ethics*. San Francisco: W. H. Freeman.

Daly, Herman E., and John B. Cobb, Jr. 1989. *For the Common Good: Redirecting the Economy Toward Community, the Environment, and a Sustainable Future*. Boston: Beacon Press.

Durning, Alan B. 1989. *Poverty and the Environment: Reversing the Downward Spiral*. Washington, D.C.: Worldwatch Institute.

Elhington, John, et al. 1990. *The Green Consumer*. New York: Penguin Books.

Foreman, David, and Dave Haywood, eds. 1987. *Ecodefense: A Field Guide to Monkey Wrenching*. 2nd ed. Tucson, Ariz.: Earth First!

Georgescu-Roegen, Nicholas. 1971. *The Entropy Law and the Economic Process*. Cambridge, Mass.: Harvard University Press.

Hall, Bob. 1990. *Environmental Politics: Lessons from the Grassroots*. Durham, N.C.: Institute for Southern Studies.

Hamrin, Robert D. 1983. *A Renewable Resource Economy*. New York: Praeger.

Henderson, Hazel. 1981. *The Politics of the Solar Age*. New York: Anchor/Doubleday.

Henning, Daniel H., and William R. Mangun. 1989. *Managing the Environmental Crisis*. Durham, N.C.: Duke University Press.

MacNeill, Jim. 1989. "Strategies for Sustainable Economic Development." *Scientific American*, September, 155–163.

Mannes, Christopher. 1990. *Radical Environmentalism*. New York: Little, Brown.

Mathews, Christopher. 1988. *Hardball: How Politics Is Played—Told by One Who Knows the Game*. New York: Summitt Books.

Montague, Peter. 1989. "What We Must Do—A Grass-Roots Offensive Against Toxics in the 1990s." *The Workbook*, vol. 14, no. 3, 90–113.

Myers, Norman. 1988. "Environment and Security." *Foreign Policy*, vol. 74, 23–41.

Ophuls, William. 1977. *Ecology and the Politics of Scarcity*. San Francisco: W. H. Freeman.

Paehlke, Robert C. 1989. *Environmentalism and the Future of Progressive Politics*. New Haven, Conn.: Yale University Press.

Pearce, David, et al. 1990. *Sustainable Development: Economics and Environment in the Third World*. London: Edward Elgar Publishing, Ltd.

Renner, Michael. 1989. *National Security: The Economic and Environmental Dimensions*. Washington, D.C.: Worldwatch Institute.

Repetto, Robert, et al. 1989. *Wasting Assets: Natural Resources in the National Income Accounts*. Washington, D.C.: World Resources Institute.

Rosenbaum, Walter A. 1985. *Environment, Politics, and Policy*. Washington, D.C.: Congressional Quarterly.

Sargoff, Mark. 1988. *The Economy of the Earth: Philosophy, Law, and the Environment*. New York: Cambridge University Press.

Tietenberg, Tom. 1988. *Environmental and Resource Economics*. 2nd ed. Glenview, Ill.: Scott, Foresman.

World Commission on the Environment and Development. 1987. *Our Common Future*. New York: Oxford University Press.

HAZARDS, RISK,

AND

HUMAN HEALTH

General Questions and Issues

1. What are common hazards that people face and what are their effects?

2. How can the risks and benefits associated with using a particular technology or product be estimated?

3. How can government or other agencies manage risks to protect the public?

4. What are the major risks that can lead to cancer and how can they be reduced?

For the first time in the history of the world, every human being is now subjected to dangerous chemicals, from the moment of conception until death.

RACHEL CARSON

A LMOST EVERYTHING WE DO and every form of technology involves some degree of risk to our health and the health of other species. In evaluating risks the key questions are whether the risks of damage from each hazard outweigh the benefits and how we can minimize the risks.

8-1 HAZARDS: TYPES AND EFFECTS

Common Hazards **Risk** is the possibility of suffering harm from a hazard. A **hazard** is a source of risk and refers to substance or action that can cause injury, disease, economic loss, or environmental damage. Most hazards come from exposure to various factors in our environment:

- *Physical hazards*—ionizing radiation (Figure 3-8, p. 44), noise (Table 6-1, p. 133), earthquakes, hurricanes, tornadoes, fires, floods (see Case Study on p. 237), and drought.

- *Chemical hazards*—harmful chemicals in air (Chapters 9 and 10), water (Chapter 11), soil (Chapter 12), and food (Chapter 14).

- *Biological hazards*—disease-causing bacteria and viruses (Figure 8-1), pollen, parasites, and attacks by hungry or vicious animals.

- *Cultural hazards*—working and living conditions, smoking, diet, drugs, excessive drinking, driving, criminal assault, unsafe sex, and poverty (Section 7-3). In the United States the most dangerous occupation is farming, followed by construction and mining.

We are mostly concerned with hazards that directly affect our own health and safety. But we are now beginning to expand this view to include hazards for us and other species at the organism, population, and community-ecosystem level, as summarized in Table 5-2, p. 104.

Chemical Hazards A **toxic substance** is a chemical that is harmful to people or other living organisms. The effects from exposure to a toxic substance may be acute or chronic. **Acute effects** are those that appear shortly after exposure, usually to a large concentration or dose over a short time. Examples are skin burns or rashes, eye irritation, chest pains, kidney damage, headache, convulsions, and death.

Effects that are delayed and usually long-lasting are called **chronic effects**. They may not appear for months to years after exposure and usually last for years. Examples are cancer (caused by exposure to *carcinogens*), lung and heart disease, birth defects (caused by exposure to *teratogens*), genetic defects (caused by exposure to *mutagens*), and nerve and behavioral disorders. Chronic effects often occur as a

Frank Lambrecht/Visuals Unlimited

Figure 8-1 Unsafe water supply in Nigeria. Transmissable diseases from drinking contaminated water is the leading killer, especially of young children, in LDCs.

result of prolonged exposure to fairly low concentrations or doses of a toxic substance. However, they may also occur as the delayed effects of short-term exposure to high doses.

Determining Toxicity Levels Determining the toxicity levels of chemicals and the harmful effects of biological organisms is difficult, costly, and controversial. It is neither ethical nor practical to use people to test toxicity. Thus, toxicity is usually determined by carrying out tests on live (*in vivo*) laboratory animals (mostly mice and rats for testing carcinogens and guinea pigs, mice, and some primates for testing harmful microorganisms), bacteria, and cell and tissue cultures.

Chemical structure and activity analyses can also be used to compare the chemical makeup and possible effects of a new chemical with the estimated effects of older chemicals with a similar chemical makeup. But often reliable data is not available for many of the older chemicals with similar structures.

There are several problems with animal testing. Except in industrial and other accidents, humans are rarely exposed to the high dose levels per unit of body weight given to test animals. Extrapolating from high dose to low dose levels is uncertain and controversial. Many scientists also question the validity of extrapolating data from test animals to humans because human physiology and metabolism are different from those of the test animals. Not all chemicals that cause cancer in one animal species also cause cancer in other species. Also animal tests take several years and cost about $500,000 or more per substance. Animal tests are coming under increasing fire from

animal rights groups and many scientists are trying to use substitute methods where possible.

There is also controversy over the effectiveness of using bacteria and cell and tissue cultures for determining harmful effects on humans. One of the most widely used bacterial tests, the Ames test, is considered to be an accurate predictor of substances that cause genetic mutations (mutagens) and is also quick (two weeks) and cheap ($1,000 to $1,500 per substance). But the accuracy of this test as a predictor of cancer-causing substances (carcinogens) is controversial. Cell and tissue culture tests have similar uncertainties, take several weeks to months, and cost about $18,000 per substance.

Another approach to toxicity testing and determining the agents causing diseases such as cancer is **epidemiology**—the study of patterns of disease or effects from toxic exposure within defined groups of people and the factors that influence these patterns. Typically the effects on people exposed to a particular toxic chemical from an industrial accident, people working under high exposure levels, or people in certain geographic areas are compared with groups of people not exposed to these conditions to see if there are statistically significant differences.

Epidemiology works best when the population exposed to the risk agent is large (such as cigarette smoking), where the harmful effects are unusual (such as a rare form of cancer), where high levels of the risk agent are present in the environment, and where a causal link has been established between the risk agent and its harmful effects.

This approach also has limitations. For many toxic agents, not enough people have been exposed to high enough levels to detect statistically significant

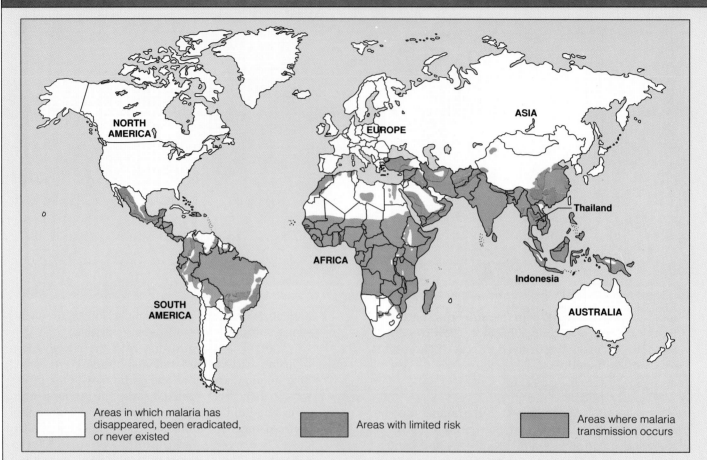

Figure 8-2 Malaria threatens half the world's population. (Data from the World Health Organization)

More than half the world's population live in malaria-prone regions in about 100 different countries in tropical and subtropical regions, especially West Africa and Southeast Asia (Figure 8-2). Malaria, spread by various species of the water-breeding *Anopheles* mosquito, afflicts 500 million people worldwide. Each year it kills at least 2.5 million (some sources say 5 million) people. At least half of its victims are children under the age of five.

Each year there are at least 100 million new cases. Even in the United States, each day an average of four people discover they have malaria. Malaria's symptoms come and go; they include fever and chills, anemia, an enlarged spleen, severe abdominal pain and headaches, extreme weakness, and greater susceptibility to other diseases.

Malaria is caused by one of four species of protozoa (one-celled organisms) of the genus *Plasmodium*. The disease is transmitted from person to person by a bite from the female of about 60 of the 400 different kinds of *Anopheles* mosquito. When an infected mosquito bites a person, *Plasmodium* parasites move into the bloodstream, multiply in the liver, and then enter blood cells to continue multiplying. When an uninfected mosquito bites an infected person, the cycle starts over (Figure 8-3). Malaria can also be transmitted when a person receives the blood of an infected donor or when a drug user shares a needle with an infected user.

differences. Also, people are exposed to many different toxic chemicals and disease-causing factors throughout their lives. Thus, it is not possible to say with much certainty that an observed epidemiological effect is caused only by exposure to a particular toxic agent or other hazardous condition. Also, epidemiology can be used only to evaluate hazards to which people have already been exposed. It is rarely useful for predicting the effects of new technologies or substances.

Thus, all of the methods we use to estimate toxicity levels have serious limitations. But they are the only practical methods we have for estimating the acute effects of toxic agents on humans.

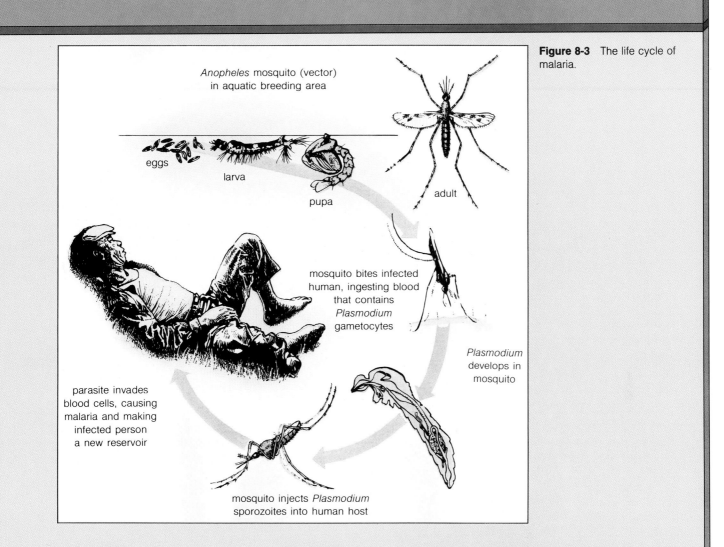

Figure 8-3 The life cycle of malaria.

Anopheles mosquito (vector)
in aquatic breeding area

eggs

larva

pupa

adult

mosquito bites infected
human, ingesting blood
that contains
Plasmodium
gametocytes

Plasmodium
develops in
mosquito

parasite invades
blood cells, causing
malaria and making
infected person
a new reservoir

mosquito injects *Plasmodium*
sporozoites into human host

Since the 1950s the spread of malaria has been reduced by draining swamplands and marshes and by spraying breeding areas with DDT, dieldrin, and other pesticides. Also drugs have been used to kill the *Plasmodium* parasites in the bloodstream.

This strategy worked for two decades, but since 1970 malaria has made a dramatic comeback in many parts of the world. Because of repeated spraying, most of the malaria-carrying species of *Anophe-les* mosquitoes have become genetically resistant to most of the insecticides used. Also the protozoa have become genetically resistant to widely used antimalarial drugs. Other factors increasing the incidence of malaria are the spread of irrigation ditches that provide new mosquito breeding grounds and reduced budgets for malaria control due to the belief that the disease was under control.

Researchers are working to develop new antimalarial drugs and vaccines and biological controls for *Anopheles* mosquitoes, but such approaches are in the early stages of development and lack adequate funding. The World Health Organization estimates that only 3% of the money spent worldwide each year on biomedical research is devoted to malaria and other tropical diseases, even though more people suffer and die worldwide from these diseases than from all others combined.

Biological Hazards: Disease, Economics, and Geography Human diseases can be broadly classified as transmissible and nontransmissible. A **transmissible disease** is caused by living organisms such as bacteria, viruses, and parasitic worms and can be spread from one person to another by air, water (Figure 8-1), food, body fluids, and in some cases by insects and other nonhuman transmitters (called vectors). Examples are malaria (see Case Study on p. 172), schistosomiasis, elephantiasis, sleeping sickness, measles, AIDS, and other sexually transmitted diseases.

A **nontransmissible disease** is not caused by living organisms and does not spread from one person to another. Examples include cardiovascular (heart and blood vessel) disorders, most cancer, diabetes, chronic respiratory diseases (bronchitis and emphysema), and malnutrition. Many of these diseases have several often unknown causes and tend to develop slowly and progressively over time.

Fortunately, major improvements in human health in LDCs can be made with primary preventive health care measures at a relatively low cost. These include providing:

- Contraceptive supplies, sex education, and family planning counseling.

- Better nutrition, prenatal care, and birth assistance for pregnant women. At least 500,000 women in LDCs die each year of mostly preventable pregnancy-related causes, compared to only 6,000 in MDCs.

- Greatly improved postnatal care (including the promotion of breast-feeding) to reduce infant mortality.

- Immunization against tetanus, measles, diphtheria, typhoid, and tuberculosis.

- Oral rehydration for diarrhea victims by feeding them a simple solution of water, salt, and sugar.

- Antibiotics for infections.

- Clean drinking water and sanitation facilities to the third of the world's population that lacks them.

Extending such primary health care to all the world's people would cost an additional $10 billion a year, one twenty-fifth as much as the world spends each year on cigarettes. In addition, relatively small expenditures on research and control of tropical diseases by MDCs and governments of LDCs would greatly reduce death rates and suffering from tropical diseases.

In most MDCs, safe water supplies, public sanitation, adequate nutrition, and immunization have nearly stamped out many transmissible diseases. In 1900 pneumonia, influenza, tuberculosis, and diarrhea and other intestinal infections were the leading causes of death in MDCs. By contrast, the four leading causes of death in MDCs today are heart disease and stroke, cancer, respiratory infections, and accidents, especially automobile accidents (see Pro/Con on p. 130).

These deaths are largely a result of environmental and lifestyle factors rather than infectious agents invading the body. Except for auto accidents, these deaths result from chronic diseases that take a long time to develop, have multiple causes, and are largely related to the area in which people live (urban or rural), their work environment, their diet, whether they smoke, how much exercise they get, and the amount of alcohol they consume.

Changing these harmful lifestyle factors could

prevent 40% to 70% of all premature deaths, one-third of all cases of acute disability, and two-thirds of all cases of chronic disability. Another area of growing concern in both MDCs and LDCs is the spread of AIDS and other sexually transmitted diseases, mostly because of ignorance, drug abuse, and failure to practice safe sex.

In 1988 about 11% of the GNP of the United States was spent on health care—an average of $2,000 per American. But more than 95% of this money was used to treat rather than prevent disease. Clearly, a much better balance between treatment (output approaches) and prevention (input approaches) is needed.

8-2 RISK ANALYSIS

Nature and Uses of Risk Analysis **Risk analysis** involves identifying hazards, evaluating the nature and severity of risks (*risk assessment*), using this and other information to determine options and make decisions about reducing or eliminating risks (*risk management*), and communicating information about risks to decision makers and the public (*risk communication*).

Risk analysis can be used to help decision makers and other interested parties to

- estimate environmental and health problems associated with an activity (such as waste disposal) and substances (such as the production and use of a particular chemical)

- estimate and compare the risks associated with using new and existing technologies (such as nuclear energy and genetic engineering)

- evaluate and compare different methods for reducing or eliminating the risk from an activity, technology, or substance

- select sites for potentially hazardous facilities such as industrial plants, power plants, landfills, incinerators, and hazardous waste dumps

- evaluate and set priorities for the management of various risks

Estimating Risks **Risk assessment** is the technical process of gathering data and making assumptions to estimate the nature, severity, and likelihood of harm to human health or the environment from a hazard (Table 5-1, p. 104). Professional risk analysts use scientific experiments, models, and statistical methods to evaluate risks. Most ordinary citizens rely on intuitive risk judgments based on past experience, media information, and perceptions about how risky something is.

Formal risk assessment is difficult, imprecise, and controversial. Probabilities based on past experience, animal and other tests, and epidemiological studies are used to estimate risks from older technologies and products. For new technologies and products,

U. S. Department of Energy

Figure 8-4 Control room of a nuclear power plant. Watching these indicators is such a boring job that government investigators have found some operators asleep and in one case found no one in the control room. Most nuclear power plant accidents have resulted primarily from unpredictable and largely uncontrollable human errors.

much more uncertain statistical probabilities, based on models rather than actual experience, must be calculated.

The more complex a technological system, the more difficult it is to make realistic calculations of risks based on statistical probabilities of the failure of equipment and people. The total reliability of any technological system is the product of two factors:

$$\frac{\text{system}}{\text{reliability (\%)}} = \frac{\text{technology}}{\text{reliability}} \times \frac{\text{human}}{\text{reliability}} \times 100$$

With careful design, quality control, maintenance, and monitoring, a high degree of technology reliability can usually be obtained in complex systems such as a nuclear power plant, space shuttle, or early warning system for nuclear attack. However, human reliability is almost always much lower than technology reliability and virtually impossible to predict; to be human is to err.

For example, suppose that the technology reliability of a system such as a nuclear power plant is 95% (0.95) and the human reliability is 65% (0.65). Then the overall system reliability is only 62% (0.95 × 0.65 = 0.62 × 100 = 62%). Even if we could increase the technology reliability to 100% (1.0), the overall system reliability would still be only 65% (1.0 × 0.65 = 0.65 × 100 = 65%).

This crucial dependence of even the most carefully designed systems on unpredictable human reliability helps explain the occurrence of extremely unlikely events that risk analysts consider almost impossible. Examples are the Three Mile Island and Chernobyl nuclear power plant accidents (see Spotlight on p. 423), the tragic (and unnecessary) explo-

sion of the space shuttle *Challenger*, and the far too frequent false alarms given by early warning defense systems on which the fate of the entire world depends.

Poor management, training, and supervision increase the chances of human errors. Maintenance workers or people who monitor warning panels in complex systems such as the control rooms of nuclear power plants (Figure 8-4) become bored and inattentive because most of the time nothing goes wrong. They may fall asleep while on duty (as has happened at several U.S. nuclear plant control rooms); they may falsify maintenance records because they believe that the system is safe without their help; they may be distracted by personal problems or illness; or they may be told by managers to take shortcuts to increase short-term profits or to make the managers look more efficient and productive.

One way to improve system reliability is to move more of the potentially fallible elements from the human side to the technical side, making the system more foolproof or "fail-safe." But chance events such as a lightning bolt can knock out automatic control systems. And no machine or computer program can replace all the skillful human actions and decisions involved in seeing that a complex system operates properly and safely. Also the parts in any automated control system are manufactured, assembled, tested, certified, and maintained by fallible human beings.

There are other problems with risk assessment. Complex processes and effects may be oversimplified or poorly understood, and risk evaluators may have overconfidence in the reliability of current scientific and technical knowledge and models. They may also fail to see how the system as a whole functions and how different components and chemicals interact. All systems have properties that cannot be determined or predicted by understanding the properties of their parts.

Risk-Benefit Analysis The key question is whether the estimated short- and long-term benefits of using a particular technology or product outweigh the estimated short- and long-term risks compared with other alternatives. One method for making such evaluations is **risk-benefit analysis**. It involves estimating the short- and long-term societal benefits and risks involved and then dividing the benefits by the risks to find a **desirability quotient**:

$$\text{desirability quotient} = \frac{\text{societal benefits}}{\text{societal risks}}$$

Assuming that accurate calculations of benefits and risks can be made (a big assumption), here are several possibilities:

1. $\text{large desirability quotient} = \frac{\text{large societal benefits}}{\text{small societal risks}}$

   Example: *X rays.* Use of ionizing radiation in the form of X rays to detect bone fractures and other medical problems has a large desirability quotient. But this is true only if X rays are not overused to protect doctors from liability suits, the dose is no larger than needed, and less harmful alternatives are not available. Other examples in this category are mining, most dams, and airplane travel.

2. $\text{very small desirability quotient} = \frac{\text{very small societal benefits}}{\text{very large societal risks}}$

   Example: *Nuclear war.* Global nuclear war has no societal benefits (except the short-term profits made by companies making weapons and weapons defense systems) and involves totally unacceptable risks to the earth's present human population and to many future generations. Global nuclear war is the single greatest threat to the human species and the earth's life-support systems for all species, as discussed in Section 10-6.

3. $\text{small desirability quotient} = \frac{\text{large societal benefits}}{\text{much larger societal risks}}$

   Example: *Coal-burning power plants* (Section 18-2) *and nuclear power plants* (Section 18-3). Nuclear and coal-burning power plants provide society with electricity, a highly desirable benefit, but many analysts contend that the short- and long-term societal risks from widespread use of these technologies outweigh the benefits. They believe that many other more economically and environmentally acceptable alternatives exist for producing electricity with less severe societal risks (see Chapter 17).

4. $\text{uncertain desirability quotient} = \frac{\text{large benefits}}{\text{large risks}}$

   Example: *Genetic engineering.* For many decades humans have selected and crossbred different genetic varieties of plants and animals to develop new varieties with certain desired qualities. Today "genetic engineers" have learned how to transfer genetic traits from one species to another to make new genetic combinations instead of waiting for nature to evolve new genetic combinations through natural selection. We are already using this biotechnology to produce new forms of life, which are then patented and sold in the marketplace. But this is a controversial new technology (see Pro/Con on p. 177).

Limitations of Risk Assessment Calculation of desirability quotients and other ways of expressing risk is extremely difficult, filled with uncertainty, and controversial. For example, many people—especially those who make their living or earn profits from the technologies involved—would disagree with the general estimates of desirability quotients just given.

Some technologies benefit one group of people (population A) while imposing a risk on another (population B). Some people making the estimates emphasize short-term risks while others put more weight on long-term risks. Which type of risk should get the most emphasis and who decides this?

Who should carry out a particular risk-benefit analysis or risk assessment? Should it be the corporation or government agency involved in developing or managing the technology, or some independent laboratory or panel of scientists? If it involves outside evaluation, who chooses the persons to do the study? Who pays the bill and thus has the potential to influence the outcome by refusing to give the lab, agency, or experts future business?

Once the study is done, who reviews the results— a government agency, independent scientists, the general public—and what influence will outside criticism have on the final decisions? Clearly, politics, economics, and value judgments that can be biased in either direction are involved at every step of the risk analysis process.

The difficulty of estimating risks does not mean that it should not be made or that it is not useful. Despite the inevitable uncertainties involved, risk assessment and risk-benefit analysis are useful ways to organize available information, identify significant hazards, focus on areas that need more research, and stimulate people to make decisions about health and environmental goals and priorities.

Scientists, politicians, and the general public who must evaluate such analyses and make decisions based on risk assessment should be aware of their serious limitations. At best they can only be expressed

The rapidly developing technology of genetic engineering excites some scientists and many business executives. They see it as a way to increase crop and livestock yields and to produce plants and livestock that have greater resistance to diseases, pests, frost, and drought and that provide greater quantities of nutrients such as proteins.

Developing bacteria that can destroy oil spills and degrade toxic wastes and new vaccines, drugs, and therapeutic hormones are other goals of those behind this new technology. Gene therapy would also be used to eliminate certain genetic diseases and other genetic afflictions. By patenting these new life forms and selling them in the marketplace, investors see genetic engineering as a source of enormous profits.

But some people are horrified by the prospect of biotechnology running amok. Most of these critics recognize that it is essentially impossible to stop the development of genetic engineering, which is already well under way, but they believe that this technology should be kept under strict control.

These critics do not believe that people have enough understanding of how nature works to be trusted with such great control over the genetic characteristics of humans and other species. Critics also fear that unregulated biotechnology could lead to the development of "superorganisms." If such organisms were released deliberately or accidentally into the environment, they could cause unpredictable, possibly harmful health and ecological effects. Most would probably be safe, but some would inevitably be dangerous.

Since many organisms, especially bacteria, are capable of rapidly reproducing and spreading to new locations, any problems they cause would be widespread. For example, genetically altered bacteria designed to clean up ocean oil spills by degrading the oil might multiply rapidly and eventually degrade the world's remaining oil supplies—including the oil in cars and trucks.

In addition to reproducing rapidly, genetically engineered organisms might also mutate and change their form and behavior. Unlike defective cars and other products, living organisms can't be recalled once they are in the environment.

The risks of this or other catastrophic events resulting from biotechnology are small. But critics contend that biotechnology is a potential source of such enormous profits that without strict controls, greed—not ecological wisdom and restraint—will take over.

Genetic scientists answer, however, that it is highly unlikely that the release of genetically engineered species would cause serious and widespread ecological problems. To have a serious effect, such organisms would have to be outstanding competitors and resistant to predation. In addition, they would have to be capable of becoming dominant in ecosystems and in the ecosphere. But critics point out that this has happened many times when we have accidentally or deliberately introduced alien organisms into biological communities.

In 1989 a committee of prominent ecologists appointed by the Ecological Society of America released a report stating that many of the assertions about the inherent safety of genetically engineered organisms vary widely with the type of organism. The committee calls for a case-by-case review of any proposed environmental releases. It also calls for carefully regulated, small-scale field tests before any bioengineered organism is put into commercial use.

This controversy illustrates the difficulty of balancing the actual and potential benefits of a technology with its actual and potential risks of harm. What do you think?

as a range of probabilities and uncertainties based on different assumptions—not the precise bottom-line numbers that decision makers want. The present uncertainty in risk analysis is so great that we need to understand that regulatory decisions using these tools are based primarily on political analysis rather than scientific analysis.

 8-3

RISK MANAGEMENT
Deciding How and What Risks to Manage   **Risk management** includes the administrative, political, and economic actions taken to decide how, and if, a particular societal risk is to be reduced to a certain level and at what cost. It is integrated with risk assessment (Figure 8-5). Risk management involves trying to answer the following questions:

■ Which of the vast number of risks facing society should be evaluated and managed with the limited funds available?

■ In what sequence or priority should the risks be evaluated and managed?

■ How reliable is the risk-benefit analysis or risk assessment carried out for each risk?

■ How much risk is acceptable? How safe is safe enough?

■ How much money will it take to reduce each risk to an acceptable level?

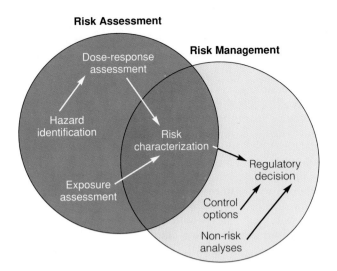

Risk Assessment

Risk Management

Dose-response assessment

Hazard identification

Risk characterization

Regulatory decision

Exposure assessment

Control options

Non-risk analyses

Figure 8-5 Summary of risk assessment and risk management. (Environmental Protection Agency)

- How much will each risk be reduced if sufficient funds are not available, as is usually the case?

- How will the risk management plan be communicated to the public, monitored, and enforced?

Risk managers must make difficult decisions involving inadequate and uncertain scientific data, potentially grave consequences for human health and the environment, and large economic effects on industry and consumers. Thus, each step in this process involves value judgments and trade-offs to find some reasonable compromise between conflicting political and economic interests.

So far most risk reduction from pollutants has focused on output or end-of-pipe pollution control techniques. Very little emphasis has been placed on reducing risks by using input or front-of-pipe methods for pollution prevention and waste reduction. Beginning with front-of-pipe pollution prevention instead of end-of-pipe pollution control is the key to risk reduction (Figure 8-6). Table 8-1 shows how such strategies can be applied at various levels of society.

Risk analysis attempts to find some politically or economically acceptable level of pollution (Figure 7-3, p. 152) or other risk. By contrast, pollution prevention aims at reducing the risk to health to the lowest possible level. If a pollutant or risky technology is eliminated or reduced to a very low level, the elaborate and uncertain system of risk assessment and standard setting and the resulting controversy and legal challenges become irrelevant.

The problem of what level of risk to accept is highly controversial. To achieve no (or zero) risk an activity or chemical would have to be banned completely, but when this happens there are usually risks associated with the substitutes. The question is which of the alternatives has the lowest risk.

Risk Perception and Communication Communicating risks to the public is a political act that is as much an art as a skill. Most of us are bad at assessing the risks from the hazards that surround us. We are risk-illiterate and full of contradictions. On the one hand we deny and shrug off high risk activities such as driving or riding in a car, not wearing seat belts, hang gliding, and exposing ourselves to the cancer-causing rays of the sun or tanning lamps to get a tan.

On the other hand we insist on zero or near zero risk from things that are quite unlikely to kill us. Some of us become almost paranoid about eating apples that might bear a trace of a pesticide, riding in a commercial airplane, or fearing that we will be killed by a burglar or mugger.

The public generally perceives that a technology or product has a greater risk than the risk estimated by experts when it:

- Is relatively new or complex (genetic engineering, nuclear power) rather than familiar (dams, automobiles).

- Is mostly involuntary (nuclear power plants, nuclear weapons, industrial pollution, food additives) instead of voluntary (smoking, drinking alcohol, driving).

- Is viewed as beneficial and necessary (cars) rather than unnecessary (CFCs and hydrocarbons as propellants in aerosol spray cans, food additives used to increase sales appeal).

- Involves a large number of deaths and injuries from a single catastrophic accident (severe nuclear power plant accident, industrial explosion, or plane crash) rather than the same or larger number of deaths spread out over a longer time (coal-burning power plants, automobiles, malnutrition in LDCs).

- Involves unfair distribution of the risks. Citizens are outraged when government officials decide to put a hazardous waste dump or incinerator in or near their neighborhood under the guise of scientific analysis. This is usually viewed as politics, not science.

- Is poorly communicated. Does the decision-making agency or company come across as trustworthy and concerned or dishonest, unconcerned, and arrogant (as Exxon was viewed after the Valdez oil spill and the Nuclear Regulatory Agency and the nuclear industry have been viewed since the Three Mile Island accident)? Does it involve the community and tell it what's going on before the real decisions are made? Does it understand, listen to, and respond to community concerns?

- Does not take into account ethical and moral concerns. Spewing out numbers and talking about cost-risk trade-offs seem very callous when the risk involves moral issues such as the

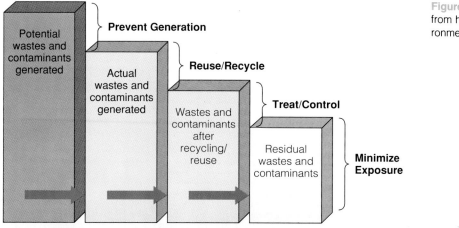

Prevent Generation

Reuse/Recycle

Treat/Control

Potential wastes and contaminants generated

Actual wastes and contaminants generated

Wastes and contaminants after recycling/reuse

Residual wastes and contaminants

Minimize Exposure

Table 8-1   Examples of Risk Reduction Strategies (Environmental Protection Agency)

Strategy	Individuals	Communities and Community Groups	Industry	Federal and State Governments
Prevent Pollutant Generation	Conserve energy	Reduce pesticide use	Substitute raw materials and redesign processes	Ban certain materials
Recycle and Reuse	Return wastes to recycling centers	Promote and operate recycling centers	Reclaim solvents	Purchase recycled products
Treat and Control	Inspect and remove asbestos	Treat water supplies	Treat hazardous waste	Mandate air and wastewater treatment standards
Reduce Residual Exposure	Avoid fishing and swimming in polluted waters	Operate clean sanitary landfills	Operate clean chemical landfills	Establish high-level radiation and toxic disposal facilities

health of people and other species and environmental quality.

People who see their lives and the lives of their children as being threatened because they live near a chemical plant or toxic-waste dump that has leaked a hazardous chemical into the nearby environment could care less that experts say the chemical is likely to kill only one out of a million people in the general population. Only a small number of these million people live near the plant or dump, as they do. To those on the risk front lines, often the poor and middle class, risk is personal, not an abstraction.

Unless risk-analysis communicators understand and take into account the perceptions, concerns, fears, interests, values, priorities, and preferences of individuals and public groups, the information they provide is likely to be ignored. Risk communicators also need to clearly point out the assumptions and uncertainties in their estimates and risk comparisons. They also must acknowledge that risk analysis is a way to help make political and economic decisions based on useful but incomplete and often controversial statistical and scientific evidence.

Some observers contend that when it comes to evaluation of large-scale, complex technologies, the public often is better at seeing the forest than the risk-benefit specialists, who look primarily at the trees. This commonsense wisdom does not usually depend on understanding or even caring about the details of scientific risk-benefit analysis. Instead, it is based on the average person's understanding that science and technology have limits and that the people responsible for potentially hazardous technological systems and products are fallible just like everyone else.

8-4 RISK FACTORS AND CANCER

What Is Cancer? Tumors are growths of cells that enlarge and reproduce at abnormal rates. Tumors are classified as benign or malignant. A **benign tumor** is one that remains within the tissue

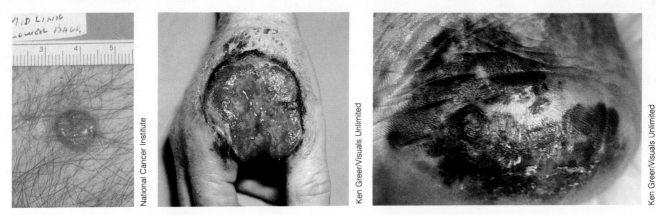

a Basal   **b** Squamous   **c** Malignant melanoma

Figure 8-7   Three types of skin cancer. These types of cancer are rising because of increased exposure to ultraviolet-B radiation. This is being caused by depletion of ozone gas in the stratosphere by chlorofluorocarbons (CFCs) and other chemicals containing chlorine or bromine atoms.

Normally nonfatal skin cancer is by far the most common form of cancer; about one in seven Americans gets it sooner or later. Cumulative exposure to ultraviolet ionizing radiation in sunlight over a number of years is the major cause of basal-cell and squamous-cell skin cancers (Figures 8-7a and 8-7b). These two types of cancer can be cured if detected early enough, although their removal may leave disfiguring scars. I have had three basal-cell cancers on my face because of too much sun exposure in my younger years. I wish I had known then what I know now.

Evidence suggests that just one severe, blistering burn as a child or teenager is enough to double the risk of contracting deadly malignant melanoma (Figure 8-7c) later in life, regardless of the skin type or the amount of cumulative exposure to the sun. This form of skin cancer of the cells that produce the skin's pigment spreads rapidly to other organs and can kill its victims. Each year it kills about 9,000 Americans and 100,000 people worldwide. The number of cases of melanoma is increasing at a rapid rate, with

white people seven to ten times more susceptible than blacks. Depletion of the ozone layer (Section 10-4) will lead to a sharp increase in all types of skin cancers.

Virtually anyone can get skin cancer but those with very fair and freckled skin run the highest risk. White Americans who spend long hours in the sun or in tanning booths (which are even more hazardous than direct exposure to the sun) greatly increase their chances of developing skin cancer and also tend to have wrinkled, dry skin by age 40. Blacks with darker skin are almost immune to sunburn but do get skin cancer, although at a rate ten times lower than whites. A dark suntan also doesn't prevent skin cancer. Outdoor workers are particularly susceptible to cancer of the exposed skin on the face, hands, and arms.

The safest thing to do is to stay out of the sun and tanning booths. Avoid direct exposure between 10 AM and 3 PM when the sun's ultraviolet rays are strongest. Sitting under an umbrella does not protect against the sun because of sunlight reflected from sand, cement, or water. Clouds are deceptive because they admit

as much as 80% of the sun's harmful ultraviolet radiation.

When you are in the sun wear tightly woven protective clothing and a wide-brimmed hat and apply a sunscreen with a protection factor of 15 or more (25 if you have light skin) to all exposed skin. Reapply sunscreen after swimming or excessive sweating. Children using a sunscreen with a protection factor of 15 any time they are in the sun from birth to age 18 decrease their chance of skin cancer by 80%.

Get to know your moles and examine your skin surface at least once a month for any changes. The warning signs are a change in the size, shape, or color of a mole or wart (the major sign of malignant melanoma, which needs to be treated quickly), sudden appearance of dark spots on the skin, or a sore that keeps oozing, bleeding, and crusting over but does not heal. You should also be on the watch for precancerous growths that appear as reddish brown spots with a scaly crust. If any of these signs are observed, you should immediately consult a doctor. What are you doing to protect your skin and your life?

where it develops and often grows slowly. A **malignant tumor** or **cancer** is one in which cells multiply uncontrollably and invade the surrounding tissue. If not detected and treated in time, many cancerous

tumors undergo **metastasis**; that is, they release malignant (cancerous) cells that travel in body fluids to various parts of the body, making treatment much more difficult. Any cell is capable of becoming cancerous, but most cancer cells are found in tissues undergoing rapid cellular division. Examples are the cells in the lungs, bone marrow, ovaries, testes, skin, and intestinal lining.

Cancers are caused by poorly understood interactions between genes and environmental factors such as ionizing radiation (Figure 3-8, p. 44) and certain chemicals and viruses. Environmental agents that can cause or promote the growth of malignant tumors or cancers are called **carcinogens**.

Only about 10% of the 70,000 chemicals in commercial use have been adequately tested to determine whether they are carcinogens. Each year about 1,000 new chemicals are introduced into the marketplace. Establishing strong evidence that a chemical or other agent does or does not cause or promote cancer in humans is difficult and costly (Section 8-1). It is virtually impossible to isolate the effects of the thousands of chemicals and other factors in the environment that people are exposed to during the 10 to 40 years that may elapse before a cancer reaches detectable size.

Cancer Incidence and Cure Rates Cancer will strike about 1 million Americans this year and one of every three Americans now living will eventually have some type of cancer (see Spotlight on p. 180). Every 66 seconds an average of one person dies from cancer in the United States.

The good news is that almost half of all Americans who get cancer can now be cured, defined as being alive and cancer-free five or more years after treatment, compared to only a 38% cure rate in 1960. Survival rates for some types of cancers now range from 66% to 88%. This has happened mostly because of a combination of early detection and improved use of surgery, radiation, and drug treatments.

Cancer Risk Factors According to the World Health Organization, environmental and lifestyle factors play a key role in causing or promoting 80% to 90% of cancers. Major sources of carcinogens are cigarette smoke (40%), dietary factors (25% to 30%), occupational exposure (10% to 15%), and environmental pollutants (5% to 10%). About 10% to 20% of cancers are believed to be caused by inherited genetic factors and some viruses.

Thus the risks of developing cancer can be greatly reduced by working and living in a less hazardous environment, not smoking or being around smokers (see Case Study on p. 182), drinking in moderation (no more than two beers or drinks a day) or not at all, adhering to a healthful diet (see Spotlight above), and shielding oneself from the sun (see Spotlight on p. 180). According to experts, 60% of all cancers could be prevented by such lifestyle changes.

SPOTLIGHT   The Prudent Diet

The National Academy of Sciences and the American Heart Association advise that the risk of certain types of cancer—lung, stomach, colon, breast, and esophagus cancer—heart disease, and diabetes can be significantly reduced by a daily diet that cuts down on certain foods and includes others. Such a diet limits

- total fat intake to 30% or less of total calories, with no more than 10% from saturated fats, and the remaining 20% divided about equally between polyunsaturated fats (like safflower oil and corn oil) and monosaturated fats (such as olive oil)

- protein (particularly meat protein) to 15% of total calories or about 171 grams (6 ounces) a day (about the amount in one hamburger)

- alcohol consumption to 15% of total caloric intake—no more than two drinks, glasses of wine, or beers a day

- cholesterol consumption to no more than 300 milligrams a day with the goal of keeping blood cholesterol levels below 200 milligrams per deciliter

- sodium intake to no more than 6 grams (about 1 teaspoon) a day to help lower blood pressure, which should not exceed 140 over 90

Also, each of us should achieve and maintain the ideal body weight for our frame size and age by a combination of diet and exercise for 20 minutes a day at least three days a week.

Our diet should include fruits (especially vitamin C rich oranges, grapefruit, and strawberries), minimally cooked orange, yellow, and green leafy vegetables such as spinach and carrots, and cabbage-family vegetables such as cauliflower, cabbage, kale, brussels sprouts, and broccoli.

It should also incorporate starches and other complex carbohydrates, 10 to 15 grams of whole-grain fiber a day (from bran and fibers in vegetables and fruits), and a daily intake of selenium not exceeding 200 micrograms. Salt-cured, nitrate-cured, and smoked ham, bacon, hot dogs, sausages, bologna, salami, corned beef, and fish should not be eaten or eaten only rarely.

Many people don't make such changes. One problem is that usually 10 to 40 years elapse between the initial cause or causes of a cancer and the appearance of detectable symptoms. For instance, healthy high school and college students and young adults have difficulty accepting the fact that their smoking, drinking, eating, and other lifestyle habits today will

Figure 8-8   Annual deaths in the United States related to tobacco use and other causes in 1989. (Data from National Center for Health Statistics)

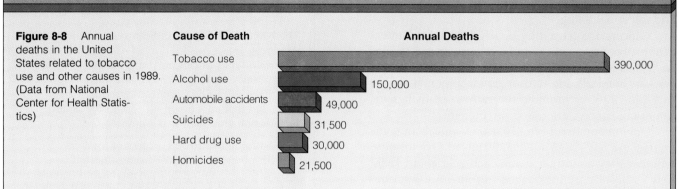

Cause of Death	Annual Deaths
Tobacco use	390,000
Alcohol use	150,000
Automobile accidents	49,000
Suicides	31,500
Hard drug use	30,000
Homicides	21,500

Smoking tobacco causes more death and suffering by far among adults than any other environmental factor. Each cigarette smoked reduces one's average life span by five and a half minutes. Worldwide, at least 3 million smokers die prematurely each year from heart disease, lung cancer, bronchitis, emphysema, other cancers, and stroke—all related to smoking.

In 1989 smoking killed about 390,000 Americans—an average of 1,068 a day. This annual death toll in the United States is equal to three fully loaded jumbo jets crashing every day with no survivors, almost eight times the number of people killed in traffic accidents each year, and seven times the number of American soldiers killed in the nine-year Vietnam War.

Nicotine is not classified as an illegal drug. Yet, it kills and harms more people each year in the United States than all illegal drugs and alcohol (the second most harmful drug), automobile accidents, suicide, and homicide combined (Figure 8-8).

Numerous studies have shown that the nicotine in tobacco is a highly addictive drug that, like heroin and cocaine, can quickly and strongly hook its victims. A British government study showed that adolescents who smoke more than one cigarette have an 85% chance of becoming smokers. An inhaled hit of nicotine takes only 5 to 10 seconds to reach the brain—twice as fast as intravenous drugs and three times faster than alcohol. The typical smoker has a 200 to 400 hit a day legalized habit.

Smokers develop tolerance to nicotine and experience withdrawal symptoms when they try to stop. Some recovering heroin addicts report they had a much harder time quitting smoking than quitting heroin. About 75% of smokers who quit start smoking again within six months, about the same relapse rate as recovering alcoholics and heroin addicts.

Smokers are not the only ones harmed by smoke from cigarettes and other tobacco products. Several studies indicate that passive smoke causes 3,000 to 15,000 premature deaths of Americans a year. According to a 1986 study by the National Research Council, nonsmoking spouses of smokers have a 30% greater chance of getting lung cancer than spouses of nonsmokers. Women exposed to passive smoke three hours or more a day appear to have a threefold increased risk for cervical cancer.

Tobacco's harmful costs to American society exceed its economic benefits to tobacco farmers and employees and stockholders of tobacco companies by more than two to one. In the United States, smoking costs society at least $52 billion (some estimate $95 billion) a year in premature death, disability, medical treat-ment, increased insurance costs, and lost productivity because of illness. These external costs amount to an average cost to society of at least $2.20 per pack of cigarettes sold.

The American Medical Association and numerous health experts have called for

- a total ban on cigarette advertising in the United States

- prohibition of the sale of cigarettes and other tobacco products to anyone under 21 with strict penalties for violations

- a ban on all cigarette vending machines

- classifying nicotine as a drug and placing the manufacture, distribution, sale, and promotion of tobacco products under the jurisdiction of the Food and Drug Administration

- eliminating all federal subsidies to U.S. tobacco farmers and tobacco companies

- taxing cigarettes at about $2.20 a pack to discourage smoking and to make smokers pay for the harmful effects of smoking now borne by society as a whole

- prohibiting elected and appointed government officials from exerting any influence on other governments to enhance the export of tobacco from the United States to other countries

be major influences on whether they will die prematurely from cancer before they reach age 50. Denial can be deadly.

Government has an important role to play in reducing the risks from the numerous hazards we are exposed to. However, when we have a choice, changing our own lifestyles is the most important way of reducing risks to our health and survival.

Though their health needs differ drastically, the rich and the poor do have one thing in common: both die unnecessarily. The rich die of heart disease and cancer, the poor of diarrhea, pneumonia, and measles. Scientific medicine could vastly reduce the mortality caused by these illnesses. Yet, half the developing world lacks medical care of any kind.

WILLIAM U. CHANDLER

DISCUSSION TOPICS

1. Considering the benefits and risks involved, do you believe that
   a. nuclear power plants should be controlled more rigidly and gradually phased out?
   b. coal-burning power plants should be controlled more rigidly and gradually phased out?
   c. genetic engineering should be prohibited?
   d. genetic engineering should be more rigidly controlled?
   In each case defend your position.

2. What are the major limitations of risk analysis? Does this mean that it is useless? Explain.

3. Explain why you agree or disagree with each of the following proposals:
   a. All advertising of cigarettes and other tobacco products should be banned.
   b. All smoking should be banned in public buildings and commercial airplanes, buses, subways, and trains.
   c. All government subsidies to tobacco farmers and the tobacco industry should be eliminated.
   d. Cigarettes should be taxed at about $2.20 a pack so that smokers—not nonsmokers—pay for the health and productivity losses now borne by society as a whole.

4. Assume you have been appointed to a technology risk-benefit-assessment board. Explain why you approve or disapprove of widespread use of each of the following:
   a. abortion pills (now used in France and China)
   b. effective sex stimulants
   c. drugs that would retard the aging process
   d. drugs that would enable people to get high but are physiologically and psychologically harmless
   e. electrical or chemical methods that would stimulate the brain to eliminate anxiety, fear, unhappiness, and aggression
   f. genetic engineering that would produce people with superior intelligence, strength, and other traits

FURTHER READINGS

Bergin, Edward J., and Ronald Grandon. 1984. *The American Survival Guide: How to Survive Your Toxic Environment.* New York: Avon.

Bernarde, Melvin A. 1989. *Our Precarious Habitat: Fifteen Years Later.* New York: John Wiley.

Cohrssen, John J., and Vincent T. Covello. 1989. *Risk Analysis: A Guide to Principles and Methods for Analyzing Health and Environmental Risks.* Springfield, Va.: National Technical Information Service.

Douglas, Mary, and Aaron Wildavsky. 1982. *Risk and Culture.* Berkeley: University of California Press.

Ehrlich, Anne, and J. Birks, eds. 1990. *Hidden Dangers.* San Francisco: Sierra Club Books.

Environmental Protection Agency. 1987. *Unfinished Business: A Comparative Assessment of Environmental Problems.* Washington, D.C.: EPA.

Freudenburg, William R. 1988. "Perceived Risk, Real Risk: Social Science and the Art of Probabilistic Risk Assessment." *Science,* vol. 242, 44–49.

Haden, Susan G. 1989. *A Citizen's Right to Know: Risk Communication and Public Policy.* Boulder, Colo.: Westview Press.

Imperato, P. J., and Greg Mitchell. 1985. *Acceptable Risks.* New York: Viking.

Krimsky, Sheldon, and Alonzo Plough. 1988. *Environmental Hazards: Communicating Risks as a Social Process.* Dover, Mass.: Auburn House.

Kupchella, Charles E. 1987. *Dimensions of Cancer.* Belmont, Calif.: Wadsworth.

Lave, Lester B. 1987. *Risk Assessment and Management.* New York: Plenum.

National Academy of Sciences. 1986. *Environment Tobacco Smoke: Measuring Exposures and Assessing Health Effects.* Washington, D.C.: National Academy Press.

National Academy of Sciences. 1989. *Diet and Health: Implications for Reducing Chronic Disease Risk.* Washington, D.C.: National Academy Press.

National Academy of Sciences. 1989. *Improving Risk Communication.* Washington, D.C.: National Academy Press.

Olsen, Steve. 1986. *Biotechnology.* Washington, D.C.: National Academy Press.

Rifkin, Jeremy. 1985. *Declaration of a Heretic.* Boston: Routledge & Kegan Paul.

Suzuki, David, and Peter Knudtson. 1989. *Genethics: The Clash Between the New Genetics and Human Values.* Cambridge, Mass.: Harvard University Press.

U.S. Department of Health and Human Services. 1986. *The Health Consequences of Involuntary Smoking: A Report of the Surgeon General.* Rockville, Md.: U.S. Department of Health and Human Services.

U.S. Department of Health and Human Services. 1988. *The Health Consequences of Smoking: Nicotine Addiction.* Washington, D.C.: Government Printing Office.

U.S. Department of Health and Human Services. 1988. *The Surgeon General's Report on Nutrition and Health.* Washington, D.C.: Government Printing Office.

Wilson, Richard, and E. A. C. Crouch. 1987. "Risk Assessment and Comparisons: An Introduction." *Science,* vol. 236, 267–270.

North America's largest copper smelter and tallest smokestack in Sudbury, Ontario, Canada.

*I am utterly convinced that most of the great environmental struggles will be either won or lost
in the 1990s, and that by the next century it will be too late to act.*
THOMAS E. LOVEJOY

CHAPTER NINE

Environ. Test
Physics Quiz
Math

AIR RESOURCES

AND

AIR POLLUTION

General Questions and Issues

1. What are the major components of the atmosphere, and how are we disrupting the earth's gaseous biogeochemical cycles?

2. What are the major types and sources of air pollutants?

3. What is smog and acid deposition?

4. What undesirable effects can air pollutants have on people, other species, and materials?

5. What legal and technological methods can be used to reduce air pollution?

I thought I saw a blue jay this morning. But the smog was so bad that it turned out to be a cardinal holding its breath.

MICHAEL J. COHEN

AKE A DEEP BREATH. About 99% of the volume of air you inhaled is gaseous nitrogen and oxygen. You also inhaled trace amounts of other gases, minute droplets of various liquids, and tiny particles of various solids. Many of these chemicals are classified as air pollutants. Most come from cars, trucks, power plants, factories, cigarettes, cleaning solvents, and other sources related to our activities. Most are related to the burning of fossil fuels, with cars responsible for at least half of the air pollution in urban areas.

You are exposed to air pollutants outdoors and indoors. Repeated exposure to trace amounts of many of these chemicals can damage lung tissue, plants, fish and other animals, buildings, metals, and other materials, as discussed in this chapter. Changes in the chemical content of the atmosphere from trace amounts of air pollutants we are adding are now altering local, regional, and global climates and increasing the amount of the sun's harmful ultraviolet radiation reaching the earth's surface. These planetary emergencies we have brought about are discussed in Chapter 10.

9-1 THE ATMOSPHERE

Our Air Resources The **atmosphere,** the thin envelope of life-sustaining gases surrounding the earth, is divided into several spherical layers, much like the successive layers of skin on an onion (Figure 9-1). About 95% of the mass of the earth's air is found in the innermost layer, known as the **troposphere,** extending only about 17 kilometers (11 miles) above the earth's surface. If the earth were an apple, this lower layer that contains the air we breathe would be no thicker than the apple's skin.

About 99% of the volume of clean, dry air in the troposphere consists of two gases: nitrogen (78%) and oxygen (21%). The remaining volume of air in the troposphere has slightly less than 1% argon and about 0.035% carbon dioxide. Air in the troposphere also holds water vapor in amounts varying from 0.01% by volume at the frigid poles to 5% in the humid tropics.

The second layer of the atmosphere, extending from about 17 to 48 kilometers (11 to 30 miles) above the earth's surface, is called the **stratosphere** (Figure 9-1). It contains small amounts of gaseous ozone (O_3) that filters out about 99% of the incoming harmful ultraviolet (UV) radiation (Figure 4-4, p. 61). This filtering action by the thin gauze of ozone in the stratosphere protects us from increased sunburn, skin cancer (see Spotlight on p. 180), eye cancer, and eye cataracts. This global sunscreen also prevents damage to some plants and aquatic organisms.

0-17 trop
17-48 strat

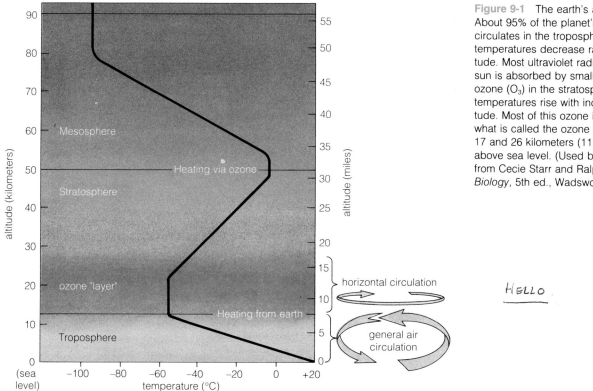

Figure 9-1 The earth's atmosphere. About 95% of the planet's mass of air circulates in the troposphere, where temperatures decrease rapidly with altitude. Most ultraviolet radiation from the sun is absorbed by small amounts of ozone (O_3) in the stratosphere, where temperatures rise with increasing altitude. Most of this ozone is found in what is called the ozone layer between 17 and 26 kilometers (11 and 16 miles) above sea level. (Used by permission from Cecie Starr and Ralph Taggart, *Biology*, 5th ed., Wadsworth, 1989)

By filtering out high-energy UV radiation, stratospheric ozone also keeps much of the oxygen in the troposphere from being converted to toxic ozone. The trace amounts of ozone that do form in the troposphere as a component of urban smog damage plants, the respiratory systems of people and other animals, and materials such as rubber.

Thus, our good health and that of many other species depends on having enough "good" ozone in the stratosphere and as little as possible "bad" ozone in the troposphere. Unfortunately, our activities are increasing ozone in the tropospheric air we must breathe and decreasing it in the stratosphere, as discussed in Section 10-4.

Disrupting the Earth's Gaseous Biogeochemical Cycles The earth's biogeochemical cycles (Section 4-4) work fine as long as we don't disrupt them by overloading them at certain points or by removing too many vital chemicals at other points. The problem is that our activities are disrupting these natural gaseous biogeochemical cycles and our impacts are growing exponentially.

We produce one-fourth as much CO_2 as nature and we are gaining as we try to burn up nature's one-time deposit of fossil fuels and clear forests as fast as possible to support a rapidly growing human population and to fuel unsustainable forms of economic growth. This has the potential to warm the

earth and alter global climate and food-producing regions (Section 10-2). We also disrupt natural energy flows and produce massive heat islands and dust domes over urban areas (Figure 6-20, p. 132).

By burning fossil fuels and using nitrogen fertilizers we are releasing three times the nitrogen oxides and gaseous ammonia into the atmosphere as natural processes in the nitrogen cycle (Figure 4-25, p. 77). In the atmosphere these nitrogen oxides are converted to nitric acid (HNO_3) that returns to the earth.

Currently we are releasing twice as much sulfur into the atmosphere as natural processes. Most of this is sulfur dioxide emitted by petroleum refining and by burning coal and oil. In the atmosphere this sulfur dioxide is converted to sulfuric acid (H_2SO_4) that returns to the earth.

These are only a few of the chemicals we are spewing into the atmosphere. Small amounts of toxic metals such as arsenic, cadmium, and lead are also circulated in the ecosphere in chemical cycles. But we now inject twice as much arsenic into the atmosphere as nature does, 7 times as much cadmium, and 17 times as much lead.

This gives you a glimpse of why many environmentalists (see quote on p. 185) believe that we may have only a few years to turn things around before we cause even larger-scale disruption of the gaseous and other chemical cycles and the energy flows that sustain life.

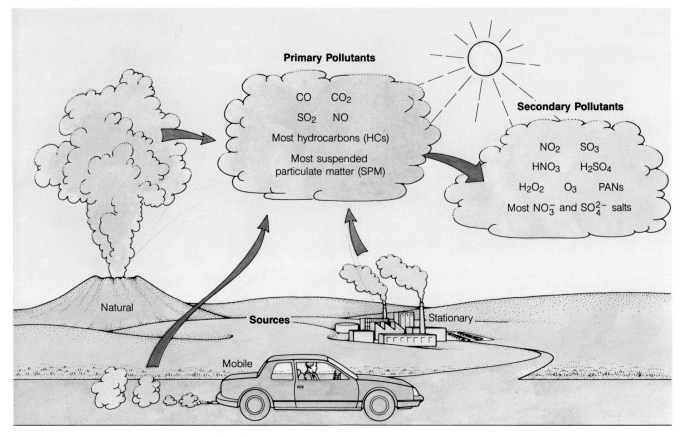

Primary Pollutants

CO   CO$_2$
SO$_2$ NO
Most hydrocarbons (HCs)
Most suspended particulate matter (SPM)

Secondary Pollutants

NO$_2$ SO$_3$
HNO$_3$ H$_2$SO$_4$
H$_2$O$_2$ O$_3$ PANs
Most NO$_3^-$ and SO$_4^{2-}$ salts

Natural

Sources

Stationary

Mobile

Figure 9-2   Primary and secondary air pollutants.

9-2 OUTDOOR AND INDOOR AIR POLLUTION

Major Types and Sources of Outdoor Air Pollution   As clean air moves across the earth's surface, it collects various chemicals produced by natural events and human activities. Once in the troposphere, these potential air pollutants mix vertically and horizontally, often reacting chemically with each other or with natural components of the atmosphere (Figure 9-1). Air movements and turbulence help dilute potential pollutants, but long-lived pollutants are transported great distances before they return to the earth's surface as solid particles, liquid droplets, or chemicals dissolved in precipitation.

Hundreds of air pollutants are found in the troposphere. However, trace amounts of nine major classes of pollutants cause most outdoor air pollution:

1. *carbon oxides*—carbon monoxide (CO) and carbon dioxide (CO$_2$)

2. *sulfur oxides*—sulfur dioxide (SO$_2$) and sulfur trioxide (SO$_3$)

3. *nitrogen oxides*—nitric oxide (NO), nitrogen dioxide (NO$_2$), and nitrous oxide (N$_2$O)

4. *volatile organic compounds (VOCs)*—hundreds of compounds such as methane (CH$_4$), benzene (C$_6$H$_6$), chlorofluorocarbons (CFCs), and bromine-containing halons

5. *suspended particulate matter (SPM)*—thousands of different types of *solid particles* such as dust (soil), soot (carbon), asbestos, and lead, arsenic, cadmium, nitrate (NO$_3^-$) and sulfate (SO$_4^{2-}$) salts, and *liquid droplets* of chemicals such as sulfuric acid (H$_2$SO$_4$), oil, PCBs, dioxins, and various pesticides

6. *photochemical oxidants*—ozone (O$_3$), PANs (peroxyacyl nitrates), hydrogen peroxide (H$_2$O$_2$), hydroxy radicals (OH), and aldehydes such as formaldehyde (CH$_2$O) formed in the atmosphere by the reaction of oxygen, nitrogen oxides, and volatile hydrocarbons under the influence of sunlight

7. *radioactive substances*—radon-222, iodine-131, strontium-90, plutonium-239, and other radioisotopes that enter the atmosphere as gases or suspended particulate matter

8. *heat*—produced when any kind of energy is transformed from one form to another, especially when fossil fuels are burned in cars, factories, homes, and power plants (Figure 6-20, p. 132)

9. *noise*—produced by motor vehicles, airplanes, trains, industrial machinery, construction machinery, lawn mowers, vacuum cleaners, food, sirens, earphones, radios, cassette players, and live concerts (Table 6-1, p. 133)

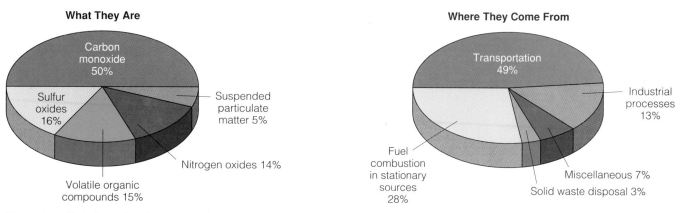

What They Are

Carbon monoxide 50%

Sulfur oxides 16%

Suspended particulate matter 5%

Nitrogen oxides 14%

Volatile organic compounds 15%

Where They Come From

Transportation 49%

Industrial processes 13%

Fuel combustion in stationary sources 28%

Miscellaneous 7%

Solid waste disposal 3%

Figure 9-3 Emissions of major outdoor air pollutants in the United States. (Data from Environmental Protection Agency)

A **primary air pollutant,** such as sulfur dioxide, directly enters the air as a result of natural events or human activities. A **secondary air pollutant,** such as sulfuric acid, is one that is formed in the air through a chemical reaction between a primary pollutant and one or more air components (Figure 9-2).

Most of the widely recognized outdoor air pollution in the United States (and other industrialized countries) comes from five groups of pollutants: carbon monoxide, nitrogen oxides, sulfur oxides, volatile organic compounds (mostly hydrocarbons), and suspended particulate matter (Figure 9-3). Other key pollutants are ozone and lead (mostly from burning leaded gasoline).

In MDCs most of these pollutants are emitted into the atmosphere from the burning of fossil fuels in power and industrial plants (*stationary sources*) and in motor vehicles (*mobile sources*). In LDCs, especially in rural areas where over half of the world's people live, most air pollution is produced by the burning of wood, dung, and crop residues in inefficient crude stoves and open fires. This adds carbon dioxide and soot to the atmosphere and forces the desperately poor to deplete forests for enough fuelwood to survive (Section 15-2).

Recently it has been recognized that trace amounts of hundreds of *unconventional outdoor air pollutants* may be a long-term threat to human health when inhaled over long periods. A study revealed that in 1988 at least 1.1 billion kilograms (2.4 billion pounds) of 320 toxic compounds, 60 of them known carcinogens, were released by industries into American skies. This does not cover perhaps an equal amount of toxic chemicals released from 178 million motor vehicles, thousands of toxic-waste dumps, and tens of thousands of small businesses such as dry cleaners and gas stations. All of these emissions are legal because in 1988 only 7 unconventional air pollutants were regulated by federal standards. The EPA estimates

that these pollutants are responsible for 2,000 excess cancer deaths a year in the United States.

Types and Sources of Indoor Air Pollution High concentrations of toxic air pollutants can also build up indoors, where most people spend 70% to 98% of their time (Figure 9-5). Indeed, scientists recently have found that the air inside many homes, schools, office buildings, factories, cars, and airliners in the United States (and presumably most MDCs) is more polluted and dangerous than outdoor air on a smoggy day. Indoor air pollution poses an especially high risk for the elderly, the very young, the sick, and factory workers who spend a large amount of time indoors.

In 1985 the EPA reported that toxic chemicals found in almost every American home are three times more likely to cause some type of cancer than outdoor air pollutants. Other air pollutants found in buildings produce dizziness, headaches, coughing, sneezing, burning eyes, and flulike symptoms in many people—a health problem called the "sick building syndrome."

An estimated one-fifth to one-third of all U.S. buildings, including the EPA headquarters, are considered "sick." Each year exposure to pollutants inside factories and businesses in the United States kills from 100,000 to 210,000 workers prematurely. According to the EPA and public health officials, cigarette smoke (see Case Study on p. 182), radioactive radon-222 gas (see Case Study on p. 190), and asbestos are the three most dangerous indoor air pollutants.

9-3 SMOG AND ACID DEPOSITION

Smog: Cars + Sunlight = Tears A mixture of dozens of primary pollutants and secondary pollutants formed when some of the primary pollutants interact under the influence of sun-

Radon-222 is a colorless, odorless, tasteless, naturally occurring radioactive gas produced by the radioactive decay of uranium-238. Small amounts of radon-producing uranium-238 are found in most soil and rock, but this isotope is much more concentrated in underground deposits of uranium, phosphate, granite, and shale rock.

When radon gas from such deposits percolates upward to the soil and is released outdoors, it disperses quickly in the atmosphere and decays to harmless levels. However, when the gas seeps or is drawn into buildings through cracks, drains, hollow concrete blocks, and drains in basements or into water in underground wells over such deposits, it can build up to high levels (Figure 9-4). Stone and other building materials obtained from radon-rich deposits can also be a source of indoor radon contamination.

Radon-222 gas quickly decays into solid particles of other radioactive elements that can be inhaled, exposing lung tissue to a large amount of ionizing radiation from alpha particles (Figure 3-8, p. 44). Smokers are especially vulnerable because the inhaled radioactive particles tend to adhere to tobacco tar deposits in the lungs and upper respiratory tract. Repeated exposure to these radioactive particles over 20 to 30 years can cause lung cancer.

Data from a sampling of indoor radon levels taken mostly in basements or crawl spaces of houses in 30 states since 1986 found harmful levels in roughly one out every four homes tested. Iowa ranked No. 1 in these radon surveys, with 71% of the homes tested exceeding safe levels. In Pennsylvania radon levels in the home of one family created a cancer risk equal to having 455,000 chest X rays a year. In 1989 the EPA reported that about 54% of 130 schools tested had dangerous levels of radon.

According to studies by the EPA and the National Research Council, prolonged exposure to high levels of radon over a 70-year lifetime causes up to 20,000 of the 136,000 lung cancer deaths each year in the United States, with 85% of these being due to a combination of radon and smoking. Radon released from water obtained from groundwater near radon-laden rock and then heated and used for showers and washing clothes may be responsible for 50 to 400 of these premature deaths.

Because radon hot spots can occur almost anywhere, it is impossible to know which buildings have unsafe levels of radon without carrying out tests. In 1988 the EPA and the U.S. Surgeon General's Office recommended that everyone living in a detached house, town house, mobile home, or first three floors of an apartment building test for radon. By 1990, however, less than 3% of U.S. households had conducted such tests.

Unsafe levels can build up easily in a superinsulated or airtight home unless the building has an air-to-air heat exchanger to change indoor air without losing much heat. Some tests also indicate higher levels in houses with electric heat. Homeowners with wells should also have their water tested for radon.

Individuals can measure radon levels in their homes or other buildings with radon detection kits that can be bought in many hardware stores and supermarkets or from mail-order firms for $10 to $50. Pick one that is EPA approved and mail them to an EPA-certified testing laboratory to get the test result.* Alpha track detectors, which must be left in place for a month or more, are the most reliable devices.

* For information see the article, ''Radon Detectors: How To Find Out If Your House Has a Radon Problem,'' in the July 1987 issue of *Consumer Reports*.

If testing reveals an unacceptable level (over four picocuries of radiation per liter of air), the EPA recommends several ways to reduce radon levels and health risks.† The first is to stop all indoor smoking or at least confine it to a well-ventilated room. The next corrective measure is natural ventilation, especially leaving basement windows partially open or crawl-space vents open. Also cracks in basement walls and floors and around pipes and joints between floors and walls should be sealed. But this does little good if foundation or basement walls are constructed of porous, hollow-core concrete blocks, unless the interior surfaces of these blocks are liberally coated with latex paint or a concrete topcoat.

Air-to-air heat exchangers ($300 to $2,500) can be installed to remove radon if radiation levels are not above 10 picocuries per liter of air. These devices also remove most other indoor air pollutants. Also large positive-ion generators costing about $400 can help in house with low to moderate levels.

For houses with serious radon gas problems, special venting systems usually have to be installed below the foundations at a cost of $700 to $2,500. To remove radon from contaminated well water, a special type of activated carbon filter can be added to holding tanks at a cost of about $1,000. Contact the state radiological health office or regional EPA office to get a list of approved contractors and avoid unscrupulous radon testing and repair firms.

In Sweden no house can be built until the lot has been tested

† A free copy of *Radon Reduction Methods* can be obtained from the Environmental Protection Agency, 401 M St. SW, Washington, D.C. 20460. A free copy of *Radon Reduction in New Construction* is available from state radiation-protection offices or the National Association of Home Builders, Attention: William Young, 15th and M Streets NW, Washington, D.C. 20005.

for radon. If the reading is high, the builder must follow government-mandated construction procedures to ensure that the house won't be contaminated.

Environmentalists urge enactment of a similar building-code program for all new construction in the United States. They also suggest that before buying a lot to build a new house, individuals should have the soil tested for radon.

Similarly, no one should buy an existing house unless it has been tested for radon by certified personnel, just as houses must now be inspected for termites. People building a new house should insist that the contractor use relatively simple construction practices that prevent harmful buildup of radon and add only $100 to $1,000 to the construction cost. This includes using foundation materials such as solid concrete blocks or poured concrete walls and installing a heat-bonded nylon mat (called Enkavent and costing $450 to $650) under the slab of a house during construction. Has the building where you live or work been tested for radon?

Figure 9-4 Sources of indoor radon-222 gas and comparable risks of exposure to various levels of this radioactive gas for a lifetime of 70 years. Levels are those in an actual living area, not a basement or crawl space where levels are much higher. Smokers have the highest risk of getting lung cancer from a combination of prolonged exposure to cigarette smoke and radon-222 gas. (Data from Environmental Protection Agency)

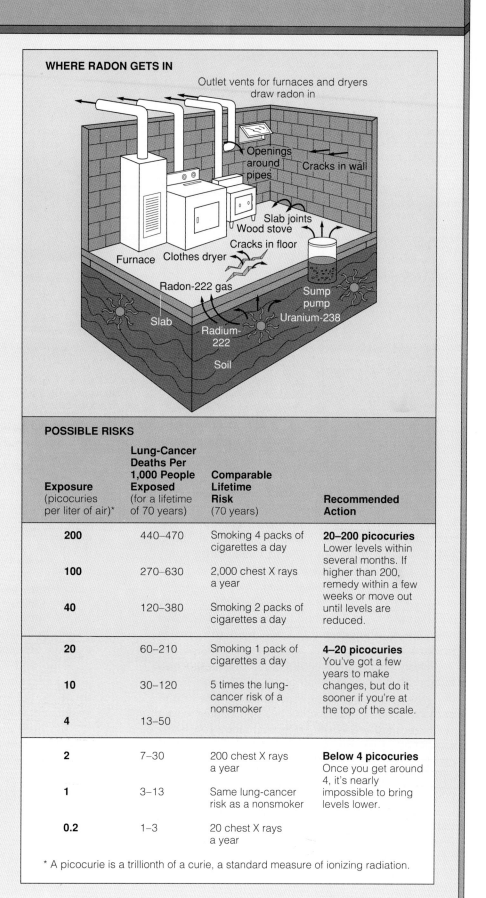

WHERE RADON GETS IN

POSSIBLE RISKS

Exposure (picocuries per liter of air)*	Lung-Cancer Deaths Per 1,000 People Exposed (for a lifetime of 70 years)	Comparable Lifetime Risk (70 years)	Recommended Action
200	440–470	Smoking 4 packs of cigarettes a day	**20–200 picocuries** Lower levels within several months. If higher than 200, remedy within a few weeks or move out until levels are reduced.
100	270–630	2,000 chest X rays a year	
40	120–380	Smoking 2 packs of cigarettes a day	
20	60–210	Smoking 1 pack of cigarettes a day	**4–20 picocuries** You've got a few years to make changes, but do it sooner if you're at the top of the scale.
10	30–120	5 times the lung-cancer risk of a nonsmoker	
4	13–50		
2	7–30	200 chest X rays a year	**Below 4 picocuries** Once you get around 4, it's nearly impossible to bring levels lower.
1	3–13	Same lung-cancer risk as a nonsmoker	
0.2	1–3	20 chest X rays a year	

* A picocurie is a trillionth of a curie, a standard measure of ionizing radiation.

Figure 9-5 Some major indoor air pollutants.

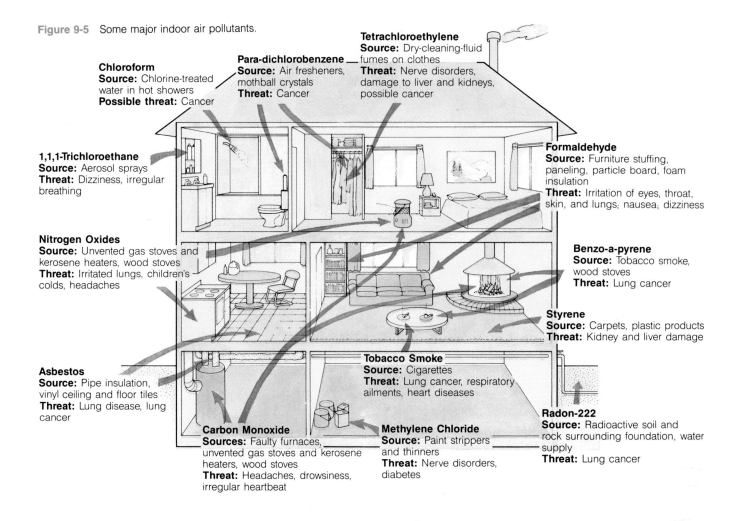

Chloroform
Source: Chlorine-treated water in hot showers
Possible threat: Cancer

Para-dichlorobenzene
Source: Air fresheners, mothball crystals
Threat: Cancer

Tetrachloroethylene
Source: Dry-cleaning-fluid fumes on clothes
Threat: Nerve disorders, damage to liver and kidneys, possible cancer

1,1,1-Trichloroethane
Source: Aerosol sprays
Threat: Dizziness, irregular breathing

Formaldehyde
Source: Furniture stuffing, paneling, particle board, foam insulation
Threat: Irritation of eyes, throat, skin, and lungs; nausea; dizziness

Nitrogen Oxides
Source: Unvented gas stoves and kerosene heaters, wood stoves
Threat: Irritated lungs, children's colds, headaches

Benzo-a-pyrene
Source: Tobacco smoke, wood stoves
Threat: Lung cancer

Styrene
Source: Carpets, plastic products
Threat: Kidney and liver damage

Asbestos
Source: Pipe insulation, vinyl ceiling and floor tiles
Threat: Lung disease, lung cancer

Carbon Monoxide
Sources: Faulty furnaces, unvented gas stoves and kerosene heaters, wood stoves
Threat: Headaches, drowsiness, irregular heartbeat

Tobacco Smoke
Source: Cigarettes
Threat: Lung cancer, respiratory ailments, heart diseases

Methylene Chloride
Source: Paint strippers and thinners
Threat: Nerve disorders, diabetes

Radon-222
Source: Radioactive soil and rock surrounding foundation, water supply
Threat: Lung cancer

light is called **photochemical smog** (Figure 9-6). Virtually all modern cities have photochemical smog, but it is much more common in those with sunny, warm, dry climates and lots of motor vehicles. Cities with serious photochemical smog include Los Angeles, Denver, Salt Lake City, Sydney, Mexico City (see Figure 6-17, p. 129), and Buenos Aires. The worst episodes of photochemical smog tend to occur in summer.

Thirty years ago cities like London, Chicago, and Pittsburgh burned large amounts of coal and heavy oil, which contain sulfur impurities, in power and industrial plants and for space heating. During winter such cities suffered from **industrial smog** consisting mostly of a mixture of sulfur dioxide, suspended droplets of sulfuric acid formed from some of the sulfur dioxide, and a variety of suspended solid particles. Today coal and heavy oil are burned only in large boilers and with reasonably good control so that industrial smog, sometimes called gray-air smog, is rarely a problem. However, this is not the case in China and in some eastern European countries, such as Poland and Czechoslovakia, where large quantities of coal are burned with inadequate controls.

Local Climate, Topography, and Smog The frequency and severity of smog in an area depend on the local climate and topography, the density of population and industry, and the major fuels used in industry, heating, and transportation. In areas with high average annual precipitation, rain and snow help cleanse the air of pollutants. Winds also help sweep pollutants away and bring in fresh air, but may transfer some pollutants to distant areas.

Hills and mountains tend to reduce the flow of air in valleys below and allow pollutant levels to build up at ground level. Buildings in cities also slow wind speed and reduce dilution and removal of pollutants.

During the day the sun warms the air near the earth's surface. Normally this heated air expands and rises, diluting low-lying pollutants and carrying them higher into the troposphere. Colder, denser air from surrounding high-pressure areas then sinks into the low-pressure area created when the hot air rises (Figure 9-7, left). This continual mixing of the air helps keep pollutants from reaching dangerous levels in the air near the ground.

Sometimes weather conditions trap a layer of dense, cool air beneath a layer of less dense, warm

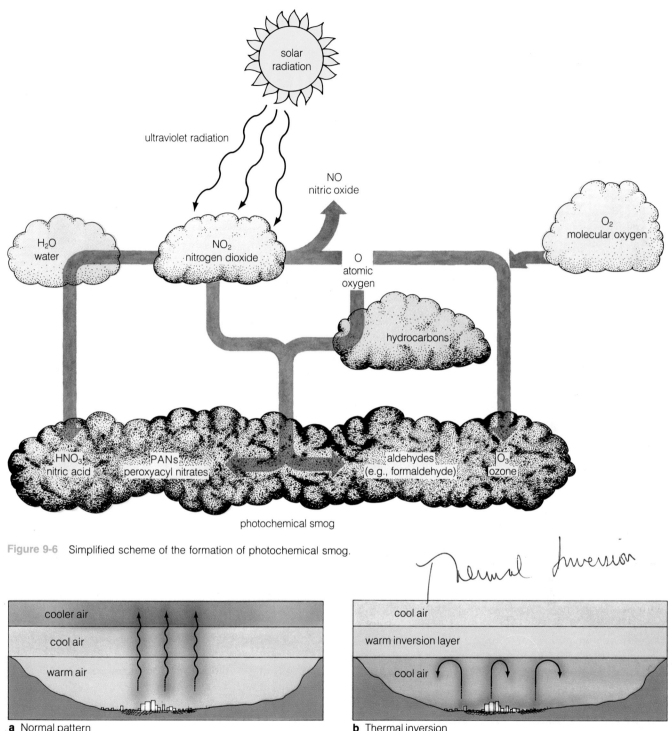

Figure 9-6 Simplified scheme of the formation of photochemical smog.

Thermal Inversion

a Normal pattern

cooler air

cool air

warm air

b Thermal inversion

cool air

warm inversion layer

cool air

Figure 9-7 Thermal inversion traps pollutants in a layer of cool air that cannot rise to carry the pollutants away.

air in an urban basin or valley. This is called a **temperature inversion, or thermal inversion** (Figure 9-7, right). In effect, a warm-air lid covers the region and prevents pollutants from escaping in upward-flowing air currents. Usually these inversions last for only a few hours, but sometimes they last for several days when a high-pressure air mass stalls over an

area. Then air pollutants at ground level build up to harmful and even lethal levels. Thermal inversions also enhance the harmful effects of urban heat islands and dust domes that build up over urban areas (Figure 6-20, p. 132).

In 1948 a lengthy thermal inversion over Donora, an industrial town in Pennsylvania, killed 20 people

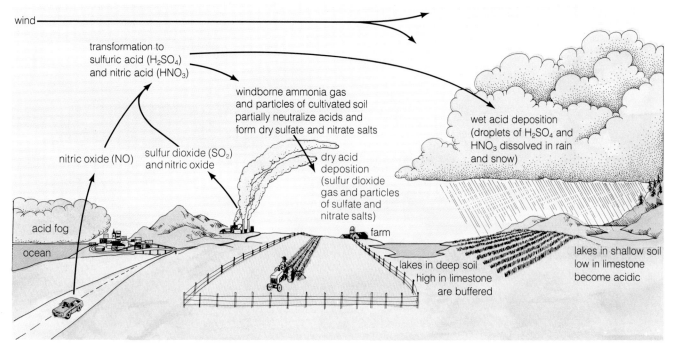

wind

transformation to
sulfuric acid (H₂SO₄)
and nitric acid (HNO₃)

windborne ammonia gas
and particles of cultivated soil
partially neutralize acids and
form dry sulfate and nitrate salts

wet acid deposition
(droplets of H₂SO₄ and
HNO₃ dissolved in rain
and snow)

nitric oxide (NO) sulfur dioxide (SO₂)
and nitric oxide

dry acid
deposition
(sulfur dioxide
gas and particles
of sulfate and
nitrate salts)

acid fog

ocean

farm

lakes in shallow soil
low in limestone
become acidic

lakes in deep soil
high in limestone
are buffered

Figure 9-8 Acid deposition.

and made 6,000 of the town's 14,000 inhabitants sick. A prolonged thermal inversion over New York City in 1963 killed 300 people and injured thousands.

Thermal inversions occur more often and last longer over towns or cities located in valleys surrounded by mountains (Donora, Pennsylvania, and Mexico City, Figure 6-17, p. 129), on the leeward sides of mountain ranges (Denver), and near coasts (New York City and Los Angeles).

A city with several million people and automobiles in an area with a sunny climate, light winds, mountains on three sides, and the ocean on the other, has the ideal conditions for photochemical smog worsened by frequent thermal inversions. This describes the Los Angeles basin. It has almost daily inversions, many of them prolonged during the summer months, 13.8 million people, 8.5 million cars, and thousands of factories. Despite having the world's toughest air pollution control program, Los Angeles is the air pollution capital of the United States.

Acid Deposition When electric power plants and industrial plants burn coal and oil, their smokestacks emit large amounts of sulfur dioxide, suspended particulate matter, and nitrogen oxides. To reduce local air pollution and meet government standards without having to add expensive air pollution control devices, power plants and industries began using tall smokestacks (see photo on p. 185) to spew pollutants above the inversion layer. As more power plants and industries began using this fairly cheap output approach to controlling local pollution in the 1960s and 1970s, pollution in downwind areas began to rise.

As emissions of sulfur dioxide and nitric oxide from stationary sources are transported long distances by winds, they form secondary pollutants such as nitrogen dioxide, nitric acid vapor, and droplets containing solutions of sulfuric acid and sulfate and nitrate salts (Figure 9-2). These chemicals descend to the earth's surface in wet form as acid rain or snow and in dry form as gases, fog, dew, or solid particles. The combination of dry deposition and wet deposition of acids and acid-forming compounds onto the surface of the earth is known as **acid deposition,** commonly called acid rain (Figure 9-8). Other contributions to acid deposition come from emissions of nitric oxide from massive numbers of automobiles in major urban areas.

Different levels of acidity and basicity of water solutions of substances are commonly expressed in terms of **pH** (Figure 9-9). A neutral solution has a pH of 7; one with a pH greater than 7 is basic, or alkaline; and one with a pH less than 7 is acidic. The lower the pH below 7, the more acidic the solution. Each whole-number decrease in pH represents a tenfold increase in acidity.

Natural precipitation varies in acidity, with an average pH of 5.0 to 5.6, but the average rain in the eastern United States is as acidic as tomato juice with a pH of 4.3. Precipitation in some areas is more than ten times as acidic with a pH of 3—as acidic as vinegar (Figure 9-9). Some cities and mountaintops downwind from cities are bathed in acid fog as acidic as lemon juice with a pH of 2.3.

Acid deposition has a number of harmful effects, especially when the pH falls below 5.1, including:

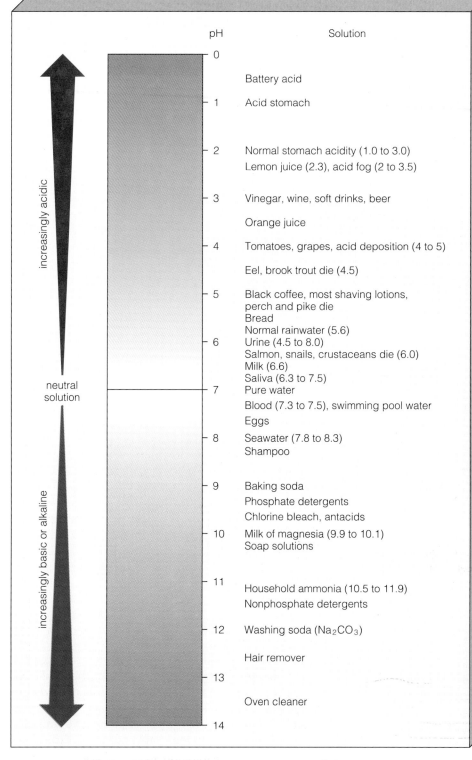

Figure 9-9 Scale of pH, used to measure acidity and alkalinity of water solutions. Values shown are approximate.

pH	Solution
0	
	Battery acid
1	Acid stomach
2	Normal stomach acidity (1.0 to 3.0)
	Lemon juice (2.3), acid fog (2 to 3.5)
3	Vinegar, wine, soft drinks, beer
	Orange juice
4	Tomatoes, grapes, acid deposition (4 to 5)
	Eel, brook trout die (4.5)
5	Black coffee, most shaving lotions, perch and pike die
	Bread
	Normal rainwater (5.6)
6	Urine (4.5 to 8.0)
	Salmon, snails, crustaceans die (6.0)
	Milk (6.6)
	Saliva (6.3 to 7.5)
7	Pure water
	Blood (7.3 to 7.5), swimming pool water
	Eggs
8	Seawater (7.8 to 8.3)
	Shampoo
9	Baking soda
	Phosphate detergents
	Chlorine bleach, antacids
10	Milk of magnesia (9.9 to 10.1)
	Soap solutions
11	Household ammonia (10.5 to 11.9)
	Nonphosphate detergents
12	Washing soda (Na_2CO_3)
	Hair remover
13	
	Oven cleaner
14	

increasingly acidic

neutral solution

increasingly basic or alkaline

- Damaging statues, buildings, metals, and car finishes.
- Killing fish, aquatic plants, and microorganisms in lakes and streams.
- Weakening or killing trees, especially conifers at high elevations, by leaching calcium, potassium, and other plant nutrients from soil (Figure 9-10).

- Damaging tree roots and killing many kinds of fish by releasing ions of aluminum, lead, mercury, and cadmium from soil and bottom sediments (Figure 9-10).

Weakening trees and making them more susceptible to attacks by diseases, insects, drought, and fungi and mosses that thrive under acidic conditions (Figure 9-10).

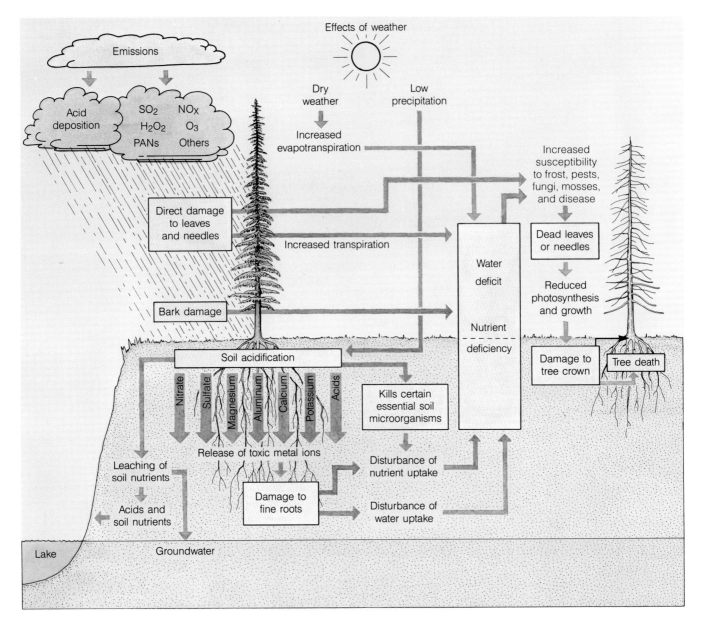

Figure 9-10 Harmful effects of air pollutants on trees.

- Stunting the growth of crops such as tomatoes, soybeans, spinach, carrots, broccoli, and cotton.
- Leaching toxic metals such as copper and lead from city and home water pipes into drinking water.
- Causing and aggravating many human respiratory diseases and leading to premature death. Dr. Philip Landrigan at New York's Mt. Sinai School of Medicine estimates that acid deposition is the third largest cause of lung disease in the United States, after smoking and indoor radon.

Acid deposition illustrates the threshold or straw-that-broke-the-camel's-back effect. Most soils, lakes, and streams contain alkaline or basic chemicals that can react with a certain amount of acids and thus neutralize them. But repeated exposure to acids year after year can deplete most of these acid-buffering chemicals. Then suddenly large numbers of trees start dying and most fish in a lake or stream die when exposed to the next year's input of acids. By this time it is 10 to 20 years too late to prevent serious damage.

Acid deposition is already a serious problem in northern and central Europe, the northeastern United States (Figure 9-11), southeastern Canada, and parts of China, Brazil, and Nigeria. It is emerging as a problem in heavily industrialized parts of Asia, Latin America, and Africa and in parts of the western United States (mostly from dry deposition).

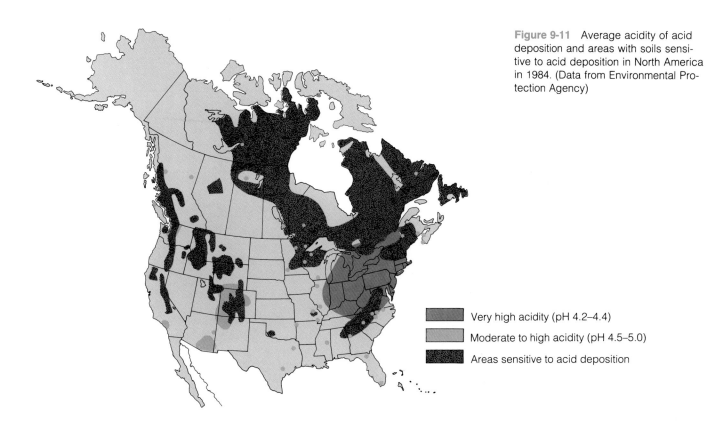

Figure 9-11 Average acidity of acid deposition and areas with soils sensitive to acid deposition in North America in 1984. (Data from Environmental Protection Agency)

Very high acidity (pH 4.2–4.4)

Moderate to high acidity (pH 4.5–5.0)

Areas sensitive to acid deposition

A large portion of the acid-producing chemicals produced in one country is exported to others by prevailing surface winds. For example, over three-fourths of the acid deposition in Norway, Switzerland, Austria, Sweden, the Netherlands, and Finland is blown to these countries from industrialized areas of western and eastern Europe.

More than half the acid deposition in heavily populated southeastern Canada and in the eastern United States originates from emissions from the heavy concentration of coal- and oil-burning power and industrial plants in seven central and upper Midwest states—Ohio, Indiana, Pennsylvania, Illinois, Missouri, West Virginia, and Tennessee (Figure 9-11). Canada produces almost twice as much sulfur dioxide per person and per unit of energy consumed as the United States and thus exports some acid deposition to the northeastern United States (see photo on p. 185). However, because U.S. production of sulfur dioxide and nitrogen oxides is so massive, Canada receives much more acid deposition from the United States than it exports across the border.

The large net flow of acid deposition from the United States to Canada is straining relations between the two countries. Canadians have become increasingly alarmed about the damage this is doing to their lakes, streams, and forests, threatening jobs for 600,000 of its citizens. So far the United States has refused to do much about this chemical assault on its neighbor, citing the need for more research. Research is very important, but it can be used by politicians as a way to avoid making difficult decisions, leading to "paralysis by analysis."

9-4 EFFECTS OF AIR POLLUTION ON LIVING ORGANISMS AND MATERIALS

Damage to Human Health Your respiratory system has a number of mechanisms that help protect you from air pollution. Hairs in your nose filter out large particles. Sticky mucus in the lining of your upper respiratory tract captures small particles and dissolves some gaseous pollutants. Automatic sneezing and coughing mechanisms expel contaminated air and mucus when your respiratory system is irritated by pollutants. Your upper respiratory tract is lined with hundreds of thousands of tiny, mucus-coated hairs, called cilia. They continually wave back and forth, transporting mucus and the pollutants they trap to your mouth, where it is either swallowed or expelled.

Years of smoking and exposure to air pollutants can overload or deteriorate these natural defenses, causing or contributing to a number of respiratory diseases such as lung cancer, chronic bronchitis, and

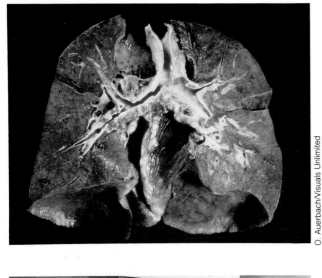

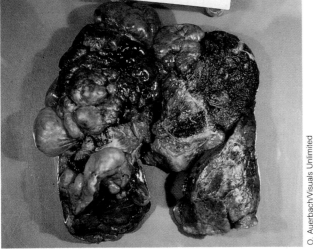

Figure 9-12 Normal appearance of human lung tissue (top) and appearance of lung taken from a person who died from emphysema (bottom).

The air pollution capital of the world may be Cubatao, an hour's drive south of São Paulo, Brazil. This city of 100,000 people lies in a coastal valley that has frequent thermal inversions. Residents call the area "the valley of death."

In this heavily industrialized city scores of plants spew thousands of tons of pollutants a day into the frequently stagnant air. More babies are born deformed there than anywhere else in Latin America.

In one recent year 13,000 of the 40,000 people living in the downtown core area suffered from respiratory disease. One resident says, "On some days if you go outside, you will vomit." The mayor refuses to live in the city.

Most residents would like to live somewhere else, but they need the jobs available in the city and cannot afford to move. The government has begun some long overdue efforts to control air pollution but has far to go. Meanwhile the poor continue to pay the price of this form of economic progress: bad health and premature death.

emphysema. Elderly people, infants, pregnant women, and persons with heart disease, asthma, or other respiratory diseases are especially vulnerable to air pollution. Recent evidence on test animals indicates that nitrogen dioxide—a common pollutant from automobile exhaust—may encourage the spread of cancer throughout the body, especially deadly melanoma (Figure 8-7c, p. 180). [handwritten: NO₂]

Fine particles are particularly hazardous to human health because they are small enough to penetrate the lung's natural defenses. They can also bring with them droplets or other particles of toxic or cancer-causing pollutants that become attached to their surfaces.

Inhaling the ozone found in photochemical smog (Figure 9-6) causes coughing, shortness of breath, nose and throat irritation and discomfort, and aggravates chronic diseases such as asthma, bronchitis, emphysema, and heart trouble. Outdoor exercise in areas where ozone levels exceed safe levels (0.12 parts

per million per hour) amplifies these effects. Many U.S. cities frequently exceed these levels, especially during warm weather.

Emphysema is an incurable condition that reduces the ability of the lungs to transfer oxygen to the blood so that the slightest exertion causes acute shortness of breath (Figure 9-12). Prolonged smoking and exposure to air pollutants can cause emphysema in anyone. However, about 2% of emphysema cases are caused by a defective gene that reduces the elasticity of the air sacs of the lungs. Anyone with this hereditary condition, for which testing is available, should not smoke and should not live or work in a highly polluted area.

The World Health Organization estimates that worldwide nearly 1 billion urban dwellers—almost one of every five people on earth—are being exposed to health hazards from air pollutants (see Spotlight above). An estimated 60% of the people living in Calcutta suffer from respiratory diseases related to air pollution.

In 1989 the EPA estimated that roughly 150 million Americans, 60% of the population, were breathing unsafe air. The congressional Office of Technology Assessment estimates that some 50,000 premature deaths occur in the United States each year due to respiratory or cardiac problems caused or aggravated by current air pollution levels. The American Lung

[handwritten: 60% unsafe air]

Figure 9-13 Injury to ponderosa pine needles from exposure to ozone and other pollutants in photochemical smog.

120,000 die air pollution

Association estimates that up to 120,000 Americans die each year as a result of air pollution.

According to the EPA air pollution costs the United States at least $110 billion annually in health care and lost work productivity. About $100 billion of this is caused by indoor air pollution—the problem we have focused on the least.

Damage to Plants Chronic exposure of leaves and needles to air pollutants can break down the waxy coating that helps prevent excessive water loss and damage from diseases, pests, drought, and frost. *Cuticle* Such exposure also interferes with photosynthesis and plant growth, reduces nutrient uptake, and causes leaves or needles to turn yellow or brown and drop off (Figure 9-13). Spruce, fir, and other coniferous trees, especially at high elevations, are highly vulnerable to the effects of air pollution because of their long life spans and the year-round exposure of their needles to polluted air.

In addition to causing direct leaf and needle damage, acid deposition can leach vital plant nutrients such as calcium, magnesium, and potassium from *leach* the soil and kill essential soil microorganisms (Figure 9-10). It also releases aluminum ions, which are normally bound to soil particles, into soil water. There they damage fine root filaments, reduce uptake of water and nutrients from the soil, and make trees more vulnerable to drought, frost, insects, fungi, mosses, and disease (Figure 9-10). This indirect damage that weakens trees is believed to be much more threatening than the direct damage from air pollution. Prolonged exposure to high levels of multiple air pollutants can kill all trees and most other vegetation in an area (Figure 9-14).

The effects of chronic exposure of trees to multiple air pollutants may not be visible for several decades. Then suddenly large numbers begin dying off because of soil nutrient depletion and increased susceptibility to pests, diseases, fungi, mosses, and drought.

Figure 9-14 Sulfur dioxide and other fumes from a copper smelter that operated for 52 years near Ducktown and Copperhill, Tennessee, killed the forest once found on this land and left a desert in its place. After decades of replanting, some vegetation has returned to the area, but recovery has been slow because of severe soil erosion.

Figure 9-15 Tree death and damage in this coniferous forest near the top of Mt. Mitchell, North Carolina, is believed to be the result of long-term exposure to multiple air pollutants, which made the trees more vulnerable to disease and drought.

waldsterben

That is what is happening to about 35% of the forested area in 28 European countries. The phenomenon, known as *waldsterben* (forest death), turns whole forests into stump-studded meadows. The four European countries with the highest percentages of their forests damaged are Czechoslovakia (71%), Greece (64%), the United Kingdom (64%), and West Germany (52%).

Similar diebacks in the United States have occurred in mostly coniferous forests on high-elevation slopes facing moving air masses. The most seriously affected areas are the Appalachian Mountains from Georgia to New England. By 1988 most spruce, fir, and other conifers atop North Carolina's Mt. Mitchell, the highest peak in the East, were dead from being bathed in ozone and acid fog for years (Figure 9-15). The soil was so acidic that new seedlings could not survive.

Table 9-1   Harmful Effects of Air Pollution on Materials

Material	Effects	Principal Air Pollutants
Stone and concrete	Surface erosion, discoloration, soiling	Sulfur dioxide, sulfuric acid, nitric acid, particulate matter
Metals	Corrosion, tarnishing, loss of strength	Sulfur dioxide, sulfuric acid, nitric acid, particulate matter, hydrogen sulfide
Ceramics and glass	Surface erosion	Hydrogen fluoride, particulate matter
Paints	Surface erosion, discoloration, soiling	Sulfur dioxide, hydrogen sulfide, ozone, particulate matter
Paper	Embrittlement, discoloration	Sulfur dioxide
Rubber	Cracking, loss of strength	Ozone
Leather	Surface deterioration, loss of strength	Sulfur dioxide
Textile fabrics	Deterioration, fading, soiling	Sulfur dioxide, nitrogen dioxide, ozone, particulate matter

Plant pathologist Robert Bruck warns that damage to mountaintop forests is an early warning that many tree species at lower elevations may soon die or be damaged by prolonged exposure to air pollution. Many scientists fear that elected officials in the United States will continue to delay establishing stricter controls on major forms of air pollution until it is too late to prevent a severe loss of valuable forest resources like that happening in Europe.

Air pollution, especially by ozone, also threatens some types of crops. In the United States it is estimated that reduced crop yields as a result of air pollution causes an economic loss of about $5.4 billion a year.

Damage to Aquatic Life   Acid deposition has a severe harmful impact on the aquatic life of freshwater lakes in areas where surrounding soils have little acid-buffering capacity (Figure 9-11). Much of the damage to aquatic life in the Northern Hemisphere is a result of *acid shock*. Acid shock is caused by the sudden runoff of large amounts of highly acidic water (along with toxic aluminum leached from the soil) into lakes when snow melts in the spring or when heavy rains follow a period of drought. The aluminum leached from the soil and lake sediment kills fish by clogging their gills.

In Norway and Sweden at least 68,000 lakes either contain no fish or have lost most of their acid-buffering capacity because of excess acidity. In Canada the Department of the Environment reports some 14,000 lakes are almost fishless, and another 150,000 are in peril because of excess acidity.

In the United States tne more than 1,000 acidified lakes 4 hectares (10 acres) or larger are concentrated in the Northeast and Upper Middle West (mostly in parts of Minnesota, Michigan, Wisconsin, and the upper Great Lakes). About one-fourth of the lakes and ponds in New York's Adirondack Mountains are too acidic to support fish. Another 20% have lost most of their acid-neutralizing capacity. About 2.7% of the nation's streams are acidified, including half the streams of the mid-Atlantic coastal plain.

Damage to Materials   Each year air pollutants cause tens of millions of dollars in damage to various materials (Table 9-1). The fallout of soot and grit on buildings, cars, and clothing requires costly cleaning. Irreplaceable marble statues, historic buildings, and stained-glass windows throughout the world have been pitted and discolored by air pollutants (Figure 9-16).

9-5   CONTROLLING AIR POLLUTION

U.S. Air Pollution Legislation   Air pollution or any other type of pollution can be controlled by laws to establish desired standards and by technology to achieve the standards. In the United States, Congress passed the Clean Air Acts of 1970 and 1977, which gave the federal government considerable power to control air pollution.

These laws required the EPA to establish **national ambient air quality standards (NAAQS)** for seven major outdoor pollutants: suspended particulate matter, sulfur oxides, carbon monoxide, nitrogen oxides, ozone, hydrocarbons, and lead. Each standard specifies the maximum allowable level, averaged over a specific time period, for a certain pollutant in outdoor (ambient) air. Each state is required to develop and enforce an implementation plan for attainment of these standards.

The EPA has also established a policy of *prevention of significant deterioration (PSD)*. It is designed to prevent a decrease in air quality in regions where the air is cleaner than required by the NAAQS for suspended particulate matter and sulfur dioxide. Otherwise, industries would move into these areas and gradually degrade air quality to the national standards for these two major pollutants.

The EPA is also required to establish *national emission standards* for less common air pollutants capable of causing serious harm to human health at low concentrations. Scientists have identified at least 600 potentially hazardous air pollutants. However, by 1988 the EPA had established emission standards for only seven hazardous air pollutants.

Part of the problem is the difficulty of getting accurate scientific data on the effects of specific pollutants on human health (Section 8-1). Economic and political pressures also hamper the EPA's work. During the 1980s Congress slashed EPA budgets and the Reagan administration reduced enforcement of air pollution laws.

Congress also set a timetable for achieving certain percentage reductions in emissions of carbon monoxide, hydrocarbons, and nitrogen oxides from motor vehicles. These standards forced automakers to build cars that emit six to eight times fewer pollutants than the cars of the late 1960s. Although significant progress has been made, a series of legally allowed extensions has pushed deadlines for complete attainment of most of these goals into the future.

Trends in U.S. Outdoor Air Quality   There is still a long way to go, but the United States has achieved remarkable progress in reducing outdoor pollution from five major pollutants. In the United States between 1975 and 1988 average outdoor concentrations of most major pollutants (carbon monoxide, ozone, sulfur dioxide, suspended particulate matter, and lead), except nitrogen oxides, dropped as a result of air pollution control laws, economic recession, and higher energy prices. Lead made the sharpest drop because of the 91% drop in the amount of lead allowed in leaded gasoline.

According to the Council on Environmental Quality, the Clean Air Act of 1970 has saved 14,000 lives and $21 billion in health, property, and other damages each year since 1970. Without the 1970 standards, emissions of the seven major outdoor pollutants would be 130% to 315% higher today.

However, from 1986 to 1988 levels of ozone and suspended particulate matter increased and nitrogen

Figure 9-16   This marble monument on a church in Surrey, England, has been damaged by exposure to acidic air pollutants.

Motor vehicles double

dioxide levels stayed the same mostly because of auto emissions. An almost doubling of the number of motor vehicles on the road travelling longer distances, relaxing fuel efficiency standards, and decreasing budgets for enforcement of air pollution laws since 1981 are just some of the symptoms of the lack of progress in controlling these pollutants.

Some 150 million Americans—three out of five—now live in areas that do not meet one or more of the health standards set by the Clean Air Act of 1970. In 1988 one out of three Americans lived in 101 major cities where average ozone levels regularly exceed safe levels. The worst was Los Angeles where ozone levels exceeded the standard for 172 days during 1988. One out of four Americans lives in 59 cities with too much carbon monoxide.

In the 1970s most Western European countries, Canada, Australia, Japan, and South Korea established automobile emissions standards similar to those in the United States, although some European countries lag behind. Brazil will have similar standards by 1997. Little, if any, attempt is made to control vehicle emissions in India, Mexico, Argentina, China, the Soviet Union, and Eastern European countries. Switzerland and Austria have the world's toughest air pollution laws.

In 1988 Mexico's ozone standard, which is *less* strict than the U.S. standard, was violated more than

300 times in Mexico City (Figure 6-17, p. 129)—nearly twice as often as in Los Angeles, with a much stricter standard. Lead levels in the blood of seven out of ten newborn babies in Mexico City exceed World Health Organization standards.

Methods of Pollution Control Once a pollution control standard has been adopted, two general approaches can be used to prevent levels from exceeding the standard. One is *input control,* which prevents or reduces the severity of the problem. The other is *output control,* which treats the symptoms. Output methods such as scrubbers on smokestacks can reduce emissions dramatically, but they are not ultimate solutions. Eventually they are overwhelmed by increases in population and industrialization (Figure 1-10, p. 14). They also create environmental problems of their own, such as the need to dispose of scrubber ash, a hazardous waste.

Input methods are usually easier and cheaper in the long run than output methods. Major input control methods for reducing the total amount of pollution of any type from reaching the environment are

- regulating population growth (Section 6-4)
- reducing unnecessary waste of metals, paper, and other matter resources through increased recycling and reuse and by designing products that last longer and are easy to repair (Section 19-6) *Recycle*
- reducing energy use
- using energy more efficiently (Section 17-1) } *obvious*
- switching from coal to natural gas, which produces less pollution and carbon dioxide when burned (Section 18-2)
- switching from fossil fuels to energy from the sun, wind, and flowing water (Chapter 17) or nuclear power (Section 18-3) *geothermal fossil fuels*
- identifying the source of pollution in a production process, eliminating it from that process, and finding a more environmentally benign substitute *Substitute*

These input methods are the most effective and least costly ways (except nuclear power) to reduce air, water, and soil pollution, including the buildup of carbon dioxide and other greenhouse gases in the atmosphere (Section 10-2). So far they have rarely been given serious consideration in national and international strategies for pollution control.

Control of Sulfur Dioxide Emissions from Stationary Sources In addition to the general input control methods mentioned above, the following approaches can lower sulfur dioxide emissions or reduce their effects:

Sulfur Dioxide Input Control Methods

1. *Burn low-sulfur coal.* Especially useful for new power and industrial plants located near deposits of such coal.

2. *Remove sulfur from coal.* Fairly inexpensive; present methods remove only 20% to 50% but new methods being tested may remove most of the sulfur.

3. *Convert coal to a gas or liquid fuel.* Low net energy yield (Figure 3-18, p. 53).

4. *Remove sulfur during combustion by fluidized-bed combustion (FBC) of coal (Figure 9-17).* Removes up to 90% of the SO_2, reduces CO_2 by 20%, and increases energy efficiency by 5%; should be commercially available for small to medium plants in the mid-1990s.

5. *Remove sulfur during combustion by limestone injection multiple burning (LIMB).* Still in the development and testing stage.

Sulfur Dioxide Output Control Methods

1. *Use smokestacks tall enough to pierce the thermal inversion layer.* Can decrease pollution near power or industrial plants but increases pollution levels in downwind areas (see photo on p. 185 and Figure 9-7)

2. *Remove pollutants after combustion by using flue gas desulfurization or scrubbing (Figure 9-18d, p. 204).* Removes up to 95% of SO_2 and 99.9% of suspended particulate matter (but not the more harmful fine particles); can be used in new plants and added to most existing large plants but is expensive. Presently most of the resulting slurry or sludge is deposited in landfills or holding ponds, but 90% of it could be converted into useful chemicals for use as fertilizers, catalysts, and construction materials.

3. *Add a tax on each unit emitted.* Encourages development of more efficient and cost-effective methods of emissions control; opposed by industry because it costs more than tall smokestacks and requires polluters to bear more of the harmful cost passed on to society (Section 7-2).

By 1985 the Soviet Union and 21 European countries had signed a treaty agreeing to reduce their annual emissions of sulfur dioxide from 1980 levels by at least 30% by 1993; 4 countries agreed to 70% cuts. While this is an important step, ecologists believe that SO_2 emissions must be cut by about 90% to prevent continuing serious ecological damage. Between 1983 and 1989 West Germany cut power plant emissions of sulfur dioxide by 90% and Switzerland and Austria reduced their emissions by more than 90%.

The United States and Great Britain refused to participate in this historic but moderate agreement, citing scientific uncertainty over the harmful effects

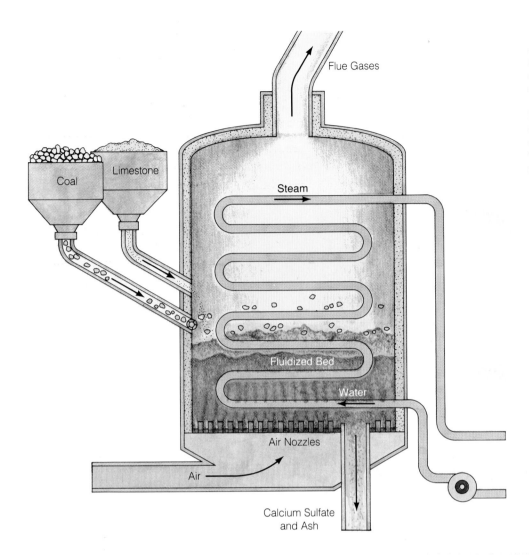

Flue Gases

Coal

Limestone

Steam

Fluidized Bed

Water

Air Nozzles

Air

Calcium Sulfate
and Ash

Figure 9-17 Fluidized-bed combustion of coal. A stream of hot air is blown into a boiler to suspend a mixture of powdered coal and crushed limestone. This removes most of the sulfur dioxide, sharply reduces emissions of nitrogen oxides, and burns coal more efficiently and cheaply than conventional combustion methods.

of sulfur dioxide. U.S. air pollution control laws encourage emissions of SO_2 and suspended particulate matter by allowing tall smokestacks (see photo on p. 185) and by not requiring coal-burning power and industrial plants built before 1972 to install effective pollution control devices. This gives corporations an incentive to keep old, polluting plants in operation rather than build new advanced plants.

Control of Emissions of Nitrogen Oxides from Stationary Sources About half the emissions of nitrogen oxides in the United States come from the burning of fossil fuels at stationary sources, primarily electric power and industrial plants. The rest comes mostly from motor vehicles.

So far little emphasis has been placed on reducing emissions of nitrogen oxides from stationary sources because control of sulfur dioxide and particulates was considered more important. Now it is clear that nitrogen oxides are a major contributor to acid deposition and that they increase tropospheric levels of ozone and other photochemical oxidants that can damage crops, trees, and materials. The following approaches can be used to decrease emissions of nitrogen oxides from stationary sources:

Input Control Methods for Nitrogen Oxides

1. *Remove nitrogen oxides during fluidized-bed combustion* (Figure 9-17). Removes 50% to 75%.

2. *Remove during combustion by limestone injection multiple burning.* Removes 50% to 60%, but is still in the development stage.

3. *Reduce by decreasing combustion temperatures.* Well-established technology that reduces production of these gases by 50% to 60%.

Output Control Methods for Nitrogen Oxides

1. *Use tall smokestacks.*

2. *Add a tax for each unit emitted.*

3. *Remove after combustion by reburning.* Removes 50% or more but is still under development for large plants.

4. *Remove after burning by reacting with isocyanic acid.* Removes up to 99% and breaks down into harmless nitrogen and water; will not be available commercially for at least ten years.

In 1988 representatives from 24 countries, including the United States, signed an agreement that would

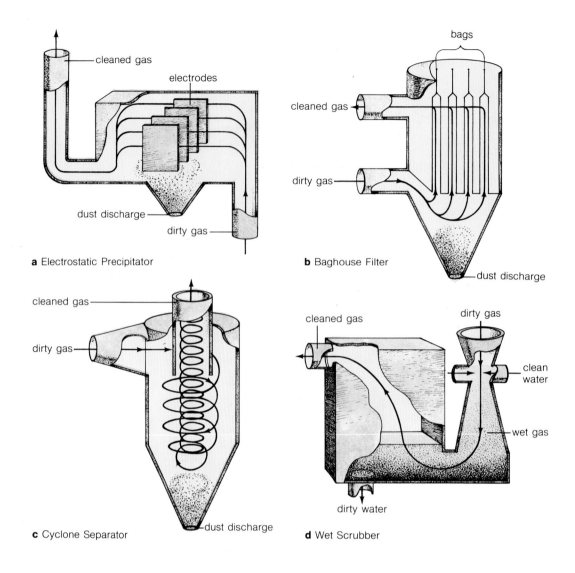

Figure 9-18 Four commonly used methods for removing particulates from the exhaust gases of electric power and industrial plants. The wet scrubber is also used to reduce sulfur dioxide emissions.

a Electrostatic Precipitator

b Baghouse Filter

c Cyclone Separator

d Wet Scrubber

freeze emissions of nitrogen oxides at 1987 levels by 1995. Twelve Western European countries agreed to cut emissions of nitrogen oxides by 30% between 1987 and 1997. Environmentalists applaud these efforts but believe that a 90% reduction in these emissions is needed to prevent continuing serious ecological damage.

Control of Particulate Matter Emissions from Stationary Sources The only input control method for suspended particulate matter is to convert coal to a gas or liquid, a method that is expensive and low in net energy yield (Figure 3-18, p. 53). The following output approaches can be used to decrease emissions of suspended particulate matter from stationary sources:

Suspended Particulate Matter Output Control Methods

1. *Use tall smokestacks.*

2. *Add a tax on each unit emitted.*

3. *Remove particulates from stack exhaust gases.* The most widely used method in electric power and

industrial plants. Several methods are in use: **(a)** electrostatic precipitators (Figure 9-18a); **(b)** baghouse filters (Figure 9-18b); **(c)** cyclone separators (Figure 9-18c); and **(d)** wet scrubbers (Figure 9-18d). Except for baghouse filters, none of these methods removes many of the more hazardous fine particles; all produce hazardous solid waste or sludge that must be disposed of safely; except for cyclone separators, all methods are expensive.

Control of Emissions from Motor Vehicles The following are other methods for decreasing emissions from motor vehicles:

Motor Vehicle Input Control Methods

1. *Rely more on mass transit, bicycles, and walking.*

2. *Shift to less-polluting automobile engines.* Examples are the stratified charge engine, engines that run on hydrogen gas (Section 17-8), or electric motors (if the additional electricity needed to charge batteries is not produced by fossil-fuel burning power plants or by expensive and po-

tentially dangerous nuclear power plants). The government could overcome the auto industry's resistance to producing such engines by specifying smog-free engines on the $5 billion of vehicles it buys each year.

3. *Shift to less-polluting fuels.* Examples are natural gas (Section 18-1), alcohols (Section 17-6), and hydrogen gas (Section 17-8).

4. *Improve fuel efficiency.* A quick and cost-effective approach (Section 17-1).

5. *Modify the internal combustion engine to reduce emissions.* Burning gasoline using a lean, or more air-rich, mixture reduces carbon monoxide and hydrocarbon emissions but increases emissions of nitrogen oxides; a new lean-burn engine that reduces emissions of nitrogen oxides by 75% to 90% may be available in about ten years.

6. *Raise annual registration fees on older, more polluting, gas-guzzling (petro-pig) cars or offer owners an incentive to retire such cars.*

7. *Add a charge on all new cars based on the amount of the major pollutants emitted by the engine according to EPA tests.* This would prod manufacturers to reduce emissions and encourage consumers to buy less-polluting cars.

8. *Give subsidies to car makers for each low-polluting, energy efficient car they sell.* This would allow consumers to pay less for this type of vehicle and much more for polluting gas guzzlers.

9. *Restrict driving in downtown areas.*

Motor Vehicle Output Control Methods

1. *Use emission control devices.* Most widely used approach; engines must be kept well tuned for such devices to work effectively; three-way catalytic converters now being developed can decrease pollutants by 90% to 95% and should be available within a few years.

2. *Require car inspections twice a year and have drivers exceeding the standards pay an emission charge based on the grams emitted per kilometer and the number of kilometers driven since the last inspection.* This would encourage drivers not to tamper with emission control devices and to keep them in good working order. Currently the emission control systems on about 60% of the U.S. car fleet have been disconnected or are not working properly.

3. *Establish emission standards for light-duty trucks* (presently not effectively regulated by U.S. air pollution control laws).

In 1989 the Bush administration proposed that standards for tail-pipe emissions from motor vehicles be regulated by "emissions averaging." This would allow cars built in the 1990s to emit more pollution than cars built in the 1980s—a step backward at a time when we should be going forward.

Control of Troposphere Ozone Levels Ozone levels in the troposphere are affected primarily by emissions of nitrogen oxides and hydrocarbons coupled with sunlight (Figure 9-6). Thus, decreasing ozone levels involves combining the input and output methods already discussed for nitrogen oxides and for motor vehicles.

It also involves decreasing hydrocarbon emissions from cars that produce half of the pollutants that cause smog and from a variety of hard-to-control sources such as oil-based paints, aerosol propellants, dry-cleaning plants, and gas stations that emit the other half.

In 1989 California's South Coast Air Quality Management District Council proposed a drastic program to reduce ozone and photochemical smog in the Los Angeles area. If approved by the state environmental agency and the EPA this plan would require

- outlawing drive-through facilities to keep vehicles from idling in lines

- substantially raising parking fees and assessing high fees for families owning more than one car to discourage automobile use and encourage car and van pooling and use of mass transit

- strictly controlling or relocating of petroleum refining, dry cleaning, auto painting, printing, baking, trash-burning plants, and other industries that release large quantities of hydrocarbons and other pollutants

- finding substitutes or banning use of aerosol propellants, paints, household cleaners, barbecue starter fluids, and other consumer products that release hydrocarbons

- gradually eliminating gasoline burning engines over two decades by converting trucks, buses, and lawnmowers to alternate fuels such as methanol, ethanol, natural gas, or electricity

- requiring gas stations to use a hydrocarbon vapor recovery system on gas pumps and sell alternative fuels

The plan will cost $2.8 to $15 billion a year and may be defeated by public opinion when residents begin to feel the pinch from such drastic changes. But proponents argue that the alternative of continuing degradation of air quality in the country's air pollution capital will cost consumers and businesses much more. Such measures are a glimpse of what most cities will have to do as people, cars, and industries proliferate. Without such changes scientists estimate that ground-level ozone pollution in U.S. urban areas could increase as much as 50% between 1990 and 2020.

The most important thing you can do to reduce air pollution and ozone depletion, slow projected global climate change, and save money is to improve energy efficiency (see Section 7 and Individuals Matter inside the back cover).

Recycle newspapers, aluminum, and other materials. Plant trees and avoid purchasing products such as Styrofoam that contain ozone-depleting chlorofluorocarbons (CFCs). If you have an auto air conditioner, have it serviced at a shop that recycles CFCs. Lobby for much stricter clean air laws and enforcement and development of international treaties to reduce ozone depletion and slow global warming (Chapter 10).

Protect yourself from most indoor air pollutants by

- Testing for radon and taking corrective measures as needed (see Case Study on p. 190).

- Installing air-to-air heat exchangers.

- Avoiding the purchase of formaldehyde products or using "low-emitting formaldehyde" or nonformaldehyde building materials.

- Reducing indoor levels of formaldehyde and several other toxic gases by using house plants such as the spider or airplane plant (the most effective), golden pathos, syngonium, and philodendron. About 20 plants can help clean the air in a typical home.

- Baking houses (especially mobile homes) out at 38°C (100°F) for three to four days and then changing the air several times.

- Changing air filters regularly, cleaning air conditioning systems, emptying humidifier water trays frequently, and not storing gasoline, solvents, or other volatile hazardous chemicals inside a home or attached garage.

- Not using room deodorizers or air fresheners.

- Not using any aerosol spray products.

- Not smoking or smoking outside or in a closed room vented to the outside.

- Attaching whole-house electrostatic air cleaners and charcoal filters to central heating and air conditioning equipment. Humidifiers, however, can load indoor air with bacteria, mildew, and viruses.

- Making sure that wood-burning stoves and fireplaces are properly installed, vented, and maintained. If you use a wood stove for heating, buy one of the newer, more energy-efficient models that greatly reduce indoor and outdoor emissions.

Controlling Toxic Emissions   The Clean Air Act of 1970 ordered the EPA to evaluate the health hazards posed by the hundreds of toxic compounds, some of them known carcinogens, emitted into the atmosphere and then set standards to control each chemical. By 1989 the EPA had issued regulations for only seven of the hundreds of potentially toxic air pollutants.

Because setting standards for each toxic pollutant is so difficult and time consuming, it has been proposed that this chemical-by-chemical approach be replaced by an industry-by-industry approach using technological standards instead of emissions standards. First, the EPA would rank industries according to the amount of toxics they emit. Then the industrial plants emitting the largest amounts would be given deadlines to reduce their emissions to certain levels using the best available technology.

Control of Indoor Air Pollution   For most people, indoor air pollution poses a much greater threat to their health than outdoor air pollution. Yet, the EPA spends $200 million a year trying to reduce outdoor air pollution and only $2 million a year on indoor air pollution.

To sharply reduce indoor air pollution it is not necessary to establish mandatory indoor air quality standards and monitor the more than 100 million homes and buildings in the United States. Instead this can be done by

- modifying building codes to prevent radon infiltration and requiring use of air-to-air heat exchangers or other devices to change indoor air at certain intervals (see Case Study on p. 190)

- removing up to 90% of the hazardous materials in furniture and building materials in new and older homes, apartments, and workplaces by baking these structures out at 38°C (100°F) for three to four days, and then using a fan to exhaust and replace the contaminated air several times

- requiring exhaust hoods or vent pipes for stoves, refrigerators, dryers, kerosene heaters, or other appliances burning natural gas or other fossil fuels

- setting emission standards for building materials that emit formaldehyde such as particle board, plywood, some types of insulation, and materials used in furniture, carpets, and carpet backing

- finding substitutes for potentially harmful chemicals in aerosols, cleaning compounds, paints, and other products used indoors and requiring all such products to have labels listing their ingredients

- requiring employers to provide safe indoor air for employees

In LDCs major reductions in respiratory illnesses would occur if governments gave rural residents and poor people in cities simple stoves that burn biofuels more efficiently (which would also reduce deforestation) and that are vented outside.

Between 1977 and 1990 conflicts between environmentalists and the automobile, utility, mining, chemical, and other industries kept U.S. lawmakers from having enough votes to pass a new air pollution control law. As this book was being written in June 1990, Congress was on the verge of passing a new clean air bill. Compromises were still being negotiated, but it is likely that the law will require SO_2 emission levels to be reduced from their 1980 levels by 50%, nitrogen oxide levels by 60%, and volatile organic compounds by 60% by the end of the year 2000. The law may also call for the emission levels of 191 hazardous air pollutants to be reduced by 80% to 90% from their 1987 levels by the end of the year 2003.

Protecting the Atmosphere Considerable progress has been made in reducing the levels of several outdoor air pollutants in the United States and many other MDCs, but much more needs to be done. And few LDCs have begun to tackle their air pollution problems, which are increasing as these countries become more urbanized and more industrialized.

As long as MDCs and LDCs rely mostly on end-of-pipe output methods for controlling air pollution, the air we breathe will eventually be overwhelmed by more people consuming more energy and other resources that emit potentially harmful chemicals into the atmosphere. Protecting this commonly shared resource will require the following major changes:

- emphasizing pollution prevention rather than pollution control in both MDCs and LDCs

- recognizing that the burning of fossil fuels is the major cause of air pollution

- integrating air pollution and energy policies, with primary emphasis on improving energy efficiency, shifting from fossil fuels to perpetual and renewable energy resources (Chapter 17), discouraging automobile use, boosting the use of public transportation, revamping transportation systems and urban design, increasing recycling and reuse, and reducing the production of all forms of waste

- developing air quality strategies based on air flows and pollution sources for an entire region instead of the current city-by-city, piecemeal approach

- controlling population growth

- recognizing that all nations and all individuals are tied together by a common fate based on protecting the earth that sustains us all

Making these changes will require major modifications in our economic systems. As long as the social costs of air pollution and other forms of pollution remain external to our economic accounting systems, industries, utilities, and individuals will have little incentive to reduce the amount of pollution they generate. Thus, internalizing the external costs of pollution is the key to protecting the atmosphere and other parts of the ecosphere from our activities (Section 7-2).

Making this fundamental shift in our economic priorities will require political involvement by individuals to overcome the built-in resistance to such changes (Section 7-5). Bringing about the needed political and economic changes from the bottom up and changing our polluting and earth-degrading lifestyles will require that we shift from our current unsustainable throwaway worldview to a sustainable-earth worldview (Section 2-3).

J. M. Stycos has observed that major social changes go through four stages:

- Phase 1: No Talk–No Do

- Phase 2: Talk–No Do

- Phase 3: Talk–Do

- Phase 4: No Talk–Do

We have reached the Talk–Do phase in some areas, but mostly we are still stuck in the Talk–No Do phase, despite more than 20 years of effort. We must now shift to phases 3 and 4.

We have precious little time to make fundamental shifts in the ways we act toward the earth and thus toward ourselves and future generations. But it is not too late, if we act now. We must heed the advice that Chief Seattle gave us more than 200 years ago: "Contaminate your bed, and you will one night suffocate in your own waste."

Turning the corner on air pollution requires moving beyond patchwork, end-of-pipe approaches to confront pollution at its sources. This will mean reorienting energy, transportation, and industrial structures toward prevention.

HILARY F. FRENCH

DISCUSSION TOPICS

1. Rising oil and natural gas prices and environmental concerns over nuclear power plants could force the United States to depend more on coal, its most plentiful fossil fuel, for producing electric power. Comment on this in terms of air pollution. Would you

favor a return to coal instead of increased use of nuclear power? Explain.

2. Evaluate the pros and cons of the statement, "Since we have not proven absolutely that anyone has died or suffered serious disease from nitrogen oxides, present federal emission standards for this pollutant should be relaxed."

3. What topographical and climate factors either increase or help decrease air pollution in your community?

4. Do you favor or oppose requiring a 50% reduction in emissions of sulfur dioxide and nitrogen oxides by fossil-fuel-burning electric power and industrial plants and a 50% reduction in emissions of nitrogen oxides by motor vehicles in the U.S. over the next ten years? Explain.

5. Should all tall smokestacks be banned? Explain.

6. Do buildings in your college or university contain asbestos? If so, what is being done about this potential health hazard?

7. Have dormitories and other buildings on your campus been tested for radon?

FURTHER READINGS

Bower, John. 1989. *The Healthy House.* New York: Lyle Stuart.

Brenner, David J. 1989. *Radon: Risk and Remedy.* Salt Lake City, Utah: W. H. Freeman.

Brookins, Douglas G. 1990. *The Indoor Radon Problem.* Irvington, N.Y.: Columbia University Press.

Environmental Protection Agency. 1988. *The Inside Story: A Guide to Indoor Air Quality.* Washington, D.C.: EPA.

French, Hilary F. 1990. *Clearing the Air: A Global Agenda.* Washington, D.C.: Worldwatch Institute.

Geller, H., et al. 1986. *Acid Rain and Energy Conservation.* Washington, D.C.: American Council for an Energy-Efficient America.

Hunter, Linda Mason. 1989. *The Healthy House: An Attic-to-Basement Guide to Toxin-Free Living.* Emmaus, Penn.: Rodale Press.

MacKenzie, James J., and Mohamed T. El-Ashry. 1988. *Ill Winds: Airborne Pollution's Toll on Trees and Crops.* Holmes, Penn.: World Resources Institute Publishing.

Mohnen, Volker A. 1988. "The Challenge of Acid Rain." *Scientific American,* vol. 259, no. 2, 30–38.

National Academy of Sciences. 1986. *Acid Deposition: Long-Term Trends.* Washington, D.C.: National Academy Press.

National Academy of Sciences. 1988. *Air Pollution, the Automobile, and Human Health.* Washington, D.C.: National Academy Press.

Office of Technology Assessment. 1989. *Catching Our Breath: Next Steps for Reducing Urban Ozone.* Washington, D.C.: Government Printing Office.

Pawlick, Thomas. 1986. *A Killing Rain: The Global Threat of Acid Precipitation.* San Francisco: Sierra Club Books.

Postel, Sandra. 1984. *Air Pollution, Acid Rain, and the Future of Forests.* Washington, D.C.: Worldwatch Institute.

Regens, James L., and Robert W. Rycroft. 1988. *The Acid Rain Controversy.* Pittsburgh, Penn.: University of Pittsburgh Press.

Wark, K., and C. F. Warner. 1986. *Air Pollution: Its Origin and Control.* 3rd ed. New York: Harper & Row.

CHAPTER TEN

CLIMATE, GLOBAL WARMING, OZONE DEPLETION, AND NUCLEAR WAR: ULTIMATE PROBLEMS

General Questions and Issues

1. What major factors determine variations in climate?

2. How can our activities cause global warming and what are some possible effects from doing this?

3. What can we do to delay and reduce possible global warming and adjust to its effects?

4. How are we depleting ozone in the stratosphere and what are some possible effects from doing this?

5. What can we do to slow down ozone depletion?

6. What effects might even limited nuclear war have on the earth's life-support systems for humans and other species?

We, humanity, have finally done it: disturbed the environment on a global scale.

THOMAS E. LOVEJOY

CLIMATE IS THE MAJOR FACTOR determining the types, distribution, and abundance of the earth's biodiversity (Figure 5-2, p. 87). It is also the major factor affecting where we can grow food, have enough water, and live and is a key natural resource supporting our economic activities. We and many types of plants and other animals also survive because a thin gauze of ozone in the stratosphere (Figure 9-1, p. 187) keeps much of the harmful ultraviolet radiation given off by the sun (Figure 4-4, p. 61) from reaching the earth's surface.

Although we have been on the earth for only an eyeblink of its existence, we are now altering the chemical content of the earth's atmosphere 10 to 100 times faster than its natural rate of change (see cartoon). Projected global warming caused by our binge of fossil fuel burning and deforestation (see photo on p. 293), ozone depletion caused by our extensive use of chlorofluorocarbons and other chemicals we could learn to do without, and the buildup of nuclear weapons are now major global environmental threats.

Largely invisible and silent, these ultimate problems will continue building until they trigger major thresholds of change. When these thresholds are crossed it will be too late to prevent the drastic, lasting, and unpredictable effects they have on the biosphere that supports us and other species. There will be no place to escape to and no place to hide from their effects. Dealing with these planetary emergencies to prevent ultimate tragedy to the world will require major changes in the way we think and act (Section 2-3) and international cooperation on an unprecedented scale.

10-1 CLIMATE: A BRIEF INTRODUCTION

Weather and Climate Every moment there are changes in temperature, barometric pressure, humidity, precipitation, sunshine (solar radiation), cloudiness, wind direction and speed, and other

A CENTURY OF ABUSE RISES TO HAUNT AN OBLIVIOUS, DECADENT SOCIETY!

REVENGE OF THE ENVIRONMENT

WITH INDUSTRIAL SELF-SERVING GREEDY AND AN APATHETIC POLLUTERS POLITICIANS DEVELOPERS PUBLIC COMING SOON!

Short-term changes conditions in the troposphere. These short-term changes in the properties of the troposphere are what we call **weather**. **Climate** is the average weather of an area. It is the general pattern of atmospheric or weather conditions, seasonal variations, and weather extremes in a region over a long period—at least 30 years. The two most important factors determining the climate of an area are its temperature with its seasonal variations and the quantity and distribution of precipitation over each year (Figure 5-3, p. 88).

Temperature + prec.

Temp + Precipitation

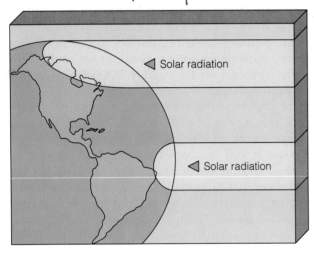

Figure 10-1 During certain times of the year the sun's rays are almost perpendicular to the earth's equatorial zones, concentrating heat on this part of the earth. The sun's rays strike the earth's poles at a slanted angle, spreading the incoming solar energy over a larger area and diluting the heat input.

Climate and Global Air Circulation Several major factors determine the patterns of global air circulation that cause the uneven patterns of average temperature and average precipitation (Figure 5-3, p. 88) and thus the climates of the world. One is variations in the amount of incoming solar energy that strikes different parts of the earth (Figure 10-1), resulting from changes in solar output and minute changes in the earth's orbit around the sun in a 22,000-year cycle.

The large input of heat at and near the equator (Figure 10-1) warms large masses of air that rise because warm air has a lower density (mass per unit of volume) than cold air. As these warm air masses rise, they spread northward and southward, carrying heat from the equator toward the poles.

At the poles the warm air cools and sinks downward because cool air is denser than warm air. These cool air masses then flow back near ground level toward the equator to fill the void left by rising warm air masses. This general global air circulation pattern in the troposphere leads to warm average temperatures near the equator, cold average temperatures near the poles, and moderate or temperate average temperatures at the middle latitudes between the two regions (Figure 5-3, p. 88). This global circulation of air and moisture is a key factor determining the types of terrestrial ecosystems found in areas with tropical, temperate, and polar climates (Figure 5-2, p. 87).

However, general average temperature and precipitation patterns vary with the seasons in all parts of the world away from the equator. These seasonal changes in climate are caused by the earth's annual orbit around the sun and its daily rotation around its tilted axis—the imaginary line connecting the North and South poles (Figure 10-2).

Figure 10-2 The seasons (shown here for the Northern Hemisphere only) are caused by variations in the amount of incoming solar energy as the earth makes its annual rotation around the sun on an axis tilted by 23.5 degrees. The tilt of the earth's axis is not fixed and varies between 22.5 and 24 degrees over thousands of years. The greater the tilt, the colder the winters and the hotter the summers.

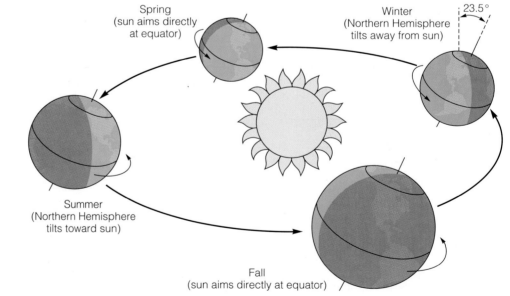

Spring (sun aims directly at equator)

Winter (Northern Hemisphere tilts away from sun)

23.5°

Summer (Northern Hemisphere tilts toward sun)

Fall (sun aims directly at equator)

Forces created on the atmosphere as the earth spins around its tilted axis also break up the general air circulation pattern from the equator to the poles and back. This creates three separate belts of moving air, or prevailing surface winds, north and south of the equator (Figure 10-3), which affect the average temperature of various areas. The general circulation of air masses in the troposphere and the prevailing surface winds also affect the distribution of precipitation over the earth.

The earth's rotation, inclination, prevailing winds, and differences in water density cause ocean currents and surface drifts that generally move parallel with the equator (Figure 10-4). Trade winds blowing almost continuously in an easterly direction toward the equator push surface ocean waters westward in the Atlantic, Pacific, and Indian oceans until these waters bounce off the nearest continent. This causes two large circular water movements, called *gyres*, that turn clockwise in the Northern Hemisphere and counterclockwise in the Southern Hemisphere (Figure 10-4). These gyres move warm waters to the north and south of the equator.

Warm and cold currents and surface drifts in the world's oceans affect the climates of nearby coastal areas. For instance, without the warm Gulf Stream, which transports 25 times more water than all the world's rivers, the climate of northwestern Europe would be more like that of the sub-Arctic. Ocean currents and drifts also help mix ocean waters and distribute the nutrients and dissolved oxygen needed by aquatic organisms.

Coastal regions have less extreme climates than areas in the interiors of continents because seawater changes temperature more slowly—warming up in summer and cooling down in winter—than land does.

Climate and the Chemical Content of the Atmosphere: The Greenhouse Effect and the Ozone Layer The chemical content of the troposphere and stratosphere is another important factor determining the earth's average temperature and thus its climates. Fairly small amounts of carbon dioxide and water vapor (mostly in clouds) and trace amounts of ozone, methane, nitrous oxide, chlorofluorocarbons, and other gases in the troposphere play a key role in this temperature regulation process.

These gases, known as **greenhouse gases,** act somewhat like the glass panes of a greenhouse or of a car left parked in the sun with its windows rolled up. They allow solar radiation to enter the earth's atmosphere. The earth's surface then absorbs much of this solar energy and degrades it to infrared radiation, or heat, which rises from the earth's surface (Figure 4-5, p. 62). Some of this heat escapes into space and some is absorbed by molecules of green-

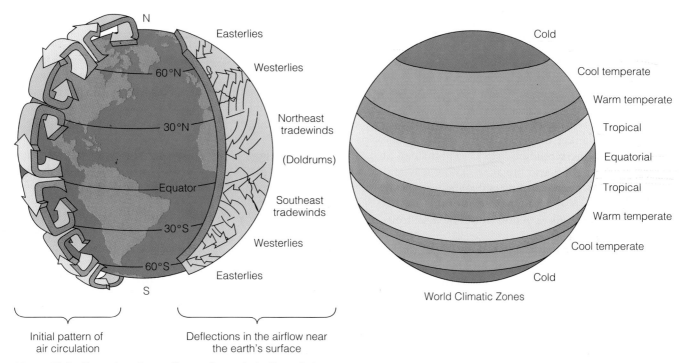

Initial pattern of air circulation

Deflections in the airflow near the earth's surface

Figure 10-3 Formation of prevailing surface winds. The twisting motion caused by the earth's rotation on its axis causes the airflow in each hemisphere to break up into three separate belts of winds. Airflow is deflected to the right in the Northern Hemisphere and to the left in the Southern Hemisphere. These prevailing winds affect the types of climate found in different areas.

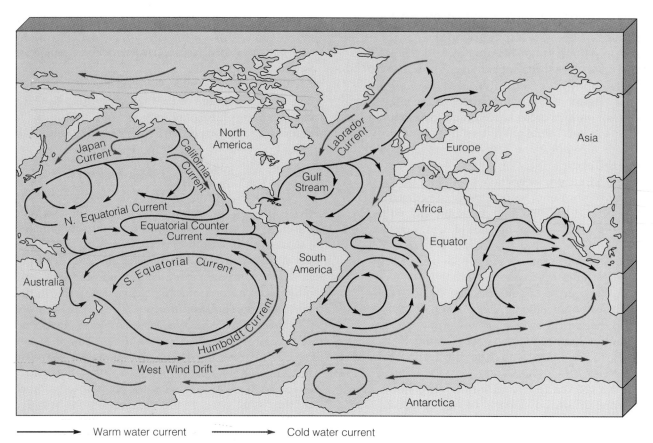

Warm water current ----→ Cold water current

Figure 10-4 Major warm and cold surface currents of the world's oceans. These water movements are produced by the earth's winds and modified by its rotational forces. They circulate water in the oceans in great surface gyres and have major effects on the climate on adjacent lands.

house gases and radiated back toward the earth's surface (Figure 10-5).

The resulting heat buildup raises the temperature of the air in the troposphere, a warming of the lower atmosphere and the earth's surface called the **greenhouse effect**, or heat lamp effect. Without our current heat-trapping blanket of gases, the earth's average surface temperature would be −18°C (0°F) instead of its current 15°C (59°F), and life as we know it would not exist. But too much warming or cooling, especially if it occurs over a few decades instead of the normal hundreds to thousands of years, would be disastrous for us and other species.

In addition to filtering out harmful ultraviolet radiation, ozone in the stratosphere affects climate. Absorption of UV radiation by ozone creates warm layers of air high in the stratosphere that prevent churning gases in the troposphere from entering the stratosphere (Figure 9-1, p. 187). This thermal cap is an important factor in determining the average temperature of the troposphere and thus the earth's current climates.

Thus, any human activities that decrease the average amount of ozone in the stratosphere and

increase the average amount of greenhouse gases in the troposphere can have far-reaching effects on climate, human health, human economic and social systems, and the health and existence of other species. This is what we are doing at a rapid and increasing rate.

Past Climate Changes Climate changes and fluctuations in levels of greenhouse gases, suspended particulate matter, and average surface temperatures lasting hundreds to thousands of years are a normal part of the earth's climatic history. They have lead to eight great ice ages over the last 700,000 years during which thick ice sheets spread southward over much of North America, Europe, and parts of Asia. The estimated coldest temperature of the last ice age (which took thousands of years to develop) is just 5°C (9°F) colder than it is today. Each of these glacial periods lasted about 100,000 years and was followed by a warmer interglacial period, such as the one we are enjoying, lasting 10,000 to 12,500 years.

The last great ice age ended about 10,000 years ago, when agriculture began. As the ice melted, average sea levels rose 100 meters (300 feet), changing

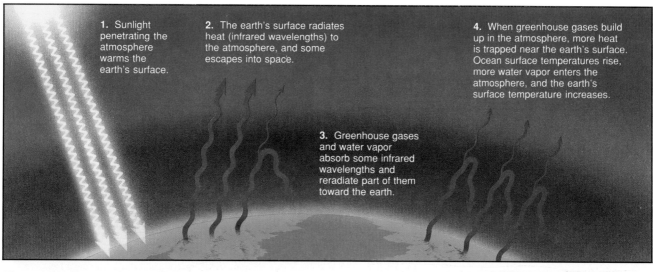

Figure 10-5 The greenhouse effect. (Adapted with permission from Cecie Starr and Ralph Taggart, *Biology: The Unity and Diversity of Life*, 5th ed., Belmont, Calif.: Wadsworth, 1989)

CO₂
Methane   nitrous
CFCs   oxide

the ecological face of the earth. The warmer climate during the present interglacial period plus agricultural knowledge has allowed agriculture to spread rapidly throughout the world and the human population to grow to 5.3 billion.

Until the Industrial Revolution began about 275 years ago, changes in the levels of greenhouse gases in the atmosphere were caused almost entirely by natural processes. Examples are variations in the sun's energy output, slight changes in the earth's orbit around the sun and the tilt of its axis, volcanic eruptions, and the shifting of ocean currents that normally take place over hundreds to thousands of years. But during the past 275 years we have released large quantities of carbon dioxide and other greenhouse gases into the troposphere. Our emissions of greenhouse gases have increased sharply since 1950 and are projected to increase even more rapidly during your lifetime as we burn most of the world's remaining fossil fuels and accelerate deforestation, industrialization, and industrialized farming.

CO₂ – most of pollution

10-2 GLOBAL WARMING FROM AN ENHANCED GREENHOUSE EFFECT

Rising Levels of Greenhouse Gases So far there is virtually unanimous agreement on two points. The greenhouse theory, first proposed in 1827 by the French mathematician Jean Baptiste Fourier, is valid, and we are putting enormous quantities of greenhouse gases (and other chemicals) into the atmosphere (Figure 10-6).

Satellite and other measurements indicate that currently carbon dioxide accounts for about 49% of the annual human-caused input of greenhouse gases, chlorofluorocarbons (CFCs) for 14%, methane for 8%, and nitrous oxide for 6%. But the latter three gases have a much greater warming effect per molecule than carbon dioxide (Figure 10-6). The United States is responsible for the largest emissions of greenhouse gases (18%), followed by European countries (13%), the Soviet Union (12%), Brazil (11%), China (7%), India (4%), Japan (4%), Canada (2%), and Mexico (1.4%).

Carbon dioxide is released when carbon or any carbon-containing compound is burned. Fossil fuels provide almost 80% of the world's energy (Figure 3-4, p. 40), cause about 75% of current CO_2 emissions, and produce most of the world's air pollution. The carbon dioxide level in the troposphere is now the highest it has been in at least 130,000 years, and the level is rising. Fossil fuel use is projected to double between 1985 and 2040. Thus, the projected global warming crisis, along with greatly increased air pollution (Chapter 9), is largely an energy crisis caused mostly by rapid, massive, and wasteful burning of the world's fossil fuels.

Deforestation, especially the wholesale clearing and burning of tropical forests (Section 15-2), is believed to account for about 20% of the increase in carbon dioxide levels. Growing forests convert CO_2 into wood through photosynthesis, but CO_2 is added to the atmosphere when forests or fuelwood is burned and when trees die and decay. As long as no more wood is harvested or burned or decays than grows back each year, there is no net increase in atmospheric CO_2 from this source. But we are

Figure 10-6 Increases in average concentrations of greenhouse gases in the atmosphere. They are projected to cause an increase in the average temperature of the troposphere. (Data from Electric Power Research Institute. Adapted by permission from Cecie Starr and Ralph Taggart, *Biology: The Unity and Diversity of Life,* 5th ed., Belmont, Calif.: Wadsworth, 1989)

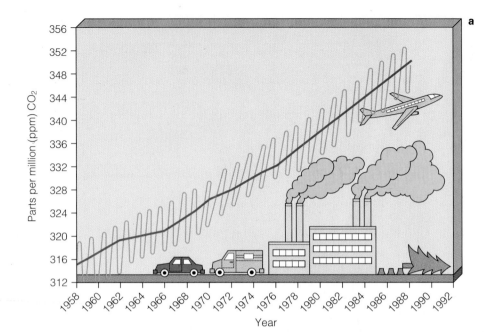

a. Carbon dioxide (CO_2) This gas is thought to be responsible for 49% of the human-caused input of greenhouse gases. Major sources are fossil-fuel burning (80%) and deforestation (20%). CO_2 remains in the atmosphere for about 500 years. Industrial countries account for about 76% of annual emissions.

b. Chlorofluorocarbons (CFCs) These gases are responsible for 14% of the human input of greenhouse gases and by 2020 will probably be responsible for about 25% of the input. CFCs also deplete ozone in the stratosphere. Major sources are leaking air conditioners and refrigerators, evaporation of industrial solvents, production of plastic foams, and propellants in aerosol spray cans (in some countries). CFCs remain in the atmosphere for 65 to 111 years, depending on the type and generally have 10,000 to 20,000 times the impact per molecule on global warming than each molecule of CO_2.

49
14
18
6

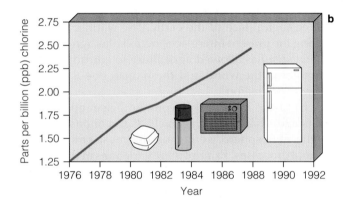

c. Methane (CH_4) This gas is responsible for about 18% of the human input of greenhouse gases. It is produced by bacteria that decompose organic matter in oxygen-poor environments. About 40% of global methane emissions come from waterlogged soils, bogs, marshes, and rice paddies. A 1°C (1.8°F) warming may increase methane emissions from these sources by 20% to 30% and amplify global warming. Other sources of methane are landfills, the guts of termites whose populations are expanding to digest the dead woody materials left after deforestation, and the digestive tracts of the billions of cattle, sheep, pigs, goats, horses, and other livestock. Some methane also leaks from coal seams, natural gas wells, pipelines, storage tanks, furnaces, dryers, and stoves. Natural sources produce an estimated one-third of the methane in the atmosphere, and human activities produce the rest. CH_4 remains in the troposphere for about 7 to 10 years and each molecule is about 25 times more effective in warming the atmosphere than a molecule of carbon dioxide.

d. Nitrous oxide (N_2O) This gas is responsible for 6% of the global warming. It is released from the breakdown of nitrogen fertilizers in soil, livestock wastes, and nitrate-contaminated groundwater, and by biomass burning. Its average stay in the troposphere is 150 years. It also depletes ozone in the stratosphere. The global warming from each molecule of this gas is about 230 times that of a CO_2 molecule.

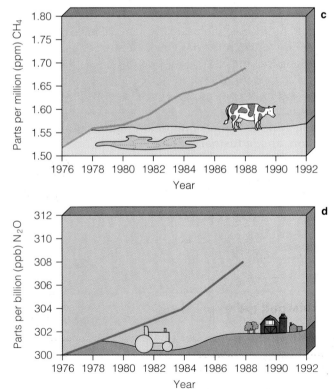

harvesting forests unsustainably (Table 1-1, p. 10, and Section 15-2). Modern farming, forestry, industries, and motor vehicles are also releasing other greenhouse gases—mostly methane, nitrous oxide, chlorofluorocarbons, and ozone formed in photochemical smog (Figure 9-6, p. 193)—into the troposphere at an accelerating rate (Figure 10-6b, c, and d).

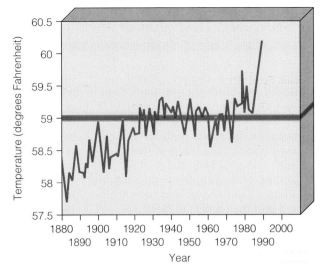

Figure 10-7 Average global temperatures, 1880–1989. Note the general increase since 1880 and the sharp rise in the 1980s, the warmest decade in the last 110 years. The baseline is the global average temperature from 1950 to 1980. (Data from National Academy of Sciences and National Center for Atmospheric Research)

Projected Global Warming Scientists accept the greenhouse theory. But they disagree widely on how large the rise in average global temperature will be, whether other factors in the climate system will counteract or amplify a temperature rise, how fast temperatures might climb, and what the effects will be on various areas. The reasons for these disagreements are uncertainty about the accuracy of the mathematical models and geological evidence used to project future changes in climate and assumptions about how rapidly we will consume fossil fuels and clear forests. Such controversy is a normal part of science (see Spotlight on p. 216).

Since 1880, when reliable measurements began, average global temperatures have risen about 0.5°C (0.9°F) (Figure 10-7). However, there is no strong evidence linking this recent warming to the greenhouse effect. The reason we don't have a "smoking gun" is that so far any temperature changes caused by an enhanced greenhouse effect have been too small to exceed normal short-term swings in average atmospheric temperatures.

The more pressing question, however, is what kind of climate is likely to develop over the next 50 years. Circumstantial evidence from the past and climatic modeling have convinced many climate experts that global warming caused by our rapidly increasing input of greenhouse gases into the atmosphere (Figure 10-6) will begin accelerating in the 1990s or in the first decade of the next century, rising above the background temperature changes (climatic noise) that presently mask this effect.

Current climatic models project that the earth's average atmospheric temperature will rise 1.5°C to 5.5°C (2.7°F to 9.9°F) over the next 50 years (by 2040) if greenhouse gases continue to rise at the current rate. By way of comparison, the typical natural variation in the earth's average temperature over periods of 100 to 200 years during the interglacial period we live in has been at most 0.5°C to 1°C (0.9°F to 1.8°F).

Because of the many uncertainties in these global climate models (see Spotlight on p. 216), their developers believe their projections are accurate within a factor of two. This means that projected global warming during the next century could be as low as 0.7°C (1.3°F) or as high as 11°C (20°F). There is about a 50% chance either way. If we keep pumping greenhouse gases into the atmosphere and continue cutting down much of the world's forests, we are flipping a coin and gambling with life as we know it on this planet.

You might be thinking, Why should we worry about a few degrees rise in the troposphere's average temperature? After all, we often have this much change between June and July, or between yesterday and today. The key point is that we are not talking about the normal swings in weather from place to place. We are talking about a projected global change in average climate in your lifetime, with various parts of the world experiencing changes well above the average.

Current models indicate that the Northern Hemisphere will warm more and faster than the Southern Hemisphere, mostly because there is so much more ocean in the south and water takes longer to warm than land. Temperatures at middle and high latitudes are projected to rise two to three times the average increase, while temperature increase in tropical areas near the equator would be less than the global average.

Projecting changes in average global temperatures is difficult enough. But this is easy compared to projecting how global temperature changes might change the climate in specific regions. About all we can hope for in this case is a series of scenarios of regional climate change based on feeding different assumptions into current and improved climatic models. Current climate models generally predict the same results on a global basis, but disagree widely on projected climate changes in different geographic regions.

The main way scientists, economists, and others project (not predict) future behavior of climate, ecological, economic, and other complex systems is to develop mathematical models that simulate such systems. Then the models are run on high speed computers. Various data and assumptions are fed into the models to make a series of projections of future behavior and effects. How well the results correspond to the real world depends on the design of the model and the accuracy of the data and assumptions used.

Another way to project how climate might change in the future is to see how it has changed in the past. Evidence about past climate change has been gathered by analyzing the chemical content and fossil evidence of climate-sensitive forms of life found in deeply buried samples of rock, sediments on the bottoms of oceans and lakes, and deep cores removed from ice sheets. Such limited and often speculative evidence can be used to test and improve computer models of the earth's climate systems.

Scientists recognize that their models of the earth's climate are crude approximations at best. Present models do not adequately include the role of cloud formation and cover on climate, interactions between the atmosphere and the oceans (which contain 50 times more CO_2 than the atmosphere), how the Greenland and Antarctic ice sheets affect climate, and how soils, forests, and other ecosystems respond to changes in atmospheric temperature. In each case these factors could dampen or amplify global warming.

For example, global warming will raise the average surface temperature of the world's oceans, which will increase the rate of evaporation of water into the atmosphere to form clouds. If there is a net increase in thick, low-level clouds that reflect sunlight back into space, this would help slow the state of global warming. On the other hand, if winds and other factors lead to a net increase in thin, high-level clouds, which act as a blanket to trap heat in the lower troposphere, the rate of global warming would increase. We don't know the net effects of such factors or how long they take to act.

Improving the models is vital. But even so we will never have the scientific certainty that decision makers want before making highly controversial decisions such as greatly slowing down the use of fossil fuels upon which much of the world's present economy depends.

Often those opposing change or wishing to delay decision making say that something should not be done until it has been scientifically "proved." This, however, misrepresents the results of science.

Scientific theories, models, and forecasts are based on mostly circumstantial and incomplete evidence and statistical probabilities, not certainties. All scientists can do is project that there is a low, medium, high, or very high chance of something happening. Such information is quite useful, but the only way to get conclusive, direct proof about a possible future event is to wait and see whether it happens.

Many climate experts believe there is already enough circumstantial evidence for us to act now to slow this warming down to a more manageable rate. This will buy us precious time to do more research, shift to less harmful practices, and adapt to a warmer earth.

They argue that if we wait for the earth's average temperature to rise to the point where it exceeds normal climatic fluctuations, it will be too late to prevent lasting and highly disruptive social, economic, and environmental changes. Waiting to act involves carrying out a gigantic experiment on ourselves and other species—a form of global Russian roulette.

Besides, since fossil fuels (especially oil) are running out and are the major causes of air pollution, water pollution, and land disruption, we need to drastically improve energy efficiency and shift to other energy sources as fast as possible, even if there were no threat from global warming. Similarly, since deforestation is one of the greatest threats to the earth's biodiversity, we should halt and reverse this form of environmental degradation regardless of whether the threat of global warming is serious or not.

Since 1945 the world's countries, mostly MDCs, have spent over $12 trillion to protect us from the possibility of nuclear war. Global warming is a much more likely and equally serious threat to national, economic, and individual security. Yet we have spent only a pittance to deal with this potentially devastating threat, while the time to act is rapidly running out. What do you think should be done?

However, one thing is clear. We now have the potential to bring about disruptive climate change at a rate 10 to 100 times faster than has occurred during the past 10,000 years. A rise of 1.5°C (2.7°F) would make the earth warmer than it has been in the last 100,000 years. A rise of 5°C (9°F) would make the earth warmer than it has been in millions of years. Such rapid and widespread changes in climate are something our species has not experienced in its brief 40,000-year existence. It will affect where we and other species can live and where we can grow food. It is guaranteed to mean nasty ecological surprises for us and for other species. Such rapid global warming would be comparable to global nuclear war (Section 10-6) in its potential to cause sudden, unpredictable, and widespread disruption of ecological, economic, and social systems.

Possible Effects on Crop Production, Ecosystems, and Biodiversity At first glance a warmer average climate might seem desirable. It could lead to lower heating bills and longer growing seasons in middle and high latitudes. Crop yields might increase 60% to 80% in some areas because more carbon dioxide in the atmosphere can increase the rate of plant photosynthesis. Increased warming of the troposphere may cause some cooling in the stratosphere, thereby slowing down reactions that destroy ozone.

But other factors could offset these effects. Air conditioning use would rise and contribute more heat to the troposphere. This would intensify and spread urban heat islands (Figure 6-20, p. 132), causing people to use even more air conditioning. Using fossil fuels to produce more electricity to run air conditioners would add more carbon dioxide and chlorofluorocarbons (used as coolants in air conditioners) to the atmosphere, accelerating global warming and ozone depletion. This would also add more nitrogen oxides and sulfur dioxide to the troposphere, increasing ground-level ozone, photochemical smog, and acid deposition.

Potential gains in crop yields from increased CO_2 levels could be wiped out by increased damage from insect pests because warmer temperatures would boost insect breeding. Higher temperatures would also increase plant aerobic respiration rates and reduce water availability. Recent evidence suggests that many plants have responded to past CO_2 increases by developing fewer of the pores they use to take in CO_2 and thus reducing their rate of photosynthesis.

Regional climate changes shift the ecological tolerance of species hundreds of kilometers horizontally and hundreds of meters vertically (Figure 5-5, p. 90), with unpredictable consequences for natural systems and crops. Past evidence and computer models indicate that climate belts would shift northward by about 161 kilometers (100 miles) for each 1°C (1.8°F) rise in the global atmospheric temperature.

Current and even improved climatic models won't be able to accurately project where such changes might occur, but the point is that there would be massive unpredictable shifts in where we could grow food. The major reason we can grow so much food today is that global and regional climates have not changed much during the past 200 years.

Having to shift the location of much of our agricultural production in only a few decades would create massive disruptions in food supplies and could lead to as many as 1 billion environmental refugees. Shifting crop production would also require huge investments in new dams, irrigation systems, and water supply distribution systems, fertilizer plants, and other parts of our agricultural systems. But since the effects of rapid climate change would be largely unpredictable we might do this only to find that food growing areas shift again if global warming accelerates or starts to drop.

Current models indicate that food production could drop in many of the world's major agricultural regions, including the U.S. midwestern grain belt (Figure 10-8), the Canadian prairie provinces, the Ukraine, and northern China, because of reduced soil moisture in the summer growing season. Parts of Africa and India and the northern reaches of the Soviet Union and Canada may get climates that could increase food production. But soils in some of these potential new food-growing areas, such as Canada and Siberia, are poor and would take centuries before they reach the productivity of current agricultural land. Meanwhile, food prices would skyrocket.

In some areas lakes, rivers, and aquifers that have nourished ecosystems, cropfields, and cities for centuries could shrink or dry up altogether, forcing entire communities and populations to migrate to areas with adequate water supplies. The Gulf Stream (Figure 10-4) might stop flowing northeastward as far as Europe, leading to a much colder climate in that part of the world.

Global warming might also speed up the bacterial decay of dead organic matter in the soil. This could lead to a rapid release of vast amounts of carbon dioxide from dry soils and methane from waterlogged wetlands and rice paddies. Huge amounts of methane tied up in hydrates in arctic tundra soils and in muds on the bottom of the Arctic Ocean could also be released if the blanket of permafrost covering tundra soils melts and the oceans warm. Because methane is such a potent greenhouse gas (Figure 10-6c), this could greatly amplify global warming.

The spread of tropical climates from the equator would bring malaria (see Case Study on p. 172 and Figure 8-2, p. 172), encephalitis, and other insect-borne diseases to formerly temperate zones. Tropical skin diseases would also spread to many areas that now have a temperate climate.

In a warmer world, highly damaging weather extremes such as prolonged heat waves and droughts would increase in frequency and intensity in many parts of the world. As the upper layers of sea water warm, hurricanes, typhoons, and tropical cyclones would become more severe in some parts of the world. For example, computer models project that giant hurricanes with 50% more destructive potential

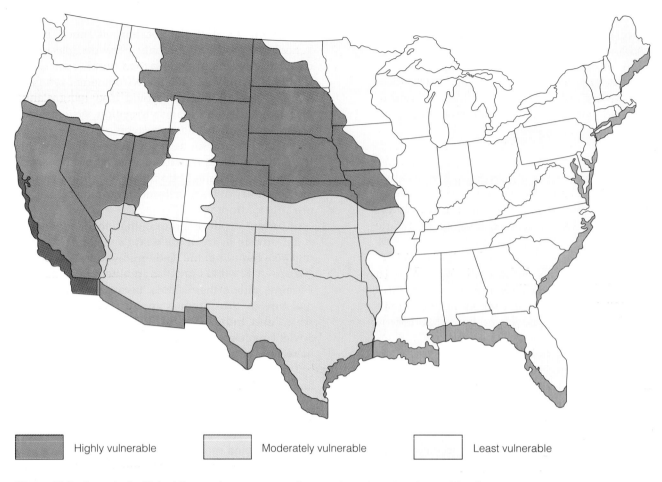

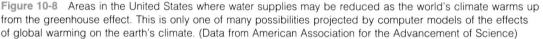

| Highly vulnerable | Moderately vulnerable | Least vulnerable |

Figure 10-8 Areas in the United States where water supplies may be reduced as the world's climate warms up from the greenhouse effect. This is only one of many possibilities projected by computer models of the effects of global warming on the earth's climate. (Data from American Association for the Advancement of Science)

than those today would hit farther north and during more months of the year. Cities such as Miami, Galveston, Atlantic City, Charleston, and Myrtle Beach (Figure 5-23, p. 98) could be devastated by the impact of such superhurricanes. Even a very slight warming of surface waters could greatly increase the intensity of hurricanes in partially enclosed ocean basins like the Gulf of Mexico and the Bay of Bengal.

Changes in regional climate brought about by global warming would be a great threat to forests, especially those in temperate climates and the northern coniferous forests in regions with a subarctic climate (Figure 5-3, p. 88). Least affected would probably be tropical rain forests, if we don't cut most of them down (Section 15-2).

As the earth warms, forest growth in temperate regions will move toward the poles and will replace open tundra and some snow and ice. However, tree species in forests can only move through the slow growth of new trees along their edges—typically, about 0.9 kilometers (0.5 miles) a year or 9 kilometers (5 miles) a decade. If climate belts move faster than this very slow migration or if migration is blocked by

cities, cropfields, highways, and other human barriers, then entire forests will wither and die. These diebacks could amplify the greenhouse effect when the decaying trees release carbon dioxide into the air. Then the increased bacterial decay of organic matter in the warmer exposed soil would release even more CO_2. Large-scale forest diebacks would also cause mass extinctions of plant and animal species that couldn't migrate to new areas. Fish would die as temperatures soared in streams and lakes and lower water levels concentrated pesticides.

Stress to trees from pests and disease microorganisms would also increase because they are able to adapt to climate change faster than trees. The number of devastating fires in drier forest areas and grasslands would increase, adding more carbon dioxide to the atmosphere. Costly efforts to plant trees may fail when many of the new trees die.

Any shifts in regional climate caused by an enhanced greenhouse effect would pose severe threats to many of the world's parks, wildlife reserves, wilderness areas, and wetlands (Chapter 15). This would accelerate the already serious and increasing loss of the earth's biodiversity (Chapter 16). Thomas Lovejoy of the Smithsonian Institution warns: "There will be no winners in this game of ecological chairs, for it will be fundamentally disruptive and destablizing, and we can anticipate hordes of environmental refugees."

Possible Effects on Sea Levels Water expands slightly when it is heated, like the fluid in a thermometer. This explains why global sea levels would rise if the oceans warm. Additional rises would occur if the higher-than-average heating at the poles causes some melting of ice sheets and glaciers.

The Greenland and Antarctic ice sheets act like enormous mirrors to cool the earth by reflecting sunlight back into space. Some scientists fear that even a small temperature rise would shrink these glaciers, allowing more sunlight to hit the earth. Global warming would amplify and cause a larger rise in average sea levels than that from the thermal expansion of water. If most of the Greenland and West Antarctic ice sheets melted, as happened during a warm period 120,000 years ago, sea levels would gradually rise as much as 6 meters (20 feet) over several hundred years.

Other scientists argue that increased warming would allow the atmosphere to carry more water vapor and increase the amount of snowfall on some glaciers, particularly the Antarctic ice sheet. If snow accumulates faster than ice is lost, the Antarctic ice sheet would grow, reflect more sunlight, and help cool the atmosphere.

Current models indicate that an increase in the average atmospheric temperature of 3°C (5°F) would raise the average global sea level by 0.2 to 1.5 meters (1 to 5 feet) over the next 50 to 100 years. If the Antarctic ice sheet grows in size because snow accumulation exceeds ice loss, the lower estimate (give or take 0.4 meter) is more likely by the year 2050. Approximately half of the world's population lives in coastal areas that would be threatened or flooded by rising seas.

Even a modest rise in average sea level would flood coastal wetlands and low-lying cities and croplands, move barrier islands further inland, and contaminate coastal aquifers with salt. A one-third meter rise would push shorelines back 30 meters (98 feet) compared to 136 meters (445 feet) for a 1.5-meter (5-foot) rise in the average sea level. Only a few of the most intensively developed resort areas along the U.S. coast have beaches wider than 30 meters at high tide. Especially hard hit would be North and South Carolina, where the slope of the shoreline is so gradual that a 0.3-meter (1-foot) rise would push the coastline back several kilometers.

A modest 1-meter (3-foot) rise would flood low-lying areas of major cities such as Shanghai, Cairo, Bangkok, and Venice and large areas of agricultural lowlands and deltas in Bangladesh, India, Egypt, and China, where much of the world's rice is grown. With a 1.5-meter (5-foot) rise many small low-lying islands like the Marshall Islands in the Pacific, the Maldives (a series of about 1,200 islands off the west coast of India), and some Caribbean nations would cease to exist, creating a multitude of environmental refugees.

Large areas of the wetlands that nourish the world's fisheries would also be destroyed (see Spotlight on p. 97). The EPA projects that a sea level rise of 1 meter (3 feet) would destroy 26% to 65% of the coastal wetlands in the United States. Even a 0.5-meter (1.6-foot) rise would drown about one-third of U.S. coastal wetlands. The salinity of streams, bays, and coastal aquifers would also increase. Tanks storing hazardous chemicals along the Gulf and Atlantic coasts would also be flooded.

Most barrier islands (Figure 5-25, p. 99) would be flooded and shifted toward the shore. Places such as the low-lying Florida Keys and the present beaches of Malibu, California, would also be covered with water.

Low-lying areas in U.S. cities such as New Orleans, New York, Atlantic City, Boston, Washington, D.C., Galveston, Charleston, and Savannah would be threatened by flooding unless billions of dollars were spent to build and maintain massive systems of dikes and levees. Even dikes and levees wouldn't save Miami (which lies at or just above sea level on swampland reclaimed from the Everglades) because it sits on a porous bed of limestone. This means that the ocean would seep in underneath the city, contaminating all fresh water supplies and making the entire area uninhabitable. A comedian joked that he was planning to buy land in Kansas because it would probably become valuable beach front property.

10-3 DEALING WITH GLOBAL WARMING

Slowing Down Global Warming We have two options for dealing with the global warming many scientists believe we have set in motion: slow it down or adjust to its effects. Many experts believe that we must do both with no time to lose.

The cures to this planetary crisis we have caused are controversial, difficult, and painful. But if the models are correct, we are in the position of a long-time alcoholic whose doctor tells him that if he doesn't stop drinking now he will die.

We and many other species can learn to live under different climate conditions, if we are given time to make the necessary changes. This is why slowing down any major climate change—either warming or cooling—caused by our activities must become the major priority of our species worldwide. Otherwise, environmental and economic security could be threatened everywhere within a single generation. Since we are emitting greenhouse gases into the atmosphere almost everywhere we must slow global warming by simultaneously using all the methods available. Ways to do this include:

Input Approaches

■ Banning all production and uses of chlorofluorocarbons and halons by 1995. This is the easiest thing we can do because we can either do without these chemicals or phase in substitutes for their essential uses. It is also the best early test of worldwide commitment to protecting the atmosphere from both global warming and ozone depletion.

■ Cutting current fossil fuel use 20% by 2000 and 50% by 2010 and 70% by 2030. The largest users of fossil fuels such as the United States and the Soviet Union should cut their use by about 35% by the year 2000.

■ Greatly improving energy efficiency. This is the quickest, cheapest, and most effective method to reduce emissions of CO_2 and other air pollutants during the next two to three decades (see Section 17-2 and Spotlight on p. 221).

■ Shifting to perpetual and renewable energy resources that do not emit CO_2 over the next 30 years (Chapter 17). Ultimately the world must move away from fossil fuels for most of its energy, even if we cut CO_2 emissions in half. Otherwise emissions would begin to rise because of increasing population and industrialization. Increased use of perpetual and renewable energy resources can cut projected U.S. CO_2 emissions 8% to 15% by the year 2000 and can virtually eliminate them by 2010.

■ Transferring energy efficiency and renewable energy technology and pollution prevention and waste reduction technology to LDCs so they can leapfrog into a new sustainable-earth age instead of following the energy- and matter-wasting and earth-depleting path of today's MDCs.

■ Increasing the use of nuclear power to produce electricity if a new generation of much safer reactors can be developed and the problem of how to store nuclear waste safely for thousands of years can be solved (Section 18-3). However, improving energy efficiency is much quicker and safer and reduces emissions of CO_2 per dollar invested 2.5 to 10 times more than nuclear power (see Pro/Con on p. 431).

■ Placing heavy taxes on gasoline and emissions fees on each unit of fossil fuel (especially coal) burned to reduce emissions of CO_2 and other air pollutants (Section 18-5). Some of the tax revenue should be used to improve the energy efficiency of dwellings and heating systems for the poor in MDCs and LDCs and to provide them with enough energy to offset higher fuel prices.

■ Sharply reducing the use of coal, which emits 60% more carbon dioxide per unit of energy produced than any other fossil fuel. Using the world's estimated coal supplies would produce at least a sixfold or eightfold increase in atmospheric carbon dioxide. But to power its industrialization program, China plans to nearly double coal use in the next decade and India plans to triple its use. MDCs must try to prevent this by helping these and other LDCs greatly improve their energy efficiency and shift from coal to perpetual and renewable energy sources.

■ Switching from coal to natural gas for producing electricity and high-temperature heat in countries, such as the United States and the Soviet Union, that have ample supplies of natural gas, which emits only half as much CO_2 per unit of energy as coal. Switching to natural gas also sharply reduces emissions of other air pollutants (Section 18-1). Because burning natural gas still emits CO_2, this is only a short-term method that helps buy time to switch to an age of energy efficiency and renewable energy. Also, a recent study indicates that methane leaking from natural gas distribution systems has such a powerful greenhouse effect that it could offset the benefits of switching from coal to natural gas.

■ Capturing methane gas emitted by landfills and using it as a fuel. Burning this methane produces carbon dioxide, but each molecule of methane reaching the atmosphere causes about 25 times more global warming than each molecule of CO_2.

■ Sharply reducing beef production to reduce the fossil-fuel inputs into agriculture and to reduce the methane produced by the animals themselves.

■ Halting unsustainable deforestation everywhere by the year 2000 (Sections 15-2 and 15-4).

■ Switching from unsustainable to sustainable agriculture (Sections 13-6 and 14-5). Worldwide, agriculture is responsible for about 14% of the greenhouse gases we emit into the atmosphere. If LDCs increase their use of unsustainable industrialized agriculture, this percentage could rise.

■ Slowing population growth (Section 6-4). If we cut greenhouse gas emissions in half and population doubles, we are back where we started.

Energy expert Amory Lovins (see Guest Essay on p. 56) says that arguments over whether global warming is happening, will happen, may not happen, may not be as severe as projected, and what its impacts will be are largely irrelevant and divert us from doing what needs to be done anyway. The reason is that the remedies just listed for slowing global warming are things we need to do *now* even if there were no threat of global warming or any other type of climate change.

Also Lovins argues that getting countries to sign treaties and agree to cut back their use of fossil fuels in time enough to reduce serious environmental effects is difficult, if not almost impossible, and very costly. Evidence for this appeared in November 1989 when representatives from 70 nations couldn't even agree to freeze their emissions of greenhouse gases at 1988 levels by the year 2005, mainly because of opposition by the United States, the Soviet Union, Japan, and China—which together account for 58% of the world's output of these gases.

In 1990 when leadership was needed President George Bush mainly called for more research rather than action to delay global warming. U.S. House of Representatives member Claudine Schneider calls this the "let's wait until the ship hits the rocks, and then figure out what to do" approach so prevalent in public policy-making today.

According to Lovins, the good news, among all the gloom-and-doom about global warming, is that improving energy efficiency is the fastest, cheapest, and surest way to sharply cut carbon dioxide emissions and emissions of most other air pollutants within two decades using existing technology (Section 17-2). This approach should also be immensely profitable, saving the world up to a trillion dollars a year—as much as the annual global military budget.

Moreover, reducing fossil fuel use by improving energy efficiency reduces all forms of pollution, helps protect biodiversity, and avoids arguments among governments about how CO_2 reductions should be divided up and enforced. This approach will also make the world's fossil fuel supplies last longer, reduce international tensions over who gets the world's dwindling oil supplies, and give us more time to phase in alternatives to fossil fuels.

Industrialized countries will have to set a better example by committing themselves to a crash program to improve energy efficiency (Section 17-2). They will also have to lead the shift from nonrenewable fossil fuels and nuclear energy to perpetual and renewable energy sources (Chapter 17).

Existing and new technologies for improving energy efficiency and using perpetual and renewable energy must also be transferred to LDCs, which on average are nearly three times less energy efficient than the average MDC. According to Lovins, this in principle could allow LDCs to expand their economies by about tenfold with no increase in energy use and avoid the dirtiest stage of the industrialization process. But instead of doing this the United States and some other industrialized countries are now exporting their least efficient energy technologies—the ones too obsolete and costly to sell at home—to LDCs.

Greatly improving energy efficiency *now* is a money-saving, life-saving, and earth-saving offer that we must not refuse. So far, however, no government has made this approach much more than a token part of its national strategy for slowing greenhouse warming, reducing dependence on oil, and reducing air and water pollution.

- Converting to sustainable-earth economies in MDCs and LDCs (see Spotlight on p. 156).

- Dismantling the global poverty trap to reduce unnecessary deaths and human suffering and environmental degradation and to help LDCs help themselves and not follow the present throwaway industrial path of today's MDCs (Section 7-3).

Output Approaches

- Developing better methods to remove carbon dioxide from the smokestack emissions of coal-burning power and industrial plants and from vehicle exhausts. Present methods remove only about 30% of the CO_2 and would at least double the cost of electricity. Eventually this approach would be overwhelmed by increased fossil fuel use. Also the recovered CO_2 must be kept out of the atmosphere, presumably by putting it in the deep ocean, spent oil and gas wells, and excavated salt caverns, or by reacting it with other substances to convert it to a solid such as limestone. The effectiveness and cost of these methods is unknown.

- Planting trees. Everyone—even students—should plant and care for at least one tree every six months. This is an important form of earth care, but we should recognize that tree planting is only a stopgap measure for slowing CO_2 emissions. Trees must be continually planted faster than they are cut down and burned or die and rot, both processes that release CO_2 into the atmosphere. To absorb the carbon dioxide we are now putting into the atmosphere each year, we would have to plant and tend an area of new forest equal to the size of Australia every year. Worldwide, this amounts to an average of 1,000 trees per person per year and 4,500 trees annually for each American citizen—18,000 trees a year for a family of four.

- Recycling CO_2 released in industrial processes.

- Removing CO_2 by photosynthesis by using tanks and ponds of marine algae or by fertilizing the oceans with iron to stimulate the growth of marine algae.

Economist William Nordhaus estimates that a 50% reduction in 1989 greenhouse-gas emissions will cost in the long run about 1% of the world's total annual economic output. At today's level of economic output this amounts to approximately $200 billion a year for about 30 years. Reducing these emissions by 20% would cost about $12 billion a year.

Adjusting to Global Warming Even if all the things just listed are done we are still likely to experience some global warming, although at a more manageable rate. Since there is a good chance that many of these things will either not be done or be done too slowly, some analysts suggest that we should also begin preparing for the effects of long-term global warming. Their suggestions include

- increasing research on the breeding of food plants that need less water and plants that can thrive in water too salty for ordinary crops

- building dikes to protect coastal areas from flooding, as the Dutch have done for hundreds of years

- moving storage tanks of hazardous materials away from coastal areas

- banning new construction on low-lying coastal areas

- storing large supplies of key foods throughout the world as insurance against disruptions in food production

- expanding existing wilderness areas, parks, and wildlife refuges northward in the Northern Hemisphere and southward in the Southern Hemisphere and creating new wildlife reserves in these areas

- developing management plans for existing parks and reserves that take into account possible climate changes

- connecting existing and new wildlife reserves by corridors that would allow mobile species to change their geographic distributions and transplanting endangered species to new areas

- wasting less water (see Section 11-3 and Individuals Matter inside the back cover)

We have known about the greenhouse effect and its possible consequences for decades. We also know what needs to be done at the international, national, local, and individual levels (see Individuals Matter on p. 223). Research must be expanded to help clear up the uncertainties that continue to exist. But to most environmentalists and many climatologists this is no excuse for doing nothing or very little now.

10-4 DEPLETION OF OZONE IN THE STRATOSPHERE

Uses of Chlorofluorocarbons and Halons

In 1974 chemists Sherwood Roland and Mario Molina theorized that human-made chlorofluorocarbons (CFCs), also known by their Du Pont trademark as Freons, were lowering the average concentration of ozone in the stratosphere and creating a global time bomb. No one suspected such a possibility when CFCs were developed in 1930.

The most widely used CFCs are CFC-11 (trichlorofluoromethane) and CFC-12 (dichlorodifluoromethane). These stable, odorless, nonflammable, nontoxic, and noncorrosive chemicals were a chemist's dream. Soon they were widely used as coolants in air conditioners and refrigerators and as propellants in aerosol spray cans. They are also used to clean electronic parts such as computer chips, as hospital sterilants, as fumigants for granaries and cargo holds, and to create the bubbles in polystyrene plastic foam (often called by its Du Pont trade name Styrofoam) used for insulation and packaging. Bromine-containing compounds, called *halons*, are also widely used, mostly in fire extinguishers. Other widely used ozone-destroying chemicals are carbon tetrachloride (used mostly as a solvent) and methyl chloroform (used as a cleaning solvent for metals and in consumer products such as correction fluid, dry cleaning sprays, spray adhesives, and other aerosols).

Since 1945 the use of the four major types of CFCs has increased sharply. Industrial countries account for 84% of CFC production, with the United States being the top producer followed by western European countries and Japan. Since 1978 the use of CFCs in aerosol cans has been banned for most uses in the United States, Canada, and most Scandinavian countries, primarily because of consumer boycotts. However, worldwide aerosols account for 25% of global CFC use. In the United States CFCs are still legally used as aerosol propellants in asthma and other medication sprays, cleaning sprays for VCRs and sewing machines, and even in canned confetti.

While waiting for the world's governments to adopt strategies for slowing global warming, each of us can take matters into our own hands.

■ Be aware of your CO_2 emissions and reduce them. In the United States each person emits an average of 16.7 metric tons (18.4 tons) of CO_2 a year, six times more than the average person emits in an LDC. Using one kilowatt-hour of electricity generated at a coal-fired power plant emits 0.9 kilogram (2 pounds) of CO_2, and burning 3.8 liters (1 gallon) of gasoline emits 8.6 kilograms (19 pounds) of CO_2. The average American car, driven 6,200 kilometers (10,000 miles), releases its own weight of carbon into the atmosphere. In the United States vehicles emit 33% of the CO_2 produced when fossil fuels are burned, explaining why driving a car is the major way you help turn up the earth's thermostat.

■ Reduce your use and unnecessary waste of energy (see Individuals Matter inside the back cover). Since use and waste of fossil fuels is the major cause of projected global warming and most other forms of pollution and environmental degradation, reducing your use and waste of energy is the most important thing you can do. Driving a car that gets at least 15 kilometers per liter (35 miles per gallon), using a carpool and mass transit, and walking or bicycling when possible are the major ways you can reduce your emissions of CO_2 and other air pollutants, and save money besides.

■ Don't use electricity to heat space or water (Section 3-6).

■ Make your house energy efficient, and heat your house and household water by using perpetual energy from the sun. Cool your home by using shade trees and available winds (Chapter 17).

■ If you can't use perpetual and renewable energy to heat your house and water, use natural gas. When burned, it produces much less carbon dioxide and other air pollutants than burning oil or than burning coal at a power plant to produce electricity for heating space and household water.

■ Plant trees and help cool the globe and your house. This must be done on a continuous basis so that the rate of planting exceeds the rate at which trees release carbon dioxide into the atmosphere when they are burned or when they die and decay. Ask your employer to sponsor a tree-planting program by buying seedlings to be planted by children from a local school. Planting trees is important, but it is not nearly as important as preventing the clearing of existing forests, improving energy efficiency, and shifting to perpetual and renewable sources of energy.

■ Use the following priorities for all items: No use unless necessary; reuse and recycle; throw away only as a last resort.

■ Buy reusable products to cut down on energy and mineral resource use, thereby reducing greenhouse-gas emissions. Use washable ceramic cups and dishes and metal tableware instead of throwaway paper and plastic. Keep a set at work. Carry a set in your car and have them filled at a fast food restaurant. Use washable wash cloths, sponges, and cloth diapers instead of throwaway paper towels and diapers. Carry your lunch in a reusable lunch box. Put your garbage out in reusable plastic containers instead of in throwaway plastic bags. Carry your groceries in your own reusable string or canvas bags.

■ Recycle items that can't be reused and buy products made from recycled materials.

■ Urge state and national legislators to sponsor bills aimed at greatly improving energy efficiency, halting the harvesting of stands of ancient forests (see front cover), and curbing emissions of greenhouse gases and other air pollutants.

■ Don't support highly unpredictable schemes such as covering the oceans with white Styrofoam chips to help reflect more energy away from the earth's surface, dumping iron into oceans to stimulate marine algae growth to remove CO_2 from the atmosphere (at a cost of about $1 billion per year), unfurling a gigantic foil-faced sun shield in space, or injecting sunlight-reflecting particulate matter into the stratosphere to cool it by exploding nuclear bombs near the earth's surface or by using aircraft or rocket systems. Some of these large-scale technological solutions may be possible in the future, but they can have harmful side effects that we can't anticipate because of our poor understanding of how the earth works.

The United States accounts for about 29% of the global consumption of CFCs. Also, Americans use six times more CFCs per person than the global average. Vehicle air conditioners account for about three-quarters of annual CFC emissions in the United States.

Ozone Layer Depletion Spray cans, discarded or leaking refrigeration and air conditioning equipment, and the production and burning of plastic foam products release CFCs into the atmosphere. Depending on the type, CFCs are so unreactive that they stay intact in the atmosphere for 22 to 111 years. This gives them plenty of time to rise slowly through the atmosphere until they reach the stratosphere. There, under the influence of high-energy UV radiation from the sun, they break down and release chlorine atoms, which speed up the breakdown of ozone into O_2 and O. Over time a single chlorine atom can convert as many as 100,000 molecules of O_3 to O_2. Although this effect was proposed in 1974, it took 15 years of interaction between scientists and politicians before countries took action to begin slowly phasing out CFCs.*

Several other stable, chlorine-containing compounds, including widely used solvents such as methyl chloroform and carbon tetrachloride, also rise into the stratosphere and destroy ozone molecules. When fire extinguishers are used their unreactive bromine-containing halon compounds enter the air and eventually reach the stratosphere where they are broken apart by UV radiation. Each of their bromine atoms destroys hundreds of times more ozone molecules than a chlorine atom. All these compounds, especially CFCs, are also greenhouse gases that contribute to global warming (Figure 10-6b).

In the 1980s researchers were surprised to find that up to 50% of the ozone in the upper stratosphere over the Antarctic is destroyed each September and October during the Antarctic spring—something not predicted by computer models of the stratosphere. During these two months of 1987 and 1989, this Antarctic ozone hole covered an area larger than the continental United States (Figure 10-9).

Measurements indicate that this large annual decrease in ozone over the South Pole is caused by the presence of large spinning vortices. Clouds of tiny ice crystals form in these vortices. The surfaces of these crystals absorb CFCs, blown in by global air circulation that moves air from the equator to the poles (Figure 10-3). This greatly increases the rate at which these chemicals destroy ozone and leads to the sharp drop in ozone over the Antarctic each fall.

After about two months the vortex breaks up and great clumps of ozone-depleted air flow northward and linger over parts of Australia, New Zealand, and the southern tips of South America and Africa for a few weeks. During this period ultraviolet levels in these areas may increase as much as 20%.

In 1988 scientists discovered that a similar but smaller ozone hole formed over the Arctic during the two-month Arctic spring, with an ozone loss of 20% to 25%. When this hole breaks up, clumps of ozone-depleted air flow southward and linger over parts of Europe and North America. This can produce a 5%

*For a fascinating account of how corporate stalling, politics, economics, and science interact, see Sharon Roan's *Ozone Crisis: The 15-Year Evolution of a Sudden Global Emergency* (New York: Wiley, 1989).

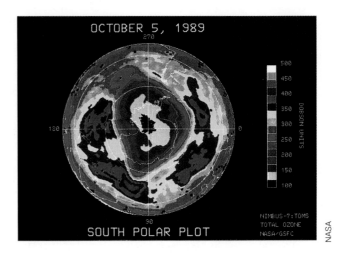

Figure 10-9 The purple shade in this image taken by the NIMBUS-7 satellite shows how far the annual ozone hole that appears during the fall in the upper stratosphere over the Antarctic had spread by October 5, 1989. This hole, where the normal ozone level has been cut in half, is ten times larger than the area of the continental United States and is caused by ozone-destroying chlorofluorocarbons we have put into the atmosphere. The lower the Dobson units shown in the scale on the right, the greater the depletion of ozone. In October 1987 the level over the Antarctic fell to an all-time low of 109 units; 350 is considered normal.

winter loss of ozone over much of the Northern Hemisphere.

In 1988 the National Aeronautics and Space Administration (NASA) released a study showing that the stratospheric ozone-depletion average over the whole year had decreased by as much as 3% over heavily populated regions of North America, Europe, and Asia since 1969 (Figure 10-10).

Unless emissions of these chemicals are cut drastically, average levels of ozone in the stratosphere could drop by 10% to 25% by 2050 or sooner, with much higher drops in certain areas (Figure 10-10).

Effects of Ozone Depletion With less ozone in the stratosphere, more biologically harmful ultraviolet B radiation will reach the earth's surface. This form of UV radiation damages DNA molecules and can cause genetic defects on the outer surfaces of plants and animals, including your skin. Each 1% loss of ozone leads to a 2% increase in the ultraviolet radiation striking the earth and a 5% to 7% increase in skin cancer, including a 1% increase in deadly malignant melanoma.

The EPA estimates that a 5% ozone depletion would cause the following effects in the United States:

■ An additional 940,000 cases annually of basal-cell and squamous-cell skin cancers (Figure 8-7a, b, p. 180), both disfiguring but usually not fatal cancers if treated in time (see Spotlight on p. 180).

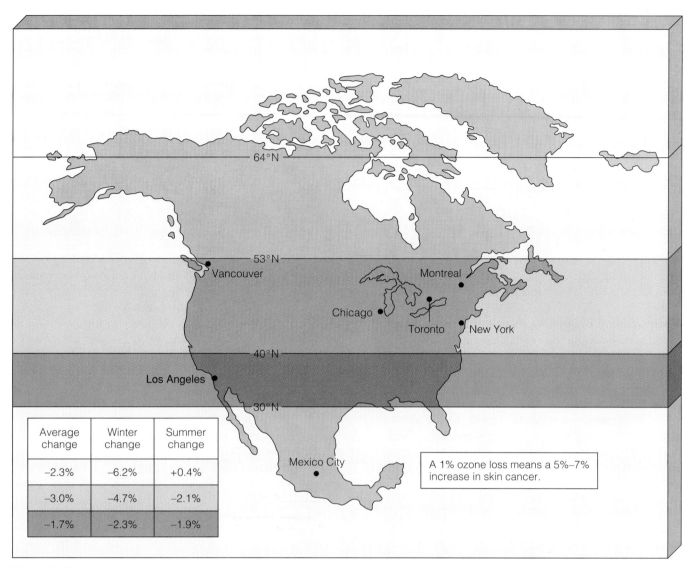

Average change	Winter change	Summer change
−2.3%	−6.2%	+0.4%
−3.0%	−4.7%	−2.1%
−1.7%	−2.3%	−1.9%

A 1% ozone loss means a 5%–7% increase in skin cancer.

Figure 10-10 Average drops in ozone levels in the stratosphere above parts of the earth between 1969 and 1986, based on data gathered from satellites and ground stations. (Data from NASA)

- An additional 30,000 cases annually of often-fatal melanoma skin cancer (Figure 8-7c, p. 180), which now kills almost 9,000 Americans each year.

- A sharp increase in eye cataracts (a clouding of the eye that causes blurred vision and eventual blindness) and severe sunburn in people and eye cancer in cattle.

- Suppression of the human immune system, which would reduce our defenses against a variety of infectious diseases, an effect similar to the AIDS virus.

- An increase in eye-burning photochemical smog, highly damaging ozone, and acid deposition in the troposphere (Chapter 9). According to the EPA, each 1% decrease in stratospheric ozone may cause a 2% increase in ozone near the ground.

- Decreased yields of important food crops such as corn, rice, soybeans, and wheat.

- Reduction in the growth of ocean phytoplankton that form the base of ocean food chains and webs and help remove carbon dioxide from the atmosphere.

- A loss of perhaps $2 billion a year from degradation of paints, plastics, and other polymer materials.

- Increased global warming.

The EPA warns that unless action is taken to halt ozone destruction, the United States alone can expect 40 million skin cancer cases and 800,000 premature deaths of people alive today and those born during the next 88 years. The Commerce Department estimates that the harmful external costs to U.S. society

" I MISS THE OZONE LAYER...."

over the next 88 years from failure to ban ozone-destroying chemicals would be $1.3 trillion; the industrial cost of the phase out would be less than $4 billion. For example, the estimated harmful costs of releasing the CFCs in a single aerosol can total $12,000. The harmful costs of CFCs released from just one automobile air conditioner during use and repair are many times this figure.

In a worst case scenario no one would be able to expose themselves to the sun (see cartoon). Cattle could only graze at dusk without eye damage. Farmers might measure their exposure to the sun in minutes, like workers in nuclear power plants.

 ## PROTECTING THE OZONE LAYER

10-5

Major Actions Models of atmospheric processes indicate that just to keep CFCs at 1987 levels would require an immediate 85% drop in total CFC emissions throughout the world. Analysts believe that the first step toward this goal should be an immediate worldwide ban on the use of CFCs in aerosol spray cans and in producing plastic foam products. Cost-effective substitutes are already available for these uses. Automotive service shops should be required to recycle CFCs from automotive air conditioners and the sale of small cans of CFCs used by consumers to charge leaky air conditioners should be banned by 1992.

The next step would be to phase out all other uses of CFCs, halons, carbon tetrachloride, and methyl chloroform by 1995. Substitute coolants in refrigeration and air conditioning will probably cost more. But compared to the potential economic and health consequences of ozone depletion, such cost increases would be minor. Although they receive little publicity compared to CFCs, the widely used solvents carbon tetrachloride and methyl chloroform contribute more

to ozone-threatening chlorine levels than all but two of the eight CFCs and halons now partially controlled by an international treaty. Substitutes are available for virtually all uses of these two chemicals.

However, we must be sure that substitutes don't contribute to atmospheric warming or cause other harmful effects. Currently, there are three major types of substitutes. One consists of chemicals outside the fluorocarbon family that can be used as cleaning and blowing agents. The other two types, useful mainly as cooling agents in refrigerators and air conditioners, are hydrofluorocarbons (HFCs) that contain no chlorine or bromine atoms and hydrochlorofluorocarbons (HCFCs) that contain fewer atoms of chlorine per molecule than conventional CFCs.

HFCs and HCFCs are decomposed more rapidly than conventional CFCs and have shorter atmospheric lifetimes of 2 to 20 years depending on the compound. But HCFCs contain some ozone-destroying chlorine atoms and both HFCs and HCFCs are still greenhouse gases. However, their ozone depletion potential is only 2% to 10% of conventional CFCs and they would contribute about 90% less per kilogram to greenhouse warming than currently used CFCs.

One HCFC, Dymel, is being marketed by Du Pont as an aerosol propellant in hair sprays, deodorants, colognes, and other products. This unnecessary use of an HCFC is incorrectly thought of by some as "environmentally friendly." But while HFC and HCFC substitutes may help make the transition away from CFCs for essential uses such as refrigeration, eventually these new chemicals will also have to be banned to halt ozone depletion.

Hopeful but Inadequate Progress Some progress has been made. In 1987, 24 nations meeting in Montreal, Canada, developed a treaty—commonly known as the "Montreal Protocol"—to reduce production of the eight most widely used and most damaging CFCs. By early 1990, 49 countries had signed this historic treaty. If carried out, it will reduce total emissions of CFCs into the atmosphere by about 35% between 1989 and 2000. According to the EPA this would prevent about 137 million cases of skin cancer, 27 million skin-cancer deaths, and 1.2 million eye cataracts.

Most scientists agree that the treaty is an important symbol of global cooperation but that it does not go far enough to prevent significant depletion of the ozone layer and global warming. Indeed by 1989 new evidence indicated that we had already destroyed as much ozone as the treaty-makers assumed we would lose by the year 2050. Since some of these chemicals are unnecessary and substitutes exist, most scientists call for phasing out all uses of ozone-depleting chemicals by 1995, as Sweden has agreed to do. Environmentalists also call for all products that contain or require CFCs, halons, or other ozone-depleting chem-

- Avoid purchasing products containing chlorofluorocarbons, halons, carbon tetrachloride, and methyl chloroform (1,1,1, trichloroethane on most ingredient labels). Such products include cleaning sprays for sewing machines, VCRs, and electronic equipment, spray-on cleaners and spot removers, bug killers and foggers, shoe polish sprays and other aerosols, and polystyrene foam insulation and packaging. Seek out the substitutes that are or will soon be available for these products.

- Don't buy CFC-containing polystyrene foam insulation. Types of insulation that don't contain CFCs are extended polystyrene (commonly called EPS or beadboard), fiberglass, rock wool, cellulose, and perlite.

- Don't buy halon or carbon dioxide fire extinguishers for home use. Instead, buy those using dry chemicals.

- Stop using all aerosol spray products, except in some medical sprays. Even those not using CFCs and HCFCs (Du Pont's Dymel) emit hydrocarbons or other propellant chemicals into the air. Use roll-on and hand-pump products instead.

- Pressure legislators to ban all uses of CFCs and halons by 1995, carbon tetrachloride and methyl chloroform by 2000, and HCFC and HFC substitutes by the year 2010.

- Pressure legislators to tax the billions of dollars in windfall profits that CFC and halon manufacturers will get from a phasedown of their use and to use these tax revenues for researching the climate, improving energy efficiency, and switching to renewable energy resources.

- As they become available, buy new refrigerators and freezers that use vacuum insulation (as

in Thermos bottles) instead of rigid-foam insulation and that use helium as a coolant instead of CFCs or HCFCs (available from Cryodynamics, 1101 Bristol Road, Mountainside, NJ 07092).

- Pressure legislators to require that all products containing or requiring CFCs, halons, or other ozone-depleting chemicals for their manufacture be clearly labelled.

- Since leaky air conditioners in cars are the single largest source of CFC emissions, if you have an auto air conditioner make sure it is not leaking. If it needs to be recharged, take it to a shop that has the equipment to recycle its CFCs.

- Don't fall for highly unpredictable schemes such as using lasers to blast CFCs out of the sky and building more supersonic commercial airplanes (SSTs), whose emissions could deplete ozone in the stratosphere.

icals for their manufacture to be clearly labelled so that consumers can consciously choose whether to use such products.

In 1989 delegates from 82 countries meeting in Helsinki, Finland, pledged in principle to phase out the use and production of the five CFCs most strongly implicated in ozone depletion by the year 2000, if substitutes are available by then. They also agreed on the need to phase out or reduce the use of other ozone-depleting substances such as halons, carbon tetrachloride, methyl chloroform, and the HCFCs now being used as substitutes for some CFCs, but took no action. According to the EPA, overall concentrations of chlorine from these ozone-destroying chemicals in the stratosphere could double, triple, or quadruple in the next century even if CFCs are completely phased out. Also, the effectiveness of any ban on ozone-depleting chemicals will depend on the willingness of large LDCs such as China and India to participate.

Even if all ozone-depleting substances were banned tomorrow, it would take about 100 years for the planet to recover from the present ozone depletion and that to come from those substances already in the atmosphere. The key question is whether MDCs and LDCs

can agree to sacrifice short-term economic gain by eliminating their use of all ozone-depleting chemicals within the next decade to protect life on earth in coming decades.

In 1975, anthropologist Margaret Mead said, "The atmosphere is the key symbol of global interdependence. If we can't solve some of our problems in the face of threats to this global commons, then I can't be very optimistic about the future of the world." Perhaps the challenges posed by global warming and ozone depletion can be a catalyst for worldwide awareness of the urgent need to get serious about sustaining the earth and learning how to deal with long-term problems that build up slowly and invisibly until they exceed threshold levels. Let's hope so and begin by sharply reducing our individual impacts on the ozone layer (see Individuals Matter above).

10-6 CLIMATE, BIODIVERSITY, AND NUCLEAR WAR

The Ultimate Ecological Catastrophe Most people believe that global nuclear war is the single greatest threat to the human species and the earth's

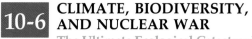

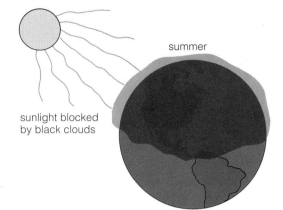

summer

sunlight blocked
by black clouds

Figure 10-11 The nuclear winter or autumn effect caused by a limited nuclear war in the Northern Hemisphere. Massive amounts of smoke, soot, dust, and other debris, lifted into the atmosphere as a result of the nuclear explosions and subsequent fires, would coalesce into huge smoke clouds. These dense clouds would cover large portions of the Northern Hemisphere, prevent 20% to 90% of all sunlight from reaching large areas, and cause a sharp drop in atmospheric temperatures for several weeks to months, depending on the size and locations of the explosions.

life-support systems for most species. The nuclear age began in August 1945, when the United States exploded a single nuclear fission atomic bomb over Hiroshima and another over Nagasaki. These blasts killed an estimated 110,000 to 140,000 people and injured tens of thousands more.

By the end of the year another 100,000 had died, mostly from exposure to radioactive fallout: dirt and debris sucked up and made radioactive by the blast and dropped back to earth near ground zero and on downwind areas hundreds and even thousands of kilometers away. Other people, exposed to nonlethal doses of ionizing radiation, developed cataracts, leukemia, and other forms of lethal cancer decades later and some are still dying today. This shows us what a small nuclear bomb can do.

Today the world's nuclear arsenals have an explosive power equal to more than 952,000 Hiroshima-type bombs or 3,333 times all the explosives detonated during World War II. Each Trident II submarine that began coming on line in 1989 carries nuclear warheads with the explosive power of 4,570 Hiroshima bombs or over 30 times all the explosives detonated in World War II. Today we live in a world with enough nuclear weapons to kill everyone on earth 67 times. By the end of this century 60 countries—one of every three in the world—will either have nuclear weapons or the knowledge and capability to build them.

In addition to the health and environmental threats, the buildup of nuclear and conventional weapons drains funds and creativity that could be used to solve most of the world's population, food,

health, resource, and environmental problems. For example, about 40% of the world's research and development expenditures and 50% of its physical scientists and engineers are devoted to developing weapons to improve our ability to kill one another.

Nuclear Winter and Nuclear Autumn Effects Some U.S. and Soviet military strategists have talked about the concept of "limited nuclear war." Since 1982, however, evaluation of previously overlooked calculations has suggested that even a limited nuclear war could kill 1 to 4 billion people—20% to 80% of the world's current population. Such an exchange would probably take place in the Northern Hemisphere, presumably between the United States and the Soviet Union, either because of direct confrontation between these two superpowers or because conflicts between other countries with nuclear weapons escalate into a global conflict involving the superpowers. If this happened, the direct effects of the explosions would kill an estimated 1 billion people, mostly in the Northern Hemisphere. Tens of millions more would suffer injuries.

Computer models indicate that within the next two years, another 1 to 3 billion might die from starvation caused by disruption of world agricultural production, first in the Northern Hemisphere and later in the Southern Hemisphere, because of what is called the *nuclear winter* or *nuclear autumn effect*.

Depending on the extent, time, and location of the explosions, average atmospheric temperatures in temperate areas would probably drop rapidly to temperatures typical of fall or early winter. This atmospheric cooling would happen because of a reverse greenhouse effect. Massive amounts of smoke, soot, dust, and other debris, lifted into the atmosphere as a result of the nuclear explosions and subsequent fires, would coalesce into huge smoke clouds. Within two weeks these dense clouds would cover large portions of the Northern Hemisphere and prevent 20% to 90% of all sunlight from reaching large areas (Figure 10-11).

The abnormally cold temperatures and reduction of sunlight would cause a sharp drop in food production in the growing season following the war. This would cause widespread starvation in the Northern Hemisphere and in African and Asian countries dependent on food imports from countries such as the United States and Canada.

Food production in the Southern Hemisphere might also be affected as the smoke clouds gradually became less dense and spread southward. Drops in temperature and sunlight would be less severe than in the Northern Hemisphere. But even a slight reduction in temperature could be disastrous to agriculture in tropical and subtropical forest areas and could lead to the extinction of numerous plant and animal species. Subtropical grasslands and savannas in Africa and South America might be the least affected of the world's ecosystems because their plants are more cold tolerant and drought resistant.

A cold and dark nuclear winter or autumn would not be the only cause of food scarcity. Plagues of rapidly producing insects and rodents—the life forms best equipped to survive nuclear war—would damage stored food and spread disease. In areas where crops could still be grown, farmers would be isolated from supplies of seeds, fertilizer, pesticides, and fuel. People hoping to subsist on seafood would find many surviving aquatic species contaminated with radioactivity, runoff from ruptured tanks of industrial liquids, and oil pouring out of damaged offshore rigs.

A limited nuclear war would also destroy 30% to 70% of the ozone layer, leading to a deadly *ultraviolet summer*. This would sharply decrease the ability to grow crops and harvest fish. It would also cause a massive increase in skin cancers and eye cataracts and impair immune systems so that already weakened survivors would die from infectious diseases.

Incineration of oil in tanks and refineries, storage tanks of hazardous chemicals, rubber tires, and other materials would produce toxic smog, which would cover much of the Northern Hemisphere. Large areas might also be assaulted with extremely acid rains. High levels of radioactivity from fallout would contaminate most remaining food and water supplies, kill many people and other warm-blooded animals, some crops, and certain coniferous trees, especially pines.

Thus, people not killed outright by the nuclear explosions would find themselves choking and freezing in a smoggy, radioactive darkness with contaminated water supplies and little chance of growing food for a year or perhaps several years. Most people who survived in underground bunkers would die slowly rather than quickly.

The computer models of the atmosphere used to make these calculations are crude and may overestimate or underestimate the effects of nuclear explosions. But scientists agree that even if the effects are less than the models suggest, they would still cause serious disruptions in regional and global climate.

It is encouraging that there is some thawing in the cold war tensions between the United States and the Soviet Union and that these countries have destroyed a small number of their nuclear missiles. But most people are unaware that only the delivery systems have been destroyed. The fissionable bomb material has merely been removed and stored for possible future use. We must go much further and abolish war, something that Kenneth Boulding thinks is now within our grasp (see Guest Essay on p. 230).

Some people have wondered whether there is intelligent life in other parts of the universe. Perhaps the real question we should ask is whether there is intelligent life on earth. If we can seriously deal with the three planetary emergencies discussed in this chapter *now*—not after decades more of discussion, research, delay, and wasteful depletion of the earth's natural capital, then the answer is a hopeful yes. Otherwise, it is a tragic no. The choice is ours. Not to decide is to decide.

Even though climate change threatens to be bigger, more irreversible, and more pervasive than other environmental problems, it's also controllable, if we act now. That much is clear.

GUS SPETH

DISCUSSION TOPICS

1. What consumption patterns and other features of your lifestyle directly add greenhouse gases to the atmosphere? Which, if any, of these things would you be willing to give up in order to slow projected global warming and reduce other forms of air pollution?

2. Explain why you agree or disagree with each of the proposals for **(a)** slowing down emissions of greenhouse gases into the atmosphere listed on pp. 220–222, and **(b)** adjusting to the effects of global warming listed on p. 222. Explain. What effects would carrying out these proposals have on your lifestyle and those of any children you might choose to have? What effects might not carrying out these actions have?

3. In 1989, U.S. Senator Albert Gore introduced a legislative package he calls the Strategic Environment Initiative (SEI), an ecological version of Ronald Reagan's Strategic Defense Initiative (SDI). Domestically the SEI would focus on improving energy efficiency, developing alternative fuels, reforestation, comprehensive recycling, and drastic cuts in ozone-depleting chemicals. It would also help LDCs obtain energy-efficient technology and develop environmentally sustainable industries and agriculture. Do you support such a bill? What things, if any, would you add? What has happened to this proposal since it was first introduced in 1989?

4. What consumption patterns and other features of your lifestyle directly and indirectly add ozone-depleting chemicals to the atmosphere? Which, if any, of these things would you be willing to give up in order to slow ozone depletion?

5. Should all uses of CFCs, halons, and other ozone-depleting chemicals be banned in the United States and worldwide? Explain. Suppose this meant that air conditioning (especially in cars and perhaps in buildings) had to be banned or became five times as expensive. Would you still support such a ban?

6. Do you think that nuclear war is preventable? How? What are you doing to help prevent nuclear war?

Kenneth E. Boulding

Kenneth E. Boulding is Distinguished Professor of Economics Emeritus and director of the Program of Research and General Social and Economic Dynamics at the Institute of Behavioral Sciences, University of Colorado at Boulder. During his long and distinguished career as an economist and social thinker he has served as president of the American Economic Association and the American Association for the Advancement of Science. He has engaged in research on peace, systems analysis, economic theory, economics and ethics, and economics and environment. He was the first leading economist to propose in the late 1960s that we move from our present throwaway or frontier economy to a sustainable-earth economy.

Perhaps the most profound change in the state of the Planet Earth in the last 100 years has been the development of more destructive forms of warfare. The first major development was aerial warfare and the bombing of cities. The second was the development of the nuclear weapon, especially the hydrogen bomb with its enormous destructive power, and the long-range missile which can deliver it to virtually any point on the earth's surface.

Something like this, however, though on a much smaller scale, has happened before in human history. A good example is the development of gunpowder and the effective cannon in the fifteenth and sixteenth centuries, which really brought the feudal system to an end and created national states the size of England or France. As long as the means of destruction were spears and arrows, there was some sense in having a castle or a city wall. With the coming of the effective cannon, these made no sense. The baron who stayed in his castle got blown up. Some attempt was made to save the city wall by building longer triangular projections from them, with cannon on them, a little reminiscent of the Strategic Defense Initiative (SDI) or "Star Wars" defense now being pursued by the United States. But this turned out to be ineffective. City walls were torn down and boulevards and castles became tourist attractions.

The nuclear weapon and the long-range missile have done for unilateral national defense precisely what the cannon did for the feudal baron and the city wall, but we haven't caught on to this yet. We delude ourselves by thinking that we have a stable deterrence. Deterrence says, "You do something nasty to me and I'll do something nasty to you."

Deterrence can work over short periods and in specific places, but it cannot be stable in the long run, otherwise it would not deter in the short run. If the probability of nuclear weapons going off were zero, that would be the same as not having them. They would deter nobody.

Deterrence always has a positive probability of breaking down. It is a fundamental principle that if there is a positive probability of anything, if we wait

FURTHER READINGS

Abrahamson, Dean E., ed. 1989. *The Challenge of Global Warming.* Covelo, Calif.: Island Press.

Dotto, Lydia. 1986. *Planet Earth in Jeopardy: Environmental Consequences of Nuclear War.* New York: Wiley.

Dotto, Lydia. 1990. *Thinking the Unthinkable: Civilization and Rapid Climate Change.* Waterloo, Ontario: Wilfrid Lanier University Press.

Ehrlich, Paul R. 1988. "The Ecology of Nuclear War." In Paul R. Ehrlich and John P. Holdren, eds., *The Cassandra Conference: Resources and the Human Predicament.* Texas Station: Texas A&M University Press.

Environmental Protection Agency. 1988. *The Potential Effects of Global Climate Change on the United States.* Washington, D.C.: Environmental Protection Agency.

Environmental Protection Agency. 1989. *Policy Options for Stabilizing Global Climate.* Washington, D.C.: Environmental Protection Agency.

Fisher, David E. 1990. *Fire & Ice: The Greenhouse Effect, Ozone Depletion & Nuclear Winter.* New York: Harper & Row.

Fishman, Albert, and Robert Kalish. 1990. *Global Alert: The Ozone Pollution Crises.* New York: Plenum.

Flavin, Christopher. 1989. *Slowing Global Warming: A Worldwide Strategy.* Washington, D.C.: Worldwatch Institute.

Greenhouse Crisis Foundation. 1990. *The Greenhouse Crisis: 101 Ways to Save the Earth.* Washington, D.C.: Greenhouse Crisis Foundation.

Gribbin, John. 1990. *Hothouse Earth: The Greenhouse Effect and Gaia.* London: Grove Weidenfeld.

Houghton, Richard A., and George M. Woodwell. 1989. "Global Climatic Change." *Scientific American,* vol. 260, no. 4, 36–44.

Jacobson, Jodi. 1990. "Holding Back the Sea." In Lester R. Brown, et al., *State of the World 1990.* Washington, D.C.: Worldwatch Institute, 79–97.

Levenson, Thomas. 1990. *Ice Time: Climate, Science, and Life on Earth.* New York: Harper & Row.

Lovins, Amory B., et al. 1989. *Least-Cost Energy: Solving the CO_2 Problem.* 2nd ed. Snowmass, Colo.: Rocky Mountain Institute.

Lyman, Francesca, et al. 1990. *The Greenhouse Trap: What We're Doing to the Atmosphere and How We Can Slow Down Global Warming.* Boston: Beacon Press.

long enough it will happen. When one reflects that the Chernobyl nuclear power plant was carefully designed *to not* go off and that nuclear weapons are carefully designed *to* go off, the possibility of even accidental nuclear war may be alarmingly high.

For the first time in human history the fate of the planet is in the hands of a very few decision makers. There is always a positive probability that one or more of them will make a fatal decision.

The only way to remove this dire threat to the planet is through the abolition of war. This too has a noticeable probability. Again there are historical parallels. We abolished dueling when it went from swords to pistols and often both parties were getting killed. We abolished slavery when it became economically and morally unacceptable.

Now the great task of the human race is to abolish national defense and transform the military into a means of production rather than a means of destruction. We have to learn that military victory is something that has disappeared from the earth, that we have to live without enemies.

The great problem of the military is that they have to have enemies in order to justify their budgets. Hence they are designed to be very ineffective at conflict management, the most important skill we need on this planet today. Peace is not something exotic and improbable. In the last 150 years we have seen the rise of stable peace among many nations, beginning perhaps in Scandinavia, spreading to North America by about 1870, now to western Europe and the Pacific, where we now have perhaps 18 nations in a great triangle from Australia to Japan to Finland who have no plans whatever to go to war with each other.

The next step is to establish stable peace between the Soviet bloc and the rest of the world. With the extraordinary things that have been happening in the Communist countries, this peace has risen above the horizon as a real possibility if the West can respond positively to it. Then, of course, the next step is to spread stable peace to the tropics and the Third World.

It is a fundamental principle that what exists must be possible. Stable peace has existed now in many nations for at least 150 years. It must clearly be possible. In that there is great hope for the human race and this incredibly beautiful planet.

Guest Essay Discussion

1. Do you think or talk about the possibility of nuclear war? Why or why not? What causes most people to largely ignore this greatest threat to their survival and the survival of their children and grandchildren?

2. The history of warfare has shown that any advance in military technology is countered by another advance in military technology and that the pace of this process is increasing. In light of this, do you believe that the United States should pour hundreds of billions of dollars into a "Star Wars" defense system against nuclear attack? Explain.

3. Do you agree that the most urgent task on this planet is to abolish war? Why? Do you believe that this is possible? Explain. If you agree that it is the most urgent task, then what are you doing as an individual to convert this possibility into reality?

MacKenzie, James J. 1989. *Breathing Easier: Taking Action on Climate Change, Air Pollution, and Energy Efficiency.* Washington, D.C.: World Resources Institute.

Makhigani, Arjon, et al. 1990. "Beyond the Montreal Protocol: Still Working on the Ozone Hole." *Technology Review,* May/June, 53–59.

McKibben, Bill. 1989. *The End of Nature.* New York: Random House.

Mintzer, Irving, et al. 1990. *Protecting the Ozone Shield: Strategies for Phasing Out CFCs During the 1990s.* Washington, D.C.: World Resources Institute.

Mintzer, Irving, and William R. Moomaw. 1990. *Escaping the Heat Trap.* Washington, D.C.: World Resources Institute.

National Academy of Sciences. 1989. *Global Environmental Change.* Washington, D.C.: National Academy Press.

National Academy of Sciences. 1989. *Ozone Depletion, Greenhouse Gases, and Climate Change.* Washington, D.C.: National Academy Press.

National Academy of Sciences. 1990. *Sea Level Change.* Washington, D.C.: National Academy Press.

Oppenheimer, Michael, and Robert H. Boyle. 1990. *Dead Heat: The Race Against the Greenhouse Effect.* New York: Basic Books.

Postel, Sandra. 1986. *Altering the Earth's Chemistry: Assessing the Earth's Risks.* Washington, D.C.: Worldwatch Institute.

Roan, Sharon L. 1989. *Ozone Crisis: The 15-Year Evolution of a Sudden Global Emergency.* New York: Wiley.

Rowland, F. Sherwood. 1989. "Chlorofluorocarbons and the Depletion of Stratospheric Ozone." *American Scientist,* vol. 77, 36–45.

Schneider, Stephen. 1989. *Global Warming: Are We Entering the Greenhouse Century?* New York: Random House.

Shea, Cynthia Pollack. 1988. *Protecting Life on Earth: Steps to Save the Ozone Layer.* Washington, D.C.: Worldwatch Institute.

Turco, R. P., et al. 1990. "Climate and Smoke: An Appraisal of Nuclear Winter." *Science,* vol. 247, 166–176.

Weiner, Jonathan. 1990. *The Next One Hundred Years: Shaping the Fate of Our Living Earth.* New York: Bantam.

CHAPTER ELEVEN

WATER RESOURCES

AND WATER POLLUTION

General Questions and Issues

1. How much usable fresh water is available for human use and how much of this supply are we using?

2. What are the major water resource problems in the world and in the United States?

3. How can water resources be managed to increase the supply and reduce unnecessary waste?

4. What are the major types and sources of water pollutants?

5. What are the major pollution problems of streams, lakes, oceans, and groundwater aquifers?

6. What legal and technological methods can be used to control water pollution?

If there is magic on this planet, it is in water.
LOREN EISLEY

W ATER IS OUR MOST abundant resource, covering about 71% of the earth's surface (Figure 5-19, p. 96). This precious film of water—most of it salt water with the remainder being fresh water—helps maintain the earth's climate, dilutes pollutants, and is essential to all life. The much smaller amount of fresh water constantly renewed by the hydrologic cycle is also a vital resource for agriculture, manufacturing, transportation, and countless other human activities.

Despite its importance, water is one of the most poorly managed resources on earth. We waste it and pollute it. We also charge too little for making it available, encouraging even greater waste and pollution of this vital renewable resource.

11-1 SUPPLY, RENEWAL, AND USE OF WATER RESOURCES

Worldwide Supply and Renewal The world's fixed supply of water in all forms (vapor, liquid, and solid) is enormous. However, about 97% of the earth's volume of water is found in the oceans and is too salty for drinking, growing crops, and most industrial uses except cooling.

The remaining 3% is fresh water. But all except 0.003% of this supply is polluted, lies too far under the earth's surface to be extracted at an affordable cost, or is locked up in glaciers, polar ice caps, atmosphere, and soil. If the world's water supply were only 100 liters (26 gallons), our usable supply of fresh water would be only about 0.003 liter (one-half teaspoon) (Figure 11-1).

Fortunately, this freshwater supply is continually collected, purified, and distributed in the *hydrologic (water) cycle* (Figure 4-29, p. 79). This natural recycling and purification process works as long as we don't pollute water faster than it is replenished, overload it with slowly degradable and nondegradable wastes, or withdraw it from slowly renewable underground deposits faster than it is replenished. Unfortunately, we are disrupting the water cycle by doing all of these things.

Surface Water and Groundwater The fresh water we use comes from two sources: surface water and groundwater (Figure 11-2). Precipitation that does not infiltrate into the ground or return to the atmosphere by evaporation or transpiration is called **surface water**. It is fresh water that is on the earth's surface in streams, lakes, wetlands, and reservoirs.

Some precipitation seeps into the ground and fills pores (spaces or cracks) in soil and rock in the earth's crust. The area where all available pores are filled by water is called the **zone of saturation**, and the water in these pores is called **groundwater** (Figure 11-3). The **water table** is the upper surface of the zone of saturation. There is 40 times as much groundwater below the earth's surface as there is in all the world's streams and lakes.

A geologic formation's ability to hold water is dependent on its porosity and permeability. **Porosity** refers to the pores in rocks, or the percentage of the rock's volume that is not occupied by the rock itself. **Permeability** is the degree to which underground rock pores are interconnected with each other and thus is a measure of the degree to which water can flow freely from one pore to another. For example, sand and coarse gravel are highly permeable, while clay has a very low permeability. Porous, water-saturated layers of rock that can yield an economically significant amount of water are called **aquifers**.

Most aquifers are recharged or replenished naturally by precipitation, which percolates downward through soil and rock in what is called **natural recharge** (Figure 11-3). Any area of land allowing water to pass through it and into an aquifer is called a **recharge area** (Figure 11-3). Groundwater moves from the recharge area through an aquifer and out to a discharge area as part of the hydrologic cycle (Figure 4-29, p. 79). Discharge areas can be wells, springs, lakes, geysers, streams, and oceans.

The direction of flow of groundwater from recharge areas to discharge areas is dependent on gravity, pressure, and friction. Normally groundwater moves from points of high elevation and pressure to points of lower elevation and pressure. This movement is quite slow, typically only a meter or so (about 3 feet) a year and rarely more than 0.3 meter (1 foot)

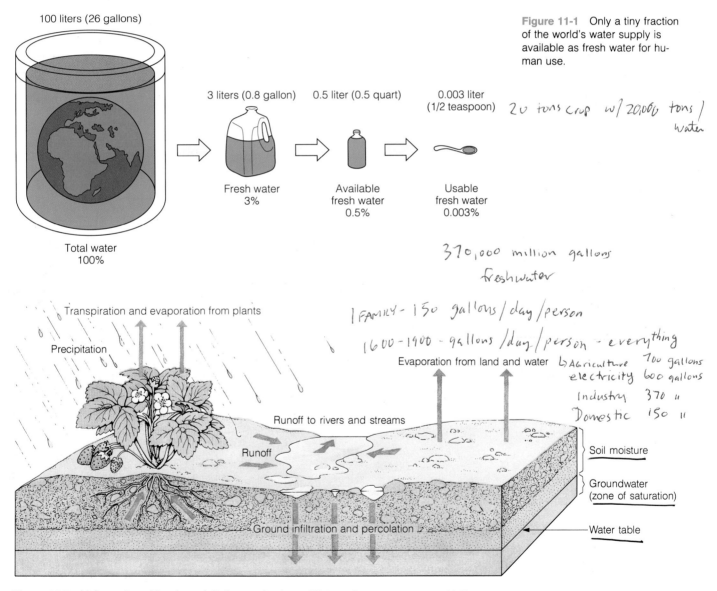

100 liters (26 gallons)

3 liters (0.8 gallon)

Fresh water
3%

0.5 liter (0.5 quart)

Available
fresh water
0.5%

0.003 liter
(1/2 teaspoon)

Usable
fresh water
0.003%

Total water
100%

Figure 11-1 Only a tiny fraction of the world's water supply is available as fresh water for human use.

20 tons crop w/ 20,000 tons / water

370,000 million gallons freshwater

1 FAMILY - 150 gallons/day/person

1600-1900 - gallons/day/person - everything
└ Agriculture 700 gallons
electricity 600 gallons
Industry 370 "
Domestic 150 "

Transpiration and evaporation from plants

Precipitation

Evaporation from land and water

Runoff to rivers and streams

Runoff

Soil moisture

Groundwater
(zone of saturation)

Ground infiltration and percolation

Water table

Figure 11-2 Major routes of local precipitation: surface runoff into surface waters, ground infiltration into aquifers, and evaporation and transpiration into the atmosphere.

a day. Thus, most aquifers are like huge, slow-moving underground lakes.

If the withdrawal rate of an aquifer exceeds its natural recharge rate, the water table around the withdrawal well is lowered, creating a waterless volume known as a cone of depression (Figure 11-4). Any pollutant discharged within the land area above the cone of depression will be pulled directly into the well and can have a devastating effect on the quality of water withdrawn from that well.

Some ancient aquifers, called "fossil aquifers," are often found deep underground, get very little recharge, and are nonrenewable resources on a human time scale. Withdrawals from these deposits amount to "water mining," eventually depleting these one-time deposits of earth capital.

World and U.S. Water Use Two common measures of human water use are withdrawal and consumption. **Water withdrawal** involves taking water from a groundwater or surface water source and transporting it to a place of use. **Water consumption** occurs when water that has been withdrawn is not available for reuse in the area from which it is withdrawn, mostly

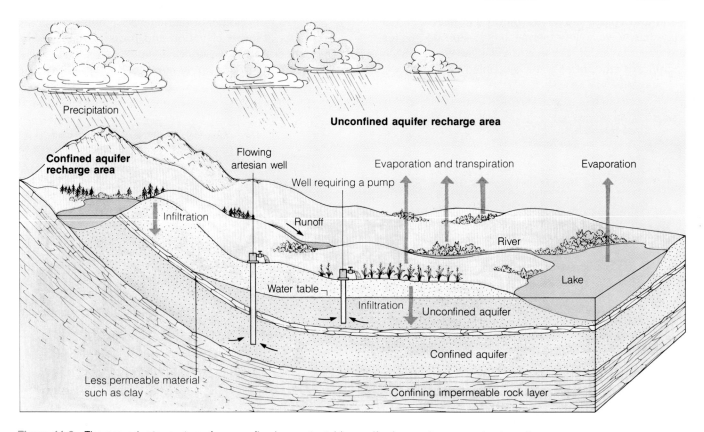

Figure 11-3 The groundwater system. An unconfined, or water table, aquifer forms when groundwater collects above a layer of rock or compacted clay through which water flows very slowly (low permeability). A confined aquifer is sandwiched between layers that have low permeability, such as clay or shale. Groundwater in this type of aquifer is confined and under pressure.

Figure 11-4 Drawdown of water table and cone of depression.

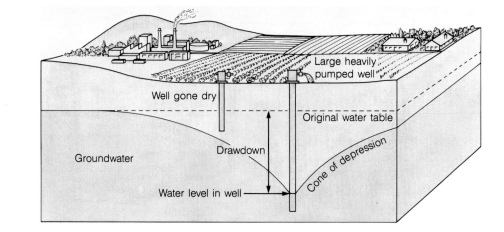

because of evaporation or transpiration into the atmosphere. In arid areas, such as much of the western United States, the major consumptive uses of water are irrigation and lawn watering.

Between 1950 and 1990 global water withdrawal increased more than threefold, largely in response to the rapid growth in population, agriculture, and industrialization. Water withdrawals are projected to at least double in the next two decades to meet the food and other resource needs of the world's rapidly growing population.

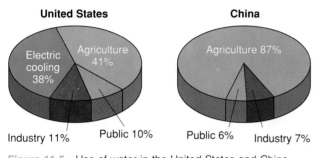

United States

Electric cooling 38%

Agriculture 41%

Industry 11%

Public 10%

China

Agriculture 87%

Public 6%

Industry 7%

Figure 11-5 Use of water in the United States and China. (Data from Worldwatch Institute and World Resources Institute)

Annual total water withdrawal varies considerably among various MDCs and LDCs, with the largest volumes drawn in decreasing order by the United States, China, India, and the Soviet Union. The United States also has the highest per capita water withdrawal in the world, averaging 2.6 million liters (687,000 gallons) per person per year. If each of the 5.3 billion people on earth had the same average per capita water withdrawal as in the United States, we would be trying to withdraw more usable fresh water than is available worldwide.

Averaged globally, about 70% of the water withdrawn each year is used to irrigate 18% of the world's cropland. The figure can reach 80% or more in some areas such as Egypt where all cropland must be irrigated and in Pakistan where 77% of the cropland is irrigated. The remaining 30% of the water we withdraw is used in industrial processing, in cooling electric power plants, and in homes and businesses (public use).

Uses of withdrawn water vary widely from one country to another (Figure 11-5). Growing food and manufacturing various products require large amounts of water (Figure 11-6), although in most cases much

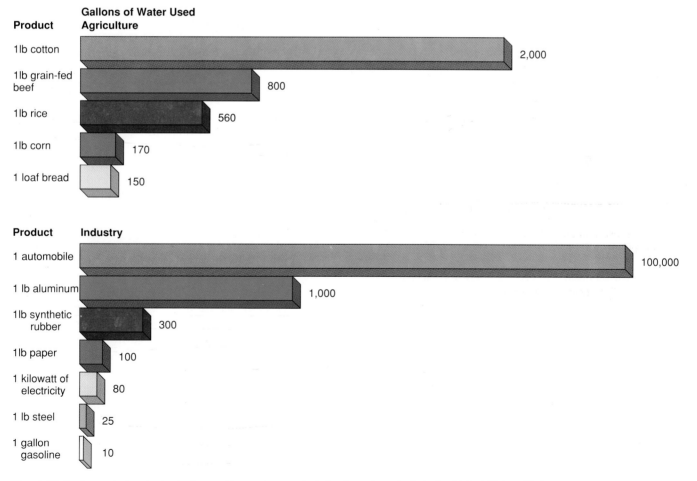

Figure 11-6 Amount of water typically used to produce various foods and products in the United States. (Data from U.S. Geological Survey)

Figure 11-7 Camp of drought refugees in Burkina Faso, West Africa, depicts the plight of some of the estimated 10 million people driven from their homes worldwide by environmental degradation.

of this water could be used more efficiently and reused.

11-2 WATER RESOURCE PROBLEMS

Too Little Water During the 1970s, major droughts affected an average of 24.4 million people per year, killed over 23,000 a year, and created large numbers of environmental refugees—a trend that continued throughout the 1980s (Figure 11-7). At least 80 arid and semiarid countries, where nearly 40% of the world's people live, have serious periodic droughts. Areas likely to face increased water shortages in the 1990s and beyond include northern Africa, parts of India, northern China, much of the Middle East, Mexico, and parts of the western United States and the central Soviet Union.

Reduced precipitation, higher than normal temperatures, or both usually trigger a drought. But rapid population growth and poor land use intensify its effects. In many LDCs large numbers of poor people have no choice but to try to survive on drought-prone land.

Drought is often viewed as a natural disaster. But often it is a human-caused disaster, the result of trying to support too many people and livestock in areas that normally have prolonged droughts.

Severe droughts will occur more frequently in some areas of the world during your lifetime unless we take action now to slow down projected global warming (Section 10-3). If we fail to deal with this problem, it is likely that irrigation systems will be poorly matched to altered rainfall patterns. This will jeopardize our ability to produce enough food for the world's growing population.

Almost 150 of the world's 214 major river systems are shared by at least two countries and 12 of these waterways are shared by five or more countries.

Together these countries contain 40% of the world's population, and they often clash over water rights. This already intense competition and conflict over water supplies between countries and within countries will escalate sharply because of increased water needs and projected global warming that could shift supplies of surface water in unpredictable ways.

For example, water-supply arguments between Egypt, Ethiopia, and the Sudan over the Nile River are escalating rapidly. Ethiopia, which controls the headwaters of 80% of the Nile's flow, has plans to divert more of this water. The drop in the water available to Egypt, where all crops must be irrigated, could lead to a dire situation. Egypt, whose population is projected to almost double from 55 million in 1990 to 102 million in 2090, is already having serious problems growing enough food. In 1989, Egypt's foreign minister warned: "The next war in our region will be over the waters of the Nile."

Competition between cities and farmers for scarce water within countries is also escalating in areas such as California (see Case Study on p. 241) and in northern China where dozens of cities, including Beijing, already face acute water shortages. Water shortages are expected in 450 of China's 644 cities by the turn of this century. In most cases the only way to lessen these problems is to shift more of a region's water from farmers, who use the most, to city dwellers whose water use is less and generally produces much greater economic benefits.

Too Much Water Some countries have enough annual precipitation but get most of it at one time of the year. In India, for example, 90% of the annual precipitation falls between June and September, the monsoon season. This downpour can cause periodic flooding, waterlogging of soils, nutrient depletion of soils, and the washing away of soils and crops.

Prolonged rains anywhere can cause streams and lakes to overflow and flood surrounding land areas. Hurricanes and cyclones can flood low-lying coastal areas.

In the 1960s the number of flood victims averaged 5.2 million a year. During the 1970s, major flood disasters affected 15.4 million people annually, killed an average of 4,700 people a year, and caused tens of billions of dollars in property damages. This trend continued throughout the 1980s. In India, for example, losses from flooding doubled in the 1980s.

Floods, like droughts, are usually called natural disasters. But, human activities have contributed to the sharp rise in flood deaths and damages since the 1960s. Cultivation of land, deforestation (Figure 1-1, p. 3), overgrazing, and mining have removed water-absorbing vegetation and soil (see Case Study on p. 237).

Urbanization also increases flooding, even with moderate rainfall. It replaces vegetation and soil with

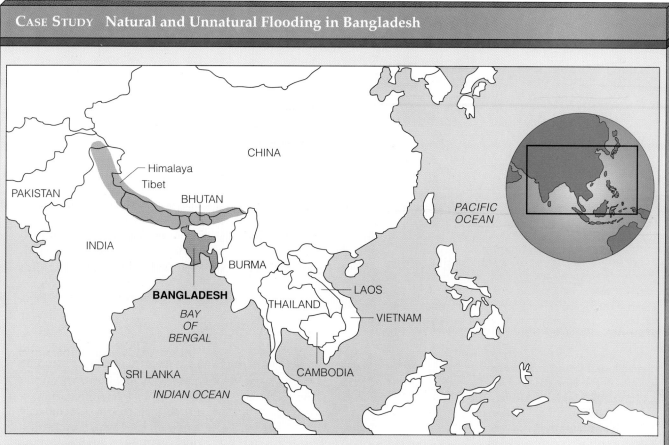

Figure 11-8   Where is Bangladesh?

Bangladesh (Figure 11-8) is one of the world's most densely populated countries. More than 115 million people—almost half the population of the United States—are packed into a country roughly the size of Wisconsin. It is also one of the world's poorest countries with an average per capita income of about $170. Eighty percent of its people can't read or write.

The country is located on a vast, low-lying delta of shifting islands of silt at the mouth of the Ganges, Brahmaputra, and Meghna rivers. Its people are accustomed to flooding after water from annual monsoon rains in the Himalayan mountain ranges of India, Nepal, Bhutan, and China flows downward through rivers to Bangladesh and into the Bay of Bengal (Figure 11-8).

Bangladesh depends on this annual flooding to grow rice, its major source of food. The annual deposit of Himalayan soil in the delta basin also helps maintain soil fertility. Thus, the people of this country are used to moderate annual flooding and need it for their survival, but massive flooding from excessive runoff from the Himalayas and from storm surges caused by cyclones in the Bay of Bengal is disastrous.

In the past, major floods occurred only once every 50 years or so, but since 1950 the number of large-scale floods has increased sharply. During the 1970s and 1980s the average interval between major floods in Bangladesh was only four years. After a flood in 1974, an estimated 300,000 people died in a famine.

In the 1980s floods have become even more severe. In 1988 a massive flood covered two-thirds of the country's land mass for several days and leveled 2 million homes after the heaviest monsoon rains in 70 years (Figure 11-9). At least 2,000 people were killed by drowning and snakebite, and 30 million people—one out of four—were left homeless. Hundreds of

thousands more contracted diseases such as cholera and typhoid fever from contaminated water and food supplies. At least a quarter of the country's crops were destroyed, costing this impoverished nation at least $1.5 billion, and many thousands died prematurely from lack of food.

Bangladesh's flooding problems begin in the Himalayan watershed. There a combination of rapid population growth, deforestation, overgrazing, and unsustainable farming on easily erodible steep mountain slopes has greatly diminished the ability of soil in this mountain watershed to absorb water. Instead of being absorbed and released slowly, water from the annual monsoon rain runs off the denuded foothills of the Himalayas north of Bangladesh's border. Then heavier than normal monsoon rains cause massive flooding in Bangladesh. This deluge of water also carries with it

(continued)

Figure 11-9 Some annual flooding in Bangladesh is necessary for growing rice, but the country now experiences more disastrous major floods because of forest clearing in other countries in the watershed of the Himalayan mountains.

the soil vital to the survival of people in the Himalayas.

In their struggle to survive, the poor in Bangladesh have cleared many of the country's coastal mangrove forests for fuelwood and for growing food. This has increased the severity of flooding because these coastal wetlands help protect the low-lying coastal areas from storm surges generated by cyclones in the Bay of Bengal. In 1970, somewhere between 200,000 and 1 million people drowned in one of these storms; no one knows the exact figure. Flood damages and deaths in areas where these forests still exist are much lower than in areas where they have been cleared.

This human-caused environmental disaster, which is likely to be repeated, causes as much damage as that wrought by a war. Rising sea levels from an enhanced greenhouse effect would put the heavily populated portion of Bangladesh that juts into the Bay of Bengal (Figure 11-8) under water. This illustrates the folly of not placing at least as much emphasis on environmental security as on military security. The severity of this problem can only be reduced if Bangladesh, Bhutan, China, India, and Nepal all agree to cooperate in reforestation efforts and flood control measures.

MDCs also have an important role to play. They must provide aid for reforestation and flood control. Equally important, they must do their part in dismantling the global poverty trap (Section 7-3) that forces the poor into unsustainable land use and makes them highly vulnerable to environmental change by having to survive on unprotected floodplains.

highways, parking lots, shopping centers, office buildings, homes, and other structures that lead to rapid runoff of rainwater. If sea levels rise during the next century as a result of projected global warming, flooding of low-lying coastal cities, wetlands, and croplands will increase dramatically.

Flood damage can be prevented or reduced. Vegetation can be replanted in disturbed areas to reduce runoff. Governments can also identify floodplains—flat areas along the banks of streams naturally subject to flooding after heavy, prolonged rains—and prohibit their use for certain types of development, as is now done in the United States. Sellers of property in these areas should be required to provide prospective buyers with information about average flood frequency.

Water in the Wrong Place In some countries the largest rivers, which carry most of the runoff, are far from agricultural and population centers where the water is needed. South America has the largest average annual runoff of any continent. But 60% of this runoff flows through the Amazon, the world's largest river, in areas far from where most people live.

In many LDCs poor people must spend a good part of their waking hours fetching water, often from polluted streams. Many women and children walk 16 to 24 kilometers (10 to 14 miles) a day, carrying heavy, water-filled jars.

Contaminated Drinking Water According to the World Health Organization, 1.5 billion people do not have a safe supply of drinking water and at least 5 million people die every year from waterborne diseases that could be prevented by improvements in drinking water supplies and sanitation. In 1980 the United Nations called for MDCs and LDCs to spend $300 billion to supply all of the world's people with clean drinking water and adequate sanitation by 1990. The $30-billion-a-year cost of this program would be roughly equal to what the world spends every 10 days for military purposes. Some progress has been made in several countries, including Indonesia, Mexico, and Ghana. But by 1990 the program had fallen far short of its goal because of lack of funding, lack

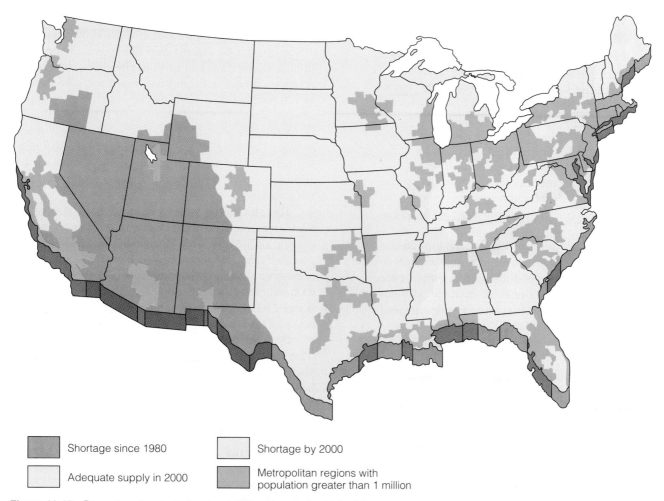

Shortage since 1980

Shortage by 2000

Adequate supply in 2000

Metropolitan regions with population greater than 1 million

Figure 11-10 Present and projected water deficit regions in the United States compared with present metropolitan regions with populations greater than 1 million. (Data from U.S. Water Resources Council and U.S. Geological Survey)

of commitment by MDCs and LDCs, and increased population.

Wastes)

In its passage through the hydrological cycle water is polluted by three major kinds of waste. One is the sediment washed from the land into surface waters by natural erosion and greatly accelerated erosion of soil from agriculture, forestry, mining, construction, and other land clearing and disturbing activities (Section 12-2). Another is organic waste from human and animal excreta and the discarded parts of harvested plants. The third is the rapidly increasing volume of a variety of hazardous chemicals produced by industrialized societies.

¹ sediment erosion
² Waste human
³) Toxic Chemicals

All three categories of these wastes are increasing because of rapid population growth, poverty, and industrialization. These forms of water pollution are discussed in the last half of this chapter.

The U.S. Situation Overall, the United States has plenty of fresh water. But much of the country's annual runoff is not in the desired place, occurs at the wrong time, or is contaminated from agricultural

and industrial activities. Most of the eastern half of the country usually has ample precipitation, while much of the western half has too little.

Many major urban centers in the United States are located in areas that don't have enough water or are projected to have water shortages by 2000, especially in the West and Midwest (Figure 11-10). These shortages could worsen if the world's climate warms up as a result of an enhanced greenhouse effect (Figure 10-8, p. 218). Because water is such a vital resource, you might find Figures 10-8 and 11-10 useful in deciding where to live in coming decades.

In many parts of the eastern United States the major water problems are flooding, inability to supply enough water to some large urban areas, and pollution of rivers, lakes, and groundwater. For example, 3 million residents of Long Island must draw all their water from an aquifer. This aquifer is becoming severely contaminated by industrial wastes, leaking septic tanks and landfills, and ocean salt water, which is drawn into the aquifer when fresh water is withdrawn faster than it is naturally recharged.

1) Floods 2) Not enough 3) Pollution

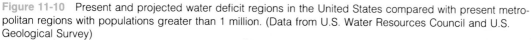

11-3 WATER RESOURCE MANAGEMENT

Methods for Managing Water Resources

One way to manage water resources is to increase the supply, mostly by building dams and reservoirs, diverting surface water from one area to another, and tapping groundwater. The other approach is to improve the efficiency of water use by decreasing unnecessary use and waste.

LDCs may or may not have enough water, but they rarely have the money needed to develop the water storage and distribution systems needed to increase their supply. Their people must settle where the water is.

In MDCs people tend to live where the climate is favorable and bring in water through expensive transfers from one watershed to another. Some settle in a desert and expect water to be brought to them at a low price. Others settle on a floodplain and expect the government to keep flood waters away.

Farms and cities in some arid areas of the U.S., India, China, and the USSR are now being supported by removing earth capital in the form of groundwater faster than it is replaced by natural resources. But this unsustainable "mining" of a slowly renewable or nonrenewable resource can only go on so long before farming must be abandoned and the populations of parched cities are reduced by migration.

Increasing the water supply in some areas is important. But this approach is eventually overwhelmed by increasing population, food production, industrialization, and unpredictable shifts in water supplies from projected global warming. Thus, it makes much more sense economically and environmentally to put the primary emphasis on increasing the efficiency of the ways we use water in agriculture and industry, and in homes (see Individuals Matter inside back cover).

Constructing Dams and Reservoirs Rainwater and water from melting snow that would otherwise be lost can be captured and stored in large reservoirs behind dams built across streams (Figure 5-29, p. 102). By controlling stream flow, dams reduce the danger of flooding in downstream areas and allow people to live on the fertile floodplains of major streams below the dam. They also provide a controllable supply of water for irrigating arid and semiarid land below the dam. Hydroelectric power plants, which use the energy of water flowing from dam reservoirs, generate more than 20% of the world's electricity. Reservoirs behind large dams can also be used for outdoor recreation such as swimming, boating, and fishing.

But the benefits of dams and reservoirs must be weighed against their costs. Dams are expensive to build and may give developers and residents in a floodplain below the dam a false sense of safety from major floods that can overwhelm the ability of a dam to control flood waters. The reservoirs fill up with silt and become useless in 40 to 200 years, depending on local climate and land-use practices. The permanent flooding of land behind dams to form reservoirs displaces people and destroys vast areas of valuable agricultural land, wildlife habitat, and scenic natural beauty. The World Bank is lending the Indian government money to build dams that will displace 1.5 million people over the next 50 years and flood large areas of land.

The storage of water behind a dam also raises the water table. The higher water table often waterlogs the soil on nearby land, decreasing its crop or forest productivity. Evaporation also increases the salinity of reservoir water by leaving salts behind, decreasing its usefulness for irrigation. The sheer weight of the water impounded in reservoirs increases the likelihood of fault movement, which cause earthquakes.

By interrupting the natural flow of a stream, a dam disrupts the migration and spawning of fish, such as salmon. By trapping silt that would normally be carried downstream, dams deprive downstream areas and estuaries of vital nutrients and decrease their productivity. Studies have also shown that large-scale dams in LDCs often tend to benefit a small minority of the well-to-do while flooding the lands and often destroying floodwater farming and fisheries that are vital to the poor majority.

Faulty construction, earthquakes, floods, landslides, sabotage, or war can cause dams to fail, taking a terrible toll in lives and property. According to a 1986 study by the Federal Emergency Management Agency, the United States has 1,900 unsafe dams in populated areas. The agency reported that the dam safety programs of most states are inadequate because of weak laws and budget cuts.

Watershed Transfers In MDCs local governments often increase the supply of fresh water in water-poor populated areas by transferring water from watersheds in water-rich areas. Two such projects are the California Water Plan, the world's largest water distribution system and water diversion from the Aral Sea in the Soviet Union (see Case Studies on pp. 241 and 242).

Tapping Groundwater In the United States about half of the drinking water (96% in rural areas and 20% in urban areas), 40% of the irrigation water, and 23% of all fresh water used is withdrawn from aquifers. Overuse of groundwater can cause or intensify several problems: aquifer depletion, subsidence (sinking of land when groundwater is withdrawn), and intrusion of salt water into aquifers (Figure 11-13). Groundwater can also become contaminated from

The California Water Plan transports water from water-rich parts of northern California to heavily populated parts of northern California and to mostly arid and semiarid, heavily populated southern California (Figure 11-11). For decades, northern and southern Californians have been feuding over how the state's water should be allocated under this plan.

People in arid southern California say they need more water from the north for growing crops and supporting Los Angeles, San Diego, and other large and growing urban areas. Opponents in the north say that sending more water south would degrade the Sacramento River, threaten fishing, and reduce the flushing action that helps clean San Francisco Bay of pollutants.

They also argue that much of the water already sent south is wasted and that an increase of only 10% in irrigation efficiency would provide enough water for domestic and industrial uses in southern California. Pointing out that agriculture accounts for only 2.5% of the state's economy while using 85% of the water withdrawn, they contend that water used in other ways contributes more to economic growth.

Conservationists believe that the government should not award new long-term water contracts that give farmers and ranchers federally subsidized water without an assessment of the long-term environmental and economic impact of using so much water for irrigating crops—especially grass

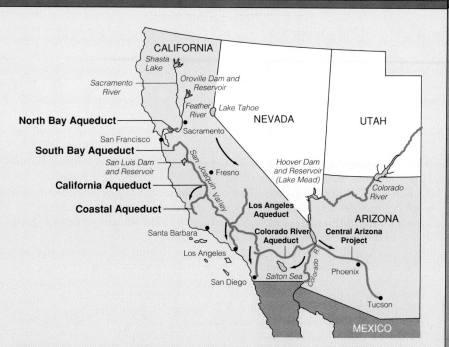

Figure 11-11 California Water Plan and Central Arizona Project for large-scale transfer of water from one watershed to another. Arrows show general direction of water flow.

for cows, rice, alfalfa, and cotton—that could be grown more cheaply in rain-fed areas.

Cities wanting more water should also be required to pay farmers to install water-saving irrigation technology and then have the right to use the water saved. But bringing about these changes is difficult because the political power of California's taxpayer-subsidized farmers greatly exceeds their contribution to the state's economy.

A related project is the federally financed $3.9 billion Central Arizona Project. It began pumping water from the Colorado River uphill to Phoenix in 1985 and is expected to deliver water to Tuc-

son by 1991 (Figure 11-11). When the first part of this project was completed in 1985, southern California, especially the arid and booming San Diego region, began losing up to one-fifth of its water, which Arizona has a legal right to divert.

If water supplies in California should drop sharply because of global warming (Figure 10-8, p. 218), water delivered by this massive distribution system would drop sharply. Most irrigated agriculture in California would have to be abandoned and much of the population of southern California might have to move to areas with more water.

---

industrial and agricultural activities, septic tanks, and other sources (Section 11-7).

Currently, about one-fourth of the groundwater withdrawn in the United States is not replenished. The major groundwater overdraft problem is in parts of the huge Ogallala Aquifer, extending under the farm belt from northern Nebraska to northwestern Texas (see Case Study on p. 242).

Aquifer depletion is also a serious problem in northern China, Mexico City, and parts of India. The most effective solution is to reduce the amount of groundwater withdrawn by wasting less irrigation water, not growing water-thirsty crops in dry areas, and developing crop strains that require less water and that are more resistant to heat stress.

When groundwater in an unconfined aquifer

Another example of large-scale watershed transfer is in Soviet Central Asia, where 90% of the cropland is irrigated because it has the driest climate in the Soviet Union. Since 1960 massive amounts of irrigation water have been diverted from the inland Aral Sea and the two rivers that replenish its water.

This has caused a regional ecological disaster, described by a Russian official "as ten times worse than the 1986 accident at the Chernobyl nuclear power plant." As the two rivers have been reduced to little more than sewers, the volume of the Aral Sea—once the world's fourth largest freshwater lake—has dropped by two-thirds, its surface area has

shrunk by 40%, and salinity levels have risen threefold.

Today virtually no fresh water reaches the Aral. Shoreline towns have been stranded inland. The area's irrigated cropland has been turned into a salt and dust desert, stretching for 322 kilometers (200 miles). All native fish species have disappeared, devastating the area's fishing industry. Winds pick up salt and dust from the dry seabed and dump it on surrounding cropland, reducing crop yields.

Local farmers have turned to herbicides and fertilizers to keep growing some crops. Many of these chemicals are leached into the water table. The area is now reporting soaring rates of hepati-

tis, typhoid fever, jaundice, intestinal infections, and cancer. It also has the highest infant mortality rate in the Soviet Union.

Salt, dust, and dried pesticide residues have also been carried by winds and deposited on towns and cropfields hundreds of kilometers away. Rains have deposited Aral salt as far west as the Black Sea and as far north as the Soviet Arctic. As the salt spreads, it kills crops and trees, wildlife, and pastureland.

Soviet officials have done little to protect the area and have kept this disaster a secret for decades. Local residents claim the government is committing genocide of the area's mostly Muslim population.

Water withdrawn from the vast Ogallala Aquifer (Figure 11-12) is used to irrigate one-fifth of all U.S. cropland in an area too dry for rainfall farming. The aquifer supports $32 billion of agricultural

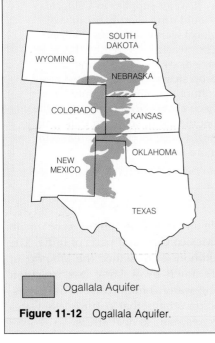

Ogallala Aquifer

Figure 11-12 Ogallala Aquifer.

production a year—mostly wheat, sorghum, cotton, corn, and 40% of the country's grain-fed beef cattle.

The Ogallala Aquifer contains a massive amount of water, but it is essentially a nonrenewable fossil aquifer with an extremely slow recharge rate. Today the overall rate of withdrawal from this aquifer is eight times its natural recharge rate.

Even higher withdrawal rates, sometimes 100 times the recharge rate, are taking place in parts of the aquifer that lie beneath Texas, New Mexico, Oklahoma, and Colorado. Water resource experts project that at the present rate of withdrawal, much of this aquifer will be dry by 2020, and much sooner in areas where it is shallow. When this water is gone, it will take thousands of years to replenish itself.

Long before this happens, however, the high cost of pumping water from rapidly dropping

water tables will force many farmers to grow crops that need much less water, such as wheat and cotton, instead of profitable but thirsty crops such as corn and sugar beets. Some farmers will have to go out of business. The amount of irrigated land already is declining in five of the seven states using this aquifer because of the high cost of pumping water from depths as great as 1,830 meters (6,000 feet).

If farmers in the Ogallala region began using water conservation measures and switched to crops with low water needs, depletion of the aquifer would be delayed. During the 1960s and 1970s, the water table in the High Plains District of Texas dropped 0.6 to 1.8 meters (2 to 6 feet) a year. However, because of improvements in irrigation efficiency, there has been little change in the water table level in this district since 1985. What do you think should be done?

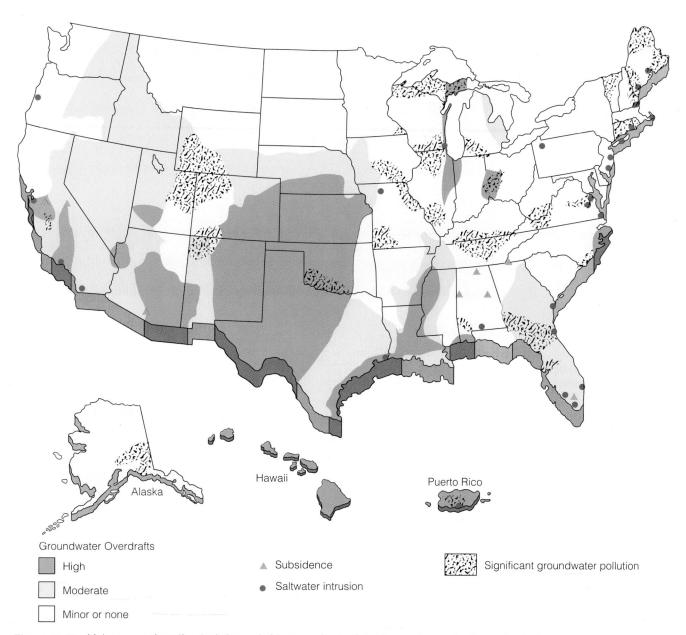

Groundwater Overdrafts

- ■ High
- ▨ Moderate
- □ Minor or none

▲ Subsidence

● Saltwater intrusion

▨ Significant groundwater pollution

Figure 11-13 Major areas of aquifer depletion, subsidence, saltwater intrusion, and groundwater contamination in the United States. (Data from U.S. Water Resources Council and U.S. Geological Survey)

(Figure 11-3) is withdrawn faster than it is replenished, relatively unconsolidated land overlying the aquifer can sink, or subside. This can damage pipelines, highways, railroad beds, and buildings (Figure 11-14).

When freshwater is withdrawn from an aquifer near a coast faster than it is recharged, salt water intrudes into the aquifer (Figure 11-15). Saltwater intrusion threatens to contaminate the drinking water of many towns and cities along the Atlantic and Gulf coasts (Figure 11-13) and in the coastal areas of Israel, Syria, and the Arabian Gulf states. Another growing problem in the United States and many other MDCs

is groundwater contamination, discussed in Section 11-7.

Desalination Removing dissolved salts from ocean water or brackish (slightly salty) groundwater is an appealing way to increase freshwater supplies. Distillation and reverse osmosis are the two most widely used desalination methods.

Distillation involves heating salt water until it evaporates and condenses as fresh water, leaving salts behind in solid form. In reverse osmosis, high pressure is used to force salt water through a thin membrane whose pores allow water molecules but

Figure 11-14 Subsidence in Winter Park, Florida, in May 1981. This large sinkhole caused $2 million in damages and swallowed parts of two businesses, part of a community swimming pool, a house, and several cars and trailers.

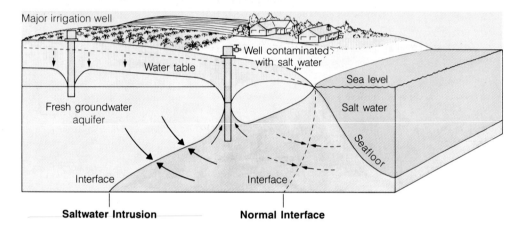

Figure 11-15 Saltwater intrusion along a coastal region. When the water table is lowered, the normal interface (dotted line) between fresh and saline groundwater moves inland (solid line).

Major irrigation well

Well contaminated
with salt water

Water table

Sea level

Fresh groundwater
aquifer

Salt water

Seafloor

Interface

Interface

Saltwater Intrusion

Normal Interface

not dissolved salts to pass through. Desalination plants in arid parts of the Middle East and North Africa produce about two-thirds of the world's desalinated water. Desalination is also used in parts of Florida.

The basic problem with these desalination methods is that they use large amounts of energy and therefore are expensive. Desalination can provide fresh water for coastal cities in arid regions, where the cost of getting fresh water by any method is high. But desalinated water will probably never be cheap enough to irrigate conventional crops or to meet much of the world's demand for fresh water, unless efficient solar-powered methods can be developed.

Another problem is that even more energy and money would be needed to pump desalinated water uphill and inland from coastal desalination plants. Also, a vast network of desalination plants would produce mountains of salt to be disposed of. The easiest and cheapest solution would be to dump the salt in the ocean near the plants. But this would increase the salt concentration and threaten food resources in estuarine waters.

Unnecessary Water Waste *An estimated 30% to 50% of the water used in the United States—the world's largest water user—is unnecessarily wasted.* A major cause of water waste in the United States is artificially low water prices kept that way by elected officials hoping to stimulate economic growth and get reelected by keeping consumers happy. Low-cost water is the only reason that farmers in Arizona can grow water-thirsty crops like alfalfa in the middle of the desert. It also allows people in Palm Springs, California, to keep their lawns and 74 golf courses green in a desert area.

Water subsidies are paid for by all taxpayers in the form of higher taxes. Because these external costs don't show up on monthly water bills, consumers have little incentive to conserve. Raising the price of water to reflect its true cost would provide powerful incentives for reducing water waste.

In the United States, the federal Bureau of Reclamation supplies one-fourth of the water used to irrigate land in the West under long-term contracts (typically 40 years) at greatly subsidized prices. For example, farmers getting subsidized water in the

Central Valley Project of California have had to pay only $50 million or 5% of the project's $931 million cost over the last 40 years. A 1982 law says that cheap federal water should only go to farms of 389 hectares (960 acres) or less. But the big growers have gotten around this by reorganizing their holdings into smaller tracts. In 1987 a single 389-hectare (960-acre) farm in the Central Valley of California received a water subsidy worth $1.8 million.

During the 1990s hundreds of these long-term water contracts will be coming up for renewal. If the Bureau of Reclamation uses this opportunity to sharply raise the price of federally subsidized water to encourage investments in improving water efficiency, many of the water supply problems in the West could be eased. Otherwise, water waste used to support unsustainable short-term economic growth will continue for another 40 years.

Outdated laws governing access and use of water resources also encourage unnecessary water waste (see Spotlight at right).

Another reason that water waste in the United States is greater than necessary is that the responsibility for water resource management in a particular watershed is divided among many state and local governments rather than being handled by one authority. For example, the Chicago metropolitan area has 349 separate water supply systems, divided among some 2,000 local units of government over a six-county area.

In sharp contrast is the regional approach to water management used in England and Wales. The British Water Act of 1973 replaced more than 1,600 separate agencies with 10 regional water authorities based on natural watershed boundaries. In this integrated approach, each water authority owns, finances, and manages all water-supply and waste-treatment facilities in its region. The responsibilities of each authority include water-pollution control, water-based recreation, land drainage and flood control, inland navigation, and inland fisheries. Each water authority is managed by a group of elected local officials and a smaller number of officials appointed by the national government.

Reducing Irrigation Losses Since irrigation accounts for 70% of water use and almost two-thirds of this water is wasted, more efficient use of even a small amount of irrigation water frees water for other uses. Most irrigation systems distribute water from a groundwater well or a surface canal by down-slope or gravity flow through unlined field ditches (Figure 11-16). This method is cheap as long as farmers in water-short areas don't have to pay the real cost of making this water available. But it provides far more water than needed for crop growth, and at least 50%

SPOTLIGHT **Water Rights in the United States**

Laws regulating water access and use differ in the eastern and western parts of the United States. In most of the East water use is based on the doctrine of **riparian rights**. Basically this system of water law gives anyone whose land adjoins a flowing stream the right to use water from the stream as long as some is left for downstream landowners. However, as population and water-intensive land uses grow, there is often not enough water to meet the needs of all the people along a stream.

In the arid and semiarid West the riparian system does not work because large amounts of water are needed in areas far from major surface water sources. In most of the West the principle of **prior appropriation** regulates water use. In this first-come, first-served approach, the first user of water from a stream establishes a legal right for continued use of the amount originally withdrawn. Some areas of the United States have a combination of riparian and prior appropriation water rights.

Prior appropriation gives later water users little access to the resource, especially during droughts, and causes unnecessary use and waste. To hold on to their rights, users must keep on withdrawing and using a certain amount of water even if they don't need it—a use-it-or-lose-it approach. This penalizes farmers who use water-conserving irrigation methods.

Most groundwater use is based on common law, which holds that subsurface water belongs to whomever owns the land above such water. This means that landowners can withdraw as much as they want to use on their land.

When many users tap the same aquifer, that aquifer becomes a common-property resource. The multiple users may remove water at a faster rate than it is replaced by natural recharge. The largest users have little incentive to conserve.

Conservationists and many economists call for a change in laws allocating rights to surface and groundwater supplies, with emphasis on *water marketing*. They believe that farmers and other users who save water through conservation or switching to less-thirsty crops should be able to sell or lease the water they save to industries and cities rather than losing their rights to this water. What do you think should be done?

---

of the water is lost by evaporation and seepage. Such overwatering without adequate drainage also decreases crop yields by waterlogging and the buildup of salts in the soil (Section 12-4).

Figure 11-16 Gravity flow systems like this one in California's San Joaquin Valley irrigate most of the world's irrigated cropland, but only about 40% to 50% of the water actually gets to the crops.

Figure 11-18 Drip irrigation greatly reduces water use and waste. A perforated pipe delivers a small volume of water close to the roots of plants in Rancho, California.

Figure 11-17 Center-pivot irrigation systems like this one in Texas can reduce water consumed by seepage and evaporation to about 30%.

Farmers could prevent seepage by placing plastic, concrete, or tile liners in irrigation canals. Lasers can also be used as a surveying aid to help level fields so that water gets distributed more evenly. Small check dams of earth and stone can be used to capture runoff from hillsides and channel this water to fields. Holding ponds can be used to store rainfall or to capture irrigation water for recycling to crops. Restoring deforested watersheds also leads to a more manageable flow of irrigation water, instead of a devastating flood (see Case Study on p. 237).

Farmers in LDCs can use inexpensive tubewells to withdraw groundwater for irrigation, watering livestock, and household use, as long as withdrawals don't exceed the natural recharge rate. In rural areas where there is no electricity, pumps for these wells can be run by photovoltaic cells powered by the sun (Section 17-3).

Many farmers served by the dwindling Ogallala Aquifer have switched from gravity-flow canal sys-

tems to center-pivot sprinkler systems (Figure 11-17), which reduce water waste from 50% or more to 30%. Some farmers are switching to low-energy precision-application sprinkler systems. These systems cut water waste to about 25% by spraying water closer to the ground and in larger droplets. They also reduce energy use and costs by 20% to 30%.

In the 1960s highly efficient trickle- or drip-irrigation systems were developed in arid Israel. A network of perforated piping, installed at or below the ground surface, releases a small volume of water close to the roots of plants (Figure 11-18). This minimizes evaporation and seepage and cuts water waste to 10% to 20%. These systems are expensive to install but are economically feasible for high-profit fruit, vegetable, and orchard crops.

Irrigation efficiency can also be improved by computer-controlled systems that monitor soil moisture and irrigate only when necessary. Farmers can switch to crop varieties that are more water efficient, drought resistant, and salt tolerant. Also, organic farming techniques produce higher crop yields per hectare and require only one quarter of the water and fertilizer used by conventional farming. Since 1950, Israel has used many of these techniques to decrease waste of irrigation water by about 84%, while expanding the country's irrigated land by 44%.

As supplies of freshwater become more scarce and cities take over much of the water formerly used for irrigation, carefully treated urban wastewater could be used for irrigation. Effluents from sewage treatment plants are rich in plant nutrients, mostly nitrates and phosphates. Presently these nutrients are often dumped into waterways where they overfertilize aquatic plant life, deplete dissolved oxygen, kill fish, and disrupt aquatic ecosystems. It makes more sense to return these nutrients to the land to fertilize trees, crops, and other vegetation. Israel is now using 35%

Use	Typical Gallons of Water Used

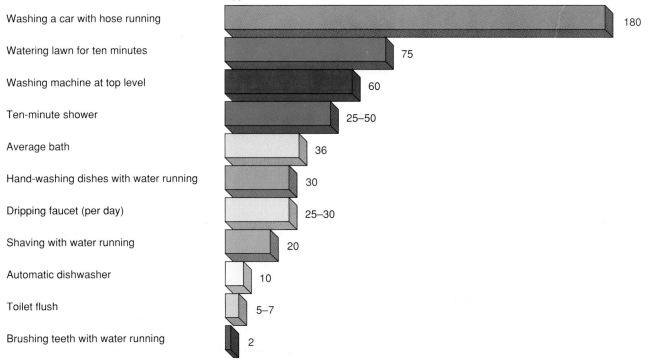

Washing a car with hose running — 180
Watering lawn for ten minutes — 75
Washing machine at top level — 60
Ten-minute shower — 25–50
Average bath — 36
Hand-washing dishes with water running — 30
Dripping faucet (per day) — 25–30
Shaving with water running — 20
Automatic dishwasher — 10
Toilet flush — 5–7
Brushing teeth with water running — 2

Figure 11-19 Some ways domestic water is wasted in the United States. (Data from American Water Works Association)

of its municipal wastewater, mostly for irrigation, and plans to reuse 80% of this flow by the year 2000.

However, as long as government-subsidized water is available at low cost, farmers have little incentive to conserve it. In most parts of the world, water is our most underpriced resource.

Wasting Less Water in Industry Manufacturing processes can either use recycled water or can be redesigned to use and waste less water. For example, to produce 0.9 metric ton (1 ton) of paper, a paper mill in Hadera, Israel, uses one-tenth as much water as most paper mills. Manufacturing aluminum from recycled scrap rather than virgin ores can reduce water needs by 97% percent.

Industry is the largest conserver of water. But the potential for water recycling in U.S. manufacturing has hardly been tapped because the cost of water to many industries is subsidized by taxpayers through federally financed water projects. A higher, more realistic price would greatly stimulate water reuse and conservation in industry.

Wasting Less Water in Homes and Businesses Flushing toilets, washing hands, and bathing account for about 78% of the water used in a typical home in the United States. However, in the arid western United States and in dry Australia, watering lawns and gardens can use up to 80% of a household's daily water expenditure. Much of this water is unneces-

sarily wasted (Figure 11-19). In 1989 Massachusetts became the first state to require builders to install toilets using no more than 6 liters (1.6 gallons) per flush.

Leaks in pipes, water mains, toilets, bathtubs, and faucets waste an estimated 20% to 35% of water withdrawn from public supplies. Because water costs so little, in most places leaking water faucets are not repaired and large quantities of water are used to clean sidewalks and streets and to irrigate lawns and golf courses. Instead of being a status symbol, a green lawn in an arid or semiarid area should be viewed as a major ecological and economic wrong and replaced with types of natural vegetation adapted to a dry climate.

Many cities offer no incentive to reduce leaks and waste. In New York City, for example, 95% of the residential units don't have water meters. Users are charged flat rates, with the average family paying less than $100 a year for virtually unlimited use of high-quality water. In Boulder, Colorado, the introduction of water meters reduced water use by more than one-third. Ways that each of us can conserve water are listed inside the back cover.

Commercially available systems can be used to purify and completely recycle wastewater from houses, apartments, and office buildings. Such a system can be leased and installed in a small shed outside a residence or building and serviced for a monthly fee about equal to that charged by most city water and

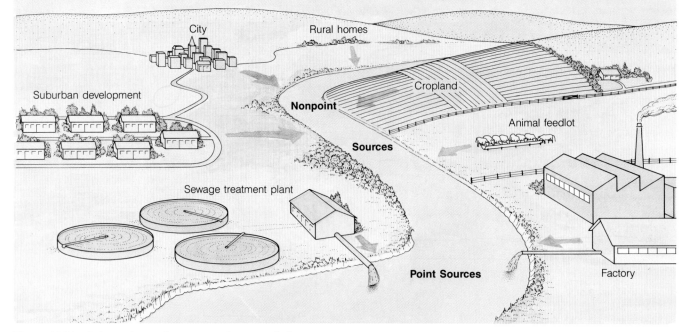

Figure 11-20 Point and nonpoint sources of water pollution.

sewer systems. In Tokyo all the water used in Mitsubishi's 60-story office building is purified for reuse by an automated recycling system.

11-4 MAJOR FORMS OF WATER POLLUTION

Major Types and Sources of Water Pollutants The following are eight common types of water pollutants:

DOWIOSRH Sowodi RH

- *Disease-causing agents*—bacteria, viruses, protozoa, and parasitic worms that enter water from domestic sewage and animal wastes. In LDCs, they are the major cause of sickness and death, prematurely killing an average of 13,700 people each day. Between 1970 and 1985, drinking water in the United States caused over 100,000 cases of disease, a larger number than in any 15-year period since 1920.

- *Oxygen-demanding wastes*—organic wastes, which can be decomposed by oxygen-consuming bacteria. Large populations of bacteria supported by these wastes can deplete water of dissolved oxygen gas. Without enough oxygen, fish and other oxygen-consuming forms of aquatic life die.

- *Water-soluble inorganic chemicals*—acids, salts, and compounds of toxic metals such as lead and mercury. Such dissolved solids can make water unfit to drink, harm fish and other aquatic life, depress crop yields, and accelerate corrosion of equipment that uses water.

- *Inorganic plant nutrients*—water-soluble nitrate and phosphate compounds that can cause excessive growth of algae and other aquatic plants, which then die and decay, depleting water of dissolved oxygen and killing fish.

- *Organic chemicals*—oil, gasoline, plastics, pesticides, cleaning solvents, detergents, and many other water-soluble and insoluble chemicals that threaten human health and harm fish and other aquatic life.

- *Sediment or suspended matter*—insoluble particles of soil and other solid inorganic and organic materials that become suspended in water and that in terms of total mass are the largest source of water pollution. Suspended particulate matter clouds the water, reduces the ability of some organisms to find food, reduces photosynthesis by aquatic plants, disrupts aquatic food webs, and carries pesticides, bacteria, and other harmful substances. Bottom sediment destroys feeding and spawning grounds of fish and clogs and fills lakes, artificial reservoirs, stream channels, and harbors.

- *Radioactive substances*—radioisotopes that are water soluble or capable of being biologically amplified to higher concentrations as they pass through food chains and webs. Ionizing radiation from such isotopes can cause birth defects, cancer, and genetic damage.

- *Heat*—excessive inputs of water that is heated when it is used mostly to cool electric power plants. The resulting increases in water temperatures lower dissolved oxygen content and make aquatic organisms more vulnerable to disease, parasites, and toxic chemicals.

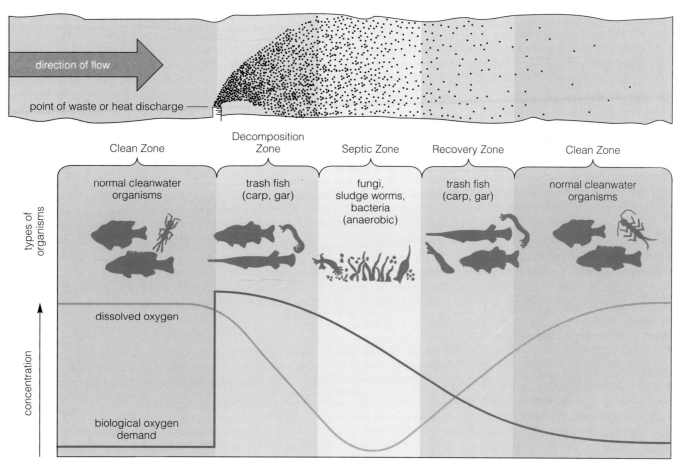

Clean Zone   Decomposition Zone   Septic Zone   Recovery Zone   Clean Zone

normal cleanwater organisms   trash fish (carp, gar)   fungi, sludge worms, bacteria (anaerobic)   trash fish (carp, gar)   normal cleanwater organisms

types of organisms

concentration

dissolved oxygen

biological oxygen demand

direction of flow

point of waste or heat discharge

time or distance downstream

Figure 11-21   The oxygen sag curve (orange) versus oxygen demand (blue). Depending on flow rates and the amount of pollutants, streams recover from oxygen-demanding wastes and heat if given enough time and if they are not overloaded.

Point and Nonpoint Sources   **Point sources** discharge pollutants at specific locations through pipes, ditches, or sewers into bodies of surface water (Figure 11-20). Examples include factories, sewage treatment plants (which remove some but not all pollutants), active and abandoned underground coal mines, offshore oil wells, and oil tankers. Because point sources are at specific places, they are fairly easy to identify, monitor, and regulate.

Nonpoint sources are big land areas that discharge pollutants into surface and underground water over a large area, and parts of the atmosphere where pollutants are deposited on surface waters (Figure 11-20). Examples include runoff into surface water and seepage into the ground from croplands, livestock feedlots, logged forests, urban and suburban lands, septic tanks, construction areas, parking lots, roadways, and acid deposition (Figure 9-8, p. 194).

Little progress has been made in the control of nonpoint water pollution because of the difficulty and expense of identifying and controlling discharges from so many diffuse sources. Controlling these inputs requires an integrated approach with emphasis on prevention by better land use (Section 12-3), soil conservation (Section 12-3), reducing resource waste (Section 19-6), controlling air pollution (Section 9-6), and regulating population growth (Section 6-4).

11-5   POLLUTION OF STREAMS AND LAKES

Streams and Oxygen-Consuming Wastes Because they flow, most streams recover rapidly from some forms of pollution, especially excess heat and degradable oxygen-demanding wastes (Figure 11-21). But this only works as long as streams are not overloaded with degradable pollutants or heat and their flow is not reduced by drought, damming, or diversion for agriculture and industries. Slowly degradable and nondegradable pollutants are not eliminated by these natural dilution and degradation processes.

The depth and width of the *oxygen sag curve* (Figure 11-21), and thus the time and distance a stream takes to recover, depend on the stream's volume, flow rate, temperature, pH, and the volume of incoming degradable wastes. Slow-flowing streams can easily be overloaded with oxygen-demanding wastes, as can normally rapid-flowing streams with reduced levels and flow rates during hot summer months or drought.

Along many streams, water for drinking is removed upstream from a city, and the city's industrial and sewage wastes are discharged downstream. The stream can then become overloaded with pollutants, as this pattern is repeated hundreds of times along the stream as it flows toward the sea.

Requiring each city to withdraw its drinking water downstream rather than upstream would dramatically improve the quality of stream water. Each city would be forced to clean up its own waste outputs rather than passing them on to downstream areas. However, this input approach for pollution control is fought by upstream users, who have the use of fairly clean water without high cleanup costs.

Poor Water Quality

Water pollution control laws enacted in the 1970s have greatly increased the number and quality of wastewater treatment plants in the United States and in many other MDCs. Laws have also required industries to reduce or eliminate point source discharges into surface waters.

Since 1972 these efforts have enabled the United States to hold the line against increased pollution of most of its streams by disease-causing agents and oxygen-demanding wastes. This is an impressive accomplishment, considering the rise in economic activity and population since 1972.

Pollution control laws have also led to improvements in dissolved oxygen content in many streams in Canada, Japan, and most western European countries since 1970. Numerous streams in the Soviet Union and in eastern Europe, however, have become more polluted with industrial wastes as industries have expanded without adequate pollution controls.

Despite progress in improving stream quality in most MDCs, large fish kills and contamination of drinking water still occur. Most of these disasters are caused by accidental or deliberate releases of toxic inorganic and organic chemicals by industries, malfunctioning sewage treatment plants, and nonpoint runoff of pesticides from cropland.

For example, in 1986 a fire at a Sandoz chemical warehouse in Switzerland released large quantities of toxic chemicals into the Rhine River, which flows through Switzerland, France, West Germany, and the Netherlands before emptying into the North Sea. These chemicals killed large numbers of aquatic life, forced temporary shutdowns of drinking water plants and commercial fishing, and set back improvements

in the river's water quality that had taken place between 1970 and 1986.

Available data indicate that pollution of streams from massive discharges of sewage and industrial wastes is a serious and growing problem in most LDCs, where waste treatment is practically nonexistent. Most of Poland's streams are severely polluted. Currently more than two-thirds of India's water resources are polluted. Of the 78 streams monitored in China, 54 are seriously polluted. In Latin America and Africa many streams are severely polluted.

Pollution Problems of Lakes and Artificial Reservoirs

Lakes and artificial reservoirs act as natural traps, collecting nutrients, suspended solids, and toxic chemicals in their growing amounts of bottom sediments. The flushing and changing of water in lakes and large artificial reservoirs can take from one to a hundred years, compared to several days to several weeks for streams.

Thus, lakes are more vulnerable than streams to contamination with plant nutrients, oil, pesticides, and toxic substances that can destroy bottom life and kill fish. Atmospheric fallout and runoff of acids into lakes is a serious problem in lakes vulnerable to acid deposition (Figure 9-8, p. 194, and Figure 9-11, p. 197).

Eutrophication of lakes is a natural process (Figure 5-28, p. 101). But the stepped-up addition of phosphates and nitrates as a result of human activities can produce in a few decades the same degree of plant nutrient enrichment that takes thousands to millions of years by natural processes. Such cultural eutrophication is a major pollution problem for shallow lakes and reservoirs, especially near urban or agricultural centers (Figure 11-22).

Overloading shallow lakes with plant nutrients (mostly nitrates and phosphates) during the summer produces dense growths of plants such as algae, water hyacinths, and duckweed. Dissolved oxygen in the surface layer of water near the shore and in the bottom layer is depleted when large masses of algae die, fall to the bottom, and are decomposed by aerobic bacteria. Then important game fish such as lake trout and smallmouth bass die of oxygen starvation, leaving the lake populated by carp and other less desirable species that need less oxygen. If excess nutrients continue to flow into a lake, the bottom water becomes foul and almost devoid of animals, as anaerobic bacteria take over and produce smelly decomposition products such as hydrogen sulfide and methane.

About one-third of the 100,000 medium to large lakes and about 85% of the large lakes near major population centers in the United States suffer from some degree of cultural eutrophication (see Case Study on p. 252).

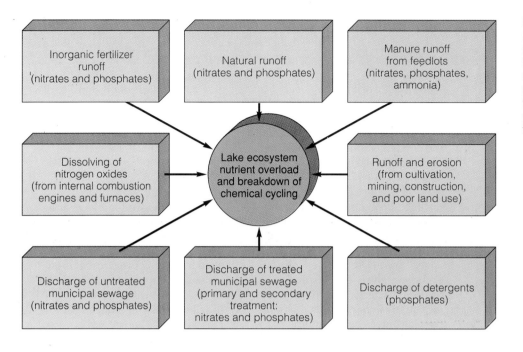

The solution to cultural eutrophication is to use input methods to reduce the flow of nutrients into lakes and reservoirs and output methods to clean up lakes suffering from excessive eutrophication. As with other forms of pollution, input approaches are the most effective and usually the cheapest in the long run.

Major input methods include advanced waste treatment, bans or limits on phosphates in household detergents and cleansers, and soil conservation and land-use control to reduce nutrient runoff. Major output methods are dredging bottom sediments to remove nutrient buildup, removing excess weeds, controlling undesirable plant growth with herbicides and algicides, and bubbling air into lakes to avoid oxygen depletion (expensive and energy intensive).

OCEAN POLLUTION

 The Ultimate Sink The oceans are the ultimate sink for much of the waste matter we produce. This is summarized in the African proverb: "Water may flow in a thousand channels, but it all returns to the sea."

In addition to natural runoff, the world's oceans receive agricultural and urban runoff, atmospheric fallout, garbage and untreated sewage from ships, and accidental oil spills from tankers and offshore oil-drilling platforms. Barges and ships also dump industrial wastes, sludge from sewage plants, and materials dredged or scraped from the bottoms of harbors and streams to maintain shipping channels into the oceans. Many of these materials are contaminated with disease-producing microorganisms and with toxic substances, including pesticides, metals such as lead and mercury, oily PCBs, and cancer-causing organic compounds.

Oceans can dilute, disperse, and degrade large amounts of sewage, sludge, oil, and some types of industrial waste, especially in deep-water areas. Marine life has also proved to be more resilient than some scientists had expected, leading these experts to suggest that it is much safer to dump much of the sewage sludge and various toxic and radioactive wastes into the deep ocean than to bury them on land or burn them in incinerators.

Other scientists dispute this idea pointing out that we know less about how the ocean works than about the moon. Using the ocean as the last large place to support our throwaway lifestyles will eventually overwhelm its dilution and renewal capacity, they add. This would delay the urgent need for pollution prevention and resource reduction while we further degraded this vital part of the earth's life-support system.

Overwhelming Coastal Areas Coastal areas, especially vital wetlands (see Spotlight on p. 97) and estuaries (see Case Study on p. 255), bear the brunt of our massive inputs of wastes into the ocean. It is no surprise that the coastal waters with the highest degree of contamination are near large coastal cities. These include Boston, New York, San Francisco, San Diego, Rio de Janeiro, Naples, Shanghai, Bombay, and Tokyo.

Regional seas such as the Baltic and Mediterranean are especially vulnerable to water pollution. This

The five interconnected Great Lakes contain at least 95% of the surface fresh water in the United States and 20% of the world's fresh water (Figure 11-23). The massive watershed around these lakes has thousands of industries and over 60 million people—one-third of Canada's population and one-eighth of the United States's population. The lakes supply drinking water for 26 million people. About 40% of U.S. industry and half of Canada's industry are located in this watershed.

Despite their enormous size, these lakes are especially vulnerable to pollution from point and nonpoint sources because less than 1% of the water entering the Great Lakes flows out to the St. Lawrence River each year. The Great Lakes also receive large quantities of acids, pesticides, and other toxic chemicals by deposition from the atmosphere—often blown in from hundreds or thousands of kilometers away.

By the 1960s, many areas of the Great Lakes were suffering from severe cultural eutrophication, massive fish kills, and contamination with bacteria and other wastes. Many bathing beaches had to be closed, and there was a sharp drop in commercial and sport fishing.

Although all five lakes were affected, the impact on Lake Erie was particularly intense because it is the shallowest of the Great Lakes and has the smallest volume of water. Also, its drainage basin is heavily industrialized and has the largest human population of any of the lakes. By 1970, the lake had lost nearly all its native fish.

Since 1972, a joint $15 billion pollution control program has been carried out by Canada and the United States for the Great Lakes. It has led to significant decreases in levels of phosphates, coliform bacteria, and many toxic industrial chemicals. Algal blooms have also decreased, and dissolved oxygen levels and sport and commercial fishing have increased. By 1988 only 8 of 516 swimming beaches around Lake Erie remained closed because of pollution.

These improvements were mainly the result of decreased point source discharges, brought about by new or upgraded sewage treatment plants and improved treatment of industrial wastes. Also, phosphate detergents, household cleaners, and water conditioners were banned or their phosphate levels were lowered in many areas of the Great Lakes drainage basin.

The most serious problem today is contamination from toxic wastes flowing into the lakes (especially Lake Erie and Lake Ontario) from land runoff, streams, and atmospheric deposition. Children under age 16 and pregnant women are advised not to eat any salmon, trout, or other fatty fish from many areas of the Great Lakes. Other people are advised not to eat such fish more than once a week.

In 1978 the United States and Canada signed a new agreement with the goal of virtual elimination of discharges of about 360 toxic chemicals. However, recent studies indicate that much of the input of toxic chemicals—more than 50% in Lake Superior—comes from the atmosphere, a source not covered by the agreement.

According to a 1989 study by the Conservation Foundation, the Great Lakes have improved in some ways but are still sick and not getting better. Solving these problems will be quite expensive and will take many years. But not dealing with the problems will cost far more.

Figure 11-23 The Great Lakes Basin. (Data from Environmental Protection Agency)

---

is because their coasts are highly populated and industrialized and they have more coastline per square kilometer than the high seas do.

Oil Pollution   Crude petroleum (oil as it comes out of the ground) and refined petroleum (fuel oil, gasoline, and other products obtained by distillation and chemical processing of crude petroleum) are accidentally or deliberately released into the environment from a number of sources. Tanker accidents and blowouts (oil escaping under high pressure from a borehole in the ocean floor) at offshore drilling rigs receive most of the publicity.

However, almost half (some experts estimate 90%) of the oil reaching the oceans comes from the land when waste oil dumped by cities, industries, and individuals onto the land or into the water ends up in streams that eventually flow into the ocean. Tanker accidents account for only 10% to 15% of the annual input of oil into the world's oceans, but concentrated spills can have severe ecological and economic impacts on coastal areas (see Case Study on p. 258).

Eventually oil reaching the ocean evaporates or is slowly degraded by bacteria. Otherwise the roughly 0.1% of the world's annual oil production that ends

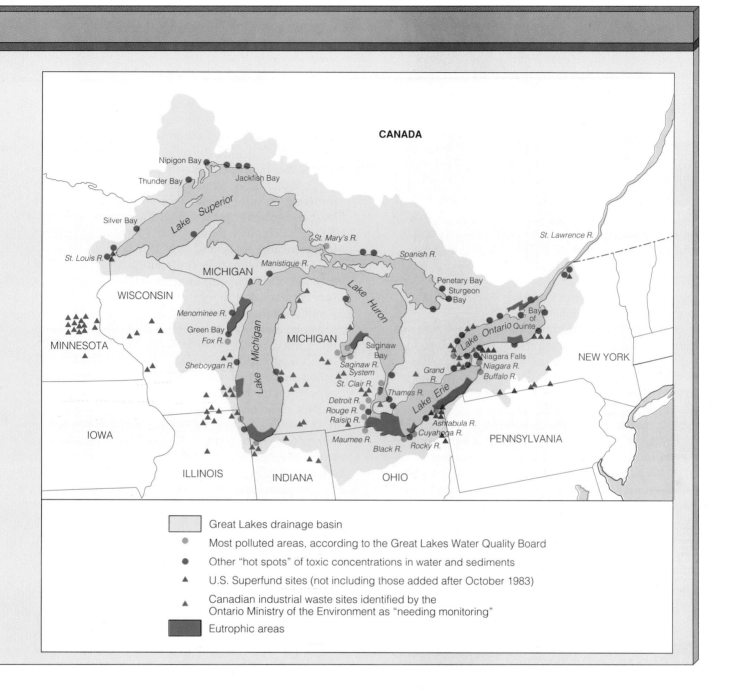

CANADA

Nipigon Bay
Thunder Bay
Jackfish Bay
Lake Superior
Silver Bay
St. Mary's R.
St. Lawrence R.
St. Louis R.
Spanish R.
MICHIGAN
Manistique R.
Penetary Bay
Sturgeon Bay
WISCONSIN
Lake Huron
Menominee R.
Green Bay
Fox R.
Lake Michigan
MICHIGAN
Saginaw Bay
Bay of Quinte
Lake Ontario
MINNESOTA
Sheboygan R.
Saginaw R. System
St. Clair R.
Grand R.
Niagara Falls
Niagara R.
Buffalo R.
NEW YORK
Detroit R.
Rouge R.
Raisin R.
Thames R.
Lake Erie
IOWA
Maumee R.
Black R.
Ashtabula R.
Cuyahoga R.
Rocky R.
PENNSYLVANIA
ILLINOIS
INDIANA
OHIO

Great Lakes drainage basin

Most polluted areas, according to the Great Lakes Water Quality Board

Other "hot spots" of toxic concentrations in water and sediments

U.S. Superfund sites (not including those added after October 1983)

Canadian industrial waste sites identified by the Ontario Ministry of the Environment as "needing monitoring"

Eutrophic areas

up in the ocean would blanket vast areas of the ocean's surface and bottom.

But while oil is evaporating and being degraded it can have a number of harmful ecological and economic effects. The effects of oil on ocean ecosystems depend on a number of factors: type of oil (crude or refined), amount released, distance of release from shore, time of year, weather conditions, average water temperature, and currents.

Volatile organic hydrocarbons in oil immediately kill a number of aquatic organisms, especially in their more vulnerable larval forms. Floating oil coats the feathers of birds (Figure 11-24), especially diving birds, and the fur of marine mammals such as seals and sea otters. This oily coating destroys the animals' natural insulation and buoyancy, and most drown or die of exposure from loss of body heat.

Heavy oil components that sink to the ocean floor or wash into estuaries can kill bottom-dwelling organisms such as crabs, oysters, mussels, and clams or make them unfit for human consumption because of their oily taste and smell. Most forms of marine life recover from exposure to large amounts of crude oil within three years. However, recovery of marine life from exposure to refined oil, especially in estuaries, may take ten years or longer. Also, the effects

Figure 11-24 A seabird coated with crude oil from an oil spill. Most of these birds die unless the oil is removed with a detergent solution. Many die even if the oil is removed.

of spills in cold waters (such as Alaska's Prince William Sound) generally last longer.

Oil slicks that wash onto beaches can have serious economic effects on coastal residents, who lose income from fishing and tourist activities. Oil-polluted beaches washed by strong waves or currents are cleaned up after about a year, but beaches in sheltered areas remain contaminated for several years. Estuaries and salt marshes suffer the most damage and cannot effectively be cleaned up. Oil cleanup is very expensive for oil companies and coastal communities.

The Valdez disaster and other oil spills are highly visible events that capture public attention and help dramatize the need for pollution prevention, which is more effective and less costly than output control. It is estimated that even with the best technology and a fast response by well-trained people, probably no more than 10% to 15% of the oil from a major spill can be recovered.

11-7 GROUNDWATER POLLUTION AND ITS CONTROL

Groundwater Contamination Groundwater is a vital source of water for drinking and irrigation in the United States and other parts of the world. Its use is expected to increase because of increasing population, irrigation, and industrialization. But this vital form of earth capital is easy to deplete because it is renewed so slowly. Also, on a human time scale, groundwater contamination can be considered permanent.

Little is known about the quality of groundwater in the United States and other parts of the world because it is very expensive to locate and test this hidden resource. Laws protecting this resource are weak and nonexistent in most countries. By 1989 only

38 of the several hundred chemicals found in U.S. groundwater were covered by federal water quality standards and routinely tested for in municipal drinking water supplies.

Results of limited testing of groundwater in the United States are alarming. In a 1982 survey the EPA found that 45% of the large public water systems served by groundwater were contaminated with synthetic organic chemicals that posed potential health threats. The EPA has documented groundwater contamination by 74 pesticides in 38 states.

Another EPA survey in 1984 found that two-thirds of the rural household wells tested violated at least one federal health standard for drinking water. The most common contaminants were nitrates from fertilizers and pesticides.

Groundwater can be contaminated from a number of point and nonpoint sources (Figure 11-26). EPA surveys indicated that by 1990 up to 58% of the 2 to 7 million underground tanks used to store gasoline, solvents, and other hazardous chemicals throughout the United States were leaking their contents into groundwater. A slow gasoline leak of just 4 liters (1 gallon) a day can seriously contaminate the water supply for 50,000 persons. Seepage of hazardous organic chemicals and toxic heavy-metal compounds from landfills (Section 19-6), thousands of abandoned hazardous-waste dumps (Section 12-5), and industrial-waste storage lagoons located above or near aquifers is also a serious problem.

Another concern is accidental leaks into aquifers from wells used to inject an estimated 57% of the country's hazardous wastes deep underground. When laws restricting the inputs of toxic chemicals into surface waters were enacted, many industries began pumping their wastes underground—a practice that attracts much less public attention than polluting more visible streams and lakes.

Laws regulating well injection are weak and poorly enforced. Reporting of the types of wastes injected is not required and no national inventory of active and abandoned wells is kept. Operators are not required to monitor nearby aquifers and are not liable for any damages from leaks once a disposal well is abandoned and plugged. Environmentalists believe that all well disposal of hazardous waste in the United States should be banned, since there are other, safer ways to deal with such wastes (Section 12-5).

When groundwater becomes contaminated, it does not cleanse itself like surface water tends to (Figure 11-21). Because groundwater flows are slow and not turbulent, contaminants are not effectively diluted and dispersed. Also there is little decomposition by aerobic bacteria because groundwater is cut off from the atmosphere's oxygen supply and has fairly small populations of aerobic and anerobic de-

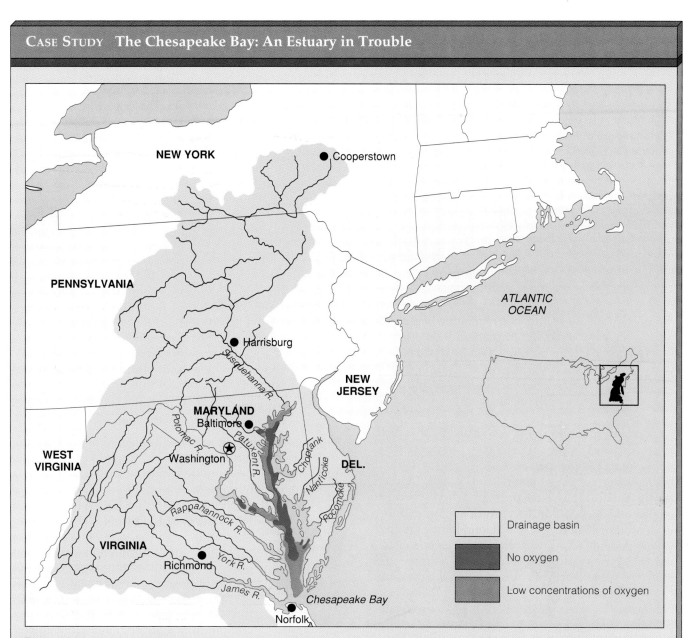

Figure 11-25   The Chesapeake Bay. The largest estuary in the United States is severely degraded as a result of water pollution from point and nonpoint sources in six states and deposition of pollutants from the atmosphere.

The Chesapeake Bay (Figure 11-25) on the East Coast is the largest estuary in the United States and one of the world's most productive. It is the single largest source of oysters in the United States and the largest producer of blue crabs in the world. The bay is also important for shipping, recreational boating, and sport fishing. Between 1940 and 1989, the number of people living close to the bay grew from 3.7 million to 13.5 million and is projected to reach 15 million by 2000.

The estuary receives wastes from point and nonpoint sources scattered throughout a huge drainage basin that includes 9 large rivers and 141 smaller streams and creeks in parts of six states. It has become a massive pollution sink because only 1% of the waste entering the bay is flushed into the Atlantic Ocean.

Levels of phosphate and nitrate plant nutrients have risen sharply in many parts of the bay, causing algal blooms and oxygen depletion. Studies have shown that point sources, primarily sewage treatment plants, contribute most of the phosphates. Nonpoint sources, mostly runoff from urban and suburban areas and agricultural activities, and acid deposition from the atmosphere are the major sources of nitrates.

(continued)

Additional pollution comes from nonpoint runoff of large quantities of pesticides from cropland and urban lawns. Point source discharge of numerous toxic wastes by industries, often in violation of their discharge permits, is also a problem.

Commercial harvests of oysters, crabs, and several commercially important fish have fallen sharply since 1960. However, populations of bluefish, menhaden, and other species that spawn in salt water and feed around algal blooms have increased.

Since 1983, over $700 million in federal and state funds have been spent on a Chesapeake Bay cleanup program that will ultimately cost several billion dollars. Between 1980 and 1987, discharges of phosphates from point sources dropped by about 20%, but there is a long way to go to reverse severe eutrophication and oxygen depletion in many areas (Figure 11-25).

Bans on phosphate-containing detergents and cleaning agents will probably have to be enacted throughout the six-state drainage basin. Forests and wetlands around the bay must also be protected from development. Halting the deterioration of this vital estuary will require the prolonged, cooperative efforts of citizens, officials, and industries.

---

composing bacteria. The cold temperature of groundwater also slows down decomposition reactions.

This means it can take hundreds to thousands of years for contaminated groundwater to cleanse itself of degradable wastes. Slowly degradable and nondegradable wastes can permanently contaminate aquifers.

Because groundwater is not visible there is little awareness and public outcry against its contamination—"out of sight, out of mind"—until wells and public water supplies must be shut down. By then slowly building pollution thresholds have been exceeded and it is too late. This explains why a number of environmentalists believe that long-lasting groundwater contamination will soon emerge as one of our most serious water resource problems as more and more threshold levels of contamination are crossed.

Control of Groundwater Pollution Groundwater pollution is much more difficult to detect and control than surface water pollution. Monitoring groundwater pollution is expensive (up to $10,000 per monitoring well), and many monitoring wells must be sunk.

Because of its location underground, pumping polluted groundwater to the surface, cleaning it up, and returning it to the aquifer is usually too expensive—$5 to $10 million or more for a single aquifer. Recent attempts to pump and treat slow-flowing contaminated aquifers show that it may take decades, even hundreds of years, of pumping before all of the contamination is forced to the surface.

Thus, preventing contamination is the only effective way to protect groundwater resources. This will require

■ banning virtually all disposal of hazardous wastes in sanitary landfills and deep wells

■ monitoring aquifers near existing sanitary and hazardous waste landfills, underground tanks, and other potential sources of groundwater contamination (Figure 11-26)

■ placing much stricter controls on the application of pesticides and fertilizers by millions of farmers and home owners

■ requiring people using private wells for drinking water to have their water tested once a year

■ establishing nationwide standards for groundwater contaminants

11-8 CONTROLLING SURFACE WATER POLLUTION

U.S. Water Pollution Legislation The Safe Drinking Water Act of 1974 requires the EPA to establish national drinking water standards, called maximum contaminant levels for any pollutants that "may" have adverse effects on human health. Environmentalists and health officials, however, have criticized the EPA for being slow in implementing this law.

By 1986, 12 years after the original legislation was passed, the EPA had set maximum contaminant levels for only 26 of at least 700 potential pollutants found in municipal drinking water supplies. In 1986 amendments to the Safe Drinking Water Act required the EPA to step up this process. Privately owned wells for millions of individual homes in suburban and rural areas are not required to meet federal drinking water standards.

Enforcement of existing laws is spotty. A National Wildlife Federation study of municipal water systems found more than 100,000 violations of federal drinking water regulations during 1988, affecting 38 million people. The EPA's inspector general also reported that the agency was failing to enforce regulations covering 140,000 non-community water systems such as those in restaurants and hospitals, serving 36 million people.

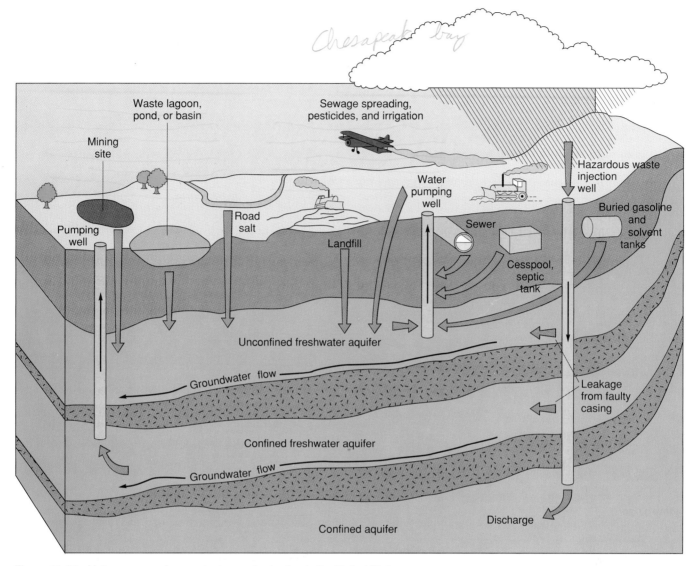

Figure 11-26 Major sources of groundwater contamination in the United States.

The Federal Water Pollution Act of 1972, renamed the Clean Water Act of 1977 when it was amended (along with amendments in 1981 and 1987), and the 1987 Water Quality Act form the basis of U.S. efforts to control pollution of the country's surface waters. The goal of these laws is to make all U.S. surface waters safe for fishing and swimming.

These acts require the EPA to establish *national effluent standards* and to set up a nationwide system for monitoring water quality. These effluent standards limit the amounts of certain conventional and toxic water pollutants that can be discharged into surface waters from factories, sewage treatment plants, and other point sources. Each point source discharger must get a permit specifying the amount of each pollutant that facility can discharge.

This has helped keep pollution of many of the country's surface waters from rising despite increased population and industrialization. However, in 1989 the EPA reported that at least 10% of the country's

lakes, rivers, estuaries, and coastal areas are contaminated by toxic chemicals that make them dangerous for aquatic life.

By 1989, 87% of the country's publicly owned sewage treatment plants complied with effluent limits set by the Clean Water Act, and about 80% of all industrial dischargers were officially in compliance with their discharge permits. But studies by the General Accounting Office have shown that most industries sometimes violate their permits.

Also 500 cities ranging from Boston to Key West, Florida, have failed to meet federal standards for sewage treatment plants. In 1989, 34 East coast cities were not doing anything more to their sewage than screening out large floating objects and discharging the rest into coastal waters.

Nonpoint Source Pollution Although most U.S. surface waters have not declined in quality since 1970, they also have not improved. The primary reason has

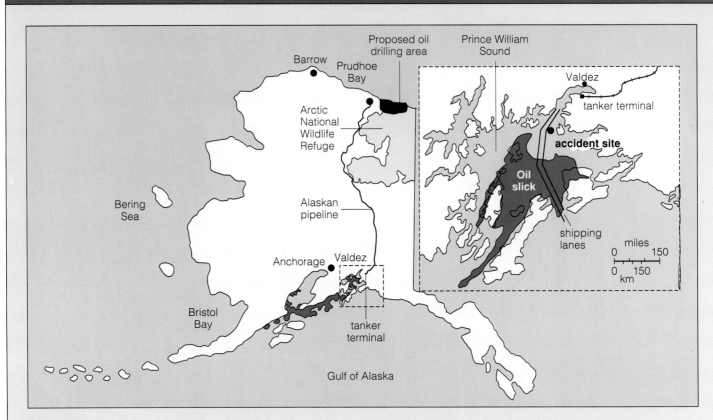

Figure 11-27   Site of the oil spill in Alaska's Prince William Sound by the tanker *Exxon Valdez* on March 24, 1989.

Crude oil extracted from fields in Alaska's North Slope near Prudhoe Bay is carried by pipeline to the port of Valdez and then shipped by tanker to the West Coast (Figure 11-27). Just after midnight on March 24, 1989, the *Exxon Valdez* tanker, more than three football fields long, went off course in a 16-kilometer-(10-mile)-wide channel in Prince William Sound near Valdez and hit submerged rocks on a reef. About 42 million liters (11 million gallons) of oil—22% of its cargo—gushed from several gashes in the hull, creating the worst oil spill in U.S. waters and spoiling one of America's most beautiful and richest wildlife areas.

In 1990 an administrative judge and the National Transportation and Safety Board found the captain of the tanker guilty of drinking before sailing and of negligently leaving the bridge before the accident and turning the ship over to an inexperienced and fatigued third mate. Since 1984 the captain had been arrested for drunken driving three times and had lost his license to drive a car, but Exxon officials still kept him in charge of one of their largest tankers carrying a $20 million cargo.

The rapidly spreading oil slick

been the absence until recently of any national strategy for controlling water pollution from nonpoint sources.

The leading nonpoint source of water pollution is agriculture. Farmers can sharply reduce fertilizer runoff into surface waters and leaching into aquifers in several ways. They should avoid using excessive amounts of fertilizer and use none on steeply sloped land. They can use slow-release fertilizers and periodically plant fields with soybeans or other nitrogen-fixing plants to reduce the need for fertilizer. Farmers should also be required to have buffer zones of permanent vegetation between cultivated fields and nearby surface water.

Similarly, farmers can reduce pesticide runoff and leaching by applying no more pesticide than is needed and by applying it only when needed. They can reduce the need for pesticides by using biological methods of pest control or integrated pest management (Section 14-5). Use of commercial inorganic fertilizers and pesticides on golf courses, yards, and public lands would also have to be sharply reduced.

Livestock growers can control runoff and infiltration by animal wastes from feedlots and barnyards using several methods. They should control animal density, plant buffers, and not locate feedlots on land sloping toward nearby surface water. Diverting runoff of animal wastes into detention basins would allow

coated and officially killed 36,471 birds (including 151 bald eagles), 1,016 sea otters, 30 seals, and unknown numbers of fish. It also jeopardized the area's $100 million-a-year fishing industry and oiled more than 5,100 kilometers (3,200 miles) of shoreline. The final toll on wildlife will never be known because most of the animals killed sank and decomposed without being counted. The good news is that these animal populations are expected to recover.

Some biologists predict recovery from this spill in three to five years. But others believe that recovery will take at least a decade because the spill took place in an enclosed body of cold water where biological decomposition of oil is slow.

In the early 1970s, conservationists predicted that a large, damaging spill might occur in these treacherous waters frequented by icebergs, submerged reefs, and violent storms. Conservationists urged that Alaskan oil be brought to the lower 48 states by pipeline over land to reduce potential damage.

But officials of Alyeska, a company formed by the seven oil companies extracting oil from Alaska's North Slope, said that a pipeline would take too long to build and that a large spill was "highly unlikely." They assured Congress that they would be at the scene of any accident within

five hours and have enough equipment and trained people to clean up any spill. The oil companies won when the 49-to-49 tie vote in the U.S. Senate was broken by Vice President Spiro Agnew under orders from President Richard M. Nixon.

When the Valdez spill occurred, Alyeska and Exxon officials did not have enough equipment and personnel and did too little too late. There is not enough equipment or trained personnel at any place in the world to deal with a spill of this size. But Alyeska clearly did not have the ability to deal with spills even half this size.

To its credit, Exxon mounted a massive cleanup effort and promptly established a claims program, which no law required.

This $2 billion accident and an unmeasurable loss of wildlife could probably have been prevented by not allowing a captain with a history of alcohol abuse to command an oil tanker and by spending $22.4 million to have the tanker built with a protective second hull to help keep oil from reaching the ocean after an accident. With profits of $5.3 billion in 1988, Exxon is one of the few companies in the world capable of paying the costs of such a spill. By the time all the claims and legal expenses are included, the costs of the spill could total $4 billion.

In the early 1970s, Interior Sec-

retary Rogers Morton told Congress that all oil tankers using Alaskan waters would have double hulls, which are on virtually all merchant ships—except oil tankers. Later, under pressure from oil companies, this requirement was dropped.

According to Jay Hair, president of the National Wildlife Federation: "This is a classic example of corporate greed. Big oil, big lies. Big lie number one was, 'Don't worry, be happy, be happy, nothing's going to happen at Valdez.' Big lie number two was, 'We're doing such a good job with the environment at the North Slope we ought to be allowed into the Arctic National Wildlife Refuge, the finest Arctic sanctuary for wildlife in the world, and Bristol Bay.'"

Others must also share the blame for this tragedy. State officials had been lax in monitoring Alyeska, and the Coast Guard did not effectively monitor tanker traffic because of inadequate radar equipment and personnel. American consumers must also share some of the blame. Their unnecessarily wasteful use of oil and gasoline (Section 17-2) is the driving force for trying to find more domestic oil without adequate environmental safeguards. This spill also highlighted the importance of pollution prevention because no spill of this magnitude can be effectively cleaned up.

this nutrient-rich water to be pumped and applied as fertilizer to cropland or forestland.

Critical watersheds should also be reforested. In addition to reducing water pollution from sediment, this would reduce soil erosion and the severity of flooding, and help slow projected global warming and loss of the earth's precious biodiversity.

Point Source Pollution: Wastewater Treatment In many LDCs and some parts of MDCs, sewage and waterborne industrial wastes from point sources are not treated. Instead, most are discharged into the nearest waterway or into **wastewater lagoons**—large ponds where air, sunlight, and microorganisms break

down wastes, allow solids to settle out, and kill some disease-causing bacteria. Water typically remains in a lagoon for 30 days. Then it is treated with chlorine and pumped out for use by a city or farms.

In MDCs, most wastes from point sources are purified to varying degrees. In rural and suburban areas with suitable soils, sewage from each house is usually discharged into a **septic tank** (Figure 11-28). It traps greases and large solids and discharges the remaining wastes over a large drainage field. As these wastes percolate downward, the soil filters out some potential pollutants and soil bacteria decompose biodegradable materials. About 24% of all homes in the United States are served by septic tanks.

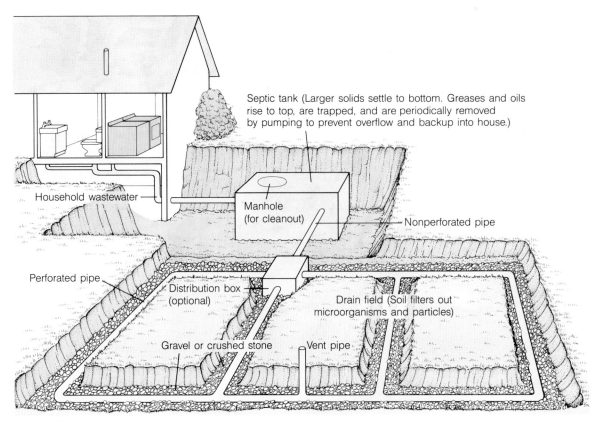

Figure 11-28 Septic tank system used for disposal of domestic sewage and wastewater in rural and suburban areas.

Labels in figure:
Septic tank (Larger solids settle to bottom. Greases and oils rise to top, are trapped, and are periodically removed by pumping to prevent overflow and backup into house.)

Household wastewater

Manhole (for cleanout)

Nonperforated pipe

Perforated pipe

Distribution box (optional)

Drain field (Soil filters out microorganisms and particles)

Gravel or crushed stone

Vent pipe

In urban areas in MDCs most waterborne wastes from homes, businesses, factories, and storm runoff flow through a network of sewer pipes to wastewater treatment plants.

Many cities have separate lines for storm water runoff, but in other areas (such as parts of Boston), lines for these two systems are combined because it is cheaper. When combined sewer systems overflow, they discharge untreated sewage directly into surface waters. When sewage reaches a treatment plant, it can undergo up to three levels of purification, depending on the type of plant and the degree of purity desired. **Primary sewage treatment** is a mechanical process that uses screens to filter out debris such as sticks, stones, and rags. Then suspended solids settle out as sludge in a settling tank (Figure 11-29). Improved primary treatment using chemically treated polymers does a better job of removing suspended solids. **Secondary sewage treatment** is a biological process that uses aerobic bacteria as a first step to remove up to 90% of degradable, oxygen-demanding organic wastes (Figure 11-30).

In the United States, combined primary and secondary treatment must be used in all communities served by wastewater treatment plants. Combined primary and secondary treatment, however, still leaves about 3% to 5% of the oxygen-demanding wastes, 3% of the suspended solids, 50% of the nitrogen

(mostly as nitrates), 70% of the phosphorus (mostly as phosphates), and 30% of most toxic metal compounds and synthetic organic chemicals in the wastewater discharged from the plant. Virtually none of any long-lived radioactive isotopes or persistent organic substances such as pesticides are removed by these two processes.

Advanced sewage treatment is a series of specialized chemical and physical processes that lower the quantity of specific pollutants still left after primary and secondary treatment (Figure 11-31). Types of advanced treatment vary depending on the contaminants in specific communities and industries. Except in Sweden, Denmark, and Norway, advanced treatment is rarely used because the plants cost twice as much to build and four times as much to operate as secondary plants.

Scientists are also evaluating solar-powered aquatics: purifying contaminated water in outdoor greenhouse-covered canals filled with purifying aquatic plants such as bulrushes or water hyacinths. In LDCs and in many areas in MDCs, using low-tech, natural ecosystems to treat wastewater may be the cheapest and the best way to purify wastewater (see Case Study on p. 263).

Before water is discharged from a sewage treatment plant, it is disinfected to remove water coloration and kill disease-carrying bacteria and some, but not

3 to 5% oxygen wastes

3% solids

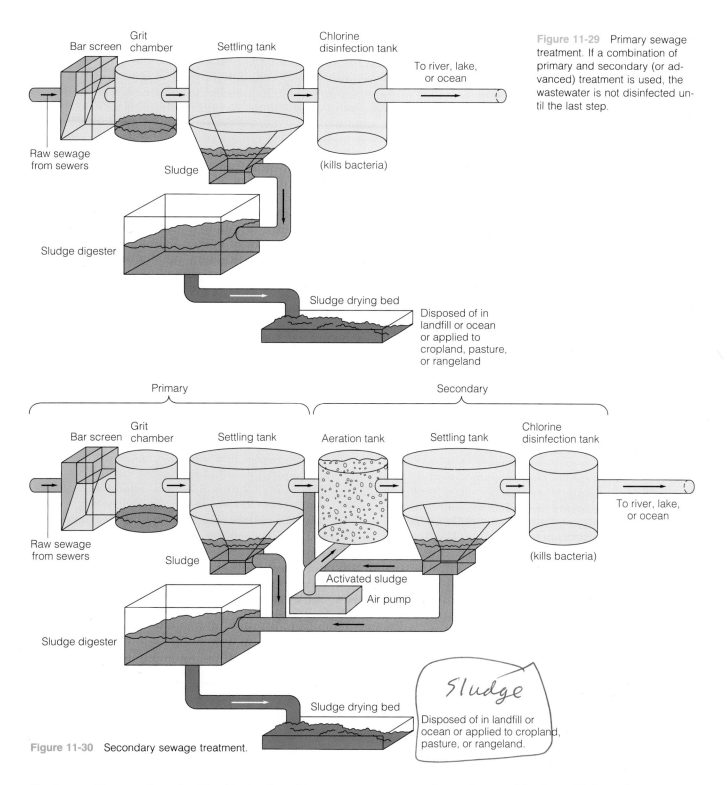

Grit
Bar screen | chamber | Settling tank | Chlorine disinfection tank

To river, lake, or ocean

Figure 11-29 Primary sewage treatment. If a combination of primary and secondary (or advanced) treatment is used, the wastewater is not disinfected until the last step.

Raw sewage from sewers

Sludge

(kills bacteria)

Sludge digester

Sludge drying bed

Disposed of in landfill or ocean or applied to cropland, pasture, or rangeland

Primary

Secondary

Grit
Bar screen | chamber | Settling tank | Aeration tank | Settling tank | Chlorine disinfection tank

Raw sewage from sewers

Sludge

To river, lake, or ocean

(kills bacteria)

Activated sludge

Air pump

Sludge digester

Sludge

Sludge drying bed

Disposed of in landfill or ocean or applied to cropland, pasture, or rangeland.

Figure 11-30 Secondary sewage treatment.

all, viruses. The usual method is chlorination. However, chlorine reacts with organic materials in the wastewater or in surface water to form small amounts of chlorinated hydrocarbons, some of which are known carcinogens. Several other disinfectants, such as ozone and UV light, are being used in some places, but are more expensive than chlorination.

Without expensive advanced treatment, effluents from primary and secondary sewage treatment plants contain enough nitrates and phosphates to contribute

to accelerated eutrophication of lakes, slow-moving streams, and coastal waters. Conventional sewage treatment has helped reduce pollution of surface waters. But environmentalists point out that it is a limited and flawed output approach that is eventually overwhelmed by more people producing more wastes.

Sewage treatment produces a gooey sludge that is difficult to dispose of and toxic. About 42% of this sludge is dumped in landfills where it can contaminate groundwater. About 6% is dumped into the ocean,

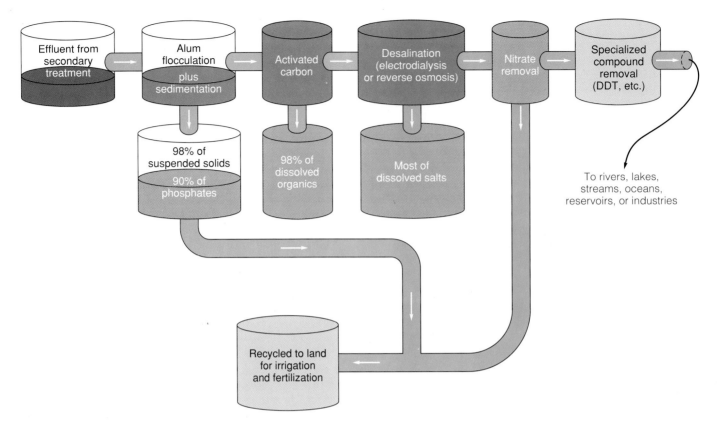

Figure 11-31 Advanced sewage treatment. This diagram shows several different types of advanced treatment. Often only one or two of these processes are used to remove specific pollutants in a particular area.

a process in which we transfer water pollution from one part of the hydrosphere to another. The 21% of the sludge that is incinerated pollutes the air with traces of toxic chemicals and the resulting highly toxic residue must be disposed of, usually in landfills where it can pollute groundwater.

A better alternative is to return the nitrate and phosphate plant nutrients in sewage plant effluent and sludge to the land as fertilizers and soil conditioners produced by composting sludge. This is now being done for more than 25% of all municipal sludge and sewage plant effluent in the United States and 48% of the amount produced in Denmark. Before it is applied, sludge can be heated to kill harmful bacteria and treated to remove heavy toxic metals and organic chemicals.

Untreated sludge can be applied to land not used for crops or livestock or to land where groundwater is already contaminated or is not used as a source of drinking water. Examples include forests, surface-mined land, golf courses, lawns, cemeteries, and highway medians.

Present sewage treatment is just not cost effective compared to pollution prevention. Without massive federal subsidies communities cannot afford to build and operate secondary sewage treatment plants. Once they are built the huge waste treatment industry is difficult to change and little money is available for

pollution prevention and development of better methods for treating sewage (see Case Study on p. 263).

Protecting Coastal Waters Major suggestions for preventing excessive pollution of coastal waters include:

Input Methods

- Eliminating the discharge of toxic pollutants into coastal waters from both industrial facilities and municipal sewage treatment plants.

- Eliminating all discharges of raw sewage from sewer line overflows.

- Promoting water conservation in homes and industries to reduce the flow to sewage treatment plants and hence the danger of overflow (see Individuals Matter inside the back cover).

- Banning all ocean dumping of sewage sludge and hazardous dredged materials.

- Enacting and enforcing laws and land-use practices to sharply reduce runoff from nonpoint sources in coastal areas.

- Protecting coastal waters that are already clean by not allowing harmful forms of development.

- Protecting sensitive marine areas from all forms of development by designating them as ocean sanctuaries, much like protected wilderness areas on land.

- Regulating the types and density of coastal development to minimize its environmental impact and eliminating subsidies and tax incentives that encourage harmful coastal development.

- Instituting a national energy policy based on energy efficiency and renewable energy resources to reduce our dependence on oil.

- Prohibiting oil drilling in ecologically sensitive offshore and nearshore areas.

- Collecting and reprocessing used oils and greases for reuse.

- Requiring new and existing oil tankers to have double hulls to lessen chances of severe leaks.

- Strictly regulating the construction and operation of oil tankers, offshore oil rigs, and oil refineries and greatly increasing the financial liability of oil companies for cleaning up oil spills to encourage pollution prevention.

- Routing tankers as far as possible from sensitive coastal areas and having Coast Guard vessels guide tankers out of all harbors and enclosed sounds and bays.

- Preventing the discharge of plastic items from vessels, sewers, and other sources.

- Sharply reducing the use of disposable plastic items (Section 19-6).

- Instituting a nationwide program to collect and safely dispose of household hazardous wastes such as pesticides, antifreeze, and household cleaning solvents that are often poured down the drain.

Output Approaches

- Adopting a nationwide tracking program to ensure that medical waste is safely disposed of.

- Greatly improving oil spill cleanup capabilities. Small spills can be effectively dealt with by quick response from teams of well-trained people supplied with enough equipment. However, according to a 1990 report by the Office of Technology Assessment, there is little chance that large spills can be effectively cleaned up. In such cases, an ounce of prevention is worth many pounds of cure.

- Upgrading all coastal sewage treatment plants to at least the degree required for inland waters (secondary treatment) or by developing new improved methods for sewage treatment.

Future Water Quality Goals Serious pollution of the hydrosphere is preventable and potentially reversible. But this requires that we shift from output pollution control to input pollution prevention. Otherwise, even the best output approaches are eventually overwhelmed by increases in population and industrialization.

CASE STUDY Working with Nature to Purify Sewage

In Lima, Peru, most of the sewage produced by its 7 million, mostly poor, people is discharged untreated into the Pacific Ocean. In poor areas this sewage runs to the ocean through a series of open ditches in which children often swim despite warnings from their parents.

The country cannot afford to build expensive waste treatment plants. However, there are now experiments in which the sewage in some areas is channeled into holding ponds, where solids fall to the bottom. Bacteria then decompose many of the wastes. After 20 to 30 days the water is safe to use. Some of it is used to irrigate corn used to feed cattle and some is pumped to other ponds where the remaining nutrients are used to raise fish.

Arcata, California, has initiated what many consider to be one of the most innovative and effective wastewater treatment systems in the country. In this coastal northern California town of 15,000, some 63 hectares (154 acres) of wetlands have been created between the town and adjacent bay in an area that was once a dump. The newly created marshes act as an inexpensive, low-tech waste treatment plant by removing nitrogen and phosphorous plant nutrients, degrading organic wastes, and filtering out toxic materials.

Instead of using expensive, high-tech waste treatment plants that only partially clean up wastewater, Arcata's beautifully simple approach is based on working with nature. First, the city's wastewater passes through oxidation ponds, where wastes settle out and are partially broken down. Then the water flows on to be further filtered and cleansed in marshes. The water then moves on to irrigate and nourish other wetlands and some is pumped into the bay, where oysters are thriving.

The marshes are also an Audubon bird sanctuary and provide habitats for thousands of otters, sea birds, and marine animals. This refuge, which draws thousands of birdwatchers each year, has no smell other than that of salt air.

Compared to sewage treatment plants, this highly effective approach is cheap and easy to maintain. However, it does require more land than the conventional approach.

This will require developing an integrated approach to managing water resources and water pollution, including consideration of urbanization, population growth and control, energy use, mining, agriculture, fisheries, forests, soil conservation, and wildlife. Pollution prevention and waste reduction

- Use commercial inorganic fertilizers, pesticides, detergents, bleaches, and other chemicals only if necessary and then in the smallest amounts possible.

- Use less harmful substances for most household cleaners (see Individuals Matter inside back cover).

- Use low-phosphate, phosphate-free, or biodegradable dishwashing liquid, laundry detergent, and shampoo.

- Don't use water fresheners in toilets.

- Contact your local Department of Health about how to dispose of household hazardous and medical wastes. Don't pour products containing harmful chemicals such as pesticides, paints, solvents, oil, and cleaning agents down the drain. The amount of old motor oil improperly disposed of by Americans each year is equivalent to 24 *Exxon Valdez* oil spills. Oil from one oil change can pollute 3.8 million liters (1 million gallons) of water.

- Join with others to encourage your local Department of Health or other agency to organize community wide Household Hazardous-Waste Collection Days.

- Recycle old motor oil and antifreeze at a gas station that has an oil recycling program. If such a program is not available, pressure local officials to start one.

- Landscape with a variety of water-conserving groundcover plants and energy-conserving trees adapted to the climate where you live rather than lawns that require watering, fertilizers, and pesticides.

- If you have to irrigate plants in yards or gardens, use drip irrigation (Figure 11-18).

- Use manure or compost to fertilize garden and yard plants (Section 12-3).

- Use biological methods or integrated pest management to control garden, yard, and household pests (Section 14-5).

- Buy organically grown produce or grow some of your own.

- Use and waste less water (see Individuals Matter inside the back cover).

- If you have a septic tank, monitor it yearly and have it cleaned out every three to five years by a reputable contractor so that it won't contribute to groundwater pollution. Do not use septic tank cleaners. They contain toxic chemicals that can kill bacteria important to sewage decomposition in the septic system and that can contaminate groundwater if the system malfunctions.

- If you are a boater, don't dump trash overboard and only discharge boat sewage into regulated onshore facilities.

- Support ecological land-use planning in your local community.

- Get to know your local bodies of water and form community watchdog groups to help monitor, protect, and restore them.

- Support tougher water pollution control laws and their enforcement at the local, state, and federal levels based on the goals listed in this section.

from all sources and minimizing waste of water, energy, and other resources must be emphasized.

Otherwise, we will continue to unnecessarily waste resources and shift environmental problems from one part of the environment to another. Ultimately this form of environmental musical chairs will fail as population and industrialization continue to grow.

Doing this will require us to truly accept the fact that the environment that we now treat as separate parts—air, water, soil, life—is an interconnected whole. This means we must organize our efforts to sustain the earth on the basis of the earth's watersheds, airsheds, and ecosystems instead of the neat geopolitical lines we draw on maps. This will require unprecedented cooperation between communities, states, and countries.

Each of us can contribute to bringing about this drastic change in the way we view and act in the world by reducing our contributions to water pollution and water waste (see Individuals Matter above).

The good news is that nature can restore polluted watersheds and airsheds, but not until we stop overwhelming them with pollutants. Trend is not destiny.

Water is more critical than energy. We have alternative sources of energy. But with water, there is no other choice.

EUGENE ODUM

DISCUSSION TOPICS

1. How do human activities increase the harmful effects of prolonged drought? How can these effects be reduced?

2. How do human activities contribute to flooding? How can these effects be reduced?

3. In your community:
   a. What are the major sources of the water supply?
   b. How is water use divided among agricultural, industrial, power-plant cooling, and public uses? Who are the biggest consumers of water?
   c. What has happened to water prices during the past 20 years? Are they too low to encourage water conservation and reuse?
   d. What water supply problems are projected?
   e. How is water being wasted?
   f. How is drinking water treated?
   g. Has drinking water been analyzed recently for the presence of synthetic organic chemicals, especially chlorinated hydrocarbons? If so, were any found and are they being removed?
   h. What are the major nonpoint sources of contamination of surface water and groundwater?

4. Explain why dams and reservoirs may lead to more flood damage than would have occurred if they had not been built. Should all proposed large dam and reservoir projects be scrapped? What criteria would you use in determining desirable projects?

5. Should the price of water for all uses in the United States be increased sharply to encourage water conservation? Explain. What effects might this have on the economy, on you, on the poor, on the environment?

6. List ten major ways to conserve water on a personal level. Which, if any, of these practices do you now use or intend to use (see Individuals Matter inside the back cover)?

7. Explain how a stream can cleanse itself of oxygen-demanding wastes. Under what conditions will this natural cleansing system fail?

8. Explain why you agree or disagree with the idea that we should deliberately dump most of our wastes in the ocean because it is a vast sink for diluting, dispersing, and degrading wastes, and if it becomes polluted, we can get food from other sources. Explain why banning ocean dumping alone will not stop ocean pollution.

9. Should the injection of hazardous wastes into deep underground wells be banned? Explain.

FURTHER READINGS

Ashworth, William. 1982. *Nor Any Drop To Drink*. New York: Summit Books.

Ashworth, William. 1986. *The Late, Great Lakes: An Environmental History*. New York: Alfred A. Knopf.

Borgese, Elisabeth Mann. 1986. *The Future of the Oceans*. New York: Harvest House.

Bullock, David K. 1989. *The Wasted Ocean*. New York: Lyons & Burford.

Center for Marine Conservation. 1989. *The Exxon Valdez Oil Spill: A Management Analysis*. Washington, D.C.: Center for Marine Conservation.

Conservation Foundation. 1989. *Great Lakes: Great Legacy*. Washington, D.C.: Conservation Foundation.

Davidson, Art. 1990. *In Wake of the Exxon Valdez*. San Francisco: Sierra Club Books.

El-Ashry, Mohamed, and Diana C. Gibbons. 1988. *Water and Arid Lands of the Western United States*. Washington, D.C.: World Resources Institute.

Gabler, Raymond. 1988. *Is Your Water Safe to Drink?* New York: Consumer Reports Books.

Goldsmith, Edward, and Nicholas Hidyard, eds. 1986. *The Social and Environmental Effects of Large Dams*. (3 vols.). New York: John Wiley.

Gottlieb, Robert. 1988. *A Life of Its Own: The Politics and Power of Water*. New York: Harcourt Brace Jovanovich.

Hansen, Nancy R., et al. 1988. *Controlling Nonpoint-Source Water Pollution*. New York: National Audubon Society and the Consumer Foundation.

Hodgson, Bryan. 1990. "Alaska's Big Spill—Can the Wilderness Heal?" *National Geographic*, Jan., 5–43.

Ives, J. D., and B. Messeric. 1989. *The Himalayan Dilemma: Reconciling Development and Conservation*. London: Routledge.

Jorgensen, Eric P., ed. 1989. *The Poisoned Well: New Strategies for Groundwater Protection*. Covelo, Calif.: Island Press.

King, Jonathan. 1985. *Troubled Water: The Poisoning of America's Drinking Water*. Emmaus, Penn.: Rodale Press.

Meybeck, Michael, et al., eds. 1990. *Global Freshwater Quality*. Cambridge, Mass.: Basil Blackwell.

National Academy of Sciences. 1985. *Oil in the Sea*. Washington, D.C.: National Academy Press.

National Academy of Sciences. 1986. *Drinking Water and Health*. Washington, D.C.: National Academy Press.

Natural Resources Defense Council. 1989. *Ebb Tide for Pollution: Actions for Cleaning Up Coastal Waters*. Washington, D.C.: Natural Resources Defense Council.

Office of Technology Assessment. 1984. *Protecting the Nation's Groundwater from Contamination*. Washington, D.C.: Government Printing Office.

Office of Technology Assessment. 1987. *Wastes in Marine Environments*. Washington, D.C.: Government Printing Office.

Office of Technology Assessment. 1990. *Coping with an Oiled Sea: An Analysis of Oil Spill Response Technologies*. Washington, D.C.: Government Printing Office.

Postel, Sandra. 1985. *Conserving Water: The Untapped Alternative*. Washington, D.C.: Worldwatch Institute.

Postel, Sandra. 1989. *Water for Agriculture: Facing the Limits*. Washington, D.C.: Worldwatch Institute.

Pringle, Laurence. 1982. *Water—The Next Great Resource Battle*. New York: Macmillan.

Reisner, Marc. 1988. "The Next Water War: Cities Versus Agriculture." *Issues in Science and Technology*, Winter, 98–102.

Rivière, J. W. Maurits. 1989. "Threats to the World's Water." *Scientific American*, Sept., 80–94.

Rocky Mountain Institute. 1988. *Catalog of Water-Efficient Technologies for the Urban/Residential Sector*. Old Snowmass, Colo.: Rocky Mountain Institute.

Simon, Anne W. 1985. *Neptune's Revenge: The Ocean of Tomorrow*. New York: Franklin Watts.

Wijkman, Anders, and Lloyd Timberlake. 1984. *Natural Disasters: Acts of God or Acts of Man?* Washington, D.C.: Earthscan.

CHAPTER TWELVE

SOIL RESOURCES

AND HAZARDOUS WASTE

General Questions and Issues

1. What are the major components and types of soil, and what properties make a soil best suited for growing crops?

2. How serious is the problem of soil erosion in the world and in the United States?

3. How can we reduce erosion and nutrient depletion in topsoil?

4. How is soil degraded by excessive salt buildup (salinization) and waterlogging?

5. How are soil and other parts of the environment degraded by hazardous chemicals?

Below that thin layer comprising the delicate organism known as the soil is a planet as lifeless as the moon.

G. Y. JACKS AND R. O. WHYTE

 NLESS YOU ARE A FARMER you probably think of soil as dirt—something you don't want on your hands, clothes, or carpet. You are acutely aware of your need for air and water, but you may be unaware that your life and that of other organisms depend on soil—especially the upper portion known as topsoil.

The nutrients in the food you eat come from soil. A more accurate version of the saying that "all flesh is grass" is "all flesh is soil nutrients." Soil also provides you with wood, paper, cotton, gravel, and many other vital materials and helps purify the water you drink.

As long as soil is held in place by vegetation, it stores water and releases it in a nourishing trickle instead of a devastating flood. Soil's decomposer organisms recycle the key chemicals we and most other forms of life need. Bacteria in soil decompose degradable forms of garbage you throw away, although this process takes decades to hundreds of years in today's compacted, oxygen-deficient landfills. Soil is truly the base of life and civilization.

Yet, since the beginning of agriculture we have abused this vital, potentially renewable resource. Today we are abusing soil more than ever. Each of us must become involved in protecting the life-giving resource we call soil. In saving the soil, we save ourselves and other forms of life.

 SOIL: COMPONENTS, TYPES, AND PROPERTIES

Soil Layers and Components Pick up a handful of soil and notice how it feels and looks. The **soil** you hold in your hand is a complex mixture of inorganic materials (clay, silt, pebbles, and sand), decaying organic matter, water, air, and billions of living organisms.

The components of mature soils are arranged in a series of zones called **soil horizons** (Figure 12-1). Each horizon has a distinct texture and composition that vary with different types of soils. A cross-sectional view of the horizons in a soil is called a *soil profile*. Most mature soils have at least three of the six possible horizons, but some new or poorly developed soils don't have horizons.

The top layer, *surface-litter layer,* or *O-horizon,* consists mostly of freshly fallen and partially decomposed leaves, twigs, animal waste, fungi, and other organic materials. Most often it is usually brown to black in color. The underlying *topsoil layer,* or *A-horizon,* is usually a porous mixture of partially decomposed organic matter (humus), living organisms, and some inorganic mineral particles. Normally it is darker and looser than deeper layers. The roots of most plants and most of a soil's organic matter are concentrated in these two upper soil layers (Figure 12-1).

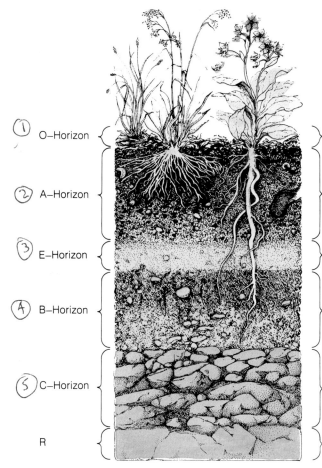

① O–Horizon

Surface litter:
Freshly fallen leaves and organic debris and partially decomposed organic matter

Iron + Al + Water + heat

② A–Horizon

Topsoil: *A*
Partially decomposed organic matter (humus), plant roots, living organisms, and some inorganic minerals

③ E–Horizon

Zone of leaching:
Area through which dissolved or suspended materials move downward

thin humus in rain forest = iron aluminum.

④ B–Horizon

Subsoil:
Unique colors and often an accumulation of iron, aluminum, and humic compounds, and clay leached down from above layers

⑤ C–Horizon

Parent material:
Partially broken-down inorganic materials

R

Bedrock:
Impenetrable layer, except for fractures

Aquifers, oils

The two top layers of most well-developed soils are also teeming with bacteria, fungi, earthworms, and small insects. These layers are also home for larger, burrowing animals such as moles and gophers. All these soil organisms interact in complex food webs (Figure 12-2). Most are bacteria and other decomposer microorganisms, with billions found in every handful of soil. They partially or completely break down some of the complex inorganic and organic compounds in the upper layers of soil into simpler, nutrient compounds that dissolve in soil water. Soil water carrying these dissolved nutrients is drawn up by the roots of plants and transported through stems and into leaves (Figure 12-3).

Some organic compounds in the two upper layers are broken down slowly and form a dark-colored mixture of partially decayed plant and animal matter called **humus**. Much of the humus is not soluble in water and it remains in the topsoil layer. Humus helps retain water and water-soluble plant nutrients so they can be taken up by plant roots. A fertile soil, useful for growing high yields of crops, has a thick topsoil layer containing a high content of humus. When soil is eroded, it is the vital surface litter and topsoil layers that are lost.

The color of the topsoil layer tells us a lot about how useful a soil is for growing crops. For example, dark brown or black topsoil has a large amount of organic matter and is nitrogen rich. Gray, bright yellow, or red topsoils are low in organic matter and will require nitrogen fertilizer to increase their fertility.

The B- and C-horizons contain most of a soil's inorganic matter. Most of this is broken-down rock in the form of varying mixtures of sand, silt, clay, and gravel. These lie on a base of bedrock (Figure 12-1).

The spaces, or pores, between the solid organic and inorganic particles in the upper and lower soil layers contain varying amounts of two other key inorganic components: air (mostly nitrogen and oxygen gas) and water. The oxygen gas, highly concentrated in the topsoil, is used by the cells in plant roots to carry out cellular respiration.

Some of the rain falling on the soil surface percolates downward through the soil layers and occupies many of the pores. As this water seeps downward, it dissolves and picks up various soil components in upper layers and carries them to lower layers—a process called **leaching.** Most materials leached from upper layers accumulate in the B-horizon, if one has developed.

Major Types of Soil Mature soils in different ecosystems of the world vary widely in color, content, pore space, acidity (pH), and depth. These differences can be used to classify soils throughout the world

CCP AD

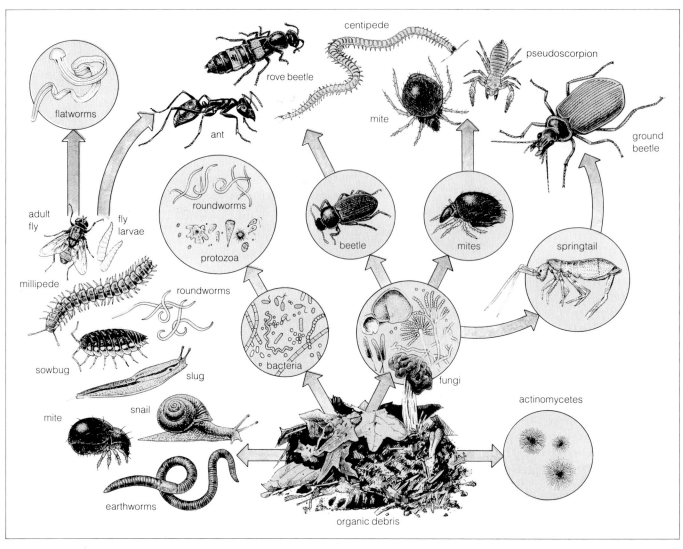

Figure 12-2 Greatly simplified food web of living organisms found in soil.

into ten major types, or orders. Five important soil orders are mollisols, alfisols, spodosols, oxisols, and aridisols, each with a distinct soil profile (Figure 12-4). Most of the world's crops are grown on grassland mollisols and on alfisols exposed when deciduous forests are cleared.

Soil Texture and Porosity Soils vary in their content of clay (very fine particles), silt (fine particles), sand (medium-sized particles), and gravel (coarse to very coarse particles). The relative amounts of the different sizes and types of particles determine **soil texture**. Figure 12-5 shows how soils can be grouped into textural classes according to clay, silt, and sand content.

Soil texture helps determine **soil porosity**—a measure of the volume of pores per volume of soil and the average distances between these spaces. A soil with a high porosity can hold more water and air than one with a lower porosity. The average pore size determines **soil permeability:** the rate at which

water and air move from upper to lower soil layers. Soil porosity is also influenced by soil structure: how the particles that make up a soil are organized and clumped together.

Soils fall into three broad textural classes: loams, sandy, and clay (Figure 12-5). Loams are the best soils for growing most crops because they retain a large amount of water that is not held too tightly for plant roots to absorb.

Sandy soils are easy to work and have less pore space per volume of soil (lower porosity) than other soils. However, sandy soils have a high permeability because their pores are larger than those in most other soils. This is why water flows rapidly through sandy soils. They are useful for growing irrigated crops or those without large water requirements, such as peanuts and strawberries.

The particles in clay soils are very small and easily compacted. When these soils get wet they form large, dense clumps, explaining why wet clay is so easy to mold into bricks and pottery. Clay soils have more

Sandy, loam, clay

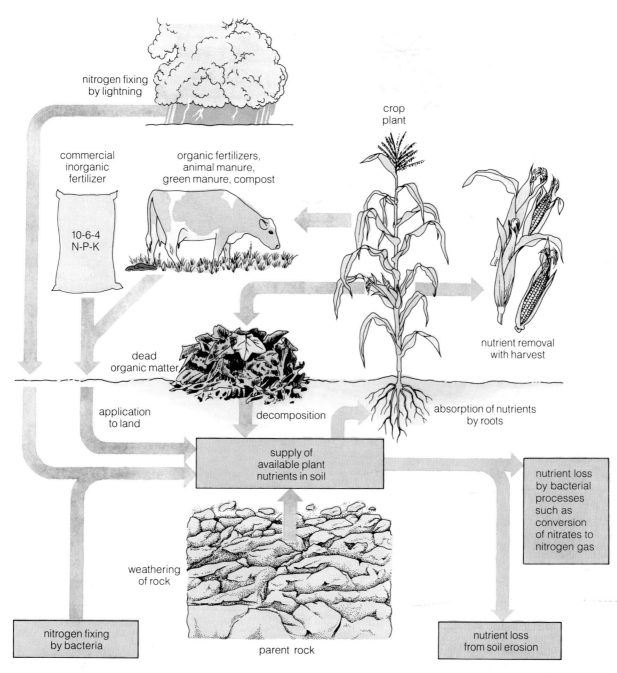

Figure 12-3 Addition and loss of plant nutrients in soils.

pore space per volume and a greater water-holding capacity than sandy soils. But the pore spaces are so small that these soils have a low permeability. Because little water can infiltrate to lower levels, the upper layers of these soils can easily become too waterlogged to grow most crops.

To get a general idea of a soil's texture, take a small amount of topsoil, moisten it, and rub it between your fingers and thumb. A gritty feel means that it contains a lot of sand. A sticky feel means that it has a high clay content, and you should be able to role it into a clump. Silt-laden soil feels smooth like flour. A loam topsoil (Figure 12-5), best suited for plant

growth, has a texture between these extremes. It has a crumbly, spongy feeling, and many of its particles are clumped loosely together.

Soil Acidity (pH) The acidity or alkalinity of a soil is another factor determining the types of crops it can support. Scientists use pH as a simple measure of the degree of acidity or alkalinity of a solution (Figure 9-9, p. 195).

Soils vary in acidity (Figure 12-4) and crops vary in the pH ranges they can tolerate. For example, wheat, spinach, peas, corn, and tomatoes grow best in slightly acidic soils; potatoes and berries do best

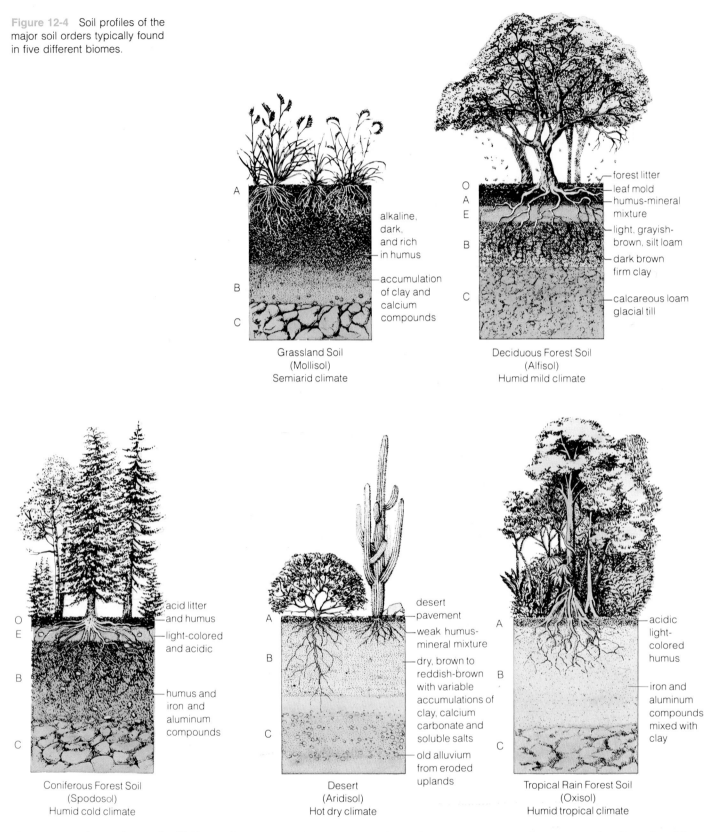

Figure 12-4 Soil profiles of the major soil orders typically found in five different biomes.

Grassland Soil
(Mollisol)
Semiarid climate

- A — alkaline, dark, and rich in humus
- B — accumulation of clay and calcium compounds
- C

Deciduous Forest Soil
(Alfisol)
Humid mild climate

- O — forest litter
- A — leaf mold
- E — humus-mineral mixture
- B — light, grayish-brown, silt loam
- B — dark brown firm clay
- C — calcareous loam glacial till

Coniferous Forest Soil
(Spodosol)
Humid cold climate

- O — acid litter and humus
- E — light-colored and acidic
- B — humus and iron and aluminum compounds
- C

Desert
(Aridisol)
Hot dry climate

- A — desert pavement
- weak humus-mineral mixture
- B — dry, brown to reddish-brown with variable accumulations of clay, calcium carbonate and soluble salts
- C — old alluvium from eroded uplands

Tropical Rain Forest Soil
(Oxisol)
Humid tropical climate

- A — acidic light-colored humus
- B — iron and aluminum compounds mixed with clay
- C

in very acidic soils; and alfalfa and asparagus in neutral soils.

When soils are too acidic for the desired crops, the acids can be partially neutralized by an alkaline substance such as lime. But adding lime speeds up the undesirable decomposition of organic matter in the soil, so manure or another organic fertilizer should also be added to maintain soil fertility.

In areas of low rainfall, such as the semiarid regions in the western and southwestern United States (Figure 11-10, p. 239), calcium and other alkaline compounds are not leached away. Soils in the

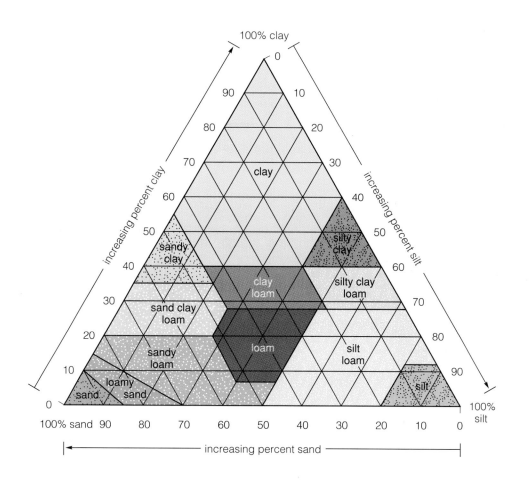

Figure 12-6 Water has eroded vital topsoil from the irrigated cropland in Arizona.

Figure 12-7 Wind eroding soil from farmland in Stevens County, Kansas.

region may be too alkaline (pH above 7.5) for some crops. If drainage is good, irrigation can reduce the alkalinity by leaching the alkaline compounds away. Adding sulfur, which is gradually converted to sulfuric acid by soil bacteria, is another way to reduce soil alkalinity. Soils in areas affected by acid deposition are becoming increasingly acidic (Figure 9-11, p. 197).

12-2 SOIL EROSION

Natural and Human-Accelerated Soil Erosion Soil does not stay in one place indefinitely. **Soil erosion** is the movement of soil components, especially topsoil, from one place to another. The two main forces causing soil erosion are flowing water (Figure 12-6 and Figure 1-1, p. 3) and wind (Figure 12-7).

Figure 12-8 Terraces on cropland in Bali, Indonesia, reduce soil erosion. Terraces also increase the amount of usable land in steep terrain.

Prato/Bruce Coleman Ltd.

Some soil erosion always takes place because of natural water flow and winds. But the roots of plants generally protect soil from excessive erosion. Agriculture, logging, construction, and other human activities that remove plant cover greatly accelerate the rate at which soil erodes.

Excessive erosion of topsoil reduces both the fertility and the water-holding capacity of a soil. The resulting sediment, the largest single source of water pollution, clogs irrigation ditches, navigable waterways, and reservoirs.

Soil, especially the topsoil, is classified as a slowly renewable resource because it is continually regenerated by natural processes. However, in tropical and temperate areas, the renewal of 2.54 centimeters (1 inch) of soil takes from 200 to 1,000 years, depending on climate and soil type. If the average rate of topsoil erosion exceeds the rate of topsoil formation on a piece of land, the topsoil on that land becomes a nonrenewable resource being depleted. Annual erosion rates for agricultural land throughout the world are 18 to 100 times the natural renewal rate (see Guest Essay on p. 289).

Soil erosion on forestland and rangeland is not as severe as erosion on cropland, but forest soil takes two to three times longer to restore itself than that of cropland. Construction sites usually have the highest erosion rates by far.

The World Situation Today topsoil is eroding faster than it forms on about one-third of the world's cropland. The amount of topsoil washing and blowing into the world's streams, lakes, and oceans each year would fill a train of freight cars long enough to encircle the planet 150 times. At this rate the world is losing about 7% of its topsoil from potential cropland each decade. The situation is worsening as farmers in MDCs and LDCs cultivate areas unsuited for agriculture to feed themselves and the world's growing population.

In mountainous areas, such as the Himalayas on the border between India and Tibet and the Andes near the west coast of South America, farmers have traditionally built elaborate systems of terraces (Figure 12-8). Terracing allowed them to cultivate steeply sloping land that would otherwise rapidly lose its topsoil.

Today farmers in some areas cultivate steep slopes without terraces, causing a total loss of topsoil in 10 to 40 years. Although most poor farmers know that cultivating a steep slope without terracing causes a rapid loss of topsoil, they often have too little time and too few workers to build terraces. This loss of protective vegetation and topsoil also greatly increases the intensity of flooding in the lowland areas of watersheds, as has happened in Bangladesh (see Case Study on p. 237).

Since the beginning of agriculture, people in tropical forests have successfully used slash-and-burn, shifting cultivation (Figure 2-4, p. 24) to provide food for relatively small populations. In recent decades, however, growing population and poverty have caused farmers in many tropical forest areas to reduce the fallow period of their fields to as little as 2 years, instead of the 10 to 30 years needed to allow the soil to regain its fertility. The result has been a sharp increase in the rate of topsoil erosion and nutrient depletion.

According to the UN Environment Program about 35% of the earth's land surface—on which about one-fifth of the world's people try to survive—is classified as arid or semiarid desert (Figure 5-2, p. 87). In drier parts of the world, desert areas are increasing at an alarming rate from a combination of natural processes and human activities.

The conversion of productive rangeland (uncultivated land used for animal grazing), rain-fed cropland, or irrigated cropland to desertlike land with a drop in agricultural productivity of 10% or more is called **desertification.**

Moderate desertification causes a 10% to 25% drop in productivity, and severe desertification causes a 25% to 50% drop. Very severe desertification causes a drop of 50% or more and usually results in formation of massive gullies and sand dunes.

Moderate desertification can go unrecognized. For example, overgrazing has reduced the productivity of much of the grasslands of the western United States. Yet, most citizens in these areas do not realize they live in an area moderately desertified by human activity.

Most desertification occurs naturally near the edges of existing deserts. It is caused by dehydration of the top layers of soil during prolonged drought and increased evaporation because of hot temperatures and high winds.

However, natural desertification is greatly accelerated by practices that leave topsoil vulnerable to erosion by water and wind:

- overgrazing of rangeland as a result of too much livestock on too little land area (the major cause of desertification)

- improper soil and water resource management that leads to increased erosion, salinization, and waterlogging of soil

- cultivation of land with unsuitable terrain or soils

- deforestation and surface mining without adequate replanting

- soil compaction by farm machinery, cattle hooves, and the impact of raindrops on denuded soil surfaces

These destructive practices are intensified by rapid population growth, high population densities, poverty, and poor land management. The consequences of desertification include intensified drought and famine, declining living standards, and swelling numbers of environmental refugees whose degraded land can no longer keep them alive.

The world's rural people affected by desertification rose from 57 million people in 1977 to 230 million in 1989. By the end of this century 350 million more people may be affected by desertification.

It is estimated that about 810 million hectares (2 billion acres)—an area the size of Brazil and 12 times the size of Texas—have become desertified during the past 50 years (Figure 12-9). Each year an estimated 6 million hectares (15 million acres)—an area the size of West Virginia—of new desert are formed. The areas most affected are in sub-Saharan Africa (between North Africa's barren Sahara desert and the plant-rich land to its south), eastern and southern Africa, Australia, the western United States, southern South America, and much of south-central Asia.

In 1977 government representatives from around the world gathered at Nairobi for the United Nations Conference on Desertification. Out of this conference came a plan of action to halt or sharply reduce the spread of desertification and to reclaim desertified land.

(continued)

Overgrazing and poor logging practices also cause heavy losses of topsoil. Intense grazing has turned many areas of North Africa from grassland to desert (see Spotlight above). In Africa, in the last three decades, soil erosion has increased twentyfold. Once-forested hills in many LDCs have been stripped bare of trees by poor people for firewood and by timber companies for use in MDCs. Because new trees are seldom planted in LDCs, the topsoil quickly erodes away.

In MDCs, where large-scale industrialized agriculture is practiced, many farmers have replaced traditional soil conservation practices with massive inputs of commercial inorganic fertilizers and irrigation water. But the tenfold increase in fertilizer use and the tripling of the world's irrigated cropland between 1950 and 1990 have only temporarily masked the effects of erosion and nutrient depletion.

Commercial inorganic fertilizer is not a complete substitute for naturally fertile topsoil; it merely hides for a time the gradual depletion of this vital resource. Nor is irrigation a long-term solution. Repeated irrigation of cropland without sufficient drainage eventually decreases or destroys its crop productivity as a result of waterlogging and salt buildup (Section 12-5).

Severe erosion accelerated by human activities is most widespread in India, China, the Soviet Union, and the United States, which together account for over half the world's food production and contain almost half the world's people.

biomass underground > above ground

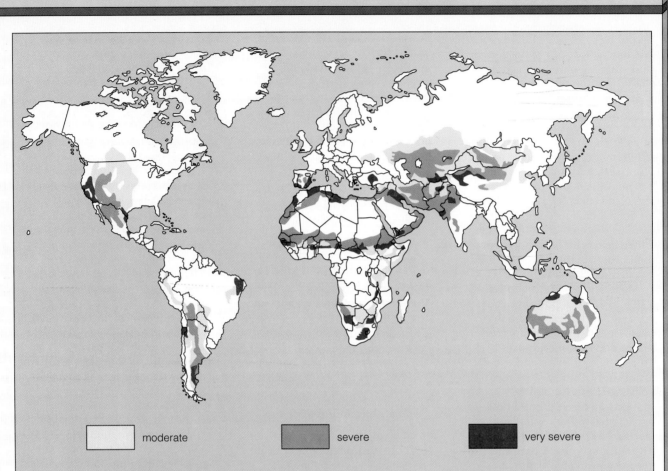

| | moderate | | severe | | very severe |

Figure 12-9 Desertification of arid and semiarid lands. (Data from UN Environmental Program and Harold E. Dregnue)

The total cost of such prevention and rehabilitation would be about $141 billion, only five and one-half times the estimated $26 billion annual loss in agricultural productivity from desertified land.

Thus, once this potential productivity is restored, the costs of the program could be recouped in five to ten years.

However, by 1990 little had been done because of inadequate

efforts and funding; only about one-tenth of the amount needed had been provided. What do you think should be done?

The U.S. Situation According to surveys by the Soil Conservation Service, about one-third of the original topsoil on U.S. croplands in use today has been washed or blown into rivers, lakes, and oceans. Surveys also show that the average rate of erosion on cultivated land in the United States is about seven times the rate of natural soil formation.

But an average national soil erosion rate masks much higher erosion in heavily farmed regions, especially the corn belt and Great Plains. Some of the country's most productive agricultural lands, such as those in Iowa, have lost about half their topsoil. In California the erosion rate is 80 times faster than the rate of natural soil formation.

Enough topsoil erodes away each day in the

United States to fill a line of dump trucks 5,600 kilometers (3,500 miles) long. Two-thirds of this soil comes from less than one-fourth of the country's cropland. The plant nutrient losses from this erosion of earth capital are worth at least $18 billion a year. Erosion also causes at least $4 billion a year in damages when silt, plant nutrients, and pesticides are carried into streams, lakes, and reservoirs.

At present, soil conservation is practiced on only about half of all U.S. farmland and on less than half of the country's most erodible cropland. Increased soil conservation is particularly important in the fertile midwestern plains, which are subject to high rates of erosion from high winds and prolonged drought (see Case Study on p. 275).

Figure 12-10 The Dust Bowl, the Great Plains area where a combination of prolonged severe drought and poor soil-conservation practices led to massive erosion of topsoil in the 1930s.

The Great Plains of the United States stretch through ten states, from Texas through Montana and the Dakotas. The region is normally dry and very windy and experiences long, severe droughts.

Before settlers began grazing livestock and planting crops in the 1870s, the extensive root systems of prairie grasses held the topsoil of these mollisol soils in place (Figure 12-4). When the land was planted in crops, the perennial grasses were replaced by annual crops with less extensive root systems.

In addition, the land was plowed up after each harvest and left bare part of the year. Overgrazing also destroyed large areas of grass, leaving the ground bare. The stage was set for crop failures during prolonged droughts, followed by severe wind erosion.

The droughts arrived in 1890 and 1910 and again with even greater severity between 1926 and 1934. In 1934, hot, dry windstorms created dust clouds thick enough to cause darkness at midday in some areas. The danger of breathing this dust-laden air was revealed by the dead rabbits and birds left in its wake.

During May 1934 the entire eastern half of the United States was blanketed with a massive dust cloud of topsoil blown off the Great Plains from as far as 2,400 kilometers (1,500 miles) away. Ships 322 kilometers (200 miles) out in the Atlantic Ocean received deposits of midwestern topsoil. These events gave a portion of the Great Plains a tragic new name: the Dust Bowl (Figure 12-10).

An area of cropland equal in size to the combined areas of Connecticut and Maryland was destroyed, and additional cropland equal in area to New Mexico was severely damaged. Thousands of displaced farm families from Oklahoma, Texas, Kansas, and other states migrated to California or to the industrial cities of the Midwest and East. Most found no jobs because the country was in the midst of the Great Depression.

In May 1934 Hugh Bennett of the U.S. Department of Agriculture (USDA) addressed a congressional hearing in Washington, pleading for new programs to protect the country's topsoil. Lawmakers took action when dust blown from the Great Plains began seeping into the hearing room.

In 1935 the United States established the Soil Conservation Service (SCS) as part of the Department of Agriculture. With Bennett as its first head, the SCS began promoting good conservation practices in the Great Plains and later in every state. Soil conservation districts were established throughout the country, and farmers and ranchers were given technical assistance in setting up soil conservation programs.

These efforts, however, did not completely stop human-accelerated erosion in the Great Plains. The basic problem is that the climate of much of the region makes it better suited for grazing than for farming.

In 1975 the Council of Agricultural Science and Technology warned that severe drought could again create a dust bowl in the Great Plains. So far, these warnings have been ignored.

Great Plains farmers, many of them debt-ridden, have continued to stave off bankruptcy by minimizing expenditures for soil conservation. Depletion of the Ogallala Aquifer is also threatening crop production and cattle raising in parts of the Dust Bowl area (see Case Study on p. 242). If projected global warming makes this region even drier (Figure 10-8, p. 218), farming will have to be abandoned. What do you think should be done about this situation?

12-3 SOIL CONSERVATION AND LAND-USE CONTROL

Conservation Tillage The practice of **soil conservation** involves using various methods to reduce soil erosion, to prevent depletion of soil nutrients, and to restore nutrients already lost by erosion, leaching, and excessive crop harvesting (Figure 12-3). Most methods used to control soil erosion involve keeping the soil covered with vegetation.

In **conventional-tillage farming,** the land is plowed, disked several times, and smoothed to make a planting surface. If plowed in the fall so that crops can be planted in the spring, the soil is left bare during the winter and early spring months—a practice that makes it vulnerable to erosion.

To lower labor costs, save energy, and reduce erosion, an increasing number of U.S. farmers are using **conservation-tillage farming,** also known as

Figure 12-11 No-till farming, Lancaster County, Pennsylvania. A specially designed machine plants seeds and adds fertilizers and weed killers at the same time with almost no disturbance of the soil.

Figure 12-12 On this gently sloping land in Illinois, contoured rows planted with alternating crops (strip cropping) reduce soil erosion.

minimum-tillage or no-till farming, depending on the degree to which the soil is disturbed. Farmers using this method disturb the soil as little as possible in planting crops.

For the minimum-tillage method, special tillers break up and loosen the subsurface soil without turning over the topsoil, previous crop residues, or any cover vegetation. For no-till farming, special planting machines inject seeds, fertilizers, and weed killers (herbicides) into slits made in the unplowed soil (Figure 12-11).

In addition to reducing soil erosion, conservation tillage reduces fuel and tillage costs, water loss from the soil, and soil compaction. It also increases the number of crops that can be grown during a season (multiple cropping). Yields are as high as or higher than yields from conventional tillage. Depending on the soil type, this approach can be used for three to seven years before more extensive soil cultivation is needed to prevent crop yields from declining. But conservation tillage is no cure-all. It requires increased use of herbicides to control weeds that compete with crops for soil nutrients (Chapter 14).

Conservation tillage is now used on about one-third of U.S. croplands and is projected to be used on over half by 2000. The USDA estimates that using conservation tillage on 80% of U.S. cropland would reduce soil erosion by at least half. So far the practice is not widely used in other parts of the world.

Contour Farming, Terracing, Strip Cropping, and Alley Cropping Soil erosion can be reduced 30% to 50% on gently sloping land by means of **contour farming**—plowing and planting across, rather than up and down, the sloped contour of the land (Figure 12-12). Each row planted at a right angle to the slope of the land acts as a small dam to help hold soil and slow the runoff of water.

Terracing can be used on steeper slopes. The slope is converted into a series of broad, nearly level terraces with short vertical drops from one to another (Figure 12-8). Some of the water running down the vegetated slope is retained by each terrace. Thus, terracing provides water for crops at all levels and decreases soil erosion by reducing the amount and speed of water runoff. In areas of high rainfall, diversion ditches must be built behind each terrace to permit adequate drainage.

In **strip cropping,** a series of rows of one crop, such as corn or soybeans, is planted in a wide strip; then the next strip is planted with a cover crop, such as alfalfa, which completely covers the soil and thus reduces erosion (Figure 12-12). The alternating strips of row crops and cover crops reduce water runoff and help prevent the spread of pests and plant diseases from one strip to another. They also help restore soil fertility if nitrogen-rich legumes such as soybeans or alfalfa are planted in some of the strips.

Erosion can also be reduced by **alley cropping** in which crops are planted in alleys between hedgerows of trees or shrubs that can be used as sources of fruits and fuelwood (Figure 12-13). The hedgerow trimmings can be used as mulch (green manure) for the crops and fodder for livestock.

Gully Reclamation and Windbreaks Water runoff quickly creates gullies in sloping land not covered by vegetation (Figure 1-1, p. 3). Such land can be restored by **gully reclamation**. Small gullies can be seeded with quick-growing plants such as oats, barley, and wheat to reduce erosion. In deeper gullies, small dams can be built to collect silt and gradually fill in the channels. Rapidly growing shrubs, vines, and trees can also be planted to stabilize the soil. Channels can be built to divert water away from the gully and prevent further erosion.

Figure 12-13 Alley cropping in Peru. Several crops are planted together in strips or alleys between trees and shrubs. The trees provide shade that reduces water loss by evaporation, helping retain soil moisture by releasing it slowly.

Figure 12-14 Windbreaks, or shelterbelts, reduce erosion on this farm in South Dakota. They also reduce wind damage, help hold soil moisture in place, supply some wood for fuel, and provide a habitat for wildlife.

Erosion caused by exposure of cultivated lands to high winds can be reduced by **windbreaks,** or **shelterbelts,** long rows of trees planted to partially block wind (Figure 12-14). They are especially effective if land not under cultivation is kept covered with vegetation. Windbreaks also provide habitats for birds, pest-eating and pollinating insects, and other animals. Unfortunately, many of the windbreaks planted in the upper Great Plains following the Dust Bowl disaster of the 1930s have been destroyed to make way for large irrigation systems (Figure 11-17, p. 246) and farm machinery.

Land-Use Classification and Control An obvious land-use approach to reducing erosion is to prohibit the planting of crops or the clearing of vegetation on marginal land. Such land is highly erodible because of a steep slope, shallow soil structure, high winds, drought, or other factors.

The most widely used approach for controlling land use is **zoning,** in which various parcels of land are designated for certain uses. Zoning can be used to protect areas from certain types of development and to control growth. Zoning is useful, but it can be influenced or modified by developers because local governments depend on property taxes for revenue. Thus, zoning often favors high-priced housing and factories, hotels, and other businesses rather than the protection of farmland and natural areas such as forests, grasslands, and wetlands.

Since World War II, the typical pattern of suburban housing development in the United States has been to bulldoze a tract of woods or farmland and build rows of houses, each standard house on a standard lot (Figure 12-15). By removing most vegetation, this approach increases soil erosion during and after construction. Someone noted that the United States is where they cut down the trees and eliminate

Figure 12-15 A typical suburban housing tract in McHenry, Illinois. Note that all trees have been removed. Such areas have very high rates of soil erosion during and after construction. Many suburban housing and other developments are built on land that was once used to grow crops.

most of the wildlife in an area and then name the streets and developments after them—Oak Lane, Cedar Drive, Pheasant Run, Fox Fields.

In hot climates these houses don't have natural cooling from trees. This greatly increases the use of air conditioning, raises electricity bills, and contributes to global warming and urban heat islands (Figure 6-20, p. 132).

In recent years, builders have made increased use of a new pattern, known as cluster development. Houses, town houses, condominiums, and garden apartments are built on only part of the tract. The rest of the area is left as open space, either in its natural state or modified for recreation (Figure 12-16). This approach helps reduce soil erosion by preserving medium-size blocks of open space and natural vegetation.

Every day an average of 6,900 hectares (17,000 acres) of farmland in the United States are destroyed

creek

marsh

cluster

creek

cluster

pond

Undeveloped Land   Typical Housing Development   Cluster Housing Development

Figure 12-16 Conventional and planned unit developments as they would appear if constructed on the same land area. The plan on the right sharply reduces soil erosion, provides wildlife habitats and outdoor recreational opportunities, and gives residents a more beautiful place to live.

or degraded, primarily due to erosion, but also from urban sprawl (Figure 12-15). To encourage wise land use and reduce erosion the Soil Conservation Service has set up a classification system for land and established almost 3,000 soil and water conservation districts. It also surveys soil in these districts and classifies soils according to type and quality.

The SCS relies on voluntary compliance with its guidelines through the local and state soil and water conservation districts and provides technical and economic assistance through the local district offices. This means that American soil conservation policy lacks teeth. The soil conservation associations that regulate the districts are made up of officials who generally represent the private interests of local farmers and ranchers. They are under intense peer pressure to make land use decisions based on short-term economic gains that can have harmful long-term environmental and economic impacts.

According to the General Accounting Office, this voluntary compliance and lack of an integrated regional and national program help to explain the fact that although the federal government has spent about $18 billion on soil erosion control efforts, these programs have had only minimal success in reducing soil erosion. The National Wildlife Federation observes that despite four decades of soil erosion control, the United States experienced 35% more soil erosion in 1981 than in the dust bowl days of the 1930s.

State and federal governments can take any of the following measures to protect cropland, forestland, wetlands, and other nonurban lands near expanding urban areas from degradation and ecologically unsound development:

- Give tax breaks to landowners who agree to use land only for specified purposes such as agriculture (with emphasis on soil conservation), wilderness, wildlife habitat, and nondestructive forms of recreation. Such agreements are called conservation easements.

- Tax land on the basis of its use as agricultural land or forestland rather than its fair market value based on its economically highest potential use. This prevents farmers and other landowners from being forced to sell land to developers to pay their tax bills.

- Purchase and protect ecologically valuable land. Such purchases (land trusts) can be made by private groups such as the Nature Conservancy, the Audubon Society, and local and regional nonprofit, tax-exempt, charitable organizations, as well as by public agencies.

- Purchase land development rights that restrict the way land can be used (for example, to preserve prime farmland near cities from development).

- Assign a limited number of transferable development rights to a given area of land.

- Require environmental impact analysis for proposed private and public projects such as roads, industrial parks, shopping centers, and suburban developments; cancel harmful projects unless they are revised to minimize harmful environmental impacts, including excessive soil erosion.

- Give subsidies to farmers for taking highly erodible cropland out of production or eliminate subsidies for farmers who farm such land or who convert wetlands to cropland, as is being done in the 1985 Farm Act.

Maintaining and Restoring Soil Fertility Organic fertilizers and commercial inorganic fertilizers can be applied to soil to partially restore and maintain plant nutrients lost by erosion, leaching, and crop harvesting and to increase crop yields (Figure 12-3). Three major types of **organic fertilizer** are animal manure, green manure, and compost. **Animal manure** includes the dung and urine of cattle, horses, poultry, and other farm animals. In some LDCs such as China and

South Korea human manure, sometimes called night soil, is used to fertilize crops.

Application of animal manure improves soil structure, increases organic nitrogen content, and stimulates the growth and reproduction of soil bacteria and fungi. It is particularly useful on crops of corn, cotton, potatoes, and cabbage.

Despite its effectiveness, the use of animal manure in the United States has decreased. One reason is that separate farms for growing crops and animals have replaced most mixed animal- and crop-farming operations. Animal manure is available at feedlots near urban areas, but transporting it to distant rural crop-growing areas usually costs too much. In addition, tractors and other motorized farm machinery have replaced horses and other draft animals that naturally added manure to the soil.

Green manure is fresh or growing green vegetation plowed into the soil to increase the organic matter and humus available to the next crop. It may consist of weeds in an uncultivated field, grasses and clover in a field previously used for pasture, or legumes such as alfalfa or soybeans grown for use as fertilizer to build up soil nitrogen.

Compost is a rich natural fertilizer and soil conditioner; farmers and home owners produce it by piling up alternating layers of carbohydrate-rich plant wastes (such as cuttings and leaves), animal manure, and topsoil (Figure 12-17). This mixture provides a home for microorganisms that aid the decomposition of the plant and manure layers.

Today, especially in the United States and other industrialized countries, farmers partially restore and maintain soil fertility by applying **commercial inorganic fertilizers.** The most common plant nutrients in these products are nitrogen (as ammonium ions, nitrate ions, or urea), phosphorus (as phosphate ions), and potassium (as potassium ions). Other plant nutrients may also be present in low or trace amounts. Farmers can have their soil and harvested crops chemically analyzed to determine the mix of nutrients that should be added.

Inorganic commercial fertilizers are easily transported, stored, and applied. Throughout the world their use increased about tenfold between 1950 and 1990. By 1990, the additional food they helped produce fed one of every three persons in the world. Without this input world food output would plummet by an estimated 40%.

Commercial inorganic fertilizers, however, have some disadvantages. They do not add humus to the soil. Unless animal manure and green manure are added to the soil along with commercial inorganic fertilizers, the soil's content of organic matter and thus its ability to hold water will decrease. If not supplemented by organic fertilizers, inorganic fertilizers cause the soil to become compacted and less

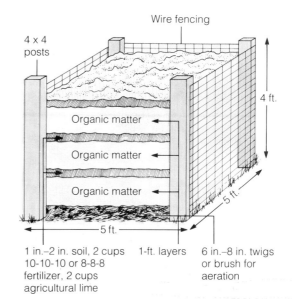

Figure 12-17 A simple home compost bin can be used to produce a mulch for garden and yard plants. A layer or two of cat litter or alfalfa meal can be used to cut down on odors. Leave a depression at the top center of the pile to collect rainwater. Turn the pile over every month or so and cover it with a tarp during winter months.

suitable for crop growth. By decreasing its porosity, inorganic fertilizers also lower the oxygen content of soil and prevent added fertilizer from being taken up as efficiently. In addition, most commercial fertilizers do not contain many of the nutrients needed in trace amounts by plants.

Water pollution is another problem caused by the widespread use of commercial inorganic fertilizers, especially on sloped land near streams and lakes. Some of the plant nutrients in the fertilizers are washed into nearby bodies of surface water, where the resulting cultural eutrophication causes excessive growth of algae, oxygen depletion, and fish kills. Rainwater seeping through soil can leach nitrates in commercial fertilizers into groundwater. High levels of nitrate ions make drinking water drawn from wells toxic, especially for infants.

A third method for preventing depletion of soil nutrients is **crop rotation**. Crops such as corn, tobacco, and cotton remove large amounts of nutrients (especially nitrogen) from the soil and can deplete the topsoil of nutrients if planted on the same land several years in a row. Farmers using crop rotation plant areas or strips with corn, tobacco, and cotton one year. The next year they plant the same areas with legumes, whose root nodules (Figure 4-26, p. 78) add nitrogen to the soil, or other crops such as oats, barley, rye, or sorghum. This method helps restore soil nutrients and reduces erosion by keeping the soil covered with vegetation. Varying the types of crops planted from year to year also reduces infestation by insects, weeds (especially if the land is planted in

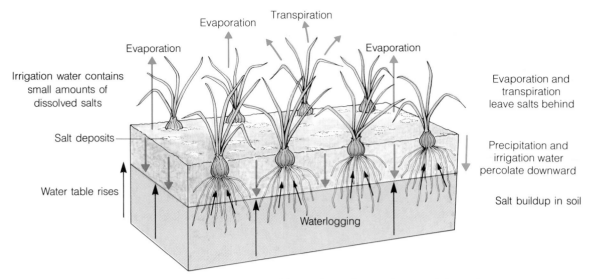

Figure 12-18 Salinization and waterlogging of soil on irrigated land without adequate drainage lead to decreased crop yields.

sorghum, which releases natural herbicides), and plant diseases.

Concern about soil erosion should not be limited to farmers. At least 40% of soil erosion in the United States is caused by timber cutting, mining, and urban development carried out without proper regard for soil conservation.

12-4 SOIL CONTAMINATION BY EXCESS SALTS AND WATER

Salinization About 18% of the world's cropland is now irrigated, producing about one-third of the world's food. Irrigated cropland is projected to at least double by 2020.

Irrigation can increase crop yields to two to three times the yields on the same area of land watered only by rain, but irrigation also has some harmful side effects. As irrigation water flows over and through the ground, it dissolves salts, increasing the salinity of the water. In dry climates much of the water in this saline solution is lost to the atmosphere by evaporation, leaving behind high concentrations of salts such as sodium chloride in the topsoil. The accumulation of salts in soils is called **salinization** (Figure 12-18). Salt buildup stunts crop growth, decreases yields, and eventually kills crop plants, making the land unproductive (Figure 12-19).

It is estimated that salinization is reducing yields on about one-fourth of the world's irrigated cropland. In Egypt, where virtually all cropland is irrigated, half is salinized enough to reduce yields. Worldwide, it is projected that 50% to 65% of all currently irrigated cropland will suffer reduced productivity from excess soil salinity by the year 2000.

One way to reduce salinization is to flush salts out of the soil by applying much more irrigation water

than is needed for crop growth. But this increases pumping and crop production costs and wastes enormous amounts of water. Another method is to pump groundwater from a central well and apply it with a sprinkler system that pivots around the well (Figure 11-17, p. 246). This method maintains downward drainage and is especially effective at preventing salinization, but at least 30% of the water is consumed by evaporation. Also, groundwater in unconfined aquifers eventually becomes too saline for irrigation and other human uses unless expensive drainage systems are installed.

Once topsoil has become heavily salinized, the farmer can renew it by taking the land out of production for two to five years, installing an underground network of perforated drainage pipe, and flushing the soil with large quantities of low-salt water. But this scheme is very expensive and only slows down the buildup of soil salinity; it does not stop the process. Flushing salts from the soil also increases the salinity of irrigation water delivered to farmers further downstream, unless the saline water can be drained into evaporation ponds rather than returned to the stream or canal.

In the Indian state of Uttar Pradesh, farmers are rehabilitating tracts of salinized land by planting a saline-tolerant tree that lowers the water table by taking up water through its roots. Tube-wells can also be used to lower water tables. But they are useful only in areas with an adequate groundwater supply.

Waterlogging A problem often accompanying soil salinity in dry regions is **waterlogging** (Figure 12-18). To keep salts from accumulating and destroying fragile root systems, farmers often apply heavy amounts of irrigation water to leach salts deeper into the soil. If drainage isn't provided, water accumulating un-

Senior project
Schoolwork
Vacation
Interview

Soil Conservation Service

derground gradually raises the water table. Saline water envelops the roots of plants and kills them.

Waterlogging is a particularly serious problem in the heavily irrigated San Joaquin Valley in California, where soils contain a clay layer with a low permeability to water. Worldwide, at least one-tenth of all irrigated land suffers from waterlogging, and the problem is getting worse. *1/10 waterlogging*

12-5 SOIL AND WATER CONTAMINATION BY HAZARDOUS WASTES

Soil and Groundwater Pollution: A Growing Threat
Soil is vulnerable to pollution from several sources. Because soil and groundwater are connected (Figure 11-3, p. 234), hazardous chemicals reaching the soil have the potential to contaminate groundwater (Figure 11-26, p. 257) and food. Harmful air pollutants settle out on soils (Figure 9-8, p. 194). Some of these pollutants are leached into groundwater supplies, and some are blown back into the air as suspended particulate matter when soil erodes (Figure 12-7). Another source of soil and groundwater contamination is hazardous wastes that are buried in landfills, injected into deep wells, or placed in surface impoundments (Figure 11-26, p. 257).

Traces of hazardous chemicals are now in the groundwater supplies, rain, and food, and in the blood and fat tissue of nearly every person in the United States, most MDCs, and many LDCs. The long-term results of this global chemical experiment on the human race are unknown. According to the EPA, 19,762 U.S. industrial plants alone released 2.1 million metric tons (2.3 million tons) of toxic chemicals into the air, surface waters, landfills, and deep underground wells in 1988. Virtually all of these releases were legal because they involved toxic chemicals not covered under present U.S. pollution control laws. To this we must add toxic chemicals discarded by households and small businesses, and vast amounts of pesticides applied to croplands, lawns, golf courses, and gardens, which end up in the surface waters and groundwater.

What Is Hazardous Waste? According to the Environmental Protection Agency, **hazardous waste** is any solid, liquid, or containerized gas that has one or more of the following properties:

- *Ignitability*—a waste that catches fire easily. Examples are waste oils, most used organic solvents, and PCBs from leaking or burning electrical transformers.

- *Corrosivity*—a highly acidic (low pH) or highly alkaline (high pH) waste, or one that corrodes steel easily.

- *Reactivity*—a highly unstable waste that can cause explosions or toxic fumes or vapors. Examples are liquid wastes from TNT operations and used cyanide solvents.

- *Toxicity*—a waste in which hazardous concentrations of toxic materials can leach out. Examples are toxic cadmium, lead, mercury compounds, and highly toxic dioxins and furans. In 1984 Congress ordered the EPA to also include chemicals that can cause cancer, genetic mutations, and birth defects in humans in this category. By 1990 this had not been done.

The EPA definition of hazardous wastes does not include radioactive wastes, hazardous materials discarded by households, mining wastes, oil and gas drilling wastes, cement kiln dust, municipal incinerator ash, and wastes from thousands of small businesses and factories that generate less than 100 kilograms (220 pounds) of hazardous waste per month.

Environmentalists contend that these hazardous wastes have not been included mostly because of lobbying by the industries involved. They urge that these categories be designated as forms of hazardous waste so they can be controlled under existing hazardous-waste laws.

In 1977, residents of a suburb of Niagara Falls, New York, discovered that "out of sight, out of mind" did not apply to them. Hazardous industrial waste buried decades earlier bubbled to the surface, found its way into groundwater, and ended up in backyards and basements.

Between 1942 and 1953, Hooker Chemicals and Plastics Corporation dumped almost 20,000 metric tons (22,000 tons) of highly toxic and cancer-causing chemical wastes (mostly in steel drums) into an old canal excavation, known as the Love Canal, named for its builder William Love. In 1953 Hooker Chemicals covered the dump site with clay and topsoil and sold the site to the Niagara Falls school board for one dollar. The deed specified that the company would have no future liability for any injury or property damage caused by the dump's contents.

An elementary school, playing fields, and a housing project, eventually containing 949 homes,

were built in the ten-square block Love Canal area. Residents began complaining to city officials in 1976 about chemical smells and chemical burns received by children playing in the canal, but these complaints were ignored. In 1977 chemicals began leaking from the badly corroded steel drums into storm sewers, gardens, and basements of homes next to the canal.

Informal health surveys conducted by alarmed residents, led by Lois Gibbs (see Individuals Matter on p. 286), revealed an unusually high incidence of birth defects, miscarriages, assorted cancers, and nerve, respiratory, and kidney disorders among people who lived near the canal. Complaints to local officials had little effect.

Continued pressure from residents and unfavorable publicity eventually led state officials to conduct a preliminary health survey and tests. They found that pregnant women in one area near the canal had a miscarriage rate

four times higher than normal. They also found that the air, water, and soil of the canal area and the basements of nearby houses were contaminated with a number of toxic and carcinogenic chemicals.

In 1978 the state closed the school, permanently relocated the 238 families whose homes were closest to the dump, and fenced off the area around the canal. On May 21, 1980, after protests from the outraged 711 families still living fairly close to the landfill, President Jimmy Carter declared Love Canal a federal disaster area and had these families relocated. Federal and New York state funds were then used to buy the homes of those who wanted to move permanently.

Since that time, the school and the 238 homes within a block and a half of the canal have been torn down and the state has purchased 570 of the remaining homes. About 45 families have remained in the desolate neighborhood, unwilling or unable to sell their

Growing Concern   There was little concern over hazardous waste in the United States and most parts of the world until 1977. Then it was discovered that hazardous chemicals leaking from an abandoned waste dump had contaminated a suburban development known as Love Canal, located in Niagara Falls, New York (see Case Study above).

Hazardous-Waste Production: Present and Past
The total quantity of hazardous wastes produced throughout the world or even in one country is impossible to estimate accurately. For the United States, estimates range from 240 million metric tons (264 million tons) by the EPA to 509 million metric tons (400 million tons) by the Office of Technology Assessment. Using the lower EPA estimate, there is about 1 metric ton (1 ton) of hazardous wastes generated per American each year.

With either of the estimates, it is clear that the United States leads the world in total and per capita hazardous-waste production. If the hazardous wastes produced each year in the United States were stacked end-to-end in 55-gallon drums, they would stretch to the moon and back.

About 95% of this waste is generated and either stored or treated on-site by large companies—chemical producers, petroleum refineries, and manufacturers. Most of these wastes are injected into deep wells or stored in surface impoundments (Figure 11-26, p. 257). The remaining 5% is handled by commercial facilities that take care of hazardous waste generated by others. Most of these wastes end up in landfills.

A serious problem facing the United States and most industrialized countries is what to do with thousands of dumps like the one at Love Canal where in the past, large quantities of hazardous wastes were disposed of in an unregulated manner (Figure 12-20). Even with adequate funding, effective cleanup is difficult because officials don't know what chemicals have been dumped and where all the sites are located.

What Happens to Hazardous Waste?   Most hazardous waste produced in the United States is disposed of in the land by deep-well injection, surface impoundments, and landfills. An estimated 56% of the hazardous wastes produced in the United States is

houses and move out of the area.

The dump site has been covered with a clay cap and surrounded by a drain system that pumps leaking wastes to a new treatment plant. After taking ten years and spending $275 million for relocation and cleanup, the EPA declared two-thirds of the area safe and renamed it Sunrise City. In June 1990, the EPA proposed a sale of the 236 remaining dilapidated and boarded-up houses at bargain prices.

Lois Gibbs, a former Love Canal resident turned environmental activist (see Individuals Matter on p. 286), says it would be "criminal to send people back in there" and filed a lawsuit to prevent the sale. She and other environmentalists point out that the dump has not been cleaned up but only capped and fitted with a drainage system. To them, selling these houses sends a message from the government that "Chemical Dumps Make Good Homes for Poor and Middle-Class Americans Who Can't Afford to Buy Homes in Safer Neighborhoods."

According to a 1989 study by the Office of Technology Assessment, about three-fourths of hazardous waste cleanups are likely to fail over the long term. Thus, sooner or later, hazardous chemicals are likely to flow from the dump as they did in the past, making Sunrise City a rerun of the earlier Love Canal chemical time bomb.

No conclusive study has been made to determine the long-term effects of exposure to hazardous chemicals on former Love Canal residents. All studies made so far have been criticized on scientific grounds. In 1988 an informal survey was made of families that once lived in a group of ten houses next to the canal. All but one had some cancer cases; there were also two suicides and three cases of birth defects among grandchildren.

The psychological damage to evacuated families is enormous. For the rest of their lives they will wonder whether a disorder will strike and will worry about the possible effects of the chemicals on their children and grandchildren.

In 1985 former Love Canal residents received payments from a 1983 out-of-court settlement from Occidental Chemical Corporation (which bought Hooker Chemicals in 1968), the city of Niagara Falls, and the Niagara Falls school board. The payments ranged from $2,000 to $400,000 for claims of injuries ranging from persistent rashes and migraine headaches to cancers and severe mental retardation.

In 1979 the EPA filed a suit against Occidental Chemical to recover cleanup costs. In 1988 a U.S. district court ruled that Occidental must pay the cleanup costs, but the company is appealing this ruling.

The Love Canal incident is a vivid reminder that we can never really throw anything away, that wastes don't stay put, and that preventing pollution is much safer and cheaper than cleaning it up.

---

injected into wells theoretically drilled beneath the aquifers tapped for drinking water and irrigation. Ideally, a well is drilled through an impervious layer of rock that separates an aquifer from a dry porous layer of rock below (Figure 11-3, p. 234).

Many environmentalists consider this poorly monitored and regulated practice a major threat to vital groundwater supplies. Wastes can spill or leak at the surface and leach into groundwater. Well pipe casings can corrode and allow wastes to escape into groundwater. Wastes can reach aquifers from inadequate or leaking seals when the well casing passes through the impervious layer of rock. Wastes can also migrate from the porous layer of rock where they are deposited to aquifers through existing fractures or new ones caused by earthquakes.

An estimated one-third of the country's hazardous wastes are deposited into ponds or pits whose bottoms are sealed with a plastic liner. The solids in these wastes settle to the bottom and accumulate, while water and other volatile compounds evaporate into the atmosphere. If the plastic bottom is well sealed and doesn't leak and evaporation exceeds the input, surface impoundments can receive wastes until

$\frac{1}{3}$ ponds

Figure 12-20 Leaking barrels in a toxic-waste dump. Most of the barrels are unlabeled, so we have little knowledge of what chemicals are being released into the environment. Such dumps are now illegal in the United States and many MDCs, but there are tens of thousands of older dumps and some illegal dumping still occurs.

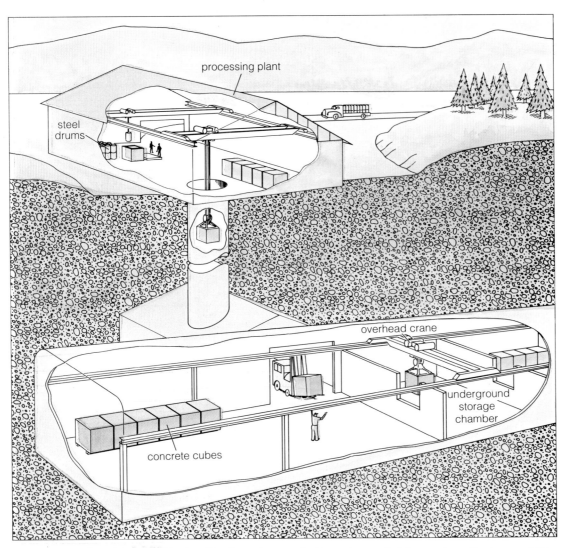

Figure 12-21 Swedish method for handling hazardous waste. Hazardous materials are placed in drums, which are stored in concrete cubes. The cubes are then placed in an underground vault.

the depressions are filled with toxic sludge. Then a new pit or pond can be dug.

But there are problems. Pond liners can develop holes and inadequate seals can allow wastes to percolate into groundwater. Most experts consider it only a matter of time before this happens. Major storms or hurricanes can cause overflows. Volatile compounds, such as hazardous organic solvents, can evaporate into the atmosphere and eventually return to the earth to contaminate surface and groundwater in other locations.

About 5% of hazardous wastes produced in the United States are concentrated, put into drums, and buried in specially designed *secured landfills*. The inside of the pit in a secure landfill is covered with two plastic liners and its bottom is covered with a layer of impervious clay. Pipes running through the bottom of the pit collect the leachate, material that leaks out of the barrels, and pump it to the surface. Wells are drilled to monitor the groundwater aquifer below the landfill. Sweden goes further and buries

its concentrated hazardous wastes in underground vaults (Figure 12-21).

Ideally, such landfills should be located in a geologically and environmentally secure place that is carefully monitored for leaks. In 1983, however, the Office of Technology Assessment concluded that sooner or later even the best-designed, secured landfill will leak hazardous chemicals into nearby surface water and groundwater.

How Should Hazardous Waste Be Controlled and Managed? There are three basic ways of dealing with hazardous waste, as outlined by the National Academy of Sciences: **(1)** waste prevention by waste reduction, recycling, and reuse, **(2)** conversion to less hazardous or nonhazardous material, and **(3)** perpetual storage (Figure 12-22). The first and most desirable method is an input, or waste prevention, approach. Its goal is to reduce the amount of waste produced by modifying industrial or other processes and by

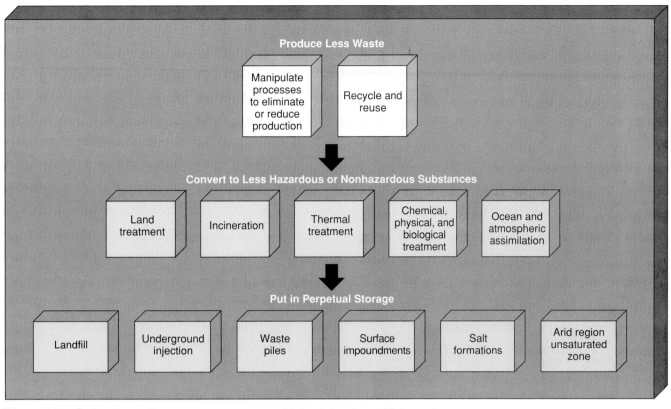

Produce Less Waste

Manipulate processes to eliminate or reduce production

Recycle and reuse

Convert to Less Hazardous or Nonhazardous Substances

Land treatment

Incineration

Thermal treatment

Chemical, physical, and biological treatment

Ocean and atmospheric assimilation

Put in Perpetual Storage

Landfill

Underground injection

Waste piles

Surface impoundments

Salt formations

Arid region unsaturated zone

Figure 12-22 Options for dealing with hazardous waste. (National Academy of Sciences)

reusing or recycling the hazardous wastes that are produced. So far no country has implemented an effective program for achieving this goal, but countries such as Denmark, the Netherlands, West Germany, and Sweden are far ahead of the United States in controlling and managing hazardous waste.

The EPA estimates that at least 20% of these materials could be recycled, reused, or exchanged so that one industry's waste becomes another's raw material. Presently, however, only about 5% of such materials are managed in this manner.

Some firms have found that waste reduction and pollution prevention save them money. Since 1975 the Minnesota Mining and Manufacturing Company (3M), which makes 60,000 different products in 100 manufacturing plants, has had a program that by 1990 had cut its waste production by two-thirds and saved over $900 million. Less hazardous raw materials can be used. Manufacturing processes can be redesigned so that they produce little, if any, hazardous wastes and those that are produced are recycled or converted into harmless chemicals on-site. Also the tracking and control of hazardous materials can be improved.

However, most firms have little incentive to reduce their output of waste because waste management makes up only about 0.1% of the total value of the products they ship. Placing a tax on each unit of hazardous waste generated would provide enough money to support a strong program for reducing, recycling, and reusing hazardous waste.

The second phase of a hazardous-waste management program is to convert any waste remaining after waste reduction, recycling, and reuse to less hazardous or nonhazardous materials (Figure 12-22). Conversion methods include spreading degradable wastes on the land, burning them on land or at sea in specially designed incinerators, thermally decomposing them, trying to cultivate microorganisms that will degrade specific chemicals, or treating them chemically or physically.

The Netherlands incinerates about half its hazardous waste. The EPA estimates that about 60% of all U.S. hazardous waste could be incinerated. With proper air pollution controls and highly trained personnel, incineration is potentially the safest method of disposal for most types of hazardous waste. But it is also the most expensive method. The ash that is left and which must be disposed of often contains toxic metals, and the gaseous and particulate combustion products emitted can be health hazards if not controlled. Another problem is that not all hazardous wastes are combustible.

Most citizens vigorously oppose locating a hazardous waste incinerator anywhere near their community, summarized by the "not-in-my-backyard" (NIMBY) slogan (see Individuals Matter on p. 286). Often elected officials and companies locate these

One highly effective frontline activist is Lois Gibbs, a housewife who organized residents of the Love Canal development near Niagara Falls, New York, when they discovered they were living near a leaking toxic-waste dump (see Case Study on p. 282). She and her neighbors eventually created enough media attention to have the state and federal governments move the families out and buy their houses.

She then went on to form the Citizens' Clearing House for Hazardous Waste, an organization that provides information and help for citizens' groups organizing to prevent hazardous-waste dumps, landfills, and incinerators from being located in their local areas. By late 1989 it had 9,000 member groups and was growing at a rate of nearly 100% a year.

She and other grass-roots activists are not swayed by the waste management industry's highly technical risk-assessment and cost-benefit studies, slick glossy publications, promises to use state of the art technology, expressions of their concern for the environment, and arguments that protesting citizens are holding up progress. Based on her experience Lois Gibbs says, "Don't listen to them. It's all BS. The simple truth is, they're trying to kill you, to sacrifice you and your family just so they can make money."

Her guidelines for earth citizens are:

- Don't compromise our children's futures by cutting deals with polluters and regulators. Environmental justice cannot be bought or sold.

- Hold polluters directly and personally accountable. There is no escaping the fact that what they are doing is wrong and they must be called to account for their actions.

- Don't fall for the argument voiced by industry that protesters against hazardous-waste dumps and incinerators are holding up progress because we have to deal with the hazardous wastes we produce. Instead, recognize that the best way to deal with pollution is not to let it happen in the first place.

- Oppose all hazardous-waste landfills and incinerators to sharply raise the cost of dealing with these materials. Only in this way will waste producers and elected officials get serious about waste reduction and pollution prevention instead of talking about it while spending little money or effort to do it. Our real goal should be "Not in Anyone's Backyard (NIABY), or "Not on Planet Earth" (NOPE).

facilities in areas where the poor and minority groups live to reduce the effectiveness of citizen opposition. Three-fourths of the hazardous-waste landfills in the southeastern United States are in low-income, black neighborhoods, and more than half of all black and Hispanic Americans live in communities with at least one toxic-waste dump.

Some countries have been incinerating hazardous waste at sea in incinerator ships. But because of lack of controls and fear of contaminating ocean ecosystems, most MDCs (including the United States) have agreed to end at-sea incineration of hazardous wastes by 1994.

Denmark, which relies almost exclusively on groundwater for drinking water, has the most comprehensive and effective program for detoxifying most of its hazardous waste. Each municipality has at least one facility that accepts paints, solvents and other hazardous wastes from households. Toxic waste from industries is delivered to 21 transfer stations scattered throughout the country. All waste is then transferred to a large treatment facility in the town of Nyborg on the island of Fyn near the country's geographic center. There about 75% of the waste is detoxified and the rest is buried in a carefully designed and monitored landfill.

The third phase of waste management involves concentrating and placing any waste left after detoxification in containers and storing them in specially designed *secured landfills* or *underground vaults* (Figure 12-21). However, some experts argue that eventually even the best-designed landfill or vault will leak and threaten groundwater supplies.

There is also growing concern about accidents during some of the 500,000 shipments of hazardous wastes in the United States each year. Between 1980 and 1988 there were 11,048 toxic-chemical accidents, causing 309 deaths, 11,000 injuries, and evacuation of 500,000 people. Most communities do not have the equipment and trained personnel to deal adequately with most types of hazardous-waste spills.

As the costs of hazardous waste disposal have risen, waste disposal firms in the United States and several other industrialized nations have shipped hazardous wastes to other countries, especially LDCs in Asia, Africa, and Latin America (see Spotlight on p. 287).

U.S. Hazardous-Waste Legislation In 1976 Congress passed the Resource Conservation and Recovery Act (RCRA, pronounced "rick-ra"), amending it in 1984. This law requires the EPA to identify hazardous wastes and set standards for their management, and provides guidelines and financial aid to establish state waste management programs. The law also requires all firms that store, treat, or dispose of more than 100 kilograms (220 pounds) of hazardous wastes per month to have a permit stating how such wastes are to be managed.

To reduce illegal dumping, hazardous-waste producers granted disposal permits by the EPA must use a "cradle-to-grave" manifest system to keep track of waste transferred from point of origin to approved off-site disposal facilities. EPA administrators, however, point out that this requirement is impossible to enforce effectively. The EPA and state regulatory agencies do not have enough personnel to review the documentation of more than 750,000 hazardous-waste generators and 15,000 haulers each year, let alone to verify them and prosecute offenders. If caught, however, violators are subject to large fines.

Some environmentalists argue that fines are not enough. They believe that people who deliberately or through negligence illegally release harmful chemicals into the environment should be subject to jail terms because such environmental crimes kill or damage the health of many people for decades.

Facilities that treat, store, or dispose of hazardous wastes must have EPA permits and follow EPA procedures for handling such waste. By 1990 there were 21 commercial hazardous waste landfills, which take wastes from anyone for a fee. There were also 35 non-commercial landfills runs by individual companies to handle their own wastes.

Operators of EPA-licensed hazardous waste landfills must prevent leakage, continually monitor the quality of groundwater around the sites, and report any contamination to the EPA. When one of these landfills reaches its capacity and is closed, the operators must cover it with a leakproof cap, monitor the nearby groundwater for 30 years, and be financially responsible for cleanup and damages from leaks for 30 years. Environmentalists consider this a serious weakness in the law because most landfills will probably begin leaking after this period.

Another loophole exempts "recycled" chemical wastes from control. This includes hazardous chemical wastes burned in industrial boilers, industrial furnaces, aggregate kilns, and cement kilns, which are not required to meet the stringent permit requirements and emission standards required for EPA-licensed hazardous waste incinerators. This practice pollutes the air with toxic metals, dioxins, furans, and other hazardous chemicals. It also produces fly ash laced with hazardous chemicals, which does not have to be disposed of in EPA-licensed hazardous waste landfills.

The 1980 Comprehensive Environmental Response, Compensation and Liability Act, known as the Superfund program, and amendments in 1986 established a $10.1 billion fund, financed jointly by federal and state governments and taxes on chemical and petrochemical industries. The money is to be used for the cleanup of abandoned or inactive haz-

Superfund — 1986 - $10 billion

---

ardous-waste dump sites and leaking underground tanks that are threats to human health and the environment. The EPA is authorized to collect fines and sue the owners of abandoned sites and tanks (if they can be found and held responsible) to recover up to three times the cleanup costs.

Here are some ways you can help protect the soil.

- Establish a soil conservation program for the land around your home.

- If you are building a home, save all the trees possible and have the contractor set up barriers to catch any soil eroded during construction. Also require the contractor to disturb as little soil as possible and to save and replace any topsoil removed instead of hauling it off and selling it. Plant any disturbed area with fast-growing native groundcover (preferably not grass) immediately after construction is completed.

- Landscape the area not used for gardening with a mix of wildflowers, herbs (for cooking and to repel insects), low-growing groundcover, small bushes, and other forms of vegetation natural to the area. This biologically diverse type of yard saves water, energy, and money and reduces infestation of mosquitoes and other damaging insects by providing a diversity of habitats for their natural predators (Section 14-5).

- Set up a compost bin (Figure 12-17) and use it to produce mulch and soil conditioner for yard and garden plants.

- Use organic methods (no commercial fertilizers or pesticides) for growing vegetables and maintaining your yard. This involves using organic fertilizers (mulch, green manure, and animal manure) and biological and cultural control of pests (Section 14-5).

To reduce your inputs of hazardous waste into the environment

- Use pesticides and other hazardous chemicals only when absolutely necessary and in the smallest amount possible.

- Use rechargeable batteries.

- Use less hazardous (and usually cheaper) cleaning products (see Individuals Matter inside the back cover).

- Do not flush hazardous chemicals down the toilet, pour them down the drain, bury them, or dump them down storm drains.* Consult your local health department or environmental agency for safe disposal methods. Find out if they have set up hazardous waste collection days or a center that accepts these wastes. If not, organize efforts to do this.

- Insist that local, state, and federal elected officials establish and strictly enforce laws and policies that sharply reduce soil erosion and contamination and emphasize pollution prevention and waste reduction.

*See *Household Hazardous Waste Wheel*, Environmental Hazards Management Institute, P.O. Box 932, Durham, NH 03824 ($3.75), and *Earth Wise Household Inventory Worksheet*, P.O. Box 682, Belmar, NJ 07719 ($2).

In 1989, the EPA estimated that there are more than 32,000 sites in the United States containing potentially hazardous wastes, with this number increasing at a rate of about 2,500 a year. The General Accounting Office estimates that there are 130,000 to 425,000 sites.

By January 1990 the EPA had placed 1,081 sites on a priority cleanup list because of their threat to nearby populations. The priority list is expected to grow by about 180 sites a year. Many of these sites are located over major aquifers and pose a serious threat to groundwater.

By early 1990 only 50 sites had been declared clean and only 24 had been removed from the priority list. In 1985 the Office of Technology Assessment estimated that the final list may include at least 10,000 sites, with cleanup costs amounting to as much as $500 billion over the next 50 years. Cleanup funds provided by taxes on industries that generate waste, has amounted to about $1 billion a year—far short of the need.

The EPA cleanup has lagged because of a lack of money and technical expertise plus unrealistic expectations by the public. By 1989 only $1.6 billion of the $2.6 billion Superfund money spent by the EPA was being used for actual site studies and cleanups. The rest was used for administrative, management, and litigation costs. Because of personnel freezes, much of Superfund spending has gone to outside consultants and experts who make more money by dragging the process out.

Cleanup procedures vary from site to site and cannot be done quickly. Often contaminated groundwater must be pumped out and purified for decades. According to a 1989 report by the Office of Technology Assessment, about 75% of the cleanups are unlikely to work over the long term.

In 1984 Congress amended the 1976 Resource Conservation and Recovery Act to make it national policy to minimize or eliminate land disposal of 450 regulated hazardous wastes by May 1990 unless the EPA has determined that it is an acceptable or the only feasible approach for a particular hazardous material. Even then, each chemical is to be treated to the fullest extent possible to reduce its toxicity before land disposal of any type is allowed.

If enforced, this policy represents a much more

David Pimentel

David Pimentel is professor of insect ecology and agricultural sciences in the College of Agriculture and Life Sciences at Cornell University. He has chaired the Board on Environmental Studies in the National Academy of Sciences (1979–1981); the Panel on Soil and Land Degradation, Office of Technology Assessment (1978–1981), and the Biomass Panel, U.S. Department of Energy (1976–1985). He has published over 300 scientific papers and 12 books on environmental topics including land degradation, agricultural pollution and energy use, biomass energy, and pesticides. He was one of the first ecologists to employ an interdisciplinary, holistic approach to investigating complex environmental problems.

At a time when the world's human population is rapidly expanding and its need for more land to produce food, fiber, and fuelwood is also escalating, valuable land is being degraded through erosion and other means at an alarming rate. Soil degradation is of great concern because soil reformation is extremely slow. Under tropical and temperate agricultural conditions, an average of 500 years (with a range of 220 to 1,000 years) are required for the renewal of 2.5 cm (1 inch) of soil—a renewal rate of about 1 metric ton (t) of topsoil per hectare (ha) of land per year (1t/ha per year). Worldwide annual erosion rates for agricultural land are 18 to 100 times this natural renewal rate.

Erosion rates vary in different regions because of topography, rainfall, wind intensity, and the type of agricultural practices used. In China, for example, the average annual soil loss is reported to be about 40t/ha while the U.S. average is 18t/ha. In states like Iowa and Missouri, however, annual soil erosion averages are greater than 35t/ha.

Worldwide, about 6 million hectares (about the size of West Virginia) of land are abandoned for crop production each year because of high erosion rates plus waterlogging of soils, salinization, and other forms of soil degradation. In addition, according to the UN Environment Program, each year crop productivity becomes uneconomical on about 20 million hectares (about the size of Nebraska) because soil quality has been severely degraded.

Soil erosion also occurs in forest land but is not as severe as that in the more exposed soil of agricultural land. However, soil erosion in managed forests is a major concern because the soil reformation rate in forests is about two to three times longer than that in agricultural land. To compound this erosion problem, at least 20 million hectares (the size of Maine and Florida combined) of forest are being cleared each year throughout the world. More than half of this is being used to compensate for loss of agricultural land caused by erosion. Average soil erosion per hectare increases when trees are removed and the land is planted with crops.

The effects of agriculture and forestry are interrelated in many other ways. Large-scale removal of forests without adequate replanting reduces fuelwood supplies and forces the poor in LDCs to substitute crop residue and manure for fuelwood. When these plant and animal wastes are burned instead of being returned to the land as ground cover and organic fertilizer, erosion is intensified and productivity of the land is decreased. These factors, in turn, increase pressure to convert more forest land into agricultural land, further intensifying soil erosion.

(continued)

ecologically sound approach to dealing with hazardous wastes. However, phasing out land disposal is hampered by a shortage of facilities to treat and handle hazardous wastes in safer ways, an inexperienced EPA staff with rapid turnover, lack of funds, strong political pressure from large waste management companies (many of which are run by former EPA officials), and too little emphasis on waste reduction, recycling, and reuse.

In 1990 several environmental groups filed a lawsuit against the EPA, charging it with not carrying out this law while giving lip service to pollution prevention. Instead of treatment, EPA regulations issued in 1990 would allow industries to dilute haz-

ardous wastes by mixing them with other wastes and then injecting the mixture into deep wells. According to environmentalists, this violates the 1984 RCRA amendments by allowing dilution to replace treatment of hazardous wastes before land disposal is allowed.

Because soil is the base of life, we must demand that soil abuse halt and be replaced with soil healing and soil protection so that these vital, slowly renewable resources are used sustainably. Each of us has a role to play (see Individuals Matter on p. 288).

Civilization can survive the exhaustion of oil reserves, but not the continuing wholesale loss of topsoil.

LESTER R. BROWN

One reason that soil erosion does not receive high priority among many governments and farmers is that it usually occurs at such a slow rate that its cumulative effects may take decades to become apparent. For example, the removal of 1 millimeter (¹⁄₂₅ inch) of soil is so small that it goes undetected. But the accumulated soil loss at this rate over a 25-year period would amount to 25 mm (1 inch)—an amount that would take about 500 years to replace by natural processes.

Although reduced soil depth is a serious concern because it is cumulative, other factors associated with erosion also reduce productivity. These are losses of water, organic matter, and soil nutrients. Water is the major limiting factor for all natural and agricultural plants and trees. When some of the vegetation on land is removed, most water is lost to remaining plants because it runs off rapidly and does not penetrate the soil. In addition, soil erosion reduces the water-holding capacity of soil because it removes organic matter and fine soil particles that hold water. When this happens, water infiltration into soil can be reduced as much as 90%.

Organic matter in soil plays an important role in holding water and in decreasing removal of plant nutrients. Thus, it is not surprising that a 50% reduction of soil organic matter on a plot of land has been found to reduce corn yields as much as 25%. When soil erodes, there is also a loss of vital plant nutrients such as nitrogen, phosphorus, potassium, and calcium. With U.S. annual cropland erosion rates of about 18t/ha, estimates are that about half of the 45 million metric tons (49.5 million tons) of commercial fertilizers that are applied annually are replacing soil nutrients lost by erosion. This use of fertilizers substantially adds to the cost of crop production.

Some analysts who are unaware of the numerous and complex effects of soil erosion have falsely concluded that the damages are relatively minor. For example, they report that an average soil loss in the United States of 18t/ha per year causes an annual reduction in crop productivity of only 0.1% to 0.5%. However, we need to consider all the ecological effects caused by erosion, including a reduction in soil depth, reduced water availability for crops, and reduction in soil organic matter and nutrients. When this is done agronomists and ecologists report a 15% to 30% reduction in crop productivity—a key factor in increased levels of costly fertilizer and declining yields on some land despite high levels of fertilization. Because fertilizers are not a substitute for fertile soil, they can only be applied up to certain levels before crop yields begin to decline.

Reduced agricultural productivity is only one of the effects and costs of soil erosion. In the United States, water runoff is responsible for transporting about 3 billion metric tons (3.3 billion tons) of sediment each year to waterways in the 48 contiguous states. About 60% of these sediments come from agricultural lands.

Estimates show that off-site damages to U.S. water storage capacity, wildlife, and navigable waterways from these sediments cost an estimated $6 billion each year. Dredging sediments from U.S. streams, harbors, and reservoirs alone costs about $570 million each year. About 25% of new water storage capacity in U.S. reservoirs is built solely to compensate for sediment buildup.

When soil sediments that include pesticides and other agricultural chemicals are carried into streams, lakes, and reservoirs, fish production is adversely affected. These contaminated sediments interfere with fish spawning, increase predation on fish, and destroy fisheries in estuarine and coastal areas.

Increased erosion and water runoff on mountain slopes flood agricultural land in the valleys below, further decreasing agricultural productivity. Eroded land also does not hold water very well, further decreasing crop productivity. This effect is magnified in the 80 countries (with nearly 40% of the world's population) that experience frequent droughts. The rapid growth in the world's population, accompanied by the need for more crops and a projected doubling of water needs in the next 20 years, will only intensify water shortages, particularly if soil erosion is not contained.

Thus, soil erosion is one of the world's critical problems and if not slowed will seriously reduce agricultural and forestry production and degrade the quality of aquatic ecosystems. Solutions are not particularly difficult but are often not implemented because erosion occurs so gradually that we fail to acknowledge its cumulative impact until damage is irreversible. Many farmers have also been conditioned to believe that losses in soil fertility can be remedied by applying increasingly higher levels of fertilizer.

The principal method of controlling soil erosion and its accompanying runoff of sediment is to maintain adequate vegetative coverage on soils by various methods discussed in Section 12-3. These methods are also cost effective in preventing erosion, especially when off-site costs of erosion are included. Scientists, policymakers, and agriculturists need to work together to implement soil and water conservation practices before world soils lose most of their productivity.

Guest Essay Discussion

1. Some analysts contend that average soil erosion rates in the United States and the world are low and that this problem has been overblown by environmentalists and can easily be solved by improved agricultural technology such as no-till cultivation and increased use of commercial inorganic fertilizers. Do you agree or disagree with this position? Explain.

2. What specific things do you believe elected officials should do to decrease soil erosion and the resulting sediment water pollution in the United States?

DISCUSSION TOPICS

1. Why should everyone, not just farmers, be concerned with soil conservation?

2. Describe briefly the Dust Bowl phenomenon of the 1930s and explain how and where it could happen again. How would you try to prevent a recurrence?

3. Survey your campus to evaluate the use of natural vegetation and other methods to reduce soil erosion and enhance natural biological diversity.

4. Would you oppose locating a hazardous-waste landfill or incinerator in your community? Explain. If you oppose both of these alternatives, how would you propose that the hazardous waste generated in your community and state be managed?

5. Give your reasons for agreeing or disagreeing with each of the following proposals for dealing with hazardous waste:
   a. Reduce the production of hazardous waste; encourage this and recycling and reuse of hazardous materials by levying a tax or fee on producers for each unit of waste generated.
   b. Ban all land disposal of hazardous waste to encourage recycling, reuse, and treatment and to protect groundwater from contamination.
   c. Provide low-interest loans, tax breaks, and other financial incentives to encourage industries that produce hazardous waste to recycle, reuse, treat, destroy, and reduce generation of such waste.
   d. Ban the shipment of hazardous waste from the United States to any other country.
   e. Ban the shipment of hazardous waste from one state to another.

6. What hazardous wastes are produced at your college or university? What happens to these wastes?

FURTHER READINGS

Batie, Sandra S. 1983. *Soil Erosion: Crisis in America's Croplands?* Washington, D.C.: Conservation Foundation.

Brady, Nyle C. 1974. *The Nature and Properties of Soils.* New York: Macmillan.

Brown, Lester R., and Edward C. Wolf. 1984. *Soil Erosion: Quiet Crisis in the World Economy.* Washington, D.C.: Worldwatch Institute.

Caplan, Ruth. 1990. *Our Earth, Ourselves: The Action-Oriented Guide to Help You Protect and Preserve Our Environment.* New York: Bantam Books.

Commoner, Barry. 1990. *Making Peace with the Planet.* New York: Pantheon.

Dale, Tom, and V. G. Carter. 1955. *Topsoil and Civilization.* Norman: University of Oklahoma Press.

Davis, Charles E., and James P. Lester, eds. 1988. *Dimensions of Hazardous Waste Politics and Policies.* Westport, Conn.: Greenwood.

Enterprise for Education. 1989. *Hazardous Waste from Homes.* Santa Monica, Calif.: Enterprise for Education.

Environmental Defense Fund. 1985. *To Burn or Not to Burn.* New York: Environmental Defense Fund.

Environmental Protection Agency. 1987. *The Hazardous Waste System.* Washington, D.C.: EPA.

Epstein, Samuel S., et al. 1982. *Hazardous Waste in America.* San Francisco: Sierra Club Books.

General Accounting Office. 1983. *Agriculture's Soil Conservation Programs Miss Full Potential in the Fight Against Soil Erosion.* Washington, D.C.: General Accounting Office.

Gibbs, Lois. 1982. *The Love Canal: My Story.* Albany, N.Y.: State University of New York Press.

Grainger, Alan. 1983. *Desertification: How People Make Deserts, How People Can Stop and Why They Don't.* Washington, D.C.: Earthscan.

Hirschorn, Joel S. 1988. "Cutting Production of Hazardous Waste." *Technology Review,* Apr., 52–61.

Huisingh, Donald, et al. 1986. *Proven Profits from Pollution Prevention: Case Studies in Resource Conservation and Waste Reduction.* Washington, D.C.: Institute for Local Self-Reliance.

Hunter, Linda M. 1989. *The Healthy Home.* Emmaus, Penn.: Rodale Press.

Lester, James P., and Ann O. Bowman, eds. 1983. *The Politics of Hazardous Waste Management.* Durham, N.C.: Duke University Press.

Muir, Warren, and Joanna Underwood. 1987. *Promoting Hazardous Waste Reduction.* New York: INFORM, Inc.

National Academy of Sciences. 1986. *Soil Conservation.* (2 vols). Washington, D.C.: National Academy Press.

Office of Technology Assessment. 1986. *Serious Reduction of Hazardous Waste.* Washington, D.C.: Government Printing Office.

Office of Technology Assessment. 1987. *From Pollution to Prevention: A Progress Report on Waste Reduction.* Washington, D.C.: Government Printing Office.

Piasecki, Bruce, and Gary Davis. 1987. *America's Future in Toxic Waste Management: Lessons from Europe.* New York: Quorum Books.

Postel, Sandra. 1987. *Defusing the Toxics Threat: Controlling Pesticides and Industrial Waste.* Washington, D.C.: Worldwatch Institute.

Segel, Edward, et al. 1985. *The Toxic Substances Dilemma: A Plan for Citizen Action.* Washington, D.C.: National Wildlife Federation.

Sophen, C. D., and J. V. Baird. 1982. *Soils and Soil Management.* Reston, Va.: Reston Publishing.

Water Pollution Control Federation. 1989. *Household Hazardous Waste: What You Should and Shouldn't Do.* Alexandria, Va.: Water Pollution Control Federation.

Whelan, Elisabeth M. 1985. *Toxic Terror.* Ottawa, Ill.: Jameson Books.

Wilson, G. F., et al. 1986. *The Soul of the Soil: A Guide to Ecological Soil Management.* 2nd ed. Quebec, Canada: Gaia Services.

The cutting and burning of an area of tropical forest in Brazil's Amazon Basin for growing crops.

It is the responsibility of all who are alive today to accept the trusteeship of wildlife and to hand on to posterity, as a source of wonder and interest, knowledge, and enjoyment, the entire wealth of diverse animals and plants. This generation has no right by selfishness, wanton or intentional destruction, or neglect, to rob future generations of this rich heritage. Extermination of other creatures is a disgrace to humankind.

WORLD WILDLIFE CHARTER

CHAPTER THIRTEEN

FOOD RESOURCES

Industrialized

General Questions and Issues

1. How is food produced throughout the world?

2. What are the world's major food problems?

3. Can increasing crop yields and cultivating more land solve the world's major food problems?

4. What government policies can increase food production?

5. What can giving food aid and redistributing land to the poor do to help solve world food problems?

6. How much food can we get from catching more fish and cultivating fish in aquaculture farms?

7. How can agricultural systems in MDCs and LDCs be designed to be ecologically and economically sustainable?

Hunger is a curious thing: At first it is with you all the time, working and sleeping and in your dreams, and your belly cries out insistently, and there is a gnawing and a pain as if your very vitals were being devoured, and you must stop it at any cost. . . . Then the pain is no longer sharp, but dull, and this too is with you always.

KAMALA MARKANDAYA

A GRICULTURE USES MORE of the earth's land, water, plant, animal, and energy resources and causes more pollution and environmental degradation than any other human activity. By 2020 the world's population is expected to reach at least 8 billion. To feed these people, we must produce as much food during the next 30 years as we have produced since the dawn of agriculture about 10,000 years ago.

Producing enough food to feed the world's population, however, is only one of a number of complex, interrelated food resource problems. Another major problem is food quality—eating food with enough proteins, vitamins, and minerals to avoid malnutrition. We must also have enough storage facilities to keep food from rotting or being eaten by pests after it is harvested. An adequate transportation and retail outlet system must be available to distribute and sell food throughout each country and the world.

There are more hungry people in the world today than at any time in human history, and their numbers are growing. Poverty is the leading cause of this hunger and premature death, resulting from lack of food quantity and quality. Making sure the poor have enough land or income to grow or buy enough food is the key to reducing deaths from malnutrition and to helping the poor escape the global poverty trap (Section 7-3). Farmers must also have economic incentives to grow enough food to meet the world's needs. Finally, the world's agricultural systems must be managed in sustainable ways to minimize the harmful environmental impacts on the soil, air, water, and wildlife of producing and distributing food.

13-1 WORLD AGRICULTURAL SYSTEMS: HOW IS FOOD PRODUCED?

Plants and Animals That Feed the World Although about 80,000 species of plants are edible, only about 30 crops feed the world and fewer than 20 produce 90% of our food. Four crops—wheat, rice, corn, and potato—make up more of the world's total food production than all others combined. These four and most other crops we depend upon are annuals. This means that each year we must disturb the soil and plant new seeds. The rest of the food people eat is mainly fish, meat, and animal products such as milk, eggs, and cheese obtained largely from eight species of domesticated livestock.

Two out of three people in the world are forced to survive on a primarily vegetarian diet. Meat and animal products are too expensive for these people, because of the loss of usable energy when an animal trophic level is added to a food chain (Figure 4-18, p. 72).

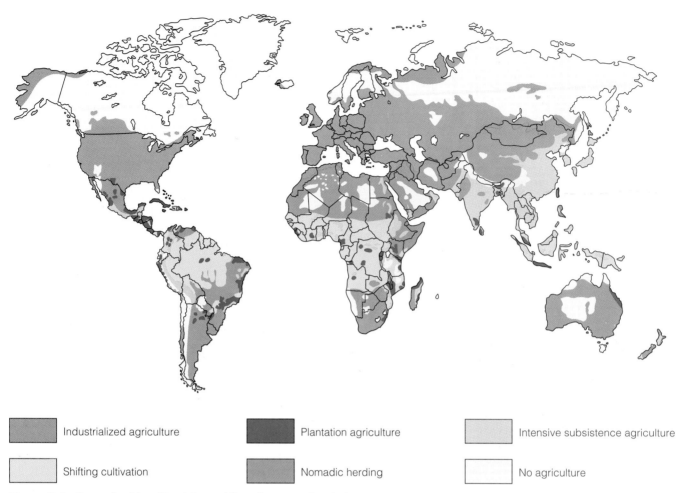

| Industrialized agriculture | Plantation agriculture | Intensive subsistence agriculture |
| Shifting cultivation | Nomadic herding | No agriculture |

Figure 13-1 Generalized location of the world's major types of agriculture.

However, as incomes rise, people consume more grain *indirectly*, in the form of meat and products from grain-fed domesticated animals. In MDCs, almost half of the world's annual grain production (especially corn and soybeans) is fed to livestock. Also, about one-third of the world's annual fish catch is converted to fish meal and fed to livestock.

Major Types of Agriculture Two major types of agricultural systems are used to grow crops and raise livestock throughout the world: industrialized agriculture and subsistence agriculture. **Industrialized agriculture** produces large quantities of a single type of crop or livestock for sale both within the country where it is grown and to other countries. Industrialized agriculture involves supplementing solar energy with large amounts of energy from fossil fuels, mostly oil and natural gas.

Industrialized agriculture, which is practiced on about 25% of all cropland, is widely used in MDCs and since the mid-1960s has spread to parts of some LDCs (Figure 13-1). It is supplemented by **plantation agriculture,** in which specialized cash crops such as

bananas, coffee, and cacao are grown in tropical LDCs primarily for sale to MDCs.

Traditional **subsistence agriculture** produces enough crops or livestock for the farm family's survival and, in good years, enough to have some left over to sell or put aside for hard times. Subsistence farmers supplement solar energy with energy from human labor and draft animals. The three major types of subsistence agriculture are shifting cultivation of small plots in tropical forests (Figure 2-4, p. 24), intensive crop cultivation on relatively small plots of land, and nomadic herding of livestock. These forms of agriculture are practiced by about 2.6 billion people—half the people on earth—who live in rural areas in LDCs. An average of 63% of the people in LDCs work in agriculture, compared to only 10% in MDCs.

Industrialized Agriculture and Green Revolutions Crop production is increased either by cultivating more land or by getting higher yields from existing cropland. Since 1950 most of the increase in world food production has come from increasing the yield per hectare in what is called a **green revolution**. This

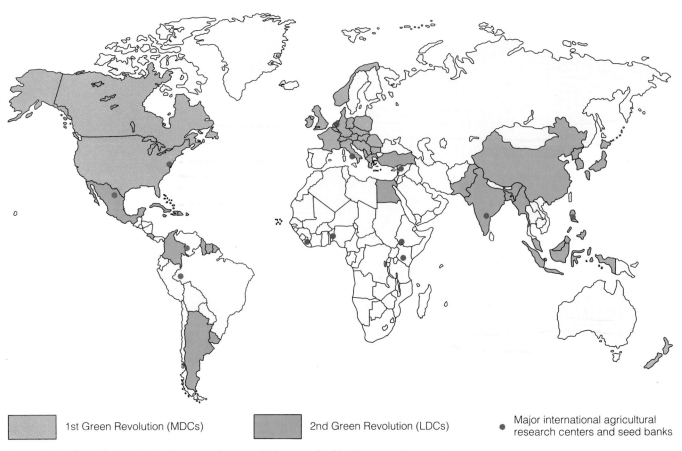

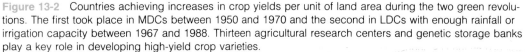

| 1st Green Revolution (MDCs) | 2nd Green Revolution (LDCs) | • Major international agricultural research centers and seed banks |

Figure 13-2 Countries achieving increases in crop yields per unit of land area during the two green revolutions. The first took place in MDCs between 1950 and 1970 and the second in LDCs with enough rainfall or irrigation capacity between 1967 and 1988. Thirteen agricultural research centers and genetic storage banks play a key role in developing high-yield crop varieties.

involves planting monocultures of scientifically bred plant varieties and applying large amounts of inorganic fertilizer, irrigation water, and pesticides.

Between 1950 and 1970 this approach led to dramatic increases in yields of major crops in the United States and most other industrialized countries, a phenomenon sometimes known as the *first green revolution* (Figure 13-2). In 1967, after 30 years of genetic research and trials, a modified version of the first green revolution began spreading to many LDCs. High-yield, fast-growing, dwarf varieties of rice and wheat, specially bred for tropical and subtropical climates, were introduced into several LDCs in what is known as the *second green revolution.*

The shorter, stronger, and stiffer stalks of the new varieties allow them to support larger heads of grain without toppling over (Figure 13-3). With large inputs of fertilizer, water, and pesticides, the wheat and rice yields of these varieties can be two to five times those of traditional varieties. The fast-growing varieties allow farmers to grow two, and even three, consecutive crops a year (multiple cropping) on the same parcel of land.

Nearly 90% of the increase in world grain output in the 1960s and about 70% of that in the 1970s were the result of the second green revolution. In the 1980s

and 1990s at least 80% of the additional production of grains is expected to be based on improved yields of existing cropland through the use of green revolution techniques.

These increases, however, depend heavily on fossil fuel inputs. Since 1950, agriculture's use of fossil fuels has increased sevenfold, the number of tractors has quadrupled, irrigated area has tripled, and fertilizer use has risen 17-fold. Agriculture now uses about one-twelfth of the world oil output.

Industrialized Agriculture in the United States
Since 1940, U.S. farmers have more than doubled crop production while cultivating about the same amount of land. They have done this through industrialized agriculture coupled with a favorable climate and fertile soils.

Less than 1% of the U.S. work force is engaged in farming. Yet the country's 2.1 million farmers— with only 650,000 working full time at farming— produce enough food to feed most of their fellow citizens better and at a lower percentage of their income than do farmers in any other country. Americans spend an average of 11% to 15% of their disposable income on food. In addition, U.S. farmers

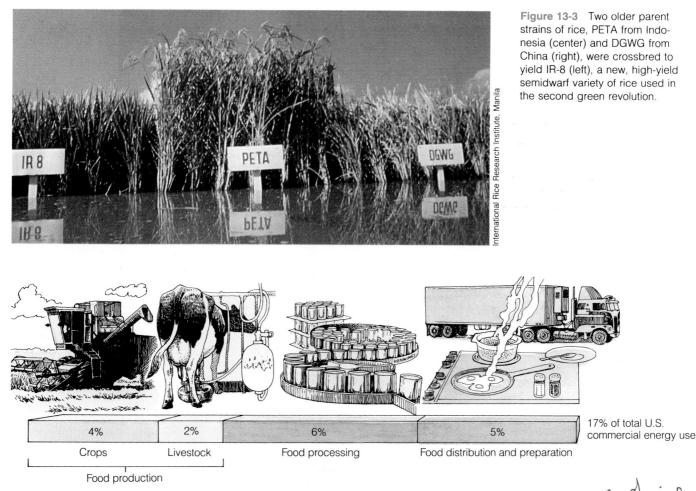

Figure 13-3 Two older parent strains of rice, PETA from Indonesia (center) and DGWG from China (right), were crossbred to yield IR-8 (left), a new, high-yield semidwarf variety of rice used in the second green revolution.

4%	2%	6%	5%	17% of total U.S. commercial energy use
Crops	Livestock	Food processing	Food distribution and preparation	

Food production

Figure 13-4 Commercial energy use by the U.S. industrialized agriculture system.

agriculture consumes 17% of all commercial energy

produce large amounts of food for export to other countries. By contrast people in much of the world spend 40% or more of their disposable income on food, and the 1 billion people making up the poorest fifth of humankind typically spend 60% to 80% of their meager income on food and still don't have an adequate diet. *40%* *60–80%*

About 23 million people are involved in the U.S. agricultural system in activities ranging from growing and processing food to selling it at the supermarket. In terms of total annual sales, the agricultural system is the biggest industry in the United States—bigger than the automotive, steel, and housing industries combined. It generates about 17% of the country's GNP and 17% of all jobs in the private sector, employing more people than any other single industry. Farmers, however, get an average of only 25 cents of every dollar spent on food in the United States, compared to 41 cents in 1950.

The amount of agricultural chemicals used to support industrialized agriculture in the United States is phenomenal. Currently, the country uses about 15% to 20% of the world's output of commercial inorganic fertilizers and 45% of the pesticides produced worldwide each year.

The gigantic American agricultural system consumes about 17% of all commercial energy used in the United States each year (Figure 13-4). Most of this energy comes from oil.

Most plant crops in the United States provide more food energy than the energy (mostly from fossil fuels) used to grow them. Raising animals for food, however, requires much more fossil-fuel energy than the animals provide as food energy.

Energy efficiency is much worse if we look at the entire U.S. food system. Counting fossil-fuel energy inputs used to grow, store, process, package, transport, refrigerate, and cook all plant and animal food, *an average of about ten units of nonrenewable fossil-fuel energy is needed to put one unit of food energy on the table—an energy loss of nine units per unit of food energy produced.* By comparison, every unit of energy from the human labor of subsistence farmers may provide ten units of food energy.

Suppose everyone in the world ate a typical American diet consisting of food produced by industrialized agriculture. If the world's known oil reserves were used only for producing this food, these reserves would be depleted in less than 12 years. This indicates that the present world population already exceeds

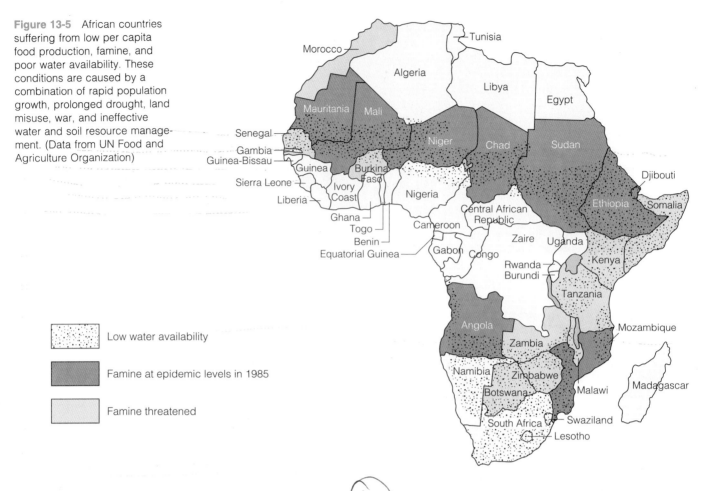

Figure 13-5 African countries suffering from low per capita food production, famine, and poor water availability. These conditions are caused by a combination of rapid population growth, prolonged drought, land misuse, war, and ineffective water and soil resource management. (Data from UN Food and Agriculture Organization)

Legend:
- Low water availability
- Famine at epidemic levels in 1985
- Famine threatened

the capacity of known supplies of arable land and petroleum to provide everyone with a U.S.-type diet using industrialized agriculture.

Examples of Subsistence Agriculture Farmers in LDCs use various forms of subsistence agriculture to grow crops on about 75% of the world's cultivated land (Figure 13-1). Many subsistence farmers imitate nature by simultaneously growing a variety of crops on the same plot. This biological diversity reduces their chances of losing most or all of their year's food supply from pests, flooding, drought, or other disasters. Common planting strategies include

- **polyvarietal cultivation,** in which a plot of land is planted with several varieties of the same crop

- **intercropping,** in which two or more different crops are grown at the same time on a plot—for example, a carbohydrate-rich grain that depletes soil nitrogen and a protein-rich legume that adds nitrogen to the soil

- **agroforestry,** a variation of intercropping in which crops and trees are planted together—for example, a grain or legume crop planted around fruit-bearing orchard trees or in rows between fast-growing trees or shrubs that can be used for fuelwood or to add nitrogen to the soil (Figure 12-13, p. 277)

- **polyculture,** a more complex form of intercropping in which a large number of different plants maturing at different times are planted together

13-2 MAJOR WORLD FOOD PROBLEMS

The Good News About Food Production World grain production expanded 2.6-fold between 1950 and 1984, and average per capita production rose by almost 40%. During the same period, average food prices adjusted for inflation dropped by 25%, and the amount of food traded in the world market quadrupled. Most of the increase in food production since 1950 came from increases in crop yields per hectare by means of improved labor-intensive subsistence agriculture in many LDCs and energy-intensive industrialized agriculture in MDCs and some LDCs (Figure 13-2).

The Bad News About Food Production The impressive improvements in world food production disguise the fact that average food production per person declined between 1950 and 1989 in 43 LDCs (22 in Africa) containing one of every seven persons on earth. Average per capita food production has been falling in Africa since 1967 and in Latin America since 1982. Thus, population growth is outstripping food production in two major regions of the world, in which 1.1 billion people live. The largest declines have occurred in Africa, where average food production per person dropped 21% between 1960 and 1989 and is projected to drop another 30% during the next 25 years (Figure 13-5). Most of the 183 nations in the world now require food imports from other nations, primarily the United States, Canada, Australia, Argentina, and France.

Another disturbing trend is that the rate of increase in world average food production per person declined during each of the past three decades. It rose 15% between 1950 and 1960, 7% between 1960 and 1970, and only 4% between 1970 and 1980. Between 1984 and 1989 average per capita food production fell by 14%. This trend is caused by a combination of population increase, a decrease in yields per unit of land area for some crops cultivated by industrialized agriculture, a levelling off or a drop in food production in some countries, unsustainable use of soil and water, and widespread drought in 1987 and 1988.

Projected changes in global climate (Section 10-2) could disrupt agricultural production, cause sharp rises in food prices, and lead to mass starvation and disease. Despite the successes of industrialized agriculture, agricultural success still depends on having favorable weather and a stable climate.

Food Quantity and Quality: Undernutrition, Malnutrition, and Overnutrition Poor people who cannot grow or buy enough food for good health and survival suffer from **undernutrition**. Survival and good health also require that people consume food containing the proper amounts of protein, carbohydrates, fats, vitamins, and minerals.

Most poor people are forced to live on a low-protein, high-starch diet of grains such as wheat, rice, or corn. As a result, they often suffer from **malnutrition,** or deficiencies of protein and other key nutrients. Many of the world's desperately poor people suffer from both undernutrition and malnutrition.

Each year 20 to 40 million people—half of them children under age five—die prematurely from undernutrition, malnutrition, or normally nonfatal infections and diseases, such as diarrhea, measles, and flu, worsened by these nutritional deficiencies. The World Health Organization estimates that diarrhea alone kills at least 5 million children under age five a year.

Adults suffering from chronic undernutrition and malnutrition are vulnerable to diseases and are too weak to work productively or think clearly. As a

---

SPOTLIGHT   Marasmus and Kwashiorkor

The two most widespread nutritional-deficiency diseases are marasmus and kwashiorkor. **Marasmus** (from the Greek "to waste away") occurs when a diet is low in both total energy (calories) and protein. Most victims of marasmus are infants in poor families in which children are not breast-fed or in which food quantity and quality are insufficient after the children are weaned.

A child suffering from marasmus typically has a bloated belly, a thin body, shriveled skin, wide eyes, and an old-looking face (Figure 1-3, p. 5). If the child is treated in time with a balanced diet, most of these effects can be reversed.

Kwashiorkor (meaning "displaced child" in a West African dialect) occurs in infants and children one to three years old who suffer from severe protein deficiency. Typically, kwashiorkor afflicts a child who no longer gets breast milk because the mother has a younger child to nurse. The displaced child's diet changes from highly nutritious breast milk to grain or sweet potatoes, which provide enough calories but not enough protein.

Children suffering from kwashiorkor have skin swollen with fluids, a bloated abdomen, lethargy, liver damage, hair loss, diarrhea, stunted growth, possible mental retardation, and irritability. If such malnutrition is not prolonged, most of the effects can be cured with a balanced diet.

---

result, their children are also underfed and malnourished (see Spotlight above). If these children survive to adulthood, many are locked in a tragic malnutrition-poverty cycle that continues these conditions in each succeeding generation and traps people in the global poverty cycle (Figure 7-5, p. 159).

Each of us must have a daily intake of small amounts of vitamins that cannot be made in the human body. Otherwise we will suffer from various effects of vitamin deficiencies. Although balanced diets, vitamin-fortified foods, and vitamin supplements have greatly reduced the number of vitamin-deficiency diseases in MDCs, millions of cases occur each year in LDCs. For example, each year more than 500,000 children in LDCs are partially or totally blinded because their diet lacks vitamin A.

Other nutritional-deficiency diseases are caused by the lack of certain minerals, such as iron and iodine. Too little iron causes anemia. Anemia causes fatigue, makes infection more likely, increases a woman's chance of dying in childbirth, and increases an infant's chances of dying from infection during its first year of life. In tropical regions of Asia, Africa, and Latin America, iron-deficiency anemia affects about 10% of the men, more than half of the children, two-thirds of the pregnant women, and about half of the other women.

Officials of the United Nations Children's Fund (UNICEF) estimate that between half and two-thirds of the worldwide annual childhood deaths from undernutrition, malnutrition, and associated infections and diseases could be prevented at an average annual cost of only $5 to $10 per child. This life-saving program would involve the following simple measures:

- immunizing against childhood diseases such as measles
- encouraging breast-feeding
- preventing dehydration from diarrhea by giving infants a solution of a fistful of sugar and a pinch of salt in a glass of water
- preventing blindness by giving people a small vitamin A capsule twice a year at a cost of about 75 cents per person
- providing family planning services to help mothers space births at least two years apart
- increasing female education with emphasis on nutrition, sterilization of drinking water, and child care

While 15% of the people in LDCs suffer from undernutrition and malnutrition, about 15% of the people in MDCs suffer from **overnutrition**. This is an excessive intake of food that can cause obesity, or excess body fat, in people who do not suffer from glandular or other disorders that promote obesity. Overnourished people exist on diets high in calories, cholesterol-containing saturated fats, salt, sugar, and processed foods, and low in unprocessed fresh vegetables, fruits, and fiber. Partly because of these dietary choices, overweight people have significantly higher than normal risks of diabetes, high blood pressure, stroke, and heart disease.

Poverty: The Geography of Hunger Today the world's massive hunger and malnutrition problem is mostly a poverty problem, not a food supply problem. If all the cereal crops currently produced in the world were divided, there would be enough to keep 6 billion people alive. In contrast, if this food were used to give everyone the typical plant and meat diet of a person in a developed country, it would support only 2.5 billion people—less than half the present world population and only one-fourth of the 14 billion

people projected by the year 2100 (Figure 1-2, p. 4). The world's supply of food, however, is not now distributed equally among the world's people, nor will it be, because of differences in soil, climate, political and economic power, and average income throughout the world.

Poverty—not lack of food production—is the chief cause of hunger, malnutrition, and premature death throughout the world. The world's 1 billion desperately poor people do not have access to land where they can grow enough food of the right kind, and they do not have the money to buy enough food of the right kind no matter how much is available. This is caused by local, national, and international forces that make up the global poverty trap (Section 7-3).

Increases in worldwide total food production and average food production per person often hide wide-spread differences in food supply and quality between and within countries. For example, about one-third of the world's hungry live in India, even though it is self-sufficient in food production.

Most of the world's hungry poor live in rural areas. In more fertile and urbanized southern Brazil, the average daily food supply per person is high. However, in Brazil's semiarid, less fertile northeastern interior, many people are severely underfed. Overall almost two out of three Brazilians suffer from malnutrition (see Case Study on p. 126).

Food is also unevenly distributed within families. In poor families the largest part of the food supply goes to men working outside the home. Children (ages 1–5) and women (especially pregnant women and nursing mothers) are the most likely to be underfed and malnourished.

MDCs also have pockets of poverty and hunger. For example, a 1985 report by a task force of doctors estimated that at least 20 million people—1 out of every 11 Americans—were hungry, mostly because of cuts in food stamps and other forms of government aid since 1980. Half of these people were children.

In 1989 the UN World Food Commission reported that progress in fighting hunger, malnutrition, and poverty in the 1970s came to a halt or was reversed in many parts of the world during the 1980s. Without a widespread increase in income and access to land, the number of chronically hungry and malnourished people in the world could increase to at least 1.5 billion by 2000—one out of every four people in the world's projected population at that time.

Environmental Effects of Producing Food Both industrialized agriculture and subsistence agriculture have a number of harmful impacts on the air, soil, and water resources that sustain all life (see Spotlight on p. 301).

Industrialized Agriculture

- Soil erosion and loss of soil fertility through poor land use, failure to practice soil conservation techniques, and too little use of organic fertilizers (Section 12-3).

- Salinization and waterlogging of heavily irrigated soils (Figure 12-18, p. 280).

- Reduction in the number and diversity of nutrient-recycling soil microorganisms due to heavy use of pesticides and commercial inorganic fertilizers and soil compaction by large tractors and other farm machinery.

- Air pollution caused by dust blown off cropland that is not kept covered with vegetation (Figure 12-7, p. 271) and from overgrazed rangeland (Section 15-5).

- Air pollution from droplets of pesticide sprayed from planes or ground sprayers and blown into the air from plants and soil.

- Air pollution caused by the extraction, processing, transportation, and combustion of massive amounts of fossil fuels used in industrialized agriculture.

- Pollution of estuaries and deep-ocean zones with oil from offshore wells and tankers (see Case Study on p. 258), and from improper disposal of oil, the main fossil fuel used in industrialized agriculture.

- Pollution of streams, lakes, and estuaries and killing of fish and shellfish from pesticide runoff.

- Depletion of groundwater aquifers by excessive withdrawals for irrigation (Figure 11-13, p. 243).

- Pollution of groundwater caused by leaching of water-soluble pesticides, nitrates from commercial inorganic fertilizers, and salts from irrigation water (Figure 11-26, p. 257).

- Overfertilization of lakes and slow-moving streams caused by runoff of nitrates and phosphates in commercial inorganic fertilizers, livestock animal wastes, and food processing wastes (Figure 11-22, p. 251). In the United States it is estimated that agriculture contributes 60% of the oxygen-demanding wastes, 64% of the suspended solids, and 76% of the phosphorus discharged into the nation's surface waters.

- Sediment pollution of surface waters caused by erosion and runoff from farm fields and animal feedlots.

- Loss of genetic diversity of plants caused by clearing biologically diverse grasslands and forests and replacing them with monocultures of single crop varieties (Figure 5-12, p. 92). Expansion of agriculture accounts for about 85% of the worldwide destruction of at least 202,000 square kilometers (78,000 square miles) of forests each year.

- Endangered and extinct animal wildlife from loss of habitat when grasslands and forests are cleared and wetlands are drained for farming (Section 15-2).

- Depletion and extinction of commercially important species of fish caused by overfishing.

- Threats to human health from nitrates in drinking water and pesticides in drinking water, food, and the atmosphere.

Subsistence Agriculture

- Soil erosion and rapid loss of soil fertility caused by clearing and cultivating steep mountain highlands without terracing, using shifting cultivation in tropical forests without leaving the land fallow long enough to restore soil fertility, overgrazing of rangeland (Section 15-5), and deforesting to provide cropland or fuelwood (Section 15-2).

- Increased frequency and severity of flooding in lowlands when mountainsides are deforested (see Case Study on p. 237).

- Desertification caused by cultivation of marginal land with unsuitable soil or terrain, overgrazing, deforestation, and failure to use soil conservation techniques (see Spotlight on p. 273).

- Air pollution caused by dust blown from cropland not kept covered with vegetation and from overgrazed rangeland.

- Sediment pollution of surface waters caused by erosion and runoff from farm fields and overgrazed rangeland.

- Endangered and extinct animal wildlife caused by loss of habitat when grasslands and forests are cleared for farming.

- Threats to human health from flooding intensified by poor land use and from human and animal wastes discharged or washed into irrigation ditches and sources of drinking water.

13-3 METHODS OF INCREASING WORLD FOOD PRODUCTION

Increasing Crop Yields Agricultural experts expect most future increases in crop production to come from increased yields per hectare on existing cropland and from expansion of green revolution technology to other parts of the world. There are two basic ways to do this. One is to alter natural environments to fit the needs of crops. If soil is deficient in plant nutrients, we add fertilizer. If an area is too dry, we add irrigation water. When pests attack crops, we spray them with pesticides. However, these inputs cannot be increased indefinitely. Eventually, crop yields begin levelling off as we reach the maximum ability of plants to absorb these inputs. Also, at some point the cost of using the inputs to increase crop yields exceeds the market value of the extra food produced.

The other approach is to genetically alter plants and livestock animals so they better fit their environments. Agricultural scientists are working to create new green revolutions—or gene revolutions—by using genetic engineering and other forms of biotechnology. Over the next 20 to 40 years they hope to breed high-yield plant strains that have greater resistance to insects and disease, thrive on less fertilizer, make their own nitrogen fertilizer like legumes (Figure 4-26, p. 78), do well in slightly salty soils, withstand drought, and make more efficient use of solar energy during photosynthesis.

If even a small fraction of this research and development is successful, the world could experience rapid and enormous increases in crop production before the middle of the next century. But some analysts point to several factors that have limited the spread and long-term success of the green revolutions.

- Without massive doses of fertilizer and water, green revolution crop varieties produce yields no higher and often lower than those from traditional strains. Biotechnologists acknowledge that without good soil and water, new genetically engineered crop strains will fail.

- So far plants have been far less responsive to genetic engineering than animals. The entire process from gene transfer to widespread planting of a new crop variety can take as long as conventional crossbreeding (typically 5 to 15 years).

- Areas without enough rainfall or irrigation water or with poor soils cannot benefit from the new varieties; that is why the second green revolution has not spread to many arid and semiarid areas (Figure 13-2).

- Increasingly greater and thus more expensive inputs of fertilizer, water, and pesticides eventually produce little or no increase in crop yields as the J-shaped curve of crop productivity reaches limits and is converted to an S-shaped curve. Scientists hope to overcome this limitation by developing new and improved varieties through crossbreeding and genetic engineering.

- Without careful land use and environmental controls (Section 12-3), degradation of water and soil can limit the long-term ecological and economic sustainability of green revolutions.

- The cost of genetically engineered crop strains is too high for most of the world's subsistence farmers in LDCs.

- The massive and increasing loss of the earth's biological diversity (Chapter 16) limits the future success of crossbreeding and genetic engineering. Geneticists can crossbreed varieties of plant and animal life (Figure 13-3) and genetic engineers can move genes from one organism to another, but they need the genetic materials found in the earth's existing plants and animals to do so.

- The loss of genetic diversity caused when a diverse mixture of natural varieties is replaced with monoculture crops limits the ability of plant scientists to use crossbreeding or genetic engineering to develop new strains for future green and gene revolutions. By the year 2000, the Food and Agriculture Organization estimates that two-thirds of all seed planted in LDCs will be of uniform strains. This botanical uniformity increases the vulnerability of food crops to pests and diseases.

To help preserve genetic variety, some of the world's native plants and native strains of food crops are being collected and stored in 13 genetic storage banks and agricultural research centers around the world (Figure 13-2). But these banks contain only a small portion of the world's known and potential varieties of agricultural crops and other plants, and many plant species cannot be successfully stored. Ecologists and plant scientists argue that the best way to preserve the genetic diversity of the world's plant and animal species is to protect large areas of representative ecosystems throughout the world from agriculture and other forms of development.

Cultivating More Land Despite our technological cleverness only about 22% of the earth's land surface is suitable for modern agriculture. The rest is covered by ice, is too dry or too wet, is too hot or too cold, is too steep, or has unsuitable soils to grow crops.

Currently, we are growing crops on land equivalent to the combined land areas of the United States and half of Canada. This amounts to about half of the world's potential cropland. The other half is currently used for grazing land or is covered by forests.

Some agricultural experts have suggested that farmers could more than double the world's cropland by clearing tropical forests and irrigating arid lands, mostly in Africa, South America, and Australia (Figure 13-6). Others believe only a small portion of these lands can be cultivated because most are too dry or too remote or lack productive soils.

Even if more cropland were developed, much of it would be on marginal land that will require large and expensive inputs of fertilizer, water, and energy. Also, these possible increases in cropland would not offset the projected loss of almost one-third of today's cultivated cropland from erosion, overgrazing, waterlogging, salinization, mining, and urbanization.

Location, Soil, and Insects as Limiting Factors About 83% of the world's potential new cropland is in the remote rain forests of the Amazon and Orinoco river basins in South America and in Africa's rain forests (Figure 5-15, p. 93). Most of the land is located in just two countries, Brazil and Zaire.

Cultivation would require massive capital and energy investments to clear the land and to transport the harvested crops to distant populated areas. The resulting deforestation would greatly increase soil erosion. It would also reduce the world's precious biological diversity by eliminating vast numbers of unique plant and animal species found only in these ecosystems.

Tropical rain forests have plentiful rainfall and long or continuous growing seasons. However, their soils often are not suitable for intensive cultivation. About 90% of the plant nutrient supply is in ground litter and vegetation above the ground rather than in the soil (Figure 12-4, p. 270).

Nearly 75% of the Amazon basin, roughly one-third of the world's potential new cropland, has highly acidic and infertile soils. In addition, an estimated 5% to 15% of tropical soils (4% of those in the Amazon basin), if cleared, would bake under the tropical sun into a brick-hard surface called laterite, useless for farming.

Some tropical soils can produce up to three crops of grain per year if massive quantities of fertilizer are applied at the right time, but costs are high. Rapid runoff and leaching of soil nutrients are also problems. The warm temperatures, high moisture, and year-round growing season also support large populations of pests and diseases that can devastate monoculture crops. Massive doses of pesticides could be used, but the same conditions that favor crop growth also favor rapid development of genetic resistance in pest species (Section 14-3).

In Africa potential cropland larger in area than the United States cannot be used for farming or livestock grazing because it is infested by 22 species of the tsetse fly, whose bite can give both people and

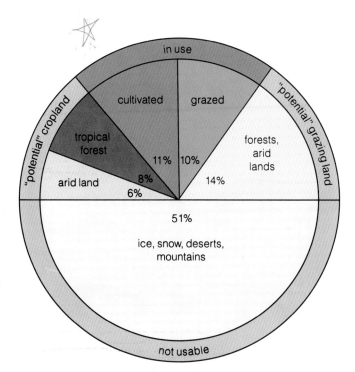

Figure 13-6 Classification of the earth's land. Theoretically, we could double in size the world's cropland by clearing tropical forests and irrigating arid lands. But converting this marginal land to cropland would destroy valuable forest resources, cause serious environmental problems, and usually not be cost effective.

livestock incurable sleeping sickness. A $120 million eradication program has been proposed, but many scientists doubt it can succeed.

Researchers hope to develop new methods of intensive cultivation in tropical areas. Some scientists, however, argue that it makes more ecological and economic sense not to use intensive cultivation in the tropics. Instead, farmers should use shifting cultivation with fallow periods long enough to restore soil fertility. Scientists also recommend plantation cultivation of rubber trees, oil palms, and banana trees, which are adapted to tropical climates and soils.

Water as a Limiting Factor Much of the world's potentially arable land lies in dry areas, where water shortages limit crop growth, especially in Africa (Figure 13-5). Large-scale irrigation in these areas would be very expensive, requiring large inputs of fossil fuel to pump water long distances. Irrigation systems would deplete many groundwater supplies and require constant and expensive maintenance to prevent seepage, salinization, and waterlogging.

There are signs that irrigation limits are being reached in land now under cultivation. Between 1950 and 1980 the world's irrigated cropland almost tripled, increasing the average irrigated area per person by 52%. However, during the 1980s growth in irrigated area slowed dramatically and fell behind population

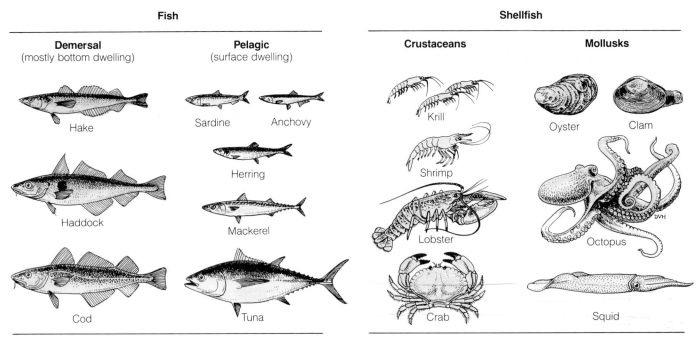

Fish		Shellfish	
Demersal (mostly bottom dwelling)	**Pelagic** (surface dwelling)	**Crustaceans**	**Mollusks**

Hake

Sardine   Anchovy

Krill

Oyster   Clam

Haddock

Herring

Shrimp

Cod

Mackerel

Lobster

Octopus

Tuna

Crab

Squid

Figure 13-7 Some major types of commercially harvested marine fish and shellfish.

growth, leading to an 8% drop in the irrigated area per person.

Do the Poor Benefit? Whether present and future green or gene revolutions reduce hunger among the world's poor depends on how new technology is applied. In LDCs the major resource available to agriculture is human labor. When green revolution techniques are used to increase yields of labor-intensive subsistence agriculture on existing or new cropland in countries with equitable land distribution, the poor benefit, as has occurred in China.

Most poor farmers, however, don't have enough land, money, or credit to buy the seed, fertilizer, irrigation water, pesticides, equipment, and fuel that the new plant varieties need. This means that the second green revolution has bypassed more than 1 billion poor people in LDCs.

Switching to industrialized agriculture makes LDCs heavily dependent on large, MDC-based multinational companies for expensive supplies, increasing the LDCs' foreign debts. It also makes their agricultural and economic systems more vulnerable to collapse from increases in oil and fertilizer prices and reduces their rates of economic growth by diverting much of their capital to pay for imported oil and other agricultural inputs. Finally, mechanization displaces many farm workers, thus increasing rural-to-urban migration and overburdening the cities (Section 6-3).

Unconventional Foods and Perennial Crops Some analysts recommend greatly increased cultivation of various nontraditional plants in LDCs to supplement

or replace traditional foods such as wheat, rice, and corn. One little-known crop is the winged bean, a protein-rich legume presently used extensively only in New Guinea and Southeast Asia. Its edible winged pods, leaves, tendrils, and seeds contain as much protein as soybeans and its edible roots contain more than four times the protein of potatoes. Indeed, this plant yields so many different edible parts that it has been called a "supermarket on a stalk."

Scientists have identified dozens of other plants that could be used as sources of food. The problem is getting farmers to cultivate such crops and convincing consumers to try new foods.

Most crops we depend on are tropical annuals. Each year the land is cleared of all vegetation, dug up, and planted with the seeds of annuals. This gives these plants a head start on weeds and other potential pests. However, David Pimentel (see Guest Essay on p. 289) and other plant scientists believe we should rely more on perennial crops, especially grains that supply 80% of all food produced worldwide. This would eliminate the need to till soil each year, greatly reducing the amount of fossil fuel, draft animal, and human energy used in agriculture each year. It would also conserve water and reduce soil erosion and sediment water pollution.

13-4 CATCHING MORE FISH AND FISH FARMING

The World's Fisheries Only 3% of the food we eat is supplied by the world's marine and freshwater fisheries. But we get an average of 24%

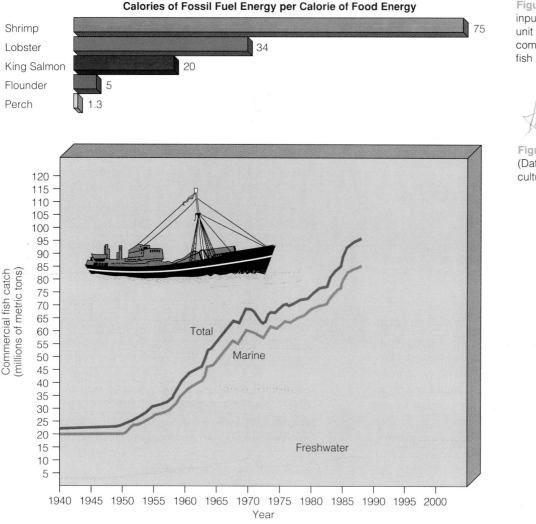

Calories of Fossil Fuel Energy per Calorie of Food Energy

Shrimp	75
Lobster	34
King Salmon	20
Flounder	5
Perch	1.3

Figure 13-8 Average energy input needed to produce one unit of food energy from some commercially desirable types of fish and shellfish.

Figure 13-9 World fish catch. (Data from UN Food and Agriculture Organization)

Commercial fish catch (millions of metric tons)

Total

Marine

Freshwater

Year

of the animal protein in our food directly from fish and shellfish and another 5% indirectly from fish meal fed to livestock. The animal protein we get from these sources is more than that from beef, twice as much as from eggs, and three times the amount from poultry. In most Asian coastal and island countries, fish and shellfish supply 30% to 90% of the animal protein eaten by people.

About 87% of the annual commercial catch of fish and shellfish comes from the ocean and the rest from fresh water. About 99% of the world marine catch is taken from plankton-rich waters within 370 kilometers (200 nautical miles) of the coast. Most of this catch comes from estuaries and upwellings where deep, nutrient-rich waters are swept up to the surface. However, this vital coastal zone is being disrupted and polluted at an alarming rate (see Spotlight on p. 97).

Only about 40 of the world's 30,000 known species of fish are harvested in large quantities. Some of the important species of fish in marine habitats are shown in Figure 13-7. Almost half of the world's commercial marine catch is taken by only five countries: Japan (16% of the catch), the USSR (13%), China (7%), the United States (6%), and Chile (6%).

More than 90% of the fish and shellfish we consume is obtained by using small and large fishing boats to hunt and gather these resources over a large area. Because 30% to 40% of the operating costs of fishing boats is spent on fuel, energy inputs for each unit of food energy obtained from most marine species are enormous (Figure 13-8).

Trends in the World Fish Catch Between 1950 and 1970, the annual commercial fish catch increased over threefold (Figure 13-9). This increase was larger than that of any other human food source during the same period. But world population also grew. This meant that between 1970 and 1986 the average fish catch per person declined in spite of slight increases in the annual harvest (Figure 13-10). Because of overfishing, pollution, population growth, and increased demand, the average catch per person is projected to drop back to the 1960 level by the year 2000.

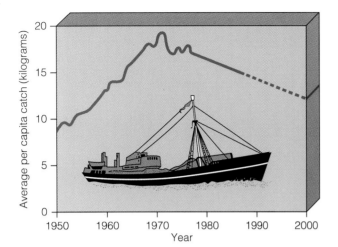

Figure 13-10 Average per capita world fish catch declined in most years since 1970. It's projected to drop further by the end of this century. (Data from United Nations and Worldwatch Institute)

Figure 13-11 Harvesting silver carp in an aquaculture farm in China.

Overfishing occurs when so many fish are taken that too little breeding stock is left to prevent a drop in numbers. Overfishing rarely causes biological extinction because commercial fishing becomes unprofitable before that point. Instead, prolonged overfishing leads to **commercial extinction,** the point at which the stock of a species is so low that it's no longer profitable to hunt and gather the remaining individuals in a specific fishery. Fishing fleets then move to a new species or to a new region, hoping that the overfished species will eventually recover.

By the early 1980s overfishing had depleted stocks of 42 valuable fisheries. Examples include cod and herring in the North Atlantic, salmon and the Alaska king crab in the northwest Pacific, and Peruvian anchovy in the southeast Pacific (see Case Study on p. 307).

The fishing industry is trying to combat the decreasing catches with more efficient fishing techniques. For example, in the Northern Pacific, more than 1,500 vessels from Japan, South Korea, Taiwan, Poland, and the Soviet Union are using driftnets to capture huge quantities of edible and inedible fish, seabirds, fur seals, dolphins, small whales, and any form of marine life that becomes entangled in the fine mesh of the nets. Each night during the two-month season, these boats set out enough nets to more than circle the world. Environmentalists have tried, without success so far, to have this method, which leads to massive biological depletion of marine life, banned.

Aquaculture **Aquaculture,** in which fish and shellfish are raised in enclosed structures for all or part of their lives, supplies about 10% of the world's commercial fish harvest. There are two major types of aquaculture. **Fish farming** involves cultivating fish in a controlled environment, usually a pond, and harvesting them when they reach the desired size (Figure 13-11). **Fish ranching** involves holding species in captivity for the first few years of their lives, releasing them, and then harvesting the adults when they return to spawn. Ranching is useful for anadromous species, such as salmon and ocean trout, which after birth move from fresh water to the ocean and then back to fresh water to spawn.

Almost three-fourths of the world's annual aquaculture catch comes from 71 LDCs. Species cultivated in LDCs include carp (Figure 13-11), tilapia, milkfish, clams, and oysters, which feed low in food webs on phytoplankton and other forms of aquatic plants. These are usually raised in small freshwater ponds or underwater cages. Aquaculture supplies 60% of the fish eaten in Israel, 40% in China, and 22% in Indonesia.

In MDCs aquaculture is used mostly to raise expensive fish and shellfish and to stock lakes and streams with game fish. This benefits anglers who fish for sport and is highly profitable for aquaculture farmers and companies. But it does little to increase food and protein supplies for the poor. In the United States, fish farms supply 40% of the oysters and most of the catfish, crawfish, and rainbow trout consumed as food.

Aquaculture has a number of advantages. It can produce high yields per unit of area. Large amounts of fuel are not required, so yields and profits are not closely tied to the price of oil, as they are in commercial fishing.

One problem, however, is that the fish can be killed by pesticide runoff from nearby croplands. Bacterial and viral infections of aquatic species can also limit aquaculture yields. Without adequate pol-

In 1953, Peru began fishing for anchovy in nutrient-rich upwellings off its western coast. The size of the fishing fleet increased rapidly. Factories were built to convert the small fish to fish meal for sale to MDCs for use as livestock feed. Between 1965 and 1971, harvests of the Peruvian anchovy made up about 20% of the world's annual commercial fish catch (Figure 13-12).

Between 1971 and 1978, however, the Peruvian anchovy became commercially extinct, only beginning a slight recovery in 1983. The collapse of this fishery is an example of how biology, geography, economics, and politics interact and often clash in fishery management.

At unpredictable intervals the productivity of the upwellings off the coast of Peru drops sharply because of a natural weather change, called the El Niño-Southern Oscillation, or ENSO, which warms the normally cool water of the Humboldt Current flowing along Peru's coast. The numbers of anchovy, other fish, seabirds, and marine mammals in food webs based on phytoplankton then drop sharply.

UN Food and Agriculture Organization biologists warned that during seven of the eight years between 1964 and 1971 the anchovy harvest exceeded the estimated sustainable yield. Peruvian fishery officials ignored these warnings. Peru's fishing industry was financed largely by short-term loans that had to be paid off. Government officials decided to risk the collapse of the fishery to pay off these loans and avoid putting thousands of people out of work. They also believed that a slight drop in the anchovy catch would be beneficial: it would cause shortages and raise the price of fish meal.

Disaster struck in 1972, when a strong ENSO arrived. The anchovy population, already at dangerously low levels due to overfishing, could not recover from the effects of the ENSO. By putting short-term economics above biology, Peru lost a major source of income and jobs and had to increase its foreign debt.

The country has made some economic recovery by harvesting the Peruvian sardine, which took over the niche once occupied by the anchovy. The catches of mackerel, bonita, and hake have also increased. Since 1983 the Peruvian anchovy fishery has been making a slight recovery.

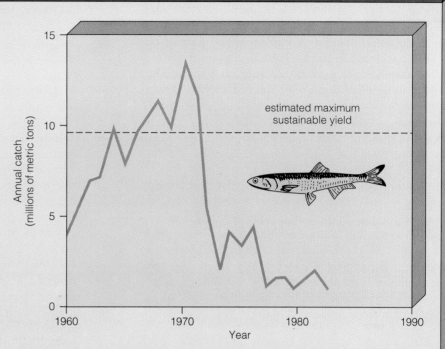

Figure 13-12   Peruvian anchovy catch, showing the combined effects of overfishing and periodic changes in climate. (Data from UN Food and Agriculture Organization)

lution control, waste outputs from shrimp farming and other large-scale aquaculture operations can also pollute nearby surface water and groundwater.

Can the Annual Catch Be Increased Significantly? Some scientists believe that the world's annual commercial catch of fish and shellfish to be at least 100 million metric tons a year. One hopeful sign was the signing of the 1982 United Nations Convention on the Law of the Sea by 159 countries. This treaty gives all coastal countries the legal right to control fishing by their own fishing fleets and by foreign ships within 364 kilometers (200 nautical miles) of their coasts. If enforced, this treaty can reduce overfishing. But 22 countries, including the Soviet Union, the United States, West Germany, and the United Kingdom, have refused to sign or ratify this treaty.

The world fish catch can also be expanded by harvesting more squid, octopus, Antarctic krill, and other unconventional species. Greatly expanded harvesting of shrimplike krill, however, could lead to sharp declines in the populations of certain whales and other species dependent on krill (Figure 4-19, p. 73). Also, food scientists haven't been able to process krill into foods that taste good enough for people to eat. Currently, krill are used to make livestock feed.

Additional increases could be brought about by a sharp decrease in the one-fifth of the annual catch now wasted, mainly from throwing back potentially useful fish taken along with desired species. More refrigerated storage at sea to prevent spoilage would also increase the catch. Experts project that annual freshwater and saltwater aquaculture production could be increased at least threefold, and possibly sixfold, between 1986 and 2000.

Other fishery experts believe that further increases in the annual marine catch are limited by overfishing and by pollution and destruction of estuaries and aquaculture ponds. Another factor that may limit the commercial fish catch from the world's oceans is the projected rise in the price of oil—and thus of boat fuel—between 1995 and 2015. Unless more seafood is produced by aquaculture, consumers may find seafood prices too high.

13-5 MAKING FOOD PRODUCTION PROFITABLE, GIVING FOOD AID, AND DISTRIBUTING LAND TO THE POOR

Government Agricultural Policies Agriculture is an especially risky business. Whether a farmer has a good or bad year is determined by factors over which the farmer has little control—weather, crop prices, crop pests and disease, interest rates, and the global market. Partly because of this and the need to have a reliable supply of food to prevent political unrest, most governments provide various forms of assistance to farmers.

Governments can influence crop and livestock prices, and thus the supply of food, in several ways.

- Keep food prices artificially low. This makes consumers happy but can decrease food production by reducing profits for farmers.

- Give farmers subsidies to keep them in business and encourage them to increase food production.

- Eliminate price controls and subsidies, allowing market competition to determine food prices and thus the amount of food produced.

Governments in many LDCs keep food prices in cities low to prevent political unrest. But low prices discourage farmers from producing enough to feed the country's population, and the government must use limited funds or go into debt to buy imported food. With food prices higher in rural areas than in cities, more rural people migrate to urban areas, aggravating urban problems and unemployment. These conditions increase the chances of political unrest, which the price control policy was supposed to prevent. However, low food prices are vital for the poor in LDCs who must spend 70% to 80% of their meager income on food.

Governments can stimulate crop and livestock production by guaranteeing farmers a certain minimum yearly return on their investment. However, if government price supports are too generous and the weather is good, farmers may produce more food than can be sold. Food prices and profits then drop because of the oversupply. The resulting availability of large amounts of food for export or food aid to LDCs depresses world food prices. The low prices reduce the financial incentive for farmers to increase domestic food production.

Unless even higher government subsidies are provided to prop up farm income by buying unsold crops or paying farmers not to grow crops on some of their land, a number of debt-ridden farmers go bankrupt. This is what happened in the United States during the 1980s. Excessive subsidies can also promote cultivation of marginal land, resulting in increased soil erosion, nutrient depletion, salinization, waterlogging, desertification, water pollution, and air pollution (by wind erosion).

In MDCs government price supports and other subsidies for agriculture farmers total more than $300 billion a year. This makes farmers and agribusiness executives happy. They are also popular with most consumers because they make food prices seem low. Politicians favor this approach because it increases their chances of staying in office. What most consumers don't realize is that they are paying higher prices for their food indirectly in the form of higher taxes to provide the subsidies. There is no free lunch.

These massive subsidies stimulate farmers to raise yields per hectare for the short term with little regard for the sustainability of the land. They send much more powerful messages to farmers than do the smaller subsidies typically provided for soil and water conservation.

Eliminating all price controls and agricultural subsidies and allowing market competition to determine food prices and production may sound like a great idea on paper. In practice, however, farmers and companies in the agriculture business with excessive economic and political influence use such power to avoid market competition. Without providing more aid for the poor, this would also increase the number of underfed and malnourished people.

Instead of eliminating subsidies, some agricultural experts suggest using subsidies to regulate crop production rather than crop acreage. This would reduce the pressure on growers to increase yields per hectare to qualify for help, thereby reducing soil erosion and excessive inputs of fertilizers, pesticides, and irrigation water. Also regulations should be set up and strictly enforced so that subsidies only go to small, needier farms instead of the present situation in which most of the money goes to large farms that do not need government payments to be profitable.

International Aid Between 1945 and 1985 the United States was the world's largest donor of nonmilitary foreign aid to LDCs. Since 1986, however, Japan has been the larger donors of such aid to LDCs. This aid is used mostly for agriculture and rural development, food relief, population planning, health, and economic development. Private charity organizations such as CARE and Catholic Relief Services and funds from benefit music concerts and record sales provide over $3 billion a year of additional foreign aid.

In addition to helping other countries, foreign aid stimulates economic growth and provides jobs in the donor country. For example, seventy cents of every dollar the United States gives directly to other countries is used to purchase American goods and services. Today 21 of the 50 largest buyers of U.S. farm goods are countries that once received free U.S. food.

Despite the humanitarian benefits and economic returns of such aid, the percentage of the U.S. gross national product used for nonmilitary foreign aid to LDCs has dropped from a high of 1.6% in the 1950s to only 0.20% in 1988—an average of only $30 per American. Sixteen other MDCs used a higher percentage of their GNP for nonmilitary foreign aid to LDCs in 1988. Some people call for greatly increased food relief for starving people from government and private sources, while others question the value of such aid (see Pro/Con at right).

Distributing Land to the Poor An important step in reducing world hunger, malnutrition, poverty, and land degradation is land reform. Land reform involves giving the landless rural poor in LDCs ownership or free use of enough arable land to produce at least enough food for their survival and ideally to produce a surplus for emergencies and for sale. China and Taiwan have had the most successful land reforms.

Such reform would increase agricultural productivity in LDCs and reduce the need to farm and degrade marginal land. It would also help reduce the flow of poor people to overcrowded urban areas by creating employment in rural areas.

Many of the countries with the most unequal land distribution are in Latin America, especially Guatemala and Brazil (see Case Study on p. 126). In Central America, 7% of the population owns 93% of

PRO/CON Is Food Relief Helpful or Harmful?

Most people view food relief as a humanitarian effort to prevent people from dying prematurely. However, some analysts contend that giving food to starving people in countries where population growth rates are high does more harm than good in the long run. By encouraging population growth and not helping people grow their own food, food relief condemns even greater numbers to premature death in the future.

Biologist Garrett Hardin (see Guest Essay on p. 145) has suggested that we use the concept of *lifeboat ethics* to decide which countries get food aid. He starts with the belief that there are already too many people in the lifeboat we call earth. If food aid is given to countries that are not reducing their population, this adds more people to an already overcrowded lifeboat. Sooner or later the boat will sink and kill most of the passengers.

Massive food aid can also depress local food prices, decrease food production, and stimulate mass migration from farms to already overburdened cities. It discourages the government from investing in rural agricultural development to enable the country to grow enough food for its population on a sustainable basis.

Another problem is that much food aid does not reach hunger victims. Transportation networks and storage facilities are inadequate, so that some of the food rots or is devoured by pests before it can reach the hungry. Typically, some of the food is stolen by officials and sold for personal profit. Some must often be given to officials as bribes for approving the unloading and transporting of the remaining food to the hungry.

Critics of food relief are not against foreign aid. Instead, they believe that such aid should be given to help countries control population growth, grow enough food to feed their population using sustainable agricultural methods (Section 13-7), or develop export crops to help pay for food they can't grow. Temporary food aid should be given only when there is a complete breakdown of an area's food supply because of natural disaster. What do you think?

the farmland. Most of this land is used for luxury export crops that do little to help the landless poor. Unfortunately, land reform is difficult to institute in countries where government leaders are unduly influenced by wealthy and powerful landowners.

13-6 SUSTAINABLE-EARTH AGRICULTURE

Sustainable-Earth Agricultural Systems
Industrialized agriculture has undeniably achieved substantial short-term gains in food production. However, it is based on eventually unsustainable practices. The bill for using the earth's resources unsustainably is now coming due in terms of eroded and nutrient-depleted soils, salted and waterlogged fields, over-fertilized and pesticide-poisoned surface waters, and drained and contaminated aquifers.

The key to reducing world hunger and the harmful environmental impacts of industrialized and traditional subsistence agriculture is to develop a variety of **sustainable-earth agricultural systems**. In these systems appropriate parts of existing industrialized and subsistence agricultural systems and new agricultural techniques would be combined to take advantage of local climates, soils, resources, and cultural systems.

A sustainable-earth agricultural system does not require large inputs of fossil fuels, promotes polyculture instead of monoculture, conserves topsoil and builds new topsoil with organic fertilizer, conserves irrigation water, and controls pests with little or no use of pesticides. This type of agriculture would increase food supplies in ways that minimize erosion, desertification, salinization, water pollution, and loss of biodiversity. This will require devising institutional mechanisms that will reward farmers for valuing the world's precious soil, water, and biodiversity.

The following are general guidelines for sustainable-earth agriculture:

- *Place primary emphasis on preserving and renewing the soil and on conserving water.* Long-term sustainability of the soil, not short-term agricultural productivity, must always come first.

- *Recognize that the marketplace cannot sustain agriculture because it does not assign an infinite value to the soil and water upon which all agriculture depends.* The marketplace is concerned with increasing short-term crop production even when this decreases the long-term sustainability of agriculture.

- *Stop thinking of and dealing with agriculture as an industry.* Agriculture involves living things and biological processes, whereas the materials of an industry are not alive and the processes are mechanical. A factory has a limited life, while the life of a farm is unlimited if its topsoil is properly used and maintained.

- *Adapt and design the agricultural system to the environment (soil, water, climate, and pest populations) of the region.* This means not trying to grow water-thirsty crops in arid and semiarid areas.

- *Emphasize small- to medium-scale intensive production of a diverse mix of fruit, vegetable, and fuelwood crops and livestock animals, rather than large-scale monoculture production of a single crop or livestock animal.*

- *Whenever possible, matter inputs should be obtained from locally available, renewable biological resources and used in ways that preserve their renewability.* Examples include using organic fertilizers from animal and crop wastes, planting fast-growing trees to supply fuelwood and add nitrogen to the soil, and building simple devices for capturing and storing rainwater for irrigating crops. Commercially produced inputs such as inorganic fertilizers and pesticides should be used only when needed and in the smallest amount possible.

- *Minimize use of fossil fuels; use locally available perpetual and renewable energy resources such as sun, wind, flowing water, and animal and crop wastes to perform as many functions as possible (Chapter 17).*

- *Emphasize use of biological pest control, windbreaks, crop rotation, green manure, and other methods that encourage beneficial organisms and discourage pests.*

- *Governments must develop agricultural development policies that include economic incentives to encourage farmers to grow enough food to meet the demand using sustainable-earth agricultural systems.*

In MDCs, such as the United States, a shift from large-scale industrialized agriculture to small- to medium-scale sustainable-earth agriculture is difficult. It would be strongly opposed by agribusiness companies, by successful farmers with large investments in industrialized agriculture, and by specialized farmers who are unwilling to learn the demanding managerial skills and agricultural knowledge needed to run a diversified farm.

The shift, however, could be brought about gradually over 10 to 20 years by a combination of methods:

- Greatly increasing government support of research and development of sustainable-earth agricultural methods and equipment. A 1989 study by the National Academy of Sciences urged U.S. farmers to begin shifting to low-input, sustainable agriculture. But this shift will be hindered as long as only 0.3% of the Department of Agriculture's annual research budget is used for this purpose.

- Setting up demonstration projects in each county so that farmers can see how sustainable systems work.

- Establishing training programs for farmers, county farm agents, and most Department of Agriculture personnel.

- Look at your lifestyle to find ways to reduce your unnecessary use and waste of food, fertilizers, and pesticides. Recognize that agriculture has a greater environmental impact than any other human activity (see Spotlight on p. 301).

- Eat lower on the food chain by eliminating or reducing meat consumption. This saves money and energy and can reduce your intake of fats that contribute to heart disease and other disorders (see Spotlight on p. 181). It also reduces air and water pollution and overgrazing. Growing grains, fruits, and vegetables uses 95% fewer raw materials than does meat production. If Americans reduced their meat intake by just 10%, the savings in grain and soybeans could adequately feed 60 million people. The amount of land needed to feed one typical meat eater could feed 20 people eating a balanced vegetarian diet, and each time an American switches to a pure vegetarian diet, 0.4 hectare (1 acre) of U.S. trees is saved.

- Use sustainable-earth cultivation techniques to grow some of your own food in a backyard plot, a window planter, a rooftop garden, or a cooperative community garden. Spending $31 to plant a living-room size garden can give you vegetables worth about $250—a better return than almost any financial investment you can make. In 1989 30% of the U.S. population grew fruits, vegetables, and herbs at home, with 57% of these foods grown organically.

- Fertilize your crops primarily with organic fertilizer produced in a compost bin (Figure 12-17, p. 279). This recycles most of your food and yard wastes instead of adding them to the growing problem of solid waste (Section 19-5). Use small amounts of commercial inorganic fertilizer only when supplies of certain plant nutrients are inadequate.

- Control pests by a combination of cultivation and biological methods (see Section 14-5). Use carefully selected chemical pesticides in small amounts only when absolutely necessary.

- Help reduce the use of pesticides on agricultural products by asking grocery stores to stock fresh produce and meat produced by organic methods (without the use of commercial fertilizers and pesticides). Insist that such foods have been tested to certify that they were grown by organic methods. Currently about 2% of U.S. farmers grow about 5% of the country's crops using organic methods. This $3-billion-a-year business is the fastest growing sector in U.S. agriculture because of rapidly increasing consumer demand.

- Recognize that organically grown fruits and vegetables may have a few holes, blemishes, or frayed leaves, but they taste just as good and are just as nutritious as more perfect-looking products (on which pesticides were used). Eating organically grown food is also a form of health insurance. You won't be ingesting small amounts of pesticides, which can also contaminate groundwater.

- Reduce unnecessary waste of food. An estimated 25% of all food produced in the United States is wasted; it rots in the supermarket or refrigerator or is thrown away off the plate. Put no more food on your plate than you intend to eat, ask for smaller portions in restaurants, and support programs that collect excess food from restaurants and school cafeterias to distribute to the poor and homeless.

- Exert pressure on candidates for public office and elected officials to support policies designed to develop and encourage sustainable-earth agricultural systems in the United States and throughout the world.

- Support efforts to regulate and slow down population growth (Section 6-4). Ultimately, the size of the world population will determine the need for food and the massive environmental impacts of producing this food.

- Establishing college curricula for sustainable-earth agriculture.

- Giving subsidies and tax breaks to farmers using sustainable agriculture and to agribusiness companies developing products for this type of farming.

Each of us has a role to play in bringing about a shift from unsustainable to sustainable agriculture at the local, national, and global levels (see Individuals Matter above).

The most important fact of all is not that people are dying from hunger, but that people are dying unnecessarily. . . . We have the resources to end it; we have proven solutions for ending it. . . . What is missing is the commitment.

WORLD HUNGER PROJECT

DISCUSSION TOPICS

1. What are the major advantages and disadvantages of
   a. labor-intensive subsistence agriculture?
   b. energy-intensive industrialized agriculture?
   c. sustainable-earth agriculture?

2. What specific actions should the following groups take to reduce the poverty that is the leading cause of hunger, malnutrition, and greatly increased chances of premature death for one out of five people living in LDCs:
   a. governments of LDCs?
   b. the U.S. government?
   c. private international aid organizations?
   d. individuals such as yourself?
   e. the poor?

3. Summarize the advantages and limitations of each of the following proposals for increasing world food supplies and reducing hunger over the next 30 years:
   a. cultivating more land by clearing tropical jungles and irrigating arid lands
   b. catching more fish in the open sea
   c. producing more fish and shellfish with aquaculture
   d. increasing the yield per area of cropland

4. Should price supports and other federal subsidies paid to U.S. farmers out of tax revenues be eliminated? Explain. Try to have one or more farmers discuss this problem with your class.

5. Is sending food to famine victims helpful or harmful? Explain. Are there any conditions you would attach to sending such aid? Explain.

6. Should tax breaks and subsidies be used to encourage more U.S. farmers to switch to sustainable-earth farming? Explain.

7. Do you eat meat? Would you be willing to go to a slaughterhouse to kill a cow or pig?

FURTHER READINGS

Amato, Paul R., and Sonia A. Partridge. 1989. *The New Vegetarians: Promoting Health and Protecting Life.* New York: Plenum.

Bardach, John. 1988. "Aquaculture: Moving from Craft to Industry." *Environment,* vol. 30, no. 2, 7–40.

Berry, Wendell. 1990. *Nature as Measure.* Berkeley, Calif.: North Point Press.

Brown, Lester R., and John E. Young. 1990. "Feeding the World in the Nineties." Pp. 59–78 in Lester R. Brown, et al., *State of the World 1990.* Washington, D.C.: Worldwatch Institute.

Crosson, Pierre R., and Norman J. Rosenberg. 1989. "Strategies for Agriculture." *Scientific American,* Sept., 128–135.

Dover, Michael J., and Lee M. Talbot. 1988. "Feeding the Earth: An Agroecological Solution." *Technology Review,* Feb./Mar., 27–35.

Doyle, Jack. 1985. *Altered Harvest: Agriculture, Genetics, and the Fate of the World's Food Supply.* New York: Viking.

Edwards, Clive A., et al., eds. 1990. *Sustainable Agricultural Systems.* Ankeny, Iowa: Soil and Water Conservation Society.

Ehrlich, Anne H. 1988. "Development and Agriculture." Pp. 75–100 in Paul R. Ehrlich and John P. Holdren, eds., *The Cassandra Conference: Resources and the Human Predicament.* Texas Station: Texas A&M University Press.

Granatstein, David. 1988. *Reshaping the Bottom Line: On-Farm Strategies for a Sustainable Agriculture.* Stillwater, Minn.: Land Stewardship Project.

Hamilton, Geoff. 1987. *The Complete Guide to Growing Flowers, Fruits, and Vegetables Naturally.* New York: Crown.

Jackson, Wes. 1980. *New Roots for Agriculture.* San Francisco: Friends of the Earth.

Lappé, Francis M., et al. 1988. *Betraying the National Interest.* San Francisco: Food First.

Mollison, Bill. 1988. *Permaculture: A Designer's Manual.* Davis, Calif.: AgAccess.

Molnar, Joesph J., and Henry Kinnucan, ed. 1989. *Biotechnology and the New Agricultural Revolution.* Boulder, Colo.: Westview Press.

National Academy of Sciences. 1989. *Alternative Agriculture.* Washington, D.C.: National Academy Press.

Office of Technology Assessment. 1988. *Enhancing Agriculture in Africa.* Washington, D.C.: Office of Technology Assessment.

Pimentel, David. 1987. "Down on the Farm: Genetic Engineering Meets Ecology." *Technology Review,* Jan., 24–30.

Pimentel, David, and Carl W. Hall. 1989. *Food and Natural Resources.* Orlando, Fla.: Academic Press.

Reganhold, John P., et al. 1990. "Sustainable Agriculture." *Scientific American,* June, 112–120.

Robbins, John. 1987. *Diet for a New America.* Waldpole, N.H.: Stillpoint Publishing.

Rodale Press. 1984. *The Encyclopedia of Organic Gardening.* Emmaus, Penn.: Rodale Press.

Sanchez, Pedro A., and Jose R. Benites. 1987. "Low-Input Cropping for Acid Soils of the Humid Tropics." *Science,* vol. 238, 1521–1527.

Schell, Orville. 1984. *Modern Meat: Antibiotics, Hormones, and the Pharmaceutical Farm.* New York: Random House.

Tudge, Colin. 1988. *Food Crops for the Future: Development of Plant Resources.* Oxford, UK: Blackwell.

United Nations World Food Commission. 1989. *The Global State of Hunger and Malnutrition.* New York: United Nations.

Wolf, Edward C. 1986. *Beyond the Green Revolution: New Approaches for Third World Agriculture.* Washington, D.C.: Worldwatch Institute.

CHAPTER FOURTEEN

PROTECTING FOOD RESOURCES: PESTICIDES AND PEST CONTROL

General Questions and Issues

1. What major types of pesticides are being used?

2. What are the advantages of using insecticides and herbicides?

3. What are the disadvantages of using insecticides and herbicides?

4. How is pesticide usage regulated in the United States?

5. What alternatives are there to using pesticides?

_____

A weed is a plant whose virtues have not yet been discovered.

RALPH WALDO EMERSON

A PEST IS ANY UNWANTED organism that directly or indirectly interferes with human activity. Pests compete with people for food, and some spread disease. Only about 100 of the at least 1 million cataloged insect species cause about 90% of the damage to food crops. In diverse ecosystems, their populations are kept in control by a variety of natural enemies.

Since 1945 vast fields planted with only one crop or only a few crops, as well as homes, home gardens, and lawns, have been treated with a variety of chemicals called **pesticides** (or *biocides*)—substances that can kill organisms that we consider to be undesirable. The most widely used types of pesticides are

- **herbicides** to kill weeds, unwanted plants that compete with crop plants for soil nutrients

- **insecticides** to kill insects that consume crops and food and transmit diseases to humans and livestock

- **fungicides** to kill fungi that damage crops

- **rodenticides** to kill rodents, mostly rats and mice

There is controversy over whether the harmful effects of these chemicals outweigh their benefits compared to other alternatives, as discussed in this chapter.

14-1 PESTICIDES: TYPES AND USES

The Ideal Pesticide The ideal pest-killing chemical would

- kill only the target pest
- have no short- or long-term health effects on nontarget organisms, including people *effects*
- be broken down into harmless chemicals in a fairly short time *broken down*
- prevent the development of genetic resistance in target organisms
- save money compared to making no effort to control pest species *$$*

Unfortunately, no known pest control method meets all these criteria.

Use of Pesticides Since 1945, chemists have developed many different types of synthetic organic chemicals for use as pesticides. Worldwide, about 2.3 million metric tons (2.5 million tons) of these pesticides are used each year—an average of 0.45 kilogram (1 pound) for each person on earth. About 85% of all pesticides are used in MDCs, but use in LDCs is growing rapidly and is projected to increase at least fourfold between 1985 and 2000. Global sales of

852) → MDC

Table 14-1 Major Types of Insecticides

Type	Examples	Persistence
1) Chlorinated hydrocarbons	DDT, aldrin, dieldrin, endrin, hepta-chlor, toxaphene, lindane, chlordane, kepone, mirex	High (2–15 years)
2) Organophosphates	Malathion, parathion, monocrotophos, methamidophos, methyl parathion, DDVP	Low to moderate (normally 1–12 weeks, but some can last several years)
3) Carbamates	Carbaryl, maneb, priopoxor, mexica-bate, aldicarb, aminocarb	Usually low (days to weeks)
4) Pyrethroids	Pemethrin, decamethrin	Usually low (days to weeks)

Hutchison Library

Figure 14-1 The heads of these pyrethrum daisy flowers being grown in Kenya are ground into a powder used as a commercial insecticide or converted to other pyrethroid insecticides.

pesticides have risen from $3 billion in 1970 to $19 billion in 1990.

In the United States, about 600 biologically active ingredients and 1,200 inert (presumably biologically inactive) ingredients are mixed to make some 50,000 individual pesticide products. Between 1964 and 1989 pesticide use in the United States almost tripled but has levelled off at around 373 million kilograms (820 million pounds) since 1980. At this rate, each year an average of 1.5 kilograms (3.3 pounds) of these products is used for each American.

Herbicides account for 66% of all pesticide use in the United States, insecticides 23%, and fungicides 11%. Four crops—corn, cotton, wheat, and soybeans—account for about 70% of the insecticides and 80% of the herbicides used on crops in the United States.

About 20% of the pesticides used each year in the United States are applied to lawns, gardens, parks, golf courses, and cemeteries. The average home owner in the United States applies about five times more pesticide per unit of land area than do farmers.

Major Types of Insecticides and Herbicides Most of the thousands of different insecticides used today fall into one of four classes of compounds: chlorinated hydrocarbons, organophosphates, carbamates, or pyrethroids (Table 14-1). These chemicals vary widely in their persistence, the length of time they remain active in killing insects (Table 14-1).

By the mid-1970s DDT and most other slowly degradable, chlorinated hydrocarbon insecticides shown in Table 14-1 (except lindane) were banned or severely restricted in the United States and most MDCs. However, many of these compounds are still produced in the United States and exported to other countries, mostly LDCs, where they have not been banned.

In the United States and most MDCs chlorinated hydrocarbon insecticides have been replaced by a number of more rapidly degradable pesticides, especially organophosphates and carbamates (Table 14-1). But some of these compounds (such as parathion) are more toxic to birds, people, and other mammals than the chlorinated hydrocarbon insecticides they replaced. They are also more likely to contaminate surface water and groundwater because they are water soluble, whereas chlorinated hydrocarbon insecticides are insoluble in water but soluble in fats. Furthermore, to compensate for their fairly rapid breakdown, farmers usually apply nonpersistent insecticides at regular intervals to ensure more effective insect control. This means they are often present in the environment almost continuously, like the slowly degradable pesticides they replaced.

Another source of insecticides is based on learning how wild plants, especially tropical species, produce chemical compounds that repel insects or inhibit

Table 14-2 Major Types of Herbicides

Type	Examples	Effects
Contact	Triazines such as atrazine and paraquat	Kills foliage by interfering with photosynthesis
Systemic	Phenoxy compounds such as 2,4-D, 2,4,5-T, and Silvex; substitute ureas such as diuron, norea, fenuron, and other nitrogen-containing compounds such as daminozide (Alar), glyphosate	Absorption creates excess growth hormones; plants die because they cannot obtain enough nutrients to sustain their greatly accelerated growth
Soil sterilants	Trifluralin, diphenamid, dalapon, butylate	Kills soil microorganisms essential to plant growth; most also act as systemic herbicides.

chemical compounds
pyrethrins

their feeding. There are two major types of these compounds—pyrethrins from wild chrysanthemum-type plants (Figure 14-1) and rotenoids produced by the roots of rain-forest legumes. Both types of compounds are biodegradable, effective at low doses, and cause little harm to higher animals such as birds and mammals, including humans.

Herbicides can be placed into three classes based on their effect on plants: contact herbicides, systemic herbicides, and soil sterilants (Table 14-2). Most herbicides are active for only a short time. In the United States and most MDCs the use of 2,4,5-T and Silvex has been banned.

14-2 THE CASE FOR PESTICIDES

Proponents of pesticides believe that the benefits of pesticides outweigh their harmful effects. They point out the following benefits:

- *Pesticides save lives.* Since World War II, DDT and other chlorinated hydrocarbon and organophosphate insecticides have probably prevented the premature deaths of at least 7 million people from insect-transmitted diseases such as malaria (carried by the *Anopheles* mosquito), bubonic plague (rat fleas), typhus (body lice and fleas), and sleeping sickness (tsetse fly). *saves lives.*

- *They increase food supplies and lower food costs.* Each year about 55% of the world's potential food supply is lost to pests before (35%) and after (20%) harvest. Proponents argue that without pesticides these losses would be much higher and food prices would increase (perhaps by 30% to 50% in the United States).

- *They increase profits for farmers.* In the United States 42% of the annual potential food supply is destroyed by pests before and after harvest. Pesticide companies estimate that every $1 spent on pesticides leads to an increase in crop yield worth $3 to $5 to farmers.

farmers more money

efficient

- *They work faster and better than other alternatives.* Compared to alternative methods of pest control, pesticides can control most pests quickly and at a reasonable cost, have a relatively long shelf life, are easily shipped and applied, and are safe when handled properly. When genetic resistance occurs in pest insects and weeds, farmers can usually keep them under control by using stronger doses or switching to other pesticides.

- *Safer and more effective products are continually being developed.* Pesticide company scientists are continually developing pesticides, such as pyrethroids (Table 14-1), that are safer to use and that cause less ecological damage. New herbicides are being developed that are effective at very low dosage rates. Genetic engineering also holds promise. But the research and development cost for a single pesticide has risen from $6 million in 1976 to $50 million today, explaining why pesticide prices have risen sharply.

Genetic Resistance

14-3 THE CASE AGAINST PESTICIDES

Development of Genetic Resistance The most serious drawback to using chemicals to control pests is that most pest species, especially insects, can develop genetic resistance to a chemical poison through natural selection. When an area is sprayed with a pesticide, most of the pest organisms are killed. However, a few organisms in a given population of a particular species survive because by chance they already had genes that made them resistant or immune to a specific pesticide.

Because most pest species—especially insects and disease organisms—have short generation times, a few surviving organisms can produce a large number of similarly resistant offspring in a short time. For example, the boll weevil (Figure 14-2), a major cotton pest, can produce a new generation every 21 days.

When populations of offspring of resistant parents are repeatedly sprayed with the same pesticide, each succeeding generation contains a higher percentage of resistant organisms. Thus, eventually widely used pesticides (especially insecticides) fail because of genetic resistance and usually lead to even larger populations of pest species, especially insects with large numbers of offspring and short generation times. In temperate regions most insects develop genetic resistance to a chemical poison within five to ten years and much sooner in tropical areas. Weeds and plant disease organisms also develop genetic resistance, but not as quickly as most insects.

Between 1950 and 1990 almost 500 major insect pest species have developed genetic resistance to one or more insecticides and at least 20 insect species are now apparently immune to all widely used insecticides. It is estimated that by the year 2000 virtually all major insect pest species will show some form of genetic resistance. About 80 species of the more than 500 major weed species are resistant to one or more herbicides. Because half of all pesticides applied worldwide are herbicides, genetic resistance in weeds is expected to increase significantly. Genetic resistance has also appeared in 70 species of fungi treated with fungicides and in 10 species of rodents (mostly rats) treated with rodenticides.

Because of genetic resistance, most widely used insecticides no longer protect people from insect-transmitted diseases in many parts of the world, leading to even more serious disease outbreaks. This is the major reason for the almost 40-fold increase in malaria between 1970 and 1988 in 84 tropical and subtropical countries (see Case Study on p. 172).

Killing of Natural Pest Enemies and Conversion of Minor Pests to Major Pests Most insecticides are broad-spectrum poisons that kill not only the target pest species but also a number of natural predators and parasites that may have been maintaining the pest species at a reasonable level. Without sufficient natural enemies, and with much food available, a rapidly reproducing insect pest species can make a strong comeback a few days or weeks after being initially controlled.

The use of broad-spectrum insecticides also kills off the natural enemies of many minor pests. Then their numbers can increase greatly and they become major pests.

The Pesticide Treadmill When genetic resistance develops, sales representatives from pesticide companies usually recommend more frequent applications, stronger doses, or switching to new (usually

Figure 14-2 Controlling the cotton boll weevil accounts for at least 25% of the pesticides used in the United States, but farmers are now increasing their use of natural predators to control this major pest.

more expensive) chemicals to keep the resistant species under control, rather than suggesting nonchemical alternatives. This puts farmers on an accelerating **pesticide treadmill,** in which they pay more and more for a pest control program that becomes less and less effective. For example, between 1940 and 1984 crop losses to insects in the United States increased from 7% to 13%, while there was a twelve-fold increase in the application of insecticides. Corn losses to insects more than tripled from 3.5% to 12% between 1945 and 1985, despite a thousand-fold increase in the amount of insecticides used on corn crops.

Worldwide, insects and weeds reduce crop production by about 30%, about the same as before modern pesticides were used. The development of new chemicals by pesticide companies to reduce the spread of genetic resistance lags far behind the development of such resistance, especially in insects and plant pathogens.

David Pimentel (see Guest Essay on p. 289), an expert in insect ecology, estimates that when the $1 billion to $2 billion a year, environmental, health and other external costs of pesticides are included, they end up saving U.S. farmers somewhere between nothing and $2.40 for each $1 invested. This is much lower than the pesticide companies' estimate of $3 to $5 saved for each $1 invested.

A 1989 study by Pimentel found that cutting pesticide use by 50% on forty different crops in the United States would save farmers $3.3 billion and food costs would rise only 0.5%, not the 40% to 50% projected by pesticide companies. Simple economics explains why an increasing number of farmers are

Figure 14-3 Crop duster spraying a pesticide on grapevines south of Fresno, California, with fungicide. No more than 10%, and often as little as 0.1%, of the chemical being applied reaches the target organisms. Aircraft are used to apply 60% of the pesticides used on cropland in the United States.

National Archives/EPA Documerica

reducing pesticide use and increasing their use of alternatives (Section 14-5).

10% - used 90% wasted

Mobility and Biological Amplification of Persistent Pesticides Pesticides don't stay put. No more than 10% of the pesticides applied to crops by aerial spraying (Figure 14-3) or ground spraying reaches the target pests. The remaining 90% or more ends up in the soil, air, surface water, groundwater, bottom sediment, food, and nontarget organisms, including people and even penguins in the Antarctic. Pesticide contamination of groundwater and surface waters is a serious and growing problem.   Contamination

According to pesticide expert David Pimentel, often less than 0.1% of insecticides and 5% of herbicides applied to crops reach target pests. Pesticides reaching the atmosphere, especially those applied by airplane or helicopter, can be carried long distances. Concentrations of fat-soluble, slowly degradable insecticides such as DDT and other chlorinated hydrocarbons (Table 14-1) can be biologically amplified thousands to millions of times in food chains and webs (see Spotlight on p. 318).

Short-Term Threats to Human Health from Pesticide Use and Manufacture The World Health Organization estimates that each year between 400,000 and 2 million people are poisoned by pesticides and 5,000 to 20,600 of them die. At least half of those poisoned and 75% of those killed are farm workers. In LDCs, where educational levels are low, warnings are few and pesticide regulation-and-control methods are often lax or nonexistent. The actual number of pesticide-related illnesses among farm workers in the United States and throughout the world is probably greatly underestimated because of poor records, lack of doctors and reporting in rural areas, and faulty diagnoses.   400,000 → 2 million poisoned

Each year at least 45,000 Americans, mostly children, are treated and about 50 die from pesticide poisonings, mostly caused by unsafe use or storage of pesticides in and around the home. In the United States, pesticides are the second leading cause of poisoning in young children, after medicines. Accidents and unsafe practices in pesticide plants can expose workers, their families, and even the general public to harmful levels of pesticides or chemicals used in their manufacture (see Case Study above).

???

Long-Term Threats to Human Health Traces of almost 500 of the 700 active ingredients used in pesticides in the United States show up in the food most people eat. The results of this long-term worldwide experiment, with people involuntarily playing

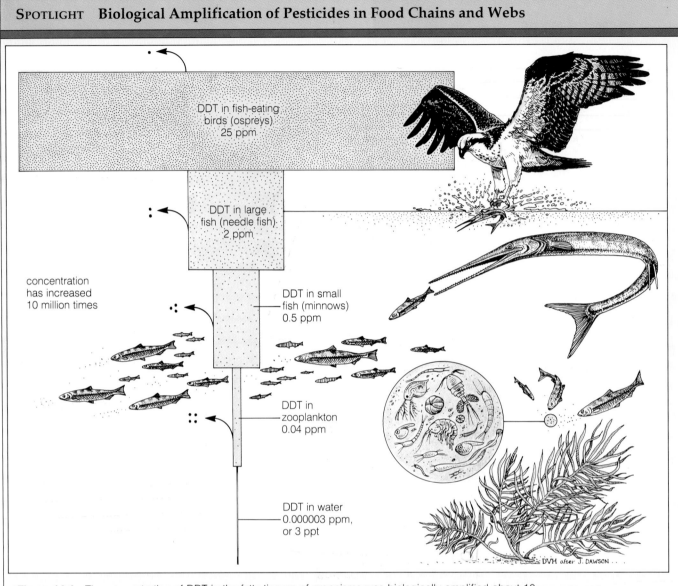

DDT in fish-eating
birds (ospreys)
25 ppm

DDT in large
fish (needle fish)
2 ppm

concentration
has increased
10 million times

DDT in small
fish (minnows)
0.5 ppm

DDT in
zooplankton
0.04 ppm

DDT in water
0.000003 ppm,
or 3 ppt

DVH after J. DAWSON

Figure 14-4 The concentration of DDT in the fatty tissues of organisms was biologically amplified about 10 million times in this food chain of an estuary adjacent to Long Island Sound near New York City. Dots represent DDT and arrows show small losses of DDT through respiration and excretion.

the role of guinea pigs, may never be known because it is almost impossible to determine that a specific chemical caused a particular cancer or other harmful effect (Section 8-1).

In 1987 the National Academy of Sciences reported that the active ingredients in 90% of all fungicides, 60% of all herbicides, and 30% of all insecticides in use in the United States may cause cancer in humans. According to the worst-case estimate in this study, exposure to pesticides in food causes 4,000 to 20,000 cases of cancer a year in the United States. In 1987 the EPA ranked pesticide residues in foods as the third most serious environmental health threat in the United States (after worker exposure and indoor radon) in terms of cancer risk.

14-4 PESTICIDE REGULATION IN THE UNITED STATES

Is the Public Adequately Protected? Because of the potentially harmful effects of pesticides on wildlife and people, Congress passed the Federal Insecticide, Fungicide, and Rodenticide Act (FIFRA) in 1972. This law, which was amended in 1975, 1978, and 1988, requires that all commercial pesticides be approved for general or restricted use by the Environmental Protection Agency.

Approval is based primarily on an evaluation of the safety of the 600 chemicals that pesticide companies designate as biologically active ingredients. These data are submitted to the EPA by the companies seeking approval.

Biological amplification occurs when the concentrations of certain chemicals in the fatty tissues of organisms increase at each successive trophic level in a food chain or web. Some synthetic chemicals, such as the pesticide DDT, PCBs, some radioactive materials, and some toxic mercury and lead compounds have the three properties needed for biological amplification. They are insoluble in water, soluble in fat, and are slowly degraded or not degraded by natural processes.

Figure 14-4 shows the biological amplification of DDT in a five-step food chain of an estuary ecosys-

Hans Reinhard/Bruce Coleman Ltd.

tem. High concentrations of DDT or other slowly degraded, fat-soluble chemicals can reduce populations of species in several ways. It can directly kill the organisms, reduce their ability to reproduce, or weaken them so that they are more vulnerable to diseases, parasites, and predators.

During the 1950s and 1960s, populations of ospreys, cormorants, eastern and California brown pelicans, and bald eagles declined drastically. These birds feed mostly on fish at the top of long aquatic food chains and webs and thus ingest large quantities of biologically amplified DDT found in their prey.

Populations of predatory birds such as prairie falcons, sparrow hawks, Bermuda petrels, and peregrine falcons (Figure 14-5) also dropped when they ate animal prey containing DDT. These birds control populations of rabbits, ground squirrels, and other crop-damaging small mammals. Only about 1,000 peregrine falcons are

Figure 14-5 The peregrine falcon is endangered in the United States, mostly because of exposure to DDT. The insecticide caused many young birds to die before hatching because their eggshells were too thin to protect them.

left in the lower 48 states. Most of them were bred in captivity and then released into the wild.

Research has shown that these population declines occurred because DDE, a chemical produced by the breakdown of DDT, accumulated in the bodies of the affected bird species. This chemical reduces the amount of calcium in the shells of their eggs. As a result, the shells are so thin that many of them break, and the unhatched chicks die.

Since the U.S. ban on DDT in 1972, populations of most of these bird species have made a comeback. In 1980, however, it was discovered that levels of DDT and other banned pesticides were rising in some areas and in some species such as the peregrine falcon and the osprey.

These species may be picking up biologically amplified DDT and other chlorinated hydrocarbon insecticides in Latin American countries, where the birds live during winter. In those countries the use of such chemicals is still legal. Illegal use of DDT and other banned pesticides in the United States may also play a role.

Since 1972 → ban 40 pesticides

Since 1972 the EPA has used this law to ban the use, except for emergency situations, of over 40 pesticides because of their potential hazards to human health. The banned chemicals include most chlorinated hydrocarbon insecticides, several carbamates and organophosphates, and several herbicides such as 2,4,5-T and Silvex.

However, according to a 1988 report by the National Academy of Sciences, federal laws regulating the use of pesticides in the United States are inadequate and poorly enforced by the Food and Drug Administration (FDA) and the EPA (see Spotlight on p. 320). According to another National Academy of Science study, by 1984 only 10% of the 600 active ingredients and none of the 1,200 inert ingredients

used in U.S. pesticide products had been tested well enough to determine their potential for producing cancer, genetic mutations, birth defects, and neurological damage in humans.

14-5 ALTERNATIVE METHODS OF INSECT CONTROL

Modifying Cultivation Procedures Opponents of the widespread use of pesticides argue that there are many safer, and in the long-run cheaper and more effective, alternatives to the use of pesticides by farmers and home owners. For centuries farmers have used cultivation methods that discourage or inhibit pests. Examples are

Numerous studies by the National Academy of Sciences and the General Accounting Office have shown that the weakest and most poorly enforced U.S. environmental law in the United States is the Federal Insecticide, Fungicide, and Rodenticide Act (FIFRA) of 1972 and its subsequent amendments.

This act required the EPA to reevaluate the 700 active ingredients approved for use in pesticide products before 1972 to determine whether any of these substances caused cancer, birth defects, or other health risks. The EPA was supposed to complete this analysis by 1975. However, by 1989 the EPA had developed data on only 192 of these chemicals and had completed its review on only two of them. In 1987 Congress extended the deadline for completing this review to 1997. The EPA claims that Congress has not appropriated enough money for it to do the job.

According to the National Academy of Sciences, up to 98% of the potential risk of developing cancer from pesticide residues on food grown in the United States would be eliminated if the EPA set the same stricter standards for pesticides registered for use before 1972 as they have for those registered after 1972.

It has also become clear that many of the 1,200 so-called inert or biologically inactive ingredients in pesticide products are in fact biologically active and can cause harm to people and some forms of wildlife. So far, most of these chemicals have not been tested and none of them have been banned, mostly because they are not covered by the present pesticide law.

This law also allows the EPA to leave inadequately tested pesticides on the market and to license new chemicals without full health and safety data. It also gives the EPA unlimited time to remove a chemical when its health and environmental effects are shown to outweigh its economic benefits. The appeals and procedures built into the law often allow a dangerous chemical to remain on the market for up to ten years.

The EPA can immediately cancel the use of a chemical on an emergency basis. Until 1990, however, the law required the EPA to use its already severely limited funds to compensate pesticide manufacturers for their remaining inventory and for all the costs of storing and disposing of the banned pesticide. This provision made it very difficult for the EPA to cancel a chemical quickly. For just one chemical, the cost of compensation could amount to more than the agency's entire pesticide budget for a single year. Usually, therefore, the only economically feasible solution has been for the EPA to allow to be sold existing stocks of a chemical that should be banned immediately.

After 20 years of pressure from environmentalists, Congress passed several new amendments to the federal pesticide law in 1988. One of these amendments shifts some, but not all, of the costs of banning and disposing of banned pesticides from the EPA to companies making the chemicals.

Environmentalists consider the 1988 amendments better than nothing. But they point out that the law still has numerous weak-

---

- *crop rotation,* in which the types of crops planted in fields are changed from year to year so that populations of pests that attack a particular crop don't have time to multiply to uncontrollable sizes

- *planting rows of hedges or trees in and around crop fields* to act as barriers to invasions by insect pests, provide habitats for their natural enemies, and serve as windbreaks to reduce soil erosion (Figure 12-14, p. 277)

- *adjusting planting times* to ensure that most major insect pests starve to death before the crop is available, or are consumed by their natural predators

- *growing crops in areas where their major pests do not exist*

- *switching from monocultures to modernized versions of intercropping, agroforestry, and polyculture* that use plant diversity to help control pests

- *disposing of diseased or infected plants*

Unfortunately, to increase profits, qualify for government subsidies, and in some cases to avoid bankruptcy, many farmers in MDCs such as the United States have abandoned these cultivation methods.

Artificial Selection, Crossbreeding, and Genetic Engineering Varieties of plants and animals that are genetically resistant to certain pest insects, fungi, and diseases can be developed. New varieties usually take a long time (10 to 20 years) to develop by conventional methods and are costly.

But insect pests and plant diseases can develop new strains that attack the once-resistant varieties, forcing scientists to continually develop new resistant strains. Genetic engineering techniques are now being used to develop resistant crops (Figure 14-6) and animals more rapidly (see Pro/Con on p. 177).

Biological Control Various natural predators (Figure 14-7), parasites, and pathogens (disease-causing bacteria and viruses) can be introduced or imported to

nesses and loopholes. One loophole allows the sale in the United States of a number of insecticide products containing as much as 15% DDT by weight, classified as an impurity. These products, along with others illegally smuggled into the United States (mostly from Mexico), are believed to be responsible for increases in DDT levels in some vulnerable forms of wildlife and on some fruits and vegetables grown and sold in the United States (especially in California).

Also, this law is the only major environmental statute that does not allow citizens to sue violators. Citizen lawsuits are an essential tool to assure government compliance with a law. Environmentalists consider it a human tragedy that the U.S. pesticide control statute is the nation's weakest and most poorly enforced environmental law since the EPA considers pesticide residues in food the third most serious environmental health threat in the country in terms of cancer risk. This is a testimony to the economic and political power of pesticide producers and the unwillingness of citizens to elect people who will protect their health.

Each year the Food and Drug Administration inspectors check less than 1% (about 12,000 samples) of domestic and imported food for pesticide contamination. The FDA's turnaround for food analysis is so long that about half of contaminated foods have been sold and eaten by the time the contamination is detected. Even when contaminated food is found, the growers and importers are rarely penalized.

Pesticide companies can make and export to other countries pesticides that have been banned in the United States or that have not been submitted to the EPA for approval. The United States leads the world in pesticide exports, followed by West Germany and the United Kingdom. About one-fourth of the 227,000 metric tons (250,000 tons) of pesticides exported by U.S. companies each year are banned from use in the United States. In what environmentalists call a *circle of poison*, residues of some of these banned chemicals return to the United States as pesticide residues on imported coffee, fruits, and vegetables.

Environmentalists have pressured Congress to halt export of banned pesticides and bar imports of food treated with these chemicals. They believe that is morally wrong for the United States to export to other countries pesticides that we have determined to be a serious risk to human health.

President George Bush opposes such a ban. He argues that such a decision should be a two-way decision between the United States and the importing country, that unilateral bans will not work, and, because the chemicals can be purchased from other countries anyway, why should the United States lose this business?

What, if anything, do you think should be done to provide more protection for the public from contamination of food and drinking water by traces of numerous pesticides?

regulate the populations of specific pests. Worldwide, more than 300 biological pest control projects have been successful, especially in China and the Soviet Union. In Nigeria, crop-spraying planes release parasitic wasps instead of pesticides in a biological assault on the cassava mealbug. These farmers get a $178 return for every $1 they spend on the wasps.

A bacterial agent (*Bacillus thuringiensis*) is a registered pesticide sold commercially as a dry powder. By choosing which of the thousands of strains of this microorganism to use, companies selling this biological agent can tailor their products to combat a variety of pests. Sales are soaring and by 1990 there were 42 of these and other biological pesticides on the market.

A gene from *Bacillus thuringiensis* has been transferred to cotton plants. These plants then produce a protein that disrupts the digestive system of pests. Insects that bite the plant die within a few hours.

Researchers have also sealed insect nematodes, parasitic worms that kill many household and agricultural insect pests, in a gelatinous capsule that also contains a specific insect attractant, a feeding stimu-

Figure 14-6 Use of genetic engineering to reduce pest damage. Both of these tomato plants were exposed to destructive caterpillars. The foliage on the normal tomato plant on the left has been almost completely eaten, while the genetically engineered plant on the right shows few signs of damage.

lant, and food for the nematodes. When a target insect eats the tiny capsule, the nematodes enter its body, feed on its tissues, and kill it within two days.

Figure 14-7 Biological control of pests. An adult convergent ladybug is consuming an aphid (left). The wolf spider, like most spiders, is harmless to humans and plays a major role in keeping insects in check (right).

Other examples of biological control include the use of

- *Guard dogs* to protect livestock from predators. Guard dogs are more effective and cost less than erecting fences and shooting, trapping, and poisoning predators, which sometimes kill nontarget organisms, including people.

- *Ducks* to devour insects and slugs. However, ducks sometimes damage vegetables, especially leafy greens, and should be kept out of these parts of gardens.

- *Geese* for weeding orchards, eating fallen and rotting fruit (often a source of pest problems), and controlling grass in gardens and nurseries. Geese also warn of approaching predators or people by honking loudly.

- *Chickens* to control insects and weeds and to increase the nitrogen content of the soil in orchards or in gardens after plants have become well established.

- *Birds* to eat insects. Farmers and home owners can provide habitats and nesting sites that attract woodpeckers, purple martins, chickadees, barn swallows, nuthatches, and other insect-eating species.

- *Spiders* to eat insects (Figure 14-7). Spiders are insects' worst enemies, devouring enough bugs worldwide in a single day to outweigh the entire human population. Most spiders, except the brown recluse and the black widow, are harmless to humans.

- *Allelopathic plants* that naturally produce chemicals that are toxic to their weed competitors or that repel or poison their insect pests. For example, certain varieties of barley, wheat, rye, sorghum, and Sudan grass can be grown in gardens or orchard trees to suppress weeds. Plant combinations that help protect against various insect pests include cassavas and beans, potatoes and mustard greens, and a mixture of

sunflowers, maize, oats, and sesame. Peppermint can be planted around houses to repel ants and to be used as a natural mouth freshener (pull off leaf, wash it, and chew it) and a cooking spice.

Biological control has a number of advantages. Normally it only affects the target species and is nontoxic to other species, including people. Once a population of natural predators or parasites is established, control of pest species is often self-perpetuating. Development of genetic resistance is minimized because both pest and predator species usually undergo natural selection to maintain a stable interaction (co-evolution). In the United States, biological control has saved farmers an average of $25 for every $1 invested, compared to the estimated maximum of $2.40 saved for every $1 invested in pesticides.

No method of pest control, however, is perfect. Typically, 10 to 20 years of research may be required to understand how a particular pest interacts with its various enemies and to determine the best biological control agent. Mass production of biological agents is often difficult, and farmers find that they are slower to act and harder to apply than pesticides.

A potential problem with biological control is that biological agents must be protected from pesticides sprayed in nearby fields. Also, there is a chance that some can later also become pests; others (such as praying mantises) may also devour other beneficial insects. In addition, some pest organisms can develop genetic resistance to viruses and bacterial agents used for biological control.

Insect Sterilization Males of some insect pest species can be raised in the laboratory and sterilized by radiation or chemicals, then released in large numbers in an infested area to mate unsuccessfully with fertile wild females. If sterile males outnumber fertile males by 10 to 1, a pest species in a given area can be eradicated in about four generations, provided rein-

Figure 14-8 A lemon infested with red scale mites. Pheromones are now being used to help control populations of red scale mites.

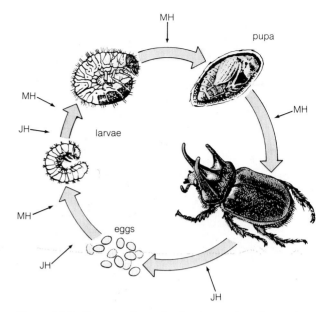

Figure 14-9 For normal growth, development, and reproduction, certain juvenile hormones (JH) and molting hormones (MH) must be present at genetically determined stages in the typical life cycle of an insect. If applied at the right time, synthetic hormones can be used to disrupt the life cycle of insect pests.

festation does not occur. This technique works best if the females mate only once, if the infested area is isolated so that it can't be periodically repopulated with nonsterilized males, and if the insect pest population has already been reduced to a fairly low level by weather, pesticides, or other factors. Success is also increased if only the sexiest—the loudest, fastest, and largest—males are sterilized.

Major problems with this approach include ensuring that sterile males are not overwhelmed numerically by nonsterile males, knowing the mating times and behavior of each target insect, preventing reinfestation with new nonsterilized males, and high costs.

Insect Sex Attractants In many insect species, when a virgin female is ready to mate she releases a minute amount (typically about one-millionth of a gram) of a species-specific chemical sex attractant called a *pheromone.* Pheromones are extracted from an insect pest species or synthesized in the laboratory. They are then used in minute amounts to lure pests into traps containing toxic chemicals or to attract natural predators of insect pests into crop fields.

Research indicates that instead of using pheromones to trap pests, it is more effective to use them to lure the pests' natural predators into fields and gardens. Sex attractants are now commercially available for use against 30 major pests (Figure 14-8).

These chemicals work on only one species, are effective in trace amounts, have little chance of causing genetic resistance, and are not harmful to non-target species. However, it is costly and time consuming to identify, isolate, and produce the specific sex attractant for each pest or natural predator species.

Pheromones have also failed for some pests because only adults are drawn to the traps; for most species, the juvenile forms—such as caterpillars—do most of the damage.

Insect Hormones Hormones are chemicals, produced in an organism's cells, that travel through the bloodstream and control various aspects of the organism's growth and development. Each step in the life cycle of a typical insect is regulated by the timely release of juvenile hormones (JH) and molting hormones (MH) (Figure 14-9).

These chemicals can be extracted from insects or synthesized in the laboratory. When applied at certain stages in an insect's life cycle (Figure 14-9), they produce abnormalities that cause the insect to die before it can reach maturity and reproduce (Figure 14-10).

They have the same advantages as sex attractants, but they take weeks to kill an insect, are often ineffective with a large infestation, and sometimes break down before they can act. Also, they must be applied at exactly the right time in the life cycle of the target insect. They sometimes affect natural predators of the target insect species and other nonpest species and can kill crustaceans if they get into aquatic ecosystems. Like sex attractants, they are difficult and costly to produce.

Irradiation of Foods Exposing certain foods to various levels of radiation is being touted by the nuclear industry and the food industry as a means of killing insects and preventing them from reproducing in certain foods after harvest, extending the

Figure 14-10 Chemical hormones can prevent insects from maturing completely and make it impossible for them to reproduce. Compare (left) a stunted and (right) a normal tobacco hornworm. The stunted hornworm was fed a compound that prevents its larvae from producing molting hormone (MH).

Agricultural Research Service/USDA

shelf life of some perishable foods, and destroying parasitic worms (such as trichinae) and bacteria (such as salmonellae that each year kill 2,000 Americans).

In 1986 the FDA approved use of low doses of ionizing radiation on spices, fruits, vegetables, and fresh pork, and it may soon be approved for use on poultry and seafood. Irradiated foods are already sold in 33 countries, including the Soviet Union, Japan, Canada, Brazil, Israel, and many west European countries.

A food does not become radioactive when it is irradiated, just as being exposed to X rays does not make the body radioactive. There is controversy, however, over irradiating food (see Pro/Con above).

Integrated Pest Management Pest control is basically an ecological problem, not a chemical problem. That is why using large quantities of broad-spectrum chemical poisons to kill and control pest populations eventually fails and ends up costing more than it is worth. As biologist Thomas Eisner puts it: "Bugs are

not going to inherit the earth. They own it now. So we might as well make peace with the landlord."

The solution is to replace this ecologically and economically unsustainable chemical approach with an ecological approach. An increasing number of pest control experts believe that in most cases the best way to control crop pests is a carefully designed **integrated pest management (IPM)** program. In this approach each crop and its pests are evaluated as an ecological system. Then a pest control program is developed that uses a variety of cultivation, biological, and chemical methods in proper sequence and timing.

The overall aim of integrated pest management is not eradication but keeping pest populations just below the size at which they cause economic loss (Figure 14-11). Fields are carefully monitored to check whether pests have reached an economically damaging level. When such a level is reached, farmers first use biological and cultivation controls. Small amounts of pesticides are applied only when absolutely necessary, and a variety of chemicals are used to retard development of genetic resistance. This approach allows farmers to escape from the pesticide treadmill and to minimize the hazards to human health, wildlife, and the environment from the widespread use of chemical pesticides.

China, Brazil, Indonesia, and the United States have led the world in the use of this approach, especially to protect cotton and soybeans. These experiences have shown that a well-designed integrated pest management program can

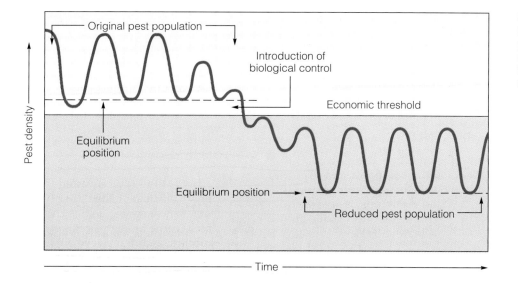

Figure 14-11 The goal of biological control and integrated pest management is to keep each pest population just below the size at which it causes economic loss.

- reduce inputs of fertilizer and irrigation water
- reduce preharvest pest-induced crop losses by 50%
- reduce pesticide use and control costs by 50% to 90%
- increase crop yields and reduce crop production costs

However, there are some drawbacks to integrated pest management. It requires expert knowledge about each pest-crop situation and is slower acting and more labor-intensive (but this creates jobs) than the use of conventional pesticides. Methods developed for a given crop in one area may not be applicable to another area with slightly different growing conditions. Although long-term costs are typically lower than those using conventional pesticides, initial costs may be higher. To date, the major focus has been on insects and pathogens, with weeds being neglected.

Switching to integrated pest management on a large scale in the United States is very difficult. First, it is strongly opposed by the politically and economically powerful agricultural chemical companies, who would suffer from a sharp drop in pesticide sales. They see little profit to be made from most alternative pest control methods, except insect sex attractants, hormones, and patented genetically engineered strains of plants and animals that have increased resistance to pests.

Second, farmers get most of their information about pest control from pesticide sales people. They also get information from U.S. Department of Agriculture county farm agents, who have supported pesticide use for decades and rarely have adequate training in the design and use of integrated pest management. The small number of integrated pest management advisors and consultants are overwhelmed by the army of pesticide sales representatives.

Third, integrated pest management methods will have to be developed and introduced to farmers by federal and state agencies because pesticide companies see little profit in this approach. However, currently only about 2% of the Department of Agriculture's budget is spent on integrated pest management.

Environmentalists urge the USDA to promote integrated pest management by

- adding a 2% sales tax on pesticides and using all of these revenues to greatly expand the federal budget for integrated pest management
- setting up a federally supported demonstration project on at least one farm in every county
- training Department of Agriculture field personnel and all county farm agents in integrated pest management so they can help farmers use this alternative
- providing federal and state subsidies and perhaps government-backed crop-loss insurance to farmers who use integrated pest management or other approved alternatives to pesticides
- gradually phasing out federal and state subsidies to farmers who depend almost entirely on pesticides once effective integrated pest management methods have been developed for major pest species

Indonesia has led the way in the no-chemical pesticide revolution. In 1986 the Indonesian government banned the use of 57 pesticides on rice and launched a nationwide program to switch to integrated pest management. By 1988 extension workers had trained over 31,000 farmers in integrated pest management techniques. This should have far-reaching benefits for Indonesia and serve as a model for other rice-growing countries.

- Ask the president and members of Congress to significantly strengthen the Federal Insecticide, Fungicide, and Rodenticide Act in order to better protect human health and the environment from the harmful effects of pesticides.

- Use pesticides in your home and on your yard or garden only when absolutely necessary, and use them in the smallest amount possible.

- Dispose of any unused pesticides in a safe manner (contact your local health department or environmental agency for safe disposal methods).

- Fix all leaking pipes and faucets, since they provide moisture, which attracts ants and roaches. This also conserves water and saves you money.

Use the following natural alternatives to pesticides for controlling common household pests:*

- *Ants.* Make sure firewood and tree branches are not in contact with the house; caulk common entry points such as windowsills, door thresholds, and baseboards (also saves energy and money). Keep ants out by planting peppermint at en-

* For further information on safe control of insect pests, contact the National Coalition Against the Misuse of Pesticides, 530 7th St., N.E., Washington, D.C. 20003.

trances and by putting coffee grounds or crushed mint leaves around doors and windows; sprinkle cayenne, red pepper, or boric acid (with an anti-caking agent) along ant trails inside your house and wipe off counter tops with vinegar. After about four days of such treatments ants usually go somewhere else.

- *Mosquitoes.* Establish nests and houses for insect-eating birds; eliminate sources of stagnant water in or near your yard; plant basil outside windows and doors; and use screens on all doors and windows, and use a yellow light bulb outside entryways. Reduce bites by not using scented soaps and not wearing perfumes, colognes, and other scented products outdoors during mosquito season. Repel mosquitoes by rubbing a bit of vinegar on exposed skin. Don't use no-pest strips for control of flying insects, especially in bedrooms or areas where food is prepared or eaten, because most contain DDVP, an organophosphate that some environmentalists have been trying unsuccessfully to have banned. Also, don't use electric zappers to kill mosquitoes and other flying insects. These devices are noisy, waste electricity, and the light

attracts insects rather than repelling them.

- *Roaches.* Caulk or otherwise plug small cracks around wall shelves, cupboards, baseboards, pipes, sinks, and bathroom fixtures (also saves energy); eliminate folded grocery sacks and newspapers which are favorite hiding places for roaches; and don't leave out dirty dishes, food spills, dog or other indoor pet food, or uncovered food or garbage overnight. Kill roaches by sprinkling boric acid under sinks and ranges, behind refrigerators, in cabinets and closets and other dark, warm places. Repel roaches by sprinkling a mixture of bay leaves and cucumbers or a mixture of 1 cup borax, ½ cup flour, ¼ cup confectioners' sugar, and 1 cup cornmeal. Establish populations of banana spiders.

- *Mice.* Seal holes and keep areas free of food, as with roaches and ants. Use glass, metal, or sturdy plastic containers to store food. Trap mice remaining in the house by using spring-loaded traps baited with a small amount of peanut butter. Put traps in an out-of-the-way place to protect children and pets. Check traps daily and remove dead mice.

Changing the Attitudes of Consumers and Farmers
Three attitudes tend to support the widespread use of pesticides and lock us into the pesticide treadmill. First, many people believe that the only good bug is a dead bug. Second, most consumers insist on buying only perfect, unblemished fruits and vegetables, even though a few holes or frayed leaves do not significantly affect the taste, nutrition, or shelf life of such produce. Third, most people accept the argument of pesticide makers that without these chemicals there wouldn't be enough to eat and food prices would soar. However, the success of integrated pest man-

agement, cultivation controls, and other alternatives to pesticides shows that we can have an adequate, safer, and affordable food supply by using pesticides only as a last, not a first, resort. Educating farmers and consumers to change these attitudes would significantly help reduce unnecessary pesticide use and the resulting risks to human health and to wildlife (see Individuals Matter above).

We need to recognize that pest control is basically an ecological, not a chemical problem.
ROBERT L. RUDD

- *Flies.* Dispose of garbage and clean garbage cans regularly; eat or remove overripened fruit; and don't leave moist, uneaten pet food out for more than an hour. Remove dog manure and clean cat litter boxes daily. Repel flies by planting sweet basil and tansy near doorways and patios and hanging a series of polyethylene strips in front entry doors (like the ones you see on some grocery store coolers); putting a blend of equal amounts of bay leaf pieces, coarsely ground cloves, clover blossoms, and eucalyptus leaves in several small bags, mosquito netting, or other mesh material and hanging them just inside entrance doors; growing sweet basil in the kitchen; or placing sweet clover in small bags made of mosquito netting and hanging them around the room.

- *Termites.* Make sure soil around and under your home is well drained and that crawl spaces are dry and well ventilated; remove scrap wood, stumps, sawdust, cardboard, firewood, and other sources of cellulose close to your house; replace heavily damaged or rotted sills, joists, or flooring; fill voids in concrete or masonry with mortar grout. In new construction install a termite shield between the foundation and floor joists, and don't let untreated wood touch soil. Inspect for damage each year, and if infestation is discovered apply a heat lamp for 10 minutes to any infested area; nematodes (tiny parasitic worms) can also be used; for a large infestation have your house treated by a professional using one of the new termicides such as Dursban, Torpedo, or Dragnet or using nematodes. Safer treatments that may soon be available are freezing termites to death by pouring liquid nitrogen into walls and infested areas, a growth regulator that transforms young termites into soldiers instead of workers needed to feed a colony, and antibiotics that eliminate the wood-digesting microorganisms that live inside termites.

You can reduce potential health risks from pesticide residues in food by

- Buying organically grown produce that has not been grown using synthetic fertilizers, pesticides, or growth regulators. Purchase only organic produce that is certified to be free of pesticide residues by independent testing laboratories, and urge your supermarket manager to carry only organic produce that meets these standards.*

- Not buying imported produce, which generally contains more pesticide residues than domestic fruits and vegetables. Show your concern and influence supermarket buying decisions by asking managers where their produce comes from

- Not buying perfect-looking fruits and vegetables, which are more likely to contain higher levels of pesticide residues.

- Buying produce in season because it is less likely to be treated with fungicides and other chemicals to preserve its appearance during storage.

- Carefully washing and scrubbing all fresh produce in soapy water.

- Removing and not using the outer leaves of lettuce and cabbage and peeling fruits that have thick skins.

- Growing your own fruits and vegetables using organic methods.

* For a list of more than 100 sources of organically grown and processed fruits, vegetables, grains, and meats, send a self-addressed business envelope with 50 cents postage to Mail-Order Organic, The Center for Science in the Public Interest, 1501 16th St., N.W., Washington, D.C. 20036.

DISCUSSION TOPICS

1. Should DDT and other pesticides be banned from use in malaria control throughout the world? Explain. What are the alternatives?

2. Environmentalists argue that because essentially all pesticides eventually fail, their use should be phased out and farmers should be given economic incentives for switching to integrated pest management. Explain why you agree or disagree with this proposal.

3. Explain how the use of insecticides can actually increase the number of insect pest problems.

4. Debate the following resolution: Because DDT and the other banned chlorinated hydrocarbon pesticides pose no demonstrable threat to human health and have saved millions of lives, they should again be approved for use in the United States.

5. Should certain types of foods used in the United States be irradiated? Explain.

6. What changes, if any, do you believe should be made in the Federal Insecticide, Fungicide, and Rodenticide Act regulating pesticide use in the United States?

7. Should U.S. companies continue to be allowed to export pesticides, medicines, and other chemicals that have been banned or severely restricted in the United States to other countries? Explain.

8. How are bugs and weeds controlled in your yard and garden, on the grounds of your college, and in the public schools, parks, and playgrounds where you live? Consider mounting efforts to have integrated pest management and organic fertilizers used on college and public grounds. Do the same thing for your yard and garden.

FURTHER READINGS

Bosso, Christopher, 1987. *Pesticides and Politics*. Pittsburgh, Penn.: University of Pittsburgh Press.

Carson, Rachel. 1962. *Silent Spring*. Boston: Houghton Mifflin.

Dover, Michael J. 1985. *A Better Mousetrap: Improving Pest Management for Agriculture*. Washington, D.C.: World Resources Institute.

Flint, Mary L., and Robert van den Bosch. 1981. *Introduction to Integrated Pest Management*. New York: Plenum Press.

Gips, Terry. 1987. *Breaking the Pesticide Habit*. Minneapolis, Minn.: IASA.

Horn, D. J. 1988. *Ecological Approach to Pest Management*. New York: Guilford Press.

Hussey, N. W., and N. Scopes. 1986. *Biological Pest Control*. Ithaca, N.Y.: Cornell University Press.

Hynes, Patricia. 1989. *The Recurring Silent Spring*. New York: Pergamon Press.

Kovrik, Robert. 1990. "Combatting Household Pests Without Chemical Warfare." *Garbage*, March/April, 22–29.

Kurzman, Dan. 1987. *A Killing Wind: Inside Union Carbide and the Bhopal Catastrophe*. New York: McGraw-Hill.

League of Women Voters. 1989. *America's Growing Dilemma: Pesticides in Food and Water*. Washington, D.C.: League of Women Voters.

Marco, G. J., et al. 1987. *Silent Spring Revisited*. Washington, D.C.: American Chemical Society.

Metcalf, R. L., and William H. Luckmann, eds. 1982. *Introduction to Insect Pest Management*. New York: John Wiley.

Mott, Lawrie, and Karen Snyder. 1988. *Pesticide Alert: A Guide to Pesticides in Fruits and Vegetables*. San Francisco: Sierra Club Books.

National Academy of Sciences. 1986. *Pesticide Resistance: Strategies and Tactics for Management*. Washington, D.C.: National Academy Press.

Natural Resources Defense Council. 1989. *Intolerable Risk: Pesticides in Our Children's Food*. New York: Natural Resources Defense Council.

Schultz, Warren. 1989. *The Chemical-Free Lawn*. Emmaus, Penn.: Rodale Press.

van den Bosch, Robert. 1978. *The Pesticide Conspiracy*. Garden City, N.Y.: Doubleday

Weir, David. 1986. *The Bhopal Syndrome: Pesticides, Environment, and Health*. San Francisco: Sierra Club Books.

Yepsen, Roger B., Jr. 1987. *The Encyclopedia of Natural Insect and Pest Control*. Emmaus, Penn.: Rodale Press.

CHAPTER FIFTEEN

LAND RESOURCES:

FORESTS, RANGELANDS,

PARKS, AND WILDERNESS

General Questions and Issues

1. Why are the world's forests such important resources?

2. Why are tropical deforestation and fuelwood shortages serious global environmental and resource problems and how should we deal with these problems?

3. What are the major types of public lands in the United States and how are they used?

4. How should forest resources be managed and conserved?

5. Why are rangelands important and how should they be managed?

6. What problems do parks face and how should parks be managed?

7. Why is wilderness important and how much should be preserved?

We abuse land because we regard it as a commodity belonging to us. When we see land as a community to which we belong, we may begin to use it with love and respect.

ALDO LEOPOLD

SNAP YOUR FINGERS. During the second this took an area of tropical forest about the size of two football fields was destroyed (see photo on p. 292) and two more football field–size areas of these incredibly diverse ecosystems were degraded. This is happening every second of every day. This explains why most of the world's remaining tropical forests as well as ancient forests in northwestern United States (see photo on front cover) will be gone within your lifetime, unless we halt this unsustainable destruction of vital earth capital.

Tropical forests, along with rangelands, parks, and wilderness are key land and biological resources that are coming under increasing stress (Table 1-1, p. 10) as populations and economic development grow. Protecting these vital resources from degradation, using them sustainably, and healing those we have degraded is one of our most important challenges.

15-1 IMPORTANCE OF FORESTS

Commercial Importance Forests supply us with lumber for housing, biomass for fuelwood, pulp for paper, medicines, and many other products, worth over $150 billion a year. Many forestlands are also used for mining, grazing livestock, and recreation and are flooded to provide reservoirs for hydropower dams and flood control.

Potentially renewable forests cover about 34% of the earth's land surface (Figure 5-2, p. 87). In the past 30 years about half of the world's forests have been destroyed to provide cropland, rangeland, lumber, fuelwood, dam reservoirs, and urban lands. Worldwide, forests, especially tropical forests, are disappearing faster than any other terrestrial ecosystem.

In the United States forests cover about one-third of the land area in the lower 48 states. The total area of forests in the United States has not declined since 1900. But since 1950 many diverse ancient or old-growth forests (Figure 15-1 and photo on front cover) have been cut down and replaced with monoculture tree farms (Figure 5-17, p. 95) to provide lumber and pulp for making paper.

Ecological Importance Forests have many vital ecological functions. Forested watersheds act as giant sponges, absorbing, holding, and gradually releasing water that recharges springs, streams, and aquifers. Thus, they regulate the flow of water from mountain highlands to croplands and urban areas and help control soil erosion, the severity of flooding (see Case Study on p. 237), and the amount of sediment washing into rivers, lakes, and artificial reservoirs.

Forests also play an important role in the global carbon cycle (Figure 4-24, p. 76) and act as an impor-

Figure 15-1 Sunlight filtering through a redwood forest in the Muir Woods near the coast of northern California. At one time large areas of these and other ancient forests covered much of the United States. Now most redwood forests have been cleared or eliminated by climate shifts. The country's few remaining ancient forests, especially in the northwest (see photo on front cover), are also being cleared rapidly for timber. This is happening when the U.S. government is urging tropical LDCs not to cut down most of their remaining ancient forests in order to protect the earth's biodiversity and to help delay projected global warming.

tant defense against global warming. Through photosynthesis trees help remove carbon dioxide from and add oxygen to the air, which explains why the world's lush tropical forests have been called "the lungs of the earth." Thus, large-scale deforestation contributes to projected greenhouse warming (Figure 10-6, p. 214).

Forests provide habitats for a larger number of wildlife species than any other land ecosystem, making them the planet's major reservoir of biological diversity (see Spotlight on p. 7). They also help buffer noise, absorb some air pollutants, and nourish the human spirit by providing solitude and beauty.

Economists and timber company managers rarely consider the ecological benefits of forests in helping sustain the biosphere. According to one calculation, a typical tree provides $196,250 worth of ecological benefits in the form of oxygen, air pollution reduction,

soil fertility and erosion control, water recycling and humidity control, wildlife habitat, and protein for wildlife. Sold as timber, this tree is worth only about $590. The problem is that the ecological benefits of forest ecosystems are assigned little or no value in the world's market-oriented economic systems (Section 7-2).

15-2 TROPICAL DEFORESTATION AND THE FUELWOOD CRISIS

Deforestation and Degradation of Tropical Forests In 1950, 30% of the earth's landmass was covered with tropical forests, the world's oldest, richest, and most complex ecosystems. By 1975 the figure was only 12%, and by 1990 only 6% was left (see photo on p. 292). Brazil contains almost 33% of the total, mostly in the Amazon basin (see Pro/Con on p. 332), and Zaire and Indonesia both have 10%.

We have cleared tropical moist forests for timber, cattle grazing, fuelwood, mining, and farming (see photo on p. 292), and the pace of destruction is accelerating (Figures 15-2 and 15-3). In 1980 the area of tropical forest cleared annually was about 11 million hectares—about the size of Pennsylvania or Louisiana. In 1989 the area of tropical deforestation was almost twice as large—20 million hectares—equalling the size of Maine and Florida combined. If clearing and degradation continue at this current rate, almost all tropical forests will be gone or severely damaged in 30 to 50 years.

Products such as rubber tapped from trees, nuts, and fruits can be harvested from tropical forests indefinitely. And patches can be cleared and used sustainably to raise crops for small populations of people if they are not cultivated or grazed for 10 to 30 years after their nutrient-poor soils are depleted (Figure 2-4, p. 24).

But tropical forests themselves must be considered as nonrenewable resources on a human time scale and should not be subjected to the large-scale clearing and degradation now taking place. Once large areas of tropical forests are cleared their nutrient poor soils (Figure 12-4, p. 270) are depleted, baked, and washed away. The moisture they once captured and released to the atmosphere is no longer available so that areas once covered with such forests tend to become drier and more desertlike (see Spotlight on p. 273).

Why Should You Care About Tropical Forests? Conservationists and ecologists consider the present destruction and degradation of tropical forests one of the world's most serious environmental and resource problems. One reason is that these incredibly diverse forests are of immense economic and ecological im-

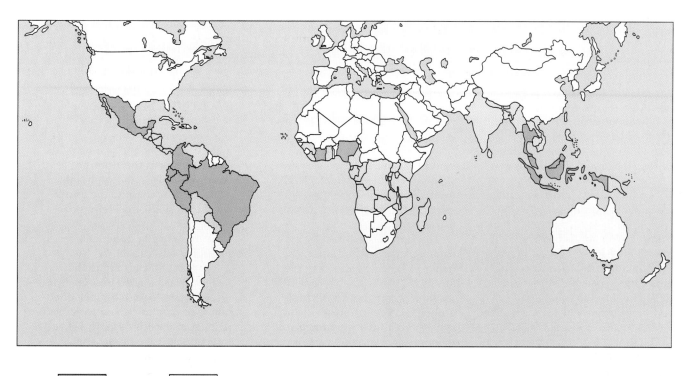

High   Moderate

Figure 15-2 Countries experiencing large annual destruction of tropical forests. (Data from UN Food and Agriculture Organization)

portance to you and everyone else on earth. They are the world's major storehouse of biological diversity. Although tropical forests occupy only about 7% of the world's land area, they provide homes for at least 50% (some estimate as much as 90%) of the earth's total stock of species (Figure 15-4, p. 334).

Tropical forests supply half of the world's annual harvest of hardwood. They also supply hundreds of food products, including coffee, spices, and tropical fruits and industrial materials such as rubber, latexes, resins, dyes, and essential oils.

The raw materials in one-fourth of the prescription and nonprescription drugs we use come from plants growing in tropical rain forests. Almost three-fourths of the 3,000 plants identified by the National Cancer Institute as containing chemicals that can fight cancer come from tropical rain forests. While you are reading this page, a plant species that can cure a type of cancer, AIDS, or some other disease that might kill you or someone you love could be wiped out forever.

Despite their immense potential, less than 1% of tropical forest species have been examined for their possible use as human resources. Destroying these forests for short-term economic gain is like throwing away an unopened present.

Tropical forests are also home for 250 million people who survive mostly by slash-and-burn and

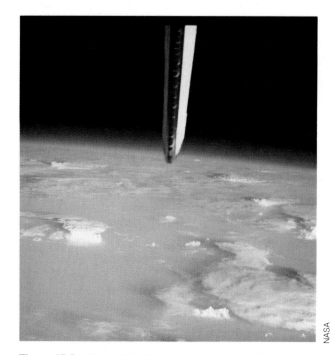

Figure 15-3 Photo taken from the space shuttle *Discovery* in September 1988 shows smoke from fires set to burn cleared areas of tropical rain forests in South America's Amazon basin. The white areas that look like clouds are plumes of smoke. The smoke shown in this photograph covers an area about three times the size of Texas. These fires may have accounted for as much as 10% of the global input of carbon dioxide to the atmosphere during 1988.

One-third of the world's tropical forests are found in South America's Amazon basin, which occupies an area equal in size to almost 90% of the land area of the continental United States (Figure 6-14, p. 127). So far only about 10% of the tropical forests in this vast greenbelt around part of the equator has been destroyed.

But destruction is occurring at a rapid rate (Figure 15-3 and photo on p. 292). Much of the development in the Amazon is financed with loans from international lending agencies and MDCs that are used to build roads that open up the forests for logging and other forms of development.

The Brazilian government's solution to rural and urban poverty is to send the landless poor off to chop down tropical forests in the Amazon basin, giving them title to the land they clear. However, the poor who move to the Amazon in a desperate search for enough land to survive rarely improve their lives. Most develop malaria and new plots must be cleared every two to three years as the soil in earlier plots is depleted.

Problems for the poor don't end there. In some areas there are violent clashes between the new immigrants and indigenous tribes who resent being driven from land they have lived on sustainably for centuries. After several years of bitter disappointment, many of the new settlers give up and leave.

The lands abandoned by agricultural squatters end up in the hands of ranchers and cash crop farmers. Thanks to government tax breaks and subsidies, these groups can profit from the land even though their operations lose money. After several years of further overexploitation, these lands are abandoned in a severely degraded state.

In the western Amazon there have been bloody confrontations between ranchers and rubber tappers who want to preserve the forests so they can continue to make a living harvesting Brazil nuts and latex from rubber trees. In 1988, Chico Mendes, a leader of the rubber tappers, was murdered for trying to stand in the way of ranchers. The rubber tappers are

living proof that the poor can make a living from the forest without destroying it.

Conservationists have mounted a global campaign to preserve 70% of the forests in the Amazon basin. But this is an uphill fight against massive economic forces concerned with short-term economic gain from depleting this treasure house of potentially renewable earth capital.

Brazilian officials argue that they must exploit the resources in these forests to help finance economic development and to help pay the interest on their massive foreign debt. They claim they are doing the same thing that the United States and other MDCs did to help finance their economic growth, including cutting down most of their native forests.

They resent the United States and environmentalists lecturing them about how they should use their resources, especially when the United States is rapidly clearing much of its remaining ancient forests (see photo on front cover) and replacing them with tree farms. During the last decade, Ha-

shifting cultivation (Figure 2-5, p. 25). About 50 million tribespeople live in these forests. These indigenous peoples, who have more knowledge about how to live sustainably in tropical forests than anyone, are seeing their land bulldozed, burned, and flooded. They are being evicted from their homelands and forced to adopt new ways that subject them to disease, hunger, and culture shock.

Tropical forests also protect watersheds and regulate water flow for farmers, who grow food for over 1 billion people in LDCs. Unless the destruction of tropical forests stops, the Environmental Policy Institute estimates that as many as 1 billion people will starve to death during the next 30 years.

Ethics demand that we protect the earth's remaining tropical forests. Regardless of the importance of tropical forests to us, sustainable-earth conservationists believe that this destruction must be stopped because it is *wrong* (Section 2-3). It could cause the premature extinction of up to 1 million plant and animal species who have as much right to exist as

we do. When we eliminate a species, it is gone for good.

Causes of Tropical Deforestation   At one level much of the destruction and degradation of the world's tropical forests is caused by gigantic projects of multinational corporations and international lending agencies that finance resource extraction, mostly to support the affluent lifestyles of people in MDCs. These projects include cattle ranches (primarily to produce raw and canned beef for export to MDCs), inefficient commercial logging and paper mills (mostly for export to MDCs), immense plantations of sugar cane and cash crops (mostly for export to MDCs), mining operations (with much of the minerals extracted exported to MDCs), and dams (often used as sources of electric power for mining and smelting operations for minerals exported to MDCs).

Governments of tropical countries feel driven to sell cash crops, timber, and minerals to MDCs at low prices to help finance economic growth and to pay

waii has cut down its remaining tropical forests faster than Brazil.

The Brazilian government is going ahead with plans to build internationally financed highways to open up larger areas of the Amazon for development, despite objections by environmentalists and other governments. Government leaders ask, "How would Americans feel if the British or some other government told them that they should not build a highway from New York to California because it would destroy their forests?"

They point out that much of the timber, beef, and mineral resources removed from tropical forests in the Amazon basin are exported to support throwaway lifestyles in MDCs. Brazilian officials also argue that the world's MDCs are the major culprits behind projected global warming, ozone depletion, and the threat of nuclear war that threaten the life-support systems for everyone on the planet (Chapter 10).

In 1988 the Brazilian government suspended new subsidies for Amazon agricultural and ranching operations. But old subsidies remain for existing ranches, which have already cost the debt-ridden government more than $2.5 billion in lost revenues. The government has also established two forests reserves and three new national parks in the Amazon region. Environmentalists applaud these actions, but fear that they are mostly window dressing to help defuse international pressures to protect much larger areas of the Amazon basin from development.

Since 1980 the Washington-based World Bank and the Inter-American Development Bank have lent Brazil more than $82 billion for development of 166 hydropower dams and reservoirs, over half of them in the Amazon basin. Most of this power will be used to support mining, smelting, and other resource extraction industries, with most of the extracted resources exported to MDCs. These dams and reservoirs will displace about 500,000 people and flood vast areas of tropical rain forests.

With funding and technical advice from the World Bank, Japan, and the European community, Brazil seeks to exploit mineral deposits and encourage settlements in a mineral-rich area of the eastern Amazon twice the size of the state of California.

Smelters in the area that will convert iron ore to pig iron will be powered by charcoal produced from wood. The cheapest way to get the charcoal is to chop down the surrounding forests and burn the trees. Environmentalists fear that this project will repeat the ecological disaster of a similar project in southeastern Brazil, where pig-iron production consumed nearly two-thirds of the area's forests. What do you think should be done about economic development in the Amazon basin?

interest on loans made by MDCs and international lending agencies controlled by MDCs. Government officials of LDCs believe this is a form of coercive economic slavery that they are powerless to control and that will leave them with few natural resources needed for their own economic development.

Since 1950 the consumption of tropical lumber has risen 15-fold, with Japan now accounting for 60% of the annual consumption. Although 68% of Japan is covered with forests, it prefers to deplete the forests of other countries. Most of the tropical hardwoods imported to Japan are processed into construction plywood, which is used to make throwaway forms for molding concrete. Other major importers of tropical hardwoods are the United States and Great Britain. The World Bank estimates that of the 33 countries that are net exporters of tropical timber, only 10 will have any timber left to export by the year 2000.

During the past 25 years Central America has lost two-thirds of its tropical forests. Much of this land has been cleared and used as rangeland to raise beef for export to the United States, Canada, and other MDCs for use in hamburgers, hot dogs, luncheon meats, chilis, stews, frozen dinners, and pet food.

For each quarter-pound hamburger made from meat imported from Central America, an area of tropical forest roughly the size of a small kitchen (55 square feet) has usually been lost. This cheaper range-fed beef reduces the cost of a pound of hamburger by a nickel but in the process helps destroy or degrade the planet's greatest storehouse of biological diversity.

After being grazed for five to ten years, tropical pastures can no longer be used for cattle. Often torrential rains and overgrazing turn the nutrient-poor soils into eroded wastelands. Ranchers then move to another area and repeat the process. This destructive "shifting ranching" is often encouraged by government tax subsidies. Tropical forests in Latin America are also being cleared and used to grow soybeans that are exported to feed cattle in western Europe.

Figure 15-4 The world's tropical forests are the planet's major storehouse of biological diversity. Two of these forests' millions of species are the red uakari monkey in Peru (left), and the keel-billed toucan in Belize (right). Most tropical forest species have specialized niches. This makes them highly vulnerable to extinction when their forest habitats are cleared or severely degraded.

Environmental and consumer groups (especially the Rainforest Action Network in San Francisco) have organized boycotts of hamburger chains buying beef imported from Central America and other tropical countries. Because of these efforts, many large chains claim they no longer buy beef from tropical countries. However, such claims are impossible to verify. In some cases cattle raised in a tropical country are imported to the United States and slaughtered there. This allows hamburger chains and other meat processors to claim they are using American beef.

Environmentalists call for a ban on all beef or beef products raised on cleared land in tropical forests. This would encourage tropical countries to raise beef on existing rangeland and to stop giving ranchers the subsidies needed to make raising beef on cleared tropical forestland profitable.

At the other end of the scale tropical forests are destroyed by the landless poor who migrate there, hoping to get enough land to survive. They are driven there by a combination of rapid population growth, the global poverty trap (Section 7-3), and ownership of much of the arable land by a few wealthy people. This mass migration of poor people to tropical forests would not be possible without the internationally financed projects that build roads and open up these usually inaccessible areas. Most of these migrants do not escape poverty and many come down with malaria (see Pro/Con on p. 332)

Reducing the Destruction and Degradation of Tropical Forests If we don't sharply reduce the destruction and degradation of the world's remaining tropical forests, we will lose one of our most important defenses against projected global warming and bring about a mass extinction of wildlife. We will also lose sources of food, fuel, new drugs that may cure AIDS and some types of cancer, and numerous raw materials. To prevent this environmentalists believe that we need to

- Establish an international ban on imports of timber, wood products, beef, or other goods that directly or indirectly destroy or degrade tropical forests.

- Provide aid and debt relief for tropical countries that ban commercial logging, cattle ranching, and other destructive uses of tropical forests and that emphasize economically and ecologically sustainable harvesting of rubber, nuts, fruits, and other renewable resources that over time can yield two times the net income derived from logging and three times that from cattle ranching.

- Set aside large areas of the world's tropical forests as reserves and parks protected from unsustainable development; participating tropical countries would act as custodians for the reserves in return for foreign aid or relief from some of their debt (debt-for-nature swaps) (see Case Study on p. 336).

- Rehabilitate degraded tropical forests and watersheds (see Case Study on p. 338).

- Provide financial incentives to villagers and village organizations for establishment of fuelwood trees and tree farms on abandoned and degraded land with suitable soil.

- Not finance the establishment of tree farms, cash crop plantations, or cattle ranches on any land now covered by tropical forests.

- Phase out and halt funding for dams, tree and crop plantations, ranches, roads, and colonization programs that threaten tropical forests.

- Include indigenous tribal peoples, women, and private local conservation organizations in the planning and execution of tropical forestry plans.

- Give indigenous people title to tropical forest lands that they and their ancestors have lived on sustainably for centuries with the stipulations that these lands cannot be developed in unsustainable ways and cannot be sold. The Colombian government is giving indigenous tribes complete control of two-thirds of the country's land area in the Amazon basin with the stipulation that they must never sell the land.

- Require an extensive environmental impact evaluation for any proposed development project in tropical forests and use internationally accepted standards for such studies.

- Prevent banks and international lending agencies from lending money for environmentally destructive projects.

- Exert political and consumer pressure on large U.S., Japanese, and British timber, paper, meat processing, and food companies now involved in destructive development projects in tropical forests. Coca-Cola recently abandoned a plan to develop a large citrus fruit plantation in Belize after pressure from the environmental organization Friends of the Earth.

- Support effective family planning and programs that attack the root causes of poverty, including unequal distribution of farmland (Section 6-4).

The Fuelwood Crisis in LDCs Almost 70% of the people in LDCs rely on biomass as their primary fuel

Figure 15-5 Women carrying fuelwood. Women, who generally do this chore, may have to walk long distances and spend much of their day seeking and carrying heavy loads of wood. Unsustainable cutting of trees for fuelwood by the poor in LDCs is one cause of deforestation. However, much more deforestation is caused by unsustainable logging of these forests with most of the timber shipped to Japan, the United States, and western European countries.

for heating and cooking. About half of this comes from burning wood or charcoal produced from wood, 33% from crop residues, and 17% from dung.

By 1985 about 1.5 billion people—almost one out of every three persons on earth—in 63 LDCs could not get enough fuelwood to meet their basic needs, or they were forced to meet their needs by consuming wood faster than it was being replenished. The UN Food and Agriculture Organization projects that by the end of this century 3 billion people in 77 LDCs will experience a fuelwood crisis.

Fuelwood scarcity leads to several harmful effects in addition to deforestation and accelerated soil erosion. It places an additional burden on the poor, especially women. Often, they must walk long distances to find and carry home bundles of fuelwood (Figure 15-5). Buying fuelwood or charcoal can take 40% of a poor family's meager income. Also, poor families who can't get enough fuelwood often burn

Bolivia is a tropical South American country (Figure 15-6). Its population of almost 9 million is growing rapidly at 2.6% a year, mostly because the average total fertility rate is 5.5 children per woman. Because the country has such a low population density, the Bolivian government believes that this fertility level, already among the highest in Latin America, is too low.

The country's average per capita GNP is less than $600 a year. Because of widespread poverty, the country's infant mortality rate is high, with one out of every nine babies dying before reaching their first birthday. The country is also saddled with a $5.7 billion foreign debt.

In 1984 biologist Thomas Lovejoy suggested that debtor nations willing to protect some of their natural resources should be eligible for discounts or credits against some of their debts. With such *debt-for-nature swaps*, a certain amount of foreign debt is cancelled in exchange for local currency investments that will improve natural resource management in the debtor country. Typically, a conservation or other organization buys a certain amount of a country's debt from a bank at a discount rate and negotiates a debt-for-nature swap.

In 1987 Conservation International, a private U.S. banking consortium, paid $100,000 to a Swiss bank to buy up $650,000 of Bolivia's national debt. In exchange for forgiveness of this part of its debt, Bolivia agreed to protect 1.5 million hectares (3.7 million acres) of tropical forest around its existing Beni Biosphere Reserve in the Amazon basin from harmful forms of development (Figure 15-6). The government was to establish maximum legal protection for the reserve and create a $250,000 fund with the interest to be used to manage the reserve.

This land is supposed to be a model of how conservation of forest and wildlife resources can be mixed with sustainable economic development (Figure 15-6). The core of this land is a virgin tropical forest to be set aside as a biological reserve. It will be surrounded by a protective buffer of savanna to be used for sustainable grazing of livestock.

Controlled commercial logging, as well as hunting and fishing by local natives, will be permitted in some of the forest in the tract (Figure 15-6). Logging will not be allowed in the mountain area above the tract to protect the area's watershed and to prevent erosion.

However, by 1990, three years after the agreement was signed, the Bolivian government had not provided legal protection for the reserve. It also waited until April 1989 to contribute only $100,000 to the reserve management fund.

Meanwhile, the Bolivian government, with the support of Conservation International, granted tree harvesting concessions to logging companies in the forested areas. Since the agreement was signed thousands of mahogany trees have been cut from the area, with most of this lumber exported to the United States.

One lesson learned from this first debt-for-nature swap is that the legislative and budget requirements should be established before the swap is made. Also such swaps need to be carefully monitored by environmental organizations to be sure that paper proposals labelled as models of sustainable development are not ways to disguise eventually unsustainable development.

Ideally, the management of the protected areas should be turned over to national or local environmental organizations in the debtor country. Debt-for-nature swaps are an excellent idea if they lead to true protection or restoration of some of the earth's remaining natural areas.

Figure 15-6 Mixing economic development and conservation in a 1.5 million hectare (3.7 million acre) tract in Bolivia. A U.S. conservation organization arranged a debt-for-nature swap to help protect this land from harmful forms of development.

dried animal dung and crop residues for cooking and heating (Figure 15-7). This keeps these natural fertilizers from reaching the soil and reduces cropland productivity, creating a vicious circle of land degradation that helps lock the poor in the global poverty trap (Section 7-3).

LDCs can reduce the severity of the fuelwood crisis by planting more fast-growing fuelwood trees such as the leucaenas, burning wood more efficiently, and switching to other fuels. Experience has shown that planting projects are most successful when local people, especially women, are involved in their plan-

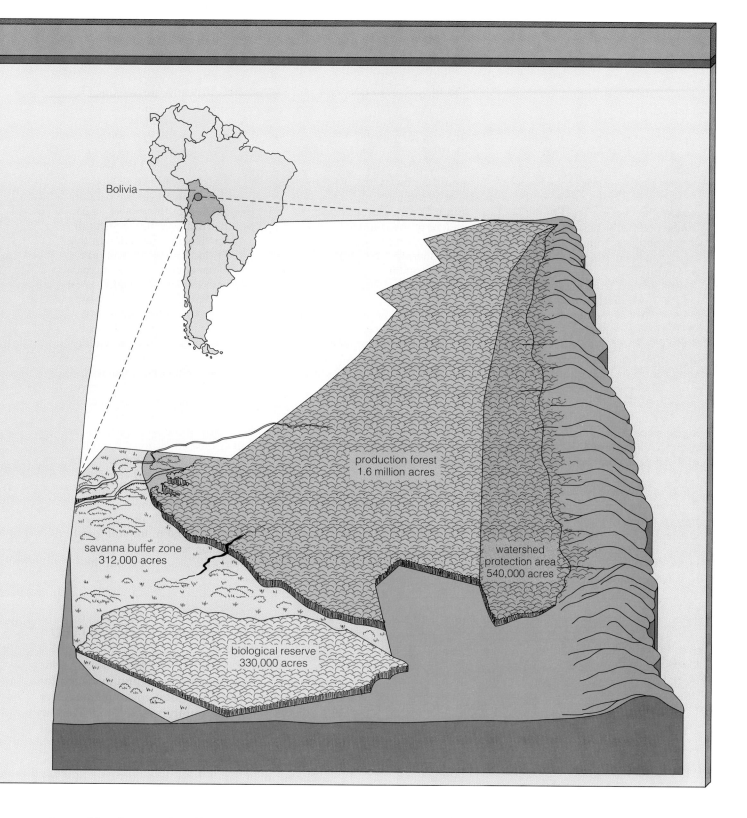

Bolivia

production forest
1.6 million acres

savanna buffer zone
312,000 acres

watershed
protection area
540,000 acres

biological reserve
330,000 acres

ning and implementation. Programs work best when village farmers own the land or are given ownership of any trees grown on common resource land. This gives them a strong incentive to plant and protect trees for their own use and for sale.

New, more efficient types of stoves must be designed to make use of locally available materials and to provide both heat and light like the open fires they replace. Villagers in Burkina Faso in West Africa have been shown how to make a stove from mud and dung that cuts wood use by 30% to 70%. It can be made by villagers in half a day at virtually no cost.

Once Costa Rica (Figure 15-8) was almost completely covered with tropical forests, containing perhaps 5% to 9% of the earth's species. However, between 1963 and 1983 politically powerful ranching families cleared much of the country's forests to graze cattle, with most of the beef exported to the United States. By 1983 only 17% of the country's original tropical forest remained and soil erosion was rampant.

This pushed large numbers of the country's rural poor deeper into the jaws of the poverty trap. Many small landholders lost their land to ranchers. And rural jobs dried up because ranching requires much less labor than crop growing. Growing numbers of landless poor migrated to expanding cities or were forced to farm fragile slopes or clear forests in ways that accelerated the country's deforestation and soil erosion.

The bright note is that Costa Rica now leads all tropical countries in efforts to protect its remaining tropical forests and restore degraded areas. In the mid-1970s, Costa Rica established a system of national parks and reserves that presently protects 12% of the country's land area. Despite widespread degradation, tiny Costa Rica is a biological "superpower," with an estimated 500,000 species of plants and animals living on a landmass just twice the size of Vermont.

Since Oscar Arias was elected president of the country in 1986, these conservation efforts have increased sharply. The country's plan is to combine conservation and sustainable economic development and to expand protected areas to 25% of the country's land by the end of this century. Even with this farsighted program, the country's forests are disappearing, as the poor clear forests to survive.

A rugged mountainous region with a tropical rain forest contains the Guanacaste National Park, which has been designated an international biosphere reserve. One of the country's most visible projects is the restoration of the dry tropical forest in this park, the world's first project of this kind.

Daniel Janzen, professor of biology at the University of Pennsylvania in Philadelphia, has helped galvanize international support for this restoration project. Janzen is a leader in the growing field devoted to the rehabilitation and restoration of degraded ecosystems. His vision is to make the nearly 40,000 local people who live near the park an integral part of the project—a concept he calls *biocultural restoration*.

By actively participating in the project, local residents will reap enormous educational, economic, and environmental benefits. Local farmers have been hired to plant large areas with tree seeds and seedlings started in Janzen's lab.

The essence of the program is to make the park a living classroom. Students in elementary schools, high schools, and universities will study the ecology of the park in the classroom and go on field trips in the park itself. There will also be educational programs for civic groups and tourists from Costa Rica and elsewhere. These visitors and activities will stimulate the local economy.

The project will also serve as a training ground in tropical forest restoration for Costa Rican and other scientists throughout the

Figure 15-7   Making fuel briquettes from cow dung in India. As fuelwood becomes scarce, more people collect and burn dung, depriving the soil of an important source of plant nutrients.

15-3 PUBLIC LANDS IN THE UNITED STATES: AN OVERVIEW

About 42% of all U.S. land consists of public lands owned jointly by all citizens and managed for them by federal, state, and local governments. Over one-third (35%) of the country's land is managed by the federal government (Figure 15-9). About 95% of this federal public land is in Alaska (73%) and in western states (22%).

These public lands have been divided by Congress into different units administered by several federal agencies. The allowed uses of these lands vary (see Spotlight on p. 341).

Federally administered public lands contain a large portion of the country's timber, grazing land, and energy resources and most of its copper, silver, asbestos, lead, molybdenum, beryllium, phosphate, and potash. Through various laws Congress has allowed private individuals and corporations to har-

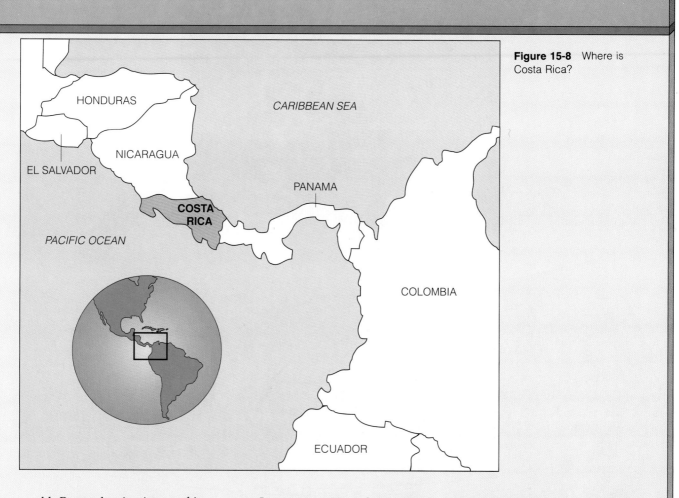

Figure 15-8 Where is Costa Rica?

world. Research scientists working on the project will give guest lectures in classrooms and lead some of the field trips.

Janzen recognizes that in 20 to 40 years the children of this area will be running the park and the local political systems. If they un-

derstand the biology and importance of their local environment, they are more likely to protect and sustain its resources.

vest or extract many of these resources—often at bargain prices. Because of the economic value of these resources, there has been a long and continuing history of conflict between various groups over management and use of these public lands.

Since the early 1900s the American conservation movement has been split into two schools of thought, the *preservationists* and the *scientific conservationists*. Preservationists emphasize protecting large areas of public lands from mining, timbering, and other forms of economic development and environmental degradation.

In contrast, scientific conservationists see public lands as resources to be used now to enhance economic growth and national strength. They believe that the government must protect these lands from degradation by managing them efficiently and scientifically for *sustainable yield* (p. 8) and *multiple use* (using publicly owned land resources such as national

forests for a mixture of human purposes including timbering, mining, recreation, grazing, hunting, construction, and water conservation). The controversy between these two groups continues today.

 15-4 FOREST MANAGEMENT AND CONSERVATION

Types of Forest Management There are two basic forest management systems: even-aged management and uneven-aged management. With **even-aged management,** trees in a given stand are maintained at about the same age and size, harvested all at once, and replanted naturally or artificially so that a new even-aged stand will grow. Growers emphasize mass production of low-quality wood with the goal of maximizing economic return on investment in as short a time as possible.

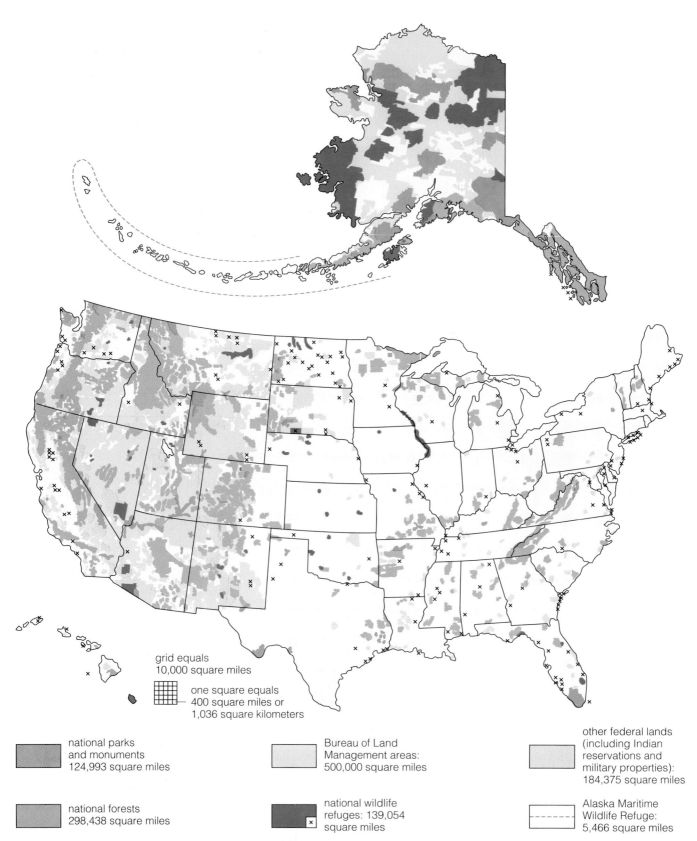

grid equals
10,000 square miles

one square equals
400 square miles or
1,036 square kilometers

national parks and monuments 124,993 square miles	Bureau of Land Management areas: 500,000 square miles	other federal lands (including Indian reservations and military properties): 184,375 square miles
national forests 298,438 square miles	national wildlife refuges: 139,054 square miles	Alaska Maritime Wildlife Refuge: 5,466 square miles

Figure 15-9 Major public lands managed by the U.S. federal government. (Data from U.S. Geological Survey)

Even-aged management begins with the cutting of all or most trees from a diverse, old-growth (see photo on front cover) or secondary forest. Then the site is replanted with an even-aged stand of a single species (monoculture) of faster-growing softwoods (Figure 5-17, p. 95). Such stands of one or only a few tree species are called **tree farms**. They need close supervision and usually require expensive inputs of fertilizers and pesticides to protect the monoculture species from diseases and insects. Once the trees

Multiple-Use Lands

- **National Forests.** The forest system includes 156 national forests and 19 national grasslands managed by the Forest Service. Excluding the 15% protected as wilderness areas, this land is supposed to be managed according to the principles of *sustained yield* and *multiple use*. The lands are used for timbering, grazing, agriculture, mining, oil and gas leasing, recreation, sport hunting, sport and commercial fishing, and conservation of watershed, soil, and wildlife resources. Off-road vehicles are usually restricted to designated routes.

- **National Resource Lands.** These lands are mostly grassland, prairie, desert, scrub forest, and other open spaces located in the western states and Alaska. They are managed by the Bureau of Land Management under the principle of *multiple use*. Emphasis is on providing a secure domestic supply of energy and strategically important nonenergy minerals and on preserving the renewability of rangelands for livestock grazing under a permit system. About 10% of these lands are being evaluated for possible designation as wilderness areas.

Moderately Restricted-Use Lands

- **National Wildlife Refuges.** This system includes 452 refuges and various ranges managed by the Fish and Wildlife Service. About 24% of this land is protected as wilderness areas. The purpose of most refuges is to protect habitats and breeding areas for waterfowl and big-game animals in order to provide a harvestable supply for hunters. A few refuges have been set aside to save specific endangered species from extinction. These lands are not officially managed under the principles of multiple use and sustained yield. Nevertheless, sport hunting, trapping, sport and commercial fishing, oil and gas development, mining (old claims only), timber cutting, livestock grazing, and farming are permitted as long as the Secretary of the Interior finds such uses compatible with the purposes of each unit.

Restricted-Use Lands

- **National Parks.** This system consists of 356 units. They include 50 major parks (mostly in the West) and 306 national recreation areas, monuments, memorials, battlefields, historic sites, parkways, trails, rivers, seashores, and lakeshores. All are managed by the National Park Service. Its management goals are to preserve scenic and unique natural landscapes, preserve and interpret the country's historic and cultural heritage, provide protected wildlife habitats, protect wilderness areas within the parks, and provide certain types of recreation. National parks may be used only for camping, hiking, sport fishing, and motorized and nonmotorized boating. Motor vehicles are permitted only on roads, and off-road vehicles are not allowed. Unlike national parks, national recreation areas may be used for sport hunting, new mining claims, and new oil and gas leasing. About 49% of the land in the National Park System is protected as wilderness areas.

- **National Wilderness Preservation System.** This system includes 474 roadless areas within the national parks, national wildlife refuges, and national forests. They are managed, respectively, by the National Park Service, the Forest Service, and the Fish and Wildlife Service. These areas are to be preserved in their essentially untouched condition "for the use and enjoyment of the American people in such a manner as will leave them unimpaired for future use and enjoyment as wilderness." Wilderness areas are open only for recreational activities such as hiking, sport fishing, camping, nonmotorized boating, and in some areas sport hunting and horseback riding. Roads, timber harvesting, grazing, mining, commercial activities, and human-made structures are prohibited, except where such activities occurred before an area's designation as wilderness. Motorized vehicles, boats, and equipment are banned except for emergency uses such as fire control and rescue operations. However, aircraft are allowed to land in Alaskan wilderness areas.

---

reach maturity, the entire stand is harvested and the area is replanted with seeds or seedlings. Genetic crossbreeding and genetic engineering can be used to improve both the quality and quantity of wood produced from tree farms.

With **uneven-aged management,** trees in a given stand are maintained at many ages and sizes to permit continuous natural regeneration. Here the goals are to sustain biological diversity, sustain long-term production of high-quality timber, provide a reasonable economic return, and allow multiple use of a forest stand. The emphasis is on selective cutting of mature trees and using clearcutting only on small patches of tree species that benefit from this type of harvesting.

Sustainable development and *ecological sustainability* have become buzzwords that are used to mean

Figure 15-10 Selective cutting of ponderosa pine in Deschutes National Forest, an old-growth forest near Sisters, Oregon.

Figure 15-11 First stage of shelterwood cutting of a stand of longleaf pine in south Alabama.

different things. To ecologists sustainable forestry means not destroying processes that help sustain the natural forest ecosystems that can produce sustainable yields of timber indefinitely. This means emphasizing uneven-growth management and not clearing and replacing diverse forests with monoculture tree farms.

To timber companies sustainable forestry means getting a sustainable yield of commercial timber in as short a time as possible. This usually means even-aged management in which diverse forest are cleared and replaced with intensely managed tree farms. To environmentalists this approach is ecologically and economically unsustainable in the long run.

Tree Harvesting The method chosen for harvesting depends on whether uneven-aged or even-aged forest management is being used. It also depends on the tree species involved, the nature of the site, and the objectives and resources of the owner.

In **selective cutting,** intermediate-aged or mature trees in an uneven-aged forest are cut singly or in small groups (Figure 15-10). This reduces crowding, encourages the growth of younger trees, and maintains an uneven-aged stand with trees of different species, ages, and sizes. Over time the stand will regenerate itself. If done properly, selective cutting helps protect the site from soil erosion and tree damage by the wind.

But selective cutting is costly unless the value of the trees removed is high. It's not useful for shade-intolerant species, which require full sunlight for seedling growth. The need to reopen roads and trails periodically for selective harvests can cause erosion of certain soils. Maintaining a good mixture of tree ages and sizes takes considerable planning and skill.

One type of selective cutting, not considered a sound forestry practice, is *high grading,* or *creaming*— removing the most valuable trees without considering the quality or distribution of the remaining trees needed for regeneration. Many loggers in tropical forests in LDCs use this destructive form of selective cutting. For example, 100 trees may be destroyed just to remove one mature mahogany tree in a tropical forest.

Some tree species do best when grown in full or moderate sunlight in forest openings or in large cleared and seeded areas. Even-aged stands or tree farms of such shade-intolerant species are usually harvested by shelterwood cutting, seed-tree cutting, or clearcutting.

Shelterwood cutting is the removal of all mature trees in an area in a series of cuttings, typically over a period of ten years. This technique can be applied to even-aged or uneven-aged stands. In the first harvest, selected mature trees, unwanted tree species, and dying, defective, and diseased trees are removed. This cut opens up the forest floor to light and leaves the best trees to cast seed and provide shelter for growing seedlings (Figure 15-11).

After a number of seedlings have taken hold, a second cutting removes more of the remaining mature trees. Some of the best mature trees are left to provide shelter for the growing young trees. After the young trees are well established, a third cutting removes the remaining mature trees and allows the even-aged stand of young trees to grow to maturity.

This method allows natural reseeding from the best seed trees and protects seedlings from being crowded out. It leaves a fairly natural-looking forest that can be used for a variety of purposes. It also helps reduce soil erosion and provides good habitat for wildlife.

Without careful planning and supervision, however, loggers may take too many trees in the initial cutting, especially the most valuable trees. Shelterwood cutting is also more costly and takes more skill and planning than clearcutting.

Figure 15-12 Seed trees left after the seed-tree cutting of a stand of longleaf pine in Alabama. About ten trees per hectare are left to reseed the area.

Figure 15-13 Forest Service clearcut in Oregon. All trees in the area are removed and the area is reseeded.

Seed-tree cutting harvests nearly all the trees on a site in one cutting, leaving a few seed-producing, wind-resistant trees uniformly distributed as a source of seed to regenerate a new crop of trees (Figure 15-12). After the new trees have become established, the seed trees are sometimes harvested.

By allowing a variety of species to grow at one time, seed-tree cutting leaves an aesthetically pleasing forest useful for recreation, deer hunting, erosion control, and wildlife conservation. Leaving the best trees for seed can also lead to genetic improvement in the new stand.

Clearcutting is the removal of all trees from a given area in a single cutting to establish a new, even-aged stand or tree farm. The clearcut area may consist of a whole stand (Figure 15-13 and photo on p. 292), a group, a strip, or a series of patches (Figure 15-14). After all trees are cut, the site is reforested naturally from seed released by the harvest, or foresters broadcast seed over the site or plant genetically superior seedlings raised in a nursery.

Clearcutting increases the volume of timber harvested per hectare, reduces road building, often permits reforesting with genetically improved stock, and shortens the time needed to establish a new stand of trees. Timber companies prefer this method because it requires much less skill and planning than other harvesting methods and usually gives them the maximum economic return.

However, large-scale clearcutting on steeply sloped land leads to severe soil erosion, sediment water pollution, flooding from melting snow and heavy rains (see Case Study on p. 237), and landslides when the exposed roots decay. The heavy logging equipment also compacts the soil and reduces its productivity. As a result, such slopes often remain barren wastelands (Figure 1-1, p. 3).

Clearcutting leaves ugly, unnatural forest openings that take decades to regenerate (Figures 15-13

Figure 15-14 Patch clearcutting and logging roads in Gifford Pinchot National Forest in Washington.

and 15-14) and reduces the recreational value of the forest. It also reduces the number and types of wildlife habitats and thus reduces biological diversity.

Protecting Forests from Diseases and Insects In a healthy, diverse forest, tree diseases and insect populations rarely get out of control and seldom destroy many trees. But a tree farm of one species has few natural defenses and is vulnerable to attack by diseases and insects.

The best and cheapest way to prevent excessive damage to trees from diseases and insects is to preserve forest biological diversity. Other methods include

- banning imported timber that might carry harmful parasites

- removing infected trees and vegetation

- clearcutting infected areas and removing or burning all debris

Figure 15-15 Surface fire in Ocala National Forest in Florida. Occasional fires like this burn deadwood, undergrowth, and litter and help prevent more destructive crown fires in some types of forests.

Figure 15-16 Destructive crown fire in Yellowstone National Park during the summer of 1988. Wildlife that can't escape are killed and wildlife habitats are destroyed. Severe erosion can occur.

- treating diseased trees with antibiotics

- developing disease-resistant tree species

- applying insecticides and fungicides (Section 14-1)

- using integrated pest management (Section 14-5)

Protecting Forests from Fires Today, scientists recognize that occasional natural fires set by lightning are an important part of the ecological cycle of many forests. Some species, especially conifers such as pines and redwoods (Figure 15-1), have thick, fire-resistant bark and benefit from occasional fires. For example, the seeds of some conifers, such as the giant sequoia and the jack pine, are released or germinate only after being exposed to intense heat.

In evaluating the effects of fire on forest ecosystems, it's important to distinguish between three kinds of forest fires: surface, crown, and ground. **Surface fires** are low-level fires that usually burn only undergrowth and leaf litter on the forest floor (Figure 15-15). These fires kill seedlings and small trees but don't kill most mature trees. Wildlife can usually escape from these fairly slow-burning fires.

In forests where ground litter accumulates rapidly, a surface fire every five years or so burns away flammable material and helps prevent more destructive crown and ground fires. Surface fires also release and recycle valuable mineral nutrients tied up in slowly decomposing litter and undergrowth, increase the activity of nitrogen-fixing bacteria, stimulate the germination of certain tree seeds, and help control diseases and insects. Some wildlife species, such as deer, moose, elk, muskrat, woodcock, and quail, depend on occasional surface fires to maintain their habitats and to provide food in the form of vegetation that sprouts after fires.

Crown fires are extremely hot fires that burn ground vegetation and tree tops (Figure 15-16). They usually occur in forests where all fire has been prevented for several decades, allowing the buildup of dead wood, leaves, and other flammable ground litter. In such forests an intense surface fire driven by a strong wind can spread to tree tops. These rapidly burning fires can destroy all vegetation, kill wildlife, and lead to accelerated erosion.

Sometimes surface fires become **ground fires,** which burn partially decayed leaves, or peat, below the ground surface. Such fires, common in northern bogs, may smoulder for days or weeks before being detected. They are very difficult to put out.

The Smokey-the-Bear educational campaign of the Forest Service and the National Advertising Council has been successful in preventing many forest fires in U.S. forests, saving many lives, and avoiding losses of billions of dollars. Ecologists, however, contend that it also has caused harm by allowing litter buildup in some forests, increasing the likelihood of highly destructive crown fires (Figure 15-16). For this reason, many fires in national parks and wilderness areas are now allowed to burn as part of the natural ecological cycle of succession and regeneration.

Prescribed surface fires in some forests can be an effective method for preventing crown fires by reducing litter buildup. They are also used to control outbreaks of tree diseases and pests. These fires are started only by well-trained personnel when weather and forest conditions are ideal for control and proper intensity of burning. Prescribed fires are also timed to keep levels of air pollution as low as possible.

Protecting Forests from Air Pollution and Global Warming Forests at high elevations and forests downwind from urban and industrial centers are exposed to a variety of air pollutants that can harm trees, especially conifers (Figure 9-15, p. 199). In addition to direct harm, prolonged exposure to multiple air pollutants makes trees much more vulnerable to drought, diseases, insects, and mosses (Figure 9-10, p. 196). The only solution is to sharply reduce emissions of the offending pollutants from coal-burning power plants, industrial plants, and cars (Section 9-6). In coming decades, an even greater threat to forests, especially temperate and boreal forests, is expected to result from projected changes in regional climate brought about by global warming (Section 10-2).

Management and Conservation of National Forests in the United States About 22% of the country's commercial forest area is located within the 156 national forests managed by the U.S. Forest Service (see Spotlight on p. 341). More than 3 million cattle and sheep graze on national forestlands each year. National forests receive more recreational visits than any other federal public lands—more than twice as many as the National Park System.

Almost half of national forestlands are open to commercial logging. They supply about 15% of the country's total annual timber harvest—enough wood to build about 1 million homes. Each year private timber companies bid for rights to cut a certain amount of timber from areas designated by the Forest Service.

The Forest Service is required by law to manage national forests according to the principles of sustained yield and multiple use. But managing a public forest for balanced multiple use is difficult and sometimes impossible. It usually involves making trade-offs that don't satisfy all the special-interest groups wanting different things from the forest.

Since 1950 there have been greatly increased demands on national forests for timber sales, development of mineral and energy resources, recreation, wilderness protection, and wildlife conservation. Because of these growing and often conflicting demands on national forest resources, the management policies of the Forest Service have been the subject of heated controversy since the 1960s. Timber company officials complain that they aren't allowed to buy and cut enough timber on public lands, especially in remaining old-growth forests in California and the Pacific Northwest (see photo on front cover). Conservationists charge that the Forest Service has made timber harvesting the dominant use in most national forests (see Pro/Con on p. 346).

Less than 10% of the ancient forests (Figure 15-1 and photo on front cover) that once stretched from California to Alaska remains, with remnants located mostly on federal lands in the Northwest. Over 80% of the total old-growth forests that remain on these public lands is slated for cutting and most could be gone by 2020. Much of this timber cut from these national forests is exported to Japan, which refuses to harvest much of its own trees. Conservationists point out that the United States is telling Brazil not to cut down its tropical forests while Americans are cutting down their remaining old-growth forests at an unprecedented rate (see Pro/Con on p. 332).

Conservationists are fighting to save most remaining ancient or old-growth stands in national and state forests in the Pacific Northwest; timber companies want to clearcut them. Loggers and sawmill operators, who don't want to lose their jobs, are caught in the middle of this struggle between conservationists and timber company owners.

Under intense pressure and lawsuits from conservationists, in 1990 the Forest Service adopted a policy that is supposed to protect about 50% of the remaining unprotected old-growth forests on public lands in the Northwest. But conservationists point out that their definition of old-growth forests includes much larger areas than the definition used by the Forest Service and the timber industry.

Recycling Wastepaper Sharply increasing the recycling of paper is a key to reducing clearance and degradation of forests, solid waste, and water and air pollution. Conservationists estimate that at least 50% of the world's wastepaper (mostly newspapers, cardboard, office paper, and computer and copier paper) could be recycled by the end of this century. But only about 25% is now recycled, although some countries such as the Netherlands, Mexico, and Japan recycle about 45%.

The United States leads the world in paper consumption and waste. Each year the average American consumes 273 kilograms (600 pounds) of paper compared to only 6 kilograms (13 pounds) per person in China. Even in MDCs such as Japan and West Germany, the average per capita consumption of paper is about half that in the United States.

During World War II, when paper drives and recycling were national priorities, the United States recycled about 45% of its wastepaper. In 1989 only about 29% was recycled. The only MDC with a lower

Conservationists are alarmed at proposals by the Forest Service to double the timber harvest on national forests between 1986 and 2030 at the urging of the Reagan and Bush administrations and the timber industry.

To accomplish this, the Forest Service has proposed the building of new roads in the national forests over the next 50 years whose overall length will be six times the length of roads in the entire interstate highway system. Conservationists charge that plans to build roads in inaccessible areas (about one-fifth of the proposed new roads) are designed to disqualify those areas from inclusion in the National Wilderness Preservation System.

Conservationists have also accused the Forest Service of poor financial management of public forests. Studies have shown that between 1975 and 1985 timber from 79 national forests was sold below the amount the Forest Serv-

ice spent on roads and preparing the trees for sale. During this period Forest Service timber sales lost $2.1 billion. Conservationists oppose these subsidies for private timber companies at taxpayers' expense. These subsidies also encourage the cutting of virgin timber and discourage the recycling of wastepaper that would sharply reduce the demand for virgin timber.

Timber industry representatives argue that such subsidies help taxpayers by keeping lumber prices down. But conservationists note that each year taxpayers already give the lumber industry tax breaks almost equal to the cost of managing the entire National Forest System.

Forestry experts and conservationists have suggested several ways to reduce exploitation of publicly owned timber resources and provide true multiple use of national forests as required by law:

- Cut the present annual harvest of timber from national forests in half instead of doubling it as proposed by the timber industry and the Forest Service.
- Keep at least 50% of remaining old-growth timber in any national forest from being cut.
- Require that timber from national forests be sold at a price that includes the costs of roads, site preparation, and site regeneration.
- Require that all timber sales in national forests yield a profit for taxpayers based on the fair market value of any timber harvested.
- Use a much larger portion of the Forest Service budget to improve management and increase timber yields of the country's privately owned commercial forestland to take pressure off the national forests.

What do you think should be done?

recycling rate is Canada, where only 16% of the paper is recycled.

Product overpackaging is a major contributor to paper use and waste. Packages inside packages and oversized containers are designed to trick consumers into thinking they're getting more for their money. Nearly $1 of every $10 spent for food in the United States goes for throwaway packaging. Junk mail also wastes enormous amounts of paper.*

Recycling the country's Sunday newspapers would save an entire forest of 500,000 trees each week. Recycling paper also saves energy because it takes 30% to 55% less energy to produce the same weight of recycled paper as making the paper from trees. The U.S. paper industry is the country's third largest consumer of energy and the largest single user of fuel oil.

Recycling paper also reduces air pollution from pulp mills by 74% to 95%, lowers water pollution by 35%, conserves large quantities of water, and saves

landfill space. Recycling paper also helps prevent groundwater contamination from the toxic ink left after paper eventually biodegrades in landfills over a 30 to 60 year period.

Recycling paper can also save money. In 1988 American Telephone and Telegraph earned more than $485,000 in revenue and saved $1.3 million in disposal costs by collecting and recycling high-grade office paper.

Requiring people to separate paper from other waste materials is a key to increased recycling. Otherwise, paper becomes so contaminated with other trash that wastepaper dealers won't buy it. Slick paper magazines and glossy newspaper and advertising supplements cause contamination and must not be included.

Textbooks and magazines with color photos and illustrations, which need a special kind of paper, cannot yet be printed effectively on recycled paper because of poor inking and other problems.

Tax subsidies and other financial incentives that make it cheaper to produce paper from trees than from recycling hinder wastepaper recycling in the United States. Widely fluctuating prices and a lack of demand for recycled paper products also make recycling wastepaper a risky financial venture.

---

*You can reduce your junk mail by about 75% by writing to Mail Preference Service, Direct Marketing Service, 11 West 42nd St., P.O. Box 3861, New York, NY 10163-3861. They will stop your name from being added to most large mailing list companies.

For example, since 1988 the supply of recycled newspapers has exceeded the capacity of U.S. paper mills to use it and the price for recycled paper has plummeted. The problem is that only 8 of 23 U.S. paper mills (and only 1 of the 40 Canadian mills) have modernized to produce acceptable newsprint from recycled paper. Most of the 11 new mills that will open by 1992 are not designed to effectively use recycled paper. Loans and tax credits to companies that invest in paper-recycling equipment could help ease this unfortunate situation that is discouraging communities from recycling paper just at a time when consumer interest in doing this has soared.

If the demand for recycled paper products increased, recycled paper would be cheaper and the price paid for wastepaper would rise. One way to increase demand is to require federal and state governments to use recycled paper products as much as possible. Half of the trash the government throws away is paper. A tax can also be added to every metric ton of virgin newsprint that is used, as is now done in Florida.

In the mid-1970s, Congress passed a law calling for federal agencies to buy as many recycled products as practical. But this law has failed because it contains so many exemptions that almost nothing has to be recycled.

Simple measures like asking teachers to instruct their students to write on both sides of the paper would also reduce unnecessary paper waste and increase environmental awareness. Conservationists call for national, state, and local policies designed to recycle half of the wastepaper in the United States by the year 2000.

 RANGELANDS

15-5 **The World's Rangeland Resources** Almost half of the earth's ice-free land is **rangeland**—land that supplies forage or vegetation (grasses, grasslike plants, and shrubs) for grazing and browsing animals. Most rangelands are grasslands in semiarid areas too dry for rain-fed cropland (Figure 5-2, p. 87). Only about 42% of the world's rangeland is used for grazing livestock. Much of the rest is too dry, cold, or remote from population centers to be grazed by large numbers of livestock animals.

Often grassland is community property, owned by inhabitants of a village or a tribe. This often discourages good care and sound management because of the tragedy of the commons (see Spotlight on p. 10).

About 29% of the total land area of the United States is rangeland. Most of this is short-grass prairies in the arid and semiarid western half of the country (Figure 5-11, p. 92). The remaining one-third of U.S. rangeland is owned by the general public and managed by the federal government, mostly by the Forest Service and the Bureau of Land Management (Figure 15-9).

Characteristics of Rangeland Vegetation Most of the grasses on rangelands have deep, complex root systems (Figure 12-4, p. 270). The multiple branches of their roots make these grasses hard to uproot, helping prevent soil erosion. When the leaf tip of most plants is eaten, the leaf stops growing. But each leaf of rangeland grass grows from its base, not its tip. When the upper half of the shoot and leaves of grass is eaten, the plant can grow back quickly. However, the lower half of the plant must remain if the plant is to survive and grow new leaves. As long as only the upper half is eaten, rangeland grass is a renewable resource that can be grazed again and again.

Rangeland Carrying Capacity and Overgrazing Each type of grassland has a herbivore **carrying capacity**—the maximum number of herbivores a given area can support without consuming the metabolic reserve needed for grass renewal. Carrying capacity is influenced by season, range condition, annual climatic conditions, past grazing use, soil type, kinds of grazing animals, and how long animals graze in an area.

Light to moderate grazing is necessary for the health of grasslands. It maintains water and nutrient cycling needed for healthy grass growth and healthy root systems, hinders soil erosion, and encourages buildup of organic soil matter.

Overgrazing occurs when too many grazing animals feed too long and exceed the carrying capacity of a grassland area. Large populations of wild herbivores can overgraze range in prolonged dry periods. But most overgrazing is caused by excessive numbers of livestock feeding too long in a particular area.

Figure 15-17 compares normally grazed and severely overgrazed grassland. Heavy overgrazing converts continuous grass cover to patches of grass and makes the soil more vulnerable to erosion, especially by wind. Then forbs (herbs other than grass) and woody shrubs such as mesquite and prickly cactus invade and take over.

Severe overgrazing combined with prolonged drought can convert potentially productive rangeland to desert (see Figure 15-18 and Spotlight on p. 273). Dune buggies, motorcycles, and other off-road vehicles also damage or destroy rangeland vegetation.

Rangeland Management The major goal of range management is to maximize livestock productivity without degrading grassland quality. The most widely used way to prevent overgrazing is to control the **stocking rate,** the number of a particular kind of animal grazing on a given area, so it doesn't exceed the carrying capacity.

Figure 15-17 Overgrazed (left) and lightly grazed (right) rangeland.

Figure 15-18 Desertification in this arid outback region of Australia was caused by cattle overgrazing the vegetation. Livestock cropped plants so severely that the vegetation died; trampling prevented the establishment of seedlings.

But determining the carrying capacity of a range site is difficult and costly. Even when the carrying capacity is known, it can change due to drought, invasions by new species, and other environmental factors.

Controlling the distribution of grazing animals over a range is the best way to prevent overgrazing and undergrazing. Ranchers can control distribution by building fences to protect degraded rangeland, rotating livestock from one grazing area to another, providing supplemental feeding at selected sites, and locating water holes and salt blocks in strategic places.

A more expensive and less widely used method of rangeland management is to suppress the growth of unwanted plants by spraying with herbicides, mechanical removal, or controlled burning. A cheaper and more effective way to remove unwanted vegetation is controlled, short-term trampling by large numbers of livestock.

Growth of desirable vegetation can be increased by seeding and applying fertilizer, but this method is usually too costly. On the other hand, reseeding is an excellent way to restore severely degraded rangeland.

Many ranchers still promote the use of poisons, trapping, and shooting to kill rabbits and rodents (such as prairie dogs), which compete with livestock for range vegetation. But this usually gives only temporary relief and is rarely worth the cost because these animals have high reproduction rates and their populations can usually recover in a short time.

For decades U.S. ranchers have shot, trapped, and poisoned predators, such as coyotes, which sometimes kill sheep and goats. However, experience has shown that killing predators is an expensive and temporary solution, one that sometimes makes matters worse.

Government agencies are required by law to manage public rangelands according to the principle of multiple use. For years, ranchers and conservationists have battled over how much ranchers should be charged for the privilege of grazing their livestock on public lands (see Pro/Con on p. 349).

15-6 PARKS: USE AND ABUSE

Threats to Parks In 1912 Congress created the U.S. National Park System and declared that national parks are to be set aside to conserve and preserve scenery, wildlife, and natural and historic objects for the use, observation, health, and pleasure of people. The parks are to be maintained in a manner that leaves them unimpaired for future generations.

Today, there are over 1,000 national parks in more than 120 countries. This is an important achievement in the global conservation movement, spurred by the development of the world's first national park system in the United States. In addition to national parks, the U.S. public has access to state, county, and city parks. Most state parks are located near urban areas and thus are used more heavily than national parks.

But these parks are increasingly threatened. In MDCs many national parks are threatened by nearby industrial development, urban growth, air and water pollution, roads, noise, invasion by alien species, and loss of natural species. Some of the most popular national parks are also threatened by overuse.

In LDCs the problems are worse. Plant and animal life in national parks is being threatened by local people who desperately need wood, cropland, and other resources. Poachers kill animals and sell their parts, such as rhino horns, elephant tusks, and furs. Park services in these countries have too little money and staff to fight these invasions, either through enforcement or public education programs. Also, most national parks in MDCs and LDCs are too small

Over 26,000 ranchers lease rights to graze on public range from the Bureau of Land Management and pay a grazing fee for this privilege. In 1981 the Reagan administration set grazing fees at about one-fifth of the average market value of federal grazing lands, a practice continued by the Bush administration in 1990.

This means that U.S. taxpayers give the 2% of ranchers with federal grazing permits subsidies amounting to about $75 million a year—the difference between the fees collected and the actual value of the grazing on this land. Each year the government also spends hundreds of millions of dollars to manage these rangelands. Overall, the government collects only about $1 from ranchers in grazing fees for every $10 spent on range management.

Conservationists call for grazing fees on public rangeland to be raised to a fair market value for use of this land. Higher fees would reduce incentives for overgrazing and provide more money for improvement of range conditions, wildlife conservation, and watershed management.

Government studies show that in 1989 about 70% of rangeland on public lands was overgrazed and in poor or fair condition. Because it is an ecological failure and economic failure (for taxpayers), conservationists believe that the current permit system should be replaced with a competitive bidding system. The primary purpose of the new system would be the long-term conservation of the public rangeland's wildlife, water, vegetation, and soil resources.

Livestock grazing would be allowed only on range in good or excellent condition and would be strictly controlled to prevent overgrazing. If the bids did not reflect the current market value of the forage, no permit would be issued. Permits would last for only 3 to 5 years. Failure to live up to the permit requirements would lead to its automatic cancellation. Ranchers with permits would share 50-50 with the government the cost of capital improvements related to livestock grazing.

Ranchers with permits fiercely oppose higher grazing fees and competitive bidding. Grazing rights on public land raise the value of their livestock animals by $1,000 to $1,500 per head. This means that a permit to graze 500 cattle on public land can be worth $500,000 to $750,000 a year to the rancher. The economic value of a permit is included in the overall worth of the ranches and can be used as collateral for a loan.

Ranchers contend that overgrazing on public rangeland is also caused by increasing numbers of elk and other grazing wildlife. The ranchers resent the fact that the government is forcing them to reduce cattle numbers to reduce overgrazing, but is not requiring game and fish departments to reduce excessive wildlife numbers.

Most ranchers who can't get a grazing permit favor open bidding for grazing rights on public land. They believe that the permit system gives politically influential ranchers an unfair economic advantage at the expense of U.S. taxpayers.

Some conservationists believe that all commercial grazing of livestock on public lands should be phased out over a 10-to-15-year period. Only about 7% of the forage consumed nationally by cattle and lambs comes from public land. Thus, phasing out livestock grazing on public lands would have little effect on the overall production and price of beef and lamb, save taxpayers money, and allow restoration of degraded public rangeland. What do you think should be done?

to sustain many of their natural species, especially larger animals.

U.S. National and State Parks   The National Park System is dominated by 50 national parks found mostly in the West (Figure 15-9). These repositories of majesty, beauty, and biological diversity have been called America's crown jewels. They are supplemented by numerous state parks.

The major problems of national and state parks stem from their spectacular success. Because of more roads, cars, and affluence, annual recreational visits to National Park System units increased 12-fold and visits to state parks 7-fold between 1950 and 1989.

Under the onslaught of people during the peak summer season, the most popular national and state parks are often overcrowded with cars and trailers and are plagued by noise, traffic jams, litter, vandalism, deteriorating trails, polluted water, drugs, and crime. Park Service rangers now spend an increasing amount of their time on law enforcement instead of resource conservation and management.

Populations of wolves, bears, and other large predators in and near various parks have dropped sharply or disappeared because of excessive hunting, poisoning by ranchers and federal officials, and the limited size of most parks. This decline has allowed populations of remaining prey species to increase sharply, destroy vegetation, and crowd out other native animal species.

The greatest danger to many parks today is from human activities in nearby areas. Wildlife and recreational values are threatened by mining, timber harvesting, grazing, coal-burning power plants, water diversion, and urban development. Within your lifetime the greatest threat to many of the world's parks may be projected shifts in regional climate caused by an enhanced greenhouse effect (Section 10-2).

Park Management: Combining Conservation and Sustainable Development Some park managers, especially in LDCs, are developing integrated management plans that combine conservation and sustainable development of the park and surrounding areas (Figure 15-6). In such a plan, the inner core and especially vulnerable areas of the park are protected from development and treated as wilderness. Controlled numbers of people are allowed to use these areas for hiking, nature study, ecological research, and other nondestructive recreational and educational activities.

In other areas controlled commercial logging, sustainable grazing by livestock, and sustainable hunting and fishing by local people are allowed. Money spent by park visitors adds to local income. By involving local people in developing park management plans, managers help them see the park as a vital resource they need to protect and sustain rather than degrade (see Case Study on p. 338).

In most cases, however, the protected inner core is too small to sustain many of its natural species. Such plans look good on paper, but often they cannot be carried out because of a lack of funds for land acquisition, enforcement, and maintenance.

In 1988 the Wilderness Society and the National Parks and Conservation Association suggested a blueprint for the future of the U.S. National Park System that included the following proposals:

- Educate the public about the urgent need to protect, mend, and expand the system.

- Establish the National Park Service as an independent agency responsible to the president and Congress. This would make it less vulnerable to the shifting political winds of the Interior Department.

- Block the mining, timbering, and other threats that are taking place near park boundaries on land managed by the Forest Service and the Bureau of Land Management.

- Acquire new parkland near threatened areas and add at least 75 new parks within the next decade. About half of the most important types of ecosystems in the United States are not protected in national parks.

- Locate most commercial park facilities (such as restaurants and shops) *outside* park boundaries.

- Raise the fees charged to private concessionaires who operate restaurants, camping, food, and recreation services inside national parks to at least 20% of their gross receipts. The present maximum return for taxpayers is only 5%, and the average is only 2.5%. Many of the large concessionaire companies have worked out contracts of up to 30 years with national park officials in which they pay the government as little as 0.75% of their gross receipts.

- Halt concessionaire ownership of facilities in national parks, which makes buying buildings back very expensive.

- Wherever feasible, place visitor parking areas outside the park areas. Use low-polluting vehicles to carry visitors to and from parking areas, and for transportation within the park.

- Greatly expand the Park Service budget for maintenance and science and conservation programs.

- Make buildings and vehicles in national and state parks educational showcases for improvements in energy efficiency and the latest developments in the use of energy from the sun, wind, flowing water, and the earth's interior heat.

- Require the Park Service and the Forest Service to develop integrated management plans so activities in nearby national forests don't degrade national parklands.

15-7 WILDERNESS PRESERVATION

How Much Wild Land Is Left? According to the Wilderness Act of 1964, **wilderness** consists of those areas "where the earth and its community of life are untrammeled by man, where man himself is a visitor who does not remain." The Wilderness Society estimates that a wilderness area should contain at least 400,000 hectares (1 million acres). Otherwise, the area can be degraded by air pollution, water pollution, and noise pollution from nearby mining, oil and natural gas drilling, timber cutting, industry, and urban development.

A 1987 survey sponsored by the Sierra Club revealed that only about 34% of the earth's land area is undeveloped wilderness in blocks of at least 400,000 hectares. About 30% of these remaining wildlands are forests. Many are in tropical forests, which are being rapidly cleared and degraded. Tundra and desert make up most of the world's remaining wildlands. Only about 20% of the undeveloped lands identified in this survey are protected by law from exploitation.

Why Preserve Wilderness? There are many reasons. We need wild places where we can experience majestic beauty and natural biological diversity. We need places where we can enhance our mental health by getting away from noise, stress, and large numbers of people. Wilderness preservationist John Muir advised:

Climb the mountains and get their good tidings. Nature's peace will flow into you as the sunshine into the trees. The winds will blow their freshness into you, and the storms their energy, while cares will drop off like autumn leaves.

Even if individuals do not use the wilderness, many want to know it is there, a feeling expressed by novelist Wallace Stegner:

Save a piece of country . . . and it does not matter in the slightest that only a few people every year will go into it. This is precisely its value . . . we simply need that wild country available to us, even if we never do more than drive to its edge and look in. For it can be a means of reassuring ourselves of our sanity as creatures, a part of the geography of hope.

Wilderness areas provide recreation for growing numbers of people. Wilderness also has important ecological values. It provides undisturbed habitats for wild plants and animals, maintains diverse biological reserves protected from degradation, and provides a laboratory in which we can discover how nature works. It is an ecological insurance policy against eliminating too much of the earth's natural biological diversity.

But to sustainable-earth conservationists the most important reason for protecting and expanding the world's wilderness areas is based on ethical grounds. Wilderness should be preserved because the wild species it contains have a right to exist without human interference (Section 2-3).

U.S. Wilderness Preservation System In the United States, preservationists have been trying to keep wild areas from being developed since 1900. Mostly they have fought a losing battle. It was not until 1964 that Congress passed the Wilderness Act. It allows the government to protect undeveloped tracts of public land from development as part of the National Wilderness Preservation System (see Spotlight on p. 341).

Only 3.9% of U.S. land area is protected as wilderness, with almost two-thirds of this in Alaska. Only 1.8% of the land area of the lower 48 states is protected in the wilderness system. Of the 413 wilderness areas there, only 4 consist of more than 400,000 hectares.

There remain almost 40.5 million hectares (100 million acres) of public lands that could qualify for designation as wilderness. Conservationists believe that all of this land should be protected as wilderness and that a massive effort should be mounted to rehabilitate other lands to enlarge existing wilderness areas and thus allow them to sustain themselves. But resource developers lobby elected officials and government agencies to build roads in these areas so that they can't be designated as wilderness.

In addition to setting aside more wilderness areas, conservationists believe that many areas need to be reclaimed as wilderness. This would be done by closing some roads in national forests and other public lands, restoring and rehabilitating native ecosystems, and reintroducing native wildlife species to create sustainable preserves of 400,000 hectares (1 million acres) or more.

Use and Abuse of Wilderness Areas Popular wilderness areas, especially in California, North Carolina, and Minnesota, are visited by so many people their wildness is threatened. Fragile vegetation is damaged, soil is eroded from trails and campsites, water is polluted from bathing and dishwashing, and litter is scattered along trails. Instead of quiet and solitude, visitors sometimes face the noise and congestion they are trying to escape.

Wilderness areas are also being degraded by air, water, and noise pollution from nearby grazing, logging, oil and gas drilling, factories, power plants, and urban areas. Projected global warming from the greenhouse effect is expected to be the biggest threat to wilderness, parks, forests, rangelands, croplands, estuaries, and inland wetlands during your lifetime (Section 10-2).

Wilderness Management To protect the most popular areas from damage, wilderness managers have had to limit the number of people hiking or camping at any one time. They have also designated areas where camping is allowed. Managers have increased the number of wilderness rangers to patrol vulnerable areas and enlisted volunteers to pick up trash discarded by thoughtless users.

Historian and wilderness expert Roderick Nash suggests that wilderness areas be divided into three categories. The easily accessible, popular areas would be intensively managed and have trails, bridges, hikers' huts, outhouses, assigned campsites, and extensive ranger patrols. Large, remote wilderness areas would not be intensively managed. They would be used only by people who get a permit by demonstrating their wilderness skills. A third category would consist of large, biologically unique areas. They would be left undisturbed as gene pools of plant and animal species, with no human entry allowed.

National Wild and Scenic Rivers System In 1968 Congress passed the National Wild and Scenic Rivers Act. It allows rivers and river segments with outstanding scenic, recreational, geological, wildlife, historical, or cultural values to be protected in the National Wild and Scenic Rivers System. The only activities allowed are camping, swimming, nonmotorized boating, sport hunting, and sport and commercial fishing. New mining claims, however, are permitted in some areas.

Conservationists have urged Congress to add 1,500 additional eligible river segments to the system by the year 2000. If this goal is achieved, about 2% of the country's unique rivers would be protected from further development. Conservationists also urge that a permanent federal administrative body be

- If you or your loved ones own forested land, develop a management and conservation plan for the sustainable use of these resources.

- Plant trees on a regular basis. Try to plant and care for a tree for every book you purchase.

- Cut down on the use of wood and paper products you don't really need, recycle paper products, and buy recycled paper. Also, pressure your office or school to start a paper recycling program. Try to buy recycled paper with a C-1 or C-1+ rating since it contains the highest percentage of consumer wastepaper that has actually been sold, used, discarded, collected, deinked, and recycled. Recycled paper with the lowest rating of C-4 is made only from paper mill scraps and similar sources and contains no recycled consumer wastepaper.

- Recycle aluminum cans or better yet switch to reusable glass beverage containers (Section 19-6). Recognize that using aluminum cans or other throwaway aluminum products encourages the development of more aluminum mines and smelters in the Amazon basin, powered by dams. These dams flood large areas of tropical forests and displace indigenous people from their lands.

- Reduce your consumption of beef, some of which is produced by clearing tropical forests.

- Don't buy tropical hardwoods. Claims that wood comes from sustainable sources are almost impossible to verify. If you must buy such hardwoods, look for the Good-Wood Seal given by Friends of the Earth.

- Contribute time, money, or both to organizations devoted to forest conservation, especially of the world's rapidly disappearing tropical forests (see Appendix 1).

- Physically help protect forests, especially ancient forests, from being cleared or degraded (see Case Study on p. 32 and Pro/Con on p. 167).

- Don't buy a cut Christmas tree or use a plastic one. Instead, buy a living tree in a tub, which can be replanted after the holidays.

- Help rehabilitate or restore a degraded area of forest or grassland near where you live.

- If you visit a park or wilderness area, don't harm or remove anything, and when you leave, don't leave anything behind.

- Support efforts to protect large areas of the world's remaining undeveloped lands and wild rivers as wilderness. In the United States, lobby elected officials to protect all undeveloped public lands and wild and scenic river segments in the lower 48 states from development.

- Pressure banks and international lending agencies not to lend money for environmentally destructive projects.

- Don't buy products produced by companies involved in destructive development projects in tropical forests. Information on such companies can be obtained from the Rainforest Action Network and Friends of the Earth (see Appendix 1).

- Lobby elected officials to protect ancient forests on public lands, reduce timber harvests in national forests, require that taxpayers get a fair return on all timber harvesting and livestock grazing on public lands, eliminate subsidies for cutting virgin timber, and adopt a national program for recycling half of the wastepaper in the United States by the year 2000.

- Support efforts to protect, expand, and mend the National Park System and get a fair return from concessionaires operating park services.

established to manage the Wild and Scenic Rivers System.

Sustaining existing forests, rangelands, wilderness, and parks and rehabilitating those that we have degraded is an urgent task. It will cost a great deal of money and require strong support from the public and changes in individual lifestyles (see Individuals Matter above). But it will cost our civilization much more if we do not protect these resources from degradation and destruction and help heal those we have wounded.

Forests precede civilizations, deserts follow them.

FRANCOIS-AUGUSTE-RENÉ DE CHATEAUBRIAND

DISCUSSION TOPICS

1. Explain why you agree or disagree with each of the proposals listed on pages 334–335 concerning protection of the world's tropical forests.

2. Should private companies cutting timber from national forests continue to be subsidized by federal payments for reforestation and for building and maintaining access roads? Explain.

3. Should all cutting, except carefully controlled selective cutting, on remaining ancient forests in U.S. national forestlands be banned? Explain.

4. Should exports of timber cut from U.S. national forests and other public lands be banned? Explain.

5. Should fees for grazing on public rangelands in the United States be **(a)** eliminated and replaced with a

competitive bidding system, **(b)** increased to the point where they equal the fair market value estimated by the Bureau of Land Management and the Forest Service? Explain your answers.

6. Should trail bikes, dune buggies, and other off-road vehicles be banned from public rangeland to reduce damage to vegetation and soil? Explain.

7. Explain why you agree or disagree with each of the proposals listed on page 350 concerning the U.S. National Park System.

8. Should more wilderness areas and wild and scenic rivers be preserved in the United States, especially in the lower 48 states? Explain.

FURTHER READINGS

Allin, Craig W. 1982. *The Politics of Wilderness Preservation.* Westport, Conn.: Greenwood Press.

Anderson, Anthony B., ed. 1990. *Alternatives to Deforestation: Steps Toward Sustainable Use of the Amazon Rain Forest.* Irvington, N.Y.: Columbia University Press.

Caufield, Catherine. 1985. *In the Rainforest.* New York: Alfred A. Knopf.

Chase, Alston. 1986. *Playing God in Yellowstone: The Destruction of America's First National Park.* New York: Atlantic Monthly Press.

Clawson, Marion. 1983. *The Federal Lands Revisited.* Washington, D.C.: Resources for the Future.

Clay, David. 1986. *Timber and the Forest Service.* Lawrence: University of Kansas Press.

Collard, Andree, and Joyce Contrucci. 1989. *Rape of the Wild: Man's Violence Against Animals and the Earth.* Bloomington: Indiana University Press.

Eckholm, Erik, et al. 1984. *Fuelwood: The Energy Crisis That Won't Go Away.* Washington, D.C.: Earthscan.

Ervin, Keith. 1989. *Fragile Majesty: The Battle for North America's Last Great Forest.* Seattle, Wash.: The Mountaineers.

Fritz, Edward. 1983. *Sterile Forest: The Case Against Clearcutting.* Austin, Tex.: Eakin Press.

Frome, Michael. 1974. *The Battle for the Wilderness.* New York: Praeger.

Frome, Michael. 1983. *The Forest Service.* Boulder, Colo.: Westview Press.

Goodland, Robert, ed. 1990. *Race to Save the Tropics: Ecology and Economics for a Sustainable Future.* Covelo, Calif.: Island Press.

Gradwohl, Judith, and Russell Greenberg. 1988. *Saving the Tropical Forests.* Covelo, Calif.: Island Press.

Hartzog, George B., Jr. 1988. *Battling for the National Parks.* New York: Moyer Bell.

Hazlewood, Peter T. 1989. *Cutting Our Losses: Policy Reform to Sustain Tropical Forest Resources.* Washington, D.C.: World Resources Institute.

Head, Catherine, and Robert Heinzman. 1989. *Lessons of the Rainforest.* San Francisco: Sierra Club Books.

Heady, H. F. 1975. *Rangeland Management.* New York: McGraw-Hill.

Hecht, Susanna, and Alexander Cockburn. 1989. *The Fate of the Forest: Developers, Destroyers, and Defenders of the Amazon.* New York: Verso (Routledge, Chapman, and Hall).

Hendee, John, et al., eds. 1977. *Principles of Wilderness Management.* Washington, D.C.: Government Printing Office.

Jacobs, Marius. 1988. *The Tropical Rain Forest: A First Encounter.* New York: Springer-Verlag.

Kelly, David. 1988. *Secrets of the Old Growth Forest.* Layton, Utah: Gibbs Smith.

Leopold, Aldo. 1949. *A Sand County Almanac.* New York: Oxford University Press.

Maser, Chris. 1988. *The Redesigned Forest.* San Pedro, Calif.: R. & E. Miles.

Myers, Norman. 1984. *The Primary Source: Tropical Forests and Our Future.* New York: W. W. Norton.

Nash, Roderick. 1982. *Wilderness and the American Mind.* 3rd ed. New Haven, Conn.: Yale University Press.

National Parks and Conservation Association. 1988. *Blueprint for National Parks* (9 vols.). Washington, D.C.: National Parks and Conservation Association.

Norse, Elliot A. 1990. *Ancient Forests of the Northwest.* Washington, D.C.: Island Press.

O'Toole, Randal. 1987. *Reforming the Forest Service.* Covelo, Calif.: Island Press.

Page, Diana. 1989. "Debt-for-Nature Swaps: Experience Gained, Lessons Learned." *International Environmental Affairs*, vol. 1, no. 4, 275–288.

Postel, Sandra, and Lori Heise. 1988. *Reforesting the Earth.* Washington, D.C.: Worldwatch Institute.

Repetto, Robert. 1988. *The Forest for the Trees? Government Policies and the Misuse of Forest Resources.* Washington, D.C.: World Resources Institute.

Repetto, Robert. 1990. "Deforestation in the Tropics." *Scientific American*, vol. 262, no. 4, 36–42.

Robinson, Gordon. 1987. *The Forest and the Trees: A Guide to Excellent Forestry.* Covelo, Calif.: Island Press.

Runte, Alfred. 1987. *National Parks: The American Experience.* 2nd ed. Lincoln: University of Nebraska Press.

Shanks, Bernard. 1984. *This Land Is Your Land.* San Francisco: Sierra Club Books.

Sierra Club. 1982. *Our Public Lands: An Introduction to the Agencies and Issues.* San Francisco: Sierra Club Books.

Simon, David J., ed. 1988. *Our Common Lands: Defending the National Parks.* Washington, D.C.: Island Press.

Society of American Foresters. 1981. *Choices in Silviculture for American Forests.* Washington, D.C.: Society of American Foresters.

Stoddard, Charles H., and Glenn M. Stoddard. 1987. *Essentials of Forestry Practice.* 4th ed. New York: John Wiley.

Wilcove, David S. 1988. *National Forests: Policies for the Future*, vols. 1 & 2. Washington, D.C.: Wilderness Society.

Wilderness Society. 1988. *Ancient Forests: A Threatened Heritage.* Washington, D.C.: The Wilderness Society.

Wilson, Edward O. 1989. "Threats to Biodiversity." *Scientific American*, Sept., 108–116.

Wuerthner, George. 1989. *Yellowstone and the Fires of Change.* Salt Lake City, Utah: Dream Garden Press.

Zaslowsky, Dyan, and The Wilderness Society. 1986. *These American Lands.* New York: Henry Holt.

CHAPTER SIXTEEN

WILD PLANT AND ANIMAL RESOURCES

General Questions and Issues

1. Why are wild species of plants and animals important to us and to the ecosphere?

2. What human activities and natural traits cause wild species to become depleted, endangered, and extinct?

3. How can endangered and threatened wild species be protected from premature extinction caused by human activities?

4. How can populations of large game be managed to have enough animals available for sport hunting without endangering the long-term survival of the species?

5. How can populations of species of freshwater and marine fish be managed to have enough available for commercial and sport fishing without endangering their long-term survival?

The mass of extinctions which the Earth is currently facing is a threat to civilization second only to the threat of thermal nuclear war.

NATIONAL ACADEMY OF SCIENCES

I N THE EARLY 1850S, Alexander Wilson, a prominent ornithologist, watched a single migrating flock of passenger pigeons darken the sky for over four hours. He estimated that this flock consisted of more than 2 billion birds and was 386 kilometers (240 miles) long and 1.6 kilometers (1 mile) wide.

By 1914 the passenger pigeon (Figure 16-1) had disappeared forever. How could the species that was once the most abundant bird in North America become extinct in only a few decades?

The answer is people. The major reasons for the extinction of this species were uncontrolled commercial hunting and loss of habitat and food supplies as forests were cleared for farms and cities.

Passenger pigeons were good to eat and were widely used for fertilizer. They were easy to kill because they flew in gigantic flocks and nested in long narrow colonies. People captured one pigeon alive and tied it to a perch called a stool. Soon a curious flock landed beside this "stool pigeon." They were then shot or trapped by nets that might contain more than 1,000 birds.

Beginning in 1858, the massive killing of passenger pigeons became a big business. Shotguns, fire, traps, artillery, and even dynamite were used. Birds were also suffocated by burning grass or sulfur below their roosts. Live birds were used as targets in shooting galleries. In 1878 one professional pigeon trapper made $60,000 by killing 3 million birds at their nesting grounds near Petoskey, Michigan.

By the early 1880s, commercial hunting ceased because only several thousand birds were left. Recovery of the species was essentially impossible because these birds laid only one egg per nest. Many of the remaining birds died from infectious disease and from severe storms during their annual fall migration to Central and South America.

By 1896 the last major breeding colony had vanished, and by 1900 only a few small, scattered flocks were left. In 1914 the last known passenger pigeon on earth, a hen named Martha after Martha Washington, died in the Cincinnati Zoo. Her stuffed body is now on view at the National Museum of Natural History in Washington, D.C. (Figure 16-1).

Sooner or later all species become extinct. However, we have become a major factor in the premature extinction of an increasing number of species. Every hour about four species become extinct because of our activities. Most mobile species that remain are forced to retreat into smaller and smaller sanctuaries in response to our relentless march across the globe.

Over the next few decades, our activities are expected to hasten the mass extinction of as many as 1 million species. The planet has not suffered a loss of this magnitude since the end of the age of dino-

Figure 16-1 The extinct passenger pigeon. The last known passenger pigeon died in the Cincinnati Zoo in 1914.

Figure 16-2 The nine-banded armadillo is used in research to find a cure for leprosy.

saurs, some 65 million years ago. Reducing this premature loss of the earth's biological diversity and restoring species that we have helped deplete are planetary emergencies that we must deal with now.

16-1 WHY PRESERVE WILD PLANT AND ANIMAL SPECIES?

Economic and Medical Importance Wild species that are actually or potentially useful to people are called **wildlife resources**. They are potentially renewable resources, if not driven to extinction or near extinction by our activities.

Most of the plants that supply 90% of the world's food today were domesticated from wild plants found in the tropics. Existing wild plant species, most of them still unclassified and unevaluated, will be needed by agricultural scientists and genetic engineers to develop new crop strains (Section 13-3). Wild animal species are a largely untapped source of food.

About 75% of the world's population relies on plants or plant extracts as sources of medicines. Roughly half of the prescription and nonprescription drugs used in the world, and 25% of those used in the United States today, have active ingredients extracted from wild organisms. Only about 5,000 of the world's estimated 250,000 to 300,000 plant species have been studied thoroughly for their possible medical uses. Many wild animal species are used to test drugs, vaccines, chemical toxicity, and surgical procedures and to increase our understanding of human health and disease (Section 8-1 and Figure 16-2).

However, animal rights advocates are protesting the use of animals in medical and biological research and teaching. Bacteria, cell and tissue cultures, and computer models are now used as alternatives for the use of animals (Section 8-1). But scientists argue

that some animal testing is vital for the welfare of people, pets, and livestock.

Aesthetic and Recreational Importance Wild plants and animals are a source of beauty, wonder, joy, and recreational pleasure for large numbers of people. Wild **game species** provide recreation in the form of hunting and fishing. Each year, almost one of every two Americans and 84% of the Canadian population participate in birdwatching, photographing, or other nondestructive forms of outdoor recreational activity involving wildlife.

Wildlife tourism is important to the economy of some LDCs, such as Kenya. One wildlife economist estimated that one male lion living to seven years of age in Kenya leads to $515,000 of expenditures by tourists. If the lion were killed for its skin, it would be worth only about $1,000.

Ecological Importance Wild species supply us and other species with food from the soil and the sea, recycle nutrients essential to agriculture, and help produce and maintain fertile soil. They also produce and maintain oxygen and other gases in the atmosphere, moderate the earth's climate, help regulate water supplies, and store solar energy as chemical energy in food, wood, and fossil fuels. Moreover, they filter and detoxify poisonous substances, decompose wastes, control most potential crop pests and carriers of disease, and make up a vast gene pool of biological diversity from which we and other species can draw.

Ethical Importance So far, the reasons given for preserving wildlife are based on the actual or potential usefulness of wild species as resources for people. Many ecologists and conservationists believe that wild species will continue to disappear at an alarming rate until we replace this *human-centered (anthropocentric)*

355

Figure 16-3 California condor in captivity at the Los Angeles Zoo. None of these birds are left in the wild. About 39 are being bred in zoos, with the hope of eventually returning some to the wild.

Figure 16-4 This endangered shallowtail butterfly was almost pushed to extinction in Great Britain, but is now hanging on, mostly in protected nature reserves.

view of wildlife and the environment either with a *life-centered (biocentric)* view or with an *ecosystem-centered (ecocentric)* view (Section 2-3).

According to the biocentric worldview, each wild species has an inherent right to exist, or at least the right to struggle to exist, equal to that of any other species. Thus, it is ethically wrong for us to hasten the extinction of any species. Some go further and believe that each individual wild creature—not just a species—has a right to survive without human interference, just as each human being has the right to survive.

Some distinguish between the survival rights of plants and those of animals. The poet Alan Watts once commented that he was a vegetarian "because cows scream louder than carrots." Many people make ethical distinctions among various types of animals. For instance, they think little about killing a fly, mosquito, cockroach, or sewer rat, or about catching and killing fish they don't eat. Unless they are strict vegetarians, they also think little about having others kill cattle, calves, lambs, and chickens in slaughterhouses to provide them with meat, leather, and other animal products. The same people, however, might deplore the killing of game animals such as deer, squirrels, or rabbits for sport or for food.

The ecocentric worldview stresses the importance of maintaining biodiversity by preserving or by not degrading entire ecosystems, rather than focusing only on individual species or an individual organism. It recognizes that saving wildlife means saving the places where they live. This view is based on Aldo Leopold's ethical principle that something is right when it tends to maintain the earth's life-support systems for us and other species, and wrong when it tends otherwise.

HOW SPECIES BECOME DEPLETED AND EXTINCT

Extinction of Species Today Extinction is a natural process (Section 5-4). However, since agriculture began about 10,000 years ago, the rate of species extinction has increased sharply as human settlements have expanded worldwide. The underlying causes of our current and projected gigantic extinction of wildlife are increases in the human population and affluence (Figure 1-11, p. 14), and the global poverty trap (Section 7-3).

Biologist Edward O. Wilson estimated that by 1990 at least 100 species per day were being lost, mostly because of our activities. Wilson, Norman Myers (see Guest Essay on p. 374), and other conservationists warn that if deforestation (especially of tropical forests), desertification, and destruction of wetlands and coral reefs continue at their present rates, then at least 500,000 and perhaps 1 million species will become extinct because of human activities between 1975 and 2000. If this massive loss of biodiversity is not slowed, by the middle of the next century we will have caused the loss of at least one-quarter, and conceivably one-half, of the earth's 5 million to 30 million species.

Animal extinctions get the most publicity. But plant extinctions are more important ecologically because most animal species depend directly or indirectly on plants for food. It is estimated that by the year 2000, from 16% to 25% of all plant species may become extinct because of our activities.

There are important differences between the present mass extinction and those in the past. First, the present "extinction spasm" is being caused by us. Second, it is taking place in only a few decades rather than over several million years. Such extinction cannot

Figure 16-5 There were an estimated 250,000 American bald eagles when this bird became the national symbol in 1782. During the late 1960s and early 1970s the number of American bald eagles in the lower 48 states declined because of loss of habitat, illegal hunting, and reproductive failure caused by pesticides in their primary diet of fish. Federal protection has led to recovery in many areas. In 1989 there were about 35,300 bald eagles in the wild, with 5,300 in the lower 48 states and about 30,000 in Alaska.

Figure 16-6 The whooping crane, shown in its winter refuge in Texas, is an endangered species in North America. Its low reproduction rates and fixed migration pattern make this species vulnerable to extinction. Mostly because of illegal shooting and loss of habitat, the number of whooping cranes in the wild dropped to only 16 by 1941. Because of a $5-million-a-year habitat protection and captive breeding program directed by the U.S. Fish and Wildlife Service, about 217 birds survive today, including more than 170 in the wild.

be balanced by speciation because it takes between 2,000 and 100,000 generations for new species to evolve. Third, plant species are disappearing as rapidly as animal species, thus threatening many animal species that otherwise would not become extinct at this time.

Endangered and Threatened Species Today Species heading toward extinction can be classified as either endangered or threatened. An **endangered species** is one having so few individual survivors that the species could soon become extinct over all or most of its natural range. Examples are the white rhinoceros in Africa (100 left), the California condor (Figure 16-3) in the United States (none left in the wild), the giant panda in central China (1,000 left), the snow leopard in central Asia (2,500 left), and the rare shallowtail butterfly (Figure 16-4).

A **threatened species** is still abundant in its natural range but is declining in numbers and is likely to become endangered. Examples are the bald eagle (Figure 16-5) and the grizzly bear.

Habitat Loss and Disturbance The greatest threat to most wild species is destruction, fragmentation, and degradation of their habitats. We build cities and suburbs, clear forests, drain and fill wetlands, plow up grasslands and plant them with crops, and disturb the land by mining. Such disruption of natural communities threatens wild species by destroying migration routes, breeding areas, and food sources. Deforestation, especially of tropical forests (Section 15-2),

Figure 16-7 Island species are especially vulnerable to extinction. The endangered *Symphonia* clings to life on the island of Madagascar, where 90% of the original vegetation has been destroyed.

is the single greatest cause of the decline in global biological diversity by habitat loss and degradation.

In the United States, forests have been reduced by 33%, wetlands by 50%, and tall-grass prairies by 98%. Furthermore, much of the remaining wildlife habitat is being fragmented and polluted at an alarming rate. Loss or degradation of habitat is the key factor in the extinction of American bird species such as the heath hen and the near extinction of Atwater's prairie chicken, the whooping crane (Figure 16-6), and the California condor (Figure 16-3).

Many rare and threatened plant and animal species live in vulnerable, specialized habitats such as islands (Figure 16-7) or single trees in tropical forests.

When European explorers arrived in North America in the late 1400s, various tribes of Native Americans depended heavily on bison for survival. The meat was their staple diet. The skin was used for tepees, moccasins, and clothes. The gut made their bowstrings, and the horns their spoons. Even the dried feces, called buffalo chips by English-speaking settlers, were used for fuel.

In 1500, before European settlers came to North America, between 60 and 125 million grass-eating American bison roamed the plains, prairies, and woodlands over most of the continent (Figure 16-8). Their numbers were so large that they were thought to be inexhaustible. By 1906, however, the once-massive range of the American bison was reduced to a tiny area, and the species was nearly driven to extinction, mostly because of overhunting and loss of habitat.

As settlers moved west after the Civil War, the sustainable balance between Native Americans and bison was upset. Plains Native Americans traded bison skins to settlers for steel knives and firearms and began killing bison in larger numbers.

But much more severe depletion of this resource was caused by other factors. First, as railroads spread westward in the late 1860s, railroad companies hired professional bison hunters to supply construction crews with meat. The well-known railroad bison hunter "Buffalo Bill" Cody killed an estimated 4,280 bison in only 18 months—surely a world record. Passengers also gunned down bison from train windows purely for the "joy" of killing, leaving the carcasses to rot.

As farmers settled the plains, they shot bison because the animals destroyed crops. Ranchers killed them because they competed with cattle and sheep for grass and knocked over fences, telegraph poles, and sod houses.

An army of commercial hunters shot millions of bison for their hides and for their tongues, which were considered a delicacy. Instead of being eaten, however, most of the meat was left to rot. "Bone pickers" then collected the bleached bones that whitened the prairies and shipped them east for use as fertilizer.

A final major factor in the near extinction of the bison occurred after the Civil War. The U.S. Army killed millions of bison to subdue plains tribes of Native Americans and take over their lands by killing off their major source of food. Between 1870 and 1875, at least 2.5 million bison were slaughtered each year.

By 1890 only one herd of about 1 million bison was left. Commercial hunters and skinners descended on this herd, and by 1892 only 85 bison were left. These were given refuge in Yellowstone National Park and protected by an 1893 law against the killing of wild animals in national parks.

In 1905 sixteen people formed the American Bison Society to protect and rebuild the captive population of the animal. In the early 1900s the federal government established the National Bison Range near Missoula, Montana. Since then, captive herds on this federal land and other herds, mostly on privately owned land scattered throughout the West, have been protected by law.

Today there are about 75,000 bison in the United States—one-fifth of them on the National Bison Range. Some captive bison are crossbred with cattle to produce hybrids, called beefalo. They have a tasty meat, grow faster, and are easier to raise than cattle, and need no expensive grain feed.

Human alterations of ecosystems fragment wildlife habitats into patches, which are often too small to support the minimum number of individuals needed to sustain a population.

Many species of insect-eating, migratory songbirds in North America are being threatened with extinction. Their summer habitats in North America and their winter habitats in the tropical forests of Mexico, Central America, South America, and the Caribbean islands are being destroyed, degraded, or fragmented.

Commercial Hunting and Poaching   There are three major types of hunting: subsistence, sport, and commercial. The killing of animals to provide enough food for survival is called **subsistence hunting. Sport hunting** is the hunting of animals for recreation and in some cases for food. **Commercial hunting** involves killing animals for profit from sale of their furs or other parts. Illegal commercial hunting or fishing is called **poaching**.

Today, subsistence hunting has declined sharply in most parts of the world because of the decrease in hunting-and-gathering societies. Sport hunting is now closely regulated in most countries. Game species are endangered by such hunting only when protective regulations do not exist or are not enforced. No animal in the United States, for instance, has become extinct or endangered because of regulated sport hunting.

Legal and illegal commercial hunting has led to the extinction or near extinction of many animal species in the past (see Case Study above). This

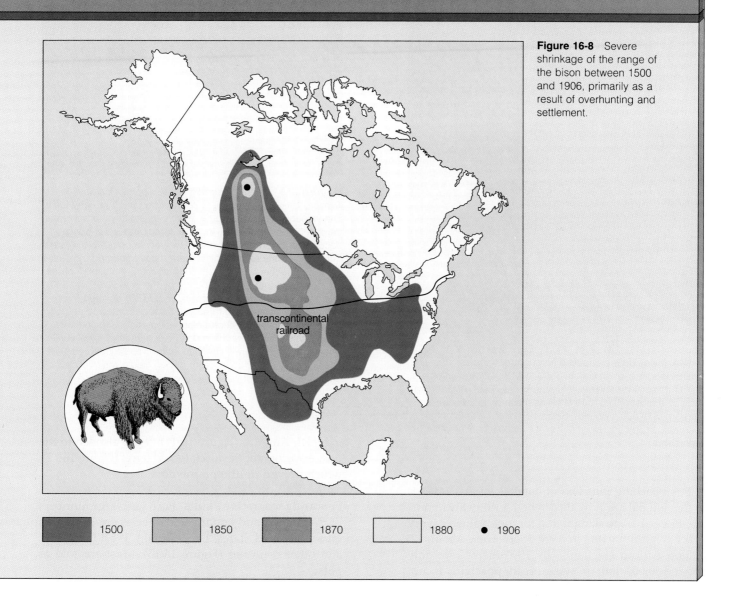

Figure 16-8 Severe shrinkage of the range of the bison between 1500 and 1906, primarily as a result of overhunting and settlement.

transcontinental railroad

| ■ 1500 | ■ 1850 | ■ 1870 | □ 1880 | ● 1906 |

continues today. The jaguar, tiger, snow leopard, and cheetah are hunted for their furs. It is not surprising that Bengal tigers face extinction, since a coat made from their fur sells for $100,000 in Tokyo. A mountain gorilla is worth $150,000; an ocelot skin, $40,000; an imperial Amazon macaw, $30,000; a snow leopard, $14,000; a leopard, $8,500; rhinoceros horn, $12,500 per pound; and tiger meat, $130 per pound.

Poaching is also increasing in the United States, especially in the western half of the country. A poached gyrofalcon sells for $120,000; a bighorn sheep, $45,000; a large saguaro cactus (Figure 4-6, p. 63), $15,000; a peregrine falcon (Figure 14-5, p. 319), $10,000; a polar bear, $6,000; a grizzly bear, $5,000; an elk calf, $5,000; a mountain goat, $3,500; and a bald eagle, $2,500. Most poachers are not caught.

There are more police officers in New York City than wildlife protection officers in the entire United States.

Even if caught, the economic incentive for a poacher far outweighs the risk of paying a small fine, and the much smaller risk of serving time in jail. As more of the world's species become endangered, their economic value and the demand for them on the black market rise sharply, hastening their extinction.

Rhinoceros (Figure 4-34, p. 84), one of the world's oldest mammals, are hunted illegally for their horns. These horns, made of the same material as your fingernails, are converted into ornate daggers in the Middle East (Figure 16-9) and ground into powder and used as alleged aphrodisiacs and in various medicines in China and other parts of Asia. Because

Figure 16-9 In parts of the Middle East, such as North Yemen, rhino horns are carved into dagger handles. These ornate daggers, which sell for $500 to $12,000, are worn as a sign of masculinity and virility. In the Middle East, Asian rhino horns, the most prized, can fetch as much as $28,600 a kilogram ($13,000 a pound). Since 1970 North Yemen has been responsible for about half of the world's consumption of rhino horns.

Figure 16-10 Vultures feeding on an elephant carcass in Tanzania. It was killed by poachers, who cut off its ivory tusks. The valuable ivory is used for jewelry, piano keys, ornamental carvings, and art objects. In 1989 Japan and Hong Kong accounted for about 75% of all raw ivory imports, with western European countries and the United States importing most of the rest. Poachers are slaughtering so many African elephants for their tusks that few may be left by the year 2000. In 1990 members of a 103-nation convention (CITES) devoted to protecting endangered and threatened species voted to ban all international trade of African elephant products. If strictly enforced, this could save the African elephant from extinction. But illegal poaching will continue, and three African countries (including South Africa, a major trader) filed a reservation against this ban that will allow them to continue trading in ivory.

of this poaching plus loss of habitat, the five species of this animal are now threatened with extinction.

Elephants are slaughtered for their valuable ivory tusks (Figure 16-10). In 1970 there were about 4.5 million African elephants. By 1990 there were only about 600,000 left. If widespread poaching is not halted, the African elephant (Figure 5-9, p. 91), could be wiped out within ten years.

Predator and Pest Control Extinction or near extinction can also occur when people attempt to exterminate pest and predator species that compete with humans for food and game. Fruit farmers exterminated the Carolina parakeet in the United States around 1914 because it fed on fruit crops. The species was easy to wipe out because when one member of a flock was shot, the rest of the birds hovered over its body, making themselves easy targets.

As animal habitats have shrunk, farmers have killed large numbers of African elephants to keep them from trampling and eating food crops. Since 1929, ranchers and government agencies have poisoned prairie dogs because sometimes horses and cattle step into the burrows and break their legs. This poisoning has killed 99% of the prairie dog population in North America (Figure 16-11). It has also led to the near extinction of the black-footed ferret (Figure 16-12), which preyed on the prairie dog.

Pets and Decorative Plants Each year large numbers of threatened and endangered species are smuggled into the United States, Great Britain, West Germany, and other countries (Figure 16-13). Most are sold as pets.

For every bird that reaches a pet shop legally or illegally, at least one other dies in transit. After purchase many of these animals are mistreated, killed, or abandoned by their owners.

Some species of exotic plants, especially orchids and cacti, are also endangered because they are gathered, often illegally (Figure 16-14). They are then sold to collectors and used to decorate houses, offices, and landscapes. A collector may pay $5,000 for a single rare orchid. Nearly one-third of the cactus species native to the United States, especially those in Texas and Arizona, are thought to be endangered because they are collected and sold for use as potted plants to nurseries, collectors, and property owners. A single prize specimen can earn cactus rustlers as much as $15,000.

Pollution and Climate Change Toxic chemicals degrade wildlife habitats, including wildlife refuges,

Figure 16-11 The Utah prairie dog is a threatened species in the United States, mostly because of widespread poisoning by ranchers and government agencies since 1929.

Figure 16-12 The black-footed ferret is one of the most endangered mammals in North America, with none left in the wild. Its near extinction occurred because most of the once abundant prairie dogs (Figure 16-11) that it ate have been eliminated. Between 1985 and 1989 the population of black-footed ferrets in captivity grew from 18 to 120.

Figure 16-13 Collectors of exotic birds may pay $10,000 for a threatened hyacinth macaw smuggled out of Brazil. Such high prices help doom such species to eventual extinction.

Figure 16-14 The black lace cactus is one of the many U.S. plants that are endangered, mostly because of development and collectors.

and kill some plants and animals. Slowly degradable pesticides, especially DDT and dieldrin, have caused populations of some bird species to decline (see Spotlight on p. 318).

Wildlife in even the best-protected and best-managed wildlife reserves throughout the world may be depleted in a few decades because of climatic change caused by projected global warming (Section 10-2).

Introduction of Alien Species As people travel around the world, they sometimes pick up plants and animals intentionally or accidentally and introduce these species to new geographical regions. Many of these alien species have provided food, game, and beauty and have helped control pests in their new environments.

However, some alien species have no natural predators and competitors in their new habitats. This allows them to dominate their new ecosystem and reduce the populations of many native species (see Case Study on p. 362). Eventually, they can cause the extinction, near extinction, or displacement of native species (Table 16-1).

Characteristics of Extinction-Prone Species Some species have natural traits that make them more

Figure 16-15 The fast-growing water hyacinth was introduced into Florida from Latin America in 1884. Since then this plant, which can double its population in only two weeks, has taken over waterways in Florida and other southeastern states.

Figure 16-16 The Florida manatee, or sea cow, feeds on aquatic weeds and could help control the growth and spread of the water hyacinth in Florida. But these gentle and playful mammals are threatened with extinction because many have died after becoming entangled in fishing nets and being struck by the propellers of power boats. Only about 1,200 are left in Florida, Georgia, and the Carolinas.

The fast-growing water hyacinth is native to Central and South America. In 1884 a woman took one of these plants from an exhibition in New Orleans and planted it in her backyard in Florida. Within ten years the plant, which can double its population in two weeks, was a public menace.

Unchecked by natural enemies and thriving on Florida's nutrient-rich waters, water hyacinths rap-idly displaced native plants. They also clogged boat traffic in many ponds, streams, canals, and rivers in Florida and in other parts of the southeastern United States (Figure 16-15).

Since 1898 mechanical harvest-ers and a variety of herbicides have been used to keep the plant in check, with little success. Large numbers of Florida manatees, or sea cows (Figure 16-16), can con-trol the growth and spread of water hyacinths in inland waters more effectively than mechanical or chemical methods. But these gentle and playful herbivores are threatened with extinction, mostly from being slashed by powerboat propellers, becoming entangled in fishing gear, or being hit on the head by oars.

In recent years scientists have introduced other alien species that feed on water hyacinths to help control its spread. They include a weevil imported from Argentina, a water snail from Puerto Rico, and the grass carp, a fish brought in from the Soviet Union. These spe-cies can help, but the water snail and grass carp also feed on other desirable aquatic plants.

There is some good news in this story. Preliminary research in-dicates that the water hyacinth can be used in several beneficial ways. They can be introduced in sewage treatment lagoons to ab-sorb toxic chemicals. They can be converted by fermentation to a biogas fuel similar to natural gas, added as a mineral and protein supplement to cattle feed, and ap-plied to the soil as fertilizer. They can also be used to clean up pol-luted ponds and lakes—if their population size can be kept under control.

vulnerable than others to premature extinction (Table 16-2). Each animal species has a critical population density and size, below which survival may be im-possible because males and females have a hard time finding each other. Once the population reaches its critical size, it continues to decline even if the species is protected, because its death rate exceeds its birth rate. The remaining small population can easily be wiped out by fire, flood, landslide, disease, or some other catastrophic event.

Table 16-1 Damage Caused by Plants and Animals Imported into the United States

Name	Origin	Mode of Transport	Type of Damage
Mammals			
European wild boar	Russia	Intentionally imported (1912), escaped captivity	Destruction of habitat by rooting; crop damage
Nutria (cat-sized rodent)	Argentina	Intentionally imported, escaped captivity (1940)	Alteration of marsh ecology; damage to levees and earth dams; crop destruction
Birds			
European starling	Europe	Intentionally released (1890)	Competition with native songbirds; crop damage; transmission of swine diseases; airport interference
House sparrow	England	Intentionally released by Brooklyn Institute (1853)	Crop damage; displacement of native songbirds
Fish			
Carp	Germany	Intentionally released (1877)	Displacement of native fish; uprooting of water plants with loss of waterfowl populations
Sea lamprey	North Atlantic Ocean	Entered via Welland Canal (1829)	Destruction of lake trout, lake whitefish, and sturgeon in Great Lakes
Walking catfish	Thailand	Imported into Florida	Destruction of bass, bluegill, and other fish
Insects			
Argentine fire ant	Argentina	Probably entered via coffee shipments from Brazil (1918)	Crop damage; destruction of native ant species
Camphor scale insect	Japan	Accidentally imported on nursery stock (1920s)	Damage to nearly 200 species of plants in Louisiana, Texas, and Alabama
Japanese beetle	Japan	Accidentally imported on irises or azaleas (1911)	Defoliation of more than 250 species of trees and other plants, including many of commercial importance
Plants			
Water hyacinth	Central America	Intentionally introduced (1884)	Clogging waterways; shading out other aquatic vegetation
Chestnut blight (fungus)	Asia	Accidentally imported on nursery plants (1900)	Destruction of nearly all eastern American chestnut trees; disturbance of forest ecology
Dutch elm disease, *Cerastomella ulmi* (fungus)	Europe	Accidentally imported on infected elm timber used for veneers (1930)	Destruction of millions of elms; disturbance of forest ecology

From *Biological Conservation* by David W. Ehrenfeld. Copyright © 1970 by Holt, Rinehart and Winston, Inc. Modified and reprinted by permission.

16-3 PROTECTING WILD SPECIES FROM EXTINCTION

Methods for Protecting and Managing Wildlife There are three basic approaches to wildlife conservation and management.

1. *The Species Approach:* Protect endangered species by identifying them, giving them legal protection, preserving and managing their critical habits, propagating species in captivity, and reintroducing species in suitable habitats.

2. *The Ecosystem Approach:* Preserve balanced populations of species in their native habitats, establish legally protected wilderness areas and wildlife reserves, and eliminate alien species from an area.

Table 16-2 Characteristics of Extinction-Prone Species

Characteristic	Examples
Low reproduction rate	Blue whale, polar bear, California condor, Andean condor, passenger pigeon, giant panda, whooping crane
Specialized feeding habits	Everglades kite (eats apple snail of southern Florida), blue whale (krill in polar upwelling areas), black-footed ferret (prairie dogs and pocket gophers), giant panda (bamboo), Australian koala (certain types of Eucalyptus leaves)
Feed at high trophic levels	Bengal tiger, bald eagle, Andean condor, timber wolf
Large size	Bengal tiger, African lion, elephant, Javan rhinoceros, American bison, giant panda, grizzly bear
Limited or specialized nesting or breeding areas	Kirtland's warbler (nests only in 6- to 15-year-old jack pine trees), whooping crane (depends on marshes for food and nesting), orangutan (now found only on islands of Sumatra and Borneo), green sea turtle (lays eggs on only a few beaches), bald eagle (prefers habitat of forested shorelines), nightingale wren (nests and breeds only on Barro Colorado Island, Panama)
Found in only one place or region	Woodland caribou, elephant seal, Cooke's kokio, and many unique island species
Fixed migratory patterns	Blue whale, Kirtland's warbler, Bachman's warbler, whooping crane
Preys on livestock or people	Timber wolf, some crocodiles
Certain behavioral patterns	Passenger pigeon and white-crowned pigeon (nest in large colonies), redheaded woodpecker (flies in front of cars), Carolina parakeet (when one bird is shot, rest of flock hovers over body), key deer (forages for cigarette butts along highways—it's a "nicotine addict")

3. *The Wildlife Management Approach:* Manage species, mostly game species, for sustained yield by using laws to regulate hunting, establishing harvest quotas, and developing population management plans.

The Species Approach: Treaties and Laws Several international treaties and conventions help protect wild species. One of the most far-reaching treaties is the 1975 Convention on International Trade in Endangered Species (CITES), developed by the International Union for the Conservation of Nature and Natural Resources (IUCN) and administered by the UN Environment Program. This treaty, now signed by 96 countries, lists 675 species that cannot be commercially traded as live specimens or wildlife products because they are endangered or threatened.

But enforcement of this treaty is spotty, and convicted violators often pay only small fines. In 1979, for example, a Hong Kong fur dealer illegally imported 319 Ethiopian cheetah skins valued at $43,900. The dealer was caught, but was fined only $1,540. Also much of the $1- to $2-billion-a-year illegal trade in wildlife and wildlife products goes on in countries, such as Singapore, that have not signed the treaty.

The United States controls imports and exports of endangered wildlife and wildlife products with two important laws. One is the Lacey Act of 1900, which prohibits transporting live or dead wild animals or their parts across state borders without a federal permit. The other law is the Endangered Species Act of 1973, including amendments in 1982 and 1988 (see Spotlight on p. 365).

The Species Approach: Wildlife Refuges By 1989 the National Wildlife Refuge System had 452 refuges (Figure 15-9, p. 340). About 85% of the area included in these refuges is in Alaska.

Over three-fourths of the refuges are wetlands for protection of migratory waterfowl. Most of the species on the U.S. endangered list have habitats in the refuge system and some refuges have been set aside for specific endangered species. These have helped the key deer, the brown pelican of southern Florida (Figure 16-18), and the trumpeter swan to recover. Conservationists complain that there has been too little emphasis on establishing refuges for endangered plants.

Congress has not established guidelines (such as multiple use or sustained yield) for management of

The Endangered Species Act of 1973 is one of the world's toughest environmental laws. This act makes it illegal for the United States to import or to carry on trade in any product made from an endangered species unless it is used for an approved scientific purpose or to enhance the survival of the species.

To make control more effective, all commercial shipments of wildlife and wildlife products must enter or leave the country through one of nine designated ports. But many illegal shipments of wildlife slip by. The 60 Fish and Wildlife Service inspectors are able to physically examine only about one-fourth of the 90,000 shipments that enter and leave the United States each year (Figure 16-17). Permits have been falsified and some government inspectors have been bribed. Even if caught, many violators are not prosecuted, and convicted violators often pay only a small fine.

The law also provides protection for endangered and threatened species in the United States and abroad. It authorizes the National Marine Fisheries Service (NMFS) to identify and list endangered and threatened marine species. The Fish and Wildlife Service (FWS) identifies and lists all other endangered and threatened species. These species cannot be hunted, killed, collected, or injured in the United States.

Any decision by either agency to add or remove a species from the list must be based only on biological grounds without economic considerations. The act also prohibits federal agencies from carrying out, funding, or authorizing projects that would jeopardize endangered or threatened species or

Figure 16-17   Confiscated products derived from endangered species. Because of a lack of funds and too few inspectors, probably no more than one-tenth of the illegal wildlife trade in the United States is discovered. The situation is much worse in most other countries.

destroy or modify their critical habitats.

Between 1970 and 1989, the number of species found only in the United States that have been placed on the official endangered and threatened list increased from 92 to 563. Another 508 species found in other parts of the world are also on the list.

Once a species is listed as endangered or threatened in the United States, the FWS or the NMFS is supposed to prepare a plan to help it recover. However, because of a lack of funds, recovery plans have been developed and approved for only about 51% of the endangered or threatened species native to the United States, and half of these plans exist only on paper. Only a handful of species have recovered sufficiently to be removed from protection.

The current annual federal

budget for endangered species is equal to the cost of about 25 Army bulldozers. This helps explain why it will take the Fish and Wildlife Service 20 years to evaluate the species presently under consideration for listing. Many species will probably disappear before they can be protected.

In 1990 President Bush's interior secretary (who is responsible for wildlife protection) proposed that the Endangered Species Act be weakened. He suggested that economic factors be included in listing endangered and threatened species and in carrying out federally funded projects that threaten the critical habitats of endangered or threatened species. This trial balloon by the Bush administration was vigorously opposed by outraged conservationists and many members of Congress. What do you think should be done?

the National Wildlife Refuge System, as it has for other public lands. As a result, the Fish and Wildlife Service has allowed many refuges to be used for hunting, fishing, trapping, timber cutting, grazing, farming, oil and gas development, mining, military

exercises, and recreational activities. By 1988 more than 60% of the refuges were open to hunting and almost 50% were open to fishing.

Development of oil, gas, and mineral resources can destroy or degrade wildlife habitats in refuges

Figure 16-18 The first national wildlife refuge was set up off the coast of Florida in 1903 to protect the brown pelican from extinction. In the 1960s this species was threatened with extinction when exposure to DDT and other persistent pesticides in the fish it eats caused reproductive losses. Now it is making a comeback.

Figure 16-19 The Arabian oryx barely escaped extinction in 1969 after being overhunted in the deserts of the Middle East. Captive breeding programs in zoos in Arizona and California have been successful in saving this antelope species from extinction. Some have been reintroduced into the wild in the Middle East.

through road building, well and pipeline construction, oil and gas leaks, and pits filled with brine or drilling muds (see Pro/Con on p. 367).

Pollution is also a problem in a number of refuges. A 1986 study by the Fish and Wildlife Service estimated that one in five federal refuges is contaminated with toxic chemicals. Most of this pollution comes from old toxic-waste dump sites and runoff from nearby agricultural land. For example, massive waterfowl deaths in the Kesterson National Wildlife Refuge in California's San Joaquin Valley in 1982 have been blamed on runoff of selenium-tainted irrigation water.

The Species Approach: Gene Banks, Botanical Gardens, and Zoos

Botanists preserve genetic information and endangered plant species by storing their seeds in gene banks—refrigerated environments with low humidity. Gene banks of most known and many potential varieties of agricultural crops and other plants now exist throughout the world (Figure 13-2, p. 296). Scientists have urged that many more be established, especially in LDCs. But some species can't be preserved in gene banks, and maintaining gene banks is very expensive.

The world's 1,500 botanical gardens also help preserve some of the genetic diversity found in the wild. However, the gardens have too little storage capacity and too little money to maintain all of the world's threatened plants.

Zoos and animal research centers are increasingly being used to preserve a representative number of individuals of critically endangered animal species. Two techniques for preserving such species are egg pulling and captive breeding.

Egg pulling involves collecting eggs produced in the wild by the remaining breeding pairs of a critically endangered bird species and hatching them in zoos or research centers. In 1983 scientists began an egg-pulling program to help save the critically endangered California condor (Figure 16-3).

For *captive breeding* some or all of the individuals of a critically endangered species still in the wild are captured and placed in zoos or research centers to breed in captivity. Scientists hope that after several decades of captive breeding and egg pulling, the captive population of an endangered species will be large enough that some individuals can be successfully reintroduced into protected wild habitats.

Captive breeding programs at zoos in Phoenix, San Diego, and Los Angeles saved the nearly extinct Arabian oryx (Figure 16-19). This large antelope species once lived throughout the Middle East. However, by the early 1970s, it had disappeared from the wild after being hunted by people using jeeps, helicopters, rifles, and machine guns. Since 1980 small numbers of these animals bred in captivity have been returned to the wild in protected habitats in the Middle East. Critically endangered U.S. species now being bred in captivity include the California condor (Figure 16-3) and the black-footed ferret (Figure 16-12).

Keeping populations of endangered animal species in zoos and research centers is limited by lack of space and money. The captive population of each species must number 100 to 500 to avoid extinction through accident, disease, or loss of genetic variability through inbreeding. Moreover, caring for and breeding captive animals is very expensive.

Because of a lack of space and money, the world's zoos now contain only 20 endangered species of

The Arctic National Wildlife Refuge on Alaska's North Slope is the second largest in the system (Figure 16-20), covering an area the size of South Carolina. Its coastal plain, the most biologically productive part of the refuge, is the only stretch of Alaska's Arctic coast that has not been opened to oil and gas development.

However, energy companies have asked Congress to open 607,000 hectares (1.5 million acres) of the coastal plain of the refuge to drilling for oil and natural gas. In 1987 the secretary of the interior joined forces with energy developers in this request. They argued that the area might contain oil and natural gas deposits that would reduce U.S. reliance on foreign oil.

Conservationists oppose this plan and want Congress to designate the entire coastal plain as wilderness. They point to Interior Department estimates that there is only a 19% chance of finding any economically recoverable large deposits of oil in this area. They do not believe that it's worth degrading a priceless and irreplaceable wildlife resource for the remote possibility of providing the U.S. with a six-month supply of oil.

Officials of oil companies claim they have developed Alaska's Prudhoe Bay oil fields without significant harm to wildlife. But the 1989

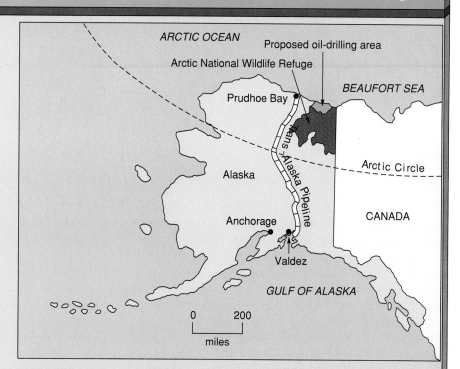

Figure 16-20   Proposed oil-drilling area in the Arctic National Wildlife Refuge. (Data from U.S. Fish and Wildlife Service)

massive oil spill from the *Exxon Valdez* tanker in Alaska's Prince William Sound cast serious doubt on such claims (see Case Study on p. 258).

Also, according to a study leaked from the Fish and Wildlife Service in 1988, oil drilling at Prudhoe Bay has caused much more air and water pollution than was estimated before drilling began in 1972. According to this study, oil

development in the coastal plain could cause the loss of 20% to 40% of the area's caribou herd, 25% to 50% of the musk oxen still left, 50% or more of the wolverines, and 50% of the snow geese that winter in this area. Do you think that oil and gas development should be allowed in the Arctic National Wildlife Refuge? Relate this to your own use and waste of oil and gasoline.

---

animals with populations of 100 or more individuals. Probably no more than 900 species of endangered animals could be protected and bred in zoos and research centers. It is doubtful that the more than $6 billion needed to take care of these animals for 20 years will be available.

Because of limited money and trained personnel, only a few of the world's endangered and threatened species can be saved by treaties, laws, wildlife refuges, and zoos. This means that wildlife experts must decide which species out of thousands of candidates should be saved. Many experts suggest that the limited funds for preserving threatened and endangered wildlife be concentrated on those species that **(1)** have the best chance for survival, **(2)** have the most ecological value

to an ecosystem, and **(3)** are potentially useful for agriculture, medicine, or industry.

The Ecosystem Approach: Protecting Habitats Most wildlife biologists believe that the best way to prevent the loss of wild species is to establish and maintain a worldwide system of reserves, parks, and other protected areas. The system would consist of at least 10% of the world's land area and include several areas of each of the world's major types of ecosystems.

This ecosystem approach would prevent many species from becoming endangered by human activities and would also be cheaper than managing endangered species one by one. An international fund to help LDCs protect and manage biosphere

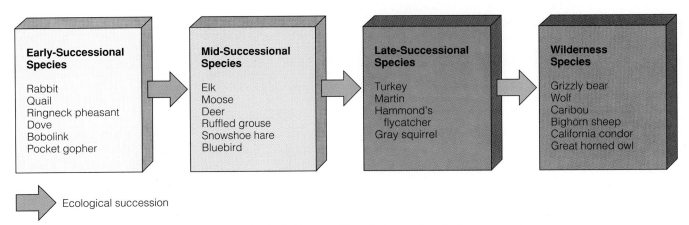

Early-Successional Species	Mid-Successional Species	Late-Successional Species	Wilderness Species
Rabbit	Elk	Turkey	Grizzly bear
Quail	Moose	Martin	Wolf
Ringneck pheasant	Deer	Hammond's	Caribou
Dove	Ruffled grouse	flycatcher	Bighorn sheep
Bobolink	Snowshoe hare	Gray squirrel	California condor
Pocket gopher	Bluebird		Great horned owl

 Ecological succession

Figure 16-21 Preferences of some wildlife species for habitats at different stages of ecological succession.

reserves would cost $100 million a year—about what the world spends on arms every 90 minutes.

In the United States, conservationists urge Congress to pass an Endangered Ecosystems Act as an important step toward preserving the country's biodiversity. Such an act would also require environmental impact studies to assess the effects of any federal activity on biological diversity.

In 1980 the IUCN, the UN Environment Program, and the World Wildlife Fund developed the *World Conservation Strategy*, a long-range plan for conserving the world's biological resources. The goals of this plan are:

■ Maintain essential ecological processes and life-support systems on which human survival and economic activities depend.

■ Preserve species diversity and genetic diversity.

■ Ensure that any use of species and ecosystems is sustainable.

By 1988, 40 countries had planned or established national conservation programs. The United States has not established such a program. If MDCs provide enough money and scientific assistance, this conservation strategy offers hope for preserving much of the world's vanishing biological and genetic diversity.

16-4 WILDLIFE MANAGEMENT

Management Approaches Wildlife management is the manipulation of wildlife populations (especially game species) and habitats for their welfare and for human benefit, the preservation of endangered and threatened wild species, and wildlife law enforcement.

The first step in wildlife management is to decide which species or groups of species are to be managed in a particular area. This is a source of much contro-

versy. Ecologists stress preservation of biological diversity. Wildlife conservationists are concerned about endangered species. Bird watchers want the greatest diversity of bird species. Hunters want large populations of game species for harvest each year during hunting season. In the United States, most wildlife management is devoted to the production of harvestable surpluses of game animals and game birds.

After goals have been set, the wildlife manager must develop a management plan. Ideally, the plan should be based on principles of ecological succession (Section 5-5), wildlife population dynamics (Section 5-4), and an understanding of the cover, food, water, space, and other habitat requirements of each species to be managed. The manager must also consider the number of potential hunters, their success rates, and the regulations available to prevent excessive harvesting.

This information is difficult, expensive, and time consuming to get. Often it is not available or reliable. That is why wildlife management is as much an art as a science. In practice it involves much guesswork and trial and error. Management plans must also be adapted to political pressures from conflicting groups and to budget constraints.

Manipulation of Habitat Vegetation and Water Supplies Wildlife managers can encourage the growth of plant species that are the preferred food and cover for a particular animal species by controlling the ecological succession of vegetation in various areas (Figure 5-37, p. 110).

Animal wildlife species can be classified into four types according to the stage of ecological succession at which they are most likely to be found: wilderness, late-successional, mid-successional, and early-successional (Figure 16-21). **Wilderness species** flourish only in fairly undisturbed, mature vegetational communities, such as large areas of old-growth forests, tundra, grasslands, and deserts. Their survival depends largely on the establishment of large state and national wilderness areas and wildlife refuges.

Late-successional species need old-growth and mature forest habitats to produce the food and cover on which they depend. These animals require the establishment and protection of moderate-sized, old-growth forest refuges (see photo on front cover).

Mid-successional species are found around abandoned croplands and partially open areas. Such areas are created by the logging of small stands of timber, controlled burning, and clearing of vegetation for roads, firebreaks, oil and gas pipelines, and electrical transmission lines.

Such openings of the forest canopy promote the growth of vegetation favored as food by mid-successional mammal and bird species. It also increases the amount of edge habitat, where two communities such as a forest and field come together. This transition zone allows animals such as deer to feed on vegetation in clearings and quickly escape to cover in the nearby forest. **Early-successional species** find food and cover in weedy pioneer plants. These plants invade an area that has been cleared of vegetation for human activities and then abandoned, as well as areas devastated by mining, fires, volcanic lava, and glaciers.

Various types of habitat improvement can be used to attract and encourage the population growth of a desired species. Improvement techniques include artificial seeding, transplanting certain types of vegetation, building artificial nests, and setting prescribed burns. Wildlife managers often create or improve ponds and lakes in wildlife refuges to provide water, food, and habitat for waterfowl and other wild animals.

Population Management by Controlled Sport Hunting The United States and most MDCs use sport hunting laws to manage populations of game animals. These laws

- require hunters to have a license
- allow hunting only during certain months of the year to protect animals during mating season
- allow hunters to use only certain types of hunting equipment, such as bows and arrows, shotguns, and rifles, and specify on which types of game each can be used
- set limits on the size, number, and sex of animals that can be killed and on the number of hunters allowed in a game refuge

But close control of sport hunting is often not possible. Accurate data on game populations may not exist and may cost too much to get. People in communities near hunting areas, who benefit from money spent by hunters, may push to have hunting quotas raised. On the other hand, some individuals and conservation groups are opposed to sport hunting and exert political pressure to have it banned or sharply curtailed.

Management of Migratory Waterfowl In North America, migratory waterfowl such as ducks, geese, and swans nest in Canada during the summer. During the fall hunting season they migrate to the United States and Central America along generally fixed routes called **flyways** (Figure 16-22).

Canada, the United States, and Mexico have signed agreements to prevent habitat destruction and overhunting of migratory waterfowl. However, since 1979 the estimated breeding populations of ducks in North America have been declining. An estimated 64 million ducks migrated southward across North America during the fall of 1989, down from an average of 92 million a year during the 1970s. The major reasons for this decrease are prolonged drought in key breeding areas and degradation and destruction of wetland and grassland breeding habitats by farmers.

The remaining wetlands are used by dense flocks of ducks and geese. This crowding makes them more vulnerable to diseases and predators such as skunks, foxes, coyotes, minks, raccoons, and hunters. Waterfowl in wetlands near croplands are also exposed to pollution from pesticides and other chemicals in the irrigation runoff they drink.

Wildlife officials manage waterfowl by regulating hunting, protecting existing habitats, and developing new habitats. More than 75% of the federal wildlife refuges in the United States are wetlands used for migratory birds. Other waterfowl refuges have been established by local and state agencies and private conservation groups such as Ducks Unlimited, the Audubon Society, and the Nature Conservancy.

Building artificial nesting sites, ponds, and nesting islands is another method of establishing protected habitats for breeding populations of waterfowl. Solar-powered electric fences are being used in some areas to keep predators away from nesting waterfowl.

In 1986 the United States and Canada agreed on a plan to spend $1.5 billion over a 16-year period, with the goal of almost doubling the continental duck breeding population. The key elements in this program will be the purchase, improvement, and protection of an additional waterfowl habitat in five priority areas.

Since 1934 the Migratory Bird Hunting and Conservation Stamp Act has required waterfowl hunters to buy a duck stamp each season they hunt. Revenue from these sales goes into a fund to buy land and easements for the benefit of waterfowl.

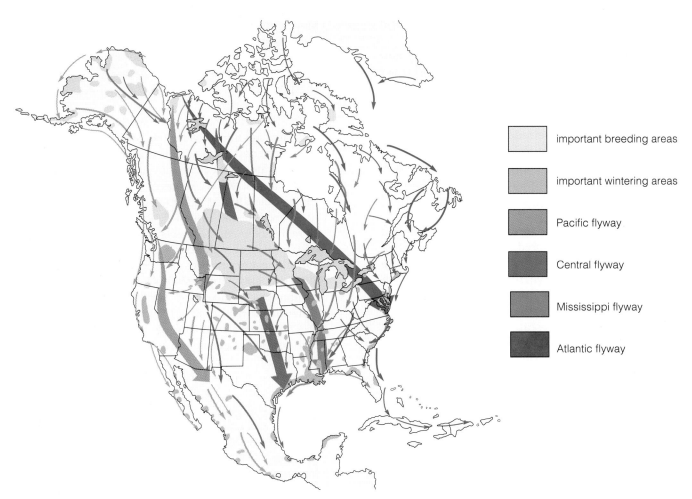

Figure 16-22 Major breeding and wintering areas and fall migration flyways used by migratory waterfowl in North America.

legend:
- important breeding areas
- important wintering areas
- Pacific flyway
- Central flyway
- Mississippi flyway
- Atlantic flyway

FISHERY MANAGEMENT

16-5

Freshwater Fishery Management The goals of freshwater fish management are to encourage the growth of populations of desirable commercial and sport fish species and to reduce or eliminate populations of less desirable species. Several techniques are used:

- regulating the timing and length of fishing seasons
- establishing the minimum-size fish that can be taken
- setting catch quotas
- requiring that commercial fish nets have a large enough mesh to ensure that young fish are not harvested
- building reservoirs and farm ponds and stocking them with game fish
- fertilizing nutrient-poor lakes and ponds with commercial fertilizer, fish meal, and animal wastes

- protecting and creating spawning sites and cover spaces
- protecting habitats from buildup of sediment and other forms of pollution and removing debris
- preventing excessive growth of aquatic plants to prevent oxygen depletion
- using small dams to control water flow
- controlling predators, parasites, and diseases by habitat improvement, breeding genetically resistant fish varieties, and using antibiotics and disinfectants
- using hatcheries to restock ponds, lakes, and streams with species such as trout and salmon

Marine Fishery Management The history of the world's commercial marine fishing and whaling industry is an excellent example of the tragedy of the commons—the overexploitation of a potentially renewable resource (see Spotlight on p. 10). As a result, many species of commercially valuable fish and whales

The blue whale is the world's largest animal. Fully grown, it's more than 30 meters (100 feet) long and weighs 136 metric tons (150 tons). The heart of an adult is as big as a Volkswagen "Bug" and some of its arteries are big enough for a child to swim through. Its brain weighs four times more than yours and this mammal shows signs of great intelligence.

Blue whales spend about eight months of the year in Antarctic waters. There they find an abundant supply of shrimplike krill, which they filter from seawater (Figure 4-19, p. 73). During the winter months they migrate to warmer waters, where their young are born.

Once an estimated 200,000 blue whales roamed the Antarctic waters. Today the species has been hunted to near biological extinction for its oil, meat, and bone (Figure 16-24).

This decline was caused by a combination of prolonged overfishing and certain natural traits of the blue whale. Their huge size made them easy to spot. They were caught in large numbers because they grouped together in their Antarctic feeding grounds. Also, they take 25 years to mature sexually and have only one offspring every 2 to 5 years. This low reproduction rate makes it hard for the species to recover once its population has been reduced to a low level.

Blue whales haven't been hunted commercially since 1964 and are classified as an endangered species. Despite this protection, some marine experts believe that not enough blue whales are left for the species to recover. Less than 1,000 blue whales may be left today. Surveys in the 1980s by the International Whaling Commission counted only about 450 blue whales in Antarctic waters. Within your lifetime the blue whale could disappear forever.

Figure 16-23   The whaling industry has pushed most of the dozen or so species of great whales to the brink of extinction through overharvesting. This photograph shows pilot whales, which are not now endangered, being butchered in the Faro Islands in the Baltic Sea between the coasts of Sweden and the Soviet Union. After continued protests since the 1960s, the International Whaling Commission finally banned commercial whaling in 1986, but only until 1992.

in international and coastal waters have been overfished to the point of commercial extinction (see Case Study above). At that point the stock of a species is so low that it's no longer profitable to hunt and gather the remaining individuals in a specific fishery.

Managers of marine fisheries can use several techniques to prevent commercial extinction and allow depleted stocks to recover. Fishery commissions, councils, and advisory bodies with representatives from countries using a fishery can be established. They can set annual quotas for harvesting fish and marine mammals and establish rules for dividing the allowable annual catch among the countries participating in the fishery.

These groups may also limit fishing seasons and regulate the type of fishing gear that can be used to harvest a particular species. Fishing techniques such as dynamiting and poisoning are outlawed. Fishery commissions may also enact size limits that make it illegal to keep fish above or below a certain size, usually the average length of the particular fish species when it first reproduces.

As voluntary associations, however, fishery commissions don't have any legal authority to compel member states to follow their rules. Nor can they compel all countries fishing in a region to join the commission and submit to its rules.

International and national laws have been used to extend the offshore fishing zone of coastal countries to 370 kilometers (200 nautical miles or 230 statute miles) from their shores. Foreign fishing vessels can take certain quotas of fish within such zones, called *exclusive economic zones*, only with government permission.

Decline of the Whaling Industry   In 1900 an estimated 4.4 million whales swam the ocean. Today only about 1 million are left (Figure 16-23). Overharvesting has caused a sharp drop in the populations of almost every whale species of commercial value

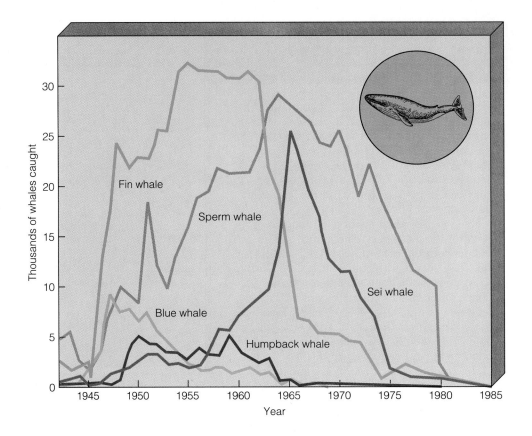

Figure 16-24 Whale harvests, showing the signs of overharvesting. (Data from International Whaling Commission)

(Figure 16-24). The populations of 8 of the 11 major species of whales once hunted by the whaling industry have been reduced to commercial extinction. This devastation has happened because of the tragedy of the commons and because whales are more vulnerable to biological extinction than fish species (see Case Study on p. 371).

In 1946 the International Whaling Commission (IWC) was established to regulate the whaling industry. Since 1949 the IWC has set annual quotas to prevent overfishing and commercial extinction. However, these quotas often were based on inadequate scientific information or were ignored by whaling countries. Without any powers of enforcement, the IWC has been unable to stop the decline of most whale species.

In 1970 the United States stopped all commercial whaling and banned all imports of whale products into the country, mostly because of pressure from conservationists and the general public. Since then conservation groups and the governments of many countries, including the United States, have called for a permanent ban on all commercial whaling.

After years of meetings and delays, the IWC established a five-year halt on commercial whaling, beginning in 1986 and ending in 1991. However, Japan, Norway, and Iceland have continued to harvest several hundred whales each year for "scientific" purposes, despite being refused permits to do this

by the IWC. Without continuing worldwide pressure from individuals and conservation organizations, large-scale commercial whaling may resume in 1992 when the present moratorium ends.

Individual Action We are all involved, at least indirectly, in the destruction of wildlife any time we buy or drive a car, build a house, consume almost anything, and waste electricity, paper, water, or any other resource. All these activities contribute to the destruction or degradation of wildlife habitats or to the killing of one or more individuals of some plant or animal species.

Modifying our consumption habits is a key goal in protecting wildlife, the environment, and ourselves (see Individuals Matter on p. 373). This also involves supporting efforts to reduce projected global warming and ozone depletion—two of the greatest threats to the earth's wildlife and the human species (Section 10-4).

Biological diversity must be treated as a global resource to be indexed and used, and above all preserved.

E. O. Wilson

- Improve the habitat on a patch of the earth in your immediate environment such as backyards, abandoned city lots, campus areas, and streams clogged with debris.

- Develop a wildlife protection and management plan for any land that you own.

- Don't buy furs, ivory, or other products made from wild animals or animals raised on fur farms where they are treated inhumanely. Non-animal substitutes are available for such products.

- Consider reducing or eliminating your consumption of meat and not using products made from leather or other materials obtained from domesticated animals raised with little regard for their well being.

- Don't buy brands of tuna fish in which thousands of dolphins are caught along with the tuna.

- Support efforts to ensure that test animals are treated humanely and to reduce their use to a minimum.

- If you have a dog or cat as a pet, have it spayed or neutered. Each year U.S. pounds and animal shelters have to kill about 15 million unwanted dogs and cats because of pet overpopulation. This is 75 times the number of dogs and cats killed each year in the United States for research and teaching purposes. Most of these test animals are obtained from animal shelters where they would have to be killed anyway because of our throwaway attitude toward pets.

- Leave wild animals in the wild. Consider not buying exotic birds, fish, and other pets imported from tropical and other areas. Typically, one or more animals die for each one that reaches a pet shop. Also, most of these pets die prematurely because they are killed, abandoned, or cared for improperly by their owners.

- Reduce habitat destruction and degradation by recycling paper, cans, plastics, and other household items. Better yet, reuse items and sharply reduce your use of throwaway items (Section 19-6).

- Support efforts to sharply reduce the destruction and degradation of tropical forests that are responsible for the largest number of wildlife extinctions (Section 15-2 and Guest Essay on p. 374).

- Pressure elected officials to pass laws requiring much larger fines and longer prison sentences for wildlife poachers and to provide more funds and personnel for wildlife protection.

- Pressure Congress to pass a national biological diversity act and to develop a national conservation program as part of the World Conservation Strategy.

- Encourage the development of an international treaty to preserve biological diversity.

DISCUSSION TOPICS

1. Discuss your gut-level reaction to this statement: "It doesn't really matter that the passenger pigeon is extinct and the blue whale, whooping crane, California condor, rhinoceros, grizzly bear, and a number of other plant and animal species are endangered mostly because of human activities." Be honest about your reaction, and give arguments for your position.

2. Make a log of your own consumption and use of food and other products for a single day. Relate your consumption to the increased destruction of wildlife and wildlife habitats in the United States and in tropical forests.

3. a. Do you accept the ethical position that each species has the inherent right to survive without human interference, regardless of whether it serves any useful purpose for humans? Explain.
   b. Do you believe that each individual of an animal species has an inherent right to survive? Explain. Would you extend such rights to individual plants and microorganisms? Explain.

4. Do you believe that the use of animals, mostly mice and rats, to test new drugs and vaccines and the toxicity of chemicals should be banned? Explain. What are the alternatives? Should animals be used to test cosmetics?

5. Use Table 16-2 to predict a species that may soon be endangered. What, if anything, is being done for this species? What pressures is it being subjected to? Try to work up a plan for protecting it.

6. Make a survey of your campus and local community to identify examples of habitat destruction or degradation that have had harmful effects on the populations of various wild plant and animal species. Develop a management plan for the rehabilitation of these habitats and wildlife.

7. Are you for or against sport hunting? Explain.

FURTHER READINGS

Credlund, Arthur G. 1983. *Whales and Whaling.* New York: Seven Hills Books.

Disilvestro, Roger L. 1989. *The Endangered Kingdom: The Struggle to Save America's Wildlife.* New York: Wiley.

Norman Myers

Norman Myers is an international consultant in environment and development with emphasis on conservation of wildlife species and tropical forests. He has served as a consultant for many development agencies and research organizations, including the U.S. National Academy of Sciences, the World Bank, the Organization for Economic Cooperation and Development, various UN agencies, and the World Resources Institute. Among his recent publications (see Further Readings) are The Sinking Ark *(1979),* Conversion of Tropical Moist Forests *(1980),* A Wealth of Wild Species *(1983),* The Primary Source *(1984), and* The Gaia Atlas of Planet Management *(1985).*

Tropical forests still cover an area roughly equivalent to the "lower 48" United States. Climatic and biological data suggest they could have once covered an area at least twice as large. So we have already lost half of them, mostly in the recent past. Worse, remote-sensing surveys show that we are now destroying the forests at a rate of at least 1.25% a year, and we are grossly degrading them at a rate of at least another 1.25% a year—and both rates are accelerating rapidly.

Unless we act now to halt this loss, within just another few decades at most, there could be little left, except perhaps a block in central Africa and another in the western part of the Amazon basin. Even these remnants may not survive the combined pressures of population growth and land hunger beyond the middle of the next century.

This means that we are imposing one of the most broad-scale and impoverishing impacts on the biosphere that it has ever suffered throughout its 4 billion years of existence. Tropical forests are the greatest celebration of nature to appear on the face of the planet since the first flickerings of life. They are exceptionally complex ecologically, and they are remarkably rich biotically. Although they now account for only 7% of the earth's land surface, they still are home for half, and perhaps three-quarters or more, of all the earth's species of plant and animal life. Thus, elimination of these forests is by far the leading factor in the mass extinction of species that appears likely over the next few decades.

Already, we are certainly losing several species every day because of clearing and degradation of tropical forests. The time will surely come, and come soon, when we shall be losing many thousands every year. The implications are profound, whether they be scientific, aesthetic, ethical—or simply economic. In medicine alone, we benefit from myriad drugs and pharmaceuticals derived from tropical forest plants. The commercial value of these products worldwide can be reckoned at $20 billion each year.

By way of example, the rosy periwinkle from Madagascar's tropical forests has produced two potent drugs against Hodgkin's disease, leukemia, and other blood cancers. Madagascar has—or used to have—at

Dunlap, Thomas R. 1988. *Saving America's Wildlife*. Princeton, N.J.: Princeton University Press.

Durrell, Lee. 1986. *State of the Ark: An Atlas of Conservation in Action*. Garden City, N.Y.: Doubleday.

Ehrlich, Paul, and Anne Ehrlich. 1981. *Extinction*. New York: Random House.

Elliot, David K. 1986. *Dynamics of Extinction*. New York: John Wiley.

Gilbert, Frederick F., and Donald G. Dodds. 1987. *The Philosophy and Practice of Wildlife Management*. Malabar, Fla.: Robert E. Krieger.

International Union for Conservation of Nature and Natural Resources. 1980. *World Conservation Strategy*. New York: Unipub.

International Union for Conservation of Nature and Natural Resources. 1985. *Implementing the World Conservation Strategy*. Gland, Switzerland: IUCN.

Kaufmann, Les, and Kenneth Mallory, eds. 1986. *The Last Extinction*. Cambridge, Mass.: MIT Press.

Koopowitz, Harold, and Hilary Kaye. 1983. *Plant Extinctions: A Global Crisis*. Washington, D.C.: Stone Wall Press.

Lackey, R. T., and L. A. Nielson. 1980. *Fisheries Management*. New York: John Wiley.

Leopold, Aldo. 1933. *Game Management*. New York: Charles Scribner's Sons.

Livingston, John A. 1981. *The Fallacy of Wildlife Conservation*. Toronto: McClelland and Stewart.

Luoma, Jon. 1987. *A Crowded Ark: The Role of Zoos in Wildlife Conservation*. Boston: Houghton Mifflin.

McNeely, Jeffery A. 1989. *Economics and Biological Diversity*. New York: Pinter Publishers.

McNeely, Jeffery A., et al. 1990. *Conserving the World's Biological Diversity*. Washington, D.C.: World Resources Institute.

Myers, Norman. 1983. *A Wealth of Wild Species: Storehouse for Human Welfare*. Boulder, Colo.: Westview Press.

least 8,000 plant species, of which more than 7,000 could be found nowhere else. Today Madagascar has lost 93% of its virgin tropical forest. The U.S. National Cancer Institute estimates that there could be many plants in tropical forests with potential against various cancers—provided pharmacologists can get to them before they are eliminated by chain saws and bulldozers.

We benefit in still other ways from tropical forests. Elimination of these forests disrupts certain critical environmental services, notably their famous "sponge effect" by which they soak up rainfall during the wet season and then release it in regular amounts throughout the dry season. When tree cover is removed and this watershed function is impaired, the result is a yearly regime of floods followed by droughts, which destroys property and reduces agricultural production. There is also concern that if tropical deforestation becomes widespread enough, it could trigger local, regional, or even global changes in climate. Such climatic upheavals would affect the lives of billions of people, if not the whole of humankind.

All this raises important questions about our role in the biosphere and our relations with the natural world around us. As we proceed on our disruptive way in tropical forests, we—that is, political leaders and the general public alike—give scarcely a moment's thought to what we are doing. We are deciding the fate of the world's tropical forests unwittingly, yet effectively and increasingly.

The resulting shift in evolution's course, stemming from the elimination of tropical forests, will rank as one of the greatest biological upheavals since the dawn of life. It will equal, in scale and significance, the development of aerobic respiration, the emergence of flowering plants, and the arrival of limbed animals, all of which took place over eons of time. But whereas these were enriching disruptions in the course of life on this planet, the loss of biotic diversity associated with tropical forest destruction will be almost entirely an impoverishing phenomenon brought about entirely by human actions. And it will all have occurred within the twinkling of a geologic eye.

In short, our intervention in tropical forests should be viewed as one of the most challenging problems that humankind has ever encountered. After all, we are the first species ever to be able to look upon nature's work and to decide whether we should consciously eliminate it or leave much of it untouched.

So the decline of tropical forest is one of the great sleeper issues of our time. Yet, we can still save much of these forests and the species they contain. Should we not consider ourselves fortunate that we alone among all generations are being given the chance to preserve tropical forests as the most exuberant expression of nature in the biosphere—and thereby to support the right to life of many of our fellow species and their capacity to undergo further evolution without human interference?

Guest Essay Discussion

1. What obligation, if any, do you as an individual have to preserve a significant portion of the world's remaining tropical forests?

2. Should MDCs provide most of the money to preserve remaining tropical forests in LDCs? Explain.

3. What can you do to help preserve some of the world's tropical forests? Which, if any, of these actions do you plan to carry out?

Nash, Roderick F. 1988. *The Rights of Nature: A History of Environmental Ethics.* Madison: University of Wisconsin Press.

National Audubon Society. Annual. *Audubon Wildlife Report.* New York: National Audubon Society.

Office of Technology Assessment. 1989. *Oil Production in the Arctic National Wildlife Refuge.* Washington, D.C.: Government Printing Office.

Prescott-Allen, Robert, and Christine Prescott-Allen. 1982. *What's Wildlife Worth?* Washington, D.C.: Earthscan.

Pringle, Laurence. 1989. *The Animal Rights Controversy.* New York: Harcourt Brace Jovanovich.

Reagan, Tom. 1983. *The Case for Animal Rights.* Berkeley: University of California Press.

Reid, Walter V. C., and Kenton R. Miller. 1989. *Keeping Options Alive: The Scientific Basis for Conserving Biodiversity.* Washington, D.C.: World Resources Institute.

Roe, Frank G. 1970. *The North American Buffalo.* Toronto: University of Toronto Press.

Shaw, J. H. 1985. *Introduction to Wildlife Management.* New York: McGraw-Hill.

Soulé, Michael E., and Bruce Wilcox, eds. 1980. *Conservation Biology.* Sunderland, Mass.: Sinauer Associates.

Tudge, Colin. 1988. *The Environment of Life.* New York: Oxford University Press.

Wallace, David Rains. 1987. *Life in the Balance.* New York: Harcourt Brace Jovanovich.

Western, David, and Mary Pearl, eds. 1989. *Conservation for the Twenty-First Century.* New York: Oxford University Press.

Wilson, E. O. 1984. *Biophilia.* Cambridge, Mass.: Harvard University Press.

Wilson, E. O., ed. 1988. *Biodiversity.* Washington, D.C.: National Academy Press.

Wolf, Edward C. 1987. *On the Brink of Extinction: Conserving the Diversity of Life.* Washington, D.C.: Worldwatch Institute.

Wind farm in a mountain pass in California.

Our entire society rests upon—and is dependent upon—our water, our land, our forests, and our minerals. How we use these resources influences our health, security, economy, and well-being.

JOHN F. KENNEDY

CHAPTER SEVENTEEN

PERPETUAL AND RENEWABLE ENERGY RESOURCES

General Questions and Issues

1. How can we evaluate present and future energy alternatives?

2. What are the advantages and disadvantages of improving energy efficiency to reduce unnecessary energy waste?

3. What are the advantages and disadvantages of capturing and using some of the sun's direct input of solar energy for heating buildings and water and for producing electricity?

4. What are the advantages and disadvantages of using flowing water and solar energy stored as heat in water for producing electricity?

5. What are the advantages and disadvantages of using wind to produce electricity?

6. What are the advantages and disadvantages of burning plants and organic waste (biomass) for heating buildings and water, producing electricity, and for transportation (biofuels)?

7. What are the advantages and disadvantages of using geothermal energy as an energy resource?

8. What are the advantages and disadvantages of using hydrogen gas to produce electricity, to heat buildings and water, and to propel vehicles when oil runs out?

If the United States wants to save a lot of oil and money and increase national security, there are two simple ways to do it: stop driving Petropigs and stop living in energy sieves.

AMORY B. LOVINS

A MAJOR THEME of this book is that energy is the thread sustaining and integrating all life and supporting all economies. This is why it is so important that we understand the nature and implications of the two energy laws that govern all energy use. For this reason, I suggest that you review Sections 3-5, 3-6, and 3-7 before studying this and the next chapter.

How and what types of energy we use are the major factors determining how much we abuse the life-support systems for us and other species. Our current dependence on fossil fuels is the major cause of air and water pollution, land disruption, and projected global warming. Most of these and other external costs are not included in the market prices of fossil fuels and electricity. This undervaluing of these key resources gives consumers and industries little incentive to conserve these nonrenewable fuels. Most analysts agree that the era of cheap oil is coming to an end. This means we must find substitutes for the oil that now supports the economies of industrialized countries and many LDCs. Others argue that to reduce the threat of projected global warming and air and water pollution, we must reduce our current use of all fossil fuels 50% by 2010, and 70% by 2030 (Section 10-3).

What is our best option for reducing dependence on oil and other fossil fuels? Cut out unnecessary energy waste by improving energy efficiency (see Guest Essay on p. 56). What is our next best energy option? Here there is disagreement.

Some say get more of the energy we need from the sun, wind, flowing water, biomass, hydrogen gas, and heat stored in the earth's interior (see Guest Essay on p. 56). These energy choices, based on using the earth's perpetual and renewable energy resources, are evaluated in this chapter.

Others say we should burn more coal and synthetic liquid and gaseous fuels made from coal. Some believe natural gas is the answer. Others think nuclear power is the answer. These choices, based on using more of the earth's nonrenewable resources, are evaluated in the next chapter.

17-1 EVALUATING ENERGY RESOURCES

Experience has shown that it takes 50 to 60 years to develop and phase in new supplemental energy resources on a large scale (Figure 3-5, p. 41). In deciding which combination of energy alternatives we should use to supplement solar energy in the future, we need to plan for three time periods: the short term (1991 to 2001), the intermediate term (2001 to 2011), and the long term (2011 to 2041).

First we must decide how much we need, or want, of different kinds of energy, such as low-temperature heat, high-temperature heat, electricity, and fuels for transportation. This involves deciding what type and quality of energy can best perform each energy task (Figure 3-7, p. 42). Then we decide which energy sources can meet these needs at the lowest cost and environmental impact by answering four questions about each alternative.

1. How much will likely be available during the short term, intermediate term, and long term?

2. What is the estimated net useful-energy yield (Figure 3-18, p. 53)?

3. How much will it cost to develop, phase in, and use?

4. What are its potentially harmful environmental, social, and security impacts, and how can they be reduced?

The most important question decision makers and individuals should ask is: What energy choices will do the most to sustain the earth for us, for future generations, and for the other species living on this planet? Despite its importance, this ethical question is rarely considered by government officials, energy company executives, and most people. Changing this situation is probably the most important and difficult challenge we face (Section 2-3).

17-2 IMPROVING ENERGY EFFICIENCY: DOING MORE WITH LESS

Reducing Energy Waste: An Offer We Can't Afford to Refuse The easiest, quickest, and cheapest way to make more energy available with the least environmental impact is to reduce or eliminate unnecessary energy use and waste (Figure 3-13, p. 48). There are two general ways to do this.

1. *Reduce energy consumption by changing energy-wasting habits.* Examples include walking or riding a bicycle for short trips, using mass transit instead of cars, wearing a sweater indoors in cold weather to allow a lower thermostat setting, turning off unneeded lights, and reducing our use of throwaway items, which require energy for raw materials, manufacture, and disposal.

2. *Improve energy efficiency by using less energy to do the same amount of work.* Examples include adding more insulation to houses and buildings, keeping car engines tuned, and switching to, or developing, more energy-efficient cars, houses, heating and cooling systems, appliances, lights, and industrial processes (Figure 3-14, p. 49).

Improving energy efficiency has the highest net useful-energy yield of all energy alternatives. It reduces the environmental impacts of using energy because less of each energy resource is used to provide the same amount of energy. It adds no carbon dioxide to the atmosphere and is the best, cheapest, and quickest way to slow projected global warming by reducing wasteful use of fossil fuels and the need for costly and politically unacceptable nuclear power.

Reducing the amount of energy we use and waste makes domestic and world supplies of nonrenewable fossil fuels last longer and buys time for phasing in perpetual and renewable energy resources. It also reduces international tensions and improves national and global military and economic security by reducing dependence on oil imports and the need for military intervention to protect sources of oil, especially in the Middle East. Furthermore, it usually provides more jobs and promotes more economic growth per unit of energy than other energy alternatives.

The only serious disadvantage of improving energy efficiency is that replacing houses, industrial equipment, and cars, as they wear out, with more energy-efficient ones takes a long time. For example, replacing most buildings and industrial equipment takes several decades, and replacing the older cars on the road with new ones takes 10 to 12 years.

Improvements in energy efficiency have saved the world more than $300 billion worth of energy every year since 1973. And these improvements have only scratched the surface (see Spotlight on p. 380).

Improving Industrial Energy Efficiency Industrial processes consume 36% of the energy used in the United States. Today American industry uses 70% less energy to produce the same amount of goods as it did in 1973, but still wastes enormous amounts of energy. Since 1983 overall U.S. industrial energy efficiency has scarcely improved. Japan has the highest overall industrial energy efficiency in the world, followed closely by West Germany.

Industries that use large amounts of both high-temperature heat or steam and electricity can save energy and money by installing *cogeneration units,* which produce both of these types of energy from the same fuel source. Today industrial cogeneration in the United States supplies electricity equal to the output of 24 large (1,000-megawatt) power plants. By the year 2000, cogeneration has the potential to produce more electricity than all of the country's nuclear power plants.

About 70% of the electricity used in U.S. industry drives electric motors. Most of these motors run at

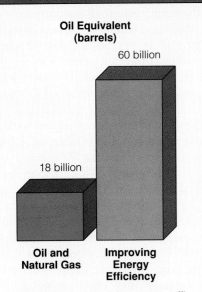

Figure 17-1 Improving energy efficiency in the United States using available technologies will produce over three times as much energy by 2020 at a lower cost than finding and developing, at a very high cost, all new oil and natural gas deposits believed to exist in the United States. (Data from Natural Resources Defense Council and Rocky Mountain Institute)

As the world's largest energy user and waster, the United States has more impact on fossil fuel depletion, oil pollution, projected global warming, acid deposition, and other forms of air pollution than any other country. At least 43% of all energy used in the United States is *unnecessarily* wasted (Figure 3-13, p. 48). This waste equals all the energy consumed by two-thirds of the world's population.

The largest untapped supplies of energy in the United States are in its energy-wasting buildings, factories, and vehicles, not in Alaska or offshore areas. This large source of energy can be found almost everywhere, can be exploited cheaply and quickly, strengthens rather than weakens the economy and national security, improves rather than damages the environment, and leaves little or no harmful wastes.

The untapped energy available by improving energy efficiency in the U.S. at a low cost is over three times that from developing remaining nonrenewable energy resources (fossil fuels and nuclear power) at a very high cost (Figure 17-1). Had the U.S. vigorously pursued a least-cost, high-energy-efficiency energy policy since 1973, instead of its high-cost, mostly fruitless search for new domestic deposits of oil, the country would have no need to import any oil today. Also, enough money would have been saved to pay off the entire national debt.

The good news is that since 1979, the United States has gotten more than seven times as much energy from improvements in energy efficiency as from all net increases in the supply of all forms of energy, with little help from federal and state governments. This reduction of energy waste has cut the country's annual energy bill by about $160 billion—about equal to the current annual national deficit. It has also reduced emissions of carbon dioxide, sulfur dioxide, and nitrogen oxides 40% below what they otherwise would have been.

The bad news is that energy efficiency in the United States is still half what it could be and has not improved much since 1985. Average gas mileage for new cars and for the entire fleet of cars is below that in most other MDCs. Most U.S. houses and buildings are still underinsulated and leaky. Electric resistance heating is the most wasteful and expensive way to heat a home (Figure 3-16, p. 51). Yet, it is installed in over half the new homes in the United

States, mostly because it saves builders money. Buyers are then left with the high heating bills.

Bringing about a low-cost, energy efficiency revolution by investing about $50 billion a year would stimulate the economy, cut carbon dioxide emissions in half, reduce urban smog and acid deposition, and save $250 billion annually between 1990 and 2000—enough to pay off the entire national debt. It would also reduce the cost of producing goods and services and make the United States more competitive in the international marketplace. For example, the United States spends about 11% of its GNP to obtain energy, while Japan uses only 5%. This gives Japanese goods an average 6% cost advantage over American goods.

Why isn't the United States pursuing an energy strategy that makes economic and environmental sense? There are several reasons. One is the political influence of companies controlling the use of nonrenewable fossil fuels and nuclear power and their emphasis on short-term profits regardless of the long-term economic and environmental consequences. Other reasons include a glut of low-cost fossil fuels, failure of elected officials to require the external costs of using fossil and nuclear fuels to be included in their market prices, and sharp cutbacks in federal support for improvements in energy efficiency and development of perpetual and renewable resources. Such shortsighted policies will continue until enough citizens demand that elected officials make improving energy efficiency and shifting to renewable energy resources the cornerstones of U.S. energy policy. What do you think should be done?

fixed speeds and voltages regardless of the tasks they perform. Adding variable-speed drives, light dimmers, and other devices that match the output of a motor or other electrical device to power needs would

eliminate the need for all existing U.S. nuclear power plants and save hundreds of millions of dollars a year.

Switching to high-efficiency lighting is another

way to save energy in industry. Industries can also use computer-controlled energy management systems to turn off lighting and equipment in nonproduction areas and make adjustments in periods of low production.

Another major way to save energy in industry is to greatly reduce the production of throwaway products. This can be done by increasing recycling and reuse and by making products that last longer and are easy to repair and recycle (Section 19-6).

For example, using recycled aluminum to produce cans and other aluminum products uses 95% less electricity than using virgin aluminum ore. Despite such enormous savings, the average world aluminum-recycling rate is only 25%. This could easily be doubled or tripled. Recycled paper requires 64% less energy than making paper from virgin wood pulp.

Improving Transportation Energy Efficiency One-fourth of the commercial energy consumed in the United States is used to transport people and goods. About one-tenth of the oil consumed in the world each day is used by American motorists on their way to and from work, 69% of them driving alone. The number one air polluter in the United States and other MDCs is the private automobile, not the coal-fired power plant. And the use of energy-saving bus and rail systems in the United States has not increased since 1979.

Today transportation consumes 63% of all oil used in the United States—up from 50% in 1973. Burning gasoline and other transportation fuels is responsible for about 35% of total U.S. emissions of carbon dioxide, and air conditioners in cars and light trucks are responsible for 75% of the country's annual CFC emissions. Thus, the best way to reduce world oil consumption and air pollution and to slow projected global warming and ozone depletion is to improve vehicle fuel efficiency (Figure 17-2), make greater use of mass transit (Figure 17-3), and haul freight more efficiently.

Between 1973 and 1990, the average fuel efficiency of new American cars doubled, and the average fuel efficiency of all the cars on the road increased by 54% (Figure 17-2). During this period these fuel-efficiency improvements saved American consumers about $285 billion in fuel costs. This is an important gain, but it is well below the 13 to 14 kilometers per liter (30 to 33 miles per gallon) fuel efficiency of new cars and the 9 to 11 kilometers per liter (22 to 25 mpg) fuel efficiency of car fleets in Japan and western Europe.

According to the U.S. Office of Technology Assessment, new cars produced in the United States could easily average between 16 to 23 kilometers per liter (38 to 55 mpg) by 1995, with only $50 to $90 added to the cost of a car. By the year 2000, new U.S.

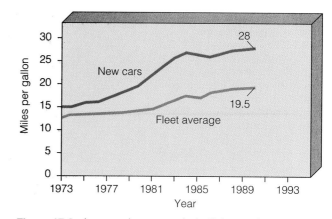

Figure 17-2 Increase in average fuel efficiency of new cars and the entire fleet of cars in the United States between 1973 and 1989. (Data from U.S. Department of Energy and Environmental Protection Agency)

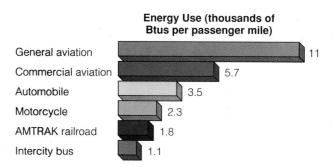
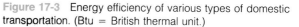

Figure 17-3 Energy efficiency of various types of domestic transportation. (Btu = British thermal unit.)

cars could average 22 to 33 kilometers per liter (51 to 78 mpg) for an additional cost of $120 to $330 a car. Also, the fuel efficiency of new light trucks—pickup trucks, minivans, and 4-wheel-drive sport vehicles—could be increased from 8.5 kilometers per liter (19.5 mpg) to 14 kilometers per liter (33 mpg) during the 1990s. Making these improvements in fuel efficiency would eliminate the need to import any oil.

However, since 1988 the three leading American car companies have sharply reduced their research and development on small, more fuel-efficient cars and have persuaded elected officials not to raise fuel-efficiency standards. The major reasons for this are the higher profits to be made on larger cars and declining consumer interest in improved fuel efficiency because of the temporary oil glut of the 1980s (see Case Study on p. 8).

Meanwhile, Japan and some western European countries have increased their research in this area. They want to have fuel-efficient cars ready when the oil crisis that is expected to occur sometime during the next two decades replaces the current oil glut.

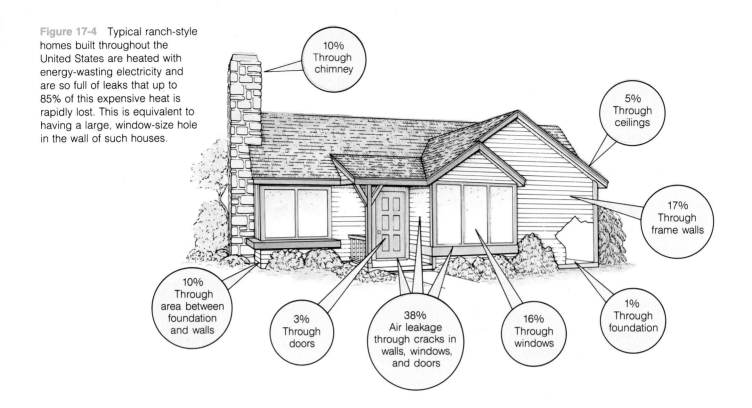

Figure 17-4 Typical ranch-style homes built throughout the United States are heated with energy-wasting electricity and are so full of leaks that up to 85% of this expensive heat is rapidly lost. This is equivalent to having a large, window-size hole in the wall of such houses.

The 1989 Chevrolet Geo Metro built in Japan by Suzuki has a fuel efficiency of 25 kilometers per liter (58 mpg), and the Honda CRX HF gets 24 kilometers per liter (56 mpg). Prototype models built by Volvo, Volkswagen, Toyota, Peugeot, and Renault get 30 to 53 kilometers per liter (71 to 124 mpg).

Another way to make the world's diminishing supply of oil last longer is to shift more freight from trucks and airplanes to trains and ships (Figure 17-3). Manufacturers can increase the energy efficiency of new transport trucks by 50% by improving their aerodynamic design and using turbocharged diesel engines and radial tires. Truck companies can reduce waste by not allowing trucks to return empty after reaching their destination (but not carrying food on the initial run and garbage or toxic materials on the return run, as is now happening in some cases). The energy efficiency of today's commercial jet aircraft fleet could be doubled by improved designs.

Improving the Energy Efficiency of Commercial and Residential Buildings Sweden and South Korea have the world's toughest standards for energy efficiency in homes and other buildings. For example, the average home in Sweden, the world's leader in energy efficiency, consumes about one-third as much energy as an average American home of the same size (Figure 17-4).

A monument to energy waste is the 110-story, twin-towered World Trade Center in Manhattan, which uses as much electricity as a city of 100,000 persons. Windows in its walls of glass cannot be opened to take advantage of natural warming and cooling. Its heating and cooling systems must run around the clock, chiefly to take away heat from its inefficient lighting.

By contrast, Atlanta's 17-story Georgia Power Company building uses 60% less energy than conventional office buildings. The largest surface of the building is oriented to capture solar energy. Each floor extends over the one below, allowing heating by the low winter sun and blocking out the higher summer sun to reduce air conditioning costs. Energy-efficient lights focus on desks rather than illuminating entire rooms. Employees working at unusual hours use an adjoining 3-story building so that the larger structure doesn't have to be heated or cooled when few people are at work.

With existing technology, the United States could save 40% to 60% of the energy used in existing buildings and 70% to 90% of the energy used in new buildings (see Spotlight on p. 384). Building a *super-insulated house* is the best way to improve the efficiency of residential space heating and cooling and save on lifetime energy costs (Figure 17-5). Such a house is heavily insulated and made extremely airtight. Heat from direct solar gain, people, and appliances warms the house, requiring little or no auxiliary heating. An air-to-air heat exchanger prevents buildup of humidity and indoor air pollution.

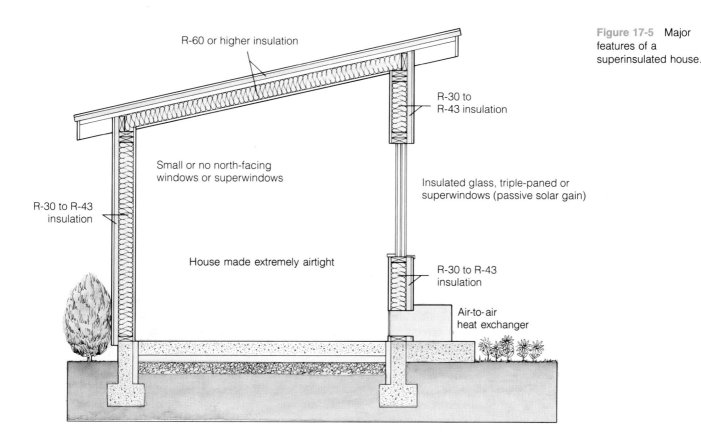

Figure 17-5 Major features of a superinsulated house.

R-60 or higher insulation

R-30 to R-43 insulation

Small or no north-facing windows or superwindows

Insulated glass, triple-paned or superwindows (passive solar gain)

R-30 to R-43 insulation

House made extremely airtight

R-30 to R-43 insulation

Air-to-air heat exchanger

Most home buyers look only at the initial price, not the more meaningful lifetime cost. A superinsulated house costs about 5% more to build than a conventional house. But this extra cost is paid back by energy savings within five years and can save a home owner $50,000 to $100,000 over a 40-year period. Yet sadly, this type of house accounts for less than 1% of new home construction in the United States, mostly because of a lack of consumer demand.

To keep the initial cost for buyers down, developers routinely construct inefficient buildings and stock them with inefficient heating and cooling systems and appliances. In the long run, such houses cost buyers 40% to 50% more when lifetime costs are considered. Also, builders of rental housing have little incentive to pay slightly more to make them energy efficient, when renters pay the fuel and electricity bills. Requiring all buildings to meet higher energy-efficiency standards and requiring a building's estimated annual and lifetime costs to be revealed to buyers would help correct these problems.

Recently engineer Michael Sykes developed an Enertia house that is heated and cooled by the earth's thermal inertia and solar energy without the need of a conventional heating or cooling system. The house's massive walls, ceilings, and floors are made of logs impregnated with salt, which allows them to soak up and dissipate heat in the summer and radiate heat in the winter. The log inner shell of the house is surrounded by a second log outer envelope that acts as convection loop to regulate temperatures without the use of fans.

Many energy-saving features can be added to existing homes, a process called *retrofitting*. Simply increasing insulation above ceilings can drastically reduce heating and cooling loads. The home owner usually recovers initial costs in two to six years and then saves money each year. Caulking and weatherstripping around windows, doors, pipes, vents, ducts, and wires save energy and money quickly.

One-third of the heat in U.S. homes and buildings escapes through closed windows (Figure 17-6)—an energy loss equal to the energy found in the oil flowing through the Alaskan pipeline every year. During hot weather these windows also let in large amounts of heat, greatly increasing the use of air conditioning. This loss and gain of heat occurs because a single-pane glass window has an insulating value of only R-1. The R-value of a material is a measure of its resistance to heat flow and thus its insulating ability. Even double-glazed windows have an insulating value of only R-2, and a typical triple-glazed window has an insulating value of R-4 to R-6.

Two U.S. firms now sell "superinsulating" R-10 to R-12 windows, about the insulating value of a normal outside wall (R-11), that pay for themselves in two to four years. If everyone in the United States used these windows, it would save more oil and

Energy experts Amory and Hunter Lovins have built a large, passively heated, superinsulated, partially earth-sheltered home and office combination in Old Snowmass, Colorado, where winter temperatures can drop to −40°C (−40°F). This structure, which also houses the research center for the Rocky Mountain Institute, gets all of its heat from the sun, uses one-tenth the normal amount of electricity, and uses less than half the normal amount of water. Electricity bills run about $30 a month, while those of conventional structures of this size in the same area run $330 to $1,000 per month.

In energy-efficient houses of the near future, microprocessors will monitor indoor temperatures, sunlight angles, and the location of people and will then send heat or cooled air where it is needed. Some will automatically open and close windows and insulated shutters to take advantage of solar energy and breezes and to reduce heat loss from windows at night and on cloudy days.

Researchers are working on "smart windows" that automatically change electronically from clear, which allows sunlight and

heat in on cold days, to reflective, which diverts sunlight away when the house gets too hot. Superinsulating windows (R-10 to R-12), already available, mean that a house can have as many windows as the owner wants in any climate without much heat loss. Insulating windows of R-12 or better should be available in the near future. Thinner insulation material will allow roofs to be insulated to R-100 and walls to R-43, far higher than today's best superinsulated houses (Figure 17-5).

Small-scale cogeneration units that run on natural gas or LPG (liquefied petroleum gas; see Section 18-1) are already available. They can supply a home with all its space heat, hot water, and electricity needs. The units are no larger than a refrigerator and make less noise than a dishwasher. Except for an occasional change of oil filters and spark plugs, they are nearly maintenance-free. Typically, this home-sized power and heating plant will pay for itself in four to five years.

Soon home owners may be able to get all the electricity they need from rolls of solar cells attached like shingles to a roof or applied

to window glass as a coating (already developed by Arco).

A West German firm (Bomin) has developed a solar-powered hydrogen gas system that can meet all energy needs of a home, apartment building, or a small village or housing development at an affordable price. A windproof panel of solar collectors automatically tracks the sun, concentrating sunlight at a fixed focus to temperatures as high as 500°C (932°F). The collected heat is stored in metal powders (hydrides) that release hydrogen gas, which burns cleanly to provide energy for cooking, electricity, heating, and cooling. This system, with an energy efficiency greater than that of electric power plants, should be available within a few years.

In 1989 Albers Technologies Corporation of Arizona patented a home air conditioner that uses water, not CFCs or HCFCs, as a coolant, draws half the electricity of a conventional unit, and costs about the same as conventional models with the same cooling capacity. A Saudi Arabian company plans to build 25,000 units a year beginning in 1992, with 20,000 units a year being exported to the United States.

natural gas each year than Alaska now supplies. The cost of saving this energy is equivalent to buying oil at $2 to $3 a barrel.

Building codes can be changed to require that all new houses use 80% less energy than conventional houses of the same size, as has been done in Davis, California (see Case Study on p. 134). Laws can require that any existing house be insulated and weatherproofed to certain standards before it can be sold, as is required in Portland, Oregon, for example.

Using the most energy-efficient appliances available can also save energy and money.* About one-

third of the electricity generated in the United States and other industrial countries is used to power household appliances.

One-fourth of the electricity produced in the United States is used for lighting—about equal to the output of 100 large power plants or half of all coal burned by the nation's electric utilities. Because conventional incandescent bulbs are only 5% efficient, their use wastes enormous amounts of energy and adds to the heat load of houses during hot weather.

Socket-type fluorescent light bulbs that use one-fourth as much electricity as conventional bulbs are now available (Figure 3-14, p. 49). Although they cost about $15 a bulb, they last about 13 times longer than conventional bulbs, save 3 times more money than they cost, and emit light indistinguishable from incandescents. Replacing a 75-watt incandescent bulb with an 18-watt compact fluorescent bulb reduces

---

* Each year the American Council for an Energy-Efficient Economy publishes a list of the most energy-efficient major appliances mass produced for the U.S. market. For a copy, send $3 to the council at 1001 Connecticut Ave., N.W., Suite 530, Washington, D.C. 20036.

Figure 17-6 Infrared photo shows heat loss around the windows, doors, roofs, and foundations (red, white, and yellow colors) of houses and stores in Plymouth, Michigan. Because of energy-inefficient design, most existing office buildings in the United States unnecessarily waste about half of the energy used to heat and cool them.

electricity consumption by 75% and prevents the generation of 0.9 metric ton (1 ton) of carbon dioxide and 11 kilograms (25 pounds) of sulfur dioxide. Over their lifetime, 15 compact fluorescent bulbs will save a homeowner about $300 to $375. Switching to these bulbs and other improved lighting equipment would save one-third of the electric energy now produced by all U.S. coal-fired plants or eliminate the need for all electricity produced by the country's 112 nuclear power plants. Sensors can also be used to turn off lights in unoccupied rooms or to dim lights whenever sunlight is available.

Residential refrigerators consume about 7% of the electricity used in the United States. If all U.S. households had the most efficient typical frost-free refrigerator now available, they would save enough electricity to eliminate the need for 18 large nuclear or coal-fired power plants. New prototype refrigerators being built in Denmark and Japan cut electricity use by another 50%.

Similar savings are possible with high-efficiency models of other energy appliances such as stoves, hot water heaters, and air conditioners. If the most energy-efficient appliances now available were installed in all U.S. homes over the next 20 years, we would save fuel equal to all the oil produced by Alaska's North Slope fields over their 25-year lifetime.

Energy expert Amory Lovins (see Guest Essay on p. 56) carries around a small briefcase that contains examples of fluorescent light bulbs, a low-flow shower head, superinsulated window glass, and other devices that save energy and money (Figure 17-7). Using these devices throughout the United States would save energy equal to the output of 200 large electric power plants. This would also save enough money to pay off the national debt, eliminate the need to import any oil, sharply reduce pollution and environ-

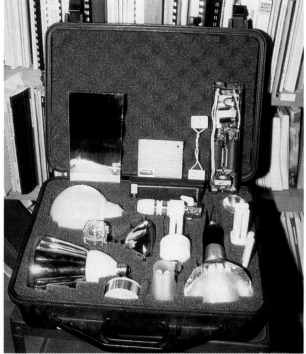

Figure 17-7 Amory Lovins's briefcase of available energy-saving lights, superinsulating window glass, water-flow restrictors, and other devices. Using these throughout the United States would save energy equal to that from 200 large electric power plants, save hundreds of billions of dollars, and sharply reduce pollution and environmental degradation.

mental degradation, slow projected global warming, and increase national and global military security.

Developing a Personal Energy Conservation Plan
Each of us can develop an individual plan for saving energy and money (see Individuals Matter inside the back cover). Four basic guidelines are:

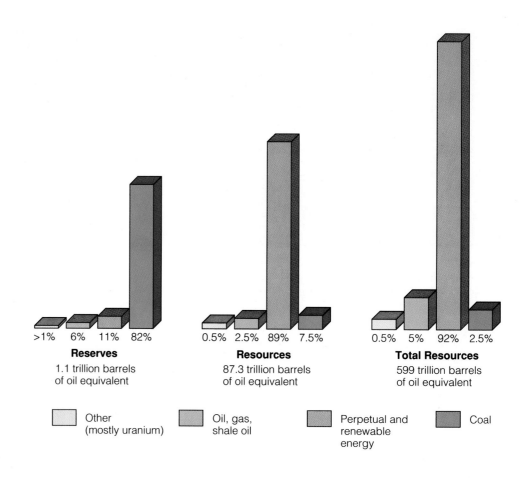

Figure 17-8 U.S. energy resource estimates. The estimated total resources (reserves plus resources) available from perpetual and renewable energy sources is more than ten times that from domestic supplies of coal, oil, natural gas, shale oil, and uranium. (Data from U.S. Department of Energy)

>1%	6%	11%	82%
Reserves			
1.1 trillion barrels of oil equivalent			

0.5%	2.5%	89%	7.5%
Resources			
87.3 trillion barrels of oil equivalent			

0.5%	5%	92%	2.5%
Total Resources			
599 trillion barrels of oil equivalent			

- Other (mostly uranium)
- Oil, gas, shale oil
- Perpetual and renewable energy
- Coal

1. Don't use electricity to heat space or water (Figure 3-7, p. 42).

2. Insulate new or existing houses heavily, caulk and weather-strip to reduce air infiltration and heat loss, and use energy-efficient windows.

3. Get as much heat and cooling as possible from natural sources—especially sun, wind, geothermal energy, and trees for windbreaks and natural shading.

4. Buy the most energy-efficient homes, lights, cars, and appliances available, and evaluate them only in terms of lifetime cost.

 DIRECT SOLAR ENERGY FOR PRODUCING HEAT AND ELECTRICITY

The Untapped Potential of Perpetual and Renewable Energy Resources The largest, mostly untapped sources of energy for all countries are perpetual and renewable energy from the sun, wind, flowing water, biomass, and the earth's internal heat (Figure 17-8). Developing these untapped resources could meet up to 80% of projected U.S. energy needs by 2010 and virtually all energy needs if coupled with improvements in energy efficiency (Figure 17-1).

Doing this would save money, eliminate the need for oil imports, produce less pollution and environmental degradation per unit of energy used, and increase economic, environmental, and military security. In the United States, geothermal power plants, wood-fired power plants, wind farms, and solar-thermal power plants can produce electricity cheaper than that generated from new nuclear power plants with far fewer subsidies from the federal government (see Guest Essay on p. 56). With a massive program to develop perpetual and renewable energy resources, these forms of energy could meet 30% to 45% of the world's projected energy demand by 2050 and bring about enough reduction in greenhouse gases to stabilize projected climate changes. The rest of this chapter evaluates the various perpetual and renewable energy resources available to us.

Passive Solar Systems for Space Heating and Cooling A **passive solar-heating system** captures sunlight directly within a structure and converts it to low-temperature heat for space heating (Figure 17-9). Insulated windows, greenhouses, and sunspaces face the sun to collect solar energy by direct gain. Thermal mass such as walls and floors of concrete, adobe, brick, stone, salt-treated logs, or tile store collected solar energy as heat and release it slowly throughout the day and night. Some designs also store heat in water-filled glass or plastic columns, black-painted barrels filled with water, and panels or cabinets containing heat-absorbing chemicals.

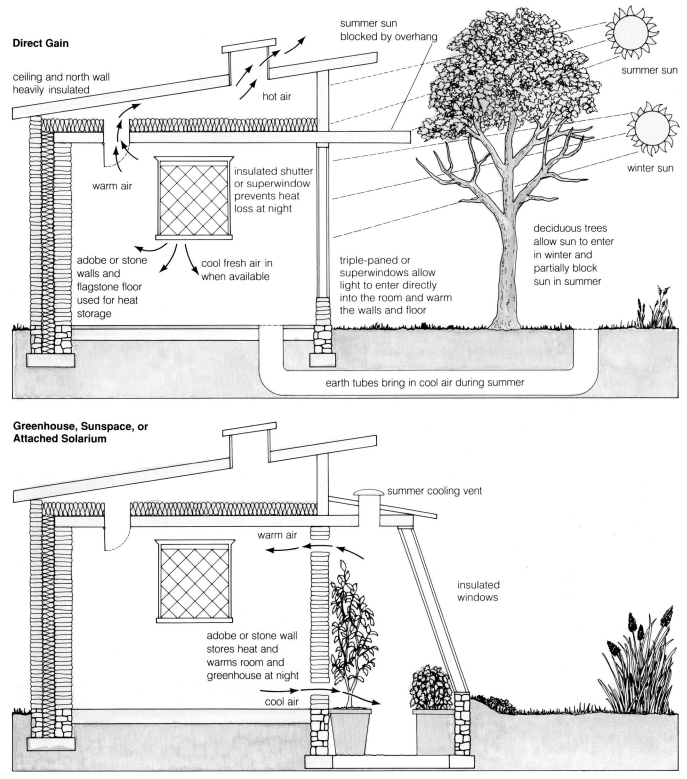

Direct Gain

ceiling and north wall heavily insulated

summer sun blocked by overhang

summer sun

hot air

winter sun

warm air

insulated shutter or superwindow prevents heat loss at night

adobe or stone walls and flagstone floor used for heat storage

cool fresh air in when available

triple-paned or superwindows allow light to enter directly into the room and warm the walls and floor

deciduous trees allow sun to enter in winter and partially block sun in summer

earth tubes bring in cool air during summer

Greenhouse, Sunspace, or Attached Solarium

summer cooling vent

warm air

insulated windows

adobe or stone wall stores heat and warms room and greenhouse at night

cool air

Figure 17-9 Three examples of passive solar design.

(continued)

Besides collecting and storing solar energy as heat, passive systems must also reduce heat loss in cold weather and heat gain in hot weather. Such structures are usually heavily insulated and caulked. Superwindows or movable, insulated shutters or curtains on windows reduce heat loss at night and on days with little sunshine.

Buildup of moisture and indoor air pollutants is minimized by an air-to-air heat exchanger, which supplies fresh air without much heat loss or gain. A small backup heating system may be used but is not necessary in a well-designed passively heated and superinsulated house in most climates. Today over 300,000 homes and 17,000 nonresidential buildings in

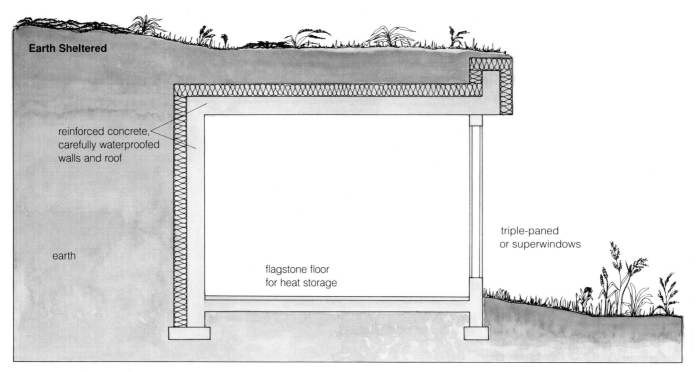

Earth Sheltered

reinforced concrete, carefully waterproofed walls and roof

earth

flagstone floor for heat storage

triple-paned or superwindows

Figure 17-9 (continued)

Pat Armstrong/Visuals Unlimited

Figure 17-10 Exterior of an earth-sheltered house in Will County, Illinois (above). The interior of the dome-shaped, earth-sheltered house, such as this one in Colorado (right), can look like that of any ordinary house. Passive solar design and skylights can provide more daylight than is found in most conventional houses. On a lifetime-cost basis, earth-sheltered houses are cheaper than conventional above-ground houses of the same size because of reduced heating and cooling costs, no exterior maintenance and painting, and low fire insurance rates. These structures also provide more privacy, quiet, and security from break-ins, fires, hurricanes, earthquakes, tornadoes, and storms than conventional homes.

Earth Systems, Inc., Durango, Colorado

the United States have passive solar designs, including earth-sheltered homes (Figure 17-10), and get 30% to 40% or more of their energy from the sun. However, this is a small number compared to the 80 million homes in the United States and the roughly 3 million new ones built each year. With technologies already available or under development, passive solar designs could provide at least 80% of a building's heating needs and at least 60% of its cooling needs with an added construction cost of 5% to 10%.

Roof-mounted passive solar water heaters (Figure 17-11) can also supply all or most of the hot water for a typical house. The most promising model, called the Copper Cricket, is produced by Sage

Active Solar Hot Water Heating System

Figure 17-11 Active and passive solar water heaters.

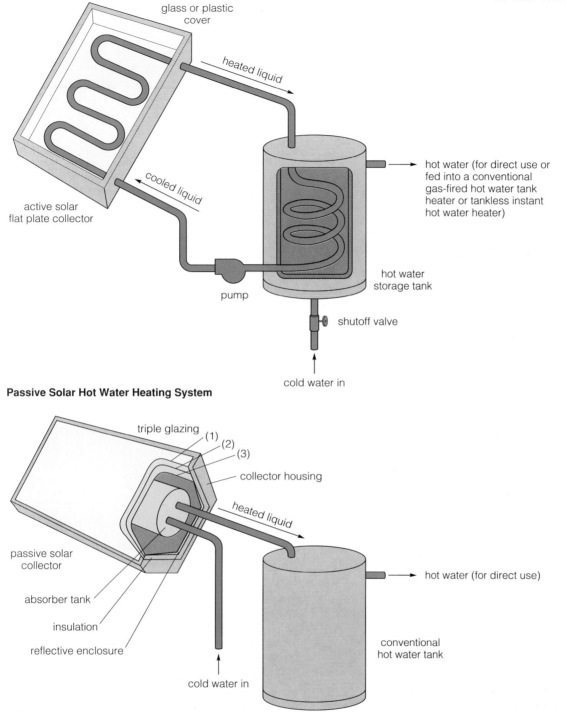

glass or plastic cover

heated liquid

active solar flat plate collector

cooled liquid

hot water (for direct use or fed into a conventional gas-fired hot water tank heater or tankless instant hot water heater)

hot water storage tank

pump

shutoff valve

cold water in

Passive Solar Hot Water Heating System

triple glazing (1) (2) (3)

collector housing

heated liquid

passive solar collector

absorber tank

insulation

reflective enclosure

hot water (for direct use)

conventional hot water tank

cold water in

Advance Company in Eugene, Oregon, and sells for $1,800.

In hot weather passive cooling can be provided by blocking the high summer sun with deciduous trees, window overhangs, or awnings (Figure 17-9). Windows and fans take advantage of breezes and keep air moving. A reflective foil sheet can be suspended in the attic to block heat from radiating down into the house.

At a depth of 3 to 6 meters (10 to 20 feet) the temperature of the earth stays about 13°C (55°F) all year long in cold northern climates and about 19°C (67°F) in warm southern climates. Earth tubes buried at this depth below ground can pipe cool and partially dehumidified air into an energy-efficient house at a cost of several dollars a summer (Figure 17-9). For a large space two to three of these geothermal cooling fields running in different directions from the house

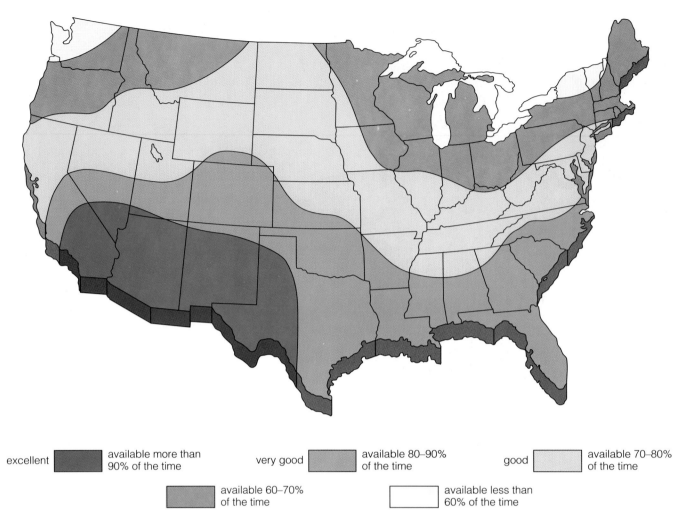

| excellent | available more than 90% of the time | very good | available 80–90% of the time | good | available 70–80% of the time |

| | available 60–70% of the time | | available less than 60% of the time |

Figure 17-12 Availability of solar energy during the day in the continental United States. (Data from U.S. Department of Energy and National Wildlife Federation)

should be installed. Then when the added heat degrades the cooling effect from one field, home owners can switch to another field.

In areas with dry climates, such as the southwestern United States, evaporative coolers can remove interior heat by evaporating water. Solar-powered air conditioners have been developed but so far are too expensive for residential use. In Reno, Nevada, some buildings stay cool throughout the hot summer without air conditioning; large, insulated tanks of water chilled by cool nighttime air keep indoor temperatures comfortable during the day.

Active Solar Systems for Heating Space and Water

In an **active solar-heating system,** specially designed collectors concentrate solar energy and store it as heat for space heating and heating water. Several connected collectors are usually mounted on a roof with an unobstructed exposure to the sun (Figure 17-11).

Solar energy, collected as heat, is transferred to water, an antifreeze solution, or air pumped through pipe inside the collector. The heated solution or air is pumped through and stored in a well-insulated tank containing water or rocks. Thermostat-controlled fans or pumps distribute this stored heat as needed, usually through conventional heating ducts. In middle and high latitudes with cold winter temperatures and moderate levels of sunlight (Figure 17-12), a small backup heating system is needed during prolonged cold or cloudy periods.

Active solar collectors can also supply hot water. Over 1 million active solar hot water systems have been installed in the United States, especially in California, Florida, and southwestern states with ample sunshine. The main barrier to their widespread use in the United States is an initial cost of $1,800 to $5,000.

Solar water heaters are used in 90% of all households in Cyprus and on most of its apartment buildings and hotels. In Israel 65% of all domestic water heating is supplied by simple active solar systems that cost less than $500 per residence. About 12% of the houses in Japan and 37% in Australia use such systems.

Comstock/© Georg Gerster

Figure 17-13 Solar furnace near Odeillo in the Pyrenees Mountains of southern France produces temperatures high enough to melt metals. A field of curved computer-driven mirrors (not shown) track the sun and reflect sunlight onto the giant parabolic collector shown in the photograph.

Luz International, Ltd., Los Angeles

Figure 17-14 The world's largest solar power facility in California's Mojave Desert produces 80 megawatts of electricity by using 405 hectares (1,000 acres) of parabolic collectors to concentrate solar energy. Sunlight is focused on water flowing through pipes in the collectors to produce steam, which is used to spin generators and produce electricity. This plant, which began operation in 1989, produces enough electricity to meet the residential needs of 115,000 people at a cost cheaper than electricity from a nuclear power plant. This type of solar power plant can be built in about a year and can be expanded as needed. By 1994 Luz International, a French-Israeli company, will have plants like this in Southern California that will produce enough electricity to meet the residential needs of 1 million people, as many as live in a city the size of Phoenix, San Antonio, or San Francisco.

Pros and Cons of Using Solar Energy for Heating Space and Water The energy supply for active or passive systems to collect solar energy for low-temperature heating of buildings is free, naturally available on sunny days, and the net useful energy yield is moderate to high. The technology is well developed and can be installed quickly. No carbon dioxide is added to the atmosphere, and environmental impacts from air pollution and water pollution are low. Land disturbance is also low because passive systems are built into structures and active solar collectors are usually placed on rooftops.

On a lifetime-cost basis, good passive solar and superinsulated design is the cheapest way to provide 40% to 100% of the space heating for a home or small building in regions with enough sunlight (Figure 17-12). Such a system usually adds 5% to 10% to the construction cost but lowers the lifetime cost of a house by 30% to 40% of the operating cost of conventional houses.

Active systems cost more than passive systems on a lifetime basis because they require more materials to build, they need more maintenance, and eventually they deteriorate and must be replaced. However, retrofitting an existing house with an active solar system is often easier than adding a passive system.

However, there are disadvantages. Higher initial costs discourage buyers not used to considering lifetime costs and buyers who move every few years. With present technology, active solar systems usually cost too much for heating most homes and small buildings. Better design and mass production techniques could change this. Many people also believe that active solar collectors, sitting on rooftops or in yards, are ugly.

Most passive solar systems require that owners open and close windows and shades to regulate heat flow and distribution, but this can be done by cheap microprocessors. Owners of passive and active solar systems also need laws that prevent others from building structures that block a user's access to sunlight. Such legislation is often opposed by builders of high-density developments.

Concentrating Solar Energy to Produce High-Temperature Heat and Electricity Huge arrays of computer-controlled mirrors can track the sun and focus sunlight on a central heat collection point atop a tall tower (Figure 17-13), or on oil-filled pipes running through the middle of curved solar collectors (Figure 17-14). This concentrated sunlight can produce tem-

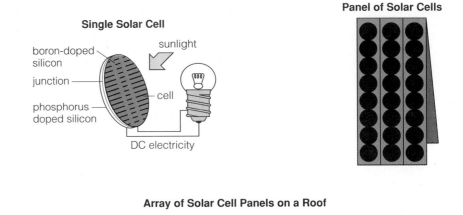

Figure 17-15 Use of photovoltaic (solar) cells to provide DC electricity for an energy-efficient home; any surplus can be sold to the local power company. Prices should be competitive sometime in the 1990s.

Single Solar Cell

boron-doped silicon

junction

phosphorus doped silicon

sunlight

cell

DC electricity

Panel of Solar Cells

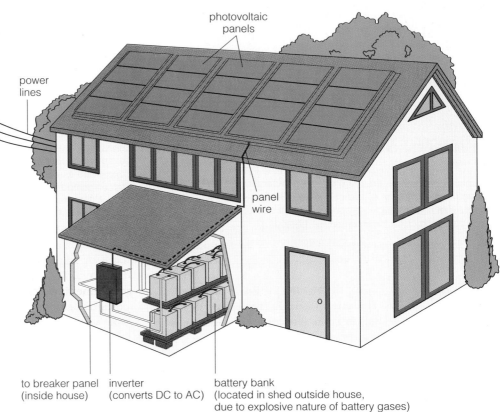

Array of Solar Cell Panels on a Roof

photovoltaic panels

power lines

panel wire

to breaker panel (inside house)

inverter (converts DC to AC)

battery bank (located in shed outside house, due to explosive nature of battery gases)

peratures high enough for industrial processes, or for making high-pressure steam to run turbines and produce electricity.

The small solar-thermal power plant shown in Figure 17-14 can produce electricity at a cost one-third less than that from a new nuclear power plant, and larger plants should be able to produce electricity as cheaply as most coal-burning power plants by 1995. Small natural gas turbines are used to run the facility when the sun is not shining. This solar-natural gas fuel mix produces less than one-sixth as much carbon dioxide per kilowatt-hour of electricity as a normal coal-fired plant.

The main use of these plants is supplying reserve power to meet daytime peak electricity loads, especially in sunny areas with large air conditioning demands. It is estimated that solar-thermal plants occupying less than 1% of the area of the Mojave Desert could supply the electricity needs of Los Angeles. Solar-thermal plants could also be used to produce hydrogen gas for use as a fuel (Section 17-8) and to convert hazardous wastes into harmless or less harmful substances.

The impact of these solar power plants on air and water is low and they can be built in only one to two years. They need large areas for solar collection but occupy one-third less land area than a coal-burning plant when the land used to extract coal is included. However, there is concern about building such structures in arid, ecologically fragile desert biomes, where there may not be enough water for use in cooling towers to recondense spent steam.

Figure 17-16 This power plant in Sacramento, California, uses solar-powered photovoltaic cells to produce electricity. The nuclear power plant in the background has been closed down.

Converting Solar Energy Directly to Electricity: Photovoltaic Cells Solar energy can be converted by **photovoltaic cells,** commonly called *solar cells,* directly into electrical energy. Most solar cells consist of layers of purified silicon, which can be made from inexpensive, abundant sand. Trace amounts of other substances (such as gallium arsenide or cadmium sulfide) are added so that the resulting semiconduction emits electrons and produces a small amount of electrical current when struck by sunlight (Figure 17-15).

Today solar cells supply electricity for at least 20,000 homes worldwide (10,000 in the United States) and for 6,000 villages in India. Most of these homes and villages are in remote areas where it costs too much to bring in electric power lines. Solar cells are also used to switch railroad tracks and to supply power for water wells, irrigation pumping, battery charging, calculators, portable laptop computers, ocean buoys, lighthouses, and offshore oil-drilling platforms in the sunny Persian Gulf.

Because the amount of electricity produced by a single solar cell is very small, many cells must be wired together in a panel to provide 30 to 100 watts of electric power (Figure 17-15). Several panels are wired together and mounted on a roof, or on a rack that tracks the sun, to produce electricity for a home or building.

Massive banks of such cells can also produce electricity at a small power plant (Figure 17-16). If the price comes down as expected, a 174 square kilometer (67 square mile) area of solar cells placed in a desert or in various sunny areas could provide all of the electricity now consumed in the United States.

The development of more energy-efficient cells and cost-effective mass-production techniques should allow solar cells to produce electricity cheaper than nuclear power plants sometime in the 1990s or shortly

after the turn of the century. In 1990 the Chronar Corporation began building a 400-kilowatt photovoltaic power plant in Davis, California, that is expected to produce electricity at a cost lower than that from a new nuclear power plant. The potential market for solar cells would be at least $100 billion a year.

Despite their enormous potential, between 1981 and 1990, the U.S. federal research and development budget for solar cells was cut by 76% and the U.S. share of the worldwide solar-cell market fell from 75% to 32%. During this same period, Japanese government expenditures in this area have tripled and Japan's share of the worldwide solar-cell market grew from 15% to 37%.

Federal and private research efforts on photovoltaics in the United States need to be increased sharply. Otherwise, the United States would lose out on a major global market and may find much of its capital being drained to pay for imports of photovoltaic cells from Japan and other countries.

Pros and Cons of Photovoltaic Cells If present projections are correct, solar cells could supply 20% to 30% of the world's electricity and half of that in the United States sometime between 2030 and 2050. This would eliminate the need to build large-scale power plants of any type and allow many existing nuclear and coal-fired power plants to be phased out.

Solar cells are reliable and quiet, have no moving parts, and should last 30 years or more if encased in glass or plastic. They can be installed quickly and easily and need little maintenance other than occasional washing to prevent dirt from blocking the sun's rays. Small or large solar-cell packages can be built and they can be easily expanded or moved as needed. Suitable locations for solar cells include deserts, marginal lands, alongside interstate highways, yards, and rooftops.

Most solar cells are made from silicon, the second most abundant element by weight in the earth's crust. They do not produce carbon dioxide during use. Air and water pollution during operation is low, air pollution from manufacture is low, and land disturbance is very low for roof-mounted systems. The net useful-energy yield is fairly high and rising with new designs.

However, solar cells do have some drawbacks. The present costs of solar-cell systems are high but are projected to become competitive in 7 to 15 years. Other problems include the opinions of some people that racks of solar cells on rooftops or in yards are unsightly; potential limits could be placed on their use by an insufficient amount of gallium or cadmium; and the possible production of moderate levels of water pollution from chemical wastes introduced through the manufacturing process in the absence of effective pollution control.

17-4 PRODUCING ELECTRICITY FROM MOVING WATER AND FROM HEAT STORED IN WATER

Types of Hydroelectric Power In *large-scale hydropower projects*, high dams are built across large rivers to create large reservoirs (Figure 5-29, p. 102). The stored water is then allowed to flow through massive pipes at controlled rates, spinning turbines and producing electricity. However, the reservoirs behind the dams eventually fill with sediment and become useless in 40 to 300 years, depending on the rate of natural and human-accelerated soil erosion from land above the dam. This means that large hydroelectric plants are nonrenewable sources of energy on a human time scale.

In *small-scale hydropower projects*, a low dam with no reservoir, or only a small one, is built across a small stream. The renewable natural water flow is used to generate electricity, but electricity production can vary with seasonal changes in stream flow.

Falling water can also be used to produce electricity in *pumped-storage hydropower systems*. Their main use is to supply extra power during times of peak electrical demand. When electricity demand is low, usually at night, pumps using electricity from a conventional power plant pump water uphill from a lake or reservoir to another reservoir at a higher elevation, usually on top of a mountain. When a power company temporarily needs more electricity than its plants can produce, water in the upper reservoir is released. On its downward trip to the lower reservoir, the water flows through turbines and generates electricity. But this is an expensive way to produce electricity. Much cheaper alternatives, such as natural gas turbines, are available. Another possibility may be the use of solar-powered pumps to raise water to the upper reservoir.

Present and Future Use of Hydropower In 1988 hydropower supplied 21% of the world's electricity and 6% of the world's total commercial energy. Hydropower supplies Norway with essentially all its electricity, Switzerland 74%, and Austria 67%.

Much of the hydropower potential of North America and Europe has been developed, but Africa has tapped only 5% of its hydropower potential, Latin America 8%, and Asia 9%. In 1988, LDCs got almost 50% of their electricity from hydropower. Between 1981 and 1995, LDCs are projected to add large hydropower plants that will produce electricity equal to that of 225 large nuclear or coal-burning power plants. Half of this new hydropower capacity will be added in Brazil, China, and India.

The United States is the world's largest producer of electricity from hydropower. Today hydropower produced at 2,029 sites supplies 10% to 14% of the electricity and 5% to 6% of all commercial energy used by the United States, with the amount varying with rain and snowfall patterns. But the large dam-building era in the United States is drawing to a close because of high construction costs and lack of suitable sites. Any new large supplies of hydroelectric power in the United States will be imported from Canada, which gets more than 70% of its electricity from hydropower.

By 1988 almost 1,500 small-scale hydroelectric power plants were operating or under construction, mostly in the West and Northeast. They were generating electricity equal to that of three large power plants. According to the U.S. Corps of Engineers, retrofitting abandoned small and medium-size hydroelectric sites and building new small-scale hydroelectric plants on suitable sites could supply the United States with electricity equal to that of 47 large power plants.

However, since 1985 the development of small-scale hydropower in the United States has fallen off sharply because of low oil prices and the loss of federal tax credits. Many projects have been opposed by local residents and conservationists because the dams reduce stream flow, disrupting aquatic life and preventing some types of recreation.

Pros and Cons of Hydropower Many LDCs have large, untapped potential hydropower sites, although many are far from where the electricity is needed. Hydropower has a moderate to high net useful energy yield and fairly low operating and maintenance costs.

Hydroelectric plants rarely need to be shut down and they produce no emissions of carbon dioxide or other air pollutants during operation. Their reservoirs have life spans two to ten times the life of coal and nuclear plants. Large dams also help control flooding and supply a regulated flow of irrigation water to areas below the dam.

Developing small-scale hydroelectric plants by rehabilitating existing dams has little environmental impact, and once rebuilt, the units have a long life. Only a few people are needed to operate them and they need little maintenance.

But hydropower has some drawbacks. Construction costs for new large-scale systems are high, and few suitable sites are left in the United States and Europe. The reservoirs of large-scale projects flood huge areas, destroy wildlife habitats, uproot people, decrease natural fertilization of prime agricultural land in river valleys below the dam, and decrease fish harvests below the dam. Without proper land-use control, large-scale projects can greatly increase soil erosion and sediment water pollution near the reservoir above the dam. This reduces the effective life of the reservoir.

By reducing stream flow, small hydroelectric projects threaten recreational activities and aquatic life, disrupt scenic rivers, and destroy wetlands. During drought periods these plants produce little if any power. Most of the electricity produced by these projects can be supplied at a lower cost and with less environmental impact by industrial cogeneration and by improving the energy efficiency of existing big dams.

Tidal, or Moon, Power Twice a day, a large volume of water flows in and out of bays and estuaries along the coast as a result of high and low tides caused by gravitational attraction of the moon. In a few places, tides flow in and out of a bay with an opening narrow enough to be obstructed by a dam with turbines to produce electricity. If the difference in water height between high and low tides is large enough, the kinetic energy in these daily tidal flows based on moon power can be used to spin turbines to produce electricity. But only about two dozen places in the world have these conditions. Currently, only two large tidal energy facilities are operating, one at La Rance, France, and the other in Canada in the Bay of Fundy. The Chinese government has built several small tidal plants.

Using tidal energy to produce electricity has several advantages. The energy source (tides from gravitational attraction) is free, operating costs are low, and the net useful-energy yield is moderate. No carbon dioxide is added to the atmosphere, air pollution is low, and little land is disturbed.

Most analysts, however, expect tidal power to make only a tiny contribution to world electricity supplies. There are few suitable sites and construction costs are high. The output of electricity varies daily with tidal flows so there must be a backup system. The dam and power plant can be damaged by storms, and metal parts are easily corroded by seawater. The disruption of normal tidal flows may also disturb aquatic life in coastal estuaries.

Wave Power The kinetic energy in ocean waves, created primarily by wind, is another potential source of energy. Japan, Norway, Great Britain, Sweden, the United States, and the Soviet Union have built small experimental plants to evaluate this form of hydropower. None of these plants has produced electricity at a competitive price, but some designs show promise.

Most analysts expect this alternative to make little contribution to world electricity production except in a few coastal areas with the right conditions. Construction costs are moderate to high and the net useful-energy yield is moderate. Equipment could be damaged or destroyed by saltwater corrosion and severe storms.

Ocean Thermal Energy Conversion Ocean water stores huge amounts of heat from the sun, especially in tropical areas. Japan and the United States have been conducting experiments to evaluate the technological and economic feasibility of using the large temperature differences between the cold deep waters and the sun-warmed surface waters of tropical oceans to produce electricity in *ocean thermal energy conversion* (OTEC) plants anchored to the bottom of tropical oceans in suitable sites.

The source of energy for OTEC is limitless at suitable sites, and a costly energy storage and backup system is not needed. No air pollution except carbon dioxide is produced during operation, and the floating power plant requires no land area. Nutrients brought up when water is pumped from the ocean bottom might be used to nourish schools of fish and shellfish.

However, most energy analysts believe that the large-scale extraction of energy from ocean thermal gradients may never compete economically with other energy alternatives. Construction costs are high—two to three times those of comparable coal-fired plants. Operating and maintenance costs are also high because of corrosion of metal parts by seawater and fouling of heat exchangers by algae and barnacles. Plants could also be damaged by hurricanes and typhoons.

Other problems include a limited number of sites and a low net useful-energy yield; possible disruption of coral reef communities and other aquatic life by pumping large volumes of deep-ocean water to the surface; and the release of large quantities of dissolved carbon dioxide into the atmosphere.

Solar Ponds A **solar pond** is a solar-energy collector consisting of at least 0.5 hectare (1 acre) of a lined cavity filled with salt water or with fresh water enclosed in black plastic bags. *Saline solar ponds* can be used to produce electricity and are usually located near inland saline seas or lakes in areas with ample sunlight. The bottom layer of water in such ponds stays on the bottom when heated because it has a higher salinity and density (mass per unit volume) than does the top layer. Heat accumulated in the bottom layer of water during daylight can be used to produce steam, which spins turbines, generating electricity.

A saline solar pond can be a naturally occurring body of salt water, such as the Dead Sea near Israel, the Salton Sea in California, or Utah's Great Salt Lake. Such ponds can also be built by excavating an area of land, lining it with plastic, and filling it with salt and water or brine. An experimental saline solar-pond power plant on the Israeli side of the Dead Sea has been operating successfully for several years. By 2000, Israel plans to build several plants around the Dead Sea to supply electricity for air conditioning and desalinating water. Several experimental saline solar ponds have been built in the United States, Australia, India, and Mexico.

Freshwater solar ponds can be used as a source of hot water and space heating. A shallow hole is dug, lined with concrete, and covered with insulation. A number of large, black plastic bags, each filled with several centimeters of water, are placed in the hole. The top of the pond is then covered with fiberglass panels, which let sunlight in and keep most of the heat stored in the water during daylight from being lost to the atmosphere. When the water in the bags has reached its peak temperature in the afternoon, a computer turns on pumps to transfer hot water from the bags to large, insulated tanks for distribution as hot water or for space heating.

Saline and freshwater solar ponds have the same advantages as OTEC systems. In addition, they have a moderate net useful-energy yield, have moderate construction and operating costs, and need little maintenance. Freshwater solar ponds can be built in almost any sunny area. They may be useful for supplying hot water and space heating for large buildings and small housing developments. Saline solar ponds are feasible in areas with moderate to ample sunlight, especially ecologically fragile deserts. Operating costs can be high because of saltwater corrosion of pipes and heat exchangers. Unless lined, the ponds can become ineffective when compounds leached from bottom sediment darken the water and reduce sunlight transmission. With adequate research and development support, solar ponds could supply 3% to 4% of U.S. energy needs by the year 2000.

17-5 PRODUCING ELECTRICITY FROM WIND

Wind Power Worldwide by 1990 there were over 20,000 wind turbines, grouped in clusters called wind farms (see photo on p. 376), which feed power to a utility grid. They produced electricity equal to that from 1.6 large (1,000 megawatt) nuclear or coal-burning power plants. Most of these are in California (15,000 machines) and Denmark, located in windy mountain passes and along coastlines.

In 1990 California wind farms produced enough electricity to meet the residential power needs of San Francisco. The state has the potential to use wind to produce electricity equal to that from 6 to 31 large power plants by the year 2000. The island of Hawaii gets about 8% of its electricity from wind, and the use of wind power is spreading to the state's other islands.

The cost of producing electricity with wind farms is about half that of a new nuclear power plant. New technological advances and mass production could reduce costs by another 25% by 1995, making wind power competitive with electricity produced by coal-fired plants. Wind power experts project that by the middle of the next century, wind power could supply more than 10% of the world's electricity and 10% to 20% of the electricity in the United States.

However, the development of this energy resource in the United States has slowed since 1986, when federal tax credits and most state tax credits for wind power were eliminated. Also, the federal budget for research and development of wind power was cut by 90% between 1981 and 1990.

Pros and Cons of Wind Power Wind power is an unlimited source of energy at favorable sites, and large wind farms can be built in only three to six months. With a moderate to fairly high net useful-energy yield, these systems emit no carbon dioxide or other air pollutants during operation, need no water for cooling, and their manufacture and use produce little water pollution. They operate 80% to 98% of the time the wind is blowing. The land occupied by wind farms can be used for grazing and other purposes, with the leases providing extra income for farmers and ranchers. Wind power has a significant cost advantage over nuclear power and should become competitive with coal in many areas sometime in the 1990s.

But wind power can be used only in areas with sufficient winds. Backup electricity from a utility company or from an energy storage system is necessary when the wind dies down, but this is not a problem at suitable sites. Backup could also be provided by linking wind farms with a solar-cell or

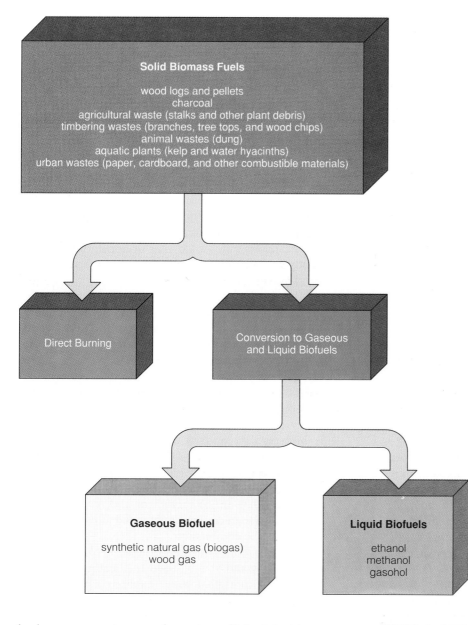

Figure 17-17 Major types of biomass fuel.

hydropower system, or by using efficient turbines fueled by natural gas.

Building wind farms in mountain passes and along shorelines can cause visual pollution. Noise and interference with local television reception have been problems with large turbines, but this can be overcome with improved design and use in isolated areas. Large wind farms might also interfere with the flight patterns of migratory birds in certain areas.

 ENERGY FROM BIOMASS

17-6 Renewable Biomass as a Versatile Fuel

Biomass is organic plant matter produced by solar energy through photosynthesis. It includes wood, agricultural wastes, and garbage. Some of this plant matter can be burned as solid fuel or converted to more convenient gaseous or liquid *biofuels* (Figure 17-17). In 1988, biomass, mostly from the burning of wood and manure to heat buildings and cook food, supplied about 15% of the world's commercial energy (4% to 5% in Canada and the United States) and about half of the energy used in LDCs.

All biomass fuels have several advantages in common. They can be used in solid, liquid, and gaseous forms for space heating, water heating, producing electricity, and propelling vehicles. Biomass is a renewable energy resource as long as trees and plants are not harvested faster than they grow back—a requirement that is not being met in most places (Section 15-2).

No net increase in atmospheric levels of carbon dioxide occurs as long as the rate of removal and burning of trees and plants and loss of below-ground organic matter does not exceed their rate of replenishment. Burning of biomass fuels adds much less

Figure 17-18 This small, 16-megawatt biomass power plant in Whitefield, New Hampshire, is operated by Thermo Electron Energy. Nearby forest residue is burned to generate electricity.

sulfur dioxide and nitric oxide to the atmosphere per unit of energy produced than the uncontrolled burning of coal and thus requires fewer pollution controls.

Biomass fuels also share some disadvantages. It takes a lot of land to grow biomass fuel. Without effective land-use controls and replanting, widespread removal of trees and plants can deplete soil nutrients and cause excessive soil erosion, water pollution, flooding, and loss of wildlife habitat. Biomass resources also have a high moisture content (15% to 95%), which lowers their net useful-energy. The added weight of the moisture makes collecting and hauling wood and other plant material fairly expensive. Each type of biomass fuel has other specific advantages and disadvantages.

Burning Wood and Wood Wastes About 80% of the people living in LDCs heat their dwellings and cook their food by burning wood or charcoal made from wood. But currently at least 1.1 billion people in LDCs cannot find or are too poor to buy enough fuelwood to meet their needs, and this number may increase to 2.5 billion by 2000 (Section 15-2).

In MDCs with adequate forests, the burning of wood, wood pellets, and wood wastes to heat homes and produce steam and electricity in industrial boilers increased rapidly during the 1970s because of price increases in heating oil and electricity. Sweden leads the world in using wood as an energy source, mostly for district heating plants.

In the United States, small wood-burning power plants located near sources of their fuel can produce electricity at about half the price of a new nuclear power plant and almost equal to that produced by burning coal (Figure 17-18). The forest products industry (mostly paper companies and lumber mills)

consumes almost two-thirds of the fuelwood used in the United States. Homes and small businesses burn the rest. The largest use of fuelwood is in New England, where wood is plentiful.

Wood has a moderate to high net useful-energy yield when collected and burned directly and efficiently near its source. But in urban areas where wood must be hauled long distances, it can cost home owners more per unit of energy produced than oil and electricity. Burning wood produces virtually no emissions of sulfur dioxide.

Harvesting and burning wood can cause accidents. Each year in the United States over 10,000 people are injured by chain saws. Also several hundred Americans are killed each year in house fires caused mostly by improperly located or poorly maintained wood stoves.

Burning fuelwood releases carbon monoxide, solid particulate matter, and unburned residues that pollute indoor and outdoor air. According to the EPA, wood burning causes as many as 820 cancer deaths a year in the United States.

This air pollution can be reduced 75% by a $100 to $250 catalytic combustion chamber in the stove or stovepipe. These units also increase the energy efficiency of a typical airtight wood stove from 55% to as high as 81% and reduce the need for chimney cleaning and the chance of chimney fires. However, these devices must be replaced every four years at a cost of about $100. Recently wood stoves have been developed that are 65% efficient and emit 90% less air pollution than conventional wood stoves, without using catalytic combustion.

Many people consider fireplaces cozy and romantic. They can also be used for heating, but they are so inefficient that they result in a net loss of energy from a house. The draft of heat and gases rising up the fireplace chimney exhausts warm air and pulls in cold air from cracks and crevices throughout a house. Fireplace inserts with glass doors and blowers help but still waste energy compared to an efficient wood-burning stove. If you must have a fireplace, shut off the room it is in from the rest of the house. Then crack a window in that room so that the fireplace won't draw much heated air from other rooms, or run a small pipe into the front of the fireplace so it can get the air it needs during combustion from outside.

In London and in South Korean cities, wood fires have been banned to reduce air pollution. Since 1990 the EPA has required all new wood stoves sold in the United States to emit at least 70% less particulate matter than earlier models. Some new stoves meet these standards by using catalytic combustion devices, others by better design or by using cleaner-burning wood pellets for fuel.

Energy Plantations One way to produce biomass fuel is to plant large numbers of fast-growing trees, shrubs, or grasses in *biomass-energy plantations* to supply fuelwood. Plantations of oil palms and varieties of Euphorbia plants, which store energy in hydrocarbon compounds (like those found in oil), can also be established. After these plants are harvested, their oil-like material can be extracted and either refined to produce gasoline or burned directly in diesel engines. Both types of energy plantations can be established on semiarid land not needed to grow crops, although lack of water can limit productivity.

This industrialized approach to biomass production usually requires heavy use of pesticides and fertilizers, which can pollute drinking supplies and harm wildlife. It also requires large areas of land.

Conversion of large areas to monoculture energy plantations also reduces biodiversity. In some areas biomass plantations might compete with food crops for prime farm land. Also, they are likely to have low or negative net useful-energy yields, as do most conventional crops grown by industrialized agricultural methods.

Burning Agricultural and Urban Wastes In agricultural areas, crop residues (the unharvested parts of food crops) and animal manure can be collected and burned (Figure 17-18) or converted to biofuels. By 1985 Hawaii was burning a residue (called *bagasse*) left after sugarcane harvesting and processing to supply almost 10% of its electricity (58% on the island of Kauai and 33% on the island of Hawaii). Other crop residues that could be burned include coconut shells, peanut and other nut hulls, and cotton stalks. Brazil gets 10% of its electricity by burning bagasse and plans to use this crop residue to produce 35% of its electricity by the year 2000.

This approach makes sense when residues are burned in small power plants located near areas where the residues are produced (Figure 17-18). Otherwise it takes too much energy to collect, dry, and transport the residues to power plants. In addition, ecologists argue that it makes more sense to use crop residues to feed livestock, retard soil erosion, and fertilize the soil.

An increasing number of cities in Japan, western Europe, and the United States have built incinerators that burn trash and use the heat released to produce electricity or to heat nearby buildings (Section 19-5). But this approach may be limited by citizen opposition because of concern about emissions of toxic gases and what to do with the resulting toxic ash. Some analysts argue that more energy is saved by composting or recycling paper (p. 345) and other organic wastes than by burning them (Section 19-6).

Converting Solid Biomass to Liquid and Gaseous Biofuels Plants, organic wastes, sewage, and other forms of solid biomass can be converted by bacteria and various chemical processes into gaseous and liquid biofuels (Figure 17-17). Examples are *biogas* (a mixture of 60% methane and 40% carbon dioxide), *liquid methanol* (methyl, or wood alcohol), and *liquid ethanol* (ethyl, or grain alcohol).

In China, anaerobic bacteria in an estimated 7 million *biogas digesters* convert organic plant and animal wastes into methane fuel for heating and cooking. After the biogas has been separated, the solid residue is used as fertilizer on food crops or, if contaminated, on nonedible crops such as trees. India has about 750,000 biogas digesters in operation, half of them built since 1986.

When they work, biogas digesters are very efficient. However, they are slow and unpredictable. Development of new, more reliable models could change this.

Methane fuel is also produced by underground decomposition of organic matter in the absence of air (anaerobic digestion) in landfills. This gas can be collected by pipes inserted into landfills, separated from other gases, and burned as a fuel. Eighty-two U.S. landfills currently recover methane, but 2,000 to 3,000 large U.S. landfills have the potential for large-scale methane recovery. Burning this gas instead of allowing it to escape into the atmosphere helps slow projected global warming because methane is 25 times more effective in causing atmospheric warming per molecule than carbon dioxide (Figure 10-6, p. 214).

Methane can also be produced by anaerobic digestion of manure produced at animal feedlots and sludge produced at sewage treatment plants (Section 11-8). Converting to methane all the manure that U.S. livestock produce each year could provide nearly 5% of the country's total natural gas consumption at the current level. But collecting and transporting manure to centralized power plants takes energy.

It is more economical to digest manure or sludge near sites where it is produced, collect the methane produced, and then burn it at farms, feedlots, or small nearby power plants. However, conservationists believe that in most cases recycling manure to the land instead of using commercial inorganic fertilizer, which requires large amounts of natural gas to produce, would probably save more natural gas than is saved by burning the manure.

Some analysts believe that methanol and ethanol can be used as liquid fuels to replace gasoline and diesel fuel when oil becomes too scarce and expensive. Both alcohols can be burned directly as fuel without requiring additives to boost octane ratings.

Currently, emphasis is on using ethanol as an automotive fuel. It can be made from sugar and grain

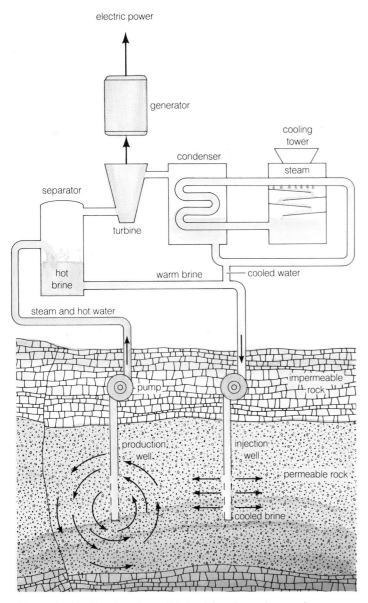

electric power

generator

cooling tower

condenser

steam

separator

turbine

hot brine

warm brine

cooled water

steam and hot water

impermeable rock

pump

production well

injection well

permeable rock

cooled brine

Figure 17-19 Tapping the earth's heat (geothermal energy) in the form of wet steam to produce electricity.

crops (sugarcane, sugar beets, sorghum, and corn) by fermentation and distillation. Pure ethanol can be burned in today's cars with little engine modification. Gasoline can also be mixed with 10% to 23% ethanol to make *gasohol*. It burns in conventional gasoline engines and is sold as super-unleaded or ethanol-enriched gasoline.

Since 1987 ethanol made by fermentation of surplus sugarcane has accounted for about half the automotive fuel consumption in Brazil. The use of ethanol helped Brazil cut its oil imports in half between 1978 and 1984 and also created an estimated 575,000 full-time jobs. But the government has spent $8 billion to prop up the country's ethanol industry.

Super-unleaded gasoline containing 90% gasoline and 10% ethanol now accounts for about 8% of gasoline sales in the United States—25% to 35% in

Illinois, Iowa, Kentucky, and Nebraska. The ethanol used in gasohol is made mostly by fermenting corn in 150 ethanol production plants built between 1980 and 1985. Excluding federal taxes, it costs about $1.60 to produce a gallon of ethanol, compared to about 50 cents for a gallon of gasoline. However, new, energy-efficient distilleries are lowering the production cost. Soon this fuel may be able to compete with other forms of unleaded gasoline without federal tax breaks, which are scheduled to expire in 1992.

Ethanol produces less carbon monoxide and nitrogen oxides per unit of energy than gasoline and is a better anti-knock fuel. However, without catalytic converters, cars burning ethanol fuels produce more aldehydes and PANs (peroxyacyl nitrates) that kill plants and cause more eye irritation than do cars burning gasoline. It also has less energy per liter than gasoline, requiring larger fuel tanks or more frequent fill-ups.

Ethanol produces less carbon dioxide per unit of energy than gasoline. But total carbon dioxide impact depends on the energy source used in the distillation process and whether the crops are grown using energy-intensive industrialized agriculture.

The distillation process to make ethanol produces large volumes of a waste material known as swill, which, if allowed to flow into waterways, kills fish and aquatic plants. Another problem is that the net useful-energy yield from producing ethanol fuel is low in older oil- or natural-gas-fueled distilleries. However, the yield is moderate at new distilleries using modern technology and powered by coal, wood, or solar energy.

Some experts are concerned that growing corn or other grains to make alcohol fuel could compete for cropland needed to grow food. It takes nine times more cropland to fuel one average U.S. automobile with ethanol for one year than it does to feed one American per year. About 40% of the entire U.S. annual harvest of corn would be needed to make enough ethanol to meet just 10% of the country's demand for automotive fuel.

Another alcohol, methanol, can be produced from wood, wood wastes, agricultural wastes (such as corn cobs), sewage sludge, garbage, coal, or natural gas at a cost of about $1.35 per gallon—more than twice the cost of producing gasoline. High concentrations of methanol corrode some metals and embrittle rubber and some plastics, but in a properly modified engine methanol burns cleanly without any problems.

A fuel of 85% methanol and 15% unleaded gasoline could reduce emissions of ozone-forming hydrocarbons 20% to 50%. Running cars on pure methanol would reduce these emissions by 85% to 95% and carbon monoxide emissions by 30% to 90%. However, cars burning pure methanol emit two to five times more formaldehyde, a suspected carcinogen, than those burning gasoline.

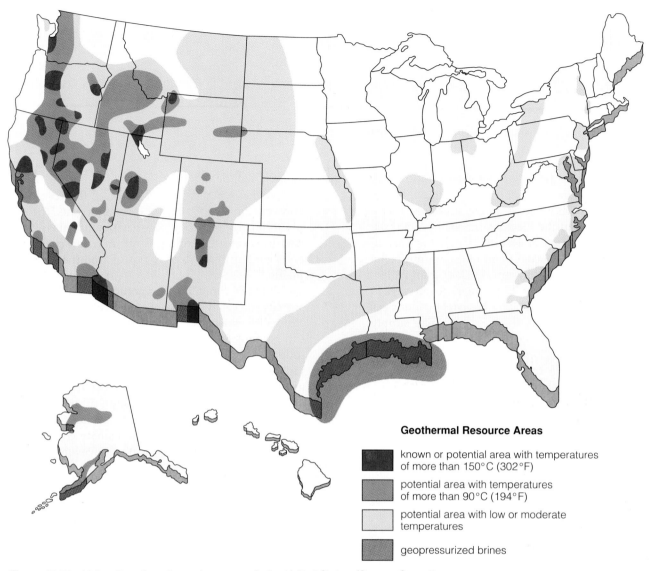

Geothermal Resource Areas

- known or potential area with temperatures of more than 150°C (302°F)
- potential area with temperatures of more than 90°C (194°F)
- potential area with low or moderate temperatures
- geopressurized brines

Figure 17-20 Major sites of geothermal resources in the United States. (Source: Council on Environmental Quality)

Methanol-powered cars emit less carbon dioxide than gasoline-powered cars. But producing the methanol from coal would double carbon dioxide emissions. Methanol also has less energy per liter than gasoline and is more flammable than gasoline.

17-7 GEOTHERMAL ENERGY

Extracting Energy from the Earth's Interior Heat contained in underground rocks and fluids is an important source of energy. At various places in the earth's crust, this **geothermal energy** from the earth's interior (Figure 4-2, p. 60) is transferred over millions of years to underground reservoirs of dry steam (steam with no water droplets), wet steam (a mixture of steam and water droplets), and hot water trapped in fractured or porous rock.

If these geothermal reservoirs are close enough to the earth's surface, wells can be drilled to extract the dry steam, wet steam trapped in rocks and fluids beneath the earth's crust, or hot water (Figure 17-19). This thermal energy can be used for space heating and to produce electricity or high-temperature heat for industrial processes.

Geothermal reservoirs can be depleted if heat is removed faster than it is renewed by natural processes. Thus, strictly speaking, geothermal resources are nonrenewable on a human time scale. However, the potential supply is so vast that it is often classified as a potentially renewable energy resource.

Currently, about 20 countries are tapping geothermal reservoirs and producing electricity equal to the output of five large electric power plants at a cost well below that of nuclear power and equal to or below that of coal in many areas. The United States accounts for 44% of the electricity generated worldwide from geothermal energy. Figure 17-20 shows that most accessible, high-temperature geothermal sites in the United States lie in the West, especially in California and the Rocky Mountain states. Iceland, Japan, and Indonesia are among the countries with the greatest potential for tapping geothermal energy.

Dry-steam reservoirs are the preferred geothermal resource, but also the rarest. A large dry-steam well near Larderello, Italy, has been producing electricity since 1904 and is a major source of power for Italy's electric railroads. Two other major dry-steam reservoirs are the Matsukawa field in Japan and the Geysers steam field about 145 kilometers (90 miles) northwest of San Francisco. Currently, the Geysers reservoir supplies more than 6% of northern California's electricity. Largely without government subsidies, this is enough to meet all the electrical needs of a city the size of San Francisco at less than half the cost of electricity from a new coal plant and one-fourth the cost of electricity from a new nuclear plant. New units can be added every 2 to 3 years (compared to 6 years for a coal plant and 10 to 12 years for a nuclear plant). By the year 2000, this reservoir may supply one-fourth of California's electricity.

Wet-steam reservoirs are more common but harder and more expensive to convert to electricity. The world's largest wet-steam power plant is in Wairaki, New Zealand. Others operate in Mexico, Japan, El Salvador, Nicaragua, and the Soviet Union. Four small-scale wet-steam demonstration plants in the western United States are producing electricity at a cost equal to paying $40 a barrel for oil.

Hot-water reservoirs are more common than dry-steam and wet-steam deposits. Almost all the homes, buildings, and food-producing greenhouses in Reykjavik, Iceland, a city with a population of about 85,000, are heated by hot water drawn from deep geothermal wells under the city. At 180 locations in the United States, mostly in the West, hot-water reservoirs have been used for years to heat homes and farm buildings and to dry crops.

A fourth potential source of nonrenewable geothermal energy and natural gas is *geopressurized zones.* These are underground reservoirs of water at a high temperature and pressure, usually trapped deep under continental shelf beds of shale or clay. With present drilling technology, they would supply geothermal energy and natural gas at a cost equal to paying $30 to $45 a barrel for oil.

There are also three types of vast, virtually perpetual sources of geothermal energy: *molten rock* (magma) found near the earth's surface; *hot dry-rock zones,* where molten rock that has penetrated the earth's crust from below heats subsurface rock to high temperatures; and low- to moderate-temperature *warm-rock reservoirs,* useful for preheating water and running geothermal heat pumps for space heating and air conditioning. According to the National Academy of Sciences, the amount of potentially recoverable energy from such reservoirs would meet U.S. energy needs at current consumption levels for 600 to 700 years.

The problem is developing methods to extract this energy economically. Several experimental projects are in progress, but so far none has been able to produce energy at a cost competitive with other energy sources.

Pros and Cons The major advantages of geothermal energy include a vast and often renewable supply of energy for areas near reservoirs, moderate net useful-energy yields for large and easily accessible reservoirs, and very small emissions of carbon dioxide. The cost of producing electricity in geothermal plants is about half the cost of generating power from new coal plants and one-fourth the cost of power from new nuclear plants.

A major limitation of geothermal energy is the scarcity of easily accessible reservoirs. Geothermal reservoirs must also be carefully managed or they can be depleted within a few decades. Without pollution control, geothermal energy production causes moderate to high air pollution from hydrogen sulfide, ammonia, and radioactive materials. It also causes moderate to high water pollution from dissolved solids (salinity) and runoff of several toxic compounds. Noise, odor, and local climate changes can also be problems. Most experts, however, consider the environmental effects of geothermal energy to be less or no greater than those of fossil fuel and nuclear power plants. In Hawaii environmentalists are fighting the construction of a large geothermal project in the only lowland tropical forest left in the United States.

17-8 HYDROGEN AS A POSSIBLE REPLACEMENT FOR OIL

Pros and Cons of Using Hydrogen Gas Some scientists have suggested that we use hydrogen gas (H_2) to fuel cars, heat homes, and provide electricity and hot water when oil and natural gas run out. Hydrogen gas does not occur in significant quantities in nature. However, it can be produced by chemical processes from nonrenewable coal or natural gas or by using heat, electricity, or perhaps sunlight to decompose fresh water or seawater (Figure 17-21).

Hydrogen gas can be burned in several ways: in a reaction with oxygen gas in a power plant, a specially designed automobile engine, a fuel cell that converts the chemical energy produced by the reaction into direct-current electricity, or a home energy system now being developed by a West German firm (see Spotlight on p. 384).

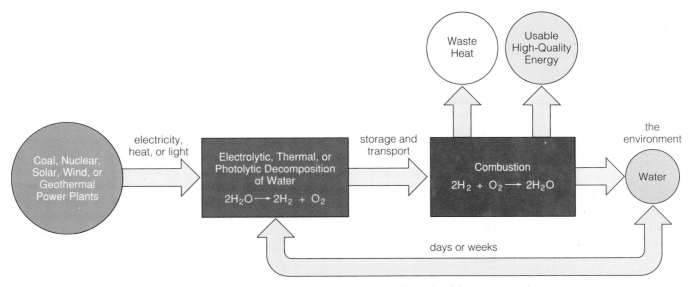

Figure 17-21 The hydrogen energy cycle. The production of hydrogen gas requires electricity, heat, or solar energy to decompose water, thus leading to a negative net useful-energy yield. However, hydrogen gas is a clean-burning fuel that can be used to replace oil and other fossil fuels.

Hydrogen burns cleanly in pure oxygen, yielding only water vapor. When burned in air, it produces only small amounts of nitrogen oxides, much less than current vehicles. Hydrogen has about 2.5 times the energy by weight of gasoline, making it an especially attractive aviation fuel. Existing natural gas pipelines could be adapted in many cases to transport hydrogen gas.

Hydrogen gas can be combined with various metals to produce solid compounds (hydrides) that can be heated to release hydrogen as needed. The hydrogen gas can then be burned as an automotive fuel or in a home heating system (see Spotlight on p. 384). Unlike gasoline, the solid metallic hydrogen compounds would not explode or burn if the tank is ruptured in an accident. Several experimental hydrogen-fueled cars are operating in the world today. The Soviet Union has flown a commercial jet partially fueled by hydrogen.

The major problem with hydrogen as a fuel is that only trace amounts of the gas occur in nature. Producing it uses high-temperature heat or electricity from another energy source, such as nuclear fission, direct solar power, or wind to decompose water. Currently it would cost about $1.40 to produce hydrogen gas with the energy found in 3.8 liters (1 gallon) of gasoline. However, the current price of gasoline does not include its numerous pollution and health costs that add at least $1 a gallon to its true cost. Thus, even today hydrogen is cheaper than gasoline and other fossil fuels when the overall societal costs are considered.

Because of the first and second energy laws (Section 3-5), hydrogen production by any method will take more energy than is released when it is burned. Thus, its net useful-energy yield will always be negative. This means that its widespread use depends on having an abundant and affordable supply of some other type of environmentally acceptable energy.

Burning hydrogen does not add carbon dioxide to the atmosphere. However, carbon dioxide would be added to the atmosphere if electricity or high-temperature heat from coal or other fossil-fuel-burning power plants were used to decompose water and produce hydrogen. Nuclear power produces about one-seventh as much carbon dioxide per kilowatt hour as coal, mainly in the processing of uranium fuel. No carbon dioxide would be added if photovoltaic cells, solar collectors, wind, geothermal, or hydroelectric power were used to produce hydrogen.

Scientists are trying to develop cells that use solar energy to split water molecules into hydrogen and oxygen gases with reasonable efficiency. Experimental solar-hydrogen plants are being built in Bavaria, West Germany, and in Saudi Arabia. In 1989 an American scientist developed a cell that uses sunlight to produce hydrogen gas from seawater.

If everything goes right, affordable commercial cells for using solar energy to produce hydrogen could become available sometime after the year 2000. According to one estimate, enough solar-produced hydrogen fuel could be produced in 2% of the world's desert area to replace all of the world's fossil fuel.

Despite the enormous potential of hydrogen as a fuel, in 1989 government research and development funding for hydrogen was only a minuscule $3 million. During this same year, the government of Japan spent $20 million on hydrogen research and development, and the West German government spent $50 million.

I am writing this book deep in the midst of some beautiful woods in Eco-Lair, a structure that Peggy, my wife and earthmate, and I designed to work with nature. This ongoing experiment is designed as a low-tech, low cost example of sustainable-earth design and living.

First, we purchased a 1954 school bus from a nearby school district for $200 and sold the tires for the same price that we paid for the bus. We built an insulated foundation, rented a crane for two hours to lift and set the gutted bus on the foundation, placed heavy insulation around the bus, and added a wooden outside frame. The interior is paneled with wood, some of it obtained from the few trees we carefully selected for removal (Figure 17-22). Most people who visit us don't know the core of the structure is a school bus unless we tell them.

We attached a solar room—a passive solar collector with double-paned conventional sliding glass windows (for ventilation)—to the entire south side of the bus structure (Figure 17-22). Thick concrete floors and filled concrete blocks on the lower half of the interior wall facing the sun absorb and slowly release solar energy collected during the day. The solar room serves as a year-round sitting and work area and contains a small kitchen with a stove and a heavily insulated refrigerator that run on liquefied petroleum gas (LPG). I plan to replace the windows with new superinsulated windows, which were not available at the time of construction.

The room collects enough solar energy to meet about 60% of the space heating needs during the cold months. The rest of the heat is provided either by a small wood stove or by a continuous loop system of water preheated by solar collectors and when necessary heated further by a tankless, instant water heater fueled by LPG (Figure 3-17, p. 52).

During sunny days active solar collectors store heat in an insulated tank, which can be a discarded conventional hot water tank that is wrapped with thick insulation. A pump connected to the water tank circulates heated water in well-insulated pipes through the instant hot water heater and from there to a heat exchanger before the water is returned to the tank. A fan transfers the heat in the water to air, which is blown through well-insulated ducts as in a conventional heating system. Some of the heat is recovered for use in the next stage when the heated water returns to the insulated tank. Indoor temperatures are controlled by a conventional thermostat. A valve on the input of the instant water heater senses water temperature and bypasses the heater when the solar heated water has a high enough temperature to provide the necessary space heating.

During the summer all hot water is supplied by the roof-mounted active solar collectors. In winter the water heated by these collectors is heated further as needed by a second tankless, instant heater fueled by LPG (Figure 3-17, p. 52). Our screw-in fluorescent light bulbs last an average of

six years and use about 60% less electricity than conventional bulbs.

For the time being, we are buying electricity from the power company. But we plan to get our electricity from roof-mounted panels of photovoltaic cells (Figure 17-15, p. 392) when their price is lower, probably sometime in the 1990s. We plan to sell back to the power company any excess power these cells produce. Present electricity bills run around $30 a month, compared with $100 or more for conventional structures of the same size.

In moderate weather cooling is provided by opening windows to capture breezes. During the hot and humid North Carolina summers additional cooling is provided by earth tubes (Figure 17-9, p. 387).

Four plastic pipes, with a diameter of 10 centimeters (4 inches), were buried about 5.5 meters (18 feet) underground, extending down a gently sloping hillside until their ends emerge some 31 meters (100 feet) away. The other ends of the tubes come up into the foundation of the bus and connect to a duct system containing a small fan with rheostat-controlled speed. When the fan is turned on, outside air at a temperature of 35°C (95°F) is drawn slowly through the buried tubes (which are surrounded by earth at about 16°C, or 60°F), entering the structure at about 22°C (72°F). This natural air conditioning costs about $1 per summer for running the fan.

Several large oak trees and other deciduous trees in front of

Taking Energy Matters into Your Own Hands

While elected officials, energy company executives, and conservationists argue over the key components of a national energy strategy, many individuals have gotten fed up and taken energy matters into their own hands. With or without tax credits, they are insulating, weatherizing, and making other changes to improve energy efficiency and save money.

Some are building passively heated and cooled solar homes. Others are building superinsulated dwellings or are adding passive or active solar heating to existing homes. Each of us can develop a personal energy strategy that not only saves money, but also improves personal and national security (see Individuals Matter inside the back cover and Spotlight above).

Figure 17-22 Eco-Lair is where I work. It is a low-tech, low-cost ongoing experiment on how to save energy and money. A south-facing solar room (left) collects solar energy passively and distributes it to a well-insulated, recycled 1954 school bus. The solar room also contains a compact, energy-efficient kitchen. Backup heat and hot water provided by solar-assisted, tankless instant water heaters fueled by LPG (Figure 3-17, p. 52). Active solar collectors are shown on the left and a passive solar water heater is shown near the ground on the right. Cooling is provided by buried tubes (earth tubes) at a cost of about $1 a summer. Electricity bills run about $30 a month. Water is conserved by the use of a water-saving toilet and showerhead and faucet aerators in an attached bathroom behind the large insulated window. The photo on the right shows the interior of the recycled school bus. The computer and desk at the far end are located where the hood and motor of the bus used to be. The large cabinet on the left folds down and serves as a double bed. The bus windows on the right can be opened as needed to allow heat collected in the attached solar room to flow into the bus space.

the solar room give us additional passive cooling during summer; in winter the trees drop their leaves, letting in the sun. A used conventional central air conditioning unit (purchased for $200) is used as a backup. It can be turned on for short periods (typically no more than 15 to 30 minutes a day) when excessive pollen or heat and humidity overwhelm our immune systems and the earth tubes. Life always involves some trade-offs.

Eco-Lair is surrounded by natural vegetation including flowers and low-level ground cover adapted to the climate of the area.

This means there is no grass to cut and no lawn mower to repair, feed with gasoline, and listen to. Plants that repel various insects have also been added, so we have few insect pest problems. The surrounding trees and other vegetation also provide habitats for various species of insect-eating birds. When ants, mice, and other creatures find their way inside we use natural alternatives to repel and control them (see Individual Matter on p. 326).

Water use has been reduced by installing water-saving faucets, a water-saving showerhead, and a

low-flush toilet. We have also experimented with a waterless composting toilet that gradually converts waste and garbage scraps to a dry, odorless powder that can be used as a soil conditioner.

Kitchen wastes are composted and recycled to the soil. Paper and bottles are carried to a local recycling center, and we are looking for a place to recycle the plastics we use. We try never to use aluminum cans and throwaway plastic bags. Extra furniture, clothes, and other items we have accumulated or salvaged over the years

(continued)

Similarly, local governments in a growing number of cities are developing successful programs to improve energy efficiency and to rely more on locally available energy resources. Across the country, towns are realizing that paying for energy is bleeding them to death economically, with 80% to 90% of the money they spend on energy leaving the local economy forever. Instead of creating local jobs and income this

money goes to wealthy Saudi Arabians, Texas oil barons, and New York investors.

Each of these individual and local initiatives are crucial political and economic actions that are bringing about change from the bottom up. Multiplied across the country, such actions can shape a sane national energy strategy with or without help from federal and state governments.

are stored in three other old school buses and recycled to family, friends, and people in need. For most household chemicals we use the substitutes shown in Individuals Matter inside the back cover.

Eco-Lair lies near the end of a narrow, 1.6-kilometer- (1-mile-) long dirt road that at times can be traversed only by a four-wheel-drive vehicle. As a result, we drive a vehicle that consumes much more gasoline than we would like. However, because I work at home I do little driving. If the technology becomes available and economically feasible in the future, we hope to purchase a vehicle that runs on hydrogen gas produced by solar photovoltaic cells that decompose water into hydrogen and oxygen gas.

Because of laziness and allergies, we get most of our food from the grocery store rather than growing it ourselves. For health and environmental reasons, we have greatly reduced our meat consumption. We should be vegetarians, but so far we have been unwilling to go quite that far.

We feel a part of the piece of land we live on and love. To us, ownership of this land means that we are ethically driven to defend and protect it from degradation. We feel that the trees, flowers, deer, squirrels, hummingbirds, songbirds, and other forms of wildlife we often see are a part of us and we are a part of them. As temporary caretakers of this small portion of the biosphere, we feel obligated to pass it on to future generations with its ecological integrity and sustainability preserved.

Most of our political activities involve thinking globally but acting locally. They include attempts to prevent an economically unnecessary nuclear power plant from opening about 24 kilometers (15 miles) away (it opened anyway), to prevent an ecologically unsound development along a nearby river that is already badly polluted (we've been successful so far), and to prevent the building of a large, conventional housing development that would double the size of the closest town (successful).

We also financially support numerous environmental and conservation organizations working at the national and global levels. We are not opposed to all forms of development, only those that are ecologically unsound and destructive.

Each year we plant several trees on our land, and I donate money to organizations to plant at least ten trees for each tree I use in writing this and other books. The publisher and I also join together in donating money to tree-planting organizations to offset the paper used in printing this book.

Working with nature gives us great joy and a sense of purpose. It also saves us money. Our attempt to work with nature is in a rural area, but people in cities can also have high-quality lifestyles that conserve resources and protect the environment (see Further Readings).

Most LDCs can also improve energy efficiency and use renewable and perpetual energy resources to meet much of their energy needs. Most renewable energy projects have short construction times and provide large numbers of jobs. Countries and communities that depend on energy from locally available renewable resources are less vulnerable to disruptions of fuel supplies and price rises, and their economies are stronger because they spend less on energy imports.

A few countries are leading the way in making the transition from the age of oil to the age of energy efficiency and renewable energy. Sweden leads the world in energy efficiency, followed by Japan. Brazil and Norway get more than half their energy from hydropower, wood, and alcohol fuel. Israel, Japan, the Philippines, and Sweden plan to rely on renewable and perpetual sources for most of their energy.

Countries that have the vision to change from an unsustainable to a sustainable energy strategy will be rewarded with increased security—not just military security but also economic, energy, and environmental security. Those that do not will experience unnecessary economic and environmental hardships and increased human suffering.

In the long run, humanity has no choice but to rely on renewable energy. No matter how abundant they seem today, eventually coal and uranium will run out. The choice before us is practical: We simply cannot afford to make more than one energy transition within the next generation.

DANIEL DEUDNEY AND CHRISTOPHER FLAVIN

DISCUSSION TOPICS

1. What are the ten most important things an individual can do to save energy in the home and in transportation (see table inside the back cover)? Which, if any, of these do you do? Which, if any, do you plan to do? When?

2. Make an energy use study of your school and use the findings to develop an energy-efficiency improvement program.

3. Should the United States institute a crash program to develop solar photovoltaic cells? Explain.

4. Explain why you agree or disagree with the ideas that the United States can get most of the electricity it needs by
   a. developing solar power plants
   b. using direct solar energy to produce electricity in photovoltaic cells
   c. building new, large hydroelectric plants
   d. building ocean thermal electric power plants
   e. building wind farms
   f. building power plants fueled by wood, crop wastes, trash, and other biomass resources
   g. tapping dry-steam, wet-steam, and hot-water geothermal deposits
   h. tapping molten rock (magma) geothermal deposits
   i. improving energy efficiency by 50%

5. Explain why you agree or disagree with the following propositions:
   a. The United States should cut average per capita energy use by at least 50% between 1991 and 2011.
   b. A mandatory energy conservation program should form the basis of any U.S. energy policy in order to help provide economic, environmental, and military security.
   c. To solve world and U.S. energy supply problems, all we need do is recycle some or most of the energy we use.

FURTHER READINGS

Also see readings for Chapter 3.

American Council for an Energy Efficient Economy. 1988. *Energy Efficiency: A New Agenda*. Washington, D.C.: American Council for an Energy Efficient Economy.

Blackburn, John O. 1987. *The Renewable Energy Alternative: How the United States and the World Can Prosper Without Nuclear Energy or Coal*. Durham, N.C.: Duke University Press.

Bleviss, Deborah Lynn. 1988. *The New Oil Crisis and Fuel Economy Technologies*. Westport, Conn.: Quorum Press.

Brower, Michael. 1990. *Coal Energy: The Renewable Solution to Global Warming*. Cambridge, Mass.: Union of Concerned Scientists.

Chiles, James R. 1990. "Tomorrow's Energy Today." *Audubon*, Jan., 59–72.

Davidson, Joel. 1987. *The New Solar Electric Home*. Ann Arbor, Mich.: Aatec Publications.

Dinga, Gustav P. 1988. "Hydrogen: The Ultimate Fuel and Energy Carrier." *Journal of Chemical Education*, vol. 65, no. 8, 688–691.

Echeverria, John, et al. 1989. *Rivers at Risk: The Concerned Citizen's Guide to Hydropower*. Covelo, Calif.: Island Press.

Flavin, Christopher. 1986. *Electricity for a Developing World: New Directions*. Washington, D.C.: Worldwatch Institute.

Flavin, Christopher. 1989. *Slowing Global Warming: A Worldwide Strategy*. Washington, D.C.: Worldwatch Institute.

Flavin, Christopher, and Alan B. Durning. 1988. *Building on Success: The Age of Energy Efficiency*. Washington, D.C.: Worldwatch Institute.

Flavin, Christopher, and Roch Piltz. 1990. *Sustainable Energy*. Washington, D.C.: Renew America.

Gever, John, et al. 1986. *Beyond Oil*. Cambridge, Mass.: Ballinger.

Gibbons, John H., et al. 1989. "Strategies for Energy Use." *Scientific American*, Sept., 136–143.

Goldenberg, Jose, et al. 1988. *Energy for a Sustainable World*. New York: Wiley.

Harnik, Peter. 1982. "How to Live Ecologically in an Apartment." *Environmental Action*, Sept. 1.

Hinkel, Kenneth M. 1990. "Wood Burning for Residential Space Heating in the United States: An Energy Efficiency Analysis." *Applied Geography*, vol. 9, 259–272.

Hubbard, H. M. 1989. "Photovoltaics Today and Tomorrow." *Science*, vol. 244, 297–304.

Lovins, Amory B. 1989. *Energy, People, and Industrialization*. Old Snowmass, Colo.: Rocky Mountain Institute.

National Academy of Sciences. 1983. *Alcohol Fuels: Options for Developing Countries*. Washington, D.C.: National Academy Press.

National Academy of Sciences. 1988. *Geothermal Energy Technology*. Washington, D.C.: National Academy Press.

Ogden, Joan M., and Robert H. Williams. 1989. *Solar Hydrogen: Moving Beyond Fossil Fuels*. Washington, D.C.: World Resources Institute.

Oppenheimer, Michael, and Robert H. Boyle. 1990. *Dead Heat: The Race Against the Greenhouse Effect*. New York: Basic Books.

Pimentel, David, et al. 1984. "Environmental and Social Costs of Biomass Energy." *BioScience*, Feb., 89–93.

Rader, Nancy, et al. 1989. *Power Surge: The Status and Near-Term Potential of Renewable Energy Technologies*. Washington, D.C.: Public Citizen.

Renner, Michael. 1988. *Rethinking the Role of the Automobile*. Washington, D.C.: Worldwatch Institute.

Rocky Mountain Institute. 1989. *Resource-Efficient Housing Guide*. Old Snowmass, Colo.: Rocky Mountain Institute.

Sawyer, Stephen W. 1986. *Renewable Energy: Progress, Prospects*. Washington, D.C.: Association of American Geographers.

Scientific American. 1990. *Energy for Planet Earth*. Entire Sept. issue.

Shea, Cynthia Pollack. 1988. *Renewable Energy: Today's Contribution, Tomorrow's Promise*. Washington, D.C.: Worldwatch Institute.

Swan, Christopher C. 1986. *Suncell: Energy, Economy, Photovoltaics*. New York: Random House.

Wade, Herb. 1983. *Building Underground: The Design and Construction Handbook for Earth-Sheltered Houses*. Emmaus, Penn.: Rodale Press.

CHAPTER EIGHTEEN

NONRENEWABLE ENERGY RESOURCES

We are an interdependent world and if we ever needed a lesson in that, we got it in the oil crisis of the 1970s.

ROBERT S. MCNAMARA

SINCE 1950 oil, coal, and natural gas have supported most of the world's economic growth. Their use is also responsible for much of the world's pollution and environmental degradation. Nuclear energy was supposed to be providing much of the world's electricity by the year 2000. But high costs (even with massive government subsidies), safety concerns, and failure to find an economically and politically acceptable solution for storing its long-lived radioactive wastes have led many countries to sharply scale back or eliminate their plans to build new nuclear power plants.

How long might various fossil fuels last? How can we reduce their environmental impact? What role should nuclear energy play in the future? What should be the energy strategy of the United States? These important and controversial issues are discussed in this chapter.

18-1 OIL AND NATURAL GAS

Conventional Crude Oil **Petroleum,** or **crude oil,** is a gooey liquid consisting mostly of hydrocarbon compounds and small amounts of compounds containing oxygen, sulfur, and nitrogen. Crude oil and natural gas are often trapped together deep within the earth's crust on land (Figure 18-1) and beneath the sea floor. The crude oil is dispersed in pores and cracks in rock.

Primary oil recovery involves drilling and pumping out the oil that flows by gravity into the bottom of the well. Thicker, slowly flowing, heavy oil is not removed. After the flowing oil has been removed, water can be injected into adjacent wells to force some of the remaining thicker crude oil into the central well and push it to the surface. This is known as **secondary oil recovery.** Usually primary and secondary recovery remove only one-third of the crude oil in a well.

For each barrel removed by primary and secondary recovery, two barrels of heavy oil are left in a typical well. As oil prices rise, it may become economical to remove about 10% of the heavy oil by **enhanced,** or **tertiary, oil recovery.** One method is to force steam into the well to soften the heavy oil so that it can be pumped to the surface. Carbon dioxide gas can also be pumped into a well to force some of the heavy oil into the well cavity for pumping to the surface.

The problem is that enhanced oil recovery is expensive. The net useful-energy yield is low because it takes energy equivalent to that in one-third of a barrel of oil to soften and pump each barrel of heavy oil to the surface. Additional energy is needed to increase the flow rate and to remove sulfur and nitrogen impurities before the heavy oil can be pumped

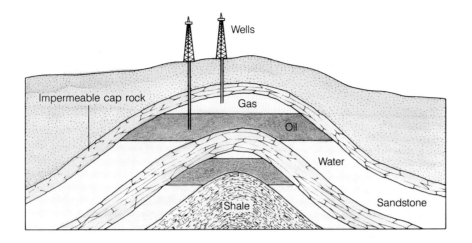

Figure 18-1 Oil and natural gas are usually found together beneath a dome of impermeable cap rock.

through a pipeline to an oil refinery. Recoverable heavy oil from known U.S. crude oil reserves could supply U.S. oil needs for only about seven years at current usage rates.

Once it is removed from a well, most crude oil is sent by pipeline to a refinery. There it is heated and distilled to separate it into gasoline, heating oil, diesel oil, asphalt, and other components. Because these components boil at different temperatures, they are removed at different levels of giant distillation columns (Figure 18-2).

Some components and products called **petrochemicals** are used as raw materials in industrial chemicals, fertilizers, pesticides, plastics, synthetic fibers, paints, medicines, and many other products. Petrochemical production accounts for about 3% of the crude oil extracted throughout the world and 7% of the oil used in the United States. This explains why the prices of many items we use go up after crude oil prices rise.

How Long Will Supplies of Conventional Crude Oil Last?

Almost two-thirds of the world's proven oil reserves are in just five countries: Saudi Arabia, Kuwait, Iran, Iraq, and the United Arab Emirates. OPEC countries have 67% of these reserves, with Saudi Arabia having 25%. Geologists believe that the Middle East also contains most of the world's undiscovered oil. This explains why OPEC is expected to have long-term control over world oil supplies and prices.

The Soviet Union is presently the world's largest oil extractor, with an annual output triple that of Saudi Arabia. The United States, the world's second largest oil extractor, has only 3% of the world's oil reserves but uses nearly 30% of the oil extracted each year. Transportation uses 63% of the 17 billion barrels of oil consumed each day in the United States. The rest is used by industry (25%), residences and commercial buildings (9%), and electric utilities (3%).

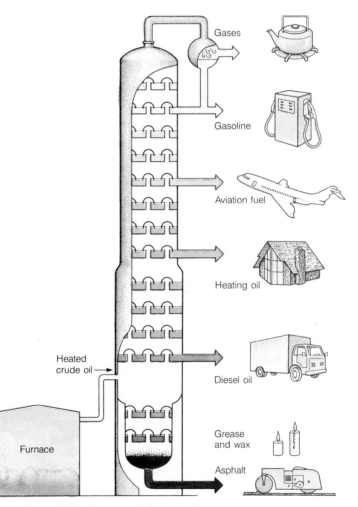

Figure 18-2 Refining of crude oil. Major components are removed at various levels, depending on their boiling points, in a giant distillation column.

Most oil occurrences in the Middle East are large and cheap to extract; most in the United States are small and more expensive to tap. Therefore, it has generally been cheaper for the United States to buy oil from other countries than to extract it from its

Figure 18-3 Major occurrences of natural gas and crude oil in the United States. The oil and natural gas industries have about 15% of the total U.S. land area under lease. (Source: Council on Environmental Quality)

own deposits. In 1989 about 48% of this oil was imported and import dependence is projected to rise (Figure 1-8, p. 8).

U.S. military intervention in the Persian Gulf by the Reagan and Bush administrations since 1985 means that U.S. oil imports from Arab OPEC countries cost $495 a barrel when the $50-billion-a-year military costs are included. Thus, Americans are buying the world's most expensive oil—costing about 25 times the average 1989 market price for oil—to meet about 8% of their consumption. They are also footing most of the bill for protecting the much larger amounts of Arab OPEC oil going to Japan and western European countries.

The Alaska oil pipeline is a much greater threat to national security than not being able to move oil by tankers through the Persian Gulf. There are now other oil suppliers and several alternate ways to get oil from the Middle East, but only one way to get oil from Alaska. Sabotage of the highly vulnerable Alaska oil pipeline could disrupt the entire American econ-

omy. The Department of Defense admits that it is impossible to protect this pipeline.

Figure 18-3 shows the locations of the major crude oil and natural gas fields in the United States. U.S. oil extraction has declined steadily since 1970 despite greatly increased exploration and test drilling.

Experts disagree over how long the world's identified and unidentified crude oil resources will last. **Reserves** are identified deposits of nonrenewable fossil fuel or minerals from which the resource can be extracted profitably at present prices using current technology. At present consumption rates, world crude oil reserves will be economically depleted in 33 years. U.S. reserves will be economically depleted by 2018 at the current consumption rate and by 2010 if oil use increases by 2% a year.

Some analysts argue that higher oil prices will stimulate the discovery and extraction of large new crude oil resources. They also believe we can extract and upgrade heavy oils from oil shale, tar sands, and enhanced recovery from existing wells.

Some believe that the earth's crust may contain 100 times more oil than usually thought. But such oil, if it exists, lies 10 kilometers (6 miles) or more below the earth's surface—about twice the depth of today's deepest wells. Most geologists do not believe this oil exists.

Other analysts argue that people making optimistic projections about future oil supplies don't understand the arithmetic and consequences of exponential growth in the use of any nonrenewable resource. Consider the following facts about the world's exponential growth in oil use, assuming that we continue to use crude oil at the current rate instead of the projected higher rates. If it were the only source,

- Saudi Arabia, with the world's largest known crude oil reserves, could supply all the world's oil needs for only ten years if it were the world's only source.

- Mexico, with the world's sixth largest crude oil reserves, could supply the world's needs for only about three years.

- The estimated crude oil reserves under Alaska's North Slope—the largest ever found in North America—would meet world demand for only six months, or U.S. demand for three years.

- The oil that oil companies have a one-in-five chance of finding by drilling in Alaska's Arctic Wildlife Refuge would meet world demand for only one month, or U.S. demand for six months.

- All estimated undiscovered, recoverable deposits of oil in the United States would meet world demand for only 1.7 years, or U.S. demand for 10 years.

- Those who believe that new discoveries will solve world oil supply problems must figure out how to discover the equivalent of a new Saudi Arabian deposit *every ten years* just to keep on using oil at the current rate.

The ultimately recoverable supply of crude oil is estimated to be three times today's proven reserves. Suppose all this new oil is found and developed—which most oil experts consider unlikely—and sold at a price of $50 to $95 a barrel, compared to the 1989 price of about $19 a barrel. About 80% of this oil would be depleted by 2073 at the current usage rate and by 2037 if oil use increased 2% a year.

We can see why most experts expect little of the world's affordable crude oil to be left by the 2059 bicentennial of the world's first oil well. Oil company executives have known this for a long time, which explains why oil companies have become diversified energy companies. To keep making money after oil runs out, these international companies now own

Figure 18-4 Sample of oil shale and the shale oil extracted from it. Major oil shale projects have now been canceled in the United States because of excessive cost.

much of the world's natural gas, coal, and uranium reserves and have bought many of the companies producing solar collectors and solar cells.

Pros and Cons of Oil Oil has been and still is fairly cheap (Figure 1-9, p. 9), can easily be transported within and between countries, and has a high net useful-energy yield (Figure 3-18, p. 53). It is also a versatile fuel that can be burned to propel vehicles, heat buildings and water, and supply high-temperature heat for industrial processes and electricity production.

Oil also has some disadvantages. Its burning releases carbon dioxide gas, which could alter global climate, and other air pollutants such as sulfur oxides and nitrogen oxides, which damage people, crops, trees, fish, and other wild species. Oil spills and leakage of toxic-drilling muds cause water pollution, and the brine solution injected into oil wells can contaminate groundwater. The crucial disadvantage of oil is that affordable supplies are expected to be depleted within 40 to 80 years.

Heavy Oil from Oil Shale **Oil shale** is a fine-grained rock (Figure 18-4) that contains varying amounts of a solid, waxy mixture of hydrocarbon compounds called **kerogen.** After being removed by surface or subsurface mining, the shale is crushed and heated to a high temperature to vaporize the kerogen (Figure 18-5). The kerogen vapor is condensed, forming a slow-flowing, dark brown, heavy oil called **shale oil.** Before shale oil can be sent by pipeline to a refinery, it must be processed to increase its flow rate and heat content and to remove sulfur, nitrogen, and other impurities.

It is estimated that the potentially recoverable heavy oil from oil shale deposits in the United States—mostly on federal lands in Colorado, Utah, and

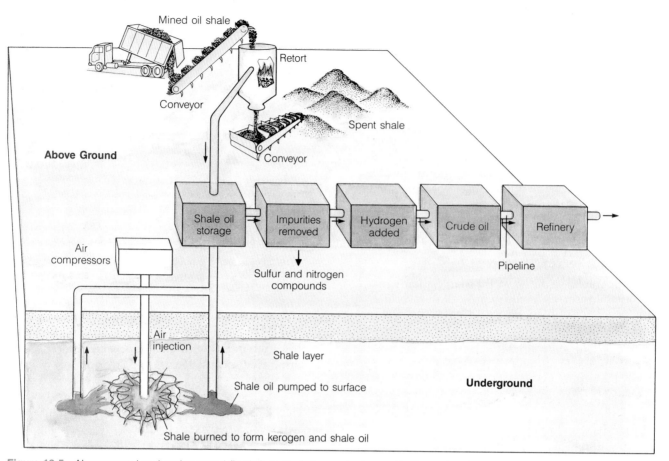

Figure 18-5 Aboveground and underground (in situ) methods for producing synthetic crude oil from oil shale.

Wyoming—could meet the country's crude oil demand for 41 years if consumption remains at the current level, and for 32 years if consumption rises 2% a year. Large oil shale deposits are also found in Canada, China, and the Soviet Union.

Environmental problems may limit shale oil production. Shale oil processing requires large amounts of water, which is scarce in the semiarid areas where the richest deposits are found. The conversion of kerogen to processed shale oil and its burning release more carbon dioxide per unit of energy than conventional oil. Nitrogen oxides and sulfur dioxide are also released. When shale is extracted aboveground, there is massive land disruption from the mining and disposal of large volumes of shale rock, which breaks up and expands like popcorn when heated. Also various salts, cancer-causing substances, and toxic metal compounds can be leached from the processed shale rock into nearby water supplies.

One way to avoid some of these environmental problems is to extract oil from shale underground, known as *in situ* (in-place) *processing* (Figure 18-5). But this method is too expensive with present technology and produces more sulfur dioxide emissions than surface processing.

The net useful-energy yield of shale oil is much lower than that of conventional oil because the energy equivalent of almost one-half of a barrel of conventional crude oil is needed to extract, process, and upgrade one barrel of shale oil. (Figure 3-18, p. 53). Also shale oil does not refine as well as crude oil and yields fewer useful products.

Heavy Oil from Tar Sand **Tar sand** (or oil sand) is a deposit of a mixture of clay, sand, water, and varying amounts of **bitumen**, a gooey, black, high-sulfur, heavy oil. Tar sand is usually removed by surface mining and heated with steam at high pressure to make the bitumen fluid enough to float to the top. The bitumen is removed and then purified and chemically upgraded into a synthetic crude oil suitable for refining (Figure 18-6). So far it is not technically or economically feasible to remove deeper deposits of tar sand by underground mining or to remove bitumen by underground extraction.

The world's largest known deposits of tar sands lie in a cold, desolate area in northern Alberta, Canada. Heavy oil in these deposits is estimated to exceed the proven oil reserves of Saudi Arabia. Other large deposits are in Venezuela, Colombia, and the Soviet Union. Smaller deposits exist in the United States, mostly in Utah. If all these were developed, they would supply all U.S. oil needs at the current usage

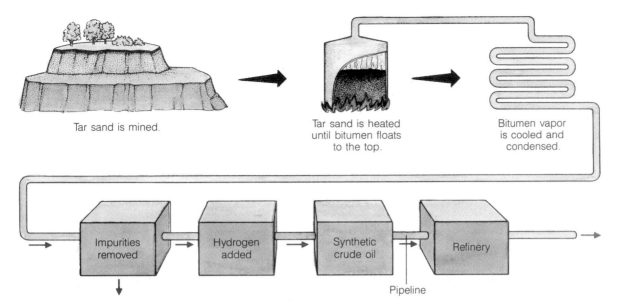

Figure 18-6 Generalized summary of how synthetic crude oil is produced from tar sand.

rate for only about three months at a price of $48 to $62 a barrel.

Since 1985 two plants have been supplying almost 12% of Canada's oil demand by extracting and processing heavy oil from tar sands at a cost of $12 to $15 a barrel—below the average world oil price between 1986 and 1989. Economically recoverable deposits of heavy oil from tar sands can supply all of Canada's projected oil needs for about 33 years at the current consumption rate. These deposits are an important source of oil for Canada, but they would meet the world's present oil needs for only about 2 years.

Producing synthetic crude oil from tar sands has several disadvantages. The net useful-energy yield is low because it takes the energy equivalent of almost one-half of a barrel of conventional oil to extract and process one barrel of bitumen and upgrade it to synthetic crude oil before it can be sent to an oil refinery. Other problems include the need for large quantities of water for processing, and the release of air and water pollutants. Upgrading bitumen to synthetic crude oil releases sulfur dioxide, hydrogen sulfide, and particulates of toxic metals.

Environmentalists charge that synthetic crude oil is produced from tar sand at a low price in Canada only because the tar sand processing plants are not required to control air pollution emissions. The plants have also created huge waste disposal ponds. Cleaning up these toxic-waste dump sites is another external cost not included in the price of crude oil produced from Canadian tar sand.

Natural Gas In its underground gaseous state, **natural gas** is a mixture of 50% to 90% methane gas and smaller amounts of heavier gaseous hydrocarbon compounds such as propane and butane. *Conventional natural gas* lies above most deposits of crude oil (Figure

Figure 18-7 Large quantities of energy are wasted when natural gas that is found with oil is sometimes burned off, as in this oil field in Saudi Arabia in the Middle East. This is done because collecting and using this natural gas costs more than it can be sold for in the oil-rich Middle East. Burning this high-quality fuel also adds carbon dioxide and other pollutants to the atmosphere, but this causes less projected global warming than allowing the methane to escape into the atmosphere.

18-1). Some of this is burned off and wasted when the primary goal is oil extraction (Figure 18-7). *Unconventional natural gas* is found by itself in other underground deposits.

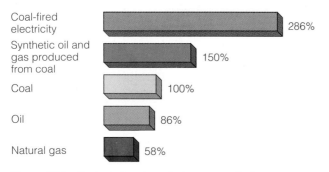

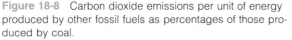

Figure 18-8 Carbon dioxide emissions per unit of energy produced by other fossil fuels as percentages of those produced by coal.

When a natural gas deposit is tapped, propane and butane gases are liquefied and removed as **liquefied petroleum gas (LPG)**. LPG is stored in pressurized tanks for use mostly in rural areas not served by natural gas pipelines. The rest of the gas (mostly methane) in the deposit is dried to remove water vapor, cleaned of hydrogen sulfide and other impurities, and pumped into pressurized pipelines for distribution.

At a very low temperature, natural gas can be converted to **liquefied natural gas (LNG)**. This highly flammable liquid form of natural gas can then be shipped to other countries in refrigerated tanker ships.

How Long Will Natural Gas Supplies Last? The Soviet Union has 40% of the world's proven reserves and is the world's largest natural gas extractor. Other countries with large proven natural gas reserves are Iran (14%), the United States (6%), Quatar (4%), Algeria (4%), Saudi Arabia (3%), and Nigeria (3%). Geologists expect to find more conventional natural gas deposits, especially in LDCs that have not been widely explored for this resource.

Most U.S. reserves of natural gas are located with the country's deposits of crude oil (Figure 18-3). About 95% of the natural gas used in the United States comes from domestic sources; the other 5% is imported by pipeline from Canada. Algeria and the Soviet Union use pipelines to supply many eastern and western European countries with natural gas and are planning more pipelines.

In 1989 about 82% of the natural gas consumed in the United States was used for space heating of residential and commercial buildings and for drying and other purposes in industry. The rest was used to produce electricity (15%) and as a vehicle fuel (3%).

Conventional supplies of natural gas should last longer than those of crude oil. Known reserves and undiscovered, economically recoverable conventional natural gas deposits in the United States are projected to last 28 years and world supplies 59 years at present consumption rates.

As the price of natural gas from conventional sources rises, it may become economical to get natural gas from unconventional sources. Such sources include coal seams, Devonian shale rock, deep underground deposits of tight sands, and deep geopressurized zones that contain natural gas dissolved in hot water. New technology for extracting gas from these resources is being developed rapidly.

In 1988 the Department of Energy estimated that technically recoverable natural gas from both conventional and unconventional sources in the lower 48 states would meet domestic needs for 50 years at current usage rates. The world's identified reserves of conventional natural gas are projected to last until 2045 at current usage rates and until 2022 if consumption rises 2% a year.

Estimated supplies of conventional natural gas, and unconventional supplies available at higher prices, would last about 200 years at the current rate and 80 years if usage rose 2% a year. If these estimates are correct, natural gas could become the most widely used fuel for space heating, industrial processes, producing electricity, and transportation.

Pros and Cons of Natural Gas Natural gas burns hotter and produces less air pollution than any fossil fuel. Burning it produces virtually no sulfur dioxide and particulate matter and only about one-sixth as much nitrogen oxides per unit of energy as burning coal, oil, and gasoline. Burning natural gas produces carbon dioxide, but the amount per unit of energy produced is much lower than that of other fossil fuels (Figure 18-8). Methane, the major component of natural gas, is a greenhouse gas that is 25 times more effective per molecule than carbon dioxide in causing global warming (Figure 10-6, p. 214). But little of the methane in the atmosphere comes from extraction and use of natural gas.

So far the price of natural gas has been low. It can be transported easily over land by pipeline and has a high net useful-energy yield (Figure 3-18, p. 53). It is a versatile fuel that can be burned cleanly and efficiently in furnaces, stoves, water heaters, dryers, boilers, incinerators, fuel cells, heat pumps, air conditioners, and refrigerators.

It can also be used as a clean-burning fuel in modified conventional and diesel engines, especially for large fleets of buses, taxis, and trucks that can be refueled in centralized facilities. Burning compressed natural gas, instead of gasoline, in retrofitted older vehicles would lower smog-forming hydrocarbon emissions by 50% to 80%; emissions would be lowered by as much as 90% in new engines designed for this fuel. Also carbon monoxide emissions would be reduced by 50% to 90%, but emissions of nitrogen oxides would increase by about 25%.

New natural-gas-burning turbines, working like jet engines, can be used to produce electricity. They

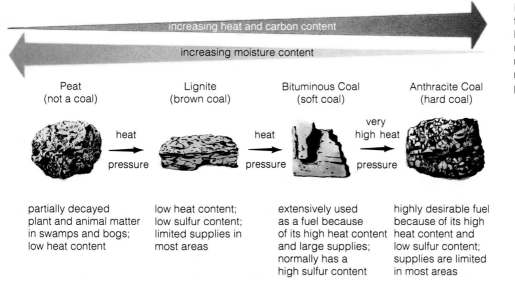

increasing heat and carbon content

increasing moisture content

Peat (not a coal)	Lignite (brown coal)	Bituminous Coal (soft coal)	Anthracite Coal (hard coal)

heat / pressure → heat / pressure → very high heat / pressure →

partially decayed plant and animal matter in swamps and bogs; low heat content

low heat content; low sulfur content; limited supplies in most areas

extensively used as a fuel because of its high heat content and large supplies; normally has a high sulfur content

highly desirable fuel because of its high heat content and low sulfur content; supplies are limited in most areas

Figure 18-9 Stages in the formation of coal over millions of years. Peat is a humus soil material. Lignite and bituminous coal are sedimentary rocks, and anthracite is a metamorphic rock (see Figure 19-2, p. 444).

cost half as much to build as a coal-fired system, are cheaper to operate, and can be put into operation within 12 to 18 months. Natural gas can also be burned cleanly and efficiently in cogenerators to produce high-temperature heat and electricity and can be burned with coal in boilers to reduce air pollution emissions.

One problem is that natural gas must be converted to liquid natural gas before it can be shipped by tanker from one country to another. Shipping LNG in refrigerated tankers is expensive and dangerous. Massive explosions could kill many people and cause much damage in urban areas near LNG loading and unloading facilities. Conversion of natural gas to LNG also reduces the net useful-energy yield by one-fourth.

If large amounts of natural gas can be extracted from nonconventional deposits at affordable prices, natural gas will be a key option for making an acceptable and orderly transition to solar and other energy options as oil is phased out over the next 50 years.

18-2 COAL

Types and Distribution Coal is a solid formed in several stages as the remains of plants are subjected to intense heat and pressure over millions of years. It is a complex mixture of organic compounds, with 30% to 98% carbon by weight plus varying amounts of water and small amounts of nitrogen and sulfur.

Three types of coal are formed at different stages: lignite, bituminous coal, and anthracite (Figure 18-9). Peat, which is the first stage of coal formation, is not a coal. It is burned in some places but has a low heat content. Low-sulfur coal produces less sulfur dioxide when burned than high-sulfur coal. The most desir-

able type of coal is anthracite because of its high heat content and low sulfur content.

About 60% of the coal extracted in the world and 70% in the United States is burned in boilers to produce steam to generate electrical power. The rest is converted to coke used to make steel and burned in boilers to produce steam used in various manufacturing processes. In 1989 coal was burned to supply 57% of the electricity generated in the United States. The rest was produced by nuclear energy (19%), natural gas (11%), hydropower (10%), and oil (3%).

Coal is the world's most abundant fossil fuel. About 68% of the world's proven coal reserves and 85% of the estimated undiscovered coal deposits are located in three countries: the United States, the USSR, and China.

Most major U.S. coalfields are located in 17 states (Figure 18-10). Anthracite, the most desirable form of coal, makes up only 2% of U.S. coal reserves. About 45% of U.S. coal reserves is high-sulfur, bituminous coal with a high fuel value. It is found mostly in the East, primarily in Kentucky, West Virginia, Pennsylvania, Ohio, and Illinois.

About 55% of U.S. coal reserves are found west of the Mississippi River. Most of these are deposits of low-sulfur bituminous and lignite coal. Unfortunately, these deposits are far from the heavily industrialized and populated East, where most coal is consumed.

Extracting Coal Surface mining is used to extract almost two-thirds of the coal used in the United States. Most surface-mined coal is removed by area strip mining or contour strip mining, depending on the terrain.

Area strip mining is used where the terrain is fairly flat. It involves stripping away the overburden and digging a cut to remove a mineral deposit, in this case coal (Figure 18-11). After the coal deposit is

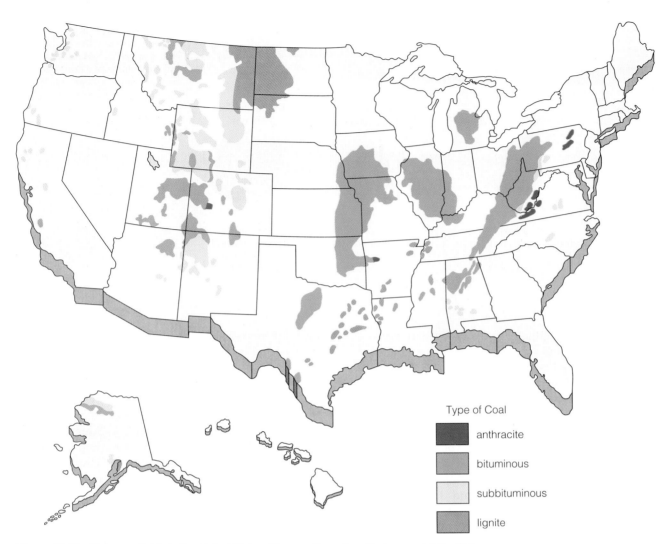

Figure 18-10 Major coalfields in the United States. (Source: Council on Environmental Quality)

Type of Coal
- anthracite
- bituminous
- subbituminous
- lignite

removed from the cut, the trench is filled with overburden. The power shovel removing coal then digs a cut parallel to the previous one. This process is repeated for the entire deposit. If the land is not restored, this type of mining leaves a wavy series of highly erodible hills of rubble called *spoil banks* (Figure 18-12).

Contour strip mining is a form of surface mining used in hilly or mountainous terrain. A power shovel cuts a series of terraces into the side of a hill or mountain. An earth mover removes the overburden, and a power shovel extracts the coal, with the overburden from each new terrace dumped onto the one below. Unless the land is restored, a wall of dirt is left in front of a highly erodible bank of soil and rock called a *highwall* (Figure 18-13). In the United States, contour strip mining is used mostly for extracting coal in the mountainous Appalachian region. If the land is not restored (Figure 18-14), this type of surface mining has a devastating impact on the land.

Subsurface mining is used to remove coal too deep to be extracted by surface mining. Miners dig a deep vertical shaft, blast subsurface tunnels and rooms to get to the deposit, and haul the coal or ore to the surface. In the *room-and-pillar method* as much as half of the coal is left in place as pillars to prevent the mine from collapsing. In the *longwall method* a narrow tunnel is created and then supported by movable metal pillars. After a cutting machine has removed the coal or ore from part of the mineral seam, the roof supports are moved forward, allowing the earth behind the supports to collapse. No tunnels are left behind after the mining operation has been completed. Sometimes giant augers are used to drill horizontally into a hillside to extract underground coal.

How Long Will Supplies Last? Identified world reserves of coal should last about 220 years at current usage and 65 years if usage rises 2% a year. The world's unidentified coal resources are projected to

Figure 18-11 Area strip mining of coal in Decker, Montana. This type of surface mining is used on flat or gently rolling terrain.

Figure 18-12 Effects of area strip mining of coal near Mulla, Colorado. Restoration of newly strip-mined areas is now required in the United States, but many previously mined areas have not been restored. In arid areas full restoration isn't possible, and enforcement of surface-mining laws is often lax.

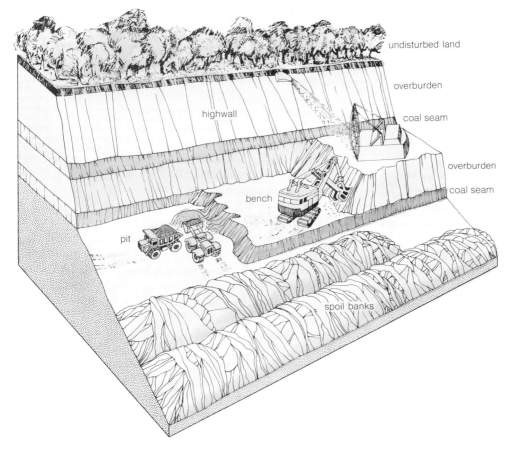

Figure 18-13 Contour strip mining of coal. This type of surface mining is used in hilly or mountainous terrain.

undisturbed land

overburden

coal seam

overburden

coal seam

highwall

bench

pit

spoil banks

last about 900 years at the current rate and 149 years if usage increases 2% a year.

Identified coal reserves in the United States should last about 300 years at the current usage rate. Unidentified U.S. coal resources could extend these supplies at the current rate for perhaps 100 years, at a much higher average cost.

Pros and Cons of Solid Coal Coal is the most abundant conventional fossil fuel in the world and

Figure 18-14 With the land returned to its original contour and grass planted to hold the soil in place, it is hard to tell that this was once a surface coal-mining site in Grantsville, Maryland.

Soil Conservation Service

in the United States. It also has a high net useful-energy yield for producing high-temperature heat for industrial processes and for generating electricity (Figure 3-18, p. 53). In countries with adequate coal supplies, burning solid coal is the cheapest way to produce high-temperature heat and electricity. But these low costs do not include requiring the best air pollution control equipment on all plants and requiring effective reclamation of all land surface mined for coal.

Since 1900 underground mining in the United States has killed more than 100,000 miners and permanently disabled at least 1 million. At least 250,000 retired U.S. miners suffer from black lung disease, a form of emphysema caused by prolonged breathing of coal dust and other particulate matter. Underground mining also causes subsidence, a depression in the earth's surface, when a mine shaft partially collapses during or after mining. Mining safety laws in most countries are much weaker than those in the United States.

Surface mining causes severe land disturbance (Figure 18-12) and soil erosion and surface-mined land in arid and semiarid areas can only be partially restored. Surface and subsurface mining of coal can cause severe pollution of nearby streams and groundwater from acids and toxic metal compounds (Figure 18-15). Once coal is mined it is expensive to move from one place to another and cannot be used in solid form as a fuel for cars and trucks.

Coal is the dirtiest fossil fuel to burn. Without expensive air pollution control devices, burning coal produces larger amounts of sulfur dioxide, nitrogen oxides, and particulate matter than other fossil fuels. These pollutants contribute to acid deposition, corrode metals, and harm trees, crops, wild animals, and people. Each year these and other air pollutants emitted when coal is burned kill about 5,000 people in the United States. They also cause 50,000 cases of

respiratory disease and several billion dollars in property damage each year. Burning coal also produces more carbon dioxide per unit of energy than other fossil fuels (Figure 18-8). This means that burning more coal to meet energy needs can accelerate projected global warming (Section 10-2). This problem by itself may prevent much of the world's coal reserves from being mined and burned.

New ways have been developed to burn coal more cleanly and efficiently. One is *fluidized-bed combustion*, which also sharply reduces emissions of sulfur dioxide and nitrogen oxides (Figure 9-17, p. 203). Successful small-scale fluidized-bed combustion plants have been built in Great Britain, Sweden, Finland, the Soviet Union, West Germany, and China. In the United States, commercial fluidized-bed combustion boilers are expected to begin replacing conventional coal boilers in the mid-1990s.

Synfuels: Converting Solid Coal to Gaseous and Liquid Fuels Besides being used in solid form, coal can also be converted to gaseous or liquid fuels, called **synfuels**. They are more useful than solid coal in heating homes and powering vehicles, and burning these fuels produces much less air pollution than burning solid coal.

Coal gasification (Figure 18-16) is the conversion of solid coal to synthetic natural gas (SNG). **Coal liquefaction** is the conversion of solid coal to a liquid hydrocarbon fuel such as methanol or synthetic gasoline. A commercial coal liquefaction plant supplies 10% of the liquid fuel used in South Africa at a cost equal to paying $35 a barrel for oil. When two new plants are completed, the country will be able to meet half of its oil needs from this source. Engineers hope to get the cost down to $25 a barrel.

Synfuels can be transported through a pipeline, burn more cleanly than solid coal, and are more versatile than solid coal. Besides being burned to

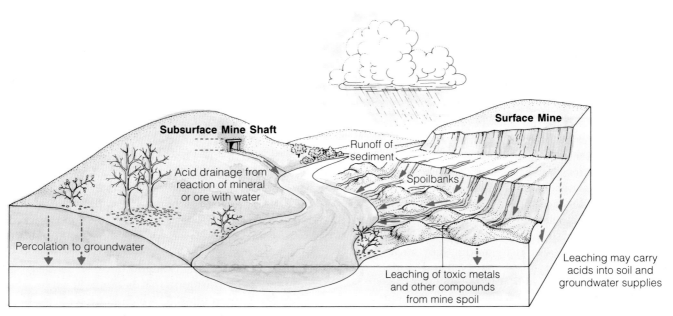

Figure 18-15 Degradation and pollution of a stream and groundwater by runoff of acids—called *acid mine drainage*—and toxic chemicals from surface- and subsurface-mining operations.

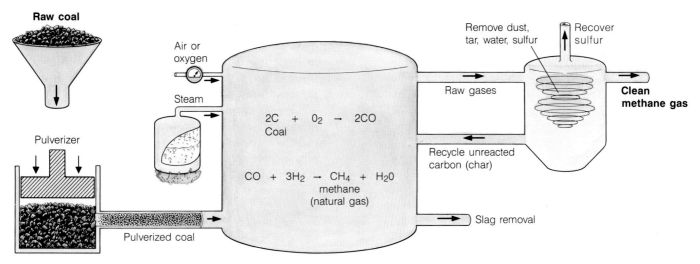

Figure 18-16 Coal gasification. Generalized view of one method for converting solid coal into synthetic natural gas (methane).

produce high-temperature heat and electricity as solid coal does, synfuels can be burned to heat houses and water and to propel vehicles.

But a synfuel plant costs much more to build and run than an equivalent coal-fired power plant fully equipped with air pollution control devices. Synfuels also have low net useful-energy yields (Figure 3-18, p. 53). The widespread use of synfuels would accelerate the depletion of world coal supplies because 30% to 40% of the energy content of coal is lost in the conversion process. It would also lead to greater land disruption from surface mining because producing a unit of energy from synfuels uses more coal than burning solid coal to provide the same amount of energy.

Producing synfuels requires huge amounts of water, and burning synfuels releases large amounts of carbon dioxide per unit of energy (Figure 18-8). Converting coal to SNG underground would solve the water problems, but currently, underground coal gasification is not competitive with conventional coal mining and aboveground coal gasification.

The major factor holding back large-scale production of synfuels in the United States is their high cost, compared to conventional oil and natural gas. Producing synfuels with current technology is the equivalent of buying oil at $38 a barrel. The U.S. Department of Energy has a goal of supporting development of new processes that will reduce the cost to $25 per barrel by 1995, but most analysts expect synfuels to play only a minor role as an energy resource in the next 30 to 50 years.

18-3 CONVENTIONAL NONRENEWABLE NUCLEAR FISSION

A Controversial Fading Dream By the end of this century, 1,800 nuclear power plants were supposed to supply 21% of the world's supplemental energy and 25% of that used in the United States. These rosy forecasts turned out to be an example of unrealistic high-tech intoxication.

By 1990, after 43 years of development and massive government subsidies, about 428 commercial nuclear reactors in 27 countries were producing only 19% of the world's electricity—equal to only about 5% of the world's supplemental energy. The percentage of the world's electricity produced by nuclear power will probably drop between 1990 and 2010 as aging nuclear plants are retired faster than new ones are built. By the year 2000, nuclear power will supply less than one-tenth of the electricity it was projected to produce.

Industrialized countries such as Japan and France, which have few fossil-fuel resources, believe that using nuclear power is the best way to reduce their dependence on imported oil. France plans to get 90% of its electricity from nuclear power by the early 1990s. However, both Japan and France already are producing more electricity than they can use (see Guest Essay on p. 56). France has been forced to sell electricity to neighboring countries at bargain prices and run its plants at partial capacity.

Since the Chernobyl nuclear accident in 1986 (see Spotlight on p. 423), many countries have scaled back or eliminated their plans to build nuclear power plants. Since 1975, no new nuclear power plants have been ordered in the United States and 108 previous orders have been canceled. In 1990 the 113 licensed nuclear plants in the United States generated almost 20% of the country's electricity. This percentage will slowly decline over the next two decades as more than 60% of the current reactors are scheduled for retirement.

What happened to nuclear power? The answer is that the nuclear industry has been crippled by high and uncertain costs of building and operating plants, billion-dollar cost overruns, frequent malfunctions, false assurances and cover-ups by government and industry officials, overproduction of electricity in some areas, poor management, and lack of public acceptance because of mistrust and concerns about safety, cost, radioactive waste disposal, and nuclear weapons proliferation. To better understand some of the problems with nuclear power, we need to know how a nuclear power plant works.

How Does a Nuclear Fission Reactor Work? When the nuclei of atoms such as uranium-235 and plutonium-239 are split by neutrons, energy is released and converted mostly to high-temperature heat in a nu-

clear fission chain reaction (Figure 3-10, p. 44). The rate at which this happens can be controlled in the nuclear fission reactor in a nuclear power plant, and the high-temperature heat released can be used to spin a turbine and produce electrical energy.

Light-water reactors (LWRs) now generate about 85% of the electricity generated worldwide (98% in the United States) by nuclear power plants. Key parts of an LWR are the core, fuel assemblies, fuel rods, control rods, moderator, and coolant (Figure 18-17). The core of an LWR typically contains about 40,000 long, thin fuel rods bundled in 180 fuel assemblies of around 200 rods each. Each fuel rod is packed with pencil-eraser-sized pellets of uranium-oxide fuel.

About 97% of the uranium in each fuel pellet is uranium-238, an isotope that is nonfissionable. The other 3% is uranium-235, which is fissionable. Uranium ore contains 97% uranium-238 by weight and only 0.7% fissionable uranium-235 (Figure 3-2, p. 38). Enrichment separates some uranium-238 from the ore, increasing the concentration of uranium-235 from 0.7% to 3% by weight. This enriched ore can be used as a fuel in a fission reactor. The uranium-235 in each fuel rod produces energy equal to that of three railroad carloads of coal over a lifetime of about three to four years.

When the fuel in the rods can no longer sustain nuclear fission, the intensely radioactive spent fuel rods are removed. If these rods are processed to remove plutonium and other very long-lived radioactive isotopes, they must be safely stored for at least 10,000 years. Otherwise, they must be stored safely for at least 240,000 years—about six times longer than our species has been around.

Control rods are made of materials such as boron or cadmium that absorb neutrons. The rods are moved in and out of the reactor core to regulate the rate of fission and the amount of power the reactor produces. All reactors place or circulate some type of material between the fuel rods and the fuel assemblies. This material, known as a moderator, slows down the neutrons emitted by the fission process so that the chain reaction can be kept going.

Three-fourths of the world's commercial reactors use ordinary water, called light water, as a moderator. Thus, the interior of most commercial nuclear reactors is somewhat like a swimming pool with a large number of movable vertical fuel rods and control rods hanging on it. The moderator in about 20% of the world's commercial reactors (50% of those in the Soviet Union, including the ill-fated Chernobyl reactor) is solid graphite, a form of carbon. Graphite-moderated reactors can also be used to produce fissionable plutonium-239 for use in nuclear weapons.

A coolant circulates through the reactor's core. It removes heat to keep fuel rods and other materials from melting and to produce steam that spins gen-

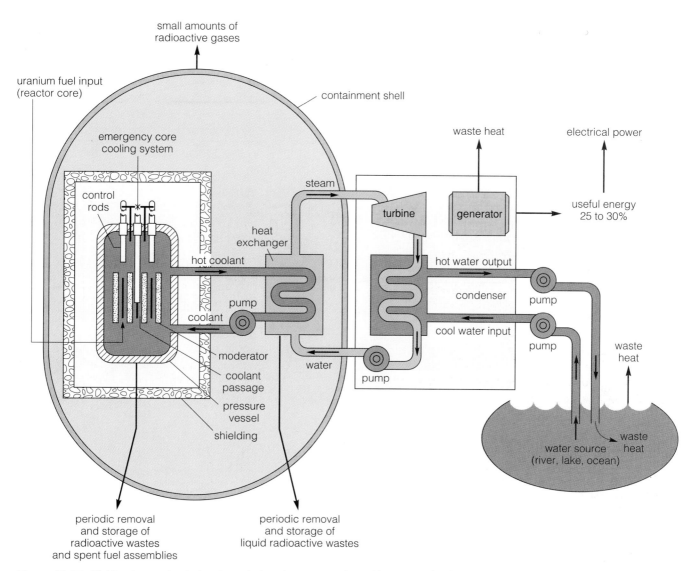

Figure 18-17 Light-water-moderated and -cooled nuclear power plant with a pressurized water reactor.

erators to produce electricity. Most water-moderated and graphite-moderated reactors use water as a coolant; a few gas-cooled reactors use an unreactive gas such as helium or argon for cooling.

A typical light-water reactor has an energy efficiency of only 25% to 30%, compared to 40% for a coal-burning plant. Graphite-moderated, gas-cooled reactors are more expensive to build and operate, but are more energy efficient (38%) than LWRs because they operate at a higher temperature.

Nuclear power plants, each with one or more reactors, are only one part of the nuclear fuel cycle necessary for using nuclear energy to produce electricity (Figure 18-18). *In evaluating the safety and economy of nuclear power, we need to look at the entire cycle—not just the nuclear plant itself.*

After about three to four years in a reactor, the concentration of fissionable uranium-235 in a fuel rod becomes too low to keep the chain reaction going, or the rod becomes damaged from exposure to ionizing

radiation. Each year about one-third of the spent fuel assemblies in a reactor are removed and stored in large, concrete-lined pools of water at the plant site.

After they have cooled for several years and lost some of their radioactivity, the spent fuel rods can be sealed in shielded, supposedly crash-proof casks and transported by truck or train to storage pools away from the reactor or to a nuclear waste repository or dump.

A third option is to send spent fuel to a fuel-reprocessing plant (Figure 18-18). There, remaining uranium-235 and plutonium-239, produced as a by-product of the fission process, are removed and sent to a fuel fabrication plant. Such plants would also handle and ship bomb-grade plutonium-239, which could be used to make nuclear weapons. Two small commercial fuel-reprocessing plants in operation (one in France and one in Great Britain) have had severe operating and economic problems. Two others are

Figure 18-18 The nuclear fuel cycle.

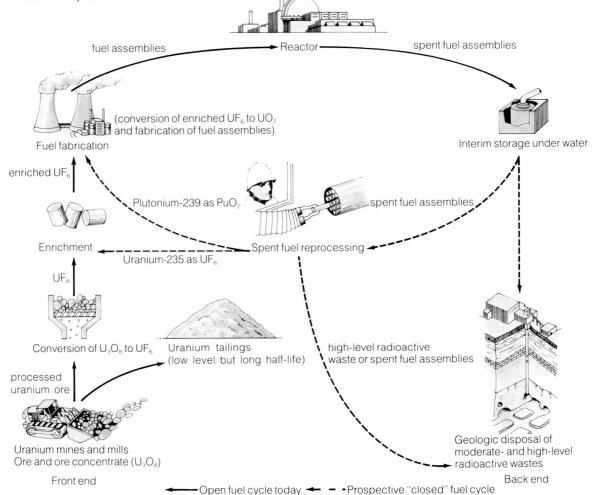

fuel assemblies → Reactor — spent fuel assemblies

(conversion of enriched UF$_6$ to UO$_2$ and fabrication of fuel assemblies)

Fuel fabrication

Interim storage under water

enriched UF$_6$

Plutonium-239 as PuO$_2$

spent fuel assemblies

Enrichment

Uranium-235 as UF$_6$

Spent fuel reprocessing

UF$_6$

Conversion of U$_3$O$_8$ to UF$_6$

Uranium tailings (low level but long half-life)

high-level radioactive waste or spent fuel assemblies

processed uranium ore

Uranium mines and mills
Ore and ore concentrate (U$_3$O$_8$)

Geologic disposal of moderate- and high-level radioactive wastes

Front end

Back end

— Open fuel cycle today ← — — Prospective "closed" fuel cycle

---

under construction—one in Japan and one in West Germany. The United States has delayed development of commercial fuel-reprocessing plants because of technical difficulties, high construction and operating costs, and adequate domestic supplies of uranium.

The fission products produced in a nuclear reactor give off radioactivity and heat, even after control rods have been inserted to stop all nuclear fission in the reactor core. To prevent a *meltdown* of the fuel rods and the core after a reactor is shut down, huge amounts of water must be kept circulating through the core. A meltdown could release massive amounts of radioactive materials into the environment.

How Safe Are Nuclear Power Plants? To greatly reduce the chances of a meltdown and other serious reactor accidents, commercial reactors in the United States (and most countries) have many safety features:

■ thick walls and concrete and steel shields surrounding the reactor vessel

■ a system for automatically inserting control rods into the core to stop fission under emergency conditions

■ a steel-reinforced concrete containment building to keep radioactive gases and materials from reaching the atmosphere after an accident (ineffective in a complete core meltdown or a massive gas explosion like the one that happened at the Chernobyl plant in 1986)

■ large filter systems and chemical sprayers inside the containment building to remove radioactive dust from the air and further reduce chances of radioactivity reaching the environment

■ systems to condense steam released from a ruptured reactor vessel and prevent pressure from rising beyond the holding power of containment building walls

■ an emergency core-cooling system to flood the core automatically with massive amounts of water within one minute to prevent meltdown of the reactor core

■ two separate power lines servicing the plant, and several diesel generators to supply backup power for the massive pumps in the emergency core-cooling system

■ X-ray inspection of key metal welds during construction and periodically after the plant goes

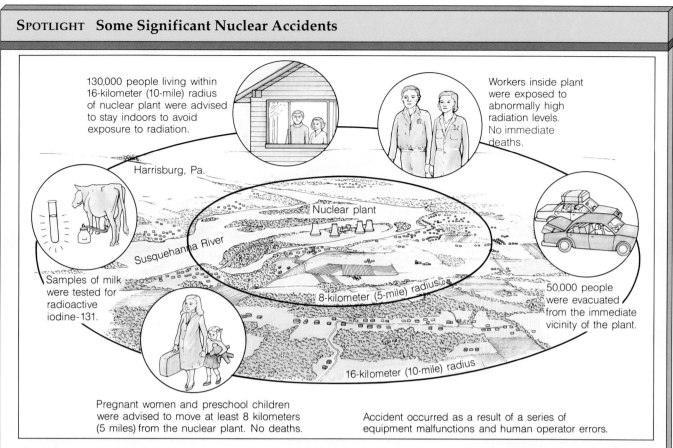

130,000 people living within 16-kilometer (10-mile) radius of nuclear plant were advised to stay indoors to avoid exposure to radiation.

Workers inside plant were exposed to abnormally high radiation levels. No immediate deaths.

Harrisburg, Pa.

Nuclear plant

Susquehanna River

Samples of milk were tested for radioactive iodine-131.

8-kilometer (5-mile) radius

50,000 people were evacuated from the immediate vicinity of the plant.

16-kilometer (10-mile) radius

Pregnant women and preschool children were advised to move at least 8 kilometers (5 miles) from the nuclear plant. No deaths.

Accident occurred as a result of a series of equipment malfunctions and human operator errors.

Figure 18-19   Three Mile Island (TMI) in eastern Pennsylvania, where a nuclear accident occurred on March 29, 1979.

- *Winter 1957.* Perhaps the worst nuclear disaster in history occurred in the Soviet Union in the southern Ural Mountains near the city of Kyshtym, then believed to be the center of plutonium production for Soviet nuclear weapons. The cause of the accident and the number of people killed and injured remain a secret. However, in 1989 Soviet officials admitted that several hundred square kilometers were contaminated with radioactivity when a tank containing radioactive wastes exploded and 10,000 people were evacuated. Today the area is deserted and sealed off, and the names of 30 towns and villages in the region have disappeared from Soviet maps.

- *October 7, 1957.* A water-cooled, graphite-moderated reactor used to produce plutonium for nuclear weapons north of Liverpool, England, caught fire as the Chernobyl nuclear plant

did 29 years later. By the time the fire was put out, 516 square kilometers (200 square miles) of countryside had been contaminated with radioactive material. Exposure to high levels of radiation caused an estimated 33 people to die prematurely from cancer.

- *March 22, 1975.* Against regulations, a maintenance worker used a candle to test for air leaks at the Brown's Ferry commercial nuclear reactor near Decatur, Alabama. It set off a fire that knocked out five emergency core-cooling systems. Although the reactor's cooling water dropped to a dangerous level, backup systems prevented any radioactive material from escaping into the environment. At the same plant, in 1978, a worker's rubber boot fell into a reactor and led to an unsuccessful search costing $2.8 million. Such incidents, caused mostly by unpredictable

human errors (Section 8-2), are common in most nuclear plants.

- *March 29, 1979.* The worst accident in the history of U.S. commercial nuclear power happened at the Three Mile Island (TMI) nuclear plant near Harrisburg, Pennsylvania (Figure 18-19). One of its two reactors lost its coolant water because of a series of mechanical failures and human operator errors not anticipated in safety studies. The reactor's core became partially uncovered. At least 70% of the core was damaged, and about 50% of it melted and fell to the bottom of the reactor. Unknown amounts of ionizing radiation escaped into the atmosphere and 144,000 people were evacuated. Investigators found that if a stuck valve had stayed opened for just another 30 to 60 minutes, there would

(continued)

have been a complete meltdown. No one is known to have died because of the accident, but its long-term health effects on workers and nearby residents are still being debated because data published on the radiation released during the accident are contradictory and incomplete.

Partial cleanup of the damaged TMI reactor will cost more than $1 billion, more than the $700 million construction cost of the reactor, and is expected to be completed in 1990 or later. Also, about $187 million of taxpayers' money has been spent by the Department of Energy on the TMI cleanup. Plant owners have also paid out $25 million to over 2,100 people who filed lawsuits for damages. The TMI cleanup is only partial. Probably in 1991 the plant will be sealed and some radioactive debris will be left in the plant for 20 to 90 years.

Confusing and misleading statements about the accident issued by Metropolitan Edison (which owned the plant) and by the Nuclear Regulatory Commission (NRC) eroded public confidence in the safety of nuclear power. Nuclear power critics contend that it is mostly luck that has prevented the TMI accident and hundreds of serious incidents since then from leading to a complete meltdown and breach of a reac-tor's containment building. U.S. nuclear industry officials claim that a catastrophic accident has not happened because the industry's multiple-backup safety systems work.

■ *April 26, 1986.* At 1:23 A.M. there were two massive explosions inside one of the four graphite-moderated, water-cooled reactors at the Chernobyl nuclear power plant north of Kiev in the Soviet Union. These blasts blew the 909-metric-ton (1,000-ton) roof off the reactor building, set the graphite core on fire, and flung radioactive debris several thousand feet into the air (Figure 18-20). Over the next several days winds carried some of these radioactive materials over parts of the Soviet Union and much of eastern and western Europe as far as 2,000 kilometers (1,250 miles) from the plant. The accident happened when engineers turned off most of the reactor's automatic safety and warning systems to keep them from interfering with an unauthorized safety experiment (Figure 18-20). It is likely that little radioactivity would have been released had the reactor been built with a strong containment dome like those found on commercial nuclear reactors in the United States and most other countries (Figure 18-17).

About 135,000 people living within 29 kilometers (18 miles) of the plant were eventually evacuated by an armada of 1,100 buses and trucks. According to Soviet officials, most of these people will never be able to return to their contaminated homes and farms.

By 1989 exposure to high levels of ionizing radiation at the accident site had killed 36 plant workers, fire fighters, and rescuers. Another 237 people were hospitalized with acute radiation sickness. Many of these people will probably die prematurely from cancer in coming years.

Soviet and Western medical experts estimate that 5,000 to 150,000 people in the Soviet Union and the rest of Europe will die prematurely over the next 70 years from cancer caused by exposure to the ionizing radiation released at Chernobyl. Thousands of others will be afflicted with thyroid tumors, cataracts, and sterility. The death toll would have been much higher if the accident had happened during the day, when people were not sheltered in houses, and if the wind had been blowing toward Kiev and its 2.4 million people. In 1988 Soviet officials revealed that the Chernobyl accident cost $14.4 billion, almost four times their original estimate of damages. The ultimate cost of

into operation to detect possible sources of leaks from corrosion

■ an automatic backup system to replace each major part of the safety system in the event of a failure

Such elaborate safety systems make a complete reactor core meltdown very unlikely. However, a partial or complete meltdown is possible through a series of equipment failures, operator errors, or both. In 1979 a reactor at the Three Mile Island plant in Pennsylvania underwent a partial meltdown because of equipment failures and operator errors (see Spotlight on p. 423).

Many studies of nuclear safety have been made since 1957 when the first commercial nuclear power plant began operating in the United States. However, there is still no officially accepted study of just how safe or unsafe these plants are and no study of the safety of the entire nuclear fuel cycle. Even if engineers can make the hardware 100% reliable, human reliability can never reach 100% (Section 8-2).

The Nuclear Regulatory Commission estimated that there is a 15% to 45% chance of a complete core

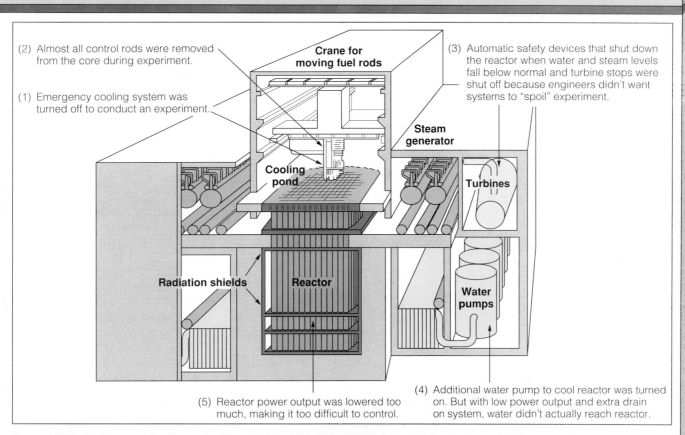

(2) Almost all control rods were removed from the core during experiment.

(1) Emergency cooling system was turned off to conduct an experiment.

Crane for moving fuel rods

(3) Automatic safety devices that shut down the reactor when water and steam levels fall below normal and turbine stops were shut off because engineers didn't want systems to "spoil" experiment.

Steam generator

Cooling pond

Turbines

Radiation shields **Reactor**

Water pumps

(5) Reactor power output was lowered too much, making it too difficult to control.

(4) Additional water pump to cool reactor was turned on. But with low power output and extra drain on system, water didn't actually reach reactor.

Figure 18-20 Major events leading to the Chernobyl nuclear power plant accident on April 26, 1986, in the Soviet Union.

the cleanup is projected to be at least $41.5 billion.

Records of radiation levels have been classified top secret, casting doubt on whether the true effects of the accident will ever be known. Some Soviet politicians and scientists charge that the accident released 20 times more radiation than the government has admitted and

killed at least 300 people. Recent measurements of radioactive contamination released could mean that 100,000 more people will have to evacuate from the area. Today the reactor is entombed in concrete and metal.

In 1987 the United States permanently shut down a Chernobyl-type military reactor

at Hanford, Washington; 54 serious safety violations had occurred at the plant during 1985 and 1986. The Chernobyl accident eroded public support for nuclear power worldwide and showed people that they need to be concerned about the safety of nuclear plants within and outside the borders of their countries.

meltdown at a U.S. reactor during the next 20 years. The commission also found that 39 U.S. reactors have an 80% chance of containment failure from a meltdown or massive gas explosion. Scientists in West Germany and Sweden project that, worldwide, there is a 70% chance of another serious core-damaging accident within the next 5.4 years.

A 1982 study by the Sandia National Laboratory estimated that a possible, but highly unlikely worst-case accident in a reactor near a large U.S. city might cause 50,000 to 100,000 immediate deaths, 10,000 to 40,000 later deaths from cancer, and $100 to $150

billion in damages. Most citizens and businesses suffering injuries or property damage from a major nuclear accident would get little if any financial reimbursement. Since the beginnings of commercial nuclear power in the 1950s, insurance companies have refused to cover more than a small part of the possible damages from an accident.

In 1957 Congress enacted the Price-Anderson Act, which limited insurance liability from a nuclear accident in the United States. In 1988 Congress extended the law for 20 years and raised the insurance liability to $7 billion—only 7% of the estimated

Some scientists believe that the long-term safe storage or disposal of high-level radioactive wastes is technically possible. Others disagree, pointing out that it is impossible to show that any method will work for the 10,000 years of failsafe storage needed for reprocessed wastes and the 240,000 years needed for unreprocessed wastes. The following are some of the proposed methods and their possible drawbacks:

1. *Bury it deep underground.* The currently favored method is to package unreprocessed spent fuel rods and bury them in a deep underground salt, granite, or other stable geological formation that is earthquake resistant and waterproof (Figure 18-21). A better method would be to reprocess the waste to remove very long-lived radioactive isotopes and convert what is left to a dry solid. The solid would then be fused with glass or a ceramic material and sealed in metal cannisters for burial. This would reduce burial time from 240,000 years to

10,000 years, but it is expensive. Some geologists question the idea of burying nuclear wastes. They argue that the drilling and tunneling to build the repository might cause water leakage and weaken resistance to earthquakes. They also contend that with present geological knowledge scientists cannot make meaningful 10,000- to 240,000-year projections about earthquake probability and paths of groundwater flows in underground storage areas.

2. *Shoot it into space or into the sun.* Costs would be very high and a launch accident, such as the explosion of the space shuttle *Challenger*, could disperse high-level radioactive wastes over large areas of the earth's surface.

3. *Bury it under the Antarctic ice sheets or the Greenland ice caps.* The long-term stability of the ice sheets is not known. They could be destabilized by heat from the wastes, and retrieval of the wastes would be difficult

or impossible if the method failed.

4. *Dump it into downward-descending, deep-ocean sediments.* The long-term stability of these sediments is unknown. Wastes could eventually be spewed out somewhere else by volcanic activity. Waste containers might leak and contaminate the ocean before being carried downward, and retrieval would probably be impossible if the method did not work.

5. *Change it into harmless or less harmful isotopes.* Presently there is no way to do this. Even if a method were developed, costs would probably be extremely high. Resulting toxic materials and low-level, but very long-lived, radioactive wastes would have to be disposed of safely.

6. *Use it in shielded batteries to run small electric generators.* Researchers claim that a wastebasket-sized battery using spent fuel could produce enough electricity to run five homes for 28 years or longer at about half the current price of

damage from a worst-case accident. Without this law the U.S. nuclear power industry would never have developed. Critics charge that the law is an unfair subsidy of the nuclear industry, and that if nuclear power plants are not safe enough to operate with adequate insurance, then they are not safe enough to operate at all.

There is also widespread lack of confidence in the NRC's ability to enforce nuclear safety. In 1989 NRC documents revealed that four out of five licensed U.S. reactors had failed to make all of the new safety changes required in 1979 after the TMI accident. None of these plants was shut down by the NRC. According to a 1987 General Accounting Office report, the NRC has also allowed plants to continue operating even after a record of repeated safety violations.

Congressional hearings in 1987 uncovered evidence that high-level NRC staff members have destroyed documents and obstructed investigations of criminal wrongdoing by utilities, suggested ways utilities can evade commission regulations, and provided utilities and their contractors with advanced

warnings of surprise inspections. Some NRC field supervisors have also harassed and intimidated lower-level NRC inspectors who cite utilities for too many violations.

In 1986 U.S. citizens learned that since the mid-1950s there had been serious disregard for the safety of workers and nearby residents in the country's nuclear weapons production facilities. Since 1957 these facilities have released huge quantities of radioactive particles into the air and dumped tons of potentially cancer-inducing radioactive waste and toxic substances into flowing creeks and leaking pits without telling local residents. The General Accounting Office and Department of Energy estimate that it will cost taxpayers $84 to $270 billion over 60 years to get these facilities cleaned up and in safe working order. Without constant pressure from citizens, Congress may not appropriate enough money to do the job.

Disposal and Storage of Radioactive Wastes Each part of the nuclear fuel cycle (Figure 18-18) for military

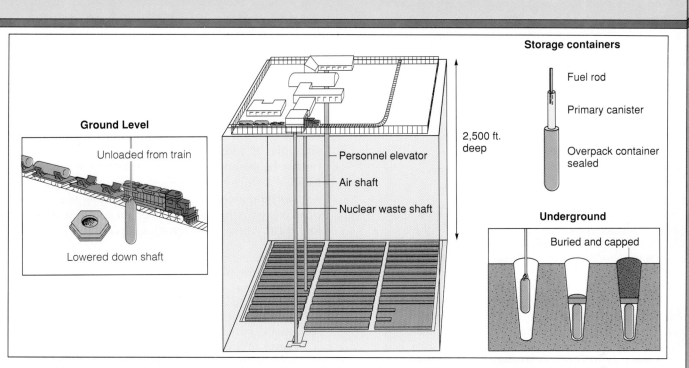

Figure 18-21 Proposed general design for deep underground permanent storage of high-level radioactive wastes from commercial nuclear power plants in the United States. (Source: U.S. Department of Energy)

electricity. But leakage could contaminate homes and communities. Dispersing high-level radioactive waste throughout a country would probably be politically unacceptable. Also this method would use only a small portion of the nuclear waste.

Critics of nuclear power are appalled that after decades there has been so little effort to solve the serious problem of what to do with nuclear waste while plunging ahead and building hundreds of nuclear reactors and weapons facilities. What do you think should be done?

and commercial nuclear reactors produces solid, liquid, and gaseous radioactive wastes. Some of these, called *low-level radioactive wastes*, give off small amounts of ionizing radiation, usually for a long time. Others are *high-level radioactive wastes*, which give off large amounts of ionizing radiation for a short time and small amounts for a long time.

From the 1940s to the 1970s, most low-level radioactive waste produced in the United States (and most other countries) was dumped into the ocean in steel drums. Since 1970 low-level radioactive wastes from military activities have been buried at government-run landfills. Three of these have been closed because of leakage.

Low-level waste materials from nuclear power plants, hospitals, universities, industries, and other producers are put in steel drums and shipped to regional landfills run by federal and state governments. By 1990 three of the six commercial landfills had been closed due to radioactive contamination of groundwater and nearby property.

In June 1990, the Nuclear Regulatory Commission caused a shockwave of opposition from environmentalists and the EPA when it proposed that most of the country's low-level radioactive waste be declared removed from federal regulation. These wastes would then be handled like ordinary trash and dumped in landfills, incinerated, reused, or recycled. According to the NRC, exposure to radiation from these unregulated wastes would kill 2,500 Americans—one out of every 100,000 citizens. The NRC contends that this is acceptable because it would save the nuclear power industry billions of dollars. This decision was made despite several recent studies showing that the hazards to humans from exposure to low-level radiation is at least 30 times higher than previously estimated.

Most high-level radioactive wastes are spent fuel rods from commercial nuclear power plants and an assortment of wastes from nuclear weapons plants. After 34 years of research and debate, scientists still don't agree on a safe method of storing these wastes (see Spotlight on p. 426). Regardless of the storage method, most U.S. citizens strongly oppose the

In 1982 Congress passed the Nuclear Waste Policy Act. It set a timetable for the Department of Energy to choose a site and build the country's first deep underground repository for storage of of high-level radioactive wastes from commercial nuclear reactors. In 1985 the Department of Energy announced plans to build the first repository, at a cost of $6 to $10 billion, based on the design shown in Figure 18-21.

The repository is supposed be built in a type of volcanic rock called tuff on federal land in the Yucca Mountain desert region, 161 kilometers (100 miles) northwest of Las Vegas, Nevada. Construction was supposed to begin in 1998 and the facility was scheduled to open by 2003.

In 1990 the Department of Energy put off the opening date to at least 2010. But it may never open. A young, active volcano is only 11 kilometers (7 miles) away, and according to DOE's own data there are 32 active earthquake faults on the site itself. Nevada ranks just behind Alaska and California in frequency of earthquakes.

Yucca Mountain's many geologic faults and its large amount of fractured rock also suggest that water flowing through the site could escape through a network of cracks. Some geologists estimate that that water carrying leached radioactive wastes could move 5 kilometers (3.1 miles) or more from the site in 400 to 500 years. This would automatically make it ineligible as a repository under current federal standards.

Since 1988 the state of Nevada has refused to give the Department of Energy permission to study the site. In 1990 the DOE asked the Justice Department to file suit against the state.

The DOE has also asked Congress to allow it to begin immediate construction of an above-ground interim storage facility for high-level radioactive wastes—a sort of halfway house for wastes awaiting permanent disposal. Critics believe that doing this would seriously undercut efforts to find a permanent storage facility. They fear the temporary facility could become the permanent site, with DOE declaring the problem solved.

Many citizens are also trying to find ways to ban shipments of radioactive wastes through their communities, but laws doing this could be overridden by the federal government.

Decommissioning Nuclear Power Plants and Weapons Facilities The useful operating life of today's nuclear power plants is hoped to be 30 to 40 years, but many plants are aging faster than expected. Because the core and many other parts contain large amounts of radioactive materials, a nuclear plant cannot be abandoned or demolished by a wrecking ball like a worn-out, coal-fired power plant.

Decommissioning nuclear power plants and nuclear weapons plants is the last step in the nuclear fuel cycle. Three ways have been proposed.

1. *Immediate dismantlement*—decontaminating and taking the reactor apart after shutdown and shipping all radioactive debris to a radioactive-waste burial facility. This promptly rids the plant site of radioactive materials and is the least expensive option. But it exposes work crews to the highest level of radiation and results in the largest volume of radioactive waste.

2. *Mothballing*—putting up a barrier and setting up a 24-hour security guard system to keep out intruders for several decades to 100 years before dismantlement. This permits short-lived radioactive isotopes to decay, which reduces the threat to dismantlement crews and the volume of contaminated waste.

3. *Entombment*—covering the reactor with reinforced concrete and putting up a barrier to keep out intruders. This allows for radioactive decay but passes a dangerous legacy to future generations.

Each method involves shutting down the plant, removing the spent fuel from the reactor core, draining all liquids, flushing all pipes, and sending all radioactive materials to an approved waste storage site yet to be built.

Worldwide, more than 20 commercial reactors (4 in the United States) have been retired and are awaiting decommissioning. Another 225 large commercial reactors (70 in the United States) will probably be retired between 2000 and 2010.

Utility company officials estimate that dismantlement of a typical large reactor should cost about $170 million and mothballing $225 million. Most analysts consider the dismantlement figure too low and put the cost at $1 to $3 billion per large reactor—roughly equal to the initial construction cost. Decommissioning costs will add to the already high price of electricity produced by nuclear fission. Politicians and nuclear industry officials in the United States and other countries may be tempted to mothball retired plants

location of a low- or high-level nuclear waste disposal facility anywhere near them (see Case Study above).

One state, California, passed a law prohibiting the licensing of any nuclear power plant until the country's nuclear waste disposal problem is solved.

and pass dismantlement costs and problems on to the next generation.

Nuclear industry scientists and government regulators are also examining the feasibility of renovating existing plants to extend their useful lives an additional 20 to 40 years. However, this will be difficult, expensive, and highly controversial.

Proliferation of Nuclear Weapons Since 1958 the United States has been giving away and selling to other countries various forms of nuclear technology. Today at least 14 other countries sell nuclear technology in the international marketplace.

For decades the U.S. government denied that the information, components, and materials used in the nuclear fuel cycle could be used to make nuclear weapons. In 1981, however, a Los Alamos National Laboratory report admitted: "There is no technical demarcation between the military and civilian reactor and there never was one"—something environmentalists had been saying for years.

Today 134 countries have signed the 1968 Nuclear-Nonproliferation Treaty. They have agreed to forego building nuclear weapons in return for help with commercial nuclear power. The International Atomic Energy Agency (IAEA) was established to monitor compliance.

But the system is not tight enough. Nuclear facilities belonging to India, Israel, South Africa, Pakistan, and other countries that have not signed the treaty are not monitored. Also nuclear facilities in countries such as China, France, Great Britain, the Soviet Union, and the United States that have signed the treaty are generally not monitored by the IAEA.

There is clear evidence that the governments of Israel, South Africa, Pakistan, and India have made almost 200 nuclear weapons, mostly by diverting weapons-grade fuel from research reactors and commercial power plants. It takes only about 10 kilograms (22 pounds) of plutonium to make a Nagasaki-sized nuclear bomb. At least seven other countries—Argentina, Brazil, Libya, Syria, Iraq, Iran, and Kuwait—are actively seeking to make nuclear weapons or to buy them from black market sources. A typical 1,000-megawatt nuclear reactor generates about 15 bombs worth of plutonium a year.

Sophisticated terrorist groups can also make a small atomic bomb by using about 2.2 kilograms (5 pounds) of plutonium or uranium-233, or about 5 kilograms (11 pounds) of uranium-235. Such a bomb could blow up a large building or a small city block and would contaminate a much larger area with radioactive materials for centuries. For example, a crude 10-kiloton nuclear weapon placed properly and detonated during working hours could topple the World Trade Center in New York City. This could easily kill more people than those killed by the atomic bomb the United States dropped on Hiroshima in 1945.

Spent reactor fuel is so highly radioactive that theft is unlikely, but plutonium separated at commercial and military reprocessing plants (Figure 18-18) is easily handled. Although plutonium shipments are heavily guarded, it could be stolen from nuclear weapons or reprocessing plants, especially by employees. Each year about 3% of the 142,000 people working in 127 U.S. nuclear weapons facilities in 23 states are fired because of drug use, mental instability, or other security risks. By the mid-1990s, hundreds of shipments of plutonium, separated at reprocessing facilities in France, Great Britain, West Germany, Japan, and India, will be traveling by land, sea, and air within and between countries.

Those who would steal plutonium need not bother to make atomic bombs. They could simply use a conventional explosive charge to disperse the plutonium into the atmosphere from atop any tall building. Dispersed in this way, 1 kilogram (2.2 pounds) of plutonium oxide powder theoretically would contaminate 8 square kilometers (3 square miles) with dangerous levels of radioactivity for several hundred thousand years.

One way to reduce the diversion of plutonium fuel from the nuclear fuel cycle is to contaminate it with other substances that make it useless as weapons material. But so far no one has come up with a way to do this, and most nuclear experts doubt that it can be done.

Soaring Costs After the United States dropped atomic bombs on Hiroshima and Nagasaki, ending World War II, the scientists who developed the bomb and the elected officials responsible for its use were determined to show the world that the peaceful uses of atomic energy would outweigh the immense harm it had done. One part of this "Atoms for Peace" program was to use nuclear power to produce electricity. American utility companies were skeptical but began ordering nuclear power plants in the late 1950s for four reasons.

1. The Atomic Energy Commission and builders of nuclear reactors projected that nuclear power would produce electricity at a very low cost compared to using coal and other alternatives.

2. The nuclear industry projected that nuclear reactors would have an 88% *capacity factor*—a measure of the time a reactor would operate each year at full power.

3. The first round of commercial reactors was built with the government paying about one-fourth of the cost and with the reactors provided to utilities at a fixed cost with no cost overruns

allowed. (The builders lost their shirts but knew they could make big profits on later rounds of plants.)

4. Congress passed the Price-Anderson Act, which protected the nuclear industry and utilities from significant liability to the general public in case of accidents.

It was an offer utility company officials could not resist. Today many wish they had.

Experience has shown that nuclear power is a very expensive way to produce electricity, even when it is heavily subsidized and enjoys partial protection from free market competition with other energy sources. According to the Department of Energy, commercial nuclear power received $1 trillion in research and development and other federal subsidies between 1952 and 1988—an average of $9 billion per reactor. Yet after almost four decades of subsidies and development, commercial nuclear reactors in the United States now deliver less of the country's energy than that provided by wood and crop wastes with hardly any subsidies.

Nuclear power plants in the United States since 1985 produce electricity at an average of 12.5 cents per kilowatt-hour—equal to buying oil at $206 per barrel. These already high costs do not include most of the costs of storing radioactive wastes and decommissioning worn-out plants.

In contrast, new coal plants with the latest air pollution control equipment produced electricity at an average cost of 5.4 cents per kilowatt-hour in 1989. Producing electricity by cogeneration costs only 5 cents per kilowatt-hour, and saving electricity by improving energy efficiency costs only about 2 cents to 4 cents per kilowatt-hour. In 1989 wind power, geothermal energy, wood-burning power plants, and combined solar thermal-natural gas power plants (Figure 17-14, p. 391) were producing electricity more cheaply than new nuclear power plants. By the year 2000 or shortly thereafter, solar photovoltaic cells are expected to be producing electricity cheaper than new nuclear plants.

Operating costs of nuclear plants have been higher than projected because U.S. pressurized water reactors operate at an average of only about 60% of their full-time, full-power capacity—far below the 88% capacity projected by proponents of nuclear power in the 1950s. The average capacity factor for PWRs in the United Kingdom is only 51% and those in Sweden 54%. Those in other countries are higher: Japan and Canada (71%), France (74%), West Germany (82%), and Switzerland (87%), mostly because of standardized design and better management.

New nuclear plants in France and Japan cost about half as much per kilowatt of power to build as those in the United States because they are better planned and use standardized designs. But France ran up an enormous $39 billion debt to finance its nuclear industry. Also France and Japan now produce more electricity than they need.

In the United States, where almost every nuclear plant has a different design, poor planning and management and stricter safety regulations since the TMI accident have increased costs and lengthened construction time. Currently, new nuclear power plants cost three times as much to build as equivalent coal-fired plants with the latest air pollution control equipment.

Banks and other lending institutions have become skeptical about financing new U.S. nuclear power plants. The Three Mile Island accident showed that utility companies could lose $1 billion or more of equipment in an hour, and at least $1 billion more in cleanup costs, even without any known harmful effects on public health. The business magazine *Forbes* has called the failure of the U.S. nuclear power program "the largest managerial disaster in U.S. business history." It involves perhaps $1 trillion in wasted investments, cost overruns, and unnecessarily high electricity costs, and production of more electricity than the country needs. No U.S. utility companies are planning the construction of any nuclear power plants because they are no longer considered cost-effective or wise investments.

Is nuclear power dead in the United States and most other MDCs? You might think so because of its high costs and massive public opposition. But powerful economic and political forces strive to maintain and expand the world's nuclear power industry (see Pro/Con on p. 431). Also, the U.S. Department of Energy and energy agencies in many other MDCs are heavily staffed with officials who continue to push for nuclear power instead of other safer and more cost-effective alternatives.

Pros and Cons of Conventional Nuclear Fission
Using nuclear fission to produce electricity has many advantages. Nuclear plants don't release particulate matter, sulfur dioxide, or nitrogen oxides into the atmosphere as do coal-fired plants. Water pollution and disruption of land are low to moderate if the entire nuclear fuel cycle operates normally. Multiple safety systems greatly decrease the likelihood of a catastrophic accident releasing deadly radioactive material into the environment.

Nuclear power also has many disadvantages. It produces electricity, which cannot be used to run vehicles without the development of affordable, long-lasting batteries to propel electric cars. Construction and operating costs for nuclear power plants in the United States and most countries are high and rising, even with massive government subsidies.

Standardized design and mass production can bring costs down, but electricity can still be produced

Since the Three Mile Island accident, the U.S. nuclear industry and utility companies have financed a massive advertising campaign by the U.S. Council for Energy Awareness. This campaign, with a $340 million budget in 1988, is designed to improve the industry's image, resell nuclear power to the American public, and downgrade the importance of solar energy, conservation, geothermal energy, wind, and hydropower as alternatives to nuclear power.

The campaign's magazine and television ads do not tell readers and viewers that the ads are paid for by the nuclear industry. Most ads use the argument that more nuclear power is needed in the United States to reduce dependence on imported oil.

The truth is that since 1979 only about 5% (3% in 1989) of the electricity in the United States has been produced by burning oil, and 95% of this is residual oil that can't be used for other purposes. Thus, *building more nuclear plants will not save the country any significant amount of domestic or imported oil.*

The nuclear industry also claims that nuclear power, unlike coal burning, does not add any carbon dioxide to the atmosphere. They argue that replacing coal-burning power plants with nuclear plants would help delay projected climate changes from an enhanced greenhouse effect. It is true that nuclear power plants don't release carbon dioxide. But the fuel cycle involved in using nuclear power does produce some carbon dioxide, mainly in the processing of uranium fuel (Figure 18-18). How-

ever, the amount of carbon dioxide produced per unit of electricity is only one-sixth that produced by a coal-burning plant.

The nuclear industry hopes to convince governments and utility companies to build hundreds of new "second-generation" plants using standardized designs. They are supposed to be safer, quicker to build (3 to 5 years), operate at full power 85% of the time, and last 60 years. Some nuclear experts believe that we can make nuclear power acceptably safe and that we have little choice but to take the risks involved in doing this (see Guest Essay on p. 438).

Nuclear advocates call these new designs, still only on drawing boards, *inherently safe.* But Robert Pollard, a former safety engineer with the NRC, points out that any scheme for fissioning atoms is inherently dangerous. You can build new reactors that are safer than existing ones, but you can't make them inherently safe.

Scientists disagree over which of the proposed new designs to pursue, and it would take at least 30 years and trillions of dollars for a new type of reactor to begin supplying 10% of U.S. electricity. According to *Nucleonics Week*, a major nuclear industry publication, "Experts are flatly unconvinced that safety has been achieved—or even substantially reduced—by the new designs."

Also, none of the new designs solve the problem of what to do with nuclear waste and the problem of the use of nuclear technology and fuel to build nuclear weapons. Indeed, these problems would become more serious as the number of nuclear plants in-

creased from a few hundred to the many thousands needed to slow global warming only a little.

If half of the U.S. use of coal burned to produce electricity was displaced by building 200 large new nuclear plants at a cost of $1.2 trillion or more, this would reduce the world's greenhouse effect by only 2%. Just to make this small dent in the carbon dioxide problem would require completing a new large nuclear reactor in the United States every 3 days for the next 37 years. To do this worldwide, we would have to build *one reactor a day* for 37 years at a total cost of $23 trillion!

Improvements in energy efficiency—especially requiring all new cars to get at least 21 kilometers per liter (50 miles per gallon) of gasoline—would save energy and result in much greater and faster reductions of carbon dioxide emissions at a fraction of the cost of building new nuclear plants. According to the Rocky Mountain Institute, if we hope to reduce carbon dioxide emissions using the least-cost methods, then investing in energy efficiency and renewable energy resources are at the top of the list and nuclear power is at the bottom (see Guest Essay on p. 56 and Spotlight on p. 221). Indeed, the full costs of nuclear power are rising, while those of perpetual and renewable energy resources are decreasing. Using the least-cost approach is not only more effective, it also frees capital for reforestation and other activities for reducing projected greenhouse warming. What do you think?

by safer methods at a cost equal to or lower than that of nuclear power. Although large-scale accidents are infrequent, a combination of mechanical failure and human errors, sabotage, or shipping accidents could again release deadly radioactive materials into the environment.

The net useful-energy yield of nuclear-generated electricity is low (Figure 3-18, p. 53). Scientists dis-

agree over how high-level radioactive wastes should be stored, and some doubt that an acceptably safe method can ever be developed. Also some carbon dioxide is released as part of the nuclear fuel cycle.

Today's military and commercial nuclear energy programs commit future generations to storing dangerous radioactive wastes for thousands of years even if nuclear fission power is abandoned tomorrow. The existence of nuclear power technology also helps spread knowledge and materials that can be used to make nuclear weapons. For these reasons, many people feel that it is unethical to use nuclear power to produce electricity.

 ## 18-4 BREEDER NUCLEAR FISSION AND NUCLEAR FUSION

Nonrenewable Breeder Nuclear Fission
At the present rate of use, the world's supply of uranium should last for at least 100 years and perhaps 200 years. However, some nuclear power proponents project a sharp rise in the use of nuclear fission to produce electricity after the year 2000. They urge the development and widespread use of breeder nuclear fission reactors that generate nuclear fuel to start up other breeders (see Guest Essay on p. 438).

Conventional fission reactors use fissionable uranium-235, which makes up only 0.7% of natural uranium ore. **Breeder nuclear fission reactors** convert nonfissionable uranium-238 into fissionable plutonium-239. Since breeders would use over 99% of the uranium in ore deposits, the world's known uranium reserves would last 1,000 years and perhaps several thousand years.

Under normal operation, a breeder reactor is considered by its proponents to be much safer than a conventional fission reactor. But if the reactor's safety system should fail, the reactor could lose some of its liquid sodium coolant. This could cause a runaway fission chain reaction and perhaps a small nuclear explosion with the force of several hundred kilograms of TNT. Such an explosion could blast open the containment building, releasing a cloud of highly radioactive gases and particulate matter. Leaks of flammable liquid sodium also can also cause fires, as as happened with all experimental breeder reactors built so far.

Since 1966, small experimental breeder reactors have been built in the the United Kingdom, the Soviet Union, West Germany, Japan, and France. In December 1986 France began operating a commercial-size breeder reactor, the Superphenix. It cost three times the original estimate to build. The little electricity it has produced is twice as expensive as that generated by France's conventional fission reactors. In 1987, shortly after the reactor began operating at full power, it began leaking liquid sodium coolant and was shut down. Repairs may be so expensive that the reactor may not be put back into operation.

Tentative plans to build full-size commercial breeders in West Germany, the Soviet Union, and the United Kingdom may be canceled because of the excessive cost of France's reactor and an excess of electric generating capacity. Also, experimental breeders built so far produce only about one-fourth of the plutonium-239 each year needed to replace their own fissionable material. If this serious problem is not solved, it would take 100 to 200 years at best for breeders to begin producing enough plutonium to fuel a significant number of other breeders. But some nuclear advocates urge us to step up research on breeder reactors because they will be needed sometime during the next century (see Guest Essay on p. 438).

Nuclear Fusion Scientists hope someday to use controlled nuclear fusion (Figure 3-11, p. 46) to provide an almost limitless source of energy for producing high-temperature heat and electricity. For 41 years research has focused on the D-T nuclear fusion reaction in which two isotopes of hydrogen—deuterium (D) and tritium (T)—fuse at about 100 million degrees, ten times as hot as the sun's interior (Figure 3-11, p. 46).

Another possibility is the D-D fusion reaction in which the nuclei of two deuterium atoms fuse together at much higher temperatures. If developed, it would run on virtually unlimited heavy water (D_2O) fuel obtained from seawater at a cost of about ten cents a gallon.

After 41 years of research, high-temperature nuclear fusion is still at the laboratory stage. Deuterium and tritium atoms have been forced together by using electromagnetic reactors the size of 12 locomotives, 120-trillion-watt laser beams, and bombardment with high-speed particles. But so far none of these approaches has produced more energy than it uses.

If researchers eventually can get more energy out than they put in, the next step is to build a small fusion reactor and scale it up to commercial size. This is considered one of the most difficult engineering problems ever undertaken. The estimated cost of a commercial fusion reactor is at least four times that of a comparable conventional fission reactor.

In 1989 two chemists announced what might be either a spectacular energy breakthrough or merely a fascinating scientific experiment—*cold nuclear fusion*. They claim to have brought about some D-D nuclear fusion at room temperature using a simple apparatus. But it is not clear whether some of this energy is coming from unexpected chemical reactions or other nonfusion processes, and their results have not been duplicated. It will probably take two to three decades to evaluate the technical and economic feasibility, if any, of cold nuclear fusion.

If everything goes right, a commercial cold fusion power plant might be built as early as 2030. But even if everything goes right, energy experts don't expect cold or high-temperature nuclear fusion to be a significant source of energy until 2100, if then. Meanwhile, several other quicker, cheaper, and safer ways can produce more electricity than we need.

18-5 DEVELOPING AN ENERGY STRATEGY FOR THE UNITED STATES

Overall Evaluation of U.S. Energy Alternatives Table 18-1 summarizes the major advantages and disadvantages of the energy alternatives discussed in this and the preceding chapter, with emphasis on their potential in the United States. Energy experts argue over these and other projections, and new data and innovations may change some information in this table. But it does provide a useful framework for making decisions based on presently available information. Four major conclusions can be drawn.

1. The best short-term, intermediate, and long-term alternatives for the United States and other countries are to improve energy efficiency and to greatly increase use of a mix of perpetual and renewable energy resources (Chapter 17).

2. Total systems for future energy alternatives in the world and the United States will probably have low to moderate net useful-energy yields and moderate to high development costs. Since there is not enough financial capital to develop all energy alternatives, projects must be chosen carefully. Otherwise, limited capital will be depleted on energy alternatives that yield too little net useful energy or prove to be economically or environmentally unacceptable.

3. We cannot and should not depend mostly on one nonrenewable energy resource like oil, coal, natural gas, or nuclear power. Instead, the world and the United States should rely more on improving energy efficiency and a mix of perpetual and renewable energy resources.

4. We should decrease dependence on using coal and nuclear power to produce electricity at large, centralized power plants. Individuals, communities, and countries should get more of their heat and electricity from locally available renewable and perpetual energy resources. This will give individuals more control over the energy they use. It would also enhance national security by eliminating large, centralized energy facilities that would be easy to knock out.

Economics and National Energy Strategy Cost is the major factor determining which commercial energy resources are widely used by consumers. Governments worldwide use three major economic and political strategies to stimulate or dampen the short- and long-term use of a particular energy resource.

1. *not attempting to control the price,* so that its use depends on open, free market competition (assuming all other alternatives also compete in the same way)

2. *keeping prices artificially low* to encourage its use and development

3. *keeping prices artificially high* to discourage its use and development

Each approach has certain advantages and disadvantages.

Free Market Competition Leaving it to the marketplace without any government interference is appealing in principle. However, a free market rarely exists in practice because business people are in favor of it for everyone but their own companies.

Most energy industry executives work hard to get control of supply, demand, and price for their particular energy resource, while urging free market competition for any competing energy resources. They try to influence elected officials and help elect those who will give their businesses the most favorable tax breaks and other government subsidies. Such favoritism distorts and unbalances the marketplace.

An equally serious problem with the open marketplace is its emphasis on today's prices to enhance short-term economic gain. This inhibits long-term development of new energy resources, which can rarely compete in their development stages without government support.

Keeping Energy Prices Artificially Low: The U.S. Strategy Many governments give tax breaks and other subsidies, pay for long-term research and development, and use price controls to keep prices for particular energy resources artificially low. This is the main approach used by the United States and the Soviet Union.

This approach encourages the development and use of energy resources getting favorable treatment. It also helps protect consumers (especially the poor) from sharp price increases, and it can help reduce inflation. Because keeping prices low is popular with consumers, this practice often helps leaders in democratic societies get reelected and helps keep leaders in nondemocratic societies from being overthrown.

But this approach also encourages waste and rapid depletion of an energy resource (such as oil) by making its price lower than it should be compared to its true value and projected long-term supply. This strategy discourages the development of energy alternatives not getting at least the same level of subsidies and price control.

Once energy industries such as the fossil-fuel and nuclear power industries get government subsidies, they usually have enough clout to maintain this support long after it becomes unproductive. And they often successfully fight efforts to provide equal or higher subsidies for the development of new energy alternatives that would allow more equal competition in the marketplace.

Table 18-1 Evaluation of Energy Alternatives for the United States (shading indicates favorable conditions)

Energy Resources	Estimated Availability			Estimated Net Useful Energy of Entire System	Projected Cost of Entire System	Actual or Potential Overall Environmental Impact of Entire System
	Short Term (1991–2001)	Intermediate Term (2001–2011)	Long Term (2011–2041)			
Nonrenewable Resources						
Fossil fuels						
Petroleum	High (with imports)	Moderate (with imports)	Low	High but decreasing	High for new domestic supplies	Moderate
Natural gas	High (with imports)	Moderate (with imports)	Moderate (with imports)	High but decreasing	High for new domestic supplies	Low
Coal	High	High	High	High but decreasing	Moderate but increasing	Very high
Oil shale	Low	Low to moderate	Low to moderate	Low to moderate	Very high	High
Tar sands	Low	Fair? (imports only)	Poor to fair (imports only)	Low	Very high	Moderate to high
Biomass (urban wastes for incineration)	Low	Moderate	Moderate	Low to fairly high	High	Moderate to high
Synthetic natural gas (SNG) from coal	Low	Low to moderate	Low to moderate	Low to moderate	High	High (increases use of coal)
Synthetic oil and alcohols from coal and organic wastes	Low	Moderate	High	Low to moderate	High	High (increases use of coal)
Nuclear energy						
Conventional fission (uranium)	Low to moderate	Low to moderate	Low to moderate	Low to moderate	Very high	Very high
Breeder fission (uranium and thorium)	None	None to low (if developed)	Moderate	Unknown, but probably moderate	Very high	Very high
Fusion (deuterium and tritium)	None	None	None to low (if developed)	Unknown, but may be high	Very high	Unknown (probably moderate to high)
Geothermal energy	Low	Low	Low	Low to moderate	Moderate to high	Moderate to high
Perpetual and Renewable Resources						
Improving energy efficiency	High	High	High	Very high	Low	Decreases impact of other sources

| Energy Resources | Estimated Availability | | | Estimated Net Useful Energy of Entire System | Projected Cost of Entire System | Actual or Potential Overall Environmental Impact of Entire System |
	Short Term (1991–2001)	Intermediate Term (2001–2011)	Long Term (2011–2041)			
Perpetual and Renewable Resources (continued)						
Water power (hydroelectricity)						
New large-scale dams and plants	Low	Low	Very low	Moderate to high	Moderate to very high	Low to moderate
Reopening abandoned small-scale plants	Moderate	Moderate	Low	High	Moderate	Low
Tidal energy	None	Very low	Very low	Unknown (probably moderate)	High	Low to moderate
Ocean thermal gradients	None	Low	Low to moderate (if developed)	Unknown (probably low to moderate)	Probably high	Unknown (probably moderate)
Solar energy						
Low-temperature heating (for homes and water)	High	High	High	Moderate to high	Moderate	Low
High-temperature heating	Low	Moderate	Moderate to high	Moderate	High initially, but probably declining fairly rapidly	Low to moderate
Photovoltaic production of electricity	Low to moderate	Moderate	High	Fairly high	High initially but declining fairly rapidly	Low
Wind energy						
Neighborhood turbines and wind farms	Low	Moderate	Moderate to high	Fairly high	Moderate	Low
Large-scale power plants	None	Very low	Probably low	Low	High	Low to moderate?
Geothermal energy (low heat flow)	Very low	Very low	Low to moderate	Low to moderate	Moderate to high	Moderate to high
Biomass (burning of wood, crop, food, and animal wastes)	Moderate	Moderate	Moderate to high	Moderate	Moderate	Variable
Biofuels (alcohols and natural gas from plants and organic wastes)	Low to moderate?	Moderate	Moderate to high	Low to fairly high	Moderate to high	Moderate to high
Hydrogen gas (from coal or water)	None	Low to moderate	Moderate	Variable	Variable	Variable, but low if produced by using solar energy

In 1984 federal tax breaks and other subsidies for the development of energy conservation and perpetual solar-based energy resources in the United States amounted to $1.7 billion. But these tax breaks were eliminated a year later and have not been restored.

By contrast, during 1984 the nuclear power industry received $15.6 billion, the oil industry $8.6 billion, the natural gas industry $4.6 billion, and the coal industry $3.4 billion in federal tax breaks and subsidies. These subsidies, unlike those for energy conservation and renewable energy, have not been eliminated. Thus, the marketplace is distorted in favor of fossil fuels and nuclear power.

Environmentalists are alarmed that an increasing share of the Department of Energy's annual budget is being used to develop nuclear weapons instead of new energy alternatives. Between 1981 and 1990, the share of the Department's budget used for making nuclear weapons and developing new ones increased from 38% to 65%. Thus, almost two-thirds of the DOE budget is actually an addition to the Defense Department's budget. Critics call for these activities to be shifted from the Department of Energy to the Department of Defense.

Most of the remaining 35% of the DOE budget is used for research and development of energy resources. Only 14% of the 1990 DOE budget was for research and development to improve energy efficiency and to develop perpetual and renewable sources of energy—amounting to an 82% cut in federal funding for these resources since 1980. The remaining 86% of the 1990 R & D budget was mostly for continued development of fossil fuels and nuclear fission and fusion. Yet, according to the Department of Energy, reserves and potential supplies of perpetual and renewable energy resources make up 92% of the total energy resources potentially available to the United States and could meet up to 80% of the country's projected energy needs by 2010 (Figure 17-8, p. 386).

Keeping Energy Prices Artificially High: The Western European Strategy Governments keep the price of an energy resource artificially high by withdrawing existing tax breaks and other subsidies or by adding taxes on its use. This encourages improvements in energy efficiency, reduces dependence on imported energy, and decreases use of an energy resource (like oil) whose future supply will be limited.

However, increasing taxes on energy use contributes to inflation and dampens economic growth. It also puts a heavy economic burden on the poor unless some of the energy tax revenues are used to help low-income families offset increased energy prices and to stimulate labor-intensive forms of economic growth such as improving energy efficiency. High

gasoline and oil import taxes have been imposed by many European governments. That is one reason why those countries use about half as much energy per person and have greater energy efficiency than the United States.

One popular myth is that higher energy prices would wipe out jobs. Actually, low energy prices increase unemployment because farmers and industries find it cheaper to substitute machines run on cheap energy for human labor. On the other hand, raising energy prices stimulates employment because building solar collectors, adding insulation, and carrying out most other forms of improving energy efficiency are labor-intensive activities.

Why the United States Has No Comprehensive Long-Term Energy Strategy After the 1973 oil embargo, Congress was prodded to pass a number of laws (see Appendix 2) to deal with the country's energy problems. Most energy experts agree, however, that these laws do not represent a comprehensive energy strategy. Indeed, analysis of the U.S. political system reveals why the United States has not been able and will probably never be able to develop a coherent energy policy.

One reason is the complexity of energy issues as revealed in this and the preceding chapter. But the major problem is that the American political process produces laws—not policies—and is not designed to deal with long-term problems (see Case Study on p. 162). Each law reflects political pressures of the moment and a maze of compromises between competing groups representing industry, conservationists, and consumers. Once a law is passed, it is difficult to repeal or modify drastically until its long-term consequences reach crisis proportions. This means that energy policy in the United States will have to be developed from the bottom up by individuals and communities taking energy matters into their own hands (see Spotlight on p. 404 and Individuals Matter inside the back cover).

A Sustainable Energy Future for the United States Citizens will have to exert intense pressure on elected officials to develop a national energy policy based on improving energy efficiency and making the transition to a mix of perpetual and sustainable energy resources. Major components of such a policy are:

■ Phasing out government subsidies for fossil fuels and nuclear energy and phasing in such subsidies for improvements in energy efficiency and greatly increased use of perpetual and renewable energy resources during the 1990s. Improvements in energy efficiency would serve as the bridge to a renewable-energy economy over the next two to three decades.

■ Adding taxes on gasoline and other fossil fuels that reflect their true costs to society, with the tax revenues used to improve energy efficiency, encourage use of perpetual and renewable energy resources, and provide energy assistance to poor and lower-middle-class Americans. Giving tax credits or government rebates for purchase of energy-efficient vehicles and adding high taxes on gas guzzlers might also help.

■ Requiring all federal and state facilities to meet the highest feasible standards for energy efficiency.

■ Buying renewable-energy systems for government facilities.

■ Requiring that all energy systems supported by government funds be based on least-cost analysis. Such analysis would be based on lifetime costs of each system and would include estimates of all major external costs. Government subsidies would be subtracted from estimated costs so that energy resources can be compared on an equal economic basis.

■ Modifying regulations so that electric utilities are required to produce electricity on a least-cost basis, can earn money for their shareholders by reducing electricity demand, and are allowed rate increases based primarily on improvements in energy efficiency. The goal of utility companies would then be to maximize production of what Amory Lovins calls energy- and money-saving "negawatts" instead of megawatts.

Energy experts estimate that implementing these policies now could save oil and money, slow projected global warming, and sharply reduce air and water pollution. This sustainable energy path would also double the percentage of energy obtained from perpetual and renewable energy resources in the United States to 15% by the year 2000 and to as high as 50% by 2020. This path to a sustainable energy future will happen only if individuals change their own energy lifestyles and elect, or keep in office, officials who pledge to support these policies.

Nuclear fission energy is safe only if a number of critical devices work as they should, if a number of people in key positions follow all their instructions, if there is no sabotage, no hijacking of the transport, if no reactor fuel processing plant or repository anywhere in the world is situated in a region of riots or guerrilla activity, and no revolution or war—even a "conventional one"—takes place in these regions. No acts of God can be permitted.

HANNES ALFVEN, NOBEL LAUREATE IN PHYSICS

DISCUSSION TOPICS

1. Explain why you agree or disagree with the ideas that the United States can get
   a. all of the oil it needs by extracting and processing heavy oil left in known oil wells
   b. all of the oil it needs by extracting and processing heavy oil from oil shale deposits
   c. all of the oil it needs by extracting heavy oil from tar sands
   d. all of the natural gas it needs from unconventional sources

2. Coal-fired power plants in the United States cause an estimated 5,000 deaths a year, mostly from atmospheric emissions of sulfur oxides, nitrogen oxides, and particulate matter. These emissions also damage many buildings and some forests and aquatic systems.
   a. Should air pollution emission standards for *all* new and existing coal-burning plants be tightened significantly? Explain.
   b. Do you favor a U.S. energy strategy based on greatly increased use of coal-burning plants to produce electricity? Explain. What are the alternatives?

3. Explain why you agree or disagree with each of the following proposals made by the nuclear power industry and currently supported by the Bush administration:
   a. The licensing time of new nuclear power plants in the United States should be halved (from an average of 12 years) so they can be built at less cost and compete more effectively with coal and other energy alternatives.
   b. A large number of new, better-designed nuclear fission power plants should be built in the United States to reduce dependence on imported oil and slow down projected global warming.
   c. Large federal subsidies (already totalling $1 trillion) should continue to be given to the commercial nuclear power industry so it does not have to compete in the open marketplace with other energy alternatives receiving no, or smaller, federal subsidies.
   d. A major program for developing the nuclear breeder fission reactor should be developed and funded largely by the federal government to conserve uranium resources and keep the United States from being dependent on other countries for uranium supplies.

4. Explain why you agree or disagree with the following propositions suggested by various energy analysts:
   a. Federal subsidies for all energy alternatives should be eliminated so that all energy choices can compete in a true free-enterprise market system.
   b. All government tax breaks and other subsidies for conventional fuels (oil, natural gas, coal), synthetic natural gas and oil, and nuclear power

Alvin M. Weinberg

Alvin M. Weinberg was a member of the group of scientists that developed the first experimental fission reactors at the University of Chicago in 1941. Since then he has been a leading figure in the development of commercial nuclear power. From 1948 to 1973, he served as director of the Oak Ridge National Laboratory. In 1974 he was director of the Office of Energy Research and Development in the Federal Energy Administration (now the Department of Energy). From 1975 to 1985, he was director of the Institute for Energy Analysis of the Oak Ridge Associated Universities, where he is now a Distinguished Fellow. He has written numerous articles and books on nuclear energy (see Further Readings) and has received many awards for his contributions to the development of nuclear energy.

There are two basically different views of the world's future. The one most popular in recent years holds that the earth's resources are limited. According to this view, nothing except drastic reduction in population, affluence, and certain types of technology can prevent severe environmental degradation.

The other view holds that as scarce materials are exhausted there will always be new, more expensive ones to take their place. According to this view, Spaceship Earth has practically infinite supplies of resources, but it will cost more and more to stay where we are as we use up those resources that are readily available.

This latter view seems to me to be the more reasonable, especially since all of our past experience has shown that as one resource becomes scarce, another takes its place. We do not use whale oil for lighting anymore, yet we have better lighting than our ancestors who burned this oil in lamps.

In the long run, humankind will have to depend on the most abundant and almost infinitely abundant elements in the earth's crust: iron, sodium, carbon, nitrogen, aluminum, oxygen, silicon, and a few others. Glass, cement, and plastics will perform many more functions than they do now. Our average standard of living will be diminished, but probably no more than by a factor of two.

Thus, in contrast to what seems to be the prevailing mood, I retain a certain basic optimism about the future. My optimism, however, is predicated on certain assumptions.

1. Technology can indeed deal with most of the effluents of this future society. Here I think I am on firm ground, for, on the whole, where technology has been given the task and been given the necessary time and funding, it has come through with very important improvements such as reducing air pollution emissions by cars. On the other hand, carbon dioxide, which is the major greenhouse gas, cannot be controlled; this may place a limit on the rate at which we burn fossil fuels.

should be removed and replaced with subsidies and tax breaks for improving energy efficiency and developing solar, wind, geothermal, and biomass energy alternatives.

c. Development of solar and wind energy should be left to private enterprise without help from the federal government, but nuclear energy and fossil fuels should continue to receive federal subsidies (present U.S. policy).

d. To solve present and future U.S. energy problems, all we need to do is find and develop more domestic supplies of conventional and unconventional oil, natural gas, and coal and increase our dependence on nuclear power (present U.S. policy).

e. The United States should not worry about heavy dependence on foreign oil imports because they improve international relations and help prevent depletion of domestic supplies (the "don't drain America first" approach).

f. A heavy federal tax should be placed on gasoline and imported oil used in the United States.

g. Between 2000 and 2020, the United States should phase out all nuclear power plants.

FURTHER READINGS

See also the readings for Chapter 3.

Burnett, W. M., and S. D. Ban. 1989. "Changing Prospects for Natural Gas in the United States." *Science*, vol. 244, 305–310.

Clark, Wilson, and Jake Page. 1983. *Energy, Vulnerability, and War*. New York: W. W. Norton.

Cohen, Bernard L. 1983. *Before It's Too Late: A Scientist's Case for Nuclear Power*. New York: Plenum Press.

2. Phosphorus, though essentially infinite in supply in the earth's crust at various locations, has no substitute. Will we be able to so revolutionize agriculture that we can eventually use the "infinite" supply of phosphorus at acceptable cost? This technological and economic question is presently unresolved, although I cannot believe it to be unresolvable.

3. All of this presupposes that we have at our disposal an inexhaustible, relatively cheap source of energy. As I and others now see the technological possibilities, there is only one energy resource we can count on—and this is *nuclear fission,* based on *breeder reactors* to extend the world's supply of fissionable uranium far into the future. This is not to say that nuclear fusion, geothermal energy, or solar energy will never be economically available. We simply do not know now that any of these will ever be available in sufficient quantity and at affordable prices. We know, however, that conventional nuclear fission and breeder reactors are already technologically feasible, and that standardized, improved, and inherently safer reactor designs already being tested or on the drawing boards should bring costs down in the future.

In opting for nuclear fission breeders—and we hardly have a choice in the matter—we assume a moral and technological burden of serious proportion. A properly operating nuclear reactor and its subsystems are environmentally a very benign energy source. In particular, a reactor emits no carbon dioxide.

The issue hangs around the words *properly operating.* Can we ensure that henceforth we shall be able to maintain the degree of intellectual responsibility, social commitment, and stability necessary to maintain this energy form so as not to cause serious harm? This is basically a moral and social question, though it does have strong technological components.

It is a Faustian bargain that we strike: In return for this essentially inexhaustible energy source, which we must have if we are to maintain ourselves at anything like our present numbers and our present state of affluence, we must commit ourselves and generations to come—essentially forever—to exercising the vigilance and discipline necessary to keep our nuclear fires well behaved.

As a nuclear technologist who has devoted his career to this quest for an essentially infinite energy source, I believe the bargain is a good one, and it may even be an inevitable one, especially if our concerns about the greenhouse effects are justified. It is essential that the full dimension and implication of this Faustian bargain be recognized, especially by the young people who will have to live with the choices that are being made on this vital issue.

Guest Essay Discussion

1. The author bases his optimism on three assumptions. Do you believe that these assumptions are reasonable? Explain. Are there any other assumptions that should be added?

2. Do you agree that we should accept the Faustian bargain of conventional and breeder nuclear fission? Explain.

3. Do you agree with the author that "we hardly have any choice" in opting for nuclear fission breeder reactors? Explain.

Flavin, Christopher. 1985. *World Oil: Coping with the Dangers of Success.* Washington, D.C.: Worldwatch Institute.

Flavin, Christopher. 1987. *Reassessing Nuclear Power: The Fallout from Chernobyl.* Washington, D.C.: Worldwatch Institute.

Flavin, Christopher. 1988. "The Case Against Reviving Nuclear Power." *World Watch,* July/Aug., 27–35.

Ford, Daniel F. 1986. *Meltdown.* New York: Simon & Schuster.

Fund for Renewable Energy and the Environment. 1987. *The Oil Rollercoaster.* Washington, D.C.: Fund for Renewable Energy and the Environment.

Glasner, David. 1986. *Politics, Prices, and Petroleum: The Political Economy of Energy.* San Francisco: Pacific Institute for Public Policy Analysis.

Golay, Michael W., and Neil E. Todreas. 1990. "Advanced Light-Water Reactors." *Scientific American,* April, 82–89.

Heede, H. Richard, et al. 1985. *The Hidden Costs of Energy.* Washington, D.C.: Center for Renewable Resources.

Hirsch, Robert L. 1987. "Impending United States Energy Crisis." *Science,* vol. 235, 1467–1473.

Hughes, Barry B., et al. 1985. *Energy in the Global Arena: Actors, Values, Policies, and Futures.* Durham, N.C.: Duke University Press.

Humphrey, Craig R., and Frederick R. Buttel. 1982. *Environment, Energy, and Society.* Belmont, Calif.: Wadsworth.

League of Women Voters Education Fund. 1985. *The Nuclear Waste Primer.* Washington D.C.: League of Women Voters.

Lidsky, Lawrence M. 1983. "The Trouble with Fusion." *Technology Review,* Oct., 32–44.

Lovins, Amory B. 1986. "The Origins of the Nuclear Power Fiasco." *Energy Policy Studies,* vol. 3, 7–34.

Lovins, Amory B., and L. Hunter Lovins. 1982. *Brittle Power: Energy Strategy for National Security*. Andover, Mass.: Brick House.

May, John. 1990. *The Greenpeace Book of the Nuclear Age*. New York: Pantheon.

Morone, Joseph G., and Edward J. Woodhouse. 1989. *The Demise of Nuclear Energy?* New Haven, Conn.: Yale University Press.

Mould, Richard F. 1988. *Chernobyl: The Real Story*. New York: Pergamon.

Murray, Raymond L. 1989. *Understanding Radioactive Waste*. 3rd ed. Columbus, Ohio: Battele Press.

National Academy of Sciences. 1990. *Energy: Production, Consumption, and Consequences*. Washington, D.C.: National Academy Press.

National Academy of Sciences. 1990. *Fuels to Drive Our Future*. Washington, D.C.: National Academy Press.

Office of Technology Assessment. 1984. *Managing the Nation's Commercial High-Level Radioactive Waste*. Washington, D.C.: Government Printing Office.

O'Hefferman, Patrick, Amory Lovins, and L. Hunter Lovins. 1984. *The First Nuclear World War*. New York: Morrow Books.

Oppenheimer, Ernest J. 1990. *Natural Gas, the Best Energy Choice*. New York: Pen & Podium.

Park, Chris C. 1989. *Chernobyl: The Long Shadow*. New York: Routledge.

Patterson, Walter C. 1984. *The Plutonium Business and the Spread of the Bomb*. San Francisco, Calif.: Sierra Club Books.

Pollock, Cynthia. 1986. *Decommissioning: Nuclear Power's Missing Link*. Washington, D.C.: Worldwatch Institute.

President's Commission on the Accident at Three Mile Island. 1979. *Report of the President's Commission on the Accident at Three Mile Island*. Washington, D.C.: Government Printing Office.

Rocky Mountain Institute. 1988. *An Energy Security Reader*. 2nd ed. Old Snowmass, Colo.: Rocky Mountain Institute.

Rosenbaum, Walter A. 1987. *Energy, Politics, and Public Policy*. 2nd ed. Washington, D.C.: Congressional Quarterly.

Schobert, Harold H. 1987. *Coal: The Energy Source of the Past and Future*. Washington, D.C.: American Chemical Society.

Sweet, William. 1989. "Chernobyl: What Really Happened." *Technology Review*, July, 43–52.

Taylor, John J. 1989. "Improved and Safer Nuclear Power." *Science*, vol. 244, 318–325.

U.S. Department of Energy. 1988. *An Analysis of Nuclear Power Operating Costs*. Washington, D.C.: Government Printing Office.

U.S. Office of Technology Assessment. 1984. *Nuclear Power in an Age of Uncertainty*. Washington, D.C.: Government Printing Office.

Watson, Robert K. 1988. *Fact Sheet On Oil and Conservation Resources*. New York: Natural Resources Defense Council.

Yergin, Daniel. 1988. "Energy Security in the 1990s." *Foreign Affairs*, Autumn.

CHAPTER NINETEEN

NONRENEWABLE MINERAL RESOURCES AND SOLID WASTE

General Questions and Issues

1. How are minerals formed and distributed?

2. How are mineral deposits found and extracted from the earth's crust?

3. What harmful environmental impacts occur from mining, processing, and using minerals?

4. How long will affordable supplies of key minerals last for the world and the United States?

5. How can we increase the supplies of key minerals?

6. How can we make supplies of key minerals last longer by reducing the production of solid waste?

Mineral resources are the building blocks on which modern society depends. Knowledge of their physical nature and origins and the web they weave between all aspects of human society and the physical earth can lay the foundations for a sustainable society.

ANN DORR

WHAT DO CARS, spoons, glasses, dishes, beverage cans, coins, electrical wiring, bricks, and sidewalks have in common? Few of us stop to think that these products and many others we use every day are made from nonrenewable minerals we have learned to extract from the earth's solid crust—the upper layer of the lithosphere (Figure 4-2, p. 60).

If LDCs are to become MDCs, the supplies of nonfuel minerals vital to industry and modern lifestyles will have to increase dramatically and we will have to greatly increase the recycling and reuse of these nonrenewable resources. We will also have to reduce the environmental impact from mining, processing, and using more of these resources.

19-1 GEOLOGIC PROCESSES AND MINERAL RESOURCES

The Earth's Internal Structure The earth is a nearly spherical planet with a diameter of about 12,800 kilometers (8,000 miles). By analyzing vibrational waves created by earthquakes, geologists have divided the earth's interior into three major spherical zones, or shells—crust, mantle, and core—according to differences in structure and composition (Figure 4-2, p. 60).

The outer zone of the earth is called the **crust.** This relatively thin shell of fairly rigid rock makes up only about 1% of the earth's volume. It consists of the *continental crust,* which underlies the continents (including the continental shelves extending into the oceans), and the *oceanic crust,* which underlies the ocean basins.

The earth's crust is composed of minerals and rocks and is the source of virtually all the nonrenewable resources we use, such as fossil fuels, metallic minerals, and nonmetallic minerals (Figure 1-7, p. 6). A **mineral** is a naturally occurring inorganic solid with a particular chemical makeup and a crystalline internal structure made up of an orderly, three-dimensional arrangement of atoms or ions. Some minerals consist of a single element, such as gold, silver, diamond (carbon), and sulfur. However, most of the more than 2,000 identified minerals occur as inorganic compounds of the ten elements that make up 99.3%, by weight, of the earth's crust (Figure 19-1). A **rock** is a combination or aggregate of minerals.

Over millions of years various geological, physical, and chemical processes have concentrated certain minerals in deposits or veins in the earth's upper crust. A deposit in which the concentration of a mineral is high enough to make its recovery profitable is called an **ore.** If the earth's mineral resources were distributed uniformly throughout the earth's crust,

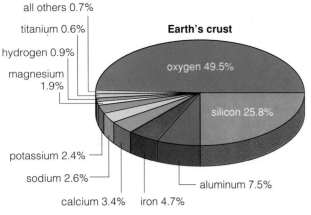

Figure 19-1 Percentage by weight of elements in the earth's crust.

their concentration would be so low that it would not be technically or economically feasible to extract them.

The earth's core is surrounded by a thick zone called the **mantle.** This largest zone of the earth's interior is rich in the elements silicon, oxygen, magnesium, and iron, which combine to form iron and magnesium silicates. Most of the mantle is solid rock, but under its rigid outermost part, there is a zone of very hot, partly melted rock that flows like soft plastic. This fluidlike portion of the mantle is called the *asthenosphere.* The **lithosphere** consists of the continental crust, oceanic crust, and the relatively rigid portion of the mantle lying above the softer asthenosphere.

The inner zone of the earth, found below the mantle beginning at a depth of about 2,900 kilometers (1,800 miles), is called the **core** and is composed mostly of iron and probably some nickel. The core has a solid inner part, surrounded by a fluidlike outer shell of molten material.

Internal Lithosphere Processes: Plate Tectonics

The lithosphere is continually being changed by internal and external processes taking place over millions of years. These processes concentrate the minerals we extract and use and also shape and alter the earth's crust to form mountains, canyons, plains, river valleys, ocean trenches and ridges, and other features of the earth's outer shell.

Internal lithosphere processes are driven by energy in the earth's core from heat released by the decay of radioactive elements. This heat is distributed within the mantle by convective currents caused by slow flows of heated rock upward and cool rock downward. These processes include volcanic eruptions, earthquakes, and the movement and fracture of rock under high temperatures and pressures.

The theory of geophysical processes that describes the formation and deformation of the lithosphere and how the continents move is called **plate tectonics.** According to this theory, the lithosphere is made up of about ten gigantic plates and a number of smaller plates that over millions of years move and rub and scrape against each other in response to convection currents in the underlying plastic portion (asthenosphere) of the mantle. In places where these convective currents reach the underside of the lithosphere, they push against and move its rigid plates at a typical speed of less than 20 centimeters (8 inches) a year.

The continents and the oceans sit on these plates. As these plates have moved over millions of years, continents have split and drifted thousands of kilometers across the earth's surface. For example, geologic evidence indicates that about 225 million years ago the continents we now call Europe, Asia, North America, and South America were combined as one continent. This gigantic continental mass, named *Pangaea,* broke up and slowly drifted apart as a result of movements of the earth's plates, and eventually the continents reached their present positions.

The movement and collision of the plates making up the lithosphere also account for most of the earth's earthquakes, volcanic eruptions, and landscape features such as volcanoes, mountain ranges, and deep oceanic trenches. These same geologic processes concentrate elements into deposits of mineral resources that are profitable to extract.

Three types of boundaries occur between the lithosphere's moving plates: divergent, convergent, and parallel. *Divergent plate boundaries* occur where molten rock material (magma) pushes up between two plates, cools, and solidifies to form new ocean floor that pushes the plates apart. Mountain chains, called *ridges,* are formed along the ocean floor where the two plates diverge.

Convergent plate boundaries form where two plates collide and one of the plates slips under the other into the mantle and melts. This leads to formation of deep trenches in the ocean floor. The area, called the *subduction zone,* where the underlying plate enters the mantle, melts, and is destroyed, is the site of most of the earth's volcanic eruptions and major earthquakes. Thus, new crust is formed at divergent plate boundaries and old crust is destroyed at convergent plate boundaries.

A *parallel boundary* is found where plates slide past one another. Along these zones, crust is neither created nor destroyed. However, this slipping of plates horizontally past one another can trigger earthquakes. For example, the San Andreas fault in California is a parallel boundary between two of the lithosphere's plates.

Some commercially important ore deposits are formed where magma intrudes into crustal rock or extrudes onto the earth's surface as lava at all three types of plate boundaries. For example, along divergent plate boundaries, magma moving upward from the mantle interacts with seawater that enters fractures in the seafloor, known as *hydrothermal* (hot water) *vents*. Through a complex series of chemical reactions, sulfide ores, made up of combinations of various metals and sulfur, form around these vents. Deposits of metallic ores can also be found at convergent plate boundaries. Some metallic elements are released when the descending plate partially melts. This metal-rich fluid is then concentrated in the magma, which rises into the crust, cools, and solidifies as metal-rich rock.

External Lithosphere Processes: Weathering and Erosion Internal lithosphere processes tend to build landscape, while external processes tend to break down and move parts of the earth's crust. These external processes include physical and chemical weathering and erosion of rock.

Physical weathering occurs when bedrock (solid rock) exposed to the atmosphere is fragmented into smaller pieces. These processes include the expansion and contraction of rock exposed to alternating hot and cold temperatures, the freezing and expansion of water within rock cracks, and the wedging apart of rocks by plant roots. *Chemical weathering* occurs mostly when rocks are decomposed by reaction with oxygen, carbon dioxide, and moisture in the atmosphere.

Physical and chemical weathering of rocks produces rock fragments called *sediments*. Some sediments are also composed of the shells and skeletons of dead organisms. Various types of *soil* are the ultimate products of weathering plus interactions with various forms of life (Figure 12-4, p. 270). Wind, flowing water, and moving ice (glaciers) remove, transport, and redeposit sediments over parts of the earth's surface by the process we call *erosion*. This geologic process, which many of our activities have accelerated (Section 12-2), can ultimately expose fresh bedrock to weathering.

Weathering and erosion also concentrate some resources in ores. For example, in some places, gold and diamonds are weathered out of their host rock and become mixed with gravel and streambeds to form sedimentary deposits called *placers*. In other cases, minerals dissolved out of rock by weathering precipitate from groundwater or seawater to form ore deposits.

The Rock Cycle The **rock cycle**, the largest and slowest of the earth's cycles, consists of a group of geologic, physical, and chemical processes that form and modify rocks and soil in the earth's crust and mantle over millions of years (Figure 19-2). This geological cycle is powered by energy from the sun

(Figure 4-5, p. 62) and the internal heat of the earth. It is also affected by water moving through the hydrologic cycle (Figure 4-29, p. 79), the other biogeochemical cycles (Section 4-4), wind created by the earth's climatic processes (Section 10-1), and the formation and deformation of the earth's crustal plates. The rocks involved in this global cycle can be classified into three major groups—igneous, sedimentary, and metamorphic—according to the way in which they are formed.

Igneous rocks, such as granite and basalt, are formed when magma (molten rock) wells up from the earth's upper mantle, cools, and solidifies. Some coarse-grained igneous rocks, such as granite, are formed below the earth's surface. Other igneous rocks are formed when lava spews forth through volcanoes or cracks in the earth's surface and solidifies to form fine-grained or glassy rocks. Igneous rocks are the most abundant type of rock and are the major source of the nonfuel minerals we use.

Exposed igneous rock is gradually disintegrated and decomposed physically and chemically by *weathering*. The resulting sediments are transported by wind, water, landslide, and ice to lower elevations. Gradually, accumulated sediments of weathered rock and the shells and skeletons of dead animals are compacted to form **sedimentary rocks.** Examples are sandstone (from compacted sand), shale, and some limestones. Highly visible sedimentary rocks form a fairly thin film over nearly three-fourths of the earth's surface.

Preexisting rocks within the lithosphere are transformed into **metamorphic rocks** by changes in temperature, pressure, and the chemical action of fluids. These changes in the chemical forms of sedimentary and igneous rocks occur over millions of years as deposits of these rocks are buried deeper and deeper in the earth's crust and are exposed to intense heat, great pressures, and in some cases chemically active fluids. Eventually, temperatures may become so high that metamorphic rocks melt to form magma. Some of the magma may then intrude into the earth's crust, cool, and solidify to form igneous rock to start the rock cycle again (Figure 19-2).

The rock cycle concentrates and disperses the earth's minerals. We know how to find and mine more than 100 minerals concentrated by the rock cycle. We convert these minerals into many everyday items we use and then discard, reuse, or recycle. When we return these materials to the rock cycle in diluted form, they are further dispersed by earth processes. Once this dispersion process has taken place, the resource may not be concentrated again for millions of years, which is why these minerals are classified on a human time scale as nonrenewable resources.

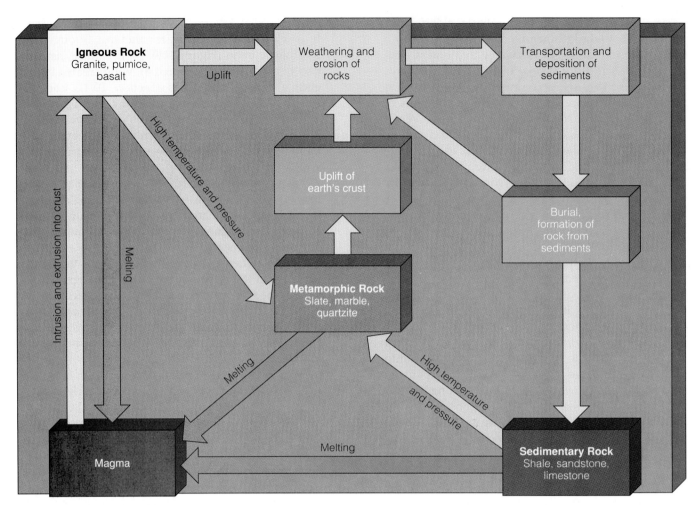

Figure 19-2 The rock cycle.

The Dynamic Earth and Time Frames The only constant thing about the earth is that change is taking place at a variety of rates. Weather changes in a matter of minutes or hours, and climate changes over decades to hundreds of years. Populations can change their size and age distribution in response to changes in environmental conditions within hours to decades. Species can respond to environmental changes and evolve into new species in thousands to millions of years. Most geological changes, except ones such as earthquakes or volcanic eruptions, take place on a time scale of millions to billions of years.

These short- and long-term changes are all interconnected in complex ways we now only partially understand and will never fully understand. A basic problem is that our exponentially growing numbers and our power to bring about major environmental changes are now measured in decades. On our short time frame we consider our species important. But on the earth's much longer time frame, we are just an unnecessary irritant that could easily wipe itself out and in the process destroy millions of other species. To survive we must now mesh our time frames of change with those that sustain the earth.

19-2 LOCATING AND EXTRACTING MINERAL RESOURCES

Finding and Mining Crustal Resource Deposits Mining companies use several methods to find promising deposits. Geological information about plate tectonics and mineral formation helps mining companies find areas for closer study. Photos taken from airplanes or images relayed by satellites sometimes reveal geological features such as rock formations, often associated with deposits of certain minerals. Other instruments on aircraft and satellites can detect deposits of minerals by their effects on the earth's magnetic or gravitational fields.

Deposits of nonfuel minerals and rocks and of coal near the earth's surface are removed by **surface mining.** Mechanized equipment strips away the overlying layer of soil and rock, known as **overburden,** and vegetation. Surface mining is used to extract about 90%, by weight, of the mineral and rock resources and more than 60%, by weight, of the coal in the United States (Section 18-2).

The type of surface mining used depends on the type of crustal resource and the local topography. In **open-pit mining,** machines dig holes and remove ore

Figure 19-3 This open pit copper mine in Bingham, Utah, is the largest human-made hole in the world. It is 4.0 kilometers (2.5 miles) in diameter and 0.8 kilometer (0.5 mile) deep. This mine produces 227,000 metric tons (250,000 tons) of copper a year along with fairly large amounts of gold, silver, and molybdenum.

deposits, such as iron and copper (Figure 19-3). This method is also used to remove sand, gravel, and building stone, such as limestone, sandstone, slate, granite, and marble.

Strip mining is surface mining in which bulldozers, power shovels, or stripping wheels remove large chunks of the earth's surface in strips. It is used mostly for removing coal (Figure 18-11, p. 417, and Figure 18-13, p. 417) and some phosphate rock (Figure 4-28, p. 79). Another form of surface mining is **dredging,** in which chain buckets and draglines scrape up sand, gravel containing placer deposits, and other surface deposits covered with water.

Some crustal resources lie so deep that surface mining is impractical. These mining deposits of metal ores and coal are removed by **subsurface mining.** In most cases, miners dig a deep vertical shaft or horizontal slits and blast tunnels and rooms. Then the resource is extracted and hauled to the surface.

Often the desired mineral in an ore makes up only a small percentage, by weight, of the rock mass that is removed by mining. This means that massive amounts of rock must be removed from the ground and processed to separate the desired mineral from the host rock. Most metals in ores are combined chemically with other elements, such as oxygen (oxide ores) or sulfur (sulfide ores). To get the desired metal, such ores must be broken down chemically through smelting (see photo on p. 184) or refining processes. These processes require large amounts of energy and produce solid, liquid, and gaseous waste products (especially sulfur dioxide and particulate matter) that must be disposed of.

19-3 ENVIRONMENTAL IMPACT OF MINING, PROCESSING, AND USING CRUSTAL RESOURCES

Overall Environmental Impact and Economics The mining, processing, and use of any nonfuel crustal resource or fuel mineral such as coal (Section 18-2) can cause land disturbance, erosion, air pollution, and water pollution (Figure 19-4). The degree to which these harmful effects are reduced depends mostly on whether their costs are included in the market prices we pay for items (Section 7-2).

Mining companies and manufacturers also have little incentive to find ways to reduce resource waste and pollution as long as they can pass many of the harmful environmental costs of their production on to society. Internalizing these external costs is a major way to reduce pollution, environmental degradation, and resource waste (Section 7-2).

Mining Impacts Mining involves removing material from the earth's crust and dumping unwanted rock and other waste materials somewhere else—usually near the mining site. Although mining uses only a small amount of the earth's surface, it has a severe local (and sometimes regional) environmental impact on the land, air, and water. Bare land, created when vegetation and topsoil are removed by surface mining, and the waste rock and other materials from mining are eroded by wind and water. Also, harmful materials run off into nearby streams, and some toxic compounds in mining wastes percolate into groundwater (Figure 18-15, p. 419). The air can be contaminated with dust and toxic substances.

Surface mining disrupts the landscape more than subsurface mining. One way to reduce the severe environmental impact of surface mining is to pass, and strictly enforce, laws that require mining companies to restore land after minerals have been removed. But when minerals are removed by surface mining in arid and semiarid regions full restoration is rarely possible. There is too little rainfall for growth of enough vegetation to prevent continuing soil erosion by rain when it does fall.

For each unit of mineral produced, subsurface mining disturbs less than one-tenth as much land as surface mining. Usually subsurface mining also produces less waste material. However, subsurface mining is more dangerous and expensive than surface mining. Roofs and walls of underground mines collapse, trapping and killing miners. Explosions of dust and natural gas kill and injure them. Prolonged inhalation of mining dust causes lung diseases. Underground fires sometimes cannot be put out. Land above underground mines caves in or subsides, causing roads to buckle, houses to crack, railroad tracks to bend, sewer lines to crack, gas mains to break, and groundwater systems to be disrupted.

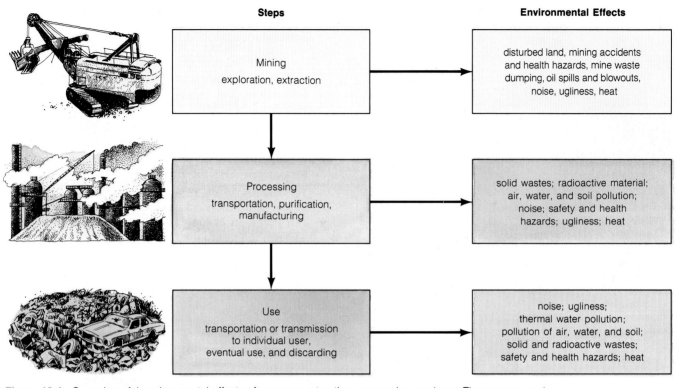

Steps		Environmental Effects
Mining exploration, extraction	→	disturbed land, mining accidents and health hazards, mine waste dumping, oil spills and blowouts, noise, ugliness, heat
Processing transportation, purification, manufacturing	→	solid wastes; radioactive material; air, water, and soil pollution; noise; safety and health hazards; ugliness; heat
Use transportation or transmission to individual user, eventual use, and discarding	→	noise; ugliness; thermal water pollution; pollution of air, water, and soil; solid and radioactive wastes; safety and health hazards; heat

Figure 19-4 Some harmful environmental effects of resource extraction, processing, and use. The energy used to carry out each step causes further pollution and environmental degradation. Most of the harm is caused by not having mining, processing, and manufacturing companies pay the full costs of the pollution and environmental degradation they cause. Instead, many of these "external" costs are passed on to society in the form of poorer health, increased health and insurance costs, and increased taxes to deal with pollution and environmental degradation. Requiring these external costs to be internalized and included in the market cost of raw materials and manufactured goods would eliminate most of these harmful effects or reduce them to more acceptable levels.

Another environmental problem of mining, especially subsurface mining, is *acid mine drainage,* which occurs when aerobic bacteria convert sulfide compounds in minerals found in mine wastes to sulfuric acid. Rainwater seeping through the mine or mine wastes transfers the acid to nearby rivers and streams (Figure 18-15, p. 419), destroying aquatic life and contaminating water supplies. Other harmful materials running off or leached from underground mines or aboveground mining wastes are iron oxides, radioactive uranium compounds, and compounds of toxic metals such as lead, arsenic, or cadmium.

Processing Impacts Processing extracted mineral deposits to remove impurities produces huge quantities of waste rock and other waste materials. Usually these wastes are piled on the ground or dumped into ponds near mining and processing sites. Unless covered and stabilized, particles of dust and toxic metals in these wastes blow into the air, and water leaches toxic and radioactive substances into nearby surface water or groundwater (Figure 18-15, p. 419).

As lower-grade ores are mined and processed, the quantity of mining waste rises sharply. Laws have

forced mining companies to reduce contamination from mining waste in the United States, but enforcement is weak. In many countries, especially LDCs, such laws don't exist.

Without effective pollution control equipment, mineral-smelting plants emit massive quantities of air pollutants that damage vegetation and soils in the surrounding area. Pollutants include sulfur dioxide, soot, and tiny particles of arsenic, cadmium, lead, and other toxic elements and compounds found in many ores. Decades of uncontrolled sulfur dioxide emissions from copper-smelting operations near Copperhill and Ducktown, Tennessee, killed all vegetation over a large area around the smelter (Figure 9-14, p. 199).

Smelting plants also cause water pollution and produce liquid and solid hazardous wastes, which must be disposed of safely or converted into less harmful substances. Workers in some smelting industries have an increased risk of cancer.

Solid Waste The United States leads the world by far in total and per capita production of nonhazardous and hazardous solid, liquid, and gaseous waste. Currently the United States produces an average of

40 metric tons (44 tons) per American of nonhazardous **solid waste**—any unwanted or discarded material that is not a liquid or a gas. With less than 5% of the world's population, the United States produces 33% of the world's trash.

About 98.5% of this nonhazardous solid waste comes from mining, oil and natural gas production, and industrial activities. Most mining waste is left piled near mine sites and can pollute the air, surface water, and groundwater. Most industrial solid waste, such as scrap metal, plastics, paper, fly ash from electrical power plants, and sludge from industrial waste treatment plants, is disposed of at the plant site where it is produced.

Municipal solid waste from homes and businesses in or near urban areas makes up the remaining 1.5% of the nonhazardous solid waste produced in the United States. The 145 metric tons (160 million tons) of municipal solid waste produced in the United States in 1990 would fill a bumper-to-bumper convoy of garbage trucks that would encircle the earth almost five times.

U.S. consumers throw away enough aluminum to rebuild the country's entire commercial airline fleet every three months. Enough glass bottles are thrown away to fill the 412-meter- (1,350-foot-) high towers of the New York World Trade Center every two weeks. If laid flat, the tires thrown away each year in the United States would circle the earth almost three times.

The average amount of municipal solid waste thrown away per person in the United States is two to five times that in most other MDCs and is growing. The United States also recycles far less than most other MDCs.

About 59% of the total weight the typical American throws away as garbage and rubbish is paper, paperboard, and yard waste from potentially renewable resources (Figure 19-5). Most of the rest are products made from glass, plastic, aluminum, iron, steel, tin, and other nonrenewable mineral resources. Only about 10% of these potentially usable resources is recycled. The rest is hauled away and dumped or burned at a cost of almost $5 billion a year.

By volume, the largest type of solid waste in landfills is paper, which makes up 55% of the volume. Plastics make up 12% by volume. Although the weight of plastics in the waste stream is growing, the volume of plastics in landfills increased only from 11% to 12% between 1970 and 1989. The reason is that plastics are crushable. By volume, the fastest growing form of solid waste in landfills is paper, with its volume increasing from 36% to 56% between 1970 and 1989.

Paper is biodegradable, but it and other biodegradable wastes break down very slowly in today's compacted and oxygen-deficient landfills. Newspapers dug up from some landfills are still readable

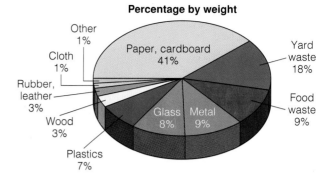

Figure 19-5 Composition by weight of urban solid waste thrown away in a typical day by each American in 1988. Because plastic containers are replacing many glass containers, plastics are expected to make up 10% of urban solid waste by the year 2000. However, by volume the largest and fastest growing source of solid waste is paper because plastics are more crushable than paper. (Data from Office of Technology Assessment and Environmental Protection Agency)

after 30 years. After 10 years, hot dogs, carrots, and chickens that have been dug up have not been degraded. New, expensive biodegradable plastics will take decades to degrade in landfills, are largely a waste of money, and discourage reduction and recycling of plastics.

Litter that people throw away is also a source of solid waste. One example is helium-filled balloons that are released into the atmosphere at sporting events and celebrations. When the helium escapes or the balloons burst, the balloons fall back to the earth as long-lasting litter. Fish, turtles, seals, whales, and other aquatic animals die when they ingest balloons that fall into oceans and lakes. This practice also suggests to children and adults that it is acceptable to litter, waste helium (a scarce resource) and the energy needed to separate helium from air, and kill wildlife (which would happen even if we switched to balloons that biodegraded within six weeks).

19-4 WILL THERE BE ENOUGH MINERAL RESOURCES?

 How Much Is There? The term **total resources** refers to the total amount of a particular material that exists on earth. It's hard to make useful estimates of total resources because the entire world has not been explored for each resource.

The U.S. Geological Survey estimates actual and potential supplies of a mineral resource by dividing the estimated total resources into two broad categories: identified and undiscovered (Figure 19-6). **Identified resources** are deposits of a particular mineral-bearing material of which the location, quantity, and quality are known or have been estimated from geological evidence and measurements.

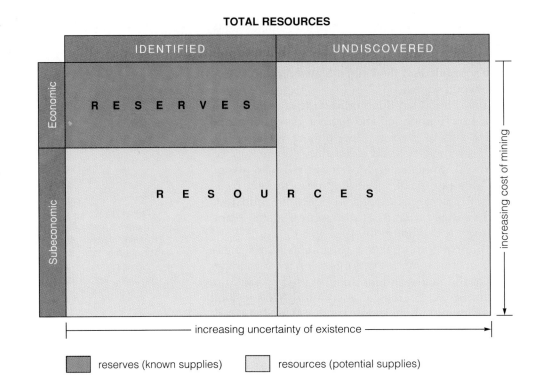

Figure 19-6 General classification of mineral resources by the U.S. Geological Survey.

IDENTIFIED

UNDISCOVERED

Economic

Subeconomic

R E S E R V E S

R E S O U R C E S

increasing cost of mining

increasing uncertainty of existence

reserves (known supplies)

resources (potential supplies)

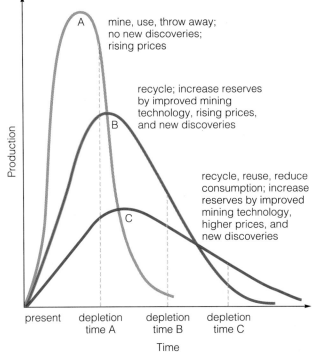

A mine, use, throw away; no new discoveries; rising prices

B recycle; increase reserves by improved mining technology, rising prices, and new discoveries

C recycle, reuse, reduce consumption; increase reserves by improved mining technology, higher prices, and new discoveries

Production

present   depletion time A   depletion time B   depletion time C

Time

Figure 19-7 Depletion curves for a nonrenewable resource, such as aluminum or copper, using three sets of assumptions. Dashed vertical lines show when 80% depletion occurs.

Undiscovered resources are potential supplies of a particular mineral. They are believed to exist on the basis of geologic knowledge and theory, though specific locations, quality, and amounts are unknown.

These two categories are subdivided into reserves and resources. **Reserves,** or **economic resources,** are

identified resources from which a usable mineral can be extracted profitably at present prices with current mining technology. **Resources** are all identified deposits, including those that can't be recovered profitably with present prices and technology, and all undiscovered resources.

Reserves are like the money you now have available to spend. Resources are the total income you expect to have during your lifetime. Your cash reserves are certain, but the resources you expect to have may or may not become available.

Most published estimates of the available supply of a particular nonrenewable mineral refer to reserves, not resources or total resources. Supplies of a mineral are usually higher than estimates of reserves. Some deposits classified as resources will be converted to reserves.

How Fast Are Supplies Being Depleted? Worldwide demand for mineral commodities is increasing exponentially due to increasing population and rising average per capita consumption (Figure 1-10, p. 14). The future supply of a nonrenewable mineral resource depends on two factors: its actual or potential supply and how rapidly the supply is being depleted.

We never completely run out of any mineral. Instead of becoming physically depleted, a mineral becomes *economically depleted* when finding, extracting, transporting, and processing the remaining lower-quality deposits cost more than they are currently worth. When this economic limit is reached, we have four choices: recycle or reuse what has already been

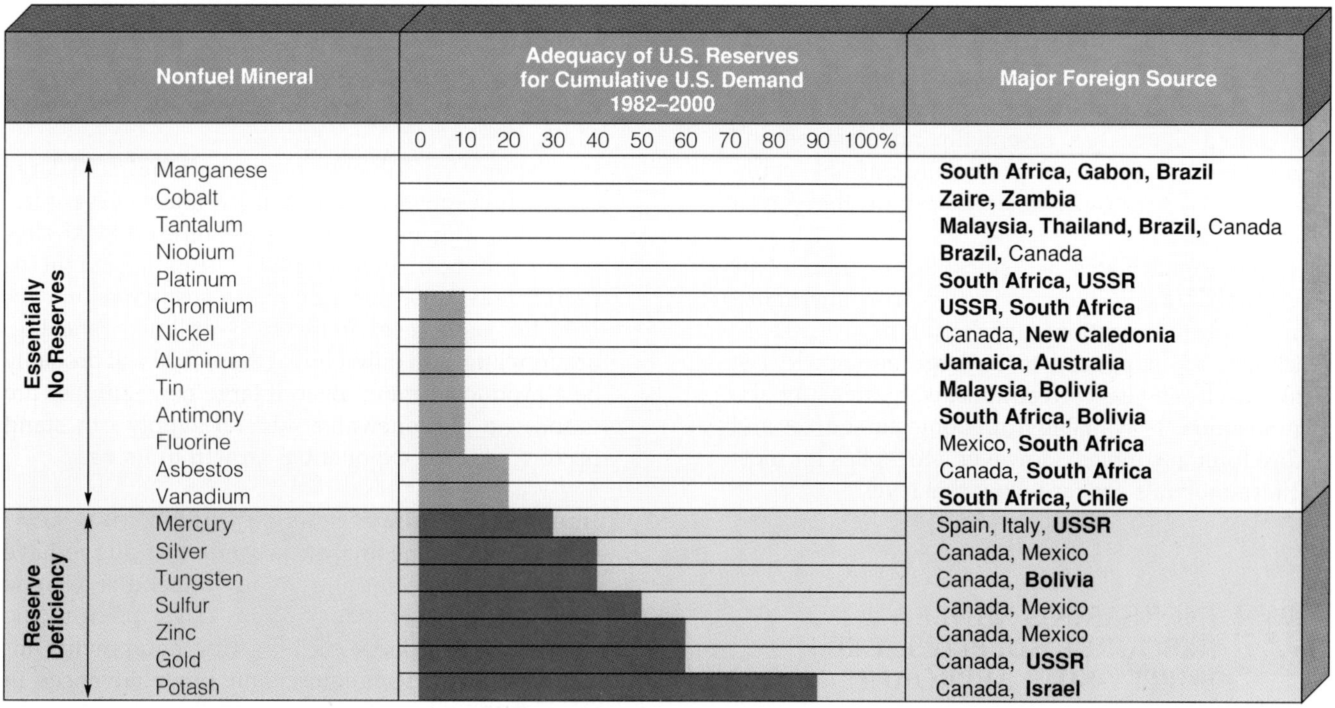

Nonfuel Mineral	Adequacy of U.S. Reserves for Cumulative U.S. Demand 1982–2000	Major Foreign Source
	0 10 20 30 40 50 60 70 80 90 100%	
Essentially No Reserves		
Manganese		South Africa, Gabon, Brazil
Cobalt		Zaire, Zambia
Tantalum		Malaysia, Thailand, Brazil, Canada
Niobium		Brazil, Canada
Platinum		South Africa, USSR
Chromium		USSR, South Africa
Nickel		Canada, New Caledonia
Aluminum		Jamaica, Australia
Tin		Malaysia, Bolivia
Antimony		South Africa, Bolivia
Fluorine		Mexico, South Africa
Asbestos		Canada, South Africa
Vanadium		South Africa, Chile
Reserve Deficiency		
Mercury		Spain, Italy, USSR
Silver		Canada, Mexico
Tungsten		Canada, Bolivia
Sulfur		Canada, Mexico
Zinc		Canada, Mexico
Gold		Canada, USSR
Potash		Canada, Israel

Figure 19-8 Estimated deficiencies of selected nonfuel mineral elements in the United States, 1982–2000, and major foreign sources of these minerals. Foreign sources subject to potential supply interruption by political, economic, or military disruption are shown in boldface. (Data from U.S. Geological Survey)

extracted, cut down on unnecessary waste of the resource, find a substitute, or do without.

Depletion time is the time it takes to use a certain portion—usually 80%—of the known reserves of a mineral at an assumed rate of use. Resource experts project depletion times and plot them on a graph by making various assumptions about the resource supply and its rate of use (Figure 19-7).

We get one estimate of depletion time by assuming that the resource is not recycled or reused, that its estimated reserves will not increase, and that its price increases over time (curve A, Figure 19-7). A longer depletion time estimate is obtained by assuming that recycling will extend the life of existing reserves, and that improved mining technology, price rises, and new discoveries will expand present reserves by some factor, say two (curve B). An even longer depletion time estimate is obtained by assuming that new discoveries will expand reserves even more, perhaps five or ten times, and that recycling, reuse, and reduced consumption will extend supplies (curve C).

Finding a substitute for a resource cancels all these curves and requires a new set of depletion curves for the new resource. Figure 19-8 shows why experts disagree over projected supplies of nonrenewable nonfuel and fuel resources. We get optimistic or pessimistic projections of the depletion time for a nonrenewable resource by making different assumptions.

Who Has the World's Nonfuel Mineral Resources? Five countries—the Soviet Union, the United States, Canada, Australia, and South Africa—supply most of the 20 minerals that make up 98%, by weight, of all nonfuel minerals consumed in the world. No industrialized country is self-sufficient in mineral resources, although the Soviet Union comes close. The United States, Japan, and western European countries are heavily dependent on imports for most of their critical nonfuel minerals.

Minerals essential to the economy of a country are called **critical minerals**, and those necessary for national defense are called **strategic minerals**. Despite its rich resource base, the United States is dependent on a steady supply of imports from 25 other countries for 24 of its 42 most critical and strategic nonfuel minerals.

Even though the United States is the world's largest producer of nonfuel mineral resources, some of these minerals are imported because they are consumed more rapidly than they can be produced from domestic supplies. Others are imported because other countries have higher-grade ore deposits that are cheaper to extract than lower-grade U.S. reserves.

The U.S. Situation Figure 19-8 shows the projected deficiency in U.S. reserves for 20 critical nonfuel minerals to the year 2000 and the major foreign sources of these minerals. Most U.S. mineral imports come from reliable and politically stable countries. There is particular concern over embargoes or sudden cutoffs of supplies of four strategic minerals—manganese, cobalt, platinum, and chromium—for which the United States has essentially no reserves and depends on imports from potentially unstable countries (South Africa, Zambia, Zaire) or the Soviet Union.

The United States has stockpiles of most of its critical and strategic minerals to cushion against short-term supply interruptions and sharp price rises. These stockpiles are supposed to be large enough to last through a three-year conventional war, after subtracting the amounts available from domestic sources and secure foreign sources. However, stockpiles for most of these minerals are far below this level.

19-5 INCREASING MINERAL RESOURCE SUPPLIES: THE SUPPLY-SIDE APPROACH

Economics and Resource Supply Geologic processes determine how a mineral resource is concentrated in the lithosphere. Economics determines what part of the total supply will be used.

According to standard economic theory, a competitive free market should control the supply and demand of goods and services. If a resource becomes scarce, its price rises. If there is an oversupply, the price falls. Some analysts believe that increased demand will raise mineral prices and stimulate new discoveries and development of more efficient mining technology. Rising prices will also make it profitable to mine ores of increasingly lower grades and stimulate the search for substitutes.

But many economists argue that this theory does not apply to nonfuel mineral resources in most MDCs. In the United States and many other MDCs, industry and government have gained so much control over supply, demand, and prices of mineral raw materials and mineral products that a competitive free market does not exist.

Because market prices of products don't reflect dwindling mineral supplies, consumers have no incentive to reduce demand soon enough to avoid economic depletion of the minerals. Low mineral prices, caused by failure to include the external costs of mining and processing them, also encourage resource waste, faster depletion, and more pollution and environmental degradation.

Finding New Land-Based Mineral Deposits Geologic exploration guided by better geologic knowledge and satellite surveys and other new techniques will increase present reserves of most minerals. However,

in MDCs and many LDCs, most of the easily accessible, high-grade deposits have already been discovered. Thus, geologists believe that most new concentrated deposits will be found in unexplored areas in LDCs.

Exploration for new resources requires a large capital investment and is a risky financial venture. Typically, if geologic theory identifies 10,000 sites where a deposit of a particular resource might be found, only 1,000 sites are worth costly exploration; only 100 justify even more costly drilling, trenching, or tunneling; and only 1 out of the 10,000 will probably be a producing mine. Even if large new supplies are found, no nonrenewable mineral supply can stand up to continued exponential growth in its use.

Improving Mining Technology and Mining Low-Grade Ore Some analysts assume that all we have to do to increase supplies of any mineral is to mine increasingly lower grades of ore. They point to the development of large earth-moving equipment, impurity removal techniques, and other advances in mining technology during the past few decades.

For example, these and other technological changes have allowed the average grade of copper ore mined in the United States to fall from about 5% copper, by weight, in 1900 to 0.4% today, with a drop in the inflation-adjusted copper price. Technological improvements also led to a 500% increase in world copper reserves between 1950 and 1980.

Since 1950 known reserves of most other minerals have also increased, despite growing consumption. Also inflation-adjusted prices for most industrial minerals have fallen. However, such price drops are not inevitable and rarely include environmental costs.

Other analysts point out that several factors limit the mining of lower-grade ores. As increasingly poorer ores are mined, energy costs increase sharply. We eventually reach a point where it costs more to mine and process such resources than they are currently worth, unless we have a virtually inexhaustible source of cheap energy. Most energy experts believe that in the future energy will neither be unlimited nor cheap and will become the limiting factor in mining increasingly lower grades of ore.

Available supplies of fresh water also may limit the supply of some mineral resources, because large amounts of water are needed to extract and process most minerals. Many areas with major mineral deposits are poorly supplied with fresh water.

Finally, exploitation of lower grades of ore may be limited by the environmental impact of waste material produced during mining and processing (Figure 19-4). At some point, land restoration and pollution control costs exceed the current value of the minerals.

Getting More Minerals from Seawater and the Ocean Floor Ocean resources are found in three areas: seawater, sediments, and deposits on the shallow continental shelf, and sediments and nodules on the deep-ocean floor. Most of the chemical elements found in seawater occur in such low concentrations that recovering them takes more energy and money than they are worth. Only magnesium, bromine, and sodium chloride are abundant enough in seawater to be extracted profitably at present prices with current technology.

Continental shelf deposits and placer deposits are already significant sources of sand, gravel, phosphates, and nine other nonfuel mineral resources. Offshore wells also supply large amounts of oil and natural gas.

The deep-ocean floor at various sites may be a future source of manganese and other metallic minerals. There are also deposits of metal sulfides of iron, manganese, copper, and zinc around hydrothermal vents found at certain locations on the deep-ocean floor. However, concentrations of metals in most of these deposits are too low to be valuable mineral resources.

Environmentalists recognize that seabed mining would probably cause less harm than mining on land. They are concerned, however, that removing seabed mineral deposits and dumping back unwanted material will stir up ocean sediments. This could destroy seafloor organisms and have unknown effects on poorly understood ocean food webs. Surface waters might also be polluted by the discharge of sediments from mining ships and rigs.

At a few sites on the deep-ocean floor, manganese-rich nodules have been found in large quantities. These cherry- to potato-size rocks contain 30% to 40% manganese, used in certain steel alloys. They also contain small amounts of other strategically important metals such as nickel, copper, and cobalt. These nodules could be sucked up from the muds of the ocean floor by pipe and carried by pipe or scooped up by a continuous cable with buckets and transported to a mining ship above.

However, most of these nodules are found in seabed sites in international waters. Development of these resources has been put off indefinitely because of squabbles between countries over who owns these resources.

Finding Substitutes Some analysts believe that even if supplies of key minerals become very expensive or scarce, human ingenuity will find substitutes. They point out that new developments by scientists are already leading to a materials revolution in which materials made of silicon and other abundant elements are being substituted for most scarce metals.

For example, ceramic materials are being used in engines, knives, scissors, batteries, and artificial limbs. Ceramics are harder, stronger, lighter, and longer-lasting than many metals. Also, they withstand enormous temperatures, and they do not corrode. Because they can burn fuel at higher temperatures than metal engines, ceramic engines can boost fuel efficiency by 30% to 40%.

High-strength plastics and composite materials, strengthened by carbon and glass fibers, are likely to transform the automobile and aerospace industries. Many of these new materials are stronger and lighter than metals. They cost less to produce because they require less energy; they don't need painting and can easily be molded into any shape.

Substitutes can probably be found for many scarce mineral resources. But finding or developing a substitute is costly and phasing it into a complex manufacturing process requires a long lead time. While an increasingly scarce mineral is being replaced, people and businesses dependent on it may suffer economic hardships as the price rises sharply.

Finding substitutes for some key materials may be extremely difficult, if not impossible. Examples are helium, phosphorus for phosphate fertilizers (Figure 4-28, p. 79), manganese for making steel, and copper for wiring motors and generators.

Also some substitutes are inferior to the minerals they replace. For example, aluminum could replace copper in electrical wiring, but the energy cost of producing aluminum is much higher than that of copper. Aluminum wiring is also more of a fire hazard than copper wiring.

19-6 WASTING RESOURCES: THE THROWAWAY APPROACH

Ways to Deal with Solid Waste There are several options for dealing with the mountains of solid waste we produce in mining, processing, manufacturing, and using resources. One is a *throwaway*, or *high-waste, approach* in which these wastes are left where they are produced, buried, or burned. The other is a *conservation*, or *low-waste, approach* based on greatly increased recycling, reuse, and waste reduction.

In the United States, about 76% of the municipal solid waste is hauled away and buried in landfills, 12% is burned in incinerators, 12% is recycled, and 1% is composted. This explains why the United States throws away more potential resources than any other country, even though waste is an outmoded concept (see Spotlight on p. 45).

Burying Solid Waste in Landfills A **sanitary landfill** is a land waste disposal site in which wastes are

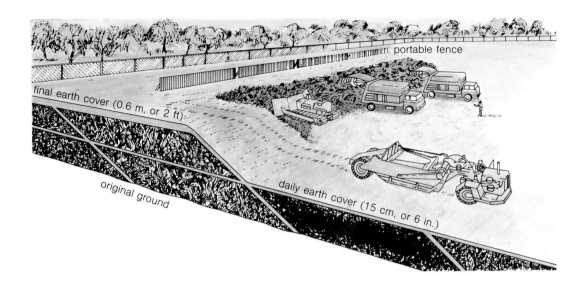

Figure 19-9 A sanitary landfill. Wastes are spread in a thin layer and then compacted with a bulldozer. A scraper (foreground) covers the wastes with a fresh layer of clay or plastic foam at the end of each day. Portable fences catch and hold windblown debris.

portable fence

final earth cover (0.6 m, or 2 ft)

original ground

daily earth cover (15 cm, or 6 in.)

spread out in thin layers, compacted, and covered with a fresh layer of clay or plastic foam each day (Figure 19-9). No open burning is allowed, odor is seldom a problem, and rodents and insects cannot thrive. Sanitary landfills should be located so as to reduce water pollution from runoff and leaching, but this is not always done.

A sanitary landfill can be put into operation quickly, has low operating costs, and can handle a massive amount of solid waste. After a landfill has been filled, the land can be graded and used as a park, a golf course, a ski hill, an athletic field, a wildlife area, or other recreation area.

However, while landfills are in operation, there is much traffic, noise, and dust. Wind can scatter litter and dust before each day's load of trash is covered with clay. That is why most people do not want a landfill nearby.

The underground anaerobic decomposition of organic wastes at landfills produces explosive methane gas and toxic hydrogen sulfide gas. These gases can get into utility lines and seep into basements several blocks away from the landfill. The buildup of these gases can cause asphyxiation or one spark can cause an explosion.

This problem can be prevented by equipping landfills with vent pipes to collect these gases so they can be burned or allowed to escape into the air. Besides saving energy, collecting and burning methane gas from all large landfills worldwide would lower atmospheric emissions of methane and help reduce projected global warming from greenhouse gases.

When rain filters through a landfill it leaches out inks, metals, and other toxic materials. Eventually the leachate seeping from the bottom of landfills has

about the same toxicity as that from a hazardous-waste landfill. This explains why contamination of groundwater and nearby surface water is a serious problem, especially for thousands of older filled and abandoned landfills.

At least 180 of the 1,081 worst hazardous-waste sites to be cleaned up by the EPA are abandoned landfills. Also at least 25%—some say 80%—of the sanitary landfills in operation today in the United States may be polluting surface water and groundwater.

Within the next five to ten years, half of existing U.S. landfills, especially in the East and Midwest, will be filled and closed. Few new landfills are being built. Either there are no acceptable sites, or construction is prevented by citizens who want their trash hauled away but don't want a landfill anywhere near them.

Some cities without enough landfill space are shipping their trash to other states or other countries, especially LDCs. But some LDCs are rebelling against garbage dumping by MDCs. Increasingly, people, businesses, and communities will have to accept responsibility for the wastes they produce instead of trying to make them somebody else's problem. This will lead to increased recycling, reuse, and waste reduction.

Burning Solid Waste Burning solid waste in incinerators kills disease-carrying organisms and reduces the amount of waste going to landfills and the need for landfill space. Although advocates claim that incineration reduces the volume of solid waste by 90%, measurements show that the actual reduction is closer to 60% to 70%.

Proponents call incineration a form of waste

reduction. But instead of reducing the total amount of waste, it puts some of it in the air as gaseous pollutants and fly ash and produces a toxic bottom ash that must be landfilled. Although the amount of material to be buried is greatly reduced, its toxicity is increased.

In *waste-to-energy incinerators*, heat released when combustible solid trash is burned can be used to run the incinerator or heat nearby buildings, or it can generate electricity. Denmark and Sweden burn about half of their solid waste to produce energy, compared to 6% in the United States.

In 1989 more than 80 waste-to-energy incinerators were in operation and 29 more were under construction in the United States. But since 1985 another 64 have been blocked, delayed, or canceled because of citizen opposition and high construction and operating costs. Of the 100 plants still in the planning stage, most face stiff opposition and will probably not be built as communities are discovering that recycling and waste reduction are cheaper, safer, and more politically acceptable alternatives.

Incineration is very expensive, and once they are built, incinerators offer very few long-term jobs. Even with advanced air pollution control devices, incinerators emit small amounts of hydrochloric acid, highly toxic dioxins and furans, and tiny particles of lead, cadmium, mercury, and other toxic substances into the atmosphere. Without continuous maintenance and good operator training and supervision, the air pollution control equipment often fails and allows the incinerator to exceed emission standards.

Incinerators also produce a residue of toxic ash. This ash, currently disposed of in leak-prone ordinary landfills, is usually contaminated with hazardous substances such as dioxins and lead, cadmium, mercury, and other toxic metals. Because the toxic ash is in the form of a powder with a large surface area, its toxic materials can be leached into groundwater much faster than that from conventional solid waste. For the same reason, you get much stronger coffee by using ground-up coffee beans in a drip coffee maker than if you used the larger raw beans. Environmentalists want the EPA to classify incinerator ash as hazardous waste and allow it to be disposed of only in landfills designed to handle hazardous waste (Section 12-5), but this has not been done.

Incinerators also waste resources that could be reused and recycled. Once they are built, incinerators hinder recycling, reuse, and waste reduction because to be economical they must be fueled with a large volume of trash every day. Environmentalists argue that no incinerator should be built unless a community has an effective plan for recycling at least 60% of its municipal solid waste.

Incinerator and landfill sites are hard to find

SPOTLIGHT Mega-Landfills

Over half of the waste hauling and disposal business in the United States is dominated by two companies—Waste Management and Browning-Ferris Industries. In 1989 they took in $3.6 billion and $2.1 billion in revenues respectively. With this sort of income, the $31 million in fines Waste Management was assessed for illegal dumping and spilling between 1981 and 1986 was paid off with just six days income as a minor cost of doing business.

But their end-of-pipe approach to the country's waste problems is threatened by widespread citizen opposition to landfills and incinerators (see Individuals Matter on p. 286) that makes recycling, reuse, and waste reduction cheaper. Fighting these battles in every community is difficult and costly. This, plus mounting legal costs and low profits, caused Browning-Ferris to announce in 1990 that it was getting out of the waste management business.

To help avoid these costly fights large waste management companies have formed an alliance with the nation's railroads to develop a few giant landfills, which can accept wastes from cities thousands of kilometers away. This way they have to fight only a few site battles. These companies have also bought up old landfills and offered to sell or lease the land as incinerator sites.

Environmentalists charge that large waste management companies have enough economic and political power to keep the country too dependent on end-of-pipe output approaches to waste management. This would hinder the shift to reuse, recycling, and waste reduction.

As evidence, environmentalists point to the revolving door relationship between the waste industry and the EPA. The agency's former prosecutor, general counsel, and former director of enforcement now work for Waste Management. William Ruckelshaus, head of the EPA during most of Ronald Reagan's last term as president, is now head of Browning-Ferris Industries. In 1989 after a breakfast meeting with the head of Waste Management, the current EPA head, William Reilly, announced he would challenge a North Carolina law giving the state the power to prevent the siting of hazardous-waste facilities.

because of citizen opposition (see Individuals Matter on p. 286). But the small number of large companies that dominate the multibillion dollar waste hauling and disposal business in the United States are seeking ways to counter this opposition (see Spotlight above).

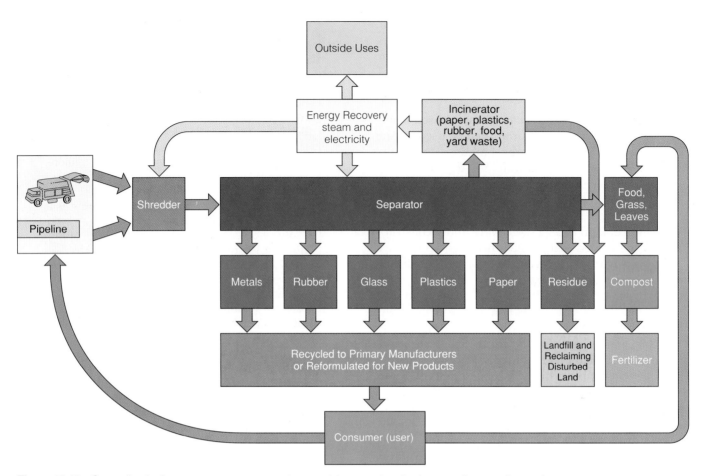

Figure 19-10 Generalized urban resource recovery system used to separate mixed wastes for recycling and burning to produce energy. Very few of this type of resource recovery plant exist today. This high-tech approach is much more expensive and wasteful of energy than having consumers separate garbage into categories for recycling. Because resource recovery plants depend on high inputs of trash to be economical, they discourage reuse and waste reduction. For the same reason, trash-to-energy incinerators discourage the recycling of paper, plastics, and other combustible items.

19-7 EXTENDING RESOURCE SUPPLIES: THE CONSERVATION APPROACH

The Low-Waste Approach Environmentalists and conservationists believe we should begin shifting from the high-waste, throwaway approach (Figure 3-19, p. 54) to a low-waste, sustainable-earth approach for dealing with nonfuel solid resources (Figure 3-20, p. 55). This would require much greater emphasis on conserving wasted solids by composting, recycling, reuse, and waste reduction, instead of dumping, burying, and burning them.

Japan has the world's most comprehensive waste management and recycling program in which 50% of its municipal solid waste is recycled. About 34% is incinerated, 16% is disposed of in landfills, and 0.2% is composted. In Japan half the paper, 55% of the glass bottles, and 66% of food and beverage cans are recycled. From their earliest school years on, Japanese children are taught about recycling and waste management. They often tour their local recycling centers and incinerators.

Composting Biodegradable solid waste from slaughterhouses and food-processing plants, kitchen and yard waste, manure from animal feedlots, and municipal sewage sludge can be mixed with soil and decomposed by aerobic bacteria to produce **compost,** a soil conditioner and fertilizer. Composting cannot be used for mixed urban waste because sorting out the glass, metals, plastic, and organic waste is too expensive.

Compost can be produced in large plants, bagged, and sold. This approach is used in many European countries, including the Netherlands, West Germany, France, Sweden, and Italy. Paper, leaves, and grass clippings can be decomposed in a backyard compost bin (Figure 12-17, p. 279) and used in gardens and flower beds. Composting yard waste would reduce the amount of solid waste in the United States by almost 18%.

Presently only 1% of the mass of solid waste in the United States is composted. But this could change because of a lack of landfill space, mandatory composting programs, and use of economic incentives to encourage composting. For example, locating a city compost heap next to a landfill would reduce the waste going to the landfill and extend its life. The compost can then be applied as landfill cover and used by public works departments for fertilizing parks, forests, roadway medians, and the grounds around public buildings.

High-Tech Resource Recovery The salvaging of usable metals, paper, plastic, and glass from municipal solid waste and selling them to manufacturing industries for recycling or reuse is called **resource recovery**. This extends the supply of minerals by reducing the amount of virgin materials that must be extracted from the earth's crust to meet demand. Recycling and reuse usually require less energy and cause less pollution and land disruption than use of virgin resources. They also cut waste disposal costs and prolong the life of landfills by reducing the volume of solid waste.

Resources can be recycled by using high- or low-technology approaches. In *high-technology resource recovery plants*, machines shred and automatically separate mixed urban waste to recover glass, iron, aluminum, and other valuable materials (Figure 19-10). These materials are then sold to manufacturing industries as raw materials for recycling. The remaining paper, plastics, and other combustible wastes are recycled or incinerated. The heat given off is used to produce steam or electricity to run the recovery plant and for sale to nearby industries or residential developments.

Currently the United States has only a few plants that recover some iron, aluminum, and glass for recycling. These plants are expensive to build and maintain. Once trash is mixed it takes a lot of money and energy to separate it. To environmentalists it makes much more sense economically to have consumers separate trash into recyclable categories before it is picked up.

Environmentalists oppose the widespread use of trash-to-energy incinerators. Unless strictly controlled and monitored, they emit hazardous air pollutants and create toxic ash. They also encourage people to keep on creating solid waste and can discourage recycling of paper, plastics, and other burnable trash. They believe that communities should use incinerators only for wastes that remain after they have established a comprehensive recycling and waste reduction program.

Low-Tech Resource Recovery With *low-technology resource recovery*, homes and businesses place various kinds of waste materials—usually glass, paper, metals, and plastics—into separate containers (Figure 19-11). Compartmentalized city collection trucks, private haulers, or volunteer recycling organizations pick up the segregated wastes and sell them to scrap dealers, compost plants, and manufacturers.

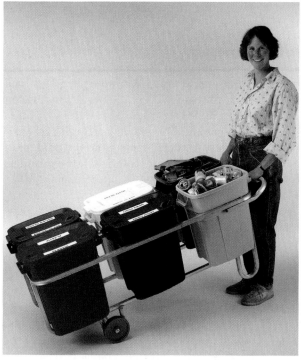

Seventh Generation

Figure 19-11 Cart for separation of paper, plastics, glass, and cans in household waste to be picked up and recycled. This low-tech approach saves more energy and money than high-tech resource recovery plants and waste-to-energy incinerators. It also does more to sensitize consumers to the need to recycle and reuse and to reduce the amount of waste they produce.

Instead of throwing garbage into one container, consumers drop these potential resources into one of several containers. Once people start doing this it takes no more time than putting garbage into one container.

A comprehensive low-technology recycling program could save 5% of annual U.S. energy use—more than the energy generated by all U.S. nuclear power plants at perhaps one-hundredth of the capital and operating costs. By contrast, burning all combustible urban solid waste in waste-to-energy plants would supply only 1% of the country's annual energy use.

The low-technology approach produces little air and water pollution, reduces litter, and has low start-up costs and moderate operating costs. It also saves more energy and provides more jobs for unskilled workers than high-technology resource recovery plants. Recycling creates three to six times more jobs per unit of material than landfilling or incineration. Another advantage is that collecting and selling aluminum cans (see Case Study on p. 456), paper (p. 345), plastics (see Case Study on p. 457), and other materials for recycling is an important source of income for many people (especially the homeless and the poor) and for volunteer service organizations.

Many communities are switching to low-tech recycling programs because it saves them lots of money. The average cost of running a weekly curbside trash collection and recycling program is $20 to $30 per 0.9 metric ton (1 ton). By comparison, it costs $40 to $60 per 0.9 metric ton (1 ton) to collect and haul trash to a landfill, and $70 to $120 per 0.9 metric ton (1 ton) to burn it.

Ideally, a community should set up one or more material recovery centers. Each center would have a reuse and repair center, a place for consumers to take household toxic wastes, a composting area, a section for handling commercial waste, and a section in which separated household paper, plastics, and aluminum, iron, and other metallic wastes are upgraded and marketed to bring in income for the community.

Beverage Container Deposit Bills Beverage container deposit laws can be used to decrease litter and encourage recycling of nonrefillable glass, metal, and plastic containers (refillable containers are discussed later in this chapter). Consumers pay a deposit (usually five to ten cents) on each beverage container they buy. The deposits are refunded when empty containers are turned in to retailers, redemption centers, or reverse vending machines, which return cash when consumers put in empty beverage cans and bottles.

Container deposit laws have been adopted in Sweden, Norway, the Netherlands, the Soviet Union, Ecuador, and parts of Australia, Canada, and Japan. Such laws have been proposed in almost every state in the United States, but have been enacted in only ten states with about one-fourth of the U.S. population. Studies by state and federal agencies show that such laws decrease litter, reduce the use of mineral resources, extend the life of landfills, increase recycling, save energy, reduce air and water pollution, and create jobs.

Environmentalists and conservationists believe a nationwide deposit law should be passed by Congress. Surveys have also shown that a national container deposit law is supported by 73% of the Americans polled. But so far such a law has been effectively opposed in all but ten states by a well-funded lobby of steel, aluminum, and glass companies, metalworkers' unions, supermarket chains, and most major brewers and soft drink bottlers.

Ad campaigns financed by Keep America Beautiful and other groups opposing deposit laws have helped prevent passage of such laws in a number of states. These groups favor litter-recycling laws, which levy a tax on industries whose products may end up as litter or in landfills. Revenues from the tax are used to set up and maintain statewide recycling centers. By 1990 seven states, containing about 14% of the U.S. population, had this type of law.

CASE STUDY   Recycling Aluminum

Recycling aluminum produces 95% less air pollution and 97% less water pollution, and requires 95% less energy than mining and processing aluminum ore. In the United States, the recycling rate for aluminum is only 29%.

In 1989 about 61% of the new aluminum beverage cans used in the United States were recycled at more than 5,000 recycling centers set up by the aluminum industry, other private interests, and local governments.

People who returned the cans got about a penny a can for their efforts, earning about $900 million. Within six weeks, the average recycled aluminum has been melted down and is back on the market as a new can.

Despite this progress, almost half the 81 billion aluminum cans produced each year in the United States are still thrown away; more aluminum is in discarded cans than most countries use for all purposes. If the aluminum cans Americans throw away in one year were laid end to end, the cans would wrap around the earth at the equator more than 164 times. Each discarded aluminum can is almost indestructible solid waste.

The electricity needed to produce one aluminum can from virgin ore would keep a 100-watt light bulb burning for 100 hours. Recycling a can takes only 5% as much energy and saves the energy equivalent of six fluid ounces of gasoline.

Environmentalists applaud aluminum recycling but believe that the best solution is to replace all throwaway aluminum, glass, and plastic beverage containers with reusable glass bottles.

Environmentalists point out that litter laws are output approaches that provide some money to clean up litter. By contrast, container deposit laws are input approaches that reward consumers who return containers and can also make people aware of the need to shift to refillable containers. This is a major reason that companies making nonrefillable containers oppose such laws.

Obstacles to Recycling in the United States Hundreds of U.S. cities are recycling 15% to 50% of the mass of their municipal solid waste. Eight states—Connecticut, Florida, Maryland, New Jersey, New York, Oregon, Pennsylvania, and Rhode Island—have comprehensive recycling programs with goals of recycling 25% to 50% of their solid waste sometime before 1995. But more than half of the states currently recycle less than 5% of their municipal solid waste.

Plastics now account for about 7% of the weight of municipal solid wastes in the United States and are the fastest-growing type of waste by weight in landfills. Most plastics used today are nondegradable or take some 200 to 400 years to degrade. When these plastics are thrown away they can harm wildlife (Figure 19-12).

Scientists have developed *photodegradable plastics* that disintegrate after a few weeks exposure to sunlight. Others are *partially biodegradable plastics* made with a combination of plastic and cornstarch. The cornstarch can be broken down in the soil by aerobic bacteria and moisture, leaving behind a fine powder of plastic. A number of companies are making partially degradable plastic garbage bags, grocery bags, six-pack yokes, and fast-food containers.

Photodegradable and biodegradable plastics make sense for products and containers likely to end up as outdoor litter. But without sunlight, photodegradable plastics take decades to hundreds of years to break down when covered in landfills. Even biodegradable plastics take decades to partially decompose in landfills because of a lack of aerobic bacteria and moisture.

We also need to know more about what degradable plastics degrade to. The residues that remain after many plastics break down could contaminate groundwater supplies. They also may be haz-

Doris Alcorn/National Marine Fisheries

Figure 19-12   Some of the plastic items we throw away harm wildlife. Here a sea lion is being slowly choked to death by a discarded yoke from a six pack of beverage cans. If you must use cans with such yokes, at least cut each of the holes before disposing of the yoke.

ardous to animals that ingest them.

To environmentalists it makes more sense to recycle plastics instead of burying slowly degradable plastics in landfills and even better sense to reduce the amount of plastic we use. Currently only 1% of all plastic wastes and 4% of plastic packaging used in the United States are recycled. Many plastic products today carry a label saying they are degradable or recyclable. While this statement is technically true, it is also highly misleading. Plastics degrade very slowly in landfills, and 99% of all plastic waste is not recycled.

One problem is that plastics rarely can be recycled to make the same products because of health and manufacturing problems. Also, plastics can't be recycled repeatedly. Another problem is that trash contains several types of plastic, and some plastic products contain several types.

With proper economic and po-

litical incentives, about 43%, by weight, of the plastic wastes produced in the United States could be recycled by the year 2000. Since most of the raw materials used to make plastics come from petroleum and natural gas, recycling would help reduce unnecessary waste of these energy resources. So far, East Germany is the only country that collects and recycles household plastics on a nationwide scale.

The $140-billion-a-year U.S. plastics industry has set up the Council for Solid Waste Solutions which runs ads promoting plastics recycling. But the main purpose of this organization is to keep us buying more plastics.

To environmentalists the best solution is simply to use much less plastic in the first place—especially throwaway items—and pressure elected officials to ban the use of plastics in products where other less harmful alternatives are available.

Several factors hinder recycling in the United States. One is that Americans have been conditioned by advertising and example to accept a throwaway lifestyle (Figure 19-13). Also, because the external costs of items are not reflected in their market prices, consumers have little incentive to recycle, reuse, or reduce their use of throwaway products.

The growth of the recycling, or secondary-materials, industry in the United States is hindered by several factors. One is that primary mining and energy industries get huge tax breaks, depletion allowances,

and other tax-supported federal subsidies to encourage them to get virgin resources out of the ground as fast as possible. In contrast, recycling industries get few tax breaks and other subsidies. The lack of large, steady markets for recycled materials also makes recycling a risky business. It is typically a boom-and-bust financial venture that attracts little investment capital.

Overcoming the Obstacles   In 1988 the EPA set a goal of recycling 25% of municipal solid waste by

Figure 19-13 Evidence of our throwaway mentality. This barge, floating off the coast of New York City, is carrying garbage nobody wants.

1992. By the year 2000, the United States could easily recycle and reuse 35% of the municipal solid waste resources it now throws away. Greatly increased recycling and reuse in the United States could be accomplished through the following measures:

- Enact a national beverage container law.

- Ban use of disposable plastic items.

- Include waste disposal costs in the price of all disposable items rather than in local taxes, so market prices of items directly reflect what it cost to dispose of them. This is now done in Florida, Massachusetts, and Minnesota.

- Provide economic incentives for recycling waste oil, plastics, tires, and CFCs used as coolants in refrigerators and air conditioners.

- Require labelling of products made with recyclable materials and the percentages used.

- Provide federal and state subsidies for secondary-materials industries and for municipal recycling and waste reduction programs. New York, North Carolina, Florida, Oregon, and Wisconsin give tax breaks to businesses that use secondary materials or buy recycling equipment.

- Decrease federal and state subsidies for primary-materials industries.

- Guarantee a large market for recycled items and stimulate the recycling industry by encouraging federal, state, and local governments to require the highest feasible percentage of recycled materials in all products they purchase. Twenty-two states have such laws, but their effectiveness varies widely.

- Use advertising and education to discourage the throwaway mentality.

- Require consumers to sort household wastes for recycling, or give them financial incentives for recycling. For example, trash separated for recycling can be picked up free, while people who don't separate their trash are charged a fee per bag or kilogram. Also, households can be pro-

vided with free reusable containers for separating recyclable trash. Ten states and many communities have laws making trash separation mandatory.

- Encourage municipal composting and backyard composting by banning the disposal of yard wastes in landfills.

Recycling 35% of our municipal solid waste is not enough. This still leaves 65% of these potential resources to be dumped in landfills or incinerated. We must strive for recycling at least 60% of our municipal solid waste.

Reusable Products Recycling is an important first step but a much more important step is reuse in which a product is used again and again in its original form. An example is glass beverage bottles that can be collected, washed, and refilled by local bottling companies.

Until 1975 most beverage containers in the United States were refillable glass. Today they make up only 15% of the market, with nonrefillable aluminum and plastic containers making up the rest. Today only ten states even have returnable bottles. We need to go back to refillable bottles, collected and filled at local bottling plants to reduce transportation and energy costs. This also creates jobs and improves the local economy.

Ecuador has a beverage container deposit fee that is 50% higher than the cost of the drink to encourage use of refillable bottles. This has been so successful that bottles as old as ten years continue to circulate. Sorting is not a problem because there are only two sizes of glass bottles allowed.

Another reusable container is the metal or plastic lunch box that most workers and school children once used. Today many people carry their lunches in paper or plastic bags that are thrown away. At work people can have their own earth-care kit consisting of a reusable glass, coffee cup, plate, knife, fork, spoon, cloth napkin, and cloth towel.

Disposable diapers made from paper and coated with plastic may make life easier, but they are also messing up life in a throwaway society that has run out of places to throw things. The 18 billion disposable diapers used each year in the United States make up about 2% of the solid waste dumped into landfills and take about 500 years to degrade. Production of disposable diapers uses trees and causes air and water pollution.

Using cloth diapers, which can be washed and reused from 80 to 200 times and then retired to become lint-free rags, keeps about 10,000 disposable diapers weighing about 1 metric ton (1.1 ton) per baby from reaching landfills. This also saves trees and money. For example, using disposable diapers during the typical time a baby needs diapers costs about $1,533. A cloth diaper service costs about $975 and washing cloth diapers at home costs about $283.

The $3.3-billion-a-year disposable diaper industry hopes to keep its business by developing biodegradable diapers. But this is not a solution. Biodegradable diapers take decades to a hundred years to break down in landfills, are still a throwaway item, and cost consumers more money than reusable cloth diapers.

However, the choice between disposable diapers and reusable cloth diapers is not clear-cut, illustrating the complexity of making environmentally sound decisions. Disposable diapers consume paper and plastic resources, end up in landfills, and produce 90 times as much solid waste as cloth diapers. But laundering cloth diapers produces 9 times as much air pollution and 10 times as much water pollution as disposables do in their lifetime. Over their lifetime, cloth diapers also consume 6 times more water and 3 times more energy than disposables. From an environmental standpoint, neither product has a clear edge, but from a financial standpoint, cloth diapers have a distinct advantage.

Other examples are reusable plastic or metal garbage cans and wastebaskets, which we should teach everyone to start calling *resource containers*. Using these containers to separate wastes for recycling and rinsing them out as needed would eliminate the need to throw any garbage away in plastic bags, which waste oil and are an unnecessary form of waste. Lining these containers with throwaway plastic bags is an expensive, unnecessary waste of matter and energy resources.

You can carry reusable baskets, canvas or plastic grocery-store bags, or string containers (Figure 19-14) when you shop for groceries or other items. Several can be folded up and kept in your pocketbook, pocket, or car. They eliminate the need to use either plastic or paper bags. This saves trees and oil and reduces pollution and environmental degradation.

Using unbleached paper coffee filters can reduce your exposure to toxic dioxins, which leach out of

Seventh Generation

Figure 19-14 An example of good earthkeeping. A reusable string bag can be used to carry groceries and avoid the use of throwaway paper and plastic grocery bags. Cloth or canvas bags can also be used.

bleached paper filters. A much better and cheaper solution is cloth filters that can be reused. After use they should be rinsed and stored in a glass of water in a refrigerator to prevent buildup of rancid coffee oils.

Reuse extends resource supplies and reduces energy use and pollution even more than recycling. Refillable glass bottles are the most energy-efficient beverage container on the market. If reusable glass bottles replaced the 80 billion throwaway beverage cans produced annually in the United States, enough energy would be saved to supply the annual electricity needs of 13 million people. Denmark has led the way by banning all nonreusable beverage containers.

Reuse is much easier if containers for products that can be packaged in reusable glass are available in only a few sizes. In Norway and Denmark, fewer than 20 sizes of reusable containers for beer and soft drinks are allowed. A popular bumper sticker reads: "Recyclers do it more than once." A much better version would be "Recyclers do it more than once, but reusers do it the most."

Waste Reduction Reducing unnecessary waste of nonrenewable mineral resources, plastics, and paper can extend supplies even more dramatically than recycling and reuse. Reducing waste generally saves more energy than recycling and reduces the environmental impacts of extracting, processing, and using resources (Figure 19-4). Table 19-1 compares the throwaway resource system of the United States, a resource recovery and recycling system, and a sustainable-earth, or low-waste, resource system. Ultimately, a sustainable-earth system is based on the

Table 19-1 Three Systems for Handling Discarded Materials

Item	For a High-Waste Throwaway System	For a Moderate-Waste Resource Recovery and Recycling System	For a Low-Waste Sustainable-Earth System
Glass bottles	Dump or bury	Grind and remelt; remanufacture; convert to building materials	Ban all nonreturnable bottles. Reuse bottles
Bimetallic "tin" cans	Dump or bury	Sort, remelt	Limit or ban production. Use returnable bottles
Aluminum cans	Dump or bury	Sort, remelt	Limit or ban production. Use returnable bottles
Cars	Dump	Sort, remelt	Sort, remelt. Tax cars lasting less than 15 years and getting less than 17 kilometers per liter (40 miles per gallon)
Metal objects	Dump or bury	Sort, remelt	Sort, remelt. Tax items lasting less than 10 years
Tires	Dump, burn, or bury	Grind and revulcanize or use in road construction; incinerate to generate heat and electricity	Recap usable tires. Tax or ban all tires not usable for at least 96,000 kilometers (60,000 miles)
Paper	Dump, burn, or bury	Incinerate to generate heat	Compost or recycle. Tax all throwaway items. Eliminate overpackaging
Plastics	Dump, burn, or bury	Incinerate to generate heat or electricity	Limit production; use returnable glass bottles instead of plastic containers; tax throwaway items and packaging
Yard wastes	Dump, burn, or bury	Incinerate to generate heat or electricity	Compost; return to soil as fertilizer; use as animal feed

principle: "If you can't recycle or reuse something, don't make it."

Manufacturers can conserve resources by using less material per product. Smaller, lighter cars, for example, save nonfuel mineral resources as well as energy. Solid-state electronic devices and microwave transmissions greatly reduce materials requirements. Optical fibers drastically reduce the demand for copper wire in telephone transmission lines.

Another low-waste approach is to make products that last longer. The economies of the United States and most industrial countries are built on the principle of planned obsolescence so that people will buy more things to stimulate the economy and raise short-term profits, even though this can eventually lead to economic and environmental grief. Many consumers can empathize with Willy Loman in Arthur Miller's *Death of a Salesman:* "Once in my life I would like to own something outright before it's broken! I'm always in a race with the junkyard."

Products should also be easy to repair. Today many items are intentionally designed to make repair impossible or too expensive. Manufacturers should adopt the principle of modular design which allows circuits in computers, television sets, and other electronic devices to be easily and quickly replaced with-

out replacing the entire item. We also need to develop remanufacturing industries that would disassemble, repair, and reassemble used and broken items.

Adopt the three Rs of earth care: Reduce, Reuse, Recycle. Think of recycling as a first and important baby step in sustaining the earth. Then move to reuse as part of environmental adolescence. Then reach environmental maturity by sharply reducing the amount of waste we produce. This means that making and using throwaway items should be considered anti-earth activities. Each of us must understand and accept our role in both creating solid waste and in conserving these vital matter resources (see Individuals Matter on p. 461).

We will always produce some wastes, but the amount we produce can be greatly reduced by not using certain products (see Individuals Matter inside the back cover) and by redesigning the ways we produce products. To do this we must force elected officials to get serious about using input approaches to prevent most waste (especially hazardous waste) from being produced or from reaching the environment. So far, less than 1% of the U.S. government's environmental spending goes for pollution prevention.

The end-of-pipe or output methods for controlling wastes that we now depend on merely move potential

- Make a conscious effort to produce less waste, mostly by not using disposable products when other alternatives are available.

- Buy refillable glass containers for beverages instead of cans or throwaway bottles.

- Use plastic or metal lunchboxes and garbage containers without plastic throwaway bags as liners. Wrap sandwiches in biodegradable wax paper, or better, put them in small, reusable plastic containers.

- Use rechargeable batteries.

- Carry groceries and other items in a reusable basket, canvas or string bag (Figure 19-14), or small cart. You could also save and reuse plastic bags from grocery and other stores. BYOC (bring your own container) is one reason why Europeans, Africans, and Asians produce so much less solid waste per person than most people in the United States. Tell store clerks and managers why you are doing this and increase their sensitivity to unnecessary waste.

- Skip the bag when you buy only a quart of milk, a loaf of bread, or anything you can carry out with your hands. Tell store clerks and managers why you are doing this.

- Use washable cloth napkins and dish towels instead of paper ones.

- Don't use throwaway paper and plastic plates, cups, eating utensils, razors, pens, lighters, and other disposable items when reusable or refillable versions are available.

- Look for and buy items made from recycled materials. A wide variety of items made from recycled paper can be ordered by catalog from Earth Care Paper, P.O. Box 335, Madison, Wisconsin 53704. A wide range of recycled and other environmentally sound items can also be bought by catalog from Seventh Generation, 10 Farell St., South Burlington, VT 05403 (800) 441-2358.

- Recycle all newspapers, glass, and aluminum.

- Buy repairable items.

- Just say no to throwaway plastic items. They are made from petrochemicals, take decades to degrade in today's landfills, can release toxins into the environment during production and when burned or degraded, hinder efforts to reduce and recycle plastics, and pose a threat to wildlife (Figure 19-12).

- Reduce the amount of junk mail you get by writing to Mail Preference Service, Direct Marketing Association, 11 West 42nd St., P.O. Box 3681, New York, NY 10163-3861, and ask them to stop your name from being sold to most large mailing list companies.

- Push for mandatory trash separation and recycling programs in your community and schools.

- Ask stores, communities, and colleges to install reverse vending machines that give you cash for each reusable or recyclable container you put in.

- Buy food items in large cans or bulk to reduce packaging.

- Choose items that have the least packaging or, better yet, no packaging. When enough people do this, producers of packaged materials will get the message.

- Don't buy helium-filled balloons, and urge elected officials and university administrators to ban balloon releases except for atmospheric research and monitoring.

- Compost your yard and food wastes and pressure local officials to set up a community composting program. Currently, an average of 15% of the solid food bought for use in U.S. households is thrown out.

- Pressure managers of businesses and schools to separate valuable ledger and computer paper for recycling.

- Ask heads of companies to switch their letterhead and paper stock to recycled products.

- Join a local environmental group and urge it to identify suppliers of recycled products in your area.

- Consider going into the recycling business as a way to make money.

- Don't litter.

pollutants from one part of the environment to another and are eventually overwhelmed by more people producing more wastes. This was summarized over a century ago by Chief Seattle: "Contaminate your bed, and you will one night suffocate in your own waste." Changing from a pollution-control culture to a pollution-prevention culture requires us to understand and live by three key principles: everything is interconnected; there is no away for the

wastes we produce; and dilution is not the solution to most pollution.

Solid wastes are only raw materials we're too stupid to use.

ARTHUR C. CLARKE

DISCUSSION TOPICS

1. What mineral resources are mined in your local area? What mining methods are used? Do local, state, or federal laws require restoration of the landscape after mining is completed? If so, how well are these laws enforced?

2. Do you believe that the United States is an overdeveloped country that uses and unnecessarily wastes too many of the world's resources relative to its population size? Explain.

3. Debate each of the following propositions:
   a. The competitive free market will control the supply and demand of mineral resources.
   b. New discoveries will provide all the raw materials we need.
   c. The ocean will supply all the mineral resources we need.
   d. We will not run out of key mineral resources because we can always mine lower-grade deposits.
   e. When a mineral resource becomes scarce, we can always find a substitute.
   f. When a nonrenewable resource becomes scarce, all we have to do is recycle it.

4. Use the second law of energy (thermodynamics) to show why the following options are usually not profitable without subsidies:
   a. extracting most minerals dissolved in seawater
   b. recycling minerals that are widely dispersed
   c. mining increasingly lower-grade deposits of minerals
   d. using inexhaustible solar energy to mine minerals
   e. continuing to mine, use, and recycle minerals at increasing rates
   f. building high-tech resource recovery plants (Figure 19-10) to separate mixed wastes for recycling

5. Explain why you support or oppose the following:
   a. eliminating all tax breaks and depletion allowances for extraction of virgin resources by mining industries
   b. passing a national beverage container deposit law
   c. requiring that all beverage containers be reusable

6. Would you favor requiring all households and businesses to sort recyclable materials for curbside pickup in separate containers? Explain.

7. Determine whether
   a. your college and your city have recycling programs
   b. your college sells soft drinks in throwaway cans or bottles
   c. your college bans release of helium-filled balloons at sporting events and other activities
   d. your state has, or is contemplating, a law requiring deposits on all beverage containers

FURTHER READINGS

Blumberg, Louis, and Robert Grottleib. 1988. *War on Waste—Can America Win Its Battle with Garbage?* Covelo, Calif.: Island Press.

Borgese, Elisabeth Mann. 1985. *The Mines of Neptune: Minerals and Metals from the Sea.* New York: Abrams.

Boyd, Susan, et al., eds. 1988. *Waste: Choices for Communities.* Washington, D.C.: CONCERN.

Clark, Joel P., and Frank R. Field III. 1985. "How Critical Are Critical Materials?" *Technology Review,* Aug./Sept., 38–46.

Connett, Paul. 1989. *Waste Management As If the Future Mattered.* Canton, N.Y.: Work on Waste.

Dorr, Ann. 1984. *Minerals—Foundations of Society.* Montgomery County, Md.: League of Women Voters of Montgomery County, Maryland.

Environmental Defense Fund. 1988. *Coming Full Circle.* New York: Environmental Defense Fund.

Environmental Protection Agency. 1989. *Solid Waste Disposal in the United States.* Washington, D.C.: Government Printing Office.

Gordon, Robert B., et al. 1988. *World Mineral Exploration: Trends and Issues.* Washington, D.C.: Resources for the Future.

Hershkowitz, Allen. 1987. "Burning Trash: How It Could Work." *Technology Review,* July, 26–34.

Huls, Jon, and Neil Seldman. 1985. *Waste to Wealth.* Washington, D.C.: Institute for Local Self-Reliance.

Institute for Local Self-Reliance. 1988. *Recycling Goals and Strategies.* Washington, D.C.: Institute for Local Self-Reliance.

Keller, Edward A. 1988. *Environmental Geology.* 5th ed. Columbus, Ohio: Merrill Publishing.

Neal, Homer A., and J. R. Schubel. 1987. *Solid Waste Management and the Environment: The Mounting Garbage and Trash Crisis.* Englewood Cliffs, N.J.: Prentice-Hall.

Office of Technology Assessment. 1985. *Strategic Materials: Technologies to Reduce U.S. Import Vulnerability.* Washington, D.C.: Government Printing Office.

Office of Technology Assessment. 1987. *Marine Minerals: Exploring Our New Ocean Frontier.* Washington, D.C.: Government Printing Office.

Office of Technology Assessment. 1989. *Facing America's Trash: What's Next for Municipal Solid Waste.* Washington, D.C.: Government Printing Office.

Pollack, Cynthia. 1987. *Mining Urban Wastes: The Potential for Recycling.* Washington, D.C.: Worldwatch Institute.

Seldman, Neil, and Bill Perkins. 1988. *Designing the Waste Stream.* Washington, D.C.: Institute for Local Self-Reliance.

Underwood, Joanna D., and Allen Hershkowitz. 1989. *Facts About U.S. Garbage Management: Problems and Practices.* New York: INFORM.

ACHIEVING A SUSTAINABLE-EARTH SOCIETY

Avoiding Some Common Traps We can read and talk about environmental and resource problems, but finally it comes down to what you and I are willing to do individually and collectively. Sustaining the earth requires each of us to make a personal commitment to live an environmentally ethical life. We must do this not because it is required by law, but because it is right. It is our responsibility to ourselves, our children and grandchildren, our neighbors, and the earth.

Start by being sure that you have not fallen into some common traps, or excuses, that lead to indifference and inaction:

- *Gloom-and-doom pessimism*—the belief that the world is doomed by nuclear war or environmental catastrophe, so we should enjoy life while we can.

- *Blind technological optimism*—the belief that human ingenuity will always be able to come up with technological advances that will solve our problems. This is the most seductive and dangerous trap. It is something that we would like to believe.

- *Fatalism*—the belief that whatever will be will be, and we have no control over our actions and the future.

- *Extrapolation to infinity*—the belief that "If I can't change the entire world quickly, I won't try to change any of it." This rationalization is reinforced by modern society's emphasis on instant gratification and quick results, with the littlest effort possible.

All of these traps represent various forms of *denial* to enable us to avoid facing up to problems and the need for change. With our present power to destroy ourselves and most other species, denial is a recipe for disaster.

Becoming Earth Citizens The good news is that we can sustain the earth and lead more meaningful and joyful lives. To do this begin with yourself by doing 12 things—the "earth-sustaining dozen."

1. **Evaluate the way you think the world works and sensitize yourself to your local environment.** Look around, experience what is going on in the environment around you, compare what is with what could and should be. Where does the water you drink and the air you breathe come from? What kind of soil is around your home? Where does your garbage go? What forms of wild plants and animals live around you? What species have become extinct in your area and what ones are threatened with extinction? What is the past history of land use in your area and what are the projected future uses of this land? What are your environmental bad habits? What is your worldview (Chapter 3)? How does your worldview influence the way you act?

2. **Become ecologically informed.** Immerse yourself in sustainable-earth thinking. Specialize in one particular area of environmental knowledge and awareness, relate this to sustainable-earth thinking, and share your knowledge and understanding with others (networking). Consider going into an environmental profession or starting or working in an environmentally responsible business, buy green products, and invest only in green companies. Everyone doesn't need to be an ecologist or a professional environmentalist, but you do need to "ecologize" your lifestyle. Keep in mind Norman Cousins's statement: "The first aim of education should not be to prepare young people for careers but to enable them to develop respect for life."

3. **Become emotionally involved in caring for the earth by experiencing nature directly and by trying to find a place that you love and must defend because you are part of it and it is part of you.** Intellectual ecological knowledge of how the world works is vitally important. But it will not be enough to bring about a change in the way you live unless it is combined with a sense of place—a feeling of oneness with and thankfulness for some piece of the earth that you truly love and respect. Poet-philosopher Gary Snyder urges us to "find our place on the planet, dig in, and take responsibility from there."

4. **Choose a simpler lifestyle by reducing resource consumption and waste and pollution production.** Do this by distinguishing between your true needs and your wants and by using

trade-offs. For every high energy, high waste, or highly polluting thing you do (buying a car, living or working in an air-conditioned building), give up a number of other things. Such a lifestyle will be less expensive and should bring you joy as you learn how to walk more gently on the earth.

5. **Focus especially on energy use and energy waste.** Recognize that energy—the integrating theme of this book—is the currency of life for us and other species. People in MDCs have achieved their higher standards of living primarily by increasing their use of energy (Figure 2-1, p. 21). At the same time, our rapid depletion of the earth's one-time deposit of fossil fuels is the major cause of most air pollution, water pollution, land degradation, and international tension over control of dwindling oil supplies. Fossil fuels also give us more energy to rapidly clear forests and degrade other ecosystems that are homes for the earth's vital biodiversity. These fuels are also used to produce petrochemicals that in turn are used to produce the plastics, pesticides, solvents, chlorofluorocarbons, and thousands of other hazardous and often slowly degradable chemicals and products we are dumping into the environment.

Thus, any effective efforts to sustain and heal the earth must be built around two things that must be carried out at the individual, local, national, and global levels within your lifetime. First, we must greatly improve energy efficiency to cut out at least 50% of the energy we are now wasting unnecessarily. Second, we must shift from lifestyles and economies built around the use of nonrenewable fossil and nuclear fuels to ones built around perpetual and renewable energy from the sun, wind, falling and flowing water, sustainable burning of biomass, and the earth's interior heat.

6. **Become more self-sustaining by trying to unhook yourself from dependence on large, centralized systems for your water, energy, food, and livelihood.** You can do this in the country or in the city. Use organic, intensive gardening techniques to grow some of your own food in a small plot, roof garden, or window box planter. Get as much of your energy as possible from renewable sources such as the sun, wind, water, or biomass.

7. **Remember that environment begins at home.** Before you start trying to convert others, begin by changing your own living patterns. If you become an earth citizen, be prepared to have everyone looking for and pointing out your own environmental sins. Your actions force people to look at what they are doing—a threatening process. People are most influenced by what we do, not by what we say.

8. **Become politically involved on local and national levels.** Start or join a local environmental group, and also join and financially support national and global environmental and conservation organizations whose causes you believe in (see Appendix 1). Work to elect sustainable-earth leaders and to influence officials once they are elected to public office.

9. **Do the little things based on thinking globally and acting locally.** Environmental problems are caused by quadrillions of small, unthinking actions by billions of people. They'll be cured by quadrillions of small actions in which you and others substitute environmentally beneficial actions for thoughtless and wasteful ones. Recycle and better yet, reuse things; don't waste energy (see inside the back cover); improve the energy efficiency of your house; don't use electricity to heat space or household water; drive a car that gets at least 17 kilometers per liter (40 miles per gallon); join a car pool; use mass transit; ride a bicycle to work; replace incandescent lights with fluorescent lights; turn off unnecessary lights; plant trees, have a compost pile; help restore a damaged part of the earth; eat lower on the food chain; grow food organically; don't waste water (see inside the back cover); choose to have no more than one, or at most, two children, and teach your children to sustain the earth; write on both sides of a piece of paper; don't discard useful clothing and other items just to be fashionable; don't buy overpackaged products; distinguish between your needs and your wants before you buy anything; buy products from and invest in companies that are working to sustain the environment.

Each of these small acts sensitizes you to earth-sustaining acts and leads to more such acts. Each of these individual actions is also a small-scale economic and political decision that, when coupled with those of others, leads to larger scale political and economic changes.

10. **Work on the big polluters and big problems, primarily through political action.** Individual actions help reduce pollution and environmental degradation, give us a sense of involvement, and help us develop a badly needed earth consciousness. Our awareness must then expand to recognize that large-scale pollution and environmental degradation are caused by industries, governments, and big agriculture driven by overemphasis on short-term economic gain that eventually leads to economic and environmental grief for us and premature extinction for many other species. The ethic of the international Greenpeace movement, for example, is "not only to personally bear witness to atrocities against life; it is to take direct nonviolent action to prevent them."

11. **Start a movement of awareness and action.** You can change the world by changing the two people next to you. For everything, big or little, that you decide to do to help sustain the earth, try to convince two others to do the same thing, and persuade them in turn to convince two others. Carrying out this doubling or exponential process only 24.5 times would convince everyone in the United States, and doing it 28.5 times would convince everyone in the world. But it is only necessary to have about 5% to 10% of the people in a community, state, country, or the world actively involved to bring about change. The national and global environmental movement is nearing this critical mass and needs your help.

12. **Don't make people feel guilty.** If you know people who are overconsuming or carrying out environmentally harmful acts, don't make them feel bad. Instead, lead by example and find the things that others are willing to do to sustain the earth. There is plenty to do, and no one can do everything. Use positive rather than negative reinforcement. We need to nurture, reassure, understand, and love one another, not threaten one another.

It is not too late. There is time to deal with the complex, interacting problems we face and to make an orderly rather than catastrophic transition to a sustainable-earth society if enough of us really care. It's not up to "them," it's up to "us." Don't wait.

Make a difference by caring. Care about the air, water, soil. Care about wild plants, wild animals, wild places. Care about people—young, old, black, white, brown—in this generation and generations to come. Let this caring be your guide for doing. Live your life caring about the earth and you will be fulfilled.

Envision the world as made up of all kinds of matter cycles and energy flows. See these life-sustaining processes as a beautiful and diverse web of interrelationships—a kaleidoscope of patterns and rhythms whose very complexity and multitude of potentials remind us that cooperation, honesty, humility, and love must be the guidelines for our behavior toward one another and the earth.

When there is no dream, the people perish.
PROVERBS 29:18

PUBLICATIONS, ENVIRONMENTAL ORGANIZATIONS, AND FEDERAL AND INTERNATIONAL AGENCIES

Publications

The following publications can help you keep well informed and up to date on environmental and resource problems. Subscription prices, which tend to change, are not given.

Ambio: A Journal of the Human Environment Royal Swedish Academy of Sciences, Box 50005, S-104 05 Stockholm, Sweden.

American Forests American Forestry Association, 1516 P St. NW, Washington, DC 20005.

Amicus Journal Natural Resources Defense Council, 122 E. 42nd St., New York, NY 10168.

Annual Review of Energy Department of Energy, Forrestal Building, 1000 Independence Ave. SW, Washington, DC 20585.

Audubon National Audubon Society, 950 Third Ave., New York, NY 10022.

Audubon Wildlife Report National Audubon Society, 950 Third Ave., New York, NY 10022. Published every two years.

Biologue National Wood Energy Association, 1730 North Lynn St., Suite 610, Arlington, VA 22209

BioScience American Institute of Biological Sciences, 730 11th St. NW, Washington, DC 20001.

Buzzworm P.O. Box 6853, Syracuse, NY 13217-7930.

The CoEvolution Quarterly P.O. Box 428, Sausalito, CA 94965.

Conservation Biology Blackwell Scientific Publications, Inc., 52 Beacon St., Boston, MA 02108.

Conservation Foundation Letter The Conservation Foundation, 1250 24th St. NW, Washington, DC 20037.

Demographic Yearbook Department of International Economic and Social Affairs, Statistical Office, United Nations Publishing Service, United Nations, NY 10017.

Earth Island Journal Earth Island Institute, 300 Broadway, Suite 28, San Francisco, CA 94133.

The Ecologist Ecosystems Ltd., 73 Molesworth St., Wadebridge, Cornway PL27 7DS, United Kingdom.

Ecology Ecological Society of America, Dr. Duncan T. Patten, Center for Environmental Studies, Arizona State University, Tempe, AZ 85281.

E Magazine P.O. Box 5098, Westport, CN 06881.

Endangered Species UPDATE School of Natural Resources, The University of Michigan, Ann Arbor, MI 48109.

Environment Heldref Publications, 4000 Albemarle St. NW, Washington, DC 20016.

Environment Abstracts Bowker A & I Publishing, 245 W. 17th St., New York, NY 10011. In most libraries.

Environmental Action 1525 New Hampshire Ave. NW, Washington, DC 20036.

Environmental Defense Letter Environmental Defense Fund, 257 Park Ave. South, New York, NY 10010.

Environmental Ethics Department of Philosophy, The University of Georgia, Athens, GA 30602.

Environmental Opportunities (Jobs) Sanford Berry, P.O. Box 969, Stowe, VT 05672.

The Environmental Professional Editorial Office, Department of Geography, University of Iowa, Iowa City, IA 52242.

EPA Journal Environmental Protection Agency, Order from Government Printing Office, Washington, DC 20402.

Everyone's Backyard Citizens' Clearinghouse for Hazardous Waste, P.O. Box 926, Arlington, VA 22216.

Family Planning Perspectives Planned Parenthood-World Population, Editorial Offices, 666 Fifth Ave., New York, NY 10019.

The Futurist World Future Society, P.O. Box 19285, Twentieth Street Station, Washington, DC 20036.

Garbage: The Practical Journal for the Environment P.O. Box 56520, Boulder, CO 80321-6520.

Greenpeace Magazine Greenpeace USA, 1436 U St. NW, Washington, DC 20009.

International Environmental Affairs University Press of New England, 17 1/2 Lebanon St., Hanover, NH 03755.

Issues in Science and Technology National Academy of Sciences, 2101 Constitution Ave. NW, Washington, DC 20077-5576.

Journal of Environmental Education Heldref Publications, 4000 Albemarle St. NW, Suite 504, Washington, DC 20016.

National Geographic National Geographic Society, P.O. Box 2895, Washington, DC 20077-9960.

National Parks and Conservation Magazine National Parks and Conservation Association, 1015 31st St. NW, Washington, DC 20007.

National Wildlife National Wildlife Federation, 1400 16th St. NW, Washington, DC 20036.

Natural Resources Journal University of New Mexico School of Law, 1117 Stanford NE, Albuquerque, NM 87131.

Nature 711 National Press Building, Washington, DC 20045.

The New Farm Rodale Research Center, Emmaus, PA 18049.

New Scientist 128 Long Acre, London, WC 2, England.

Newsline Natural Resources Defense Council, 122 E. 42nd St., New York, NY 10168.

Not Man Apart Friends of the Earth, 530 Seventh St. SE, Washington, DC 20003.

One Person's Impact P.O. Box 751, Westborough, MA 01581.

Organic Gardening & Farming Magazine Rodale Press, Inc., 33 E. Minor St., Emmaus, PA 18049.

Pollution Abstracts Cambridge Scientific Abstracts, 7200 Wisconsin Ave., Bethesda, MD 20814. Found in many libraries.

Population Bulletin Population Reference Bureau, 777 14th St. NW, Suite 800, Washington, DC 20005.

Rachel's Hazardous Waste News Environmental Research Foundation, P.O. Box 3541, Princeton, NJ 08543-3541.

Renewable Energy News Solar Vision, Inc., 7 Church Hill, Harrisville, NH 03450.

Renewable Resources 5430 Grosvenor Ln., Bethesda, MD 20814.

Rocky Mountain Institute Newsletter 1739 Snowmass Creek Road, Snowmass, CO 81654.

Science American Association for the Advancement of Science, 1333 H St. NW, Washington, DC 20005.

Science News Science Service, Inc., 1719 N St. NW, Washington, DC 20036.

Scientific American 415 Madison Ave., New York, NY 10017.

Sierra Sierra Club, 730 Polk St., San Francisco, CA 94108.

State of the States Renew America, 1001 Connecticut Ave. NW, Suite 719, Washington, DC 20036. Published annually.

State of the World Worldwatch Institute, 1776 Massachusetts Ave. NW, Washington, DC 20036. Published annually.

Statistical Yearbook Department of International Economic and Social Affairs, Statistical Office, United Nations Publishing Service, United Nations, NY 10017.

Technology Review Room E219-430, Massachusetts Institute of Technology, Cambridge, MA 02139.

Transition Laurence G. Wolf, ed., Department of Geography, University of Cincinnati, Cincinnati, OH 45221.

The Trumpeter Journal of Ecosophy P.O. Box 5883 Stn. B, Victoria, B.C., Canada V8R 6S8.

Wilderness The Wilderness Society, 1400 I St. NW, 10th Floor, Washington, DC 20005.

World Rainforest Report Rainforest Action Network, 300 Broadway, Suite 298, San Francisco, CA 94133.

World Resources World Resources Institute, 1735 New York Ave. NW, Washington, DC 20006. Published every two years.

World Watch Worldwatch Institute, 1776 Massachusetts Ave. NW, Washington, DC 20036.

Worldwatch Papers Worldwatch Institute, 1776 Massachusetts Ave. NW, Washington, DC 20036.

Yearbook of World Energy Statistics Department of International Economic and Social Affairs, Statistical Office, United Nations Publishing Service, United Nations, NY 10017.

Environmental and Resource Organizations

For a more detailed list of national, state, and local organizations, see Conservation Directory, *published annually by the National Wildlife Federation, 1400 16th St. NW, Washington, DC 20036 and* World Directory of Environmental Organizations, *published by the California Institute of Public Affairs, P.O. Box 10, Claremont, CA 91711.*

African Wildlife Foundation 1717 Massachusetts Ave. NW, Washington, DC 20036.

American Council for an Energy Efficient Economy 1001 Connecticut Ave. NW, Suite 535, Washington, DC 20013.

American Forestry Association 1516 P St. NW, Washington, DC 20005.

American Geographical Society 156 Fifth Ave., Suite 600, New York, NY 10010.

American Institute of Biological Sciences Inc. 730 11th St. NW, Washington, DC 20001.

American Solar Energy Society 2400 Central Ave., B1, Boulder, CO 80301.

American Water Resources Association 5410 Grosvenor Lane, Suite 220, Bethesda, MD 20814.

American Wildlife Association 1717 Massachusetts Ave. NW, Washington, DC 20036.

Bioregional Project (North American Bioregional Congress) Turtle Island Office, 1333 Overhulse Rd. NE, Olympia, WA 98502.

Center for Marine Conservation 1725 DeSales St. NW, Suite 500, Washington, DC 20036.

Center for Science in the Public Interest 1501 16th St. NW, Washington, DC 20036.

Citizens' Clearinghouse for Hazardous Waste P.O. Box 926, Arlington, VA 22216.

Clean Water Action 317 Pennsylvania Ave. SE, Washington, DC 20003.

Conservation Foundation 1250 42nd St. NW, Washington, DC 20076.

Council for Economic Priorities 30 Irving Place, New York, NY 10003.

Council for Solid Waste Solutions 1275 K St. NW, Suite 400, Washington, DC 20005.

Cousteau Society 930 W. 21st St., Norfolk, VA 23517.

Defenders of Wildlife 1244 19th St. NW, Washington, DC 20036.

Ducks Unlimited One Waterfowl Way, Long Grove, IL 60047.

Earth First! P.O. Box 2358, Lewiston, ME 04241.

Earth Island Institute 300 Broadway, Suite 28, San Francisco, CA 94133.

Elmwood Institute P.O. Box 5805, Berkeley, CA 94705.

Environmental Action, Inc. 1525 New Hampshire Ave. NW, Washington, DC 20036.

Environmental Defense Fund Inc. 257 Park Ave. South, New York, NY 10010.

Environmental Law Institute 1616 P St. NW, Suite 200, Washington, DC 20036.

Environmental Policy Institute 218 D St. SE, Washington, DC 20003.

Food and Agriculture Organization of the United Nations Via delle di Caracalla, Rome 00100 Italy.

Food First 1885 Mission St., San Francisco, CA 94103.

Friends of Animals P.O. Box 1244, Norwalk, CT 06856.

Friends of the Earth, Inc. 530 Seventh Ave. SE, Washington, DC 20003.

Global Greenhouse Network 1130 17th St. NW, Suite 530, Washington, DC 20036.

Global Tomorrow Coalition 1325 G Street NW, Suite 915, Washington, DC 20005.

Greenhouse Crisis Foundation 1130 17th St. NW, Suite 630, Washington, DC 20036.

Greenpeace USA, Inc. 1436 U St. NW, Washington, DC 20009.

Humane Society of the United States, Inc. 2100 L St. NW, Washington, DC 20037.

INFORM 381 Park Ave. South, New York, NY 10016.

Institute for Alternative Agriculture 9200 Edmonston Rd., Suite 117, Greenbelt, MD 20770.

Institute for Local Self-Reliance 2425 18th St. NW, Washington, DC 20009.

International Planned Parenthood Federation 105 Madison Ave., 7th Floor, New York, NY 10016.

International Union for the Conservation of Nature and Natural Resources (IUCN) Avenue du Mont Blanc, CH-1196 Gland, Switzerland (022.64 71 81).

Izaak Walton League of America 1401 Wilson Blvd., Level B, Arlington, VA 22209.

Land Institute Route 3, Salina, KS 67401.

League of Conservation Voters 2000 L St. NW, Suite 804, Washington, DC 20036. Political arm of Friends of the Earth.

League of Women Voters of the U.S. 1730 M St. NW, Washington, DC 20036.

National Audubon Society 950 Third Ave., New York, NY 10022.

National Clean Air Coalition 1400 16th St. NW, Washington, DC 20036.

National Coalition Against the Misuse of Pesticides 530 7th St. SE, Washington, DC 20003.

National Geographic Society 17th and M Sts. NW, Washington, DC 20036.

National Parks and Conservation Association 1015 31st St. NW, Washington, DC 20007.

National Recreation and Park Association 3101 Park Center Drive, 12th Floor, Alexandria, VA 22302.

National Solid Waste Management Association 1730 Rhode Island Ave. NW, Suite 100, Washington, DC 20036.

National Toxics Campaign 37 Temple Place, 4th Floor, Boston, MA 02111.

National Wildlife Federation 1400 16th St. NW, Washington, DC 20036. Research and education.

National Wood Energy Association 1730 North Lynn St., Suite 610, Arlington, VA 22209.

Natural Resources Defense Council 122 East 42nd St., New York, NY 10168.

The Nature Conservancy 1800 N. Lynn St., Arlington, VA 22209.

New Alchemy Institute 237 Hatchville Road, East Falmouth, MA 02536.

The Oceanic Society Executive Offices, 1536 16th St. NW, Washington, DC 20036.

Organization of Tropical Studies P.O. Box DM, Duke Station, Durham, NC 27706.

Permaculture Association P.O. Box 202, Orange, MA 01364.

Planet/Drum Foundation P.O. Box 31251, San Francisco, CA 94131.

Planned Parenthood Federation of America 810 Seventh Ave., New York, NY 10019.

Population Crisis Committee 1120 19th St. NW, Suite 550, Washington, DC 20036-3605.

Population Institute 110 Maryland Ave. NE, Suite 207, Washington, DC 20002.

Population Reference Bureau 777 14th St. NW, Suite 800, Washington, DC 20005.

Public Citizen 215 Pennsylvania Ave. SE, Washington, DC 20003.

Rainforest Action Network 300 Broadway, Suite 29A, San Francisco, CA 94133.

Renewable Natural Resources Foundation 5430 Grosvenor Ln., Bethesda, MD 20814.

Renew America 1001 Connecticut Ave. NW, Suite 719, Washington, DC 20036.

Resources for the Future 1616 P St. NW, Washington, DC 20036.

Rocky Mountain Institute 1739 Snowmass Creek Road, Snowmass, CO 81654.

Rodale Research Center 222 Main St., Emmaus, PA 18098.

Scientists' Institute for Public Information 355 Lexington Ave., New York, NY 10017.

Sea Shepherd Conservation Society P.O. Box 7000-S, Redondo Beach, CA 90277.

Sierra Club 730 Polk St., San Francisco, CA 94109.

Smithsonian Institution 1000 Jefferson Drive SW, Washington, DC 20560.

Social Investment Forum C.E.R.E.S. Project, 711 Atlantic Ave., Boston, MA 02111.

Society of American Foresters 5400 Grosvenor Lane, Bethesda, MD 20814.

Soil and Water Conservation Society 7515 N.E. Ankeny Road, Ankeny, IA 50021.

Student Conservation Association, Inc. P.O. Box 550, Charlestown, NH 03603.

Union of Concerned Scientists 26 Church St., Cambridge, MA 02238.

U.S. Public Interest Research Group 215 Pennsylvania Ave. SE, Washington, DC 20003.

The Wilderness Society 900 17th St. NW, Washington, DC 20006.

Work on Waste 82 Judson St., Canton, NY 13617.

World Resources Institute 1735 New York Ave. NW, Washington, DC 20006.

Worldwatch Institute 1776 Massachusetts Ave. NW, Washington, DC 20036.

World Wildlife Fund 1250 24th St. NW, Washington, DC 20037.

Zero Population Growth 1601 Connecticut Ave. NW, Washington, DC 20009.

Addresses of Federal and International Agencies

Bureau of Land Management U.S. Department of the Interior, 18th and C Sts., Room 5660, Washington, DC 20240.

Bureau of Mines 2401 E St. NW, Washington, DC 20241.

Bureau of Reclamation Washington, DC 20240.

Congressional Research Service 101 Independence Ave. SW, Washington, DC 20540.

Conservation and Renewable Energy Inquiry and Referral Service P.O. Box 8900, Silver Spring, MD 20907, (800) 523-2929.

Consumer Product Safety Commission Washington, DC 20207.

Council on Environmental Quality 722 Jackson Place NW, Washington, DC 20006.

Department of Agriculture 14th St. and Jefferson Dr. SW, Washington, DC 20250.

Department of Commerce 14th St. between Constitution Ave. and E St. NW, Washington, DC 20230.

Department of Energy Forrestal Building, 1000 Independence Ave. SW, Washington, DC 20585.

Department of Health and Human Services 200 Independence Ave. SW, Washington, DC 20585.

Department of Housing and Urban Development 451 Seventh St. SW, Washington, DC 20410.

Department of the Interior 18th and C Sts. NW, Washington, DC 20240.

Department of Transportation 400 Seventh St. SW, Washington, DC 20590.

Environmental Protection Agency 401 M St. SW, Washington, DC 20460.

Federal Energy Regulatory Commission 825 N. Capitol St. NE, Washington, DC 20426.

Fish and Wildlife Service Department of the Interior, 18th and C Sts. NW, Washington, DC 20240.

Food and Drug Administration Department of Health and Human Services, 5600 Fishers Lane, Rockville, MD 20852.

Forest Service P.O. Box 96090, Washington, DC 20013.

Geological Survey 12201 Sunrise Valley Drive, Reston, VA 22092.

Government Printing Office Washington, DC 20402.

International Whaling Commission The Red House, 135 Station Rd., Histon, Cambridge CB4 4NP England 02203 3971.

Marine Mammal Commission 1625 I St. NW, Washington, DC 20006.

National Academy of Science Washington, DC 20550.

National Aeronautics and Space Administration 400 Maryland Ave. SW, Washington, DC 20546.

National Cancer Institute 9000 Rockville Pike, Bethesda, MD 20892.

National Center for Appropriate Technology 3040 Continental Dr., Butte, MT 59701.

National Center for Atmospheric Research P.O. Box 3000, Boulder, CO 80307.

National Marine Fisheries Service U.S. Dept. of Commerce, NOAA, 1335 East-West Highway, Silver Spring, MD 20910.

National Oceanic and Atmospheric Administration Rockville, MD 20852.

National Park Service Department of the Interior, P.O. Box 37127, Washington, DC 20013.

National Science Foundation 1800 G St. NW, Washington, DC 20550

National Solar Heating and Cooling Information Center P.O. Box 1607, Rockville, MD 20850.

National Technical Information Service U.S. Department of Commerce, 5285 Port Royal Rd., Springfield, VA 22161.

Nuclear Regulatory Commission 1717 H St. NW, Washington, DC 20555.

Occupational Safety and Health Administration Department of Labor, 200 Constitution Ave. NW, Washington, DC 20210.

Office of Ocean and Coastal Resource Management 1825 Connecticut Ave., Suite 700, Washington, DC 20235.

Office of Surface Mining Reclamation and Enforcement 1951 Constitution Ave. NW, Washington, DC 20240.

Office of Technology Assessment U.S. Congress, 600 Pennsylvania Ave. SW, Washington, DC 20510.

Soil Conservation Service P.O. Box 2890, Washington, DC 20013.

Solar Energy Research Institute 6536 Cole Blvd., Golden, CO 80401.

United Nations Environment Programme New York Liaison Office, Room DC2-0803, United Nations, New York, NY 10017.

U.S. International Development Cooperation Agency 320 21st St. NW, Washington, DC 20523.

MAJOR U.S. RESOURCE CONSERVATION AND ENVIRONMENTAL LEGISLATION

General

National Environmental Policy Act of 1969 (NEPA)

International Environmental Protection Act of 1983

Energy

National Energy Act of 1978, 1980

National Appliance Energy Conservation Act of 1987

Water Quality

Water Quality Act of 1965

Water Resources Planning Act of 1965

Federal Water Pollution Control Acts of 1965, 1972

Ocean Dumping Act of 1972

Ocean Dumping Ban Act of 1988

Safe Drinking Water Act of 1974, 1984

Water Resources Development Act of 1986

Clean Water Act of 1977, 1987

Air Quality

Clean Air Act of 1963, 1965, 1970, 1977, 1990

Noise Control

Noise Control Act of 1965

Quiet Communities Act of 1978

Resources and Solid Waste Management

Solid Waste Disposal Act of 1965

Resource Recovery Act of 1970

Resource Conservation and Recovery Act of 1976

Marine Plastic Pollution Research and Control Act of 1987

Toxic Substances

Hazardous Materials Transportation Act of 1975

Toxic Substances Control Act of 1976

Resource Conservation and Recovery Act of 1976

Comprehensive Environmental Response, Compensation, and Liability (Superfund) Act of 1980, 1986

Nuclear Waste Policy Act of 1982

Pesticides

Federal Insecticide, Fungicide, and Rodenticide Control Act of 1972, 1988

Wildlife Conservation

Anadromous Fish Conservation Act of 1965

Fur Seal Act of 1966

National Wildlife Refuge System Act of 1966, 1976, 1978

Species Conservation Act of 1966, 1969

Marine Mammal Protection Act of 1972

Marine Protection, Research, and Sanctuaries Act of 1972

Endangered Species Act of 1973, 1982, 1985, 1988

Fishery Conservation and Management Act of 1976, 1978, 1982

Whale Conservation and Protection Study Act of 1976

Fish and Wildlife Improvement Act of 1978

Fish and Wildlife Conservation Act of 1980 (Nongame Act)

Land Use and Conservation

Taylor Grazing Act of 1934

Wilderness Act of 1964

Multiple Use Sustained Yield Act of 1968

Wild and Scenic Rivers Act of 1968

National Trails System Act of 1968

National Coastal Zone Management Act of 1972, 1980

Forest Reserves Management Act of 1974, 1976

Forest and Rangeland Renewable Resources Act of 1974, 1978

Federal Land Policy and Management Act of 1976

National Forest Management Act of 1976

Soil and Water Conservation Act of 1977

Surface Mining Control and Reclamation Act of 1977

Antarctic Conservation Act of 1978

Endangered American Wilderness Act of 1978

Alaskan National Interests Lands Conservation Act of 1980

Coastal Barrier Resources Act of 1982

Food Security Act of 1985

GLOSSARY

abiotic Nonliving. Compare *biotic.*

absolute resource scarcity Situation in which there are not enough actual or affordable supplies of a resource left to meet present or future demand. Compare *relative resource scarcity.*

abyssal zone Bottom zone of the ocean, consisting of deep, dark, cold water and the ocean bottom (benthos). Compare *bathyal zone, euphotic zone.*

accelerated eutrophication See *cultural eutrophication.*

acid deposition The falling of acids and acid-forming compounds from the atmosphere to the earth's surface. Acid deposition is commonly known as *acid rain,* a term that refers to only wet deposition of droplets of acids and acid-forming compounds.

acidic See *acid solution.*

acid rain See *acid deposition.*

acid solution Any water solution that has more hydrogen ions (H⁺) than hydroxide ions (OH⁻); any water solution with a pH less than 7. Compare *basic solution, neutral solution.*

active solar heating system System that uses solar collectors to capture energy from the sun and store it as heat for space heating and heating water. A liquid or air pumped through the collectors transfers the captured heat to a storage system such as an insulated water tank or rock bed. Pumps or fans then distribute the stored heat or hot water throughout a dwelling as needed. Compare *passive solar heating system.*

acute effect Harmful effect that appears shortly after exposure to a toxic substance or disease-causing organism. Compare *chronic effect.*

advanced sewage treatment Specialized chemical and physical processes that reduce the amount of specific pollutants left in wastewater after primary and secondary sewage treatment. This type of treatment is usually expensive. See also *primary sewage treatment, secondary sewage treatment.*

aerobic organism Organism that needs oxygen to stay alive. Compare *anaerobic organism.*

age structure (age distribution) Percentage of the population, or the number of people of each sex, at each age level in a population.

Agricultural Revolution Gradual shift from small, mobile hunting and gathering bands to settled agricultural communities. It began about 10,000 to 12,000 years ago.

agroforestry Planting trees and crops together.

air pollutant See *air pollution, primary air pollutant, secondary air pollutant.*

air pollution One or more chemicals or substances in high enough concentrations in the air to harm humans, other animals, vegetation, or materials. Such chemicals or physical conditions (such as excess heat or noise) are called air pollutants. See *primary air pollutant, secondary air pollutant.*

algae Simple, one-celled or many-celled plants that usually carry out photosynthesis in rivers, lakes, ponds, oceans, and other surface waters.

algal bloom Population explosion of algae in surface waters due to an increase in plant nutrients such as nitrates and phosphates.

alien species See *immigrant species.*

alkaline solution See *basic solution.*

alley cropping Planting of crops in strips with rows of trees or shrubs on each side.

alpha particle Positively charged matter consisting of two neutrons and two protons that is emitted as a form of radioactivity from the nuclei of some radioisotopes. See also *beta particle, gamma rays.*

altitude Height above sea level. Compare *latitude.*

ambient Outdoor.

anaerobic organism Organism that does not need oxygen to stay alive. Compare *aerobic organism.*

ancient forest Uncut virgin forest or old secondary forest. See *old-growth forest, secondary forest.*

animal manure Dung and urine of animals that can be used as a form of organic fertilizer. Compare *green manure.*

annual Plant that grows, sets seed, and dies in a single year. Compare *perennial.*

annual rate of natural change Annual rate at which the size of a population changes, usually expressed in percent as the difference between crude birth rate and crude death rate divided by 10.

aquaculture Growing and harvesting of fish and shellfish for human use in freshwater ponds, irrigation ditches, and lakes or in cages or fenced-in areas of coastal lagoons and estuaries. See *fish farming, fish ranching.*

aquatic Pertaining to water. Compare *terrestrial.*

aquatic ecosystem Any water-based ecosystem such as a stream, pond, lake, or ocean. Compare *biome.*

aquifer Porous, water-saturated layers of underground rock that can yield an economically significant amount of water. See *confined aquifer, unconfined aquifer.*

aquifer depletion Withdrawal of groundwater from an aquifer faster than it is recharged by precipitation.

arable land Land that can be cultivated to grow crops.

area strip mining Cutting deep trenches to remove minerals such as coal and phosphate found near the earth's surface in flat or rolling terrain. Compare *contour strip mining, open-pit surface mining.*

arid Dry. A desert or other area with an arid climate has little precipitation.

artificial reservoir See *reservoir.*

atmosphere The whole mass of air surrounding the earth. See *stratosphere, troposphere.*

atoms Minute particles that are the basic building blocks of all chemical elements and thus all matter.

autotroph See *producer.*

average life expectancy at birth See *life expectancy.*

average per capita GNP Annual gross national product (GNP) of a country divided by its total population. See *average per capita real GNP, gross national product.*

average per capita NEW Annual net economic welfare (NEW) of a country divided by its total population. See *average per capita real NEW, net economic welfare.*

average per capita real GNP Average per capita GNP adjusted for inflation.

average per capita real NEW Average per capita NEW adjusted for inflation. See *average per capita NEW, net economic welfare.*

bacteria One-celled organisms. Some transmit diseases. Most act as decomposers that break down dead organic matter into substances that dissolve in water and are used as nutrients by plants.

basic See *basic solution.*

basic solution Water solution with more hydroxide ions (OH⁻) than hydrogen ions (H⁺); water solution with a pH greater than 7. Compare *acidic solution, neutral solution.*

bathyal zone Cold, fairly dark ocean zone below the euphotic zone, in which there is some sunlight but not enough for photosynthesis. Compare *abyssal zone, benthic zone, euphotic zone.*

benign tumor A growth of cells that are reproducing at abnormal rates but remain within the tissue where it develops. Compare *cancer.*

benthic zone Bottom of a body of water. Compare *abyssal zone, bathyal zone, euphotic zone, limnetic zone, littoral zone.*

beta particle Swiftly moving electron emitted by the nucleus of a radioactive isotope. See also *alpha particle, gamma rays.*

biodegradable Material that can be broken down into simpler substances (elements and compounds) by bacteria or other decomposers. Paper and most organic wastes, such as animal manure, are biodegradable but can take decades to biodegrade in modern landfills. Compare *nonbiodegradable.*

biodiversity See *biological diversity.*

biofuel Gas or liquid fuel (such as ethyl alcohol) made from plant material (biomass).

biogeochemical cycle Natural processes that recycle nutrients in various chemical forms from the environment, to organisms, and then back to the environment. Examples are the carbon, oxygen, nitrogen, phosphorus, and hydrologic cycles.

biological amplification Increase in concentration of DDT, PCBs, and other slowly degradable, fat-soluble chemicals in organisms at successively higher trophic levels of a food chain or web.

biological community See *community.*

biological diversity Variety of different species, genetic variability among individuals within each species, and variety of ecosystems. Compare *ecological diversity, genetic diversity, species diversity.*

biological evolution See *evolution.*

biological magnification See *biological amplification.*

biological oxygen demand (BOD) Amount of dissolved oxygen needed by aerobic decomposers to break down the organic materials in a given volume of water at a certain temperature over a specified time period.

biological pest control Control of pest populations by natural predators, parasites, or disease-causing bacteria and viruses (pathogens).

biomass Total dry weight of all living organisms that can be supported at each trophic level in a food chain or web; dry weight of all organic matter in plants and animals in an ecosystem; plant materials and animal wastes used as fuel.

biome Large land (terrestrial) ecosystem such as a forest, grassland, or desert. Compare *aquatic ecosystem.*

biosphere The living and dead organisms found near the earth's surface in parts of the lithosphere, atmosphere, and hydrosphere. See also *ecosphere.*

biotic Living. Living organisms make up the biotic parts of ecosystems. Compare *abiotic.*

birth rate See *crude birth rate.*

bitumen Gooey, black, high-sulfur, heavy oil extracted from tar sand and then upgraded to synthetic fuel oil. See *tar sand.*

breeder nuclear fission reactor Nuclear fission reactor that produces more nuclear fuel than it consumes by converting nonfissionable uranium-238 into fissionable plutonium-239.

calorie Unit of energy; amount of energy needed to raise the temperature of 1 gram of water 1°C. See also *kilocalorie.*

cancer Group of more than 120 different diseases—one for most major cell types in the human body. Each type of cancer produces a tumor in which cells multiply uncontrollably and invade surrounding tissue.

capital goods Tools, machinery, equipment, factory buildings, transportation facilities, and other manufactured items made from natural resources and used to produce and distribute consumer goods and services. Compare *labor, natural resources.*

carbon cycle Cyclic movement of carbon in different chemical forms from the environment, to organisms, and then back to the environment.

carcinogen Chemical or form of high-energy radiation that can directly or indirectly cause a cancer.

carnivore Animal that feeds on other animals. Compare *herbivore, omnivore.*

carrying capacity Maximum population of a particular species that a given habitat can support over a given period of time.

cell Basic structural unit of all organisms.

cellular aerobic respiration Complex process that occurs in the cells of plants and animals in which nutrient organic molecules such as glucose ($C_6H_{12}O_6$) combine with oxygen (O_2) and produce carbon dioxide (CO_2), water (H_2O), and energy. Compare *photosynthesis.*

CFCs See *chlorofluorocarbons.*

chain reaction Series of nuclear fissions taking place within the critical mass of a fissionable isotope that release an enormous amount of energy in a short time.

chemical One of the millions of different elements and compounds found in the universe.

chemical change Interaction between chemicals in which there is a change in the chemical composition of the elements or compounds involved. Compare *physical change.*

chemical reaction See *chemical change.*

chemosynthesis Process in which certain organisms (mostly specialized bacteria) convert chemicals obtained from the environment into nutrient molecules without using sunlight. Compare *photosynthesis.*

chlorinated hydrocarbon Organic compound made up of atoms of carbon, hydrogen, and chlorine. Examples are DDT and PCBs.

chlorofluorocarbons (CFCs) Organic compounds made up of atoms of carbon, chlorine, and fluorine. An example is Freon-12 (CCl_2F_2), used as a refrigerant in refrigerators and air conditioners and in plastics such as Styrofoam. Gaseous CFCs can deplete the ozone layer when they slowly rise into the stratosphere and their chlorine atoms react with ozone molecules.

chromosome A grouping of various genes and associated proteins in plant and animal cells which carry certain types of genetic information. See *gene.*

chronic effect A harmful effect from exposure to a toxic substance or disease-causing organism that is delayed and usually long-lasting. Compare *acute effect.*

clearcutting Method of timber harvesting in which all trees in a forested area are removed in a single cutting. Compare *selective cutting, seed-tree cutting, shelterwood cutting, whole-tree harvesting.*

climate General pattern of atmospheric or weather conditions, seasonal variations, and weather extremes in a region over a long period—at least 30 years. Compare *weather.*

climax community See *mature community.*

coal Solid, combustible mixture of organic compounds with 30% to 98% carbon by weight, mixed with varying amounts of water and small amounts of compounds containing sulfur and nitrogen. It is formed in several stages as the remains of plants are subjected to intense heat and pressure over millions of years.

coal gasification Conversion of solid coal to synthetic natural gas (SNG), or a gaseous mixture that can be burned as a fuel.

coal liquefaction Conversion of solid coal to a liquid fuel such as synthetic crude oil or methanol.

coastal wetland Land along a coastline, extending inland from an estuary that is flooded with salt water all or part of the year. Examples are marshes, bays, lagoons, tidal flats, and mangrove swamps. Compare *inland wetland.*

coastal zone Relatively warm, nutrient-rich, shallow part of the ocean that extends from the high-tide mark on land to the edge of a shelflike extension of continental land masses known as the continental shelf. Compare *open sea.*

cogeneration Production of two useful forms of energy such as high-temperature heat and electricity from the same process.

commensalism An interaction between organisms of different species in which one type of organism benefits while the other type is neither helped nor harmed to any great degree. Compare *mutualism.*

commercial extinction Depletion of the population of a wild species used as a resource to a point where it is no longer profitable to harvest the species.

commercial fishing Finding and catching fish for sale. Compare *sport fishing, subsistence fishing.*

commercial hunting Killing of wild animals for profit from sale of their furs or other parts. Compare *sport hunting, subsistence hunting.*

commercial inorganic fertilizer Commercially prepared mixtures of plant nutrients such as nitrates, phosphates, and potassium applied to the soil to help restore fertility and increase crop yields. Compare *organic fertilizer.*

common property resource Resource to which people have virtually free and unmanaged access. Examples are air, fish in parts of the ocean not under the control of a coastal country, migratory birds, and the ozone content of the stratosphere. See *tragedy of the commons.*

commons See *common property resource.*

community Populations of different plants and animals living and interacting in an area at a particular time.

competition Two or more individual organisms of a single species (*intraspecific competition*) or two or more individuals of different species (*interspecific competition*) attempting to use the same scarce resources in the same ecosystem.

competitive exclusion principle No two species in the same ecosystem can occupy exactly the same ecological niche indefinitely.

compost Partially decomposed organic plant and animal matter that can be used as a soil conditioner or fertilizer.

composting Partial breakdown of organic plant and animal matter by aerobic bacteria to produce a material (compost) that can be used as a soil conditioner or fertilizer.

compound Combination of two or more different chemical elements held together by chemical bonds. Compare *element.* See *inorganic compound, organic compound.*

concentration Amount of a chemical in a particular volume or weight of air, water, soil, or other medium.

confined aquifer Groundwater between two layers of rock that have a low permeability, such as clay or shale. Compare *unconfined aquifer.*

conifer See *coniferous trees.*

coniferous trees Cone-bearing trees, mostly evergreens, that have needle-shaped or scale-like leaves. They produce wood known commercially as softwood. Compare *deciduous trees.*

conservation Use, management, and protection of resources so that they are not degraded, depleted, or wasted and are available on a sustainable basis for use by present and future generations. Methods include preservation, balanced multiple use, reducing unnecessary waste, recycling, reuse, and decreased use of resources.

conservationists People who believe that resources should be used, managed, and protected so that they will not be degraded and unnecessarily wasted and will be available to present and future generations. See *preservationists, scientific conservationists, sustainable-earth conservationists.*

conservation-tillage farming Crop cultivation in which the soil is disturbed little (*minimum-tillage farming*), or not at all (*no-till farming*), to reduce soil erosion, lower labor costs, and save energy. Compare *conventional-tillage farming.*

constancy Ability of a living system, such as a population, to maintain a certain size. Compare *inertia, resilience.*

consumer Organism that cannot produce its own food and must get it by eating or decomposing other organisms; generally divided into *primary consumers* (herbivores), *secondary consumers* (carnivores), and *microconsumers* (decomposers). In economics, one who uses economic goods.

consumption overpopulation Situation in which people use resources at such a high rate and without sufficient pollution control that significant depletion, pollution, and environmental degradation occur. Compare *people overpopulation*.

consumptive water use See *water consumption*.

continental shelf Shallow undersea land adjacent to a continent.

contour farming Plowing and planting across rather than up and down the slope of land to help retain water and reduce soil erosion.

contour strip mining Cutting a series of shelves or terraces on the side of a hill or mountain to remove a mineral such as coal from a deposit found near the earth's surface. Compare *area strip mining, open-pit surface mining*.

contraceptive Physical, chemical, or biological method used to prevent pregnancy.

conventional-tillage farming Making a planting surface by plowing land, disking it several times to break up the soil, and then smoothing the surface. Compare *conservation-tillage farming*.

core Inner zone of the earth. It has a solid inner part, surrounded by a fluidlike shell of molten material. Compare *crust, mantle*.

cost-benefit analysis Estimates and comparison of short-term and long-term costs (losses) and benefits (gains) from an economic decision. If the estimated benefits exceed the estimated costs, the decision to buy an economic good or provide a public good is considered worthwhile.

critical mass Amount of fissionable isotopes needed to sustain a nuclear fission chain reaction.

critical mineral A mineral necessary to the economy of a country. Compare *strategic mineral*.

crop rotation Planting the same field or areas of fields with different crops from year to year to reduce depletion of soil nutrients. A plant such as corn, tobacco, or cotton, which remove large amounts of nitrogen from the soil, is planted one year. The next year a legume such as soybeans, which add nitrogen to the soil, is planted.

crown fire Extremely hot forest fire that burns ground vegetation and tree tops. Compare *ground fire, surface fire*.

crude birth rate Annual number of live births per 1,000 persons in the population of a geographical area at the midpoint of a given year. Compare *crude death rate*.

crude death rate Annual number of deaths per 1,000 persons in the population of a geographical area at the midpoint of a given year. Compare *crude birth rate*.

crude oil Gooey liquid made up mostly of hydrocarbon compounds and small amounts of compounds containing oxygen, sulfur, and nitrogen. Extracted from underground deposits, it is sent to oil refineries, where it is converted to heating oil, diesel fuel, gasoline, and tar.

crust Outer zone of the solid earth. See *lithosphere*. Compare *core, mantle*.

cultural eutrophication Overnourishment of aquatic ecosystems with plant nutrients (mostly nitrates and phosphates) due to human activities such as agriculture, urbanization, and discharges from industrial plants and sewage treatment plants. See *eutrophication*.

DDT Dichlorodiphenyltrichloroethane, a chlorinated hydrocarbon that has been widely used as a pesticide.

death rate See *crude death rate*.

deciduous trees Trees such as oaks and maples that lose their leaves during part of the year. Compare *coniferous trees*.

decomposers Organisms, such as bacteria, mushrooms, and fungi, that get nutrients by breaking down organic matter in the wastes and dead bodies of other organisms into simpler chemicals. Most of these chemicals are returned to the soil and water for reuse by producers. Compare *consumers, detritivores, producers*.

deep ecology See *sustainable-earth worldview*.

deforestation Removal of trees from a forested area without adequate replanting.

degradable See *biodegradable*.

degree of urbanization Percentage of the population in the world or a country living in areas with a population of more than 2,500 people.

delta Built-up deposit of river-borne sediments at the mouth of a river.

demographic transition Hypothesis that as countries become industrialized, they have declines in death rates followed by declines in birth rates.

demography Study of characteristics and changes in the size and structure of the human population in the world or other geographical area.

depletion time How long it takes to use a certain fraction—usually 80%—of the known or estimated supply of a nonrenewable resource at an assumed rate of use. Finding and extracting the remaining 20% usually cost more than it is worth.

desalination Purification of salt water or brackish (slightly salty) water by removing dissolved salts.

desert Type of terrestrial ecosystem where evaporation exceeds precipitation and the average amount of precipitation is less than 25 centimeters (10 ten inches) a year. Such areas have little vegetation or have widely spaced, mostly low vegetation. Compare *forest, grassland*.

desertification Conversion of rangeland, rain-fed cropland, or irrigated cropland to desertlike land with a drop in agricultural productivity of 10% or more. It is usually caused by a combination of overgrazing, soil erosion, prolonged drought, and climate change.

desirability quotient A number expressing the results of risk-benefit analysis by dividing the estimate of the benefits to society of using a particular product or technology by its estimated risks. See *risk-benefit analysis*. Compare *cost-benefit analysis*.

detritivores Consumer organisms that feed on detritus, or dead organic plant and animal matter. The two major types are *detritus feeders* and *decomposers*.

detritus Dead organic plant and animal matter.

detritus feeders Organisms that injest dead organisms and their cast-off parts and organic wastes. Examples are earthworms, clams, and crabs. Compare *decomposers*.

deuterium (D: hydrogen-2) Isotope of the element hydrogen with a nucleus containing one proton and one neutron, and a mass number of 2. Compare *tritium*.

developed country See *more developed country*.

differential reproduction Ability of individuals with adaptive genetic traits to outreproduce individuals without such traits. See also *natural selection*.

dissolved oxygen (DO) content (level) Amount of oxygen gas (O_2) dissolved in a certain amount of water at a particular temperature and pressure, often expressed as a concentration in parts of oxygen per million parts of water.

diversity Variety. In biology the number of different species in an ecosystem (*species diversity*) or diversity in the genetic makeup of different species or within a single species (*genetic diversity*).

DNA (deoxyribonucleic acid) Large molecules that carry genetic information in living organisms. They are found in the cells of organisms.

doubling time Time in years that it takes for a quantity growing exponentially to double in size. It can be calculated by dividing the annual percentage growth rate into 70.

drainage basin See *watershed*.

dredge spoils Materials scraped from the bottoms of harbors and rivers to maintain shipping channels. They are often contaminated with high levels of toxic substances that have settled out of the water. See *dredging*.

dredging Type of surface mining in which materials such as sand and gravel are scooped up from seabeds. It is also used to remove sediment from streams and harbors to maintain shipping channels. See *dredge spoils*.

drip irrigation Using small tubes or pipes to deliver small amounts of irrigation water to the roots of plants.

drought Condition in which an area does not get enough water because of lower than normal precipitation, higher than normal temperatures that increase evaporation, or both.

dust dome Dome of heated air that surrounds an urban area and traps pollutants, especially suspended particulate matter. See also *urban heat island*.

early-successional species Wild animal species found in pioneer communities of plants at the early stage of ecological succession. Compare *late-successional species, midsuccessional species, wilderness species*.

ecological diversity The variety of ecosystems—forests, deserts, grasslands, oceans, streams, lakes, and other biological communities—interacting with one another and with their nonliving environment. See *biodiversity*. Compare *genetic diversity, species diversity*.

ecological land-use planning Method for deciding how land should be used by developing an integrated model that considers geological, ecological, health, and social variables.

ecological niche Description of all the physical, chemical, and biological factors that a species needs to survive, stay healthy, and reproduce in an ecosystem.

ecological succession Process in which communities of plant and animal species in a particular area are replaced over time by a series of different and usually more complex communities. See *primary ecological succession, secondary ecological succession*.

ecology Study of the interactions of living organisms with each other and with their environment; study of the structure and functions of nature.

economic decision Choosing what to do with scarce resources; deciding what goods and services to produce, how to produce them, how much to produce, and how to distribute them to people.

economic depletion Exhaustion of 80% of the estimated supply of a nonrenewable resource. Finding, extracting, and processing the remaining 20% usually cost more than it is worth; may also apply to the depletion of a potentially renewable resource, such as trees or a species of fish.

economic good Any service or material item that gives people satisfaction and whose present or ultimate supply is limited.

economic growth Increase in the real value of all final goods and services produced by an economy; an increase in real GNP. Compare *productivity*.

economic needs Types and amounts of certain economic goods—food, clothing, water, oxygen, shelter—that each of us must have to survive and to stay healthy. Compare *economic wants*. See also *poverty*.

economic resources Natural resources, capital goods, and labor used in an economy to produce material goods and services. See *capital goods, labor, natural resources*.

economics Study of how individuals and groups make decisions about what to do with scarce resources to meet their needs and wants.

economic wants Economic goods that go beyond our basic economic needs. These wants are influenced by the customs and conventions of the society we live in and by our level of affluence. Compare *economic needs*.

economy System of production, distribution, and consumption of economic goods.

ecosphere Collection of living and dead organisms (biosphere) interacting with one another and their nonliving environment (energy and chemicals) throughout the world. See also *biosphere*.

ecosystem Community of organisms interacting with one another and with the chemical and physical factors making up their environment.

efficiency Measure of how much output of energy or of a product is produced by a certain input of energy, materials, or labor. See *energy efficiency*.

egg pulling Collecting eggs produced in the wild pairs of a critically endangered species and hatching the eggs in zoos or research centers.

electron Tiny particle moving around outside the nucleus of an atom. Each electron has one unit of negative charge ($-$) and almost no mass.

element Chemical, such as hydrogen (H), iron (Fe), sodium (Na), carbon (C), nitrogen (N), or oxygen (O), whose distinctly different atoms serve as the basic building blocks of all matter. There are 92 naturally occurring elements. Another 15 have been made in laboratories. Two or more elements combine to form compounds that make up most of the world's matter. Compare *compound*.

emigration Migration of people out of one country or area to take up permanent residence in another country or area. Compare *immigration*.

endangered species Wild species with so few individual survivors that the species could soon become extinct in all or most of its natural range. Compare *threatened species*.

energy Ability to do work by moving matter or by causing a transfer of heat between two objects at different temperatures.

energy conservation Reduction or elimination of unnecessary energy use and waste. See *energy efficiency*.

energy efficiency Percentage of the total energy input that does useful work and is not converted into low-temperature, usually useless, heat in an energy conversion system or process. See *net useful energy*.

energy quality Ability of a form of energy to do useful work. High-temperature heat and the chemical energy in fossil fuels and nuclear fuels is concentrated high-quality energy. Low-quality energy such as low-temperature heat is dispersed or diluted and cannot do much useful work. See *high-quality energy, low-quality energy*.

enhanced oil recovery Removal of some of the heavy oil left in an oil well after primary and secondary recovery. Compare *primary oil recovery, secondary oil recovery*.

environment All external conditions that affect an organism or other specified system during its lifetime.

environmental degradation Depletion or destruction of a potentially renewable resource such as soil, grassland, forest, or wildlife by using it at a faster rate than it is naturally replenished. If such use continues, the resource can become nonrenewable on a human time scale or nonexistent (extinct). See also *sustainable yield*.

environmentalists People who are primarily concerned with preventing pollution and degradation of the air, water, and soil. See *conservationists*.

EPA Environmental Protection Agency. It is responsible for managing federal efforts in the United States to control air and water pollution, radiation and pesticide hazards, ecological research, and solid waste disposal.

epidemiology Study of the patterns of disease or other harmful effect from toxic exposure within defined groups of people.

epilimnion Upper layer of warm water with high levels of dissolved oxygen in a stratified lake. Compare *hypolimnion, thermocline*.

erosion See *soil erosion*.

estuarine zone Area near the coastline that consists of estuaries and coastal saltwater wetlands, extending to the edge of the continental shelf.

estuary Zone along a coastline where fresh water from rivers and streams and runoff from the land mix with seawater.

ethics What we believe to be right or wrong behavior.

euphotic zone Surface layer of an ocean, lake, or other body of water which gets enough sunlight for photosynthesis. Compare *abyssal zone, bathyal zone*.

eutrophication Physical, chemical, and biological changes that take place after a lake, estuary, or slow-flowing stream receives inputs of plant nutrients—mostly nitrates and phosphates—from natural erosion and runoff from the surrounding land basin. See also *cultural eutrophication*.

eutrophic lake Lake with a large or excessive supply of plant nutrients—mostly nitrates and phosphates. Compare *mesotrophic lake, oligotrophic lake*.

evaporation Physical change in which a liquid changes into a vapor or gas.

even-aged management Method of forest management in which trees, usually of a single species, in a given stand are maintained at about the same age and size and are harvested all at once so a new stand may grow. Compare *uneven-aged management*.

even-aged stand Forest area where all trees are about the same age. Usually, such stands contain trees of only one or two species. See *even-aged management, tree farm*. Compare *uneven-aged management, uneven-aged stand*.

evergreen plants Pines, spruces, firs, and other plants that keep some of their leaves or needles throughout the year. Compare *deciduous trees*.

evolution Changes in the genetic composition (gene pool) of a population exposed to new environmental conditions as a result of differential reproduction. Evolution can lead to the splitting of a single species into two or more different species. See also *differential reproduction, natural selection, speciation*.

exhaustible resource See *nonrenewable resource*.

exponential growth Growth in which some quantity, such as population size, increases by a constant percentage of the whole during each year or other time period; when the increase in quantity over time is plotted, this type of growth yields a curve shaped like the letter *J*. Compare *linear growth*.

external benefit Beneficial social effect of producing and using an economic good that is not included in the market price of the good. Compare *external cost, internal cost, true cost*.

external cost Harmful social effect of producing and using an economic good that is not included in the market price of the good. Compare *external benefit, internal cost, true cost*.

externalities Social benefits ("goods") and social costs ("bads") not included in the market price of an economic good. See *external benefit, external cost*. Compare *internal cost, true cost*.

extinction Complete disappearance of a species from the earth. This happens when a species cannot adapt and successfully reproduce under new environmental conditions. Compare *speciation*. See also *endangered species, threatened species*.

factors of production See *economic resources*.

family planning Providing information, clinical services, and contraceptives to help couples choose the number and spacing of children they want to have.

famine Widespread malnutrition and starvation in a particular area because of a shortage of food, usually caused by drought, war, flood, earthquake, or other catastrophic event that disrupts food production and distribution.

feedlot Confined outdoor or indoor space used to raise hundreds to thousands of domesticated livestock. Compare *rangeland*.

fertilizer Substance that adds inorganic or organic plant nutrients to soil and improves its ability to grow crops, trees, or other vegetation. See *commercial inorganic fertilizer, organic fertilizer*.

first law of ecology We can never do merely one thing. Any intrusion into nature has numerous effects, many of which are unpredictable.

first law of energy See *first law of thermodynamics*.

first law of thermodynamics (energy) In any physical or chemical change, any movement of matter from one place to another, or any change in temperature, energy is neither created nor destroyed but may be transformed from one form to another; you can't get more energy out of something than you put in; in terms of energy quantity you can't get something for nothing, or there is no free lunch. See also *second law of thermodynamics*.

fishery Concentrations of particular aquatic species suitable for commercial harvesting in a given ocean area or inland body of water.

fish farming Form of aquaculture in which fish are cultivated in a controlled pond or other environment and harvested when they reach the desired size. See also *fish ranching*.

fishing Finding and capturing a desirable species of fish or shellfish.

fish ranching Form of aquaculture in which members of an anadromous fish species such as salmon are held in captivity for the first few years of their lives, released, and then harvested as adults when they return from the ocean to their freshwater birthplace to spawn. See also *fish farming*.

fissionable isotope Isotope that can split apart when hit by a neutron or other particle moving at the right speed and thus undergo nuclear fission. Examples are uranium-235 and plutonium-239. Compare *nonfissionable isotope*.

floodplain Land along a stream that is periodically flooded when the stream overflows its banks.

fluidized-bed combustion (FBC) Process for burning coal more efficiently, cleanly, and cheaply. A stream of hot air is used to suspend a mixture of powdered coal and limestone during combustion. About 90% to 98% of the sulfur dioxide produced during combustion is removed by reaction with limestone to produce solid calcium sulfate.

flyway Generally fixed route along which waterfowl migrate from one area to another at certain seasons of the year.

food chain Series of organisms, each eating or decomposing the preceding one. Compare *food web*.

food web Complex network of many interconnected food chains and feeding interactions. Compare *food chain*.

forage Vegetation eaten by animals, especially grazing and browsing animals.

forest Terrestrial ecosystem (biome) with enough average annual precipitation (at least 76 centimeters or 30 inches) to support growth of various species of trees and smaller forms of vegetation. Compare *desert, grassland*.

fossil fuel Buried deposits of decayed plants and animals that have been converted to crude oil, coal, natural gas, or heavy oils by exposure to heat and pressure in the earth's crust over hundreds of millions of years. See *coal, crude oil, natural gas*.

Freons See *chlorofluorocarbons*.

frontier worldview See *throwaway worldview*.

fungicide Chemical used to kill fungi that damage crops.

fungus Type of decomposer; a plant without chlorophyll that gets its nourishment by breaking down the organic matter of other plants. Examples are molds, yeasts, and mushrooms.

Gaia hypothesis Proposal that the earth is alive and can be considered a gigantic superorganism.

game See *game species*.

game species Type of wild animal that people hunt or fish for fun and recreation and sometimes for food.

gamma rays High-energy, ionizing, electromagnetic radiation emitted by some radioisotopes. Like X rays, they easily penetrate body tissues.

gasohol Vehicle fuel consisting of a mixture of gasoline and ethyl or methyl alcohol—typically 10% to 23% ethanol by volume.

gene pool All genetic (hereditary) information contained in a reproducing population of a particular species.

genes The parts of various DNA molecules that control hereditary characteristics in organisms.

genetic adaptation Changes in the genetic makeup of organisms of a species that allow the species to reproduce and gain a competitive advantage under changed environmental conditions.

genetic diversity Variability in the genetic makeup among individuals within a single species. See *biodiversity*. Compare *ecological diversity, species diversity*.

geothermal energy Heat transferred from the earth's molten core to underground reservoirs of dry steam (steam with no water droplets), wet steam (a mixture of steam and water droplets), hot water, or rocks lying fairly close to the earth's surface.

GNP See *gross national product*.

grassland Terrestrial ecosystem found in regions where moderate annual average precipitation (25 to 76 centimeters, or 10 to 30 inches) is enough to support the growth of grass and small plants but not enough to support large stands of trees. Compare *desert, forest*.

greenhouse effect Trapping and buildup of heat in the atmosphere (troposphere) near the earth's surface. Some of the heat flowing back toward space from the earth's surface is absorbed by water vapor, carbon dioxide, ozone, and several other gases in the atmosphere and then reradiated back toward the earth's surface. If the atmospheric concentrations of these greenhouse gases rise, the average temperature of the lower atmosphere will gradually increase.

greenhouse gases Gases in the earth's atmosphere that cause the greenhouse effect. Examples are carbon dioxide, chlorofluorocarbons, ozone, methane, and nitrous oxide.

green manure Freshly cut or still-growing green vegetation that is plowed into the soil to increase the organic matter and humus available to support crop growth. Compare *animal manure*.

green revolution Popular term for introduction of scientifically bred or selected varieties of grain (rice, wheat, maize) that, with high enough inputs of fertilizer and water, can greatly increase crop yields.

gross national product (GNP) Total market value in current dollars of all final goods and services produced by an economy during a year. Compare *average per capita GNP, average per capita real NEW, real GNP*.

ground fire Fire which burns decayed leaves or peat deep below the ground surface. Compare *crown fire, surface fire*.

groundwater Water that sinks into the soil and is stored in slowly flowing and slowly renewed underground reservoirs called aquifers; underground water in the zone of saturation below the water table. See *confined aquifer, unconfined aquifer*. Compare *runoff, surface water*.

habitat Place or type of place where an organism or community of organisms lives and thrives.

hazard Something that can cause injury, disease, economic loss, or environmental damage.

hazardous waste Any solid, liquid, or containerized gas that can catch fire easily, is corrosive to skin tissue or metals, is unstable and can explode or release toxic fumes, or which has harmful concentrations of one or more toxic materials that can leach out. See also *toxic waste*.

heat Form of kinetic energy that flows from one body to another when there is a temperature difference between the two bodies. Heat always flows spontaneously from a hot sample of matter to a colder sample of matter. This is one way to state the second law of thermodynamics. Compare *temperature*.

heavy oil Black, high-sulfur, tarlike oil found in deposits of crude oil, tar sands, and oil shale.

herbicide Chemical that kills a plant or inhibits its growth.

herbivore Plant-eating organism. Examples are deer, sheep, grasshoppers, and zooplankton. Compare *carnivore, omnivore*.

heterotroph See *consumer*.

high-quality energy Energy that is concentrated and has great ability to perform useful work. Examples are high-temperature heat and the energy in electricity, coal, oil, gasoline, sunlight, and nuclei of uranium-235. Compare *low-quality energy*.

high-quality matter Matter that is organized and concentrated and contains a high concentration of a useful resource. Compare *low-quality matter*.

host Plant or animal upon which a parasite feeds.

humus Complex mixture of partially decomposed organic matter and inorganic compounds in topsoil. This insoluble material helps retain water and water-soluble nutrients so they can be taken up by plant roots.

hunter-gatherers People who get their food by gathering edible wild plants and other materials and by hunting wild animals and fish.

hydrocarbon Organic compound of hydrogen and carbon atoms, such as methane (CH_4).

hydroelectric power plant Structure in which the energy of falling or flowing water spins a turbine generator to produce electricity.

hydrologic cycle Biogeochemical cycle that collects, purifies, and distributes the earth's fixed supply of water from the environment, to living organisms, and back to the environment.

hydropower Electrical energy produced by falling or flowing water. See *hydroelectric power plant*.

hydrosphere All the earth's liquid water (oceans, smaller bodies of fresh water, and underground aquifers), frozen water (polar ice caps, floating ice, and frozen upper layer of soil known as permafrost), and small amounts of water vapor in the atmosphere.

hypolimnion Bottom layer of water in a stratified lake. This layer is colder and more dense than the top or epilimnion layer. Compare *epilimnion, thermocline*.

identified resources Deposits of a particular mineral-bearing material of which the location, quantity, and quality are known or have been estimated from geological evidence and measurements. Compare *total resources*.

igneous rocks Rocks formed when magma (molten rock) wells up from the earth's upper mantle, cools, and solidifies. See *rock cycle*. Compare *metamorphic rocks, sedimentary rocks*.

immature community Community at an early stage of ecological succession. It usually has a low number of species and ecological niches and cannot capture and use energy and cycle critical nutrients as efficiently as more complex, mature communities. Compare *mature community*.

immigration Migration of people into a country or area to take up permanent residence. Compare *emigration*.

indicator species Species that serve as early warnings that a community or ecosystem is being degraded. Compare *immigrant species, keystone species, native species*.

industrialized agriculture Using large inputs of energy from fossil fuels (especially oil and natural gas) to produce large quantities of crops and livestock for domestic and foreign sale. Compare *subsistence agriculture*.

Industrial Revolution Uses of new sources of energy from fossil fuels—and later, nuclear fuels—and use of new technologies to grow food and manufacture products.

industrial smog Type of air pollution consisting mostly of a mixture of sulfur dioxide, suspended droplets of sulfuric acid formed from some of the sulfur dioxide, and a variety of suspended solid particles. Compare *photochemical smog.*

inertia Ability of a living system to resist being disturbed or altered. Compare *constancy, resilience.*

infant mortality rate Annual number of deaths of infants under one year of age per 1,000 live births.

inland wetland Land away from the coast, such as a swamp, marsh, or bog, that is flooded all or part of the year with fresh water. Compare *coastal wetland.*

inorganic compound Combination of two or more elements other than those used to form organic compounds. Compare *organic compound.*

inorganic fertilizer See *commercial inorganic fertilizer.*

input pollution control Method that prevents a potential pollutant from entering the environment or that sharply reduces the amount entering the environment. Compare *output pollution control.*

insecticide Chemical designed to kill insects.

integrated pest management (IPM) Combined use of biological, chemical, and cultivation methods in proper sequence and timing to keep the size of a pest population below the size that causes economically unacceptable loss of a crop or livestock animal.

intercropping Growing two or more different crops at the same time on a plot. For example, a carbohydrate-rich grain that depletes soil nitrogen and a protein-rich legume that adds nitrogen to the soil may be intercropped. Compare *monoculture, polyculture, polyvarietal cultivation.*

intermediate goods See *capital goods.*

internal cost Direct cost paid by the producer and buyer of an economic good. Compare *external cost.*

interspecific competition Members of two or more species trying to use the same scarce resources in an ecosystem. See *competition, intraspecific competition.*

intraspecific competition Two or more individual organisms of a single species trying to use the same scarce resources in an ecosystem. See *competition, interspecific competition.*

inversion See *thermal inversion.*

invertebrates Animals that have no backbones. Compare *vertebrates.*

ion Atom or group of atoms with one or more positive (+) or negative (−) electrical charges.

ionizing radiation Fast-moving alpha or beta particles or high-energy radiation (gamma rays) emitted by radioisotopes. They have enough energy to dislodge one or more electrons from atoms they hit, forming charged ions that can react with and damage living tissue.

isotopes Two or more forms of a chemical element that have the same number of protons but different mass numbers or numbers of neutrons in their nuclei.

J-shaped curve Curve with the shape of the letter J that represents exponential growth.

kerogen Solid, waxy mixture of hydrocarbons found in oil shale rock. When the rock is heated to high temperatures, the kerogen is vaporized. The vapor is condensed and then sent to a refinery to produce gasoline, heating oil, and other products. See also *oil shale, shale oil.*

keystone species Species that play roles affecting many other organisms in an ecosystem. Compare *immigrant species, indicator species, native species.*

kilocalorie (kcal) Unit of energy equal to 1,000 calories. See *calorie.*

kilowatt (kw) Unit of electrical power equal to 1,000 watts. See *watt.*

kinetic energy Energy that matter has because of its motion and mass. Compare *potential energy.*

kwashiorkor Type of malnutrition that occurs in infants and very young children when they are weaned from mother's milk to a starchy diet low in protein. See also *marasmus.*

labor Physical and mental talents of people used to produce, distribute, and sell an economic good. Labor includes entrepreneurs, who assume the risk and responsibility of combining the resources of land, capital goods, and workers who produce an economic good. Compare *capital goods, natural resources.*

lake Large natural body of standing fresh water formed when water from precipitation, land runoff, or groundwater flow fills a depression in the earth created by glaciation, earthquake, volcanic activity, or a giant meteorite. Compare *reservoir.* See *eutrophic lake, mesotrophic lake, oligotrophic lake.*

landfill See *sanitary landfill.*

land-use planning Process for deciding the best use of each parcel of land in an area. See *ecological land-use planning.*

late-successional species Wild animal species found in moderate-size old-growth and mature forest habitats. Compare *early-successional species, mid-successional species, wilderness species.*

latitude Distance from the equator. Compare *altitude.*

law of conservation of energy See *first law of thermodynamics.*

law of conservation of matter In any ordinary physical or chemical change, matter is neither created nor destroyed but merely changed from one form to another; in physical and chemical changes existing atoms are either rearranged into different spatial patterns (physical changes) or different combinations (chemical changes).

law of energy degradation See *second law of thermodynamics.*

law of tolerance The existence, abundance, and distribution of a species are determined by whether the levels of one or more physical or chemical factors fall above or below the levels tolerated by the species. See also *tolerance limit.*

LDC See *less developed country.*

leaching Process in which various chemicals in upper layers of soil are dissolved and carried to lower layers, and in some cases to groundwater.

less developed country (LDC) Country that has low to moderate industrialization and low to moderate average GNP per person. Most LDCs are located in the tropical (or low) latitudes in Africa, Asia, and Latin America. Compare *more developed country.*

life-cycle cost Initial cost plus lifetime operating costs of an economic good.

life expectancy Average number of years a newborn infant can be expected to live.

limiting factor Single factor that limits the growth, abundance, or distribution of the population of a particular organism in an ecosystem. See *limiting factor principle.*

limiting factor principle Too much or too little of any single abiotic factor can limit or prevent growth of the populations of particular plant and animal species in an ecosystem even if all other factors are at or near the optimum range of tolerance for the species.

limnetic zone Open water surface layer of a lake, away from the shore, where there is enough sunlight for photosynthesis. Compare *benthic zone, littoral zone, profundal zone.*

linear growth Growth in which a quantity increases by some fixed amount during each unit of time. Compare *exponential growth.*

liquified natural gas (LNG) Natural gas converted to liquid form by cooling to a very low temperature.

liquefied petroleum gas (LPG) Mixture of liquefied propane and butane gas removed from a deposit of natural gas.

lithosphere Soil and rock in the earth's crust and upper mantle. See *crust, mantle.*

littoral zone Shallow waters near the shore of a body of water, in which sunlight penetrates to the bottom. Compare *benthic zone, limnetic zone, profundal zone.*

low-quality energy Energy such as low-temperature heat that is dispersed or dilute and has little ability to do useful work. Compare *high-quality energy.*

low-quality matter Matter that is disorganized, dilute, dispersed, or contains a low concentration of a useful resource. Compare *high-quality matter.*

LPG See *liquefied petroleum gas.*

macronutrient Chemical that a plant or animal needs in large amounts to stay alive and healthy. Compare *micronutrient.*

magma Molten rock.

malignant tumor See *cancer.*

malnutrition Faulty nutrition. Caused by a diet that does not supply an individual with enough proteins, essential fats, vitamins, minerals, and other nutrients needed for good health. See *kwashiorkor, marasmus.* Compare *overnutrition, undernutrition.*

mantle Zone of the earth's interior between its core and crust. See *lithosphere.* Compare *core, crust.*

manure See *animal manure, green manure.*

marasmus Nutritional-deficiency disease caused by a diet that does not have enough calories and protein to maintain good health. See *kwashiorkor, malnutrition.*

mass The amount of material in an object.

mass number Sum of the number of neutrons and the number of protons in the nucleus of an atom. It gives the approximate mass of that atom.

mass transit Buses, trains, trolleys, and other forms of transportation that carry large numbers of people.

matter Anything that has mass and occupies space; the material the world is made of.

matter quality Measure of how useful a matter resource is based on its availability and concentration. See *high-quality matter, low-quality matter.*

matter-recycling society Society which emphasizes recycling the maximum amount of all resources that can be recycled. The goal is to allow economic growth to continue without depleting matter resources and without producing excessive pollution and environmental degradation. Compare *sustainable-earth society, throwaway society.*

mature community Fairly stable, self-sustaining community in an advanced stage of ecological succession. It usually has a diverse array of species and ecological niches, captures and uses energy, and cycles critical chemicals more efficiently than simpler, immature communities. Compare *immature community.*

maximum sustainable yield See *sustainable yield.*

MDC See *more developed country.*

meltdown The melting of the core of a nuclear reactor.

mesotrophic lake Lake with a moderate supply of plant nutrients. Compare *eutrophic lake, oligotrophic lake.*

metabolic reserve Lower half of rangeland grass plants; these plants can grow back as long as this part is not consumed by herbivores.

metamorphic rocks Rocks derived from preexisting rocks within the lithosphere by changes in temperature, pressure, and the chemical action of fluids. See *rock cycle.* Compare *igneous rocks, sedimentary rocks.*

metastasis Spread of malignant (cancerous) cells from a cancer to other parts of the body.

microconsumer See *decomposer.*

micronutrient Chemical that a plant or animal needs in small, or trace, amounts to stay alive and healthy. Compare *macronutrient.*

midsuccessional species Wild species found around abandoned croplands and partially open areas at the middle stages of ecological succession. Compare *early-successional species, late-successional species, wilderness species.*

mineral Any naturally occurring inorganic solid with a particular chemical makeup and a crystalline internal structure made up of an orderly, three-dimensional array of atoms or ions.

mineral resource Nonrenewable chemical element or compound in solid form that is used by humans. Mineral resources are classified as metallic (such as iron and tin) or nonmetallic (such as sand and salt).

minimum-tillage farming See *conservation-tillage farming.*

molecule Chemical combination of two or more atoms of the same chemical element (such as O_2) or different chemical elements (such as H_2O).

monoculture Cultivation of a single crop usually on a large area of land. Compare *polyculture.*

more developed country (MDC) Country that is highly industrialized and has a high average GNP per person. Compare *less developed country.*

multiple use Principle of managing public land, such as a national forest, so it is used for a variety of purposes, such as timbering, mining, recreation, grazing, wildlife preservation, and soil and water conservation. See also *sustainable yield.*

municipal solid waste Solid materials discarded by homes and businesses in or near urban areas. See *solid waste.*

mutagen Chemical or form of radioactivity that can increase the rate of genetic mutation in a living organism. See *mutation.*

mutation An inheritable change in the DNA molecules in the genes found in the chromosomes. See *mutagen.*

mutualism An interaction between individuals of different species that benefits each individual. Compare *commensalism.*

nanoplankton Extremely small photosynthetic algae and bacteria. See also *phytoplankton.*

national ambient air quality standards (NAAQS) Maximum allowable level, averaged over a specific time period, for a certain pollutant in outdoor (ambient) air.

native species Species that normally live and thrive in a particular ecosystem. Compare *keystone species, immigrant species, indicator species.*

natural eutrophication See *eutrophication.*

natural gas Underground deposits of gases consisting of 50% to 90% methane (CH_4) and small amounts of heavier gaseous hydrocarbon compounds such as propane (C_3H_8) and butane (C_4H_{10}).

natural ionizing radiation Ionizing radiation in the environment from natural sources.

natural radioactivity Nuclear change in which unstable nuclei of atoms spontaneously shoot out "chunks" of mass, energy, or both at a fixed rate.

natural recharge Natural replenishment of an aquifer by precipitation, which percolates downward through soil and rock. See *recharge area.*

natural resources Area of the earth's solid surface, nutrients and minerals in the soil and deeper layers of the earth's crust, water, wild and domesticated plants and animals, air, and other resources produced by the earth's natural processes. Compare *capital goods, labor.*

natural selection Process by which some genes and gene combinations in a population of a species are reproduced more than others when the population is exposed to an environmental change or stress. When individual organisms in a population die off over time because they cannot tolerate a new stress, they are replaced by individuals whose genetic traits allow them to cope better with the stress. When these better-adapted individuals reproduce, they pass their adaptive traits on to their offspring. See also *evolution.*

neritic zone See *coastal zone.*

net economic welfare (NEW) Measure of annual change in quality of life in a country. It is obtained by subtracting the value of all final products and services that decrease the quality of life from a country's GNP. See *average per capita NEW.*

net energy See *net useful energy.*

net primary productivity Rate at which all the plants in an ecosystem produce net useful chemical energy. It is equal to the difference between the rate at which the plants in an ecosystem produce useful chemical energy and the rate at which they use some of this energy through cellular respiration.

net useful energy Total amount of useful energy available from an energy resource or energy system over its lifetime minus the amount of energy used (the first energy law), automatically wasted (the second energy law), and unnecessarily wasted in finding, processing, concentrating, and transporting it to users.

neutral solution Water solution containing an equal number of hydrogen ions (H^+) and hydroxide ions (OH^-); water solution with a pH of 7. Compare *acid solution, basic solution.*

neutron (n) Elementary particle in the nuclei of all atoms (except hydrogen-1). It has a relative mass of 1 and no electric charge.

NEW See *net economic welfare.*

niche See *ecological niche.*

nitrogen cycle Cyclic movement of nitrogen in different chemical forms from the environment, to organisms, and then back to the environment.

nitrogen fixation Conversion of atmospheric nitrogen gas into forms useful to plants by lightning, bacteria, and blue-green algae (cyanobacteria); it is part of the nitrogen cycle.

no-dumping principle No one has the right to dump anything into the world's air, water, and soil unless they can show that this will cause no unacceptable harm to people or other living things.

nonbiodegradable Substance that cannot be broken down in the environment by natural processes. Compare *biodegradable.*

nondegradable See *nonbiodegradable.*

nonpoint source Large land area such as crop fields and urban areas that discharge pollutants into the environment over a large area. Compare *point source.*

nonrenewable resources Resources available in a fixed amount (stock) in the earth's crust. They can be exhausted either because they are not replaced by natural processes (copper) or because they are replaced more slowly than they are used (oil and coal). Compare *perpetual resource, renewable resource.*

nontransmissible disease A disease that is not caused by living organisms and that does not spread from one person to another. Examples are most cancer, diabetes, cardiovascular disease, and malnutrition. Compare *transmissible disease.*

no-till cultivation See *conservation-tillage farming.*

nuclear change Process in which nuclei of certain isotopes spontaneously change or are forced to change into one or more different isotopes. The three major types of nuclear change are natural radioactivity, nuclear fission, and nuclear fusion. Compare *chemical change.*

nuclear energy Energy released when atomic nuclei undergo a nuclear reaction such as the spontaneous emission of radioactivity, nuclear fission, or nuclear fusion.

nuclear fission Nuclear change in which the nuclei of certain isotopes with large mass numbers (such as uranium-235 and plutonium-239) split apart into two lighter nuclei when struck by a neutron. This process also releases more neutrons and a large amount of energy. Compare *nuclear fusion.*

nuclear fusion Nuclear change in which two nuclei of isotopes of elements with a low mass number (such as hydrogen-2 and hydrogen-3) are forced together at a very high temperature until they fuse to form a heavier nucleus (such as helium-4). This process releases a large amount of energy. Compare *nuclear fission.*

nucleus Extremely tiny center of an atom, making up most of the atom's mass. It contains one or more positively charged protons and one or more neutrons with no electrical charge (except for a hydrogen-1 atom, whose nucleus has one proton and no neutrons).

nutrient Element or compound needed for the survival, growth, and reproduction of a plant or animal. See *macronutrient, micronutrient.*

ocean thermal energy conversion (OTEC) Using the large temperature differences between the cold bottom waters and the sun-warmed surface waters of tropical oceans to produce electricity.

oil See *crude oil.*

oil shale Underground formation of a fine-grained rock containing varying amounts of kerogen, a solid, waxy mixture of hydrocarbon compounds. Heating the rock to high temperatures converts the kerogen to a vapor which can be condensed to form a slow-flowing heavy oil called shale oil. See *kerogen, shale oil.*

old-growth forest Uncut forest containing massive trees that are often hundreds of years old. Examples include forests of Douglas fir, western hemlock, giant sequoia, and coastal redwoods in the western United States. Compare *secondary forest, tree farm.*

oligotrophic lake Lake with a low supply of plant nutrients. Compare *eutrophic lake, mesotrophic lake.*

omnivore Animal organism that can use both plants and other animals as food sources. Examples are pigs, rats, cockroaches, and people. Compare *carnivore, herbivore.*

open-pit surface mining Removal of materials such as stone, sand, gravel, iron, and copper by digging them out of the earth's surface and leaving a large depression or pit. See also *area strip mining, contour strip mining.*

open sea The part of an ocean that is beyond the continental shelf. Compare *coastal zone.*

ore Deposit in which the concentration of a mineral is high enough to make its recovery profitable.

organic compound Molecule that contains atoms of the element carbon, usually combined with itself and with atoms of one or more other elements such as hydrogen, oxygen, nitrogen, sulfur, phosphorus, chlorine, or fluorine. Compare *inorganic compound.*

organic farming Producing crops and livestock naturally by using organic fertilizer (manure, legumes, compost) and natural pest control (bugs that eat harmful bugs, plants that repel bugs, and environmental controls such as crop rotation) instead of using commercial inorganic fertilizers and synthetic pesticides and herbicides.

organic fertilizer Organic material such as animal manure, green manure, and compost, applied to cropland as a source of plant nutrients. Compare *commercial inorganic fertilizer.*

organism Any form of life.

output pollution control Method for reducing the level of pollution after pollutants have been produced or have entered the environment. Examples are automobile emission control devices and sewage treatment plants. Compare *input pollution control.*

overburden Layer of soil and rock overlying a mineral deposit that is removed during surface mining.

overfishing Harvesting so many fish of a species, especially immature ones, that there is not enough breeding stock left to replenish the species so that it becomes unprofitable to harvest them.

overgrazing Consumption of rangeland grass by grazing animals to the point that it cannot be renewed, or can be only slowly renewed, because of damage to the root system.

overnutrition Diet so high in calories, saturated (animal) fats, salt, sugar, and processed foods, and so low in vegetables and fruits that the consumer runs high risks of diabetes, hypertension, heart disease, and other health hazards. Compare *malnutrition, undernutrition.*

overpopulation State in which the life-support systems in the world or geographic area are impaired because people use nonrenewable and renewable resources to such an extent that the resource base is degraded or depleted, and air, water, and soil are severely polluted. See *consumption overpopulation, people overpopulation.*

oxygen cycle Cyclic movement of oxygen in different chemical forms from the environment, to organisms, and then back to the environment.

oxygen-demanding wastes Organic materials that are usually biodegraded by aerobic (oxygen-consuming) bacteria if there is enough dissolved oxygen in the water. See also *biological oxygen demand.*

ozone layer Layer of gaseous ozone (O_3) in the stratosphere that protects life on earth by filtering out harmful ultraviolet radiation from the sun.

PANs Peroxyacyl nitrates. Group of chemicals found in photochemical smog.

parasite Consumer organism that feeds on a living plant or animal, known as the host, over an extended period of time. A parasite harms the host organism but does not kill it—at least not immediately, as most other consumers do.

particulate matter Solid particles or liquid droplets suspended or carried in the air.

parts per billion (ppb) Number of parts of a chemical found in one billion parts of a particular gas, liquid, or solid mixture.

parts per million (ppm) Number of parts of a chemical found in one million parts of a particular gas, liquid, or solid.

passive solar heating system System that captures sunlight directly within a structure and converts it to low-temperature heat for space heating. Compare *active solar heating system.*

pathogen Organism that produces disease.

PCBs See *polychlorinated biphenyls.*

people overpopulation Situation in which there are more people in the world or a geographic region than available supplies of food, water, and other vital resources can support. It can also occur where the rate of population growth so exceeds the rate of economic growth or the distribution of wealth is so inequitable that a number of people are too poor to grow or buy enough food, fuel, and other important resources. Compare *consumption overpopulation.*

perennial Plant that grows from the root stock each year and does not need to be replanted. Compare *annual.*

permafrost Water permanently frozen year-round in thick underground layers of soil in tundra.

permeability The degree to which underground rock pores are interconnected with each other and thus a measure of the degree to which water can flow freely from one pore to another. Compare *porosity.*

perpetual resource Resource such as solar energy that comes from a virtually inexhaustible source on a human time scale. Compare *nonrenewable resource, renewable resource.*

persistence See *inertia.*

pest Unwanted organism that directly or indirectly interferes with human activities.

pesticide Any chemical designed to kill or inhibit the growth of an organism that people consider to be undesirable. See *fungicide, herbicide, insecticide, rodenticide.*

pesticide treadmill Situation in which the cost of using pesticides increases while their effectiveness decreases, mostly because the pest species develop genetic resistance to the pesticides.

petrochemicals Chemicals obtained by refining (distilling) crude oil. They are used as raw materials in the manufacture of most industrial chemicals, fertilizers, pesticides, plastics, synthetic fibers, paints, medicines, and many other products.

petroleum See *crude oil.*

pH Numeric value that indicates the relative acidity or alkalinity of a substance on a scale of 0 to 14, with the neutral point at 7. Acid solutions have pH values lower than 7, and basic, or alkaline, solutions have pH values greater than 7.

phosphorus cycle Cyclic movement of phosphorus in different chemical forms from the environment, to organisms, and then back to the environment.

photochemical smog Complex mixture of air pollutants produced in the atmosphere by the reaction of hydrocarbons and nitrogen oxides under the influence of sunlight. Especially harmful components include ozone, peroxyacyl nitrates (PANs), and various aldehydes. Compare *industrial smog.*

photosynthesis Complex process that takes place in cells of green plants. Radiant energy from the sun is used to combine carbon dioxide (CO_2) and water (H_2O) to produce oxygen (O_2) and simple nutrient molecules, such as glucose ($C_6H_{12}O_6$). Compare *cellular aerobic respiration, chemosynthesis.*

photovoltaic cell (solar cell) Device in which radiant (solar) energy is converted directly into electrical energy.

physical change Process that alters one or more physical properties of an element or compound without altering its chemical composition. Examples are changing the size and shape of a sample of matter (crushing ice and cutting aluminum foil) and changing a sample of matter from one physical state to another (boiling and freezing water). Compare *chemical change.*

phytoplankton Small, drifting plants, mostly algae and bacteria, found in aquatic ecosystems. See also *nanoplankton.* Compare *plankton, zooplankton.*

pioneer community First integrated set of plants, animals, and decomposers found in an area undergoing primary ecological succession. See *immature community, mature community.*

plankton Small plant organisms (phytoplankton and nanoplankton) and animal organisms (zooplankton) that float in aquatic ecosystems.

plantation agriculture Growing specialized crops such as bananas, coffee, and cacao in tropical LDCs, primarily for sale to MDCs.

plate tectonics Theory of geophysical processes that describes the formation and deformation of the lithosphere and how the continents move. It views the lithosphere as being made up of about ten major plates and a number of smaller plates that over millions of years move and rub and scrape against each other in response to convective currents in the underlying plastic portion (asthenosphere) of the earth's mantle.

point source A single identifiable source that discharges pollutants into the environment. Examples are a smokestack, sewer, ditch, or pipe. Compare *nonpoint source.*

politics Process through which individuals and groups try to influence or control the policies and actions of governments that affect the local, state, national, and international communities.

pollution A change in the physical, chemical, or biological characteristics of the air, water, or soil that can affect the health, survival, or activities of humans in an unwanted way. Some expand the term to include harmful effects on all forms of life.

pollution prevention See *input pollution control.*

polychlorinated biphenyls (PCBs) Group of 209 different toxic, oily, synthetic chlorinated hydrocarbon compounds that can be biologically amplified in food chains and webs.

polyculture Complex form of intercropping in which a large number of different plants maturing at different times are planted together. See also *intercropping*. Compare *monoculture, polyvarietal cultivation*.

polyvarietal cultivation Planting a plot of land with several varieties of the same crop. Compare *intercropping, monoculture, polyculture*.

population Group of individual organisms of the same species living within a particular area.

population crash Large number of deaths over a fairly short time brought about when the number of individuals in a population is too large to be supported by available environmental resources.

population density Number of organisms in a particular population found in a specified area.

population distribution Variation of population density over a particular geographical area. For example, a country has a high population density in its urban areas and a much lower population density in rural areas.

population dynamics Major abiotic and biotic factors that tend to increase or decrease the population size and age and sex composition of a species.

porosity The pores (cracks and spaces) in rocks or soil, or the percentage of the rock's volume not occupied by the rock itself. Compare *permeability*.

potential energy Energy stored in an object because of its position or the position of its parts. Compare *kinetic energy*.

potentially renewable resource See *renewable resource*.

poverty Inability to meet basic needs. People in different societies differ in what they consider to be basic needs.

ppb See *parts per billion*.

ppm See *parts per million*.

precipitation Water in the form of rain, sleet, hail, and snow that falls from the atmosphere onto the land and bodies of water.

predation Situation in which an organism of one species (the predator) captures and feeds on parts or all of an organism of another species (the prey).

predator Organism that captures and feeds on parts or all of an organism of another species (the prey).

predator-prey relationship Interaction between two organisms of different species in which one organism called the predator captures and feeds on parts or all of another organism called the prey.

prescribed burning Deliberate setting and careful control of surface fires in forests to help prevent more destructive crown fires and to kill off unwanted plants that compete with commercial species for plant nutrients; may also be used on grasslands. See *crown fire, ground fire, surface fire*.

preservationists People who stress the need to limit human used of parks, wilderness, estuaries, wetlands, and other types of ecosystems primarily to nondestructive recreation, education, and research. Compare *scientific conservationists, sustainable-earth conservationists*.

prey Organism that is captured and serves as a source of food for an organism of another species (the predator).

primary air pollutant Chemical that has been added directly to the air by natural events or human activities and occurs in a harmful concentration. Compare *secondary air pollutant*.

primary consumer See *herbivore*.

primary ecological succession Sequential development of communities in a bare area that has never been occupied by a community of organisms. Compare *secondary ecological succession*.

primary oil recovery Pumping out the crude oil that flows by gravity into the bottom of an oil well. Compare *enhanced oil recovery, secondary oil recovery*.

primary sewage treatment Mechanical treatment of sewage in which large solids are filtered out by screens and suspended solids settle out as sludge in a sedimentation tank. Compare *advanced sewage treatment, secondary sewage treatment*.

prime reproductive age Years between ages 20 and 29, during which most women have most of their children. Compare *reproductive age*.

principle of multiple use See *multiple use*.

prior appropriation Legal principle by which the first user of water from a stream establishes a legal right to continued use of the amount originally withdrawn. See also *riparian rights*.

producer Organism that uses solar energy (green plant) or chemical energy (some bacteria) to manufacture its own organic nutrients from inorganic nutrients. Compare *consumer, decomposer*.

productivity Measure of the output of economic goods and services produced by the input of the factors of production (natural resources, capital goods, labor). Increasing economic productivity means getting more output from less input. Compare *economic growth*.

profundal zone Deep, open-water region of a lake, a region not penetrated by sunlight. Compare *benthic zone, limnetic zone, littoral zone*.

proton (p) Positively charged particle in the nuclei of all atoms. Each proton has a relative mass of 1 and a single positive charge.

public land resources Land that is owned jointly by all citizens but is managed for them by an agency of the local, state, or federal government. Examples are state and national parks, forests, wildlife refuges, and wilderness areas.

public property resource See *public land resource*.

pyramid of energy flow Diagram representing the flow of usable, high-quality energy through each trophic level in a food chain. With each energy transfer, only a small part (typically 10%) of the usable energy entering one trophic level is transferred to the organisms at the next trophic level. See also *pyramid of energy loss*.

pyramid of energy loss Diagram showing the amount of low-quality energy, usually low-temperature heat, lost to the environment at each trophic level in a food chain. Typically 90% of the high-quality energy entering a trophic level is converted to low-quality energy and lost to the environment. See *pyramid of energy flow*.

radioactive isotope See *radioisotope*.

radioactive waste Radioactive waste products of nuclear power plants, research, medicine, weapons production, or other processes involving nuclear reactions.

radioactivity Nuclear change in which unstable nuclei of atoms spontaneously shoot out "chunks" of mass, energy, or both, at a fixed rate. The three major types of radioactivity are gamma rays, fast-moving alpha particles, and beta particles.

radioisotope Isotope of an atom whose unstable nuclei spontaneously emit one or more types of radioactivity (alpha particles, beta particles, gamma rays).

range See *rangeland*.

range condition Estimate of how close a particular area of rangeland is to its potential for producing vegetation that can be consumed by grazing or browsing animals.

rangeland Land, mostly grasslands, whose plants can provide food (forage) for grazing or browsing animals. Compare *feedlot*.

range of tolerance Range of chemical and physical conditions that must be maintained for populations of a particular species to stay alive and grow, develop, and function normally. See *law of tolerance*.

real GNP Gross national product adjusted for inflation. Compare *average per capita GNP, average per capita real GNP, gross national product*.

recharge area Area in which an aquifer is replenished with water by the downward percolation of precipitation through soil and rock.

recycling Collecting and reprocessing a resource so it can be used again. An example is collecting aluminum cans, melting them down, and using the aluminum to make new cans or other aluminum products. Compare *reuse*.

relative resource scarcity Situation in which a resource has not been depleted but there is not enough available to meet the demand. This can be caused by a war, a natural disaster, or other events that disrupt the production and distribution of a resource, or by deliberate attempts of its producers to lower production to drive prices up. Compare *absolute resource scarcity*.

renewable resource Resource that normally is replenished through natural processes. Examples are trees in forests, grasses in grasslands, wild animals, fresh surface water in lakes and streams, most deposits of groundwater, fresh air, and fertile soil. If such a resource is used faster than it is replenished, it can be depleted and converted to a nonrenewable resource. Compare *nonrenewable resource, perpetual resource*. See also *environmental degradation*.

replacement-level fertility Number of children a couple must have to replace themselves. The average for a country or the world is usually slightly higher than 2 children per couple (2.1 in the the United States and 2.5 in some LDCs) because some children die before reaching their reproductive years. See also *total fertility rate*.

reproductive age Ages 15 to 44, when most women have all their children. Compare *prime reproductive age*.

reserves (economic resources) Identified deposits of a particular resource in known locations that can be extracted profitably at present prices and with current mining technology. Compare *resources*.

reservoir Human-created body of standing fresh water often built behind a dam. Compare *lake*.

resilience Ability of a living system to restore itself to original condition after being exposed to an outside disturbance that is not too drastic. See also *constancy, inertia*.

resource Anything obtained from the environment to meet human needs and wants.

resource conservation See *conservation*.

resource recovery Salvaging usable metals, paper, and glass from solid waste and selling them to manufacturing industries for recycling.

resources Identified and unidentified deposits of a particular mineral that cannot be recovered profitably with present prices and mining technology. Some of these materials may be converted to reserves when prices rise or mining technology improves. Compare *reserves*.

respiration See *cellular aerobic respiration*.

reuse To use a product over and over again in the same form. An example is collecting, washing, and refilling glass beverage bottles. Compare *recycle*.

riparian rights System of water law that gives anyone whose land adjoins a flowing stream the right to use water from the stream as long as some is left for downstream users. Compare *prior appropriation*.

risk The probability that something undesirable will happen from deliberate or accidental exposure to a hazard. See *risk assessment, risk-benefit analysis, risk management*.

risk analysis Identifying hazards, evaluating the nature and severity of risks (*risk assessment*), using this and other information to determine options and make decisions about reducing or eliminating risks (*risk management*), and communicating information about risks to decisionmakers and the public (*risk communication*).

risk assessment Process of gathering data and making assumptions to estimate short- and long-term harmful effects on human health or the environment from exposure to hazards associated with the use of a particular product or technology. See *risk, risk analysis, risk-benefit analysis*.

risk-benefit analysis Estimate of the short- and long-term risks and benefits of using a particular product or technology. See *desirability quotient, risk*. Compare *cost-benefit analysis*.

risk communication Communicating information about risks to decisionmakers and the public. See *risk, risk analysis, risk-benefit analysis*.

risk management Using risk assessment and other information to determine options and make decisions about reducing or eliminating risks. See *risk, risk analysis, risk-benefit analysis, risk communication*.

rock Combination or aggregate of minerals. See *mineral*.

rock cycle Largest and slowest of the earth's cycles, consisting of geologic, physical, and chemical processes that form and modify rocks and soil in the earth's crust and mantle over millions of years.

rodenticide Chemical designed to kill rodents.

runoff Fresh water from precipitation and melting ice that flows on the earth's surface into nearby streams, lakes, wetlands, and artificial reservoirs. Compare *groundwater*.

rural area Geographical area in the United States with a population of less than 2,500 people. The number of people used in this definition may vary in different countries. Compare *urban area*.

salinity Amount of various salts (especially sodium chloride) dissolved in a given volume of water.

salinization Accumulation of salts in soil that can eventually make the soil unable to support plant growth.

saltwater intrusion Movement of salt water into freshwater aquifers in coastal and inland areas as groundwater is withdrawn faster than it is recharged by precipitation.

sanitary landfill Land waste disposal site in which waste is spread in thin layers, compacted, and covered with a fresh layer of soil each day.

scientific conservationists People who believe that the findings of science and technology should be used to manage resources according to the principle of multiple-use and the principle of sustainable yield so that they are available for future generations. Compare *preservationists, sustainable-earth conservationists*.

secondary air pollutant Harmful chemical formed in the atmosphere by reaction between normal air components and other air pollutants. Compare *primary air pollutant*.

secondary consumer See *carnivore*.

secondary ecological succession Sequential development of communities in an area in which natural vegetation has been removed or destroyed, but the soil or sediment is not destroyed. Compare *primary ecological succession*.

secondary forest Stands of trees resulting from secondary ecological succession. Compare *old growth forest, tree farm*.

secondary oil recovery Injection of water into an oil well after primary oil recovery to force out some of the remaining thicker crude oil. Compare *enhanced oil recovery, primary oil recovery*.

secondary sewage treatment Second step in most waste treatment systems, in which aerobic bacteria break down up to 90% of degradable, oxygen-demanding organic wastes in wastewater. This is usually done by bringing sewage and bacteria together in trickling filters or in the activated sludge process. Compare *advanced sewage treatment, primary sewage treatment*.

second law of ecology Everything is connected to and intermingled with everything else.

second law of energy See *second law of thermodynamics*.

second law of thermodynamics In any conversion of heat energy to useful work, some of the initial energy input is always degraded to a lower-quality, more-dispersed, less useful form of energy, usually low-temperature heat that flows into the environment; you can't break even in terms of energy quality. See *first law of thermodynamics*.

sediment Insoluble particles of soil and other solid inorganic and organic materials that become suspended in water and eventually fall to the bottom of a body of water.

sedimentary rocks Gradually accumulated and compacted sediments of weathered rock and shells and skeletons of dead animals. See *rock cycle*. Compare *igneous rocks, metamorphic rocks*.

seed-tree cutting Removal of nearly all trees on a site in one cutting, with a few seed-producing trees left uniformly distributed to regenerate the forest. Compare *clearcutting, selective cutting, shelterwood cutting, whole-tree harvesting*.

selective cutting Cutting of intermediate-aged, mature, or diseased trees in an uneven-aged forest stand, either singly or in small groups. This encourages the growth of younger trees and maintains an uneven-aged stand. Compare *clearcutting, seed-tree cutting, shelterwood cutting, whole-tree harvesting*.

septic tank Underground tank for treatment of wastewater from a home in rural and suburban areas. Bacteria in the tank decompose organic wastes and the sludge settles to the bottom of the tank. The effluent flows out of the tank into the ground through a field of drain pipes.

sewage sludge See *sludge*.

shade-intolerant tree species A tree species that needs lots of sunlight in the early growth stages and thrives in forest openings. Compare *shade-tolerant tree species*.

shade-tolerant tree species A tree species that can grow in dim or moderate light under the crown cover of larger trees. Compare *shade-intolerant tree species*.

shale oil Slow-flowing, dark brown, heavy oil obtained when kerogen in oil shale is vaporized at high temperatures and then condensed. Shale oil can be refined to yield gasoline, heating oil, and other petroleum products. See *kerogen, oil shale*.

shelterbelt See *windbreak*.

shelterwood cutting Removal of mature, marketable trees in an area in a series of partial cuttings to allow regeneration of a new stand under the partial shade of older trees, which are later removed. Typically this is done by making two or three cuts over a decade. Compare *clear-cutting, seed-tree cutting, selective cutting, whole-tree harvesting*.

shifting cultivation Clearing a plot of ground in a forest, especially in tropical areas, and planting crops on it for a few years (typically 2 to 5 years) until the soil is depleted of nutrients, or until the plot has been invaded by a dense growth of vegetation from the surrounding forest. Then a new plot is cleared and the process is repeated. The abandoned plot cannot successfully grow crops for 10 to 30 years. See also *slash-and-burn cultivation*.

silviculture Science and art of cultivating and managing forests to produce a renewable supply of timber.

slash-and-burn cultivation Cutting down trees and other vegetation in a patch of forest, leaving the cut vegetation on the ground to dry, and then burning it. The ashes that are left add plant nutrients to the nutrient-poor soils found in most tropical forest areas. Crops are planted between tree stumps. Plots must be abandoned after a few years (typically 2 to 5 years) because of loss of soil fertility or invasion of vegetation from the surrounding forest. See also *shifting cultivation*.

sludge Gooey solid mixture of bacteria- and virus-laden organic matter, toxic metals, synthetic organic chemicals, and solid chemicals removed from wastewater at a sewage treatment plant.

smog Originally a combination of smoke and fog, but now used to describe other mixtures of pollutants in the atmosphere. See *industrial smog, photochemical smog*.

soil Complex mixture of inorganic minerals (clay, silt, pebbles, and sand), decaying organic matter, water, air, and living organisms.

soil conservation Methods used to reduce soil erosion, to prevent depletion of soil nutrients, and to restore nutrients already lost by erosion, leaching, and excessive crop harvesting.

soil erosion Movement of soil components, especially topsoil, from one place to another usually by exposure to wind, flowing water, or both. This natural process can be greatly accelerated by human activities that remove vegetation from soil.

soil horizons Horizontal zones that make up a particular mature soil.

soil porosity Measure of the volume of pores, or spaces, and the average distances between them in a sample of soil.

soil profile Cross-sectional view of the horizons in a soil.

soil texture Relative amounts of the different types and sizes of particles in a sample of soil.

solar cell See *photovoltaic cell*.

solar collector Device for collecting radiant energy from the sun and converting it into heat. See *active solar heating system, passive solar heating system*.

solar energy Direct radiant energy from the sun. It also includes indirect forms of energy such as wind, falling or flowing water (hydropower), ocean thermal gradients, and biomass, which are produced when direct solar energy interacts with the earth.

solar pond Fairly small body of fresh water or salt water from which stored solar energy can be extracted because of temperature difference between the hot surface layer exposed to the sun during daylight and the cooler layer beneath it.

solid waste Any unwanted or discarded material that is not a liquid or a gas.

speciation Formation of new species from existing ones through natural selection in response to changes in environmental conditions; usually takes thousands to millions of years. Compare *extinction*.

species All organisms of the same kind; for organisms that reproduce sexually, a species is all organisms that can interbreed.

species diversity Number of different species and their relative abundances in a given area. See *biological diversity*. Compare *ecological diversity, genetic diversity*.

spoils Unwanted rock and other waste materials produced when a material is removed from the earth's surface or subsurface by mining.

sport fishing Finding and catching fish mostly for recreation. Compare *commercial fishing, subsistence fishing*.

sport hunting Finding and killing animals mostly for recreation. Compare *commercial hunting, subsistence hunting*.

S-shaped curve Leveling off of an exponential, J-shaped curve when a rapidly growing population exceeds the carrying capacity of its environment and ceases to grow in numbers.

stability Ability of a living system to withstand or recover from externally imposed changes or stresses. See *inertia, resilience*.

stocking rate Number of a particular kind of animal grazing on a given area of grassland.

strategic materials Fuel and nonfuel minerals vital to the industry and defense of a country. Ideally, supplies are stockpiled to cushion against supply interruptions and sharp price rises.

stratosphere Second layer of the atmosphere, extending from about 19 to 48 kilometers (12 to 30 miles) above the earth's surface. It contains small amounts of gaseous ozone (O_3), which filters out about 99% of the incoming harmful ultraviolet (UV) radiation emitted by the sun. Compare *troposphere*.

strip cropping Planting regular crops and close-growing plants, such as hay or nitrogen-fixing legumes, in alternating rows or bands to help reduce depletion of soil nutrients.

strip mining Form of surface mining in which bulldozers, power shovels, or stripping wheels remove large chunks of the earth's surface in strips. See *surface mining*. Compare *subsurface mining*.

subatomic particles Extremely small particles—electrons, protons, and neutrons—that make up the internal structure of atoms.

subsidence Sinking down of part of the earth's crust due to underground excavation, such as a coal mine, or removal of groundwater.

subsistence agriculture Supplementing solar energy with energy from human labor and draft animals to produce enough food to feed oneself and family members; in good years there may be enough food left over to sell or put aside for hard times. Compare *industrialized agriculture*.

subsistence farming See *subsistence agriculture*.

subsistence fishing Finding and catching fish to get food for survival. Compare *commercial fishing, sport fishing*.

subsistence hunting Finding and killing wild animals to get enough food and other animal material for survival. Compare *commercial hunting, sport hunting*.

subsurface mining Extraction of a metal ore or fuel resource such as coal from a deep underground deposit. Compare *surface mining*.

succession See *ecological succession*.

succulent plants Plants such as cacti that store water and produce the food they need in the thick, fleshly tissue of their green stems and branches.

superinsulated house House that is heavily insulated and extremely airtight. Typically, active or passive solar collectors are used to heat water and an air-to-air heat exchanger is used to prevent buildup of excessive moisture and indoor air pollutants.

surface fire Forest fire that burns only undergrowth and leaf litter on the forest floor. Compare *crown fire, ground fire*.

surface mining Removing soil, subsoil, and other strata and then extracting a mineral deposit found fairly close to the earth's surface. See *area strip mining, contour strip mining, open-pit surface mining*. Compare *subsurface mining*.

surface water Precipitation that does not infiltrate into the ground or return to the atmosphere by evaporation or transpiration. See *runoff*. Compare *groundwater*.

sustainable agriculture See *sustainable-earth agricultural system*.

sustainable-earth agricultural system Method of growing crops and raising livestock based on organic fertilizers, soil conservation, water conservation, biological control of pests, and minimal use of nonrenewable fossil fuel energy.

sustainable-earth conservationists People who believe that the earth's resources should be protected and sustained for the human and other species. They have a life-centered rather than a human-centered approach to managing and sustaining the earth's resources by working with nature, not wasting resources unnecessarily, and interfering with nonhuman species only to meet important human needs. Compare *preservationists, scientific conservationists*.

sustainable-earth economy Economic system in which the number of people and the quantity of goods are maintained at some constant level. This level is ecologically sustainable over time and meets at least the basic needs of all members of the population.

sustainable-earth society Society based on working with nature by recycling and reusing discarded matter, conserving matter and energy resources by reducing unnecessary waste and use, not degrading renewable resources, and by building things that are easy to recycle, reuse, and repair. Compare *matter-recycling society, throwaway society*.

sustainable-earth worldview Belief that the earth is a place with finite room and resources so that continuing population growth, production, and consumption inevitably put severe stress on natural processes that renew and maintain the resource base of air, water, and soil that support all life. To prevent environmental overload, environmental degradation, and resource depletion, people should work with nature by controlling population growth, reducing unnecessary use and waste of matter and energy resources, and not causing the premature extinction of any other species. Compare *throwaway worldview*.

sustainable yield (sustained yield) Highest rate at which a renewable resource can be used without impairing or damaging its ability to be fully renewed. This changes and is hard to determine. See also *environmental degradation*.

sustained yield See *sustainable yield*.

synfuels Synthetic gaseous and liquid fuels produced from solid coal or sources other than natural gas or crude oil.

synthetic natural gas (SNG) Gaseous fuel containing mostly methane produced from solid coal.

tailings Rock and other waste materials removed as impurities when minerals are mined and mineral deposits are processed. These materials are usually dumped on the ground or into ponds.

tar sand Swamplike deposit of a mixture of fine clay, sand, water, and variable amounts of a tar-like heavy oil known as bitumen. Bitumen can be extracted from tar sand by heating. It is then purified and upgraded to synthetic crude oil. See *bitumen*.

temperature Measure of the average speed of motion of the atoms or molecules in a substance or combination of substances at a given moment. Compare *heat*.

temperature inversion See *thermal inversion*.

teratogen Chemical which, if ingested by a pregnant female, causes malformation of the developing fetus.

terracing Planting crops on a long, steep slope that has been converted into a series of broad, nearly level terraces that follow the slope of the land to retain water and reduce soil erosion.

terrestrial Pertaining to land. Compare *aquatic*.

terrestrial ecosystem See *land ecosystem*.

tertiary (and higher) consumers Animals that feed on animal-eating animals. They feed at high trophic levels in food chains and webs. Examples are hawks, lions, bass, and sharks. Compare *carnivores, decomposers, herbivores*.

tertiary sewage treatment See *advanced sewage treatment*.

thermal enrichment Beneficial effects in an aquatic ecosystem from a rise in water temperature. Compare *thermal pollution*.

thermal inversion Layer of dense, cool air trapped under a layer of less dense warm air, thus reversing the normal situation. In a prolonged inversion, air pollution in the trapped layer may build up to harmful levels.

thermal pollution Increase in water temperature that has harmful effects on an aquatic ecosystem. See *thermal shock*. Compare *thermal enrichment*.

thermal shock A sharp change in water temperature that can kill or harm fish and other aquatic organisms. See *thermal pollution*. Compare *thermal enrichment*.

third law of ecology Any substance we produce should not interfere with any of the earth's natural biogeochemical cycles.

threatened species Wild species that is still abundant in its natural range but is likely to become endangered because of a decline in numbers. Compare *endangered species*.

threshold effect The harmful or fatal effect of a small change in environmental conditions that exceeds the limit of tolerance of an organism, population, or volume of air, water, or soil.

throwaway society Society found in most advanced industrialized countries, in which ever-increasing economic growth is sustained by maximizing the rate at which matter and energy resources are used, with little emphasis on recycling, reuse, reduction of unnecessary waste, and other forms of resource conservation. Compare *matter-recycling society, sustainable-earth society.*

throwaway worldview Belief that the earth is a place of unlimited resources. Any type of resource conservation that hampers short-term economic growth is unnecessary because if we pollute or deplete resources in one area, we will find substitutes, control the pollution through technology, and if necessary get resources from the moon and asteroids in the "new frontier" of space. Compare *sustainable-earth worldview.*

total fertility rate (TFR) Estimate of the number of live children the average woman will bear if she passes through all her childbearing years (ages 15 to 44) conforming to the age-specific fertility rates of each year.

total resources Total amount of a particular resource material that exists on earth. Compare *identified resources, reserves, resources, unidentified resources.*

toxic substance Chemical that is harmful to people or other organisms.

toxic waste Form of hazardous waste that causes death or serious injury (such as burns, respiratory diseases, cancers, or genetic mutations) to humans. See *hazardous waste.*

tragedy of the commons Depletion or degradation of a resource to which people have free and unmanaged access. An example is the depletion of commercially desirable species of fish in the open ocean beyond areas controlled by coastal countries. See *common property resource.*

transmissible disease A disease that is caused by living organisms such as bacteria, viruses, and parasitic worms and that can spread from one person to another by air, water, food, body fluids, or in some cases by insects. Compare *nontransmissible disease.*

transpiration Process by which water moves up through a living plant and is transferred to the atmosphere as water vapor from exposed parts of the plant.

tree farm Site planted with one or only a few tree species in an even-aged stand. When the stand matures, it is usually harvested by clearcutting and replanted. Normally used to grow rapidly growing tree species for fuelwood, timber, or pulpwood. See *even-aged management.* Compare *uneven-aged management, uneven-aged stand.*

tritium (T:hydrogen-3) Isotope of hydrogen with a nucleus containing one proton and two neutrons, thus having a mass number of 3. Compare *deuterium.*

trophic level All organisms that consume the same general types of food in a food chain or food web. For example, all producers belong to the first trophic level and all herbivores belong to the second trophic level in a food chain or a food web.

troposphere Innermost layer of the atmosphere. It contains about 95% of the earth's air and extends about 18 kilometers (11 miles) above the earth's surface. Compare *stratosphere.*

true cost Cost of a good when its internal costs and its short- and long-term external costs are included in its market price. Compare *external cost, internal cost.*

unconfined aquifer Collection of groundwater above a layer of rock or compacted clay through

which water flows very slowly (low permeability). Compare *confined aquifer.*

undergrazing Degradation in the quality of rangeland vegetation as a source of food for livestock or wild herbivores because of a lack of grazing. Compare *overgrazing.*

undernutrition Not taking in enough food to meet one's minimum daily energy requirement for a long enough time to cause harmful effects. Compare *malnutrition, overnutrition.*

undiscovered resources Potential supplies of a particular mineral resource, believed to exist because of geologic knowledge and theory, though specific locations, quality, and amounts are unknown. Compare *resources, reserves.*

uneven-aged management Method of forest management in which trees of different species in a given stand are maintained at many ages and sizes to permit continuous natural regeneration. Compare *even-aged management.*

uneven-aged stand Stand of trees in which there are considerable differences in the ages of individual trees. Usually, such stands have a variety of tree species. See *uneven-aged management.* Compare *even-aged stand, tree farm.*

upwelling Movement of nutrient-rich bottom water to the ocean's surface. This occurs along certain steep coastal areas where the surface layer of ocean water is pushed away from shore and replaced by cold, nutrient-rich bottom water.

urban area Geographic area with a population of 2,500 or more people. The number of people used in this definition may vary in different countries.

urban growth Rate of growth of an urban population. Compare *degree of urbanization.*

urban heat island Buildup of heat in the atmosphere above an urban area. This heat is produced by the large concentration of cars, buildings, factories, and other heat-producing activities. See also *dust dome.*

urbanization See *degree of urbanization.*

vertebrates Animals with backbones. Compare *invertebrates.*

wastewater lagoon Large pond meters deep where air, sunlight, and microorganisms break down wastes, allow solids to settle out, and kill some disease-causing bacteria. Water typically remains in a lagoon for 30 days. Then it is treated with chlorine and pumped out for use by a city or spread over cropland.

wastewater pond See *wastewater lagoon.*

water consumption Water that has been withdrawn from a groundwater or surface water source and is not available for reuse in the area from which it was withdrawn because of evaporation and transpiration. See *water withdrawal.*

water cycle See *hydrologic cycle.*

waterlogging Saturation of soil with irrigation water or excessive precipitation so that the water table rises close to the surface.

water pollution Any physical or chemical change in surface water or groundwater that can harm living organisms or make water unfit for certain uses.

watershed Land area that delivers runoff water, sediment, and dissolved substances to bodies of surface water.

water table Upper surface of the zone of saturation in which all available pores in the soil and rock in the earth's crust are filled with water.

water withdrawal Removing water from a groundwater or surface water source and transporting it to a place of use. Compare *water consumption.*

watt Unit of power, or rate at which electrical work is done. See *kilowatt.*

weather Short-term changes in the properties of the troposphere from place to place. Compare *climate.*

weathering Physical and chemical processes in which rock is gradually broken down into small bits and pieces that make up most of the soil's inorganic material by being exposed to weather and chemicals and invaded by certain organisms.

wetland Land that stays flooded all or part of the year with fresh or salt water. See *coastal wetland, inland wetland.*

whole-tree harvesting Use of machines to cut trees off at ground level or to pull entire trees from the ground and then reduce the trunks and branches to small wood chips.

wilderness Area where the earth and its community of life have not been seriously disturbed by humans and where humans are only temporary visitors.

wilderness species Wild animal species that flourish only in undisturbed mature vegetational communities such as large areas of mature forest, tundra, grassland, and desert. Compare *early-successional species, late-successional species, midsuccessional species.*

wildlife All free, undomesticated species of plants, animals, microorganisms.

wildlife conservation Activity of protecting, preserving, managing, and studying wildlife and wildlife resources.

wildlife management Manipulation of populations of wild species (especially game species) and their habitats for human benefit, the welfare of other species, and the preservation of threatened and endangered wildlife species.

wildlife resources Species of wildlife that have actual or potential economic value to people. See also *game species.*

wild species See *wildlife.*

windbreak Row of trees or hedges planted in a north-to-south direction to partially block wind flow and reduce soil erosion on cultivated land.

wind farm Cluster of small to medium-sized wind turbines in a windy area to capture wind energy and convert it to electrical energy.

work What happens when a force is used to move a sample of matter over some distance or to raise its temperature. Energy is defined as the capacity to do work.

worldview How we think the world works and what we think our role is. See *sustainable-earth worldview, throwaway worldview.*

zero population growth (ZPG) State in which the birth rate (plus immigration) equals the death rate (plus emigration) so that population of a geographical area is no longer increasing.

zone of saturation Area where all available pores in soil and rock in the earth's crust are filled with water. See *water table.*

zoning Regulating how various parcels of land can be used.

zooplankton Animal plankton. Small floating herbivores that feed on plant plankton (phytoplankton and nanoplankton). Compare *nanoplankton, phytoplankton.*

INDEX

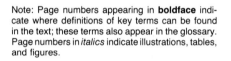

Note: Page numbers appearing in **boldface** indicate where definitions of key terms can be found in the text; these terms also appear in the glossary. Page numbers in *italics* indicate illustrations, tables, and figures.

Bathroom (65% of residential water use; 40% for toilet flushing)

- For existing toilets, reduce the amount of water used per flush by putting a tall plastic container weighted with a few stones into each tank, or buy and insert a toilet dam.

- In new houses, install water-saving toilets or, where health codes permit, waterless or composting toilets.

- Flush toilets only when necessary, using the advice found on a bathroom wall in a drought-stricken area: "If it's yellow, let it mellow—if it's brown, flush it down."

- Take short showers of less than 5 minutes instead of baths. Shower by wetting down, turning off the water while soaping up, and then rinsing off.

- Install water-saving flow restrictors on all faucets and shower heads.

- Check frequently for water leaks and repair them promptly.

- Don't keep water running while brushing teeth, shaving, or washing.

Laundry Room (15%)

- Wash only full loads; use the short cycle and fill the machine to the lowest possible water level.

- When buying a new washer, choose one that uses the least amount of water and fills up to different levels for loads of different sizes.

- Check for leaks frequently and repair all leaks promptly.

Kitchen (10%)

- Use an automatic dishwasher only for full loads; use the short cycle and let dishes air-dry to save energy.

- When washing many dishes by hand, don't let the faucet run. Instead use one filled dishpan for washing and another for rinsing.

- Keep a jug of water in the refrigerator rather than running water from a tap until it gets cold enough to drink.

- Check for leaks frequently and repair them promptly.

- Try not to use a garbage disposal or water-softening system—both are major water users. Instead, try to compost your food wastes.

Outdoors (10%, higher in arid areas)

- Don't wash your car or wash it less frequently. Wash the car from a bucket of soapy water; use the hose only for rinsing.

- Sweep walks and driveways instead of hosing them off.

- Reduce evaporation losses by watering lawns and gardens in the early morning or in the evening, rather than in the heat of midday or when windy. Better yet, landscape with native plants adapted to local average annual precipitation so that watering is not necessary.

- Use drip irrigation systems and mulch on home gardens to improve irrigation efficiency and reduce evaporation.

- To irrigate plants, install a system to capture rainwater or collect, filter, and reuse normally wasted gray water from bathtubs, showers, sinks, and the clothes washer.

Chemical	Alternative	Chemical	Alternative
Deodorant	Sprinkle baking soda on a damp wash rag and wipe skin.	General surface cleaner	Mixture of vinegar, salt, and water.
Oven cleaner	Baking soda and water paste, scouring pad.	Bleach	Baking soda or borax.
Toothpaste	Baking soda.	Mildew remover	Mix ½ cup vinegar, ½ cup borax, and warm water.
Drain cleaner	Pour ½ cup salt down drain, followed by boiling water; or pour 1 handful baking soda and ½ cup white vinegar and cover tightly for one minute.	Disinfectant and general cleaner	Mix ½ cup borax in 1 gallon of hot water.
Window cleaner	Add 2 teaspoons white vinegar to 1 quart warm water.	Furniture or floor polish	Mix ½ cup lemon juice and 1 cup vegetable or olive oil.
Toilet bowl, tub, and tile cleaner	Mix a paste of borax and water; rub on and let set one hour before scrubbing.	Carpet and rug shampoos	Sprinkle on cornstarch, baking soda, or borax and vacuum.
		Detergents and detergent boosters	Washing soda or borax and soap powder.
Floor cleaner	Add ½ cup vinegar to a bucket of hot water; sprinkle a sponge with borax for tough spots.	Fabric softener	Add 1 cup white vinegar or ¼ cup baking soda to final rinse.
		Dishwasher soap	1 part borax and 1 part washing soda.
		Pesticides (indoor and outdoor)	Use natural biological controls.